THERMOELECTRICS AND ITS ENERGY HARVESTING

MATERIALS, PREPARATION, AND CHARACTERIZATION IN THERMOELECTRICS

THERMOELECTRICS AND ITS ENERGY HARVESTING

MATERIALS, PREPARATION, AND CHARACTERIZATION IN THERMOELECTRICS

Edited by **D. M. Rowe**
OBE, DSc, PhD

CRC Press
Taylor & Francis Group
Boca Raton London New York

CRC Press is an imprint of the
Taylor & Francis Group, an **informa** business

CRC Press
Taylor & Francis Group
6000 Broken Sound Parkway NW, Suite 300
Boca Raton, FL 33487-2742

Printed in the United States of America on acid-free paper
Version Date: 20120309

International Standard Book Number: 978-1-4398-7470-7 (Hardback)

Library of Congress Cataloging-in-Publication Data

Thermoelectrics and its energy harvesting / edited by David Michael Rowe.
v. cm.
Summary: "This book includes updated theoretical considerations which provide an insight into avenues of research most likely to result in further improvements in material performance. It details the latest techniques for the preparation of thermoelectric materials employed in energy harvesting, together with advances in the thermoelectric characterisation of nanoscale material. The book reviews the use of neutron beams to investigate phonons, whose behaviour govern the lattice thermal conductivity and includes a chapter on patents"-- Provided by publisher.
Includes bibliographical references and index.
Contents: v. 1. Materials, preparation, and characterization in thermoelectrics -- v. 2. Modules, systems, and applications in thermoelectrics.
ISBN 978-1-4398-7470-7 (hardback : v. 1) -- ISBN 978-1-4398-7472-1 (hardback : v. 2)
1. Thermoelectric apparatus and appliances. I. Rowe, David Michael.

TK2950.T533 2012
621.31'243--dc23 2011052390

Visit the Taylor & Francis Web site at
http://www.taylorandfrancis.com

and the CRC Press Web site at
http://www.crcpress.com

Contents

SECTION III Commercial

Preface

During the intervening years since the publication of the *Thermoelectrics Handbook: Macro to Nano* there has been an unprecedented upsurge of activity in all aspects of thermoelectrics. This is primarily due to rapid advances in nanotechnology and its use in developing nano-architectured materials with substantially improved thermoelectric performance and a worldwide awareness of the importance of thermoelectrics in energy harvesting.

This two-volume set, collectively titled *Thermoelectrics and Its Energy Harvesting*, complements two previous volumes, the *CRC Thermoelectrics Handbook and Thermoelectrics Handbook: Macro to Nano.* It includes updated theoretical considerations which provide an insight into avenues of research most likely to result in further improvements in material and device performance.

The latest techniques for the preparation of thermoelectric materials employed in energy harvesting are detailed, together with advances in the thermoelectric characterization of nanoscale material. The use of neutron beams to investigate phonons, whose behavior governs the lattice thermal conductivity, is reviewed.

This book is comprised of 24 chapters contributed by 49 leading experts in their field. In keeping with previous Handbooks, the majority of chapters are of review type. It is presented in three sections titled General Principles and Theoretical Considerations; Materials Preparation and Measurement; and Commercial. This final section which is comprised of one chapter titled "Patent Basics Every Researcher and Engineer Should Know" covers, in depth, a topic of increasing importance, in view of the considerable commercial value of the discoveries and innovations of thermoelectricians.

As in previous CRC Thermoelectric Handbooks, I have served as an acquisition editor rather than a copy editor. Although some rephrasing has been undertaken to clarify the meaning, every attempt has been made to preserve the international and multicultural flavor of the earlier books.

Again I am indebted to my fellow thermoelectricians who have unselfishly contributed to this book and to my wife Barbara who has assisted me in dealing with the unbelievable amount of correspondence.

D. M. Rowe

Editor

David Michael Rowe is an emeritus professor in the School of Engineering, Cardiff University and research director at Babrow Thermoelectric Consultants Ltd.

He obtained a BSc in both pure mathematics and in applied physics from Swansea University, an MSc from Bristol University, and a PhD from University of Wales Institute of Science and Technology. In 1987, he was awarded a DSc by the University of Wales for "The Development of Semiconductor Alloys with Improved Thermoelectric Figures of Merit."

Professor Rowe started his career as a Harwell United Kingdom Atomic Energy Fellow researching thermoelectric materials for application in nuclear-powered cardiac pacemaker batteries. In 1966, he joined the academic staff at Cardiff University and was awarded a Personal Chair in 1995. He became a distinguished research professor in 2005 and in 2010 received the title emeritus professor.

Professor Rowe's research interests are in thermoelectric materials and their applications. In 1981, he was the first to demonstrate that the thermal conductivity of thermoelectric materials was reduced by phonon-boundary scattering in very small grain size compacted materials, and in 1988 patented the first miniature thermoelectric generator fabricated using ITC technology. Professor Rowe pioneered the thermoelectric recovery of low-temperature heat and a generator that resulted from his research was demonstrated at the Kyoto Energy Summit as an example of green technology.

Forty years of research effort is embodied in more than 350 publications, author-/coauthorship of three books, editorship of three International Conference Proceedings, and of the best-selling definitive texts—the *CRC Handbook of Thermoelectrics* (1995) and (2005). *Thermoelectrics Handbook: Macro to Nano* (2012).

He is a member of the editorial board of several international journals and serves as an expert assessor for European and U.S. funding agencies. As research director of Babrow Consultants, he is actively involved in consultancy roles for a number of national and international thermoelectric technology projects.

In 1994, Professor Rowe was elected as a Fellow of the Institute of Physics and Fellow of Institution of Electrical Engineers, and in 1995 he founded Cardiff's NEDO Laboratory of Thermoelectric Engineering and served as its director. During 1997–2001, he was president of the International Thermoelectric Society (ITS) and since 2004 has been its Secretary. On May 23, 2007, Professor Rowe received the Order of Knighthood called "The most excellent order of the British Empire" (OBE) from Her Royal Highness Queen Elizabeth II at Buckingham Palace. In 2008, he was a recipient of the 2008 Gold Prize from the International Thermoelectric Academy for contributions to thermoelectricity and in the same year was elected an Honorary Academician of the Institution of Refrigeration. In 2009, Professor Rowe was elected as president of the European Thermoelectric Society.

Contributors

L. I. Anatychuk
Institute of Thermoelectricity
Chernivtsi, Ukraine

Alexander A. Balandin
Nano-Device Laboratory
University of California, Riverside
Riverside, California

L. E. Bell
BSST LLC
Irwindale, California

Laurent Chaput
Jean Lamour Institute
University of Nancy
Vandoeuvre Les Nancy, France

Lidong Chen
State Key Laboratory of High Performance Ceramics and Superfine Microstructure
Shanghai Institute of Ceramics
Chinese Academy of Sciences
Shanghai, People's Republic of China

Mogens Christensen
Department of Chemistry
Aarhus University
Aarhus, Denmark

Anuja Datta
Department of Physics
University of South Florida
Tampa, Florida

Baoli Du
State Key Laboratory of Advanced Technology for Materials Synthesis and Processing
Wuhan University of Technology
Wuhan, People's Republic of China

Espen Flage-Larsen
Department of Physics
University of Oslo
Oslo, Norway

H. J. Goldsmid
School of Physics
University of New South Wales
New South Wales, Australia

L. M. Goncalves
Department of Industrial Electronics
University of Minho
Guimaraes, Portugal

C. Goupil
CRISMAT Laboratory
ISMRA and University of Caen
Caen, France

Dieter M. Gruen
Materials Science Division
Argonne National Laboratory
Argonne, Illinois

Robert M. Hansen
The Marbury Law Group, PLLC
Reston International Center
Reston, Virginia

Joseph P. Heremans
Department of Mechanical and Aeronautical Engineering and Department of Physics
The Ohio State University
Columbus, Ohio

Daniel Huzel
RheinMain University of Applied Sciences
Wiesbaden, Germany

Yukihiro Isoda
National Institute for Materials Science
Tsukuba, Japan

Han Li
State Key Laboratory of Advanced Technology for Materials Synthesis and Processing
Wuhan University of Technology
Wuhan, People's Republic of China

Shanghua Li
Functional Materials Division
Royal Institute of Technology
Stockholm, Sweden

Ole Martin Løvvik
SINTEF Materials and Chemistry
and
Department of Physics
University of Oslo
Oslo, Norway

Andrew F. May
Materials Science and Technology Division
Oak Ridge National Laboratory
Oak Ridge, Tennessee

Mamoun Muhammed
Functional Materials Division
Royal Institute of Technology
Stockholm, Sweden

E. Müller
Institute of Materials Research
German Aerospace Center
Köln, Germany

George S. Nolas
Department of Physics
University of South Florida
Tampa, Florida

Adrian Popescu
Department of Physics
University of South Florida
Tampa, Florida

Heiko Reith
RheinMain University of Applied Sciences
Wiesbaden, Germany

Frédéric Sansoz
School of Engineering and Materials Science
The University of Vermont
Burlington, Vermont

Matthias Schmitt
RheinMain University of Applied Sciences
Wiesbaden, Germany

W. Seifert
Institute of Physics
University Halle-Wittenberg
Halle, Germany

Li Shi
Department of Mechanical Engineering and Texas Materials Institute
The University of Texas
Austin, Texas

Xun Shi
State Key Laboratory of High Performance Ceramics and Superfine Microstructure
Shanghai Institute of Ceramics
Chinese Academy of Sciences
Shanghai, People's Republic of China

G. Jeffrey Snyder
Materials Science
California Institute of Technology
Pasadena, California

Tsunehiro Takeuchi
EcoTopoia Science Institute
Nagoya University
Nagoya, Japan

Xinfeng Tang
State Key Laboratory of Advanced Technology for Materials Synthesis and Processing
Wuhan University of Technology
Wuhan, People's Republic of China

Patrick J. Taylor
Sensors and Electron Devices Directorate
U.S. Army Research Laboratory
Adelphi, Maryland

Jean-Claude Tedenac
Institute of Molecular Chemistry and Materials
Montpellier 2 University
Montpellier, France

Janusz Tobola
Department of Condensed Matter Physics
AGH University of Science and Technology
Kraków, Poland

Muhammet S. Toprak
Functional Materials Division
Royal Institute of Technology
Stockholm, Sweden

Terry M. Tritt
Department of Physics and Astronomy
Clemson University
Clemson, South Carolina

Haruhiko Udono
Department of Electrical and Electronic Engineering
Ibaraki University
Ibaraki, Japan

Ctirad Uher
Department of Physics
University of Michigan
Ann Arbor, Michigan

Friedemann Völklein
Institute of Microtechnologies
RheinMain University of Applied Sciences
Wiesbaden, Germany

Lilia Woods
Department of Physics
University of South Florida
Tampa, Florida

Wenjie Xie
State Key Laboratory of Advanced Technology for Materials Synthesis and Processing
Wuhan University of Technology
Wuhan, People's Republic of China

Jihui Yang
Electrochemical Energy Research Lab
General Motors R&D Center
Warren, Michigan

Jiong Yang
State Key Laboratory of High Performance Ceramics and Superfine Microstructure
Shanghai Institute of Ceramics
Chinese Academy of Sciences
Shanghai, People's Republic of China

K. Zabrocki
Institute of Materials Research
German Aerospace Center
Köln, Germany

Qingjie Zhang
State Key Laboratory of Advanced Technology for Materials Synthesis and Processing
Wuhan University of Technology
Wuhan, People's Republic of China

Wenqing Zhang
State Key Laboratory of High Performance Ceramics and Superfine Microstructure
Shanghai Institute of Ceramics
Chinese Academy of Sciences
Shanghai, People's Republic of China

I

General Principles and Theoretical Considerations

1

Transverse Thermoelectric Effects and Their Application

H. J. Goldsmid
University of New South Wales

1.1 Transverse Seebeck and Peltier Effects

Thermoelectric energy conversion is usually based on the longitudinal Seebeck and Peltier effects. These effects are manifest in so-called thermocouples, each consisting of a negative and a positive thermoelement. A single thermocouple is capable of handling only a small amount of power at a very small voltage whether it is used in the generation or refrigeration mode. For this reason, practical thermoelectric devices commonly consist of thermoelectric modules made up of a great many thermocouples. Even more complex structures are required when the performance is enhanced using a cascade. We shall see that the adoption of the transverse thermoelectric effects might enable much simpler arrangements to be employed.

The transverse Seebeck and Peltier effects are illustrated in Figures 1.1 and 1.2. In the transverse Seebeck effect, a longitudinal temperature gradient, dT/dx, produces an electric field, dV/dy, in a perpendicular direction. The transverse Seebeck coefficient is defined as

$$\alpha_{yx} = \frac{dV/dy}{dT/dx} \tag{1.1}$$

The transverse Peltier coefficient appears as a heat flux in a direction perpendicular to an electric current. The transverse Peltier coefficient is defined as

$$\pi_{xy} = \frac{q_x}{i_{xy}} \tag{1.2}$$

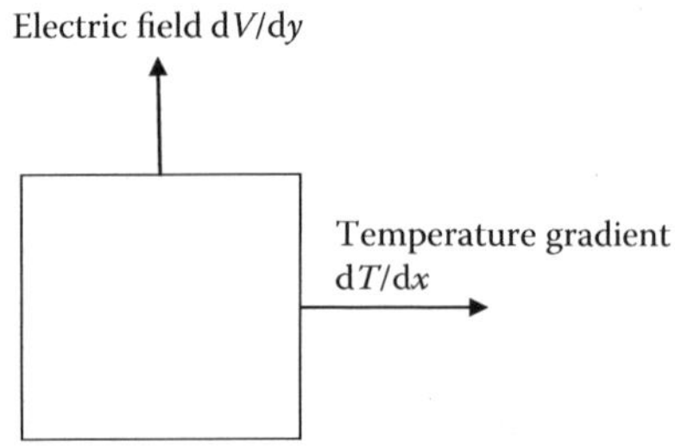

FIGURE 1.1 Transverse Seebeck effect.

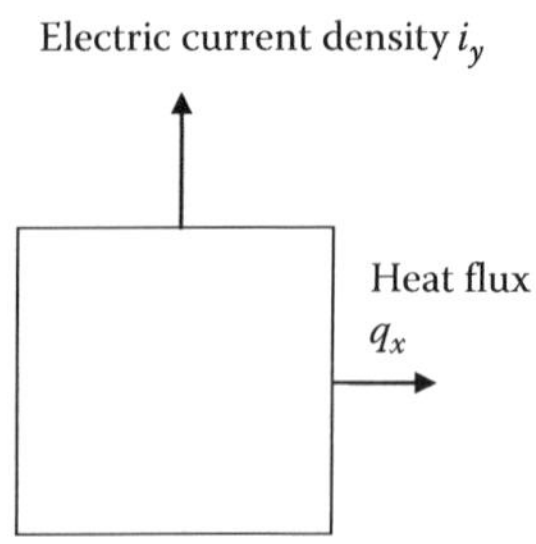

FIGURE 1.2 Transverse Peltier effect.

where q_x is the heat flux per unit area in the x-direction and i_y is the electric current density in the y-direction.

The transverse effects may be observed in any conductor in which the Seebeck coefficient is anisotropic though not for all orientations. For example, if a rectangular sample is cut from a crystal with its edges parallel to the axes, no transverse effect will be seen. In this example, the transverse effects will appear only when the edges of the specimen are at some angle to the crystal axes. There will be a particular orientation at which the effects will be largest.

1.2 Transverse Figure of Merit

The theory of energy conversion using the transverse thermoelectric effects follows the same pattern as conventional thermocouple theory.[1] There are, however, some important differences.

We suppose that the transverse device takes the form of a rectangular block as shown in Figure 1.3. We shall assume that the length in the direction of the electric flow is much greater than that in the direction of heat flow. This allows us to neglect the disturbances in the equipotentials and isothermal surfaces at the ends.

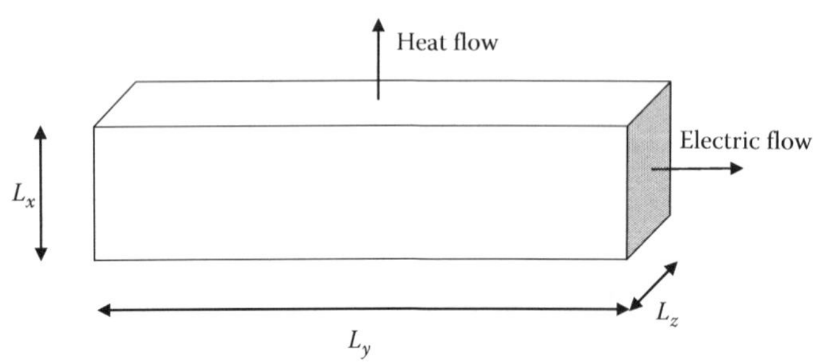

FIGURE 1.3 Transverse energy converter.

One must, of course, substitute the transverse Seebeck coefficient α_{yx} for the longitudinal Seebeck coefficient. Also, the electrical resistance is equal to $L_y\rho_y/L_xL_z$ and the thermal conductance is $L_yL_z\lambda_x/L_x$, where ρ_y is the electrical resistivity in the y-direction and λ_x is the thermal conductivity in the x-direction.

The expressions for the efficiency of a transverse generator and the coefficient of performance of a transverse refrigerator have the same form as the corresponding equations for a longitudinal device. Thus, one can define a transverse figure of merit as $\alpha_{yx}^2/\rho_y\lambda_x$ and this quantity has the same significance as Z for a conventional thermocouple.

In spite of these similarities, it is important to note that the product of electrical resistance and thermal conductance is equal to $\rho_y\lambda_x L_y^2/L_x^2$ whereas for a conventional couple this product is just equal to $\rho\lambda$. In the transverse device the product can be altered by adjusting the form factor of the specimen whereas in the longitudinal case it is invariable. This has practical consequences. Thus, the cooling power of a single thermocouple can be increased by making the cross-sectional area larger but this may raise the electric current to an impracticably high level. Peltier devices with a high cooling power are invariably modules consisting of many thermocouples. On the other hand, the cooling power of a transverse Peltier refrigerator can be increased by changing the form factor without changing the electric current. Likewise, the voltage output of a transverse thermoelectric generator can be adjusted by changing the shape of the element. These characteristics of transverse thermoelectric devices can be exploited in fast thermal detectors and infinite-staged Peltier coolers.

The response time of a thermal detector depends very strongly on its thickness.[2] For a given heat flux the output voltage of a conventional thermoelectric detector becomes very small when the length of the thermoelements is decreased but reduction of the thickness of a transverse detector does not change the voltage.

A conventional thermoelectric cascade usually consists of a large number of thermocouples in a pyramidal structure. However, a transverse Peltier cooler with an infinite number of stages can be made by continually changing the cross-sectional area in the direction of heat flow.[3] The width L_z has to vary with the distance from the heat source according to an exponential law. However, even a trapezoidal shape seems to yield a significant increase in the maximum temperature difference between the source and sink.[4]

1.3 Single-Phase Transverse Thermoelements

The transverse Seebeck and Peltier coefficients are unlikely to be very large in a conductor that has only one type of charge carrier. That is because these coefficients are a measure of the energy that is transported by the electrons or holes and, to a first approximation, this is independent of direction. For a nondegenerate extrinsic semiconductor, the Seebeck coefficient is expressed as

$$\alpha = \mp\frac{k}{e}\left[\eta - \left(r + \frac{5}{2}\right)\right]. \tag{1.3}$$

where k is Boltzmann's constant, e is the elementary charge, η is the reduced Fermi energy, and r is the exponent of energy in the scattering law.[5] Anisotropy can arise only if the parameter r is dependent on orientation. It is not sufficient for the relaxation time for carrier scattering to be anisotropic; rather it is its dependence on energy that has to be anisotropic.

Simple transport theory suggests that the limits of the parameter r are −1/2 and +3/2. Thus, it is theoretically possible for the Seebeck coefficients in perpendicular directions to differ by $2k/e$, that is about 172 μV/K. This would require acoustic-mode lattice scattering of the carriers to predominate in one direction and ionized-impurity scattering to predominate in the other. This is an unlikely situation but it is just possible that the relaxation time for lattice scattering could be very strongly anisotropic with the relaxation time for ionized-impurity scattering having some intermediate value. Then, ionized-impurity scattering might dominate in the direction in which the lattice scattering is weak.

In practice, there does not seem to be strong anisotropy of the Seebeck coefficient in any homogeneous conductor in which there is only one type of charge carrier. There can, however, be substantial anisotropy when both electrons and holes are present. Consider, for example, an intrinsic conductor in which the electron mobility is much the higher in one direction and the hole mobility much the higher in a perpendicular direction. Then in one direction the Seebeck coefficient should be equal to its partial value for the electrons and at right angles it should equal the partial value for holes. The anisotropy of the Seebeck coefficient might then amount to several hundred microvolts per degree.

Among the anisotropic conductors that might be used as transverse thermoelements we shall first discuss bismuth telluride. We know that the Seebeck coefficient of this compound is isotropic when there is only one type of carrier but we would expect it to become anisotropic in the mixed and intrinsic ranges. It is, in fact, rather difficult to produce intrinsic bismuth telluride crystals at ordinary temperatures. The compound tends to be nonstoichiometric with an excess concentration of acceptors.[6] However, from the known energy gap, the electron and hole mobilities, and the lattice conductivity in the *a*- and *c*-directions, we can calculate the anisotropy of the Seebeck coefficient for intrinsic bismuth telluride.[7] The energy gap at 300 K is equal to 5 kT and we find that the difference between the Seebeck coefficients in the two directions should amount to nearly 80 μV/K. The transport properties of an intrinsic bismuth telluride alloy are given in Table 1.1.

The transverse figure of merit Z_ϕ has its maximum value when the specimen is tilted at an angle ϕ of about 45° to the principal axes. The value of the $Z_\phi T$ is then about 0.002, which, of course, compares most unfavorably with ZT for conventional thermocouples. Thus, a Peltier cooler using the transverse effect in bismuth telluride would have a value of ΔT_{max} of <10°. The problem is the relatively high electrical resistivity that stems from the fact that the Fermi level lies about 2.5 kT within the forbidden gap.

Bismuth looks a much more promising material since it has close to zero energy gap. The properties of an undoped single crystal of bismuth[8] are shown in Table 1.2. The difference between the ratios of electron to hole mobility in the *a*- and *c*-directions is not as large as one might wish so the difference between the two Seebeck coefficients is only just over 50 μV/K. Nevertheless, the electrical resistivity is low and the optimum value of $Z_\phi T$ is over 0.02 as shown in Figure 1.4. This is still much less than unity but is probably large enough for a thin film of monocrystalline bismuth, deposited at an appropriate orientation, to act as a useful fast thermal detector. A marginally higher figure of merit seems to have been achieved in practice.[9] In fact, it appears that a thermal detector based on the transverse Seebeck effect in CdSb is useful even though this material has an even smaller figure of merit.[10]

As we shall see, much larger transverse figures of merit can be reached using two-phase systems. However, it might prove difficult to make two-phase materials with the spacing between the phases

TABLE 1.1 Thermoelectric Properties of an Intrinsic Bismuth Telluride Alloy at 300 K

Parameter	*a*-Direction	*c*-Direction
α, μV/K	0	76
ρ, Ω m	0.5×10^{-4}	1.73×10^{-4}
λ, W/m K	2.3	1.1

TABLE 1.2 Anisotropy of the Thermoelectric Parameters in Intrinsic Bi at 300 K

Parameter	*a*-Direction	*c*-Direction
α, μV/K	−50	−102
ρ, Ω m	1.1×10^{-6}	1.35×10^{-6}
λ, W/m K	10	6

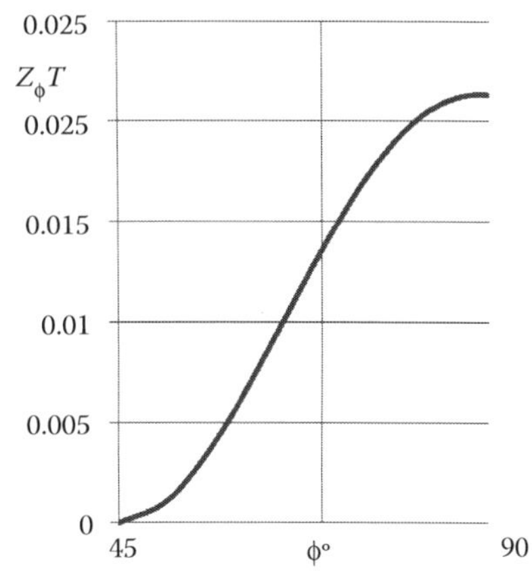

FIGURE 1.4 Dimensionless transverse figure of merit for pure Bi at 300 K as a function of the angle between the direction of current flow and the *a* axis.

small enough to allow thin-film devices to be constructed. Consequently, there may be a place for single-phase transverse thermoelements. With this in mind, it would certainly be worthwhile to search for semimetals in which the electron or hole mobility is strongly anisotropic.

It is easily shown that the anisotropy of the Seebeck coefficient of an intrinsic conductor can be expressed as

$$\alpha_1 - \alpha_2 = \frac{(c_1 - c_2)(\alpha_n - \alpha_p)}{(c_1 + 1)(c_2 + 1)} \equiv \Delta(\alpha_n - \alpha_p), \tag{1.4}$$

where α_n and α_p are the partial Seebeck coefficients of the electrons and holes. We have denoted the mobility ratio μ_{n1}/μ_{p1} by c_1 and the ratio μ_{n2}/μ_{p2} by c_2. The quantity Δ represents the anisotropy of the Seebeck coefficient as a proportion of the difference between the partial Seebeck coefficients. Since $(\alpha_p - \alpha_n)$ is about 400 μV/K for bismuth, the value of Δ for this material must be about 0.1. Clearly, there is room for improvement and there may well exist semimetals having a greater anisotropy of Seebeck coefficient than bismuth with an equally good combination of mobility, effective mass, and lattice conductivity.

1.4 Theory of Two-Phase Materials

We have seen that it is difficult if not impossible to obtain a large transverse figure of merit using single-phase material. However, one can do much better if a two-phase system is adopted.[10,11] We shall discuss a configuration in which a synthetic anisotropic material is made from a stack of layers of two different conductors 1 and 2, as shown in Figure 1.5.

The two components have different Seebeck coefficients α_1 and α_2. Ideally, the electrical resistance R_2 in the y_0-direction of one of the components will be much greater than that R_2 of the other. On the other hand, in the x_0-direction, the thermal resistance $1/K_1$ of component 1 should be much higher than that $1/K_2$ of component 2. This means that the effective Seebeck coefficient in the y_0-direction will be close to α_1 and the Seebeck coefficient in the x_0-direction will be close to α_2. The equations for the Seebeck coefficient in the two directions are

$$\alpha_{xo} = \frac{\alpha_1/K_1 + \alpha_2/K_2}{1/K_1 + 1/K_2} \tag{1.5}$$

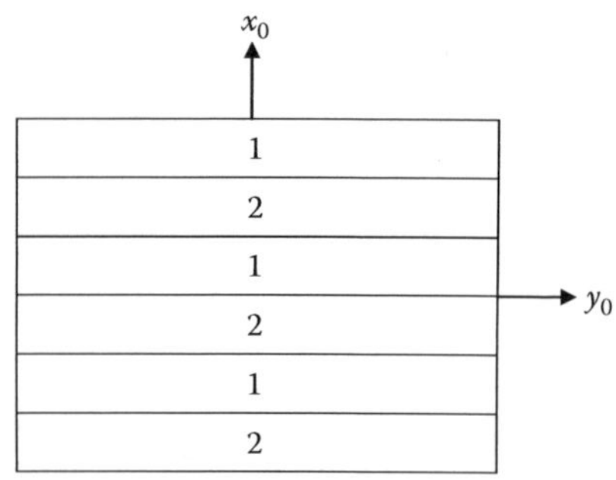

FIGURE 1.5 A synthetic anisotropic material with a layered structure.

and

$$\alpha_{y_0} = \frac{\alpha_1/R_1 + \alpha_2/R_2}{1/R_2 + 1/R_2} \tag{1.6}$$

We have stated that ideally $K_1 \gg K_2$ and $R_2 \gg R_1$ but this is only possible if the components satisfy the requirement $\sigma_1\lambda_1 \gg \sigma_2\lambda_2$. We are using the symbols σ, ρ, and λ to represent electrical conductivity, electrical resistivity, and thermal conductivity, respectively. Our aim is to find a pair of components that can be combined to form a conventional thermocouple with a high figure of merit and which at the same time satisfy the condition $\sigma_1\lambda_1 \gg \sigma_2\lambda_2$.

Having produced a layered structure, one must then cut a specimen at an angle ϕ so as to optimize the figure of merit. It is also necessary to optimize the relative thicknesses of the layers. We suppose that the layer thicknesses are d_1 and d_2 such that $d_2/d_1 = n$. Then, we find that $K_1/K_2 = \lambda_1/n\lambda_2$ and $R_1/R_2 = n\rho_1/\rho_2$.

The effective electrical resistivities in the x_0- and y_0-directions are

$$\rho_{x_0} = \frac{\rho_1 + n\rho_2}{n+1} \tag{1.7}$$

and

$$\rho_{y_0} = \frac{n+1}{1/\rho_1 + n/\rho_2} \tag{1.8}$$

The effective thermal conductivities are

$$\lambda_{x_0} = \frac{n+1}{1/\lambda_1 + n/\lambda_2} \tag{1.9}$$

and

$$\lambda_{y_0} = \frac{\lambda_1 + n\lambda_2}{n-1}(1 + Z_{12}T) \tag{1.10}$$

A comment should be made about Equation 1.10. It will be noticed that the thermal conductance in the y_0-direction does not consist merely of that for the two conductors in parallel. There is an additional factor that contains $Z_{12}T$, the conventional dimensionless figure of merit for a thermocouple made from the two components. This factor exists because of the Peltier effect associated with circulating electric currents. $Z_{12}T$ is given by

$$Z_{12}T = \frac{(\alpha_1 - \alpha_2)^2 T}{(\lambda_1 + n\lambda_2)(\rho_1 + \rho_2/n)} \tag{1.11}$$

We now have to obtain expressions for the transverse Seebeck coefficient $\alpha_{y\phi x\phi}$, the effective electrical resistivity $\rho_{x\phi x\phi}$, and the effective thermal conductivity $\lambda_{y\phi y\phi}$. These expressions are

$$\alpha_{y\phi x\phi} = (\alpha_{xo} - \alpha_{yo})\sin\phi\cos\phi, \tag{1.12}$$

$$\rho_{x\phi x\phi} = \rho_{xo}\cos^2\phi + \rho_{yo}\sin^2\phi, \tag{1.13}$$

and

$$\lambda_{y\phi y\phi} = \lambda_{xo}\sin^2\phi + \lambda_{yo}\cos^2\phi. \tag{1.14}$$

The transverse figure of merit is then

$$Z_\phi = \frac{\alpha^2_{y\phi x\phi}}{\lambda_{y\phi y\phi}\rho_{x\phi x\phi}} \tag{1.15}$$

It is not difficult to find the optimum value for the angle ϕ and the maximum figure of merit by computation but simple expressions for these quantities exist when the condition $\sigma_1\lambda_1 \gg \sigma_2\lambda_2$ is satisfied. Then, as shown by Babin et al.,[10]

$$\tan\phi_{\text{opt}} = \frac{\sqrt{n}}{n+1}\left[\frac{\rho_1\lambda_2}{\rho_2\lambda_1}(1 + Z_{12}T)\right]^{1/4} \tag{1.16}$$

and

$$Z_\phi^{\max} = z_1\frac{(1 - \alpha_2/\alpha_1)^2}{\left[1 + \sqrt{(\lambda_2\rho_2/\lambda_1\rho_1)(1 + Z_{12}T)}\right]^2} \tag{1.17}$$

It will be noticed that Equation 1.16 does not contain n, the ratio of the thicknesses of the layers and it was stated by Babin and his coworkers that the transverse figure of merit is not particularly sensitive to this quantity. They gave an approximate optimum value for n as

$$n_{\text{opt}} \simeq \left(\frac{2(\lambda_2/\lambda_1)(\rho_2/\rho_1)}{1 + (\lambda_2/\lambda_1)(\rho_2/\rho_1)}\right)^{1/2} \tag{1.18}$$

The product $\lambda\rho$ cannot exceed the value given by the Wiedemann–Franz law and it would be favorable if this law were satisfied for both components. If this were the case, the optimum value for n would be unity. As we shall see, some experimental work has made use of a good thermoelectric material like bismuth telluride as one component and a metal as the other. For this combination, n_{opt} would be equal to about 0.6. It is probably satisfactory in general to adopt more-or-less equal thicknesses for the layers.

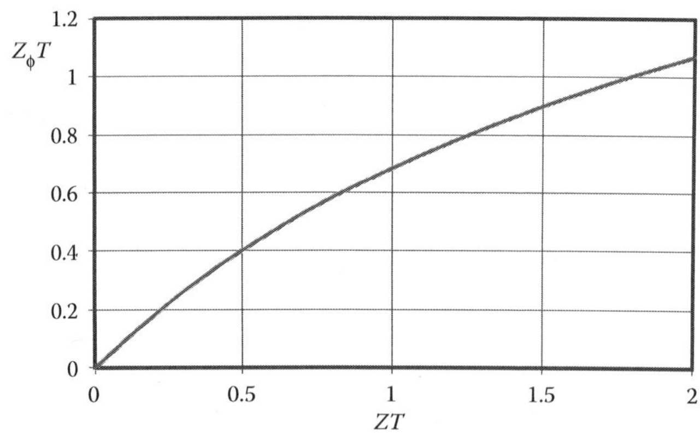

FIGURE 1.6 Plot of dimensionless transverse figure of merit against dimensionless longitudinal figure of merit for components having the same ratio of electrical to thermal conductivity.

Equation 1.16 shows that the maximum transverse figure of merit differs from the longitudinal figure of merit Z_{12} through the inclusion of the term $(1 + Z_{12}T)$. When $Z_\phi T$ is very much less than unity, this term is negligible and the two figures of merit are equal. However, when $Z_\phi T$ is of the order of unity or greater the difference between the two figures of merit is substantial as shown in Figure 1.6. In this diagram, $Z_\phi T$ is plotted against $Z_{12}T$ assuming that $\rho\lambda$ is the same for both components. We see that when $Z_{12}T$ is equal to 2, $Z_\phi T$ is only about 1. Transverse thermoelements are therefore likely to be less useful as the figure of merit of conventional thermocouples improves.

1.5 Demonstration of the Transverse Effects

The transverse thermoelectric effects were demonstrated by Korolyuk et al.[12] who made use of the semiconductor cadmium antimonide. For this material, $Z_\phi T > 0.002$ although the Seebeck coefficient displays a rather large anisotropy of 280 μV/K.

Much better figures of merit have been observed for two-phase materials. As long ago as 1978, Gudkin et al.[4] found a value of $Z_\phi T$ equal to 0.25 using a composite of bismuth–antimony telluride and bismuth. A rectilinear transverse Peltier cooler based on this composite was found to yield ΔT_{max} equal to 23°. These authors did not make an exponentially shaped cascade but they did observe that the maximum cooling effect rose to 35° when the cross section became trapezoidal with a 10:1 ratio between the areas in contact with the heat sink and the source.

Many of the experiments on the transverse effects have emphasized their application in thermal detectors. Zahner et al.[2] made their transverse thermoelement from a stack of metal foils, presumably with a very low figure of merit. Nevertheless, they were able to demonstrate the very rapid response of a thermal detector as the thickness was reduced. The response time fell from about 10 ms for a thickness of 1 mm to about 0.1 ms at a thickness of 0.1 mm.

Huber et al.[13] showed that the transverse Seebeck effect could be observed using $Bi_2Sr_2CaCu_2O_8$ but this material does not seem to have any practical use. Because of the ready availability of the components, a transverse thermoelement made from chromel and constantan by Fischer et al.[14] might be more useful but the figure of merit must be very small. Probably the most successful of recent experiments was described by Kyarad and Lengfellner[15] who made their composite from layers of bismuth telluride and lead. A bismuth telluride–lead thermocouple has only a modest longitudinal figure of merit ZT but the two components satisfy the condition $\sigma_1\lambda_1 \gg \sigma_2\lambda_2$. Thus, the transverse figure of merit has almost the same value as ZT. The transverse Peltier effect using the transverse thermoelement of Kyarad and Lengfellner produced a maximum temperature difference of 22°.

The construction of two-phase thermoelements clearly presents some problems. Thus, the method used by Kanno et al.[16] for making a device composed of bismuth and copper is of interest. Molten bismuth was forced into slots that had been cut in a block of copper. A rectangular specimen was then cut from the block at an angle of 30° to produce a transverse thermoelement with multiple layers. $Z_\phi T$ for this device was not more than 0.04 so it could hardly be used for effective Peltier cooling and it would undoubtedly be easier to make a fast thermal detector using single-phase bismuth. Nevertheless, the method of production has considerable merit and it could, perhaps, be adapted for other systems.

1.6 Selection of Materials

We have already discussed the prospects for finding improved materials for single-phase transverse thermoelements but it is apparent that only two-phase systems will yield high figures of merit. The difficulty is to find a pair of materials that form a conventional couple with a large value of ZT while simultaneously satisfying the requirement $\sigma_1\lambda_1 \gg \sigma_2\lambda_2$.

It is possible that the combination of bismuth–antimony telluride and bismuth that was used by Gudkn et al.[4] could be improved upon by using monocrystalline bismuth oriented with the layers parallel to the c axis. This is the direction in which the negative Seebeck coefficient is highest while the thermal conductivity is higher in the a-direction. Both these factors help toward a large anisotropy of the Seebeck coefficient.

Of course, the inclusion of layers of single crystal bismuth in the device means that it would be both expensive and difficult to make the structure. Consequently, the use of a bismuth telluride alloy in combination with some other material would be preferred. The combination of p-type and n-type bismuth telluride alloys, of course, has a large value of ZT but with $\sigma_1\lambda_1 \simeq \sigma_2\lambda_2$, the transverse figure of merit would be small. Thus, the second component would probably have to be a metal or semimetal. A useful material would seem to be $YbAl_{2.96}Mn_{0.04}$, which has a metal-like electrical conductivity with a Seebeck coefficient of −90 μV/K.[17] When combined with bismuth telluride the condition $\sigma_1\lambda_1 >> \sigma_2\lambda_2$ is not really satisfied but one can improve the situation by using sintered polycrystalline bismuth–antimony telluride rather than material with a preferential orientation. Furthermore, the bismuth telluride alloy can be doped so that its carrier concentration is rather less than its optimum value in a conventional couple.

1.7 Advantages of Porous Components

Despite the above steps being taken, the ratio of $\sigma_1\lambda_1$ to $\sigma_2\lambda_2$, where material 1 is $YbAl_{2.96}Mn_{0.04}$ and material 2 is bismuth–antimony telluride, is still not as large as we would like. Consequently, $Z_\phi T$ still falls well short of ZT. It has been suggested[18] that the condition $\sigma_1\lambda_1 >> \sigma_2\lambda_2$ can be met if the bismuth–antimony telluride component is porous. This has the effect of reducing $\sigma_2\lambda_2$ without changing the ratio σ_2/λ_2, at least when the porosity is not too great. At high porosities there is the possibility of heat transfer within the pores. Radiative transfer can probably be neglected but gas-filled pores could lead to a significant heat loss.[19] We define a porosity factor p as the ratio of the electrical resistivity of the porous material to that of a fully dense sample. When p is equal to 10 the relative density is about 15%. It might be impractical in any case to make the porosity factor >10.

In theory, the combination of high-porosity bismuth–antimony telluride and $YbAl_{2.96}Mn_{0.04}$ would yield a value of $Z_\phi T$ equal to about 0.7 provided that there were no heat losses across the pores. Figure 1.7 shows how $Z_\phi T$ would actually vary with the porosity-actor of the bismuth telluride alloy component if heat conduction through the gas in the pores is taken into account. It is seen that the value of $Z_\phi T$ would be only about 0.3 if both components were fully dense but even moderate porosity of the one component would enable significantly larger values to be reached. Even with a porosity factor as low as 3, $Z_\phi T$ should become as large as 0.5. This means that transverse Peltier cooling through about 60° should be possible using a specimen of constant cross section with perhaps 100° being possible with a shaped cascade.

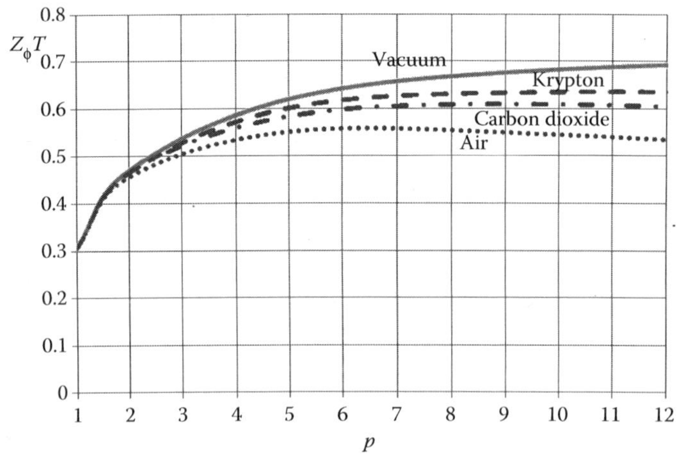

FIGURE 1.7 Plots of dimensionless transverse figure of merit for a $(Bi\text{-}Sb)_2Te_3$ and $YbAl_{2.96}Mn_{0.04}$ composite against the porosity factor of the bismuth telluride alloy component. The pores are either evacuated or filled with the gases indicated.

1.8 Possible Structures

Although a layered structure may well be the most convenient way of realizing the concept of a two-phase transverse thermoelement, it is not the only configuration that is possible. An alternative would be the use of rod-like inclusions.

Suppose that the rod shown in Figure 1.8 has electrical and thermal conductivities that are much greater than those of the surrounding material. Then, in the axial direction, electric current will flow primarily through the rod and the Seebeck coefficient will take on its value. On the other hand, in a direction perpendicular to the axis, significant temperature gradients will exist only in the surrounding material. Thus, in this direction the Seebeck coefficient of this material will dominate. In other words, the behavior will be much the same as that of a layered structure. Anatychuk[11] has shown that the rod-like inclusions need not be continuous and has suggested that a eutectic material may be suitable for transverse themoelements.

We have already pointed out that an advantage of transverse thermoelements is the capability of adjusting the electrical and thermal resistances independently. This means, for example, that the length in the direction of the electric current flow can be made large so as to make the current small and the voltage large. By selecting an appropriate value for the thickness in the heat flow direction the required

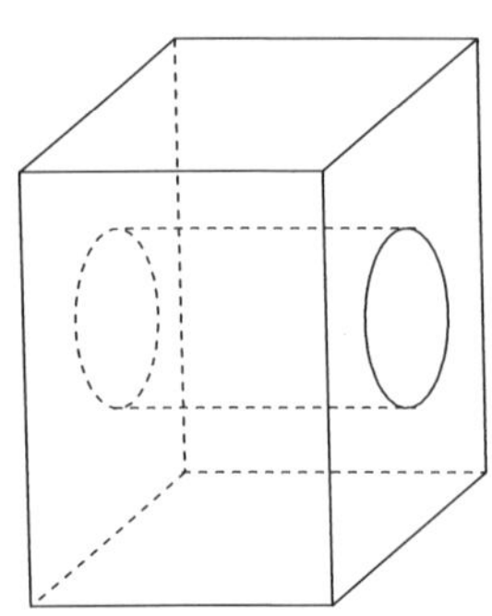

FIGURE 1.8 One cell of a two-phase material with rod-like inclusions.

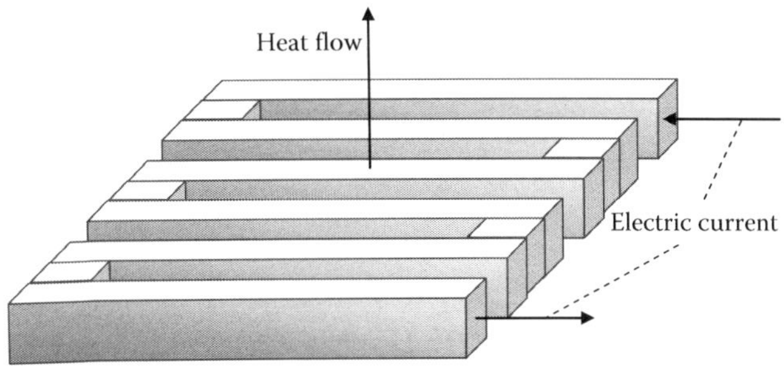

FIGURE 1.9 Serpentine structure of a transverse thermoelement.

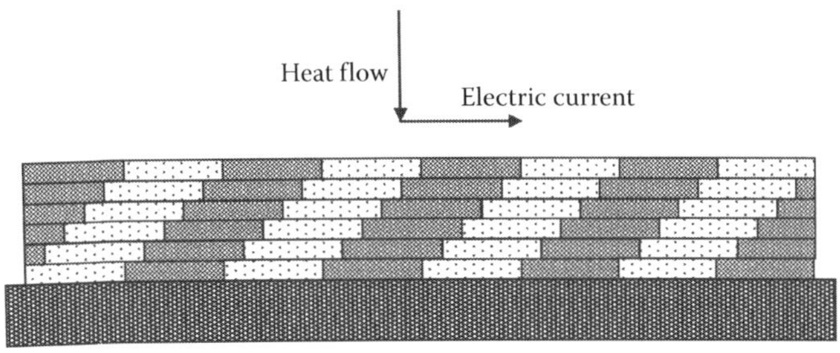

FIGURE 1.10 Multiple thin-film arrangement suitable for a fast thermal detector based on a two-phase transverse thermoelement.

thermal resistance can be achieved. A method for obtaining a large length of thermoelement in a compact space involves the use of a serpentine structure as shown in Figure 1.9. Such a structure could not be made from a single piece of material, however, since the orientation has to change as the flow direction is reversed.

It would rather be difficult to make a fast thermal detector from a bulk two-phase material. The width of each component would have to be very small and even then the preparation of a thin section would present problems. However, a two-phase transverse thermoelement consisting of thin films could take the form shown in Figure 1.10. Some compromise in the selection of components might be necessary but the device would probably be superior to any single-phase transverse thermoelement available at the present time.

1.9 Conclusions

In the past, transverse thermoelements have found little practical use. Generally, their performance has been greatly inferior to that which can be achieved using conventional thermocouples. The one application that seems to have gained acceptance is the rapid detection of thermal radiation. A fast thermal detector using the transverse Seebeck effect could be improved if a semimetal with appropriate anisotropies of the electron and hole mobilities could be found. Alternatively, a thin-film two-phase transverse thermoelement might be developed.

It is thought that a two-phase single-stage transverse Peltier cooler would now have a performance only slightly inferior to that of a conventional thermoelectric refrigerator. This means that an infinite-staged transverse device should definitely be a worthwhile proposition. However, as ZT for conventional thermocouples becomes significantly greater than unity in the future, transverse cooling devices will compare unfavorably with them due to the heat transfer effects associated with internal circulating currents.

References

1. H. J. Goldsmid, *Introduction to Thermoelectricity*, Springer, Heidelberg, pp. 7, 191, 2010.
2. T. Zahner, R. Forg, and H. Lengfellner, Transverse thermoelectric response of a tilted metallic multilayer structure, *Appl. Phys. Lett.*, **73**, 1364, 1998.
3. B. J. O'Brien and C. S. Wallace, Ettingshausen effect and thermoelectric cooling, *J. Appl. Phys.*, **29**, 1010, 1958.
4. T. S. Gudkin, E. K. Iordanishvili, and E. E. Fiskind, With reference to theory of anisotropic thermoelectric cooler, *Sov. Phys. Tech. Phys. Lett.*, **4**, 844, 1978.
5. A. F. Ioffe, *Semiconductor Thermoelements and Thermoelectric Cooling*, Infosearch, London, p. 45, 1956.
6. H. Scherrer and S. Scherrer, Bismuth telluride, antimony telluride and their solid solutions. In *CRC Handbook of Thermoelectrics*, D. M. Rowe, Ed., CRC Press, Boca Raton, FL, p. 211, 1994.
7. R. T. Delves, A. E. Bowley, D. W. Hazelden, and H. J. Goldsmid, Anisotropy of the electrical conductivity in bismuth telluride, *Proc. Phys. Soc.*, **78**, 838, 1961.
8. C. G. Gallo, B. S. Chandrasekhar, and P. H. Sutter, Transport properties of bismuth single crystals, *J. Appl. Phys.*, **34**, 144, 1963.
9. A. A. Snarskii and L. P. Bulat, Anisotropic thermoelements. In *Thermoelectrics Handbook: Macro to Nano*, D. M. Rowe, Ed., CRC Press, Boca Raton, FL, p. 45-1, 2006.
10. V. P. Babin, T. S. Gudkin, Z. M. Dashevskii, L. D. Dudkin, E. K. Iordanishvili, V. I. Kaidanov, N. V. Kolomoets, O. M. Narva, and L. S. Stil'bans, Anisotropic synthetic thermoelements and their maximum capabilities, *Sov. Phys. Semicond.*, **8**, 478, 1974.
11. L. I. Anatychuk, *Thermoelectric Power Convertors*, Institute of Thermoelectricity, Kiev, pp. 168, 172, 2005.
12. S. L. Korolyuk, I. M. Pilat, A. G. Samoilovich, V. N. Slipchenko, A. A. Snarskii, and E. F. Tsar'kov, Anisotropic thermoelement, *Sov. Phys. Semicond.*, **7**, 502, 1973.
13. W. M. Huber, S. T. Li, A. Ritzer, D. Bauerle, H. Lengfellner, and W. Prettl, Transverse Seebeck effect in $Bi_2Sr_2CaCu_2O_8$, *Appl. Phys. A*, **64**, 487, 1997.
14. K. Fischer, C. Stoiber, A. Kyarad, and H. Lengfellner, Anisotropic thermopower in tilted metallic multilayer structures, *Appl. Phys. A*, **78**, 323, 2004.
15. A. Kyarad and H. Lengfellner, Transverse Peltier effect in tilted Pb—Bi_2Te_3 multilayer structures, *Appl. Phys. Lett.*, **89**, 192103, 2006.
16. T. Kanno, S. Yotsuhashi, A. Sakai, K. Takahashi, and H. Adachi, Enhancement of transverse thermoelectric power factor in tilted Bi/Cu multilayer, *Appl. Phys. Lett.*, **94**, 061917-1, 2009.
17. T. He, J. J. Krajewski, and M. A. Subramanian, U.S. Patent 7371960 B2, 2008.
18. H. J. Goldsmid, Synthetic transverse hermoelements made from porous bismuth telluride and single crystal bismuth, *Phys. Status Solidia*, **205**, 2966, 2008.
19. H. J. Goldsmid, Porous thermoelectric materials, *Materials*, **2**, 903, 2009.

2

Thermoelectric Induction in Power Generation: Prospects and Proposals

L. I. Anatychuk
Institute of Thermoelectricity

2.1 Physical Models of Thermoelectric Power Converters

Description of thermoelectric power conversion is the basis for the physics of thermoelectricity. This description employs physical models that are commonly called "thermoelements."

Correct or appropriate choice of physical models predetermines their accuracy in the description of real processes and characteristics, as well as the efforts necessary to reach the desired results. As is known, the ability of assuming and ignoring in physical models is eventually responsible for the success of research.

The results from the research of physical models also determine the success of practical applications of thermoelectricity. Therefore, studying the prerequisites for the formation of some or other physical models of thermoelements is exceptionally important and interesting.[1]

Traditionally, thermoelectricity employs thermoelements based on the effects of origination of thermoelectromotive forces. In doing so, the thermoelements are given the shape which corresponds to the simplest manifestation of such effects. These are parallelepipeds or cylinders where one-dimensional heat and electricity flows can be arranged.

The thermocouple is the most popular and simplest model in thermoelectricity. However, the simplicity of such a model finally runs into complex technological problems. In the manufacture of modules, thermoelectric material should be first cut into elements causing its large losses, and then connected into series electric circuits. The metal–semiconductor connection is known to be the most vulnerable point of modules that entails not only technological difficulties, but also reliability reduction.

The simplicity of such initial model does not lead to simple solutions in the calculation of power converters either. The electric current flow results in the origination of new effects—the Joule, Peltier, and Thomson—and in the heat release in the contact and connection resistances. In this case, the effect of temperature dependences of the leg material properties does not allow obtaining the analytical expressions for the efficiency, electric power, and electric voltage. The situation is aggravated with regard to lateral heat exchange in the legs and current distribution in connecting elements, which makes the problem non-one-dimensional with all the ensuing complications. Similar problems also arise in the calculation of cooling modules. Such difficulties compel one to address computer methods of approximation calculation.

Hence, it follows that the initial physical model of a thermocouple is not quite good, so it is worthwhile to understand its origin.

Studies on the history of thermoelectricity[2] show that it is related to the name of Öersted, who on opening a circuit of two dissimilar materials in Seebeck's experiment, obtained a thermocouple and demonstrated the origination of thermoelectromotive forces therein. However, neither Seebeck nor Öersted set themselves the task of discovering the ways for the direct conversion of thermal energy into electric energy. They found the thermocouple model accidentally. Seebeck used the short-circuited thermocouple and the resulting magnetic field to prove the reason for the origination of the magnetic field of the Earth. He called his effect thermomagnetism not having recognized its electric nature proved by Öersted. This confirms the truth that thermoelectricity is based on a casual physical model.

At that time (the early nineteenth century), purposeful research was pursued in other fields that led to a discovery of fundamentals for power conversion processes based on the use of sufficiently general physical models. Carnot established thermoelectric cycles for a model of an ideal heat engine which resulted in the creation of physics of thermal into mechanical energy conversion and in the construction of efficient turbines and internal combustion engines. Faraday, using a model of current loop in a magnetic field, discovered the law of electromagnetic induction that gave birth to electric engineering and high-performance converters of mechanical into electric energy.

Meanwhile, in thermoelectricity, a partial thermocouple model has been used for more than 200 years. Finally, it did not contribute to progress in thermoelectricity—in the last 50 years, the efficiency of thermoelectric power conversion has increased only marginally.

2.2 Generalized Thermoelement Model

It is therefore important to consider the construction of a possible general physical model of thermoelectric power converter. Its investigation could give a better chance for extending the capabilities of thermoelectricity. Its investigation would allow describing in the general context the existing thermoelements and create conditions for devising new thermoelement types, extending the capabilities of thermoelectricity.

For this purpose, first came the full classification of thermoelectric effects.[3] Their number amounted to 63. Among them were found nearly 20 new, previously unknown effects. However, they did not result in the discovery of new, efficient thermoelectric power converters. The remaining thermoelements mostly used thermoelectromotive forces arising in thermocouples in the anisotropic thermoelements and in the Nernst–Ettingshausen thermomagnetic element (Figure 2.1).

However, together with the external electric load R, such a combination of thermoelements can be presented in the form of one generalized model which is an inhomogeneous, anisotropic, and gyrotropic medium exposed to physical fields—temperature, electric, magnetic, and force.[4]

In the presence of thermoelectric power conversion, closed currents will flow in the medium that have eddy properties.

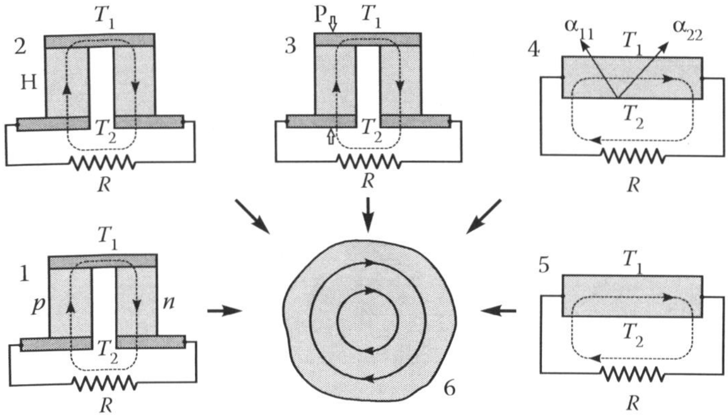

FIGURE 2.1 Generalization of typical models of thermoelements: 1, thermocouple; 2, thermocouple in a magnetic field; 3, thermocouple in a force field; 4, anisotropic thermoelement; 5, Nernst–Ettingshausen thermoelement; and 6, generalized model.

2.3 Examples of Eddy Thermoelectric Currents

Descriptions of such currents are known in thermoelectricity. They had been first excited by Seebeck in his experiments with the short-circuited thermocouples. As noted earlier, in this way, he tried to explain the nature of the magnetic field of the Earth. The followers of this idea can be found even at present. Thus, the possible origination of huge eddy currents on magma inhomogeneities that can form or affect the magnetic field of the Earth is mentioned in Ref. [5] (Figure 2.2).

The magnetic field of eddy currents also finds practical applications. It was established that such currents arise in metallurgical furnaces. Magnetic fields excited thereby give information on the technological processes occurring there (Figure 2.3).

Hirose[6] elaborated a method for the determination of electroplating quality by the EMF arising on the surfaces of bimetallic plates due to eddy currents.

In Ref. [7], this method was modified to determine the conductivity of high-resistance semiconductor epitaxial layers deposited on the low-resistance substrates.[8]

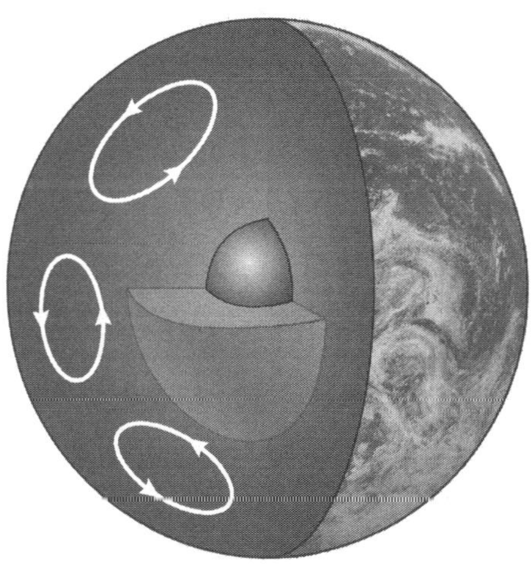

FIGURE 2.2 Eddy thermoelectric currents in the bulk of the Earth.

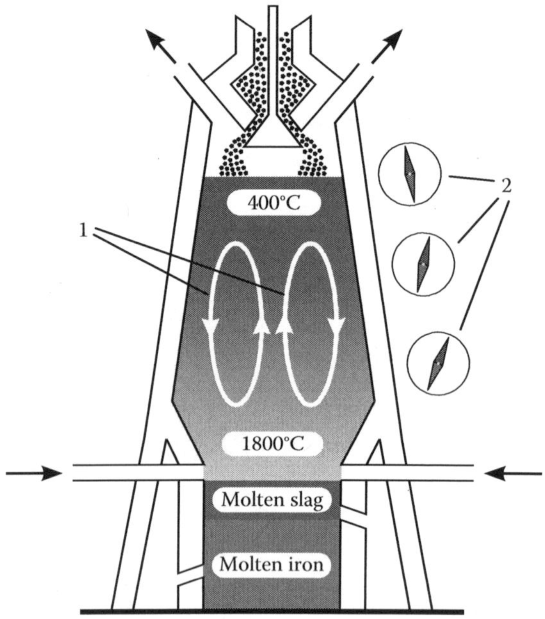

FIGURE 2.3 Eddy thermoelectric currents in a metallurgical furnace. 1, Eddy current lines; 2, magnetic field sensors.

Of decisive importance are eddy currents in the excitation of the transverse electric voltages in the artificially anisotropic-layered thermoelements.[9–14]

However, in the majority of cases, eddy thermoelectric currents are considered as parasitic, reducing the efficiency of power converters much like the Foucault currents in electric power engineering.

For example, Odelevsky[15,16] has demonstrated such efficiency reduction in thermocouple legs in the presence of regular inhomogeneities along the temperature gradient, as well as for inhomogeneities in the form of spherical inclusions. In Ref. [17] is shown the origination of parasitic eddy currents in the homogeneous anisotropic thermoelements in the dynamic modes, and in Ref. [18] in the near-contact areas of the same thermoelements.

However, the author of this work has shown[3] that eddy currents in thermoelectricity, similar to eddy currents of electromagnetic nature in electrodynamics, can be used for the description of thermoelectric power conversion in the most general form.[19]

2.4 Law of Thermoelectric Induction of Currents

Let us consider a more general model of thermoelectric power conversion as compared with that shown in Figure 2.1, making it three dimensional and supplementing with the effect on the conductive medium of the alternating electromagnetic fields. Such a model is shown in Figure 2.4.

A general description of the processes of thermoelectric power conversion can be obtained having generalized Maxwell's equations for the nonisothermal processes[20]

$$\operatorname{div} \vec{D} = 4\pi\delta, \tag{2.1}$$

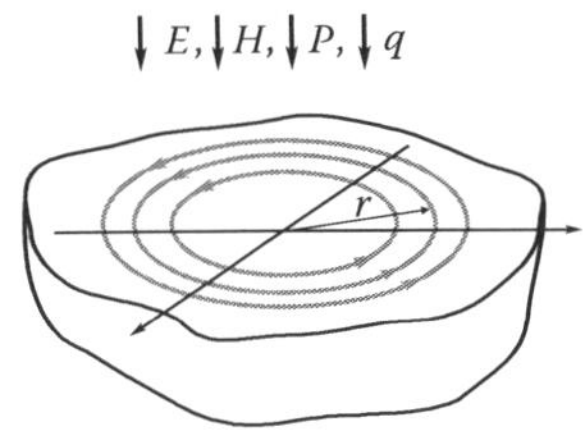

FIGURE 2.4 Generalized model of thermoelectric power conversion.

$$\operatorname{div}\vec{B} = 0, \tag{2.2}$$

$$\operatorname{rot}\vec{E} = -\frac{\partial\vec{B}}{\partial t}, \tag{2.3}$$

$$\operatorname{rot}\vec{H} = \vec{j} + \frac{\partial\vec{D}}{\partial t}, \tag{2.4}$$

using the generalized Ohm's law

$$\vec{j} = \hat{\sigma}\vec{E} - \hat{\alpha}\hat{\sigma}\nabla T, \tag{2.5}$$

where the first summand describes current caused by electric field and the second summand describes thermoelectric current caused by temperature gradient ∇T. Formulae (2.1–2.5) employ common notations used in electrodynamics and thermoelectricity.

Let us consider the induction of currents in the presence of both electromagnetic and temperature fields. For this purpose, we write Equation 2.5 as

$$\hat{\rho}\left(\operatorname{rot}\vec{H} - \frac{\partial\vec{D}}{\partial t}\right) = \vec{E} - \alpha\nabla T \tag{2.6}$$

and affect with operator rot on both sides of this equation. Then

$$\operatorname{rot}\hat{\rho}\left(\operatorname{rot}\vec{H} - \frac{\partial\vec{D}}{\partial t}\right) = \frac{\partial\vec{B}}{\partial t} - \operatorname{rot}\hat{\alpha}\nabla T. \tag{2.7}$$

From Equation 2.7 it follows that

$$\operatorname{rot}\hat{\rho}\,\vec{j}_{\text{total}} = -\frac{\partial\vec{B}}{\partial t} - \operatorname{rot}\hat{\alpha}\nabla T, \tag{2.8}$$

where $\vec{j}_{\text{total}}$ is full density of conduction and displacement currents.

Equation 2.8 is a generalization of Faraday's law of electromagnetic induction for the case of nonisothermal medium. The first term on the right-hand side of the equation describes the induction of eddy currents of electromagnetic nature, and the second term describes the induction of eddy thermoelectric currents. In the integral form, the generalized Faraday's law becomes

$$\varepsilon = \frac{\partial \Phi_C}{\partial t} - \int_C \hat{\alpha} \nabla T dl, \tag{2.9}$$

where C is an arbitrary closed loop in a conducting medium, ε is EMF induced in the loop, and Φ is magnetic flux through the loop.

In the absence of alternating electromagnetic field $(\partial \vec{B}/\partial t) = 0$ (2.9) assumes the form

$$\text{rot}\hat{\rho}\vec{j}_{\text{total}} = -\text{rot}\hat{\alpha}\nabla T. \tag{2.10}$$

Expression (2.10) is the law of thermoelectric induction of currents. Its application allows describing the effects of thermoelectric power conversion from more general positions as compared with classical ones based on the use of conventional thermoelement models.

The law of thermoelectric induction (2.10) implies the most general condition of thermoelectric power conversion in the form

$$\text{rot}\hat{\alpha}\nabla T \neq 0. \tag{2.11}$$

Let us analyze condition (2.11) in more detail, taking into account that under conditions of large temperature gradients the Seebeck tensor can be a function of not only spatial coordinates, temperature, but also temperature gradient and a magnetic field **B**

$$\hat{\alpha} = \hat{\alpha}(x_i, T(x_i), \nabla_i T, \mathbf{B}). \tag{2.12}$$

Here, index i numbers spatial coordinates. Then, condition (2.11) assumes the form

$$\begin{aligned}\text{rot}_i(\hat{\alpha}(x_i, T(x_i), \nabla_i T)\nabla T) &= e_{ikl}\frac{\partial \alpha_{lm}(x_m, T(x_m), \nabla_m T, \mathbf{B})}{\partial x_k}\frac{\partial T}{\partial x_m} \\ &= e_{ikl}\left[\left.\frac{\partial \alpha_{lm}(\mathbf{B})}{\partial x_k}\right|_{T=\text{const}} + \left.\frac{\partial \alpha_{lm}(\mathbf{B})}{\partial T}\right|_{x_k=\text{const}}\frac{\partial T}{\partial x_k}\right]\frac{\partial T}{\partial x_m} \neq 0,\end{aligned} \tag{2.13}$$

where summation is done for doubly recurrent indexes.

From condition (2.13) it follows that for thermoelectric power conversion one should use media (materials) that possess one of the following properties:

1. Anisotropy, including artificial, determined by tensor $\hat{\alpha}(x_i, T(x_i), \nabla_i T, \mathbf{B})$
2. Inhomogeneity, determined by the first summand $((\partial \alpha_{lm}(\mathbf{B}))/\partial x_k)\,|_{T=\text{const}}$ in square brackets
3. Inhomogeneity of properties on exposure to a large temperature gradient determined by square in ∇T expression

$$\left.\frac{\partial \alpha_{lm}(\mathbf{B})}{\partial T}\right|_{x_k=\text{const}}\frac{\partial T}{\partial x_k}\frac{\partial T}{\partial x_m}$$

4. Gyrotropy, created by magnetic field **B**

Some restrictions are also imposed on the temperature fields:

1. For the inhomogeneous media the direction of change in the inhomogeneity shall not coincide with the direction of temperature gradient.

2. For the anisotropic media the crystallographic axes or singled-out directions of anisotropy in materials with artificial anisotropy shall not coincide with the direction of temperature gradient.
3. For the gyrotropic medium the vector of magnetic field induction shall not coincide with temperature gradient.

A combination of these conditions allows classification of all media (materials) and physical fields leading to thermoelectric power conversion. In this case, the difference from zero of electric current density serves a peculiar indicator of the presence of thermoelectric power conversion. The result of this classification is shown in Table 2.1.

From Table 2.1, it follows that 4 out of 128 variants of such conditions cannot lead to thermoelectric power conversion. This is also helpful information which warns that in these cases, seeking new thermoelement types makes no sense. Despite this, scientific papers[21] and even a patent[22] have been found that describe such power conversions. A more thorough study proved the invalidity of these works.

Of the remaining 124 variants, only 3 are used in practice—A3, A7, and B1. They include typical thermoelements given in Figure 2.1. There are still 15 variants under study. The remaining 109 require investigation. They are blank spots on the map of thermoelectricity. And these unstudied variants possess the greatest functional opportunities, so one should expect from them, accordingly, the most interesting results in terms of practical applications.

Thus, from Table 2.1, it follows that thermoelectricity is still at the initial stages of its development. The table allows us to realize a state-of-the-art systemic analysis in thermoelectricity and to choose the most attractive lines of research corresponding to specific tasks of practical applications of thermoelectricity. On the whole, the table for thermoelectric power conversions resembles Mendeleyev's table in chemistry, where not only known variants of thermoelectric conversions are systematized, but also those that have to be studied.

Naturally, this raises the question of how to find from each table box at least one thermoelement type. The procedure for construction of power converters based on current induction is given in Figure 2.5.

Electrically conducting medium A possesses properties that correspond to one of the boxes in the table, and it is exposed to fields corresponding to this box. As a result of power conversion, closed eddy

TABLE 2.1 Thermoelectric Media and External Influences Converting Thermal Energy into Electrical Energy

rot $\mathbf{j} \neq \mathbf{0}$

External influence		Medium properties															
		1	2	3	4	5	6	7	8	9	10	11	12	13	14	15	16
		σ	$\sigma(x_i)$	σ	$\sigma(x_i)$	σ_{ik}	σ	σ_{ik}	$\sigma(x_i)$	$\sigma(x_i)$	$\sigma(x_i)$	σ_{ik}	σ	$\sigma(x_i)$	$\sigma_{ik}(x_i)$	$\sigma(x_i)$	$\sigma_{ik}(x_i)$
		α	α	$\alpha(x_i)$	$\alpha(x_i)$	α	α_{ik}	α_{ik}	α	α_{ik}	α_{ik}	$\alpha(x_i)$	$\alpha_{ik}(x_i)$	$\alpha_{ik}(x_i)$	$\alpha_{ik}(x_i)$	$\alpha_{ik}(x_i)$	$\alpha_{ik}(x_i)$
A	ΔT	0	0	■	▨	0	■	▨	0	░	░	■	░	░	░	░	░
B	$\Delta T, H$	▨	■	■	■	░	░	■	■	■	■	░	░	░	░	░	░
C	$\Delta T, P$	■	■	░	░	░	░	■	░	░	░	░	░	░	░	░	░
D	$\Delta T, E$	░	░	░	░	░	░	░	░	░	░	░	░	░	░	░	░
E	$\Delta T, H, P$	■	░	░	░	░	░	░	░	░	░	░	░	░	░	░	░
F	$\Delta T, H, E$	░	░	░	░	░	░	░	░	░	░	░	░	░	░	░	░
G	$\Delta T, H, P, E$	░	░	░	░	░	░	░	░	░	░	░	░	░	░	░	░
H	$\nabla' T$	■	░	░	░	░	░	░	░	░	░	░	░	░	░	░	░

Notations: 0, conversion is impossible; ▨, conversion is used; ■, conversion is under investigation; ░, conversion is not investigated; ΔT is temperature field, H is magnetic field, P is force field, $\nabla' T$ is high temperature gradient, σ, α are homogeneous and isotropic electrical conductivity and Seebeck coefficient, σ_{ik}, α_{ik} are homogeneous and anisotropic electrical conductivity and Seebeck coeffiicient, $\sigma(x_i)$, $\alpha(x_i)$ are inhomogeneous and isotropic electrical conductivity and Seebeck coefficient, $\sigma_{ik}(x_i)$, $\alpha_{ik}(x_i)$ are inhomogeneous and anisotropic electrical conductivity and Seebeck coefficient.

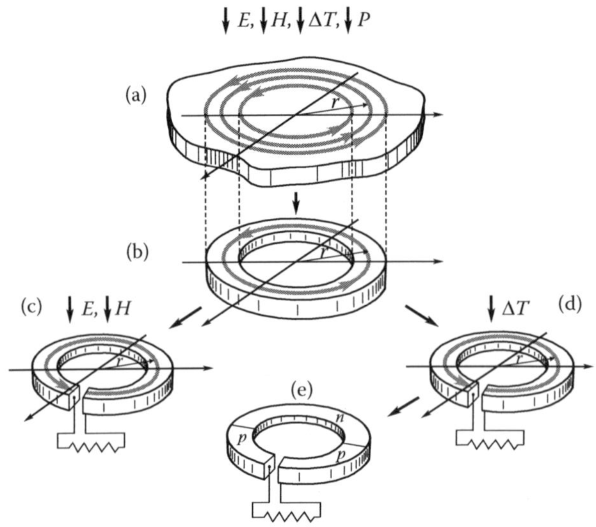

FIGURE 2.5 Method for creation of power converters based on eddy current induction. (a) Electrically conducting medium, (b) current loop, (c) power converter of electromagnetic nature, (d) thermoelectric power converter, and (e) partial case of thermocouple model.

currents are created in this medium, namely, the Foucault currents from the effect of electromagnetic field **E, H** and thermoelectric currents from the effect of temperature field ΔT. Let us mentally make cuts along current lines. Such cuts do not result in the distortion of current distributions. As a result, current loop *B* will be created. Cutting the loop along the surfaces of equal electric potential and connecting to these surfaces of the external electric load will form power converters: *C*—of electromagnetic and *D*—of thermoelectric nature. As a partial case, a loop with a pronounced inhomogeneity will bring about a model of thermocouple *E*.

Such a method demonstrates the possibility of creating thermoelements using not the traditional way based on thermoelectric effects, but a new, more general method based on the application of current induction.

Design of thermoelements on the basis of thermoelectric effects is a direct problem of thermoelectricity, where medium parameters, medium geometry, and the boundary temperature conditions are known, and temperature, current, and the electric potential distributions are sought.

The properties of thermoelements based on the induction of thermoelectric currents are found from the inverse problems of thermoelectricity, where medium parameters and current configuration are given, and temperature distribution and thermoelement configuration are sought.

2.5 Control of Spatial Distribution of Eddy Currents: Inverse Problems of Thermoelectricity

In the general case, the inverse problem of thermoelectricity is formulated as follows. In the space is assigned a singly connected area *D* filled with conductive medium, which is characterized by tensors of electric conductivity $\hat{\rho}(\mathbf{r})$, thermoEMF $\hat{\alpha}(\mathbf{r})$ that are certainly dependent on radius vector **r**. In this area, *D* a vector field of current density is assigned as a continuous function $\mathbf{f}(\mathbf{r})$

$$\mathbf{j}(\mathbf{r}) = \mathbf{f}(\mathbf{r}). \tag{2.14}$$

Thus, the problem lies in constructing a scalar field of temperatures $T(\mathbf{r})$ or vector field of temperature gradient $\nabla T(\mathbf{r})$, which necessarily leads to creation of the assigned distribution of currents (2.7) in the area *D*.

It was shown that for vector field $\hat{\alpha}(\mathbf{r})\nabla T$ two differential values are uniquely determined, namely, this field rotor and its divergence.[23,24] The well-known Helmholtz theory asserts that the problem of devising this unknown vector field is correct, provided its rotor and divergence are known. In this way, correctness of formulation of the inverse problem of thermoelectricity was proved, as well as the fact that its solution exists, if the predetermined distribution of currents satisfies the law of conservation of charge.

It was also proved that solution of the inverse problem of thermoelectricity is unique to an accuracy of arbitrary passive temperature distribution $T_p(\mathbf{r})$, incapable of exciting eddy thermoelectric currents.

In the general case of a medium with an arbitrary inhomogeneity and anisotropy of its parameters $\hat{\rho}(\mathbf{r})$, $\hat{\alpha}(\mathbf{r})$ the inverse problem of thermoelectricity lies in finding a solution of vector differential equation in second-order partial derivatives for unknown temperature distribution

$$\nabla \times \hat{\alpha}\nabla T = -\nabla \times (\hat{\rho}\mathbf{f}(\mathbf{r})) \tag{2.15}$$

According to the Helmholtz theorem, the general solution has the form

$$\nabla T(\mathbf{r}) = \hat{\alpha}\left\{\nabla\left(-\nabla\iiint_V \frac{(\hat{\rho}\mathbf{f}(\mathbf{r})) - 4\pi\delta(\mathbf{r})}{|\mathbf{r}-\mathbf{r}_m|}\,\mathrm{d}V_m\right) + \nabla\times\iiint_V \frac{\nabla\times(\hat{\rho}\mathbf{f}(\mathbf{r}))}{|\mathbf{r}-\mathbf{r}_m|}\,\mathrm{d}V_m\right] \tag{2.16}$$

To begin with, it was natural to formulate such an inverse problem which would result in finding conditions for the excitation of round-shaped eddy currents similar to currents in Faraday's experiments. The infinite area D in this case can be restricted to flat, circular-shaped samples singled out from the area by current lines.

For the anisotropic medium, the temperature distribution leading to circular-shaped eddy currents is given by the expression

$$T(x_1,x_2) = T_0 + \frac{\Delta T}{R_0^2}\left[x_1x_2\cos 2\varphi + \frac{2\kappa_{22}x_2^2 - 2\kappa_{11}x_1^2 + R_0^2(\kappa_{22}-\kappa_{11})}{2(\kappa_{22}+\kappa_{11})}\right], \tag{2.17}$$

where x_1, x_1 are axes of the Cartesian coordinates; R_0 is sample radius; φ is angle between the crystallo graphic axis and the line connecting sample center to a point where the temperature on the external loop of the sample is maximum; ΔT is maximum temperature difference; and σ_{11} and σ_{22}, α_{11} and α_{22}, κ_{11} and κ_{22} are components of electric conductivity, thermoEMF, thermal conductivity tensors, respectively (Figure 2.6a). In this case, current density in the sample is as follows:

$$j_\varphi = \frac{\sigma_{11}\sigma_{22}}{R_0^2(\sigma_{11}+\sigma_{22})}\Delta T(\sigma_{11}-\sigma_{22})2\cos 2\varphi. \tag{2.18}$$

For the inhomogeneous medium with typical semiconductor properties

$$\alpha(x_1) = B + D\frac{x_1}{G}, \quad \sigma = A\exp\left(\frac{x_1}{G}\right), \tag{2.19}$$

the sought-for temperature field is given by the expression

$$T(x_1,x_2) = T_0 - \frac{CG}{DA}x_2\exp\left(\frac{x_1}{L}\right)\left[\frac{x_1}{G}+2\right], \tag{2.20}$$

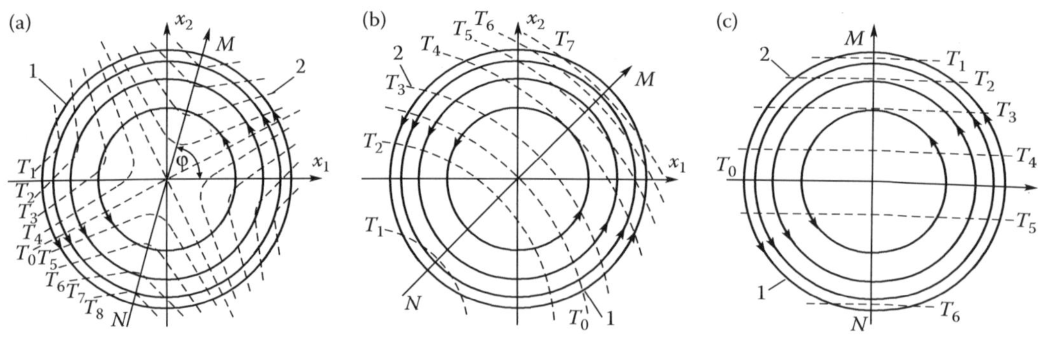

FIGURE 2.6 Solutions of the inverse problem of thermoelectricity for the circular eddy thermoelectric currents. (a) Homogeneous anisotropic medium. (b) Inhomogeneous medium. (c) Gyrotropic medium. 1, Current lines; 2, isotherms.

where *A*, *B*, *C*, *D*, *G* are constants (Figure 2.6b). The integral value of eddy current is equal to

$$I = \frac{\Delta T}{4} b \frac{DA}{CG} \frac{\exp(-R_0/G)\cos\varphi^*}{\sin\varphi[R_0/G + 2]}, \qquad (2.21)$$

where *b* is sample thickness, and φ^* is root of equation

$$\frac{R_0}{G}\cos 2\varphi^* + \cos 2\varphi^* - \frac{R_0^2}{G^2}\sin^2\varphi^*\cos\varphi^* + 2\frac{R_0}{G}\sin 2\varphi^* = 0. \qquad (2.22)$$

For a gyrotropic medium, the temperature distribution (Figure 2.6c) has the form

$$T(x) = T_0 - \frac{2C}{\sigma_s\alpha_a}x^2, \qquad (2.23)$$

where σ_s is a component of the symmetric part of electric conductivity tensor, α_a is a component of the antisymmetric part of thermoEMF tensor, and *C* is a constant.

The integral eddy current is found from the formula

$$I = 2\sigma_s\alpha_a b\Delta T. \qquad (2.24)$$

2.6 Experiments Confirming Control over the Eddy Thermoelectric Currents: The First Eddy Thermoelement

At first, these theoretical results were verified with the use of the anisotropic medium. As the anisotropic material, Czochralski-grown CdSb single crystals were used[25] with the thermoEMF anisotropy values $\alpha_{11} - \alpha_{22} = 280 \times 10^{-6}$ V/K.

From formula (2.17) and Figure 2.7a it follows that for creation of eddy currents in a sample, a series of heaters and coolers must be embedded on the periphery of the sample, to achieve the required temperature distribution. Realization of these conditions is difficult, so the unusual solution was found which is shown in Figure 2.7b. A plate of material of higher thermal conductivity (aluminum) cut along

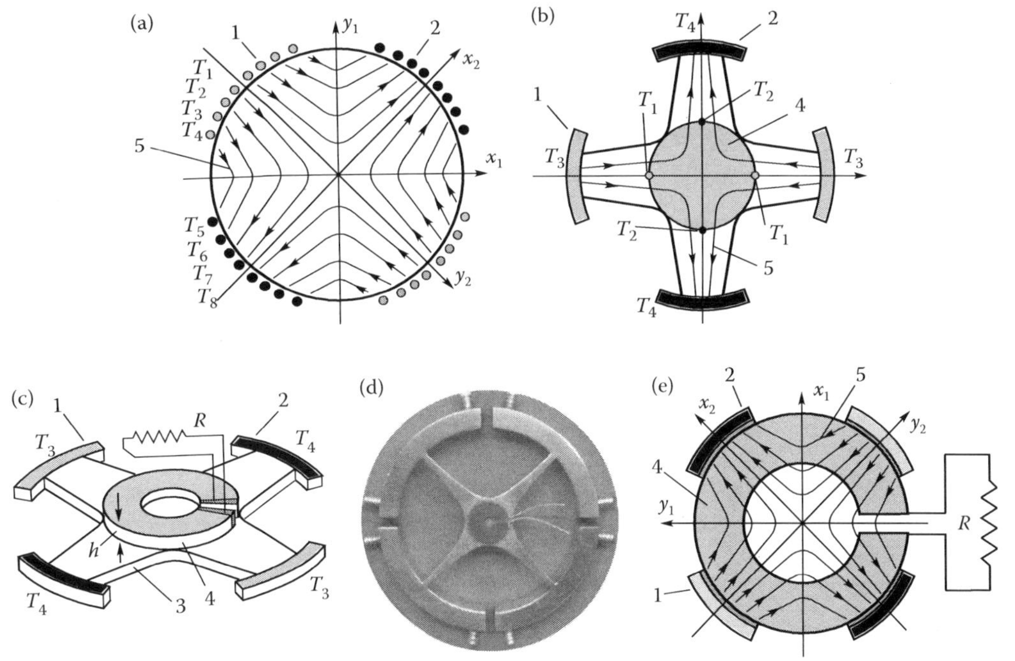

FIGURE 2.7 Excitation of circular eddy currents. (a) Conditions for creation of temperature distribution in a sample. (b) Thermally conducting plate cut along the hyperbolae. (c) Device for excitation of eddy thermoelectric currents. (d) Device appearance. (e) Currents in a sample with two heaters and two coolers. 1, Heaters; 2, coolers; 3, thermally conducting plate; 4, sample of CdSb; and 5, heat flux lines.

the hyperbolae was used. Two heaters and coolers create the required temperature distribution in the central part of the plate. At this point the plate is brought into a thermal contact with the electrically isolated sample under study, on which temperature distribution in the plate is reproduced (Figure 2.7c). The excited current was tapped to the external circuit. Its value to an accuracy of 5–10% coincided with that calculated by formula (2.18).[26]

Similar experiments were conducted on the inhomogeneous and gyrotropic samples.

Eddy currents close to circular are also excited in a simplified model, where a plurality of heat sources and sinks are replaced by two pairs of heaters and coolers (Figure 2.7e).

In fact, such a device represents the first thermoelectric power converter of a new type that is customarily called eddy. It develops the EMF which is given by the expression

$$E = \frac{\pi}{2}\Delta T(\alpha_{11} - \alpha_{22})\frac{1 - b^2/a^2}{\ln a/b}, \tag{2.25}$$

where a and b are the outer and inner radii of thermoelement. The thermoelement efficiency can be represented in a simplified form as follows:

$$\eta = \frac{\pi}{16}\frac{(\alpha_{11} - \alpha_{22})^2(\sigma_{11}\sigma_{22})\bar{T}(T_1 - T_2)}{(\sigma_{11} + \sigma_{22})\kappa T_1}r^*, \tag{2.26}$$

where $\kappa \approx \kappa_{11} \approx \kappa_{22}$ is the averaged thermal conductivity value, $r^* \approx 0.7$.

The advantages of such thermoelements include the absence of connections. An important advantage is also the possibility of creating spiral thermoelements on its basis (Figure 2.8) that correspond in their

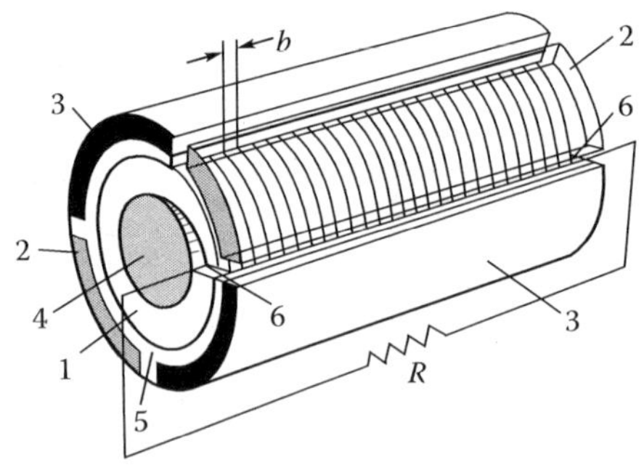

FIGURE 2.8 Thermoelectric power converter with a working medium in the form of a round-shaped anisotropic spiral. 1, Spiral of single-crystal substance; 2, heaters; 3, coolers; 4, filling agent; 5, electric insulation; and 6, electric contacts.

properties to spiral coils in electric power engineering: the larger the number of coils, the higher voltage is developed, and the larger the cross section of coils, the higher current is developed.

Using the method of eddy current excitation, nearly 20 new thermoelement types were created that enriched considerably the element basis of thermoelectricity.[12,25]

Of greatest interest among them are spiral thermoelements which will be considered in more detail.

2.7 Spiral Eddy Thermoelements

2.7.1 Anisotropic

The thermoelement shown in Figure 2.8, while confirming the possibility of creating spiral eddy power converters, nevertheless, comprises two heaters and two coolers, restricting the possibility of using them in practice. A spiral structure comprising one localized heat source (Figure 2.9a) would be more convenient. By solving the inverse problems of thermoelectricity, it was established that the best results

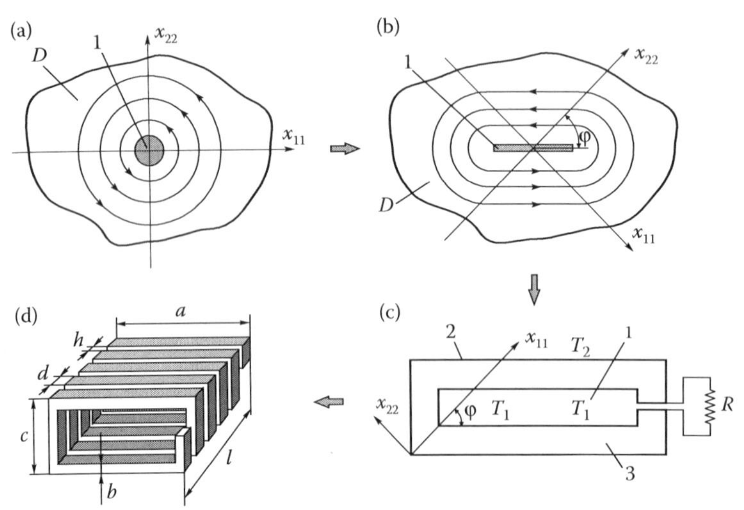

FIGURE 2.9 Stages of creating eddy spiral thermoelement with one heat source. (a) Eddy current excitation in the anisotropic medium D with heat source 1. (b) Eddy currents caused by a linear heater oriented at an angle to the crystallographic axis. (c) Eddy anisotropic thermoelement. 1, Cavity for a heater; 2, external thermostat; and 3, thermoelement working medium. (d) Spiral eddy thermoelement.

should be expected with the use of a linear heater with temperature T_1 (Figure 2.9b), oriented at an angle φ to crystallographic axis x_{22}. Finally, such a thermoelement is a rectangular of material with the anisotropic thermoEMF, having a heat source with temperature T_1 installed in its cavity. It is covered by a thermostat with temperature T_2 (Figure 2.9c).

Precise definition of temperature (Figure 2.10a) and eddy current (Figure 2.10b) distributions was found by computer methods with the use of special object-oriented programs.[27] It was established that in addition to the basic eddy currents, closed parasitic eddy currents are created in thermoelement angles (Figure 2.10c); however, their presence has a minor effect on thermoelement operation.

Based on the rectangular thermoelement (Figure 2.9c), the sought-for spiral structure was created (Figure 2.9d) with a heater inside. The exact parameter values of such power converter are determined by computer methods. However, for calculations with adequate accuracy, one can also use the analytical expressions for the thermoEMF developed by a spiral

$$E = (\alpha_{11} - \alpha_{22})(T_1 - T_2)\frac{a-c}{b}\frac{l}{d+h}\frac{a}{b}, \tag{2.27}$$

and the efficiency

$$\eta = \tfrac{1}{4}Z_a\eta_k\bar{T}\left(\frac{a-c}{b}\right)^2, \tag{2.28}$$

where $Z_a = 1/4((\Delta\alpha^2\sigma_0)/\kappa_0)$.

The formulae were obtained for the angle of $\varphi = 45°$, which corresponds to the condition of reaching maximum EMF, σ_0 and κ_0 are the averaged electric and thermal conductivity values.

Such converters made of CdSb and Bi develop elevated electric voltages. To obtain such voltages by means of conventional thermopiles, tens of thousands of thermocouples have to be connected in series.

Apart from the elevated voltages developed by spiral converters, their advantage lies in the absence of connections and, respectively, contact noises, thus creating a favorable situation for their application in the measuring technique. From the information-energy theory[28,29] it was established that Bi–Sn spiral anisotropic converters are more informative than Bi–Te thermocouple thermopiles.

Anisotropic spiral converters (sensors) were used to create microcalorimeters with their information capability considerably higher compared to known analogs.[29] Investigations for the identification of microorganisms and the effect of antibiotics on their vital functions were conducted on such devices.[30]

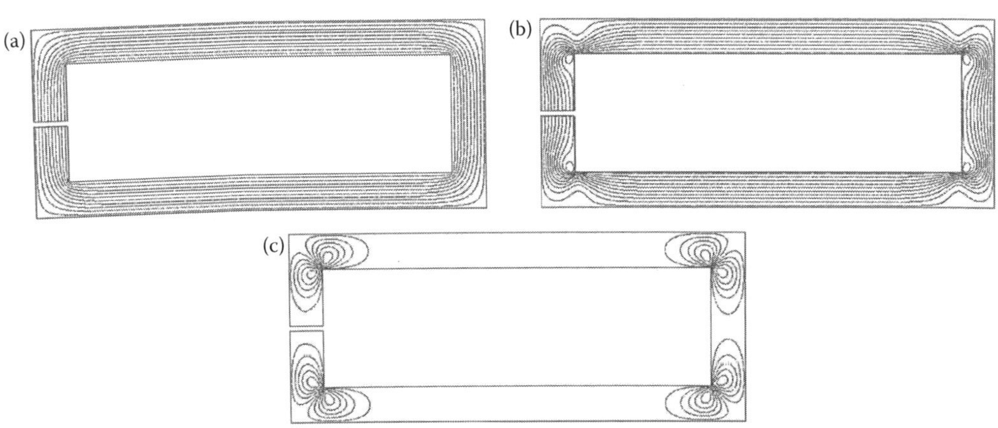

FIGURE 2.10 Distribution of temperatures (a) and eddy current (b, c) in a rectangular eddy thermoelement.

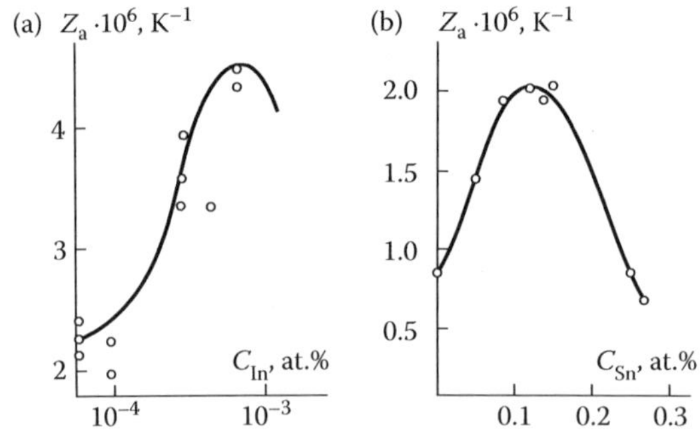

FIGURE 2.11 Optimization of Z in materials. (a) CdSb doped with In impurity. (b) Bi doped with Sn impurity.

Such information becomes increasingly relevant due to the effect of genetic engineering on the life of microorganisms. There are other successful applications of spiral sensors in laser metrology, alternating current metrology, and so on.[31,32]

Physical methods for increasing the figure of merit of materials Z_a for the anisotropic power converters are basically different from those developed for thermocouples and primarily depend on the mechanisms of origination of thermoEMF anisotropy.[33–37] Let us consider these mechanisms.

ThermoEMF anisotropy $\Delta\alpha$ is created in semiconductors in the presence of several scattering mechanisms and a sort of current carriers. However, according to estimates,[35,36] $\Delta\alpha$ in this case is low.

Low $\Delta\alpha$ anisotropy is also created due to deformation.[38]

The electron–phonon drag can cause considerable $\Delta\alpha$. However, the thermal conductivity in this case is drastically increased.[35,36]

The most efficient mechanism is the one whereby electric conductivity is provided by several sorts of current carriers with one anisotropic scattering mechanism. Optimization theory for such materials is presented in Refs. [35,36]. It implies that to reach maximum Z_a, one should choose materials of minimal thermal conductivity and introduce one type of impurity: the donor type, if the hole mobility is larger than the electron mobility; and the acceptor type, if the electron mobility is larger than the hole mobility. The results of such optimization for CdSb and Bi are given in Figure 2.11. It is obvious that Z_a is almost twice improved.

Further Z_a improvement can be achieved by suppressing phonon thermal conductivity through the introduction of isovalent impurities resulting in more complicated material structures: ZnSb–CdSb and Bi–Sb.

For power applications such converters are yet ineffective because of their low-efficiency values. More promising for such purposes can be spiral converters based on the use of eddy currents in the inhomogeneous media.

2.7.2 Inhomogeneous

Current distribution in a zone-inhomogeneous medium, obtained by the object-oriented computer method, is shown in Figure 2.12a. For tapping the eddy current to the external circuit, a central hole should be provided in the middle. Current distribution in this case has a form shown in Figure 2.12b. Spiral structure of such a loop is shown in Figure 2.12c.

For better reliability, isothermal conducting plates 3 are embedded on a semiconductor spiral 1 (Figure 2.13) with metal wire spiral 2 attached to them. On deviation of semiconductor loop integrity, current flows through a redundant metal loop. In doing so, the reliability factor MTBF increases by a factor of 10^3–10^5.

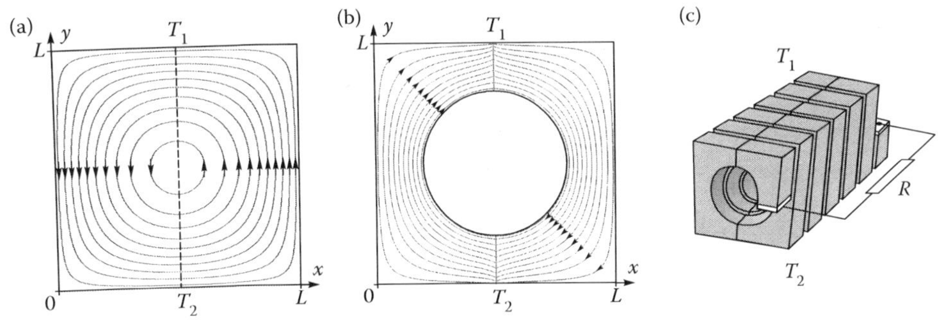

FIGURE 2.12 Zone-inhomogeneous spiral structure. (a) Current distribution in a zone-inhomogeneous medium. (b) Current distribution in the presence of a central hole. (c) Spiral inhomogeneous thermoelement.

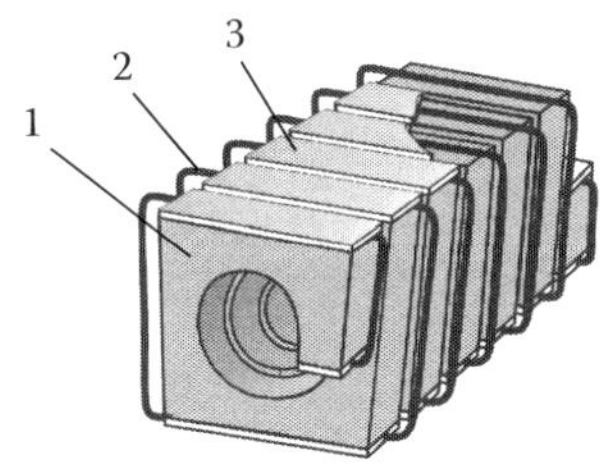

FIGURE 2.13 Inhomogeneous thermoelectric converter with a double spiral.

For efficiency increase optimal inhomogeneity distributions in the plane of each spiral loop were found (Figure 2.14).

Such materials were obtained by the extrusion method (Figure 2.15). The initial cylinder billet 1 of Bi–Te comprises four materials n_1, n_2, p_1, and p_2 with different impurity concentrations. Following extrusion in container 2, rod 3, 4 is cut by multiwire instrument to the middle 5, and then again cut 6 to join with the previous cuts 7. As a result, the required spiral 8 is formed.

Test results have confirmed that optimally inhomogeneous spirals realize the efficiency 5–10% higher than thermocouple thermopiles made of the same materials.

Thus, the obtained spiral thermoelectric power converters are more adaptable to manufacture as compared with conventional thermocouple modules and are practically waste-free as regards the thermoelectric material employed, offer higher efficiency, and are much more reliable.

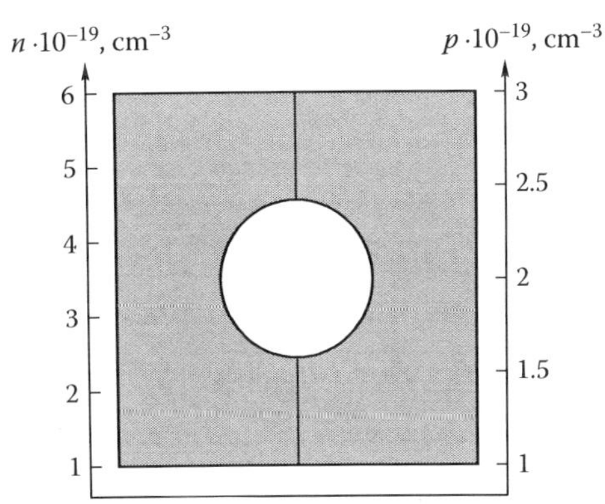

FIGURE 2.14 Optimal inhomogeneity distribution in the plane of Bi–Te spiral.

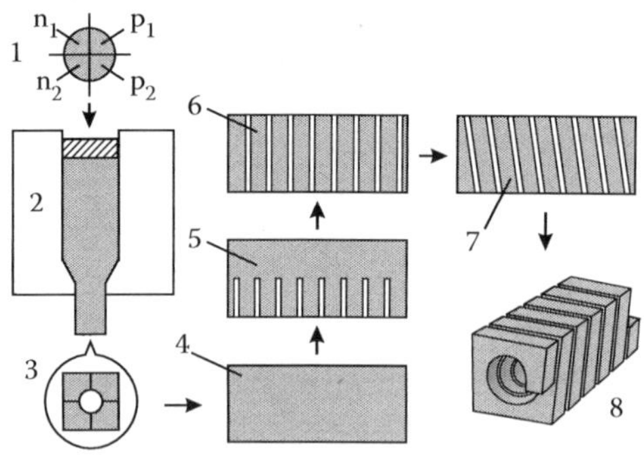

FIGURE 2.15 Schematic of obtaining a spiral inhomogeneous power converter.

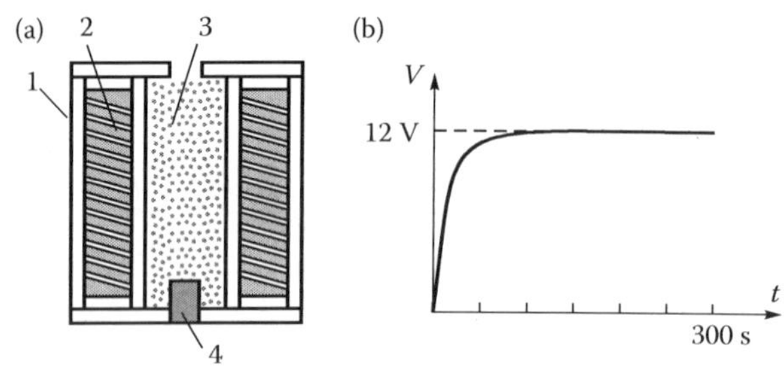

FIGURE 2.16 Shockproof generator of spiral inhomogeneous power converters. (a) Construction. (b) Dynamics of achieving the operating characteristics. 1, Casing; 2, spiral converters; 3, pyrotechnic material; and 4, shock device for generator actuation.

An important advantage of such converters is their shock resistance. Spiral structures offer increased elasticity, are junction free, and hence withstand shock loads of a factor of 5–10 higher as compared to thermocouple thermopiles. This allows their use in generators that are exposed to shock loads.

Figure 2.16 shows a variant of such a generator designed for power supply to electronic control systems in the flight path. The generator shock resistance is 20,000–80,000 *g*.

2.7.3 Gyrotropic

When creating such spiral structures, it is easy to realize a condition of heat source arrangement inside a spiral and heat sink on its external surface (Figure 2.17). Like before, the exact parameters of such power converters are determined from the distributions of temperatures, heat fluxes, electric potentials, and

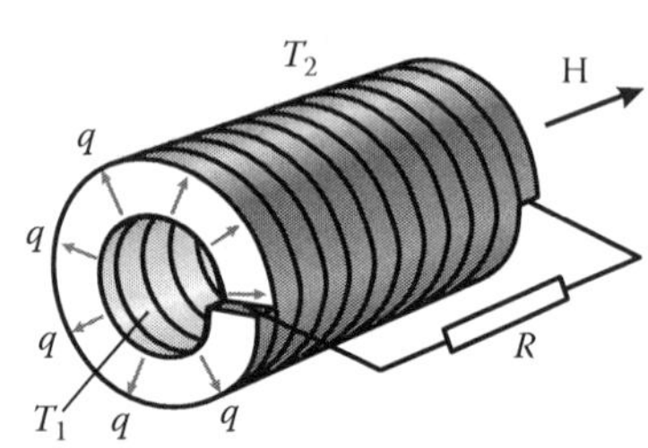

FIGURE 2.17 Schematic of spiral thermoelement in a magnetic field with a radial heat flux q.

electric currents in a spiral that are found from the object-oriented computer solutions. Approximately, spiral properties are determined by the formulae:

Developed EMF

$$E = 2\pi n Q_{\perp} B \Delta T \frac{1}{\ln(r_2/r_1)}, \tag{2.29}$$

Heat flux through the spiral

$$Q = 2\pi n h \kappa \Delta T \frac{1}{\ln(r_2/r_1)}, \tag{2.30}$$

Efficiency

$$\eta_{\max} = \frac{(Q_{\perp}B)^2}{\rho_0 \kappa}\frac{\Delta T}{4} = Z_H \frac{\Delta T}{4}, \tag{2.31}$$

where n is the number of spiral loops, $Q_{\perp}$ is the Nernst–Ettingshausen coefficient, $\Delta T = T_1 - T_2$, κ is the thermal conductivity, Z_H is the figure of merit of thermoelectric material, r_1 and r_2 are the inner and outer spiral radii, h is the spiral loop thickness, and ρ_0 is the electric resistivity.

Requirements to materials were determined whereby maximum Z_H is achieved for conduction band and valence band of different complexity.[39,40] It was established that

- For substances with two sorts of current carriers, Z_H reaches maximum in the region of intrinsic conductivity on equality of the electron and hole mobilities.
- If the electron and hole mobilities are not equal, Z_H maximum is achieved in the region of mixed conductivity on doping by acceptor impurities, if the electron mobility is higher than the hole mobility, and by donor impurities, if the electron mobility is lower than the hole mobility.

The figure of merit Z_H reaches maximum under a strong magnetic field $uB^2 > 1$. This condition is rationally realized in the presence of a large mobility u of current carriers.

To achieve a large electric conductivity, one should use materials with a large effective mass of current carriers. Combined with the large mobility, this condition can be realized in the multivalley anisotropic semiconductors.

As with the other thermoelectric power converters, one should choose materials with a minimum lattice conductivity which is typical of substances with large atomic masses.

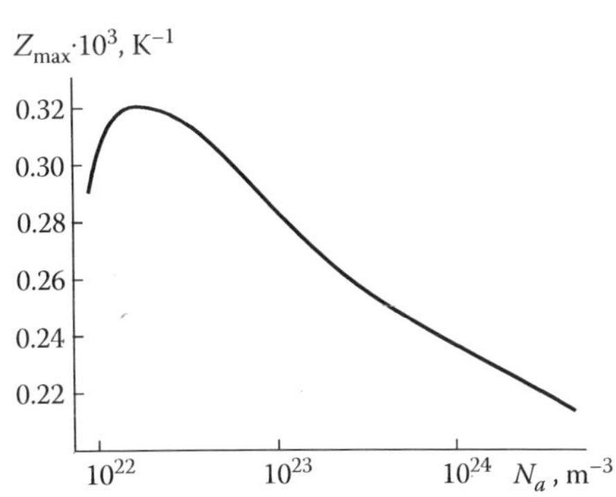

FIGURE 2.18 Dependence of maximum thermomagnetic figure of merit on the acceptor concentration.

At present, the most acceptable material for this purpose is InSb. Optimization theory for this material was developed, and an sufficiently exact one, since the scope of its current carriers is well studied. Doping conditions were determined. Figure 2.18 shows the results of InSb optimization by doping with acceptor impurities.

The use of spiral thermoelements in a magnetic field is advisable when creating multifunctional sensors with thermoelectric signal modulation.

2.8 Progress in the Artificially Anisotropic Thermoelements

In Section 2.6, it is shown that crystals have very low values of the anisotropic figure of merit Z_a. This gave impetus to seeking the ways for its improvement. One of them is the artificially anisotropic materials in the form of periodic zone-inhomogeneous structures (Figure 2.19) proposed by Geiling as far back as 1950.[9,10] The theory of thermoelements of such materials has been thoroughly studied.[11–14] The promising character of such thermoelements is emphasized by H. J. Goldsmid in Section 1.1, Chapter 1 of this book.

It was established that with the use of semiconductor thermoelectric material layers and shorting metal layers in such media, the efficiency of such artificially anisotropic thermoelements can outperform considerably the analogs made of the anisotropic crystalline materials.

However, despite their good theoretical substantiation, such thermoelements have not found practical application up to now. It is attributed to technological difficulties. The attempts of creating such structures by evaporation or consecutive electroplating eventually have not met with success.

Therefore, it is worthwhile to gain a better insight into the reasons for using such thermoelement models. Why exactly this thermoelement model was selected by Geiling is not clear from his papers. Most probably, this two-layer medium comprising a plurality of thin layers was mentally conceived such that it possesses the integral anisotropic properties along and perpendicular to the layers.

To describe the properties of such thermoelements, one generally employs the formalism of zone-inhomogeneous medium representation by the homogeneous anisotropic one. However, this approach does not reveal the essence of such thermoelement operation. In fact, as was shown in Refs. [13,14,41–43], the transverse thermoEMF in this thermoelement is not due to properties of the homogeneous anisotropic medium. It results from the combination of the transverse electric voltage components arising at excitation in the pairs of inhomogeneous zones of periodic eddy currents flowing at an oblique angle to thermoelement surface (Figure 2.19). This situation is vividly demonstrated by the electric potential distributions in thermoelements obtained by computer calculations (Figure 2.20).

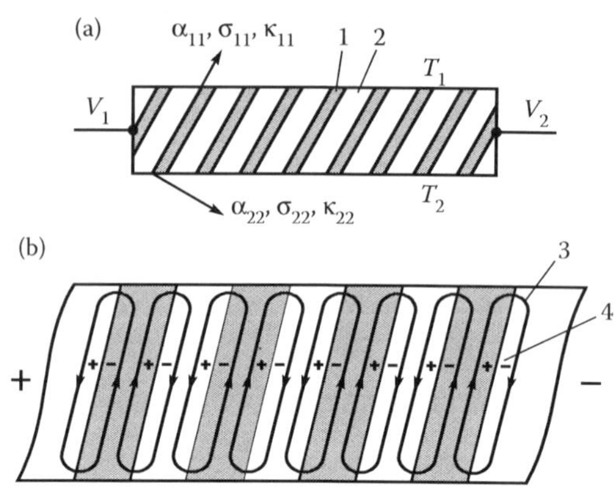

FIGURE 2.19 Layered artificially anisotropic thermoelement (a) and the distribution of eddy thermoelectric currents therein (b). 1,2 Layers of different materials; 3, eddy currents; and 4, transverse EMF due to the effect of currents.

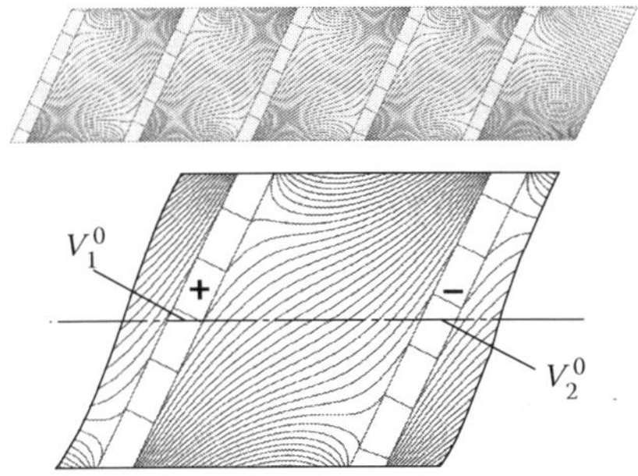

FIGURE 2.20 Example of the electric potential distribution in a zone-inhomogeneous thermoelement.

Formalism of substitution of the zone-inhomogeneous medium with eddy currents by the anisotropic medium without eddy currents can cause serious mistakes in the calculation of thermoelement parameters. Figure 2.20 shows an example of computer calculation of current distribution in a zone-inhomogeneous thermoelement, and Figure 2.21 shows the values of relative errors $E_{\perp}^{\text{theor}}/E_{\perp}^{\text{comp}}$ versus thermoelement height b and zone thickness δ_1, δ_2, where $E_{\perp}^{\text{theor}}$ are the results of calculation of the transverse EMF for the formalized anisotropic model, and $E_{\perp}^{\text{comp}}$ is the transverse EMF obtained from the model of a zone-inhomogeneous medium. It can be seen that under certain conditions, the value of error can reach 20–300%.

Thus, the key factor in the formation of the transverse thermoEMF in the artificially anisotropic zone-inhomogeneous thermoelements is a short-circuited electric current flowing at an angle to thermoelement surfaces. This opens up the prospects of seeking for other, more technologically feasible thermoelement types.

Figure 2.22 shows a model of such thermoelements consisting of a semiconductor bar with metal stripes deposited on it and arranged at an angle to its surfaces. The difference in temperatures between the upper and lower bar surfaces leads to the formation of closed currents flowing at an angle to bar surfaces, similar to currents in the Geiling thermoelement. The best results can be obtained when shorting stripes are arranged along the two lateral surfaces of the bar. The exact distributions of currents and electric potential for such thermoelement models were obtained by computer methods in Refs. [41–43]. The fabrication technique of such artificially anisotropic thermoelements is exceptionally simple and feasible.

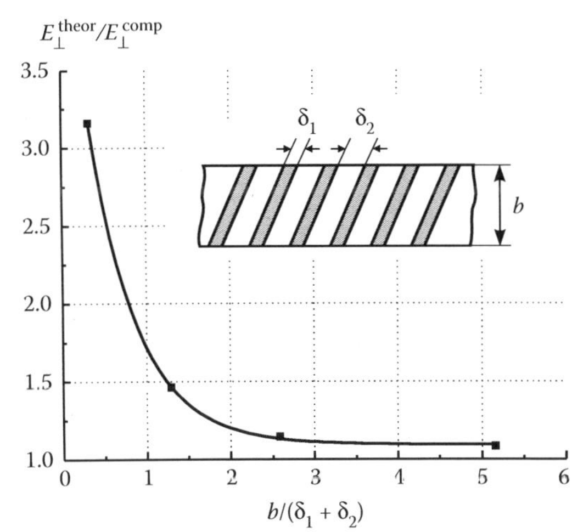

FIGURE 2.21 Errors of a formalized model of a zone-inhomogeneous thermoelement.

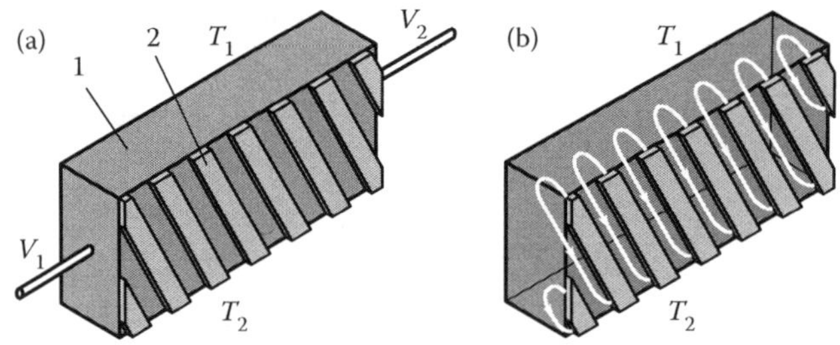

FIGURE 2.22 Zone-inhomogeneous thermoelement with a lateral arrangement of shorting stripes. (a) Thermoelement. (b) Distribution of eddy currents. 1, Semiconductor bar; 2, metal shorting stripes.

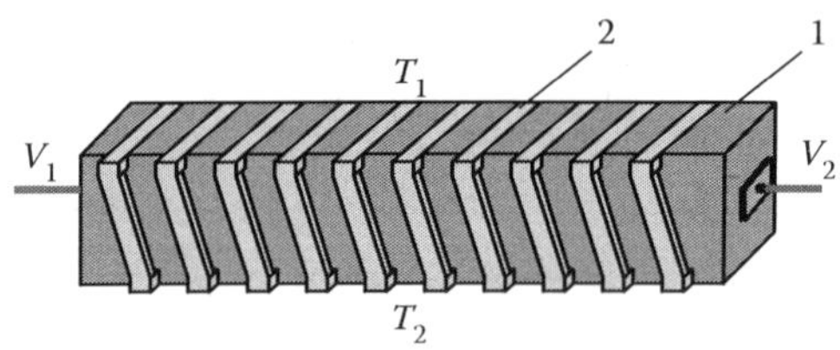

FIGURE 2.23 Short-circuited thermoelement. 1, Semiconductor bar; 2, shorting conductors.

Nevertheless, the distributions of eddy currents in such thermoelements have a complex configuration which finally leads to low thermoelement efficiency values. Closed current, which is more regular and approaching the one-dimensional distribution, is formed in the thermoelement model given in Figure 2.23. It comprises a bar of thermoelectric material and shorting conductors. The theory of this thermoelement is developed in Refs. [41–43].

Such thermoelements can employ conventional homogeneous thermoelectric materials used in the manufacture of thermoelectric modules. The efficiency of such thermoelements in the mode of transverse EMF generation reaches 10–15% from the efficiency of modules which is a factor of 10 higher than in the anisotropic thermoelements of crystalline materials.

However, the best efficiency results were obtained with the use of two mutually closing bars with connecting conductors along the cold and hot surfaces (Figure 2.24). Such a construction eliminates heat losses through closing passive conductors. As compared to the previous variant, the efficiency of such thermoelements increases by a factor of 2–2.5 on retention of adaptability to manufacture and increased

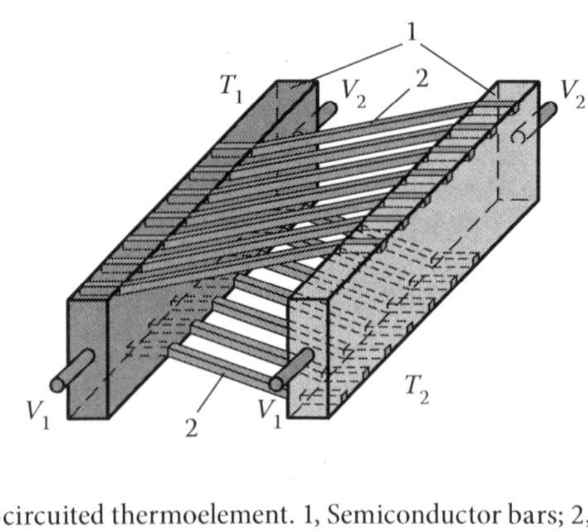

FIGURE 2.24 Double-short-circuited thermoelement. 1, Semiconductor bars; 2, shorting conductors.

reliability. Such converters expand the element basis of thermoelectricity and, accordingly, increase the potential of practical applications.

2.9 Some Consequences and Outlook

The previous sections demonstrate that the approach to constructing the models of thermoelectric power converters based on the use of eddy nature of thermoelectric current is rather efficient. Conventional thermoelements based on a direct realization of thermoelectric effects follow from this approach as particular cases. However, this approach furnishes the opportunities of creating numerous new thermoelement types, not evident in terms of thermoelectric effects.

The method for devising new thermoelement types based on the eddy current excitation and tapping to the external circuit can be further improved, for instance, by including the thermoelement electric load to the model.

Thermoelectric power conversion based on the use of large temperature gradients is represented in Table 2.1 without regard to cumulative effect of other fields. However, a general trend to microminiaturization in electronics engineering will inevitably generate a need to consider these cases as well.

Thermoelectric material science, rather well developed for a particular case of inhomogeneous medium with a sharp inhomogeneity (thermocouple), has contributed little so far to providing other thermoelement types with the optimal materials. It seems that pleasant surprises are in store on this way, such as that proposed by H.J. Goldsmid[44,45] on porous materials for the artificially anisotropic media.

The initial equations for the construction of Table 2.1 consider only the steady-state temperature fields. The dynamic modes have not been adequately studied, though it is evident that their consideration will suggest new possibilities for improving the efficiency of thermoelectric power conversion. The superposition of eddy currents of electromagnetic and thermoelectric nature is quite unknown, which can also provide new knowledge and the unusual possibilities of practical applications for thermoelectricity.

On the whole, the increasing significance of information technologies should generate growing interest in the multifunctional new thermoelement types.

References

1. Anatychuk, L.I. 2003. Physical models of thermoelements. *J. Thermoelectricity* 1:5–17.
2. Anatychuk, L.I. 1994. Seebeck or Volta? *J. Thermoelectricity* 1:9–10.
3. Anatychuk, L.I. 1998. *Thermoelectricity, Vol. 1, Physics of Thermoelectricity*. Institute of Thermoelectricity, Kyiv, Chernivtsi.
4. Anatychuk L.I. 1998. *Thermoelectricity, Vol. 1, Physics of Thermoelectricity* Institute of Thermoelectricity, Kyiv, Chernivtsi, Chapter 2, pp. 77–137.
5. Yerofeev, R.S. 2010. The influence of thermoelectric effects on tectonic processes and the Earth climate. *J. Thermoelectricity* 1:86–92.
6. Hirose, A. 1954. Mathematical theory of multidimensional thermoelectricity with some experimental proofs. *J. Inst. Electr. Jpn.* 74:1054.
7. Anatychuk, L.I. and O.J. Luste, 1973. Method of measuring of semiconductor epytaxial films specific resistance. *USSR Author Certificate SU372492.*
8. Anatychuk, L.I. and O.J. Luste, 1997. Thermoelectric eddy currents. Measuring devices on their basis. *Proceedings of the XVI International Conference on Thermoelectrics*, Dresden, pp. 741–744.
9. Geiling, L. 1950. Thermoelement, Patent FR592016.
10. Geiling, L. 1951. Das Thermoelement als Strahlungsmesser. *Z. Angew. Phys.* 3(12):467–477.
11. Babin, V.P., T.S. Gudkin, Z.M. Dashevsky, N.V. Kolomoets, and L.M. Stilbans, 1974. Artificial anisotropic thermoelements and their limit possibilities. *Fizika i Tehnika Poluprovodnikov* 8:748–753.

12. Anatychuk, L.I., R.R. Kobylanskiy, R.V. Kuz`, O.J. Luste, O.V. Nitsovich, and A.V. Pribyla, 2008. New thermoelement types. *ECT 2008, 6th European Conference on Thermoelectrics*, July 2–4, Paris, France, pp. 1–14.
13. Anatychuk, L.I., O.J. Luste, and O.V. Nitsovich, 2005. Cross-layered artificially-anisotropic thermoelement. *J. Thermoelectricity* 2:52–52.
14. Anatychuk, L.I., O.J. Luste, R.V. Kuz`, and O.V. Nitsovich, 2010. On the true reason for the origination of transverse thermoelectromotive forces in the artificially anisotropic thermoelements. *8th European Conference on Thermoelectrics, ECT 2010, Società del Casino*, Como, Italy, 22–24 September.
15. Odelevsky, V.I. 1951. Calculation of generalized conductivity of heterogeneous systems. 1. Matrix two-phase systems with inclusions. *Zhurnal technicheskoi fiziki* 21:667–677.
16. Odelevsky, V.I. 1951. Calculation of generalized conductivity of heterogeneous systems. 2. Polycrystals. *Zhurnal technicheskoi fiziki* 21:1379–1382.
17. Anatychuk, L.I., V.T. Dimitrashchuk, O.J. Luste, and Yu.S. Tsiganyuk, 1972. Eddy thermoelectric current in nonsteady state temperature field. *Izv. Vuzov. Fizika* 3:23–29.
18. Prybyla, A.V. 2010. Spiral anisotropic and anisotropic thermoelements. *J. Thermoelectricity* 4:77–86.
19. Anatychuk, L.I. 1999. Generalization of Faraday law to the process of thermoelectric energy conversion. *Proceedings of the V European Workshop on Thermoelectrics*, Pardubice, Czech Republic, pp. 114–119.
20. Maxwell, J.C. 1873. *Treatise on Electricity and Magnetism, 2 Vols*. Clarendon Press, Oxford.
21. Pilat, I.M., A.B. Belikov, L.L. Kazanskaya, and A.A. Ashcheulov, 1976. Thermoelement with anisotropic heat conductivity. *Fizika i tekhnika poluprovodnikov* 10:1019.
22. Justi, E. 1960. *Thermoelektrische Kombination insbesondere Thermosaule*, Patent DE1076210.
23. Luste, O.J. 2001. Inverse problems of mathematical theory of devising new thermoelement types. Part 1. Statement of problems and solution methods. *J. Thermoelectricity* 4:3–10.
24. Luste, O.J. 2002. Inverse problems of mathematical theory of devising new thermoelement types. Part 2. Usage of inverse problems for creation of predesigned distribution of thermoelectric eddy currents. *J. Thermoelectricity* 1:3–8.
25. Anatychuk, L.I. 2005. *Thermoelectricity, Vol. 2, Thermoelectric Power Converters, Thermoelements, Element Basis of Thermoelectricity*, Institute of Thermoelectricity, Kyiv, Chernivtsi.
26. Anatychuk, L.I. and O.J. Luste, 1966. Eddy thermoelectric currents in CdSb. *Fizika tverdogo tela* 6:2492–2494.
27. Prybyla, A.V. 2006. Specific features of thermoelectric phenomena in spiral rectangular anisotropic transverse thermoelement. *J. Thermoelectricity* 1:23–29.
28. Anatychuk, L.I., O.J. Luste, and E.L. Maslyanchuk, 1992. Informative-energetic description of thermoelectric converters. *Proceedings of XI International Conference on Thermoelectrics*, Arlington, USA, pp. 160–163.
29. Demchuk, B.N. and O.J. Luste, 2003. Informational-energetic description of thermoelectric multielement microthermopiles. *J. Thermoelectricity* 2:47–49.
30. Anatychuk, L.I., B.N. Demchuk, and O.J. Luste, 1995. Thermoelectric microcalorimetry of biological objects. *Proceedings of XIV International Conference on Thermoelectrics*, St. Petersburg, Russia, pp. 464–466.
31. Anatychuk, L.I. and A.V. Prybyla, 2008. Thermoelectric sensors of anisotropic thermoEMF materials. *XX International Conference on Photoelectronics and Night Vision Devices*, Moscow, Russia.
32. Anatychuk, L.I., O.J. Luste, and A.V. Prybyla, 2006. *Heat Flow Sensor*. Patent UA82913.
33. Nitsovich, M.V. 1959. Candidate thesis. Chernivtsi State University.
34. Anatychuk, L.I. 1965. Candidate thesis. Chernivtsi State University.
35. Samoilovich, A.G. 1966–2007. *Thermoelectric and Thermomagnetic Energy Conversion*. LKI, Moscow.
36. Samoilovich, A.G., M.V. Nitsovich, and V.M. Nitsovich, 1966. On the theory of anisotropic thermoelectric power in semiconductors. *Phys. Stat. Sol.* 16:459.

37. Anatychuk, L.I., V.D. Iskra, and O.J. Luste, 1968. *Ukrainskyi fizychnyi zhurnal* 13:1226.
38. Anatychuk, L.I. and V.D. Iskra, 1967. PiezothermoEMF of n-Si. *Fizika tekhnika poluprovodnikov* 1:1263–1265.
39. Anatychuk, L.I. 1979. *Thermoelements and Thermoelectric Devices: Handbook*. Naukova Dumka, Kyiv.
40. Luste, O.J. and Ya.G. Fedoruk, 2008. Optimization of materials for gyrotropic thermoelements. *J. Thermoelectricity* 4:20–25.
41. Anatychuk, L.I., O.J. Luste, and O.V. Nitsovich, 2007. Computer simulation of power conversion processes in a double-layer thermoelement with periodically profiled surface. *J. Thermoelectricity* 2:73–80.
42. Anatychuk, L.I., O.J. Luste, and O.V. Nitsovich, 2007. Computer research on the transverse thermo-EMF in a three-layer thermoelement with periodically profiled surface. *J. Thermoelectricity* 3:22–28.
43. Anatychuk, L.I., O.J. Luste, and O.V. Nitsovich, 2008. Temperature dependences of parameters of a double-layer thermoelement with periodically profiled surface. *J. Thermoelectricity* 2:32–40.
44. Goldsmid, H.J. 2008. Synthetic transverse thermoelements with a porous component. *J. Thermoelectricity* 1:42–48.
45. Goldsmid, H.J. 2010. Application of the transverse thermoelectric effects. *ICT2010*, May 30–June 3, Shanghai, China.

3

Thermoelectric Devices as Heat Engines: Alternative Thermodynamic Cycles

L. E. Bell
BSST LLC

3.1 Background

Thermoelectric (TE) systems are solid-state heat engines that convert heat flux induced by a temperature differential into electrical power output (Figure 3.1a). They are also capable of the reverse process, converting electric power to cooling, heating and temperature control (Figure 3.1b).

Even though such direct energy conversion is attractive for many large-scale applications, TE devices have seen very limited usage because of low efficiency, high cost, and difficulties encountered scaling to high power levels. TEs have limited application to niche markets where the positive attributes of no moving parts, ability to scale to small size, and availability of planar geometries have provided an overall system benefit.

TE materials with higher ZT are considered essential to broader use of TE technology. While an increase in ZT is very beneficial, in some important applications, other attributes of TE heat engines can be exploited to make TE systems more attractive, and ultimately, more competitive with other technologies. Each TE couple in a typical TE power generator has a volume of 0.05–0.1 mL compared to a volume of about 300–600 mL for each piston in a diesel engine, so that the TE engine unit is roughly 1/6000 that of a piston engine. The ratio is about a factor of 10 greater for typical cooling/heating system heat pumps. This very large difference suggests an evaluation of alternative cycles in which TE engines are organized in arrays that operate under localized optimal conditions as do, for example, the stages of turbine engines do. The 15–20 compressor stages and 10–15 expansion stages in turbines, each independent heat engines, operate nearly reversibly, coordinated to first compressing and then expanding a working fluid. The result is that the total turbine engine has higher efficiency than a diesel engine. Since TE engines are typically used in large arrays, it may be possible to use alternative cycles and compound cycles to improve system performance (Bell 2008).

In this chapter, improved TE performance means some combination of greater efficiency, lower cost, and smaller size, weight, or volume. To reduce complexity, to allow for comparisons among different approaches, and to better interpret results, it is assumed that all TE material properties, for example, Z, are constant over the temperature range in which the TE engines operate. Further, all parasitic losses

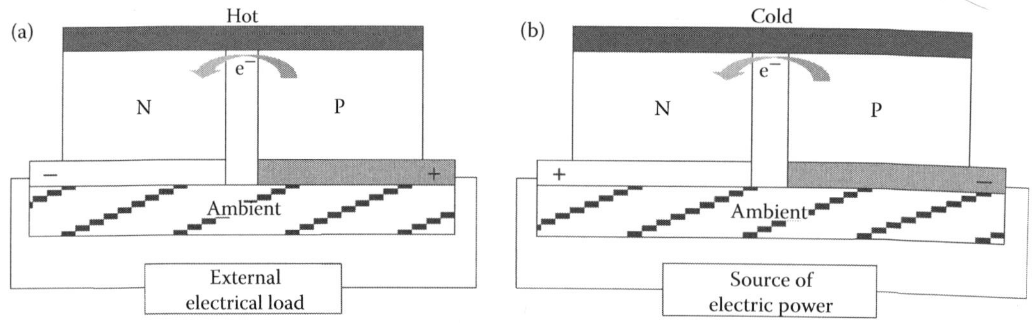

FIGURE 3.1 (a) TE heat engine. Temperature gradient produces potential difference and current flow through an external load. (b) TE heat pump. Externally apply electric current creates temperatures gradient and ability to pump thermal power from the cold to hot side.

associated with the structure of the heat engines are assumed to be zero and losses associated with electrical and thermal exchanges with external systems are assumed to be zero.

3.2 Cooling, Heating, and Temperature Control

The two cycles shown in Figure 3.2 are used in cooling, heating, and temperature control. The first, Figure 3.2a, is the traditional cycle in which all the TE engines have the same cold, T_C and hot-side, T_H operating temperatures. The fundamental equations governing performance of TE systems have been formulated based on this configuration (Goldsmid 1986). Written in terms of T_C and T_A (the average temperature), the maximum coefficient of performance (COP) at the optimum efficiency in cooling, COP_{OC}, and the maximum hot-to-cold-side temperature differential ΔT_M are:

$$COP_{OC} = \frac{T_C}{\Delta T}\left(\frac{\sqrt{ZT_A+1}-1-\Delta T/T_C}{\sqrt{ZT_A+1}+1}\right) \tag{3.1}$$

$$\Delta T_M = ZT_C\left(\frac{T_C}{2}\right) \tag{3.2}$$

TE engines that satisfy this condition are supplied as standard TE modules.

Figure 3.2b is a cascade, a well-known alternative cycle used for cooling critical sensor components to temperatures lower than obtainable with a single-stage TE engine. It is made up of standard cycles connected thermally so that all the thermal heat flux rejected from the hot side of the top-most TE engine passes directly to the cold side of the heat engine immediately below it. The thermal flux Q_{1C} from

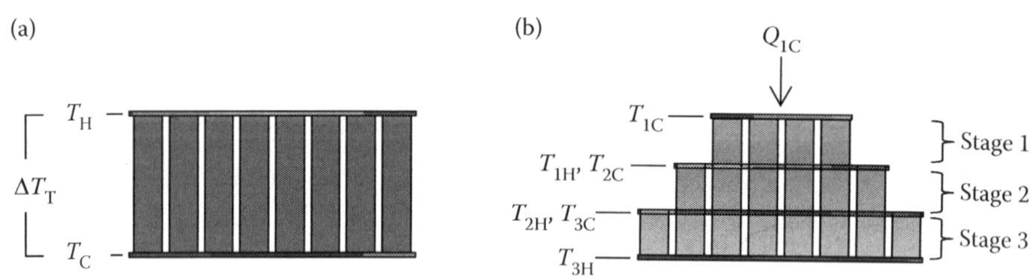

FIGURE 3.2 (a) Standard TE module. All TE engines (shaded rectangles) operate at the same hot and cold side temperatures. (b) Cascade module. TE engines are fabricated with standard TE modules connected thermally in series.

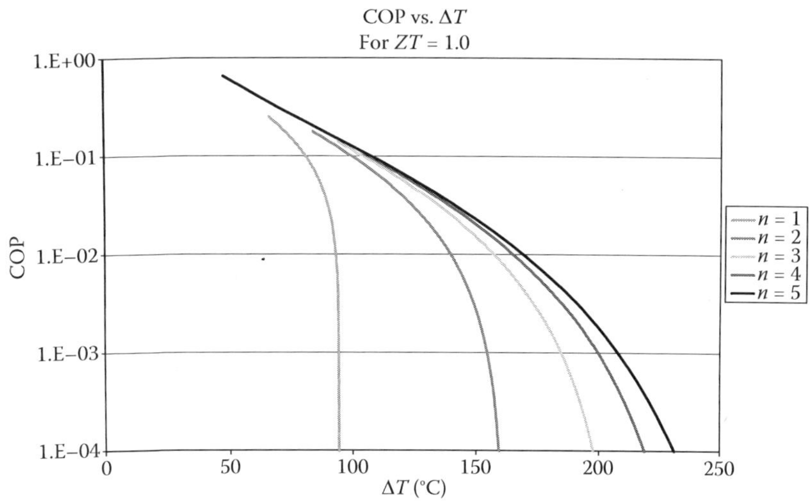

FIGURE 3.3 Ideal cascade TE systems. COP and maximum ΔT for up to 5 stages.

the cold surface at temperature T_{1C}, plus the electric power into that TE engine exit TE engine 1 at hot-side temperature T_{1H}. Thus, each heat engine performs the same as in Figure 3.2a with boundary conditions that $T_{1H} = T_{2C}$ and $Q_{1H} = Q_{2C}$. Similarly, more stages can be added (Marlow Industries 2008). Figure 3.3 presents the theoretical performance of multiple stages. For convenience, it has been assumed that for each stage the value of peak ZT is 1.0, and the ratio $\Delta T_1/T_{1C}$ is the same for each cold-side temperature T_C. In practical applications, the number of stages is limited to about 6, as unavoidable parasitic losses diminish performance gains that come from adding more stages. The characteristic of TE systems with multiple stages is an increase in maximum ΔT and very low efficiency at large ΔT as shown in Figure 3.3. In practice, the low efficiency has limited the use of cascade TE systems to applications in which efficiency is not as important as small size, ability to achieve large temperature differentials, and the reliability gains associated with solid-state operation. An example of usage is charge-coupled amplifier coolers in night vision systems.

Figure 3.4a shows a compound cycle constructed with multiple, independent heat engines. Thermal power is extracted from a working fluid entering at temperature T_A, and progressively cooling it to T_C. The heat extracted and the electrical power applied is rejected to a heat sink, usually a second working

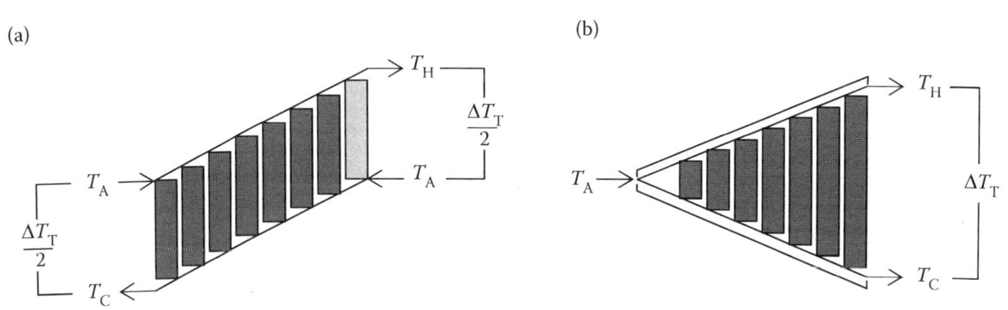

FIGURE 3.4 (a) Counterflow engine. TE cycle with thermal isolation between TE engines in the direction of working fluid flow. TE engines operate in different temperature ranges. (b) Parallel flow. TE cycle with thermal isolation between TE engines in the direction of working fluid flow. TE engines operate at progressively increasing temperature differentials.

fluid which is heated incrementally by each heat engine, and exits at temperature T_H. A critical assumption is that thermal transport between the TE engines in the direction of fluid flow is negligible so that the stages are thermally isolated from each other. These configurations can have progressive temperature change on both the hot and cold sides or a uniform temperature on one side. Also, the inlet temperatures on the hot and cold sides can have any value. These and other conditions are discussed more fully in Bell (2002b). To show the character of these cycles, the boundary conditions in Figure 3.4a are used. In a very large array of heat engines in the directions of working fluid flow, each TE engine has ~1/2 the ΔT of the heat engine in the standard cycle, so that the maximum COP for each engine is

$$\mathrm{COP}_1 = \left(\frac{T_{1C}}{\Delta T_1/2}\right)\left(\frac{\sqrt{ZT_A+1} - 1\,\Delta T_1/2T_{1C}}{\sqrt{ZT_A+1}+1}\right) \tag{3.3}$$

This leads to a factor of 2 increase in the numerator of the Carnot term for the system, and thus at least twice the efficiency of the standard cycle. Further, the factor of 2 introduced in the second TE-specific term approximately doubles the maximum possible ΔT between the hot and cold outlets. From an external reference, the measured performance is that of a conventional TE engine with a higher ZT material. Table 3.1 presents the equivalent ZT of a conventional TE system required to give the same efficiency as the compound cycle of Figure 3.4a under the stated boundary conditions. The nonlinear relationship between ZT in a system of Figure 3.4 and that of a standard cycle is due to the square root of $\sqrt{ZT_A+1}$ appearing in the efficiency Equations 3.1 and 3.3.

In the cycle shown in Figure 3.4b, both working fluids enter from the same end, so TE engines in the direction of flow are progressively exposed to larger temperature differentials. In the assumed ideal device, the TE engines where the working fluid enters have very small ΔT and so efficiency can be very high for those engines if they are designed to operate at maximum efficiency. Since $\Delta T = 0$ at the inlet and if the TE engines are constructed so that ΔT increases linearly as the working fluid moves to the right, efficiency will decrease approximately as $1/\Delta T$, and reach the total system ΔT_T at the exit. Thus, the final stage has the same ΔT_T as in a conventional system with the same T_H and T_C. The theoretical maximum efficiency is slightly greater than twice that of a standard cycle (Bell 2002b).

Both configurations shown in Figure 3.4a and b have efficiency gains over a standard cycle but differ in that the configuration in Figure 3.4a has an increased maximum ΔT while the configuration of Figure 3.4b leaves maximum ΔT unchanged. Experimental results confirm the increase in efficiency and ΔT (Diller 2002). Potential practical applications include heating, ventilating, and air conditioning (HVAC) systems with liquid working fluid, liquid working fluid thermal management systems, and industrial process recuperators.

Alternatively, the same result can be achieved by changing the temperature differential across TE engines with time so that individual engines operate in quasi-equilibrium. Cycles in which the TE engines move linearly or rotate have been analyzed (Ghoshal 1999, Bell 2003b). In concept, these cycles extract and reject thermal power across reduced temperature differentials while avoiding thermal transport

TABLE 3.1 Comparison of TE Material ZT Required for Standard Modules with the Same Performance as That of Counterflow Systems for $T_C = 300$ K and $\delta = 30$ K

ZT of Material Used in Counterflow Cycle	Material ZT Equivalent for Standard Module Cycle
0.7	1.9
1.0	3.0
1.5	5.1
2.0	9.2

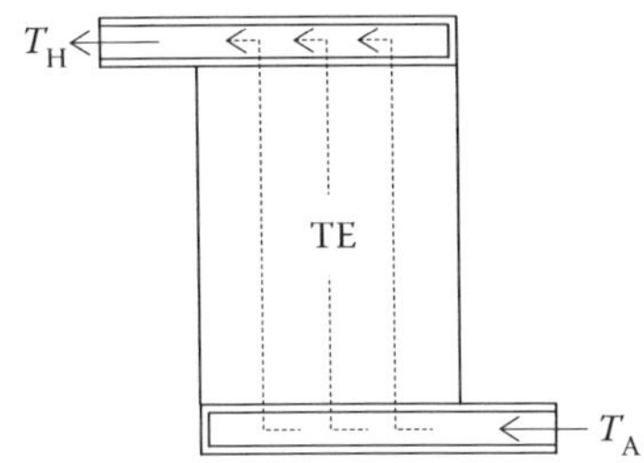

FIGURE 3.5 Convective TE engine. Convective flow of a working fluid convects a portion of the Joule heating and conductive losses that otherwise would reduce TE engine performance.

across more than one TE engine. All are calculated to have about a factor of 2 efficiency gain over that of the standard cycle under the boundary conditions of Figure 3.4a. The linear system that has been proposed for electronics cooling and rotary systems may have application in large HVAC and liquid dispenser systems.

A time-dependent, nonequilibrium cycle has been investigated which increases the maximum ΔT by about 33% for times the order of a second. The author discusses possible uses in periodic cooling of temperature-sensitive sensor components (Snyder 2002).

If a working fluid, or the TE material itself, is convected from the cold side to the hot side, as shown in Figure 3.5, then Joule heating, I^2R, within a TE element will be convected toward the hot end reducing heat lost to the cold end. Similarly, the conduction out the cold side, $-K\Delta T$ will be reduced. With suitable boundary conditions and constant transport properties, a set of equations for optimum COP and heat flows can be written that parallel those for standard heat engines (Bell 2002a), with the result that efficiency gains of 30–60% are possible. Similarly, but less readily apparent, gains in cooling of about 30% appear possible. Transient convective designs have been proposed, based on similar considerations (Echigo 1993, Tada 1996). In these designs, a working fluid is convected from the end of one TE element to the opposite end and then through the second element in a U-shaped path.

To date, both convective concepts await conclusive experimental validation of their respective theoretical models. Convective cycles appear to be more effective if the ratio of TE element length in the direction of current flow to effective cross-sectional diameter is large. This suggests more applicability to power generation than to cooling and heating applications.

3.3 Waste Heat Recovery and Primary Power Generation

One of the original and most successful uses of TE heat engines has been primary electric power generation for spacecraft. The TE generators (TEGs) developed in the United States and the former USSR dominated energy conversion for spacecraft operating beyond the effective range of sun-powered photo voltaic cells. The TE systems exhibit compact design, exceptional durability, and stable power output over decades (Bennett 2002b) (Chmielewski et al. 1992, 1994). Low efficiency of such systems has been outweighed by these positive characteristics. Recent interest in CO_2 emission reduction has led to renewed interest in TEG technology for exhaust waste heat recovery, especially in the automotive and industrial sectors. In addition, the need for continuous primary electric power production for remote communications and control system applications has also increased interest in TEGs. Generally, the cost benefit of such applications is proportional to energy conversion efficiency, so emphasis is laid on higher-ZT materials and by developing higher-efficiency thermodynamic cycles.

In the conventional TE configuration, both the hot-side heat source and cold-side heat rejection radiators operate at constant temperature, as shown in Figure 3.6a. Assuming TE properties do not vary with temperature, the maximum efficiency of the standard TE heat engine is (Angrist 1976)

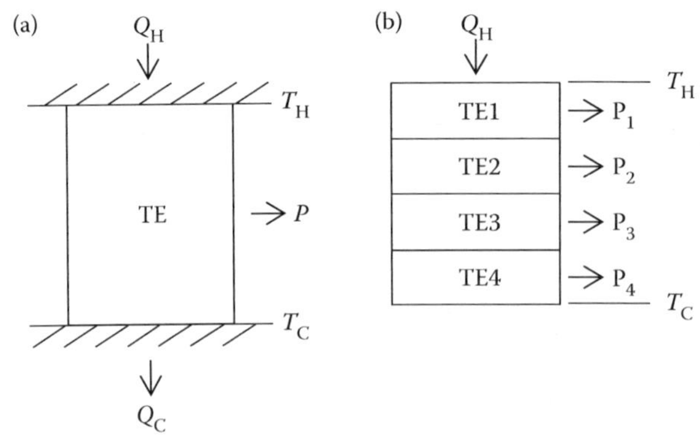

FIGURE 3.6 (a) Standard TE heat engine. Schematic of a standard TE heat engine. Under steady state conditions $Q_H = Q_C + P$ where P is the electric output. (b) Cascade TE power generator. Each stage is connected thermally to adjacent stages. Electrical power outputs P_1, P_2, P_3, P_4 are produced by each respective stage.

$$\eta = \left(\frac{\Delta T}{T_H}\right)\left(\frac{\sqrt{ZT_A + 1} - 1}{\sqrt{ZT_A + 1} + 1 - \Delta T / T_H}\right) \tag{3.4}$$

The configuration of Figure 3.6a is appropriate for small-scale solar conversion systems, most isotope generators, and at the high temperature end of systems limited by the maximum operating temperature of the TE materials used. A variant to the standard heat engine is a cascade of several TE engines as shown in Figure 3.6b. The operation parallels that discussed for cooling and heating systems, with the exception that electrical power is extracted separately from each engine stage in the cascade, thus the top hot-side engine is exposed to the maximum thermal power, and the thermal power flowing to subsequent engines is diminished by the accumulative electrical power extracted at higher temperatures. The theoretical maximum efficiency is higher than that of the standard configuration, since the added stages have lower hot-side temperatures, increasing slightly the Carnot term and the configuration-specific terms of each stage. In this chapter, peak ZT is assumed to have the same value for each stage. This convention tends to favor cascades since single stages spanning large temperature differentials have constant Z and hence lower average ZT. In *direct energy conversion* (Angrist 1976), it is noted that the parasitic losses ignored here have prevented net gains from more than two stages in production systems.

In many potential applications of commercial interest, waste heat takes the form of flowing hot exhaust gas. As thermal power is extracted, the gas cools incrementally, so that the heat extraction system operates at progressively lower ΔT in the direction of fluid flow (Cohen 2001). Figure 3.7a and b have variable–temperature heat sources and constant–temperature heat sinks (Bell 2003a). Figure 3.7a shows a cycle in which the portion of the thermal power extracted from the hot-side working fluid is between T_H and T_M, an intermediate temperature. As a convention, as well as for comparison purposes, T_M is taken as the arithmetic mean between the hot-and cold-side temperatures. For example, this cycle could be employed if limits were imposed on the peak operating temperature of the TE material. It also represents the case of a single TE module which has high thermal conduction in the direction of working fluid flow.

Figure 3.7b shows a similar cycle but with the TE material operating locally at the highest possible temperature as the fluid is cooled. It is apparent from the decreasing ΔT in the direction of fluid flow that the energy conversion efficiency becomes progressively lower. Since it is not practical to reduce the outlet temperature of the working fluid to T_C, the system has maximum average efficiency of about 1/2 that of the engine in Figure 3.6a. These cycles describe conditions in automotive waste heat recovery and some industrial waste heat and cogeneration applications (Crane 2010).

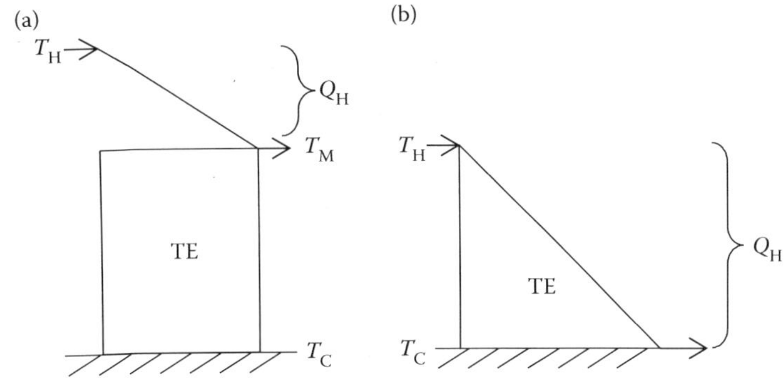

FIGURE 3.7 (a) Fixed hot, cold sides. Thermal power from a working fluid enters at T_H and exits at T_M, transferring thermal power Q_H to the TE engine. Entire hot side operates at T_M and cold side at T_C. (b) Fixed cold side. Working fluid enters at T_H and exits at T_C. Individual TE engines operate at progressively lower temperature differentials and efficiencies.

Figure 3.8a shows a cycle in which the heat rejected from the cold side is removed so that the entire TE engine operates at uniform hot and cold-side temperatures T_H and T_M. In the cycle in Figure 3.8b, the hot-side is at a uniform temperature and the cold-side is progressively heated.

Figure 3.8a and b describe conditions where the hot-side temperature is fixed or otherwise uniform and the heat rejection-side temperature varies because of coolant capacity limitations or the coolant is intentionally heated as part of recuperation in energy conversion systems. Examples are some portions of combustion-powered systems and large TE arrays for solar and isotope fission systems.

Figure 3.9a shows a system in which both the hot and heat rejection-side temperatures vary. Such applications are encountered in many practical high-capacity systems. Figure 3.9b describes a system in which the heat rejected from the cold side decreases the portion of heat input required to preheat working fluids such as combustion reactants. This system can be advantageous for primary power generation.

Convective cycles (Figure 3.5) operate as they do in heating, by employing working fluids to transport a larger portion of the thermal power from internal Joule heating to the hot side than occurs with conduction alone. Also, conductive losses to the heat rejection side are reduced. In another application,

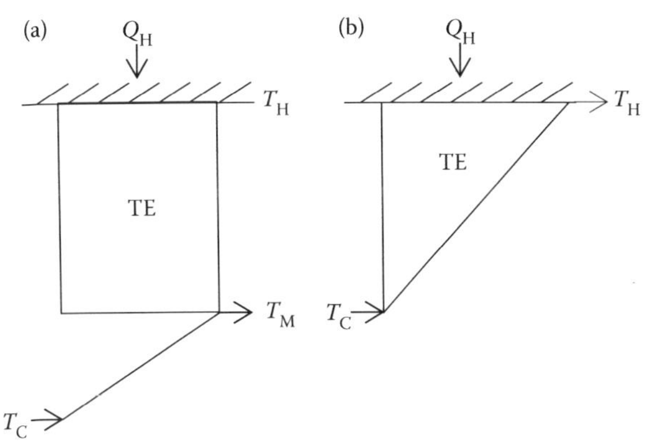

FIGURE 3.8 (a) Fixed hot, cold sides. All TE engine hot sides operate at T_H and cold sides at T_M. Working fluid enters at T_C, absorbs rejected waste thermal power and exits at T_M. (b) Fixed hot side. All hot sides of TE engines operate at T_H. Cold side working fluid enters at T_C and exists at T_H. TE engines operate at different temperature differentials.

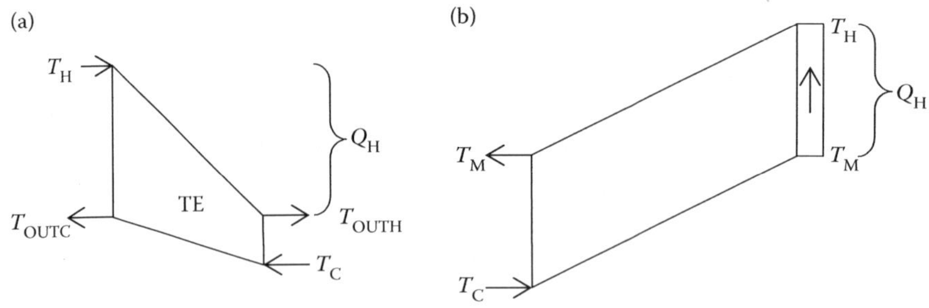

FIGURE 3.9 (a) High capacity system. Hot side working fluid enters at T_H and exits at T_{OUTH}. Cold side working fluid enters at TC and exits at T_{OUTC}. Operating conditions of TE engine hot and cold sides vary. (b) Power generator. Working fluid (reactants) enter at T_C and exhaust exits at T_M. External thermal power (or combustion) Q_H heats working fluid to T_H. Working fluid transfers Q_H thermal power to hot side of TE engines.

preheating reactants in a combustion system, the potential exists for efficiency gains, while complexity and size are reduced. An example of this concept, preheating O_2 and H_2 by this process, has been studied (Yamamoto 2006)as have systems in which a convective fluid is preheated in the TE system and expanded through a turbine generator system (Bell 2002a). A time-varying cyclic convective cycle has been studied in which porous TE elements are heated in a central zone by the combustion of working fluids (Kotyrlo 1974). By alternating the direction of reactant flow and combustion products, the hottest zone is maintained at the center junction of the TE materials and the reactants are preheated by a recuperative action during percolation through porous TE materials (Katsuki 2002).

As noted above, TE cycles can be combined as compound cycles to more fully use available thermal power and to increase efficiency. The example in Figure 3.10a is of benefit under conditions where the working fluid is convected to the hot side of TE engine to the left and a second engine uses the heated fluid as its source of thermal power. Figure 3.10b is a similar cogeneration cycle in which convection is used to preheat the working fluid of a different type of heat engine, such as a steam turbine. The convective cycles are predicted to increase system efficiency under conditions where the convected thermal power exiting at T_H can be used by a second energy conversion system. If it is assumed that all of the convected thermal power can be utilized in a cogenerator and that the loss mechanisms that degrade actual system performance can be ignored, the ideal performance of the convective TE engine portion is predicted to have efficiencies approaching Carnot. The cogenerator and the working fluid would be selected for compatibility and ability to convert efficiently the thermal power convected to the hot side

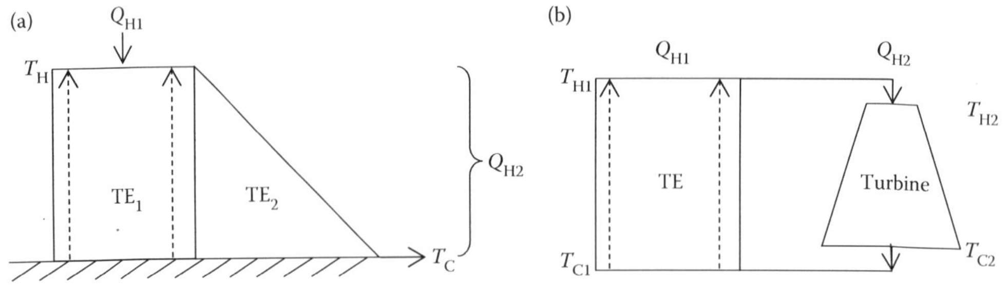

FIGURE 3.10 (a) TE cogeneration cycle. Total thermal power input is Q_{H1}. Working fluid enters at T_C at left, passes through TE, and exits at T_H absorbing thermal power Q_{H2}. The fluid delivers Q_{H2} to TE_2. (b) Combined cogeneration cycle. Working fluid, such as steam enters convective TE at T_{C1} and is heated to T_{H1}. A portion of Q_{H1} is converted to electric power output by the TE engine, another portion exits at T_C as waste heat. The remainder contained in the working fluid, Q_{H2} powers a turbine or other heat engine.

TABLE 3.2 Maximum Ideal Efficiency of Power Generation Cycles for Configurations Discussed in the Text

Configuration (Figure Number)	Efficiency (%) $ZT_H = 1$, $T_H = 600$ K, $T_C = 300$ K	$ZT_H = 2$, $T_H = 600$ K, $T_C = 300$ K	$ZT_H = 1$, $T_H = 900$ K, $T_C = 400$ K	$ZT_H = 2$, $T_H = 900$ K, $T_C = 400$ K
3.6a	8.86	13.96	9.88	15.60
3.6b	11.20	16.90	13.00	19.50
3.7a	3.04	4.79	3.56	5.61
3.7b	5.44	8.57	6.25	9.86
3.8a	4.53	7.24	5.07	8.13
3.8b	4.50	7.14	5.02	8.00
3.9b	5.10	8.03	5.79	9.13
3.10a	9.30	14.79	10.45	16.69

of the TEG. Of interest is the theoretical potential for convective TEG to increase the capacity of present, higher-efficiency power generation systems or in some cases, to slightly increase efficiency (Bell 2002a).

3.4 Summary and Conclusions

The attractive features of TE systems, solid-state function, small size, and direct energy conversion have been of sufficient importance to create niche military and commercial markets for the technology. However, low efficiency has limited wider acceptance of TE technology (Vining 2009). Under a range of operating conditions, alternate thermodynamic cycles have demonstrated improvements equivalent to at least a 100% increase in material *ZT*, and thus can make a substantial contribution to overall system efficiency. Any future increase in TE material *ZT* could make TE systems, in combination with the alternate cycles, competitive with traditional two-phase refrigerants in heating, ventilating, and air-conditioning systems. In combination with efforts to improve TE material *ZT*, alternate cycles warrant further study, development, and commercialization.

In contrast to cooling and heating, alternate cycles have not demonstrated significant performance gains over the standard cycle for power generation and waste heat recovery. Means to achieve high efficiency in systems where the thermal power is in an exhaust fluid is made especially difficult by the decrease in hot-side temperature as thermal power is extracted. Such systems have ¼ to ½ the efficiency of the standard cycle when operating under the same temperature limits. Table 3.2 summarizes the efficiency of the standard and alternate cycles. Thus, alternate cycles can provide a degree of efficiency increase in circumstances where it is not possible to apply the standard TE cycle. For example, they are being employed in developmental industrial and automotive waste heat recovery systems. For direct electric power generation, compound cycles provide some benefit, but additional gains in TE material *ZT* and cycle efficiency are needed to make TE power production efficiency competitive with alternative technologies such as conventional internal combustion engines.

References

Angrist, S. *Direct Energy Conversion*, 3 edition. Boston: Allyn and Bacon, Inc., 1976.

Bell, L. E. Alternative thermoelectric thermodynamic cycles with improved power generation efficiencies. *Proceedings of the 21st International Conference on Thermoelectrics*, pp. 558–561. Hérault, France, 2003a.

Bell, L. E. Cooling, heating, generating power, and recovering waste heat with thermoelectric systems. *Science*, 321(5895), 2008: 1457–1461.

Bell, L. E. Increased thermoelectric system thermodynamic efficiency by use of convective heat transport. *Proceedings of the 21st International Conference on Thermoelectrics*, pp. 488–499. Long Beach, 2002a.

Bell, L. E. Thermoelectric transient cooling and heating systems. United States of America Patent 6,637,210. October 28, 2003b.

Bell, L. E. Use of thermoelectric isolation to improve thermoelectric system operating efficiency. *Proceedings of the 21st International Conference on Thermoelectrics*, pp. 477–487. Long Beach, 2002b.

Bennett, G. L., Lombardo, J., Helmer, R. J., Silverman, G., Whitmore, C. W., Amos, W. R., Johnson, E. W. et al. Mission of daring: the general purpose heat source radioisotope thermoelectric generator. *4th International Energy Conversion Engineering Conference and Exhibit (IECEC)*, San Diego, CA, June 26–29, 2006.

Chmielewski, A. and Borshchevsky, A. *A Survey of Russian RTG Capabilities.* Technical Report, JPL NASA, 1992 and 1994.

Cohen, A. L. Microcombustion-based thermoelectric microgenerator. World Intellectual Property Organization. W/O 01/52332 A2, July 2001.

Crane, D. T. An introduction to system level steady-state and transient modeling and optimization of high power density thermoelectric generator devices made of segmented thermoelectric elements. *International Conference on Thermoelectrics*, pp. 561–569. Shanghai, China, 2010.

Diller, R. W. and Chang, Y. W. Experimental results confirming improved performance of systems using thermal isolation. *Proceedings of the 21st International Conference on Thermoelectrics*, pp. 548–550. Long Beach, 2002.

Echigo, R. et al. An extended analysis on thermodynamic cycle of advanced heating/cooling method by porous thermoelectric conversion device. *12th ICT.* Yokohama, Japan, 1993.

Ghoshal, U. Thermoelectric cooling with plural dynamic switching to isolate heat transport mechanisms. United States of America Patent 5,867,990. 1999.

Goldsmid, H. J. *Electronic Refrigeration.* Great Britain: Page Bros (Norwich) Limited, pp. 1–169. 1986.

Katsuki, F. et al. Temperature distributions and thermoelectric performance of a porous $FeSi_2$ element in a steady state of reciprocatory flow combustion. The Japan Institute of Metals, 2002.

Kotyrlo, G. K. and Shchegolev, G. M. Thermoelectric devices, being blown through with a substance in the direction of heat flow. NTIS U.S. Department of Commerce, April 6, 1974.

Marlow Industries. Six stage thermoelectric modules datasheet MI6030-01P. 2008.

Snyder, G. J., Fleurial, P., Caillat, T., Yang, R., and Chen G. Supercooling of Peltier cooler using a current pulse. *Journal Applied Physics*, 92(3), 2002: 1564–1569.

Tada, S., Echigo, R., and Yoshida, H. A new concept of porous thermoelectric module using a reciprocating for cooling/heating system. *15th ICT*, pp. 264–268. Pasadena, USA, 1996.

Vining, C. B. An inconvenient truth about thermoelectrics. *Nature Materials*, 8, 2009: 83–85.

Yamamoto, A., Noguchi, T., Obara, H., and Ueno, K. All-metal miniature power generation device and its application to gas combustion system. Umezono 1-1-1, Tsukuba Ibaraki 305-8568, Japan: Energy Technology Research Institute, National Institute of Advanced Industrial Science and Technology (AIST), 2006.

4

Functionally Graded Thermoelectric Generator and Cooler Elements

E. Müller
German Aerospace Center

K. Zabrocki
German Aerospace Center

C. Goupil
ISMRA and University of Caen

G. Jeffrey Snyder
California Institute of Technology

W. Seifert
University Halle-Wittenberg

4.1 Introduction

Since the discovery of semiconductor thermoelements, there has been much effort to enhance the efficiency of thermoelectric (TE) devices. Along with efforts to increase a material's figure of merit, the use of functionally graded materials (FGM) offers other ways to further improve device performance [1–40]. Graded TEs are characterized by a macroscopic gradient in their functional properties, caused by spatial variation of the composition (including doping) or microstructure. Also noncontinuously graded (i.e., segmented) elements are considered FGM since they lead to the same functional effect.

Although a TE gradient means a related spatial variation of all TE material properties, the strongest effect is generally linked to the variation of the Seebeck coefficient which is the primary material parameter of TE coupling between thermal and electrical energy transport. This is because a gradient in the Seebeck coefficient is linked to an additional, local heat release or absorption whereas a resistivity gradient will merely gradually shift the location of Joule heat release along the element. A gradient in the thermal conductivity will just cause an asymmetry in the outflow of the internally released heat to the hot and cold terminals and may thus deform the temperature profile along the element. It will tend to concentrate the temperature gradient at regions with low thermal conductivity.

Essentially, the functional effect of TE FGM is already observed in a homogeneous TE element with real, temperature-dependent material properties.

Due to the temperature-bound or explicitly position-dependent Seebeck coefficient, Peltier heat is absorbed or released inside the material. In a homogeneous but temperature-dependent material this is known as Thomson effect. In the case of an explicit Seebeck gradient, it is also referred as distributed Peltier effect or extrinsic Thomson effect [13,18,41–43].

Simulations and modeling of TE devices and materials is an important part to get guidelines how to improve them for real-life usage, see, for example, [44–55]. The central target of theoretical TE FGM studies is to elaborate recipes for optimal gradients in the TE material properties which ensure optimum operation of a TE device [5,6,21]. Ideally, it would be best to have a local criterion for optimizing global performance; currently local criteria are known for the efficiency of a thermogenerator (TEG) and the coefficient of performance of a thermoelectric cooler (TEC) [31,35]. However, global optimization requires constraints for the allowed range of temperature-dependent material properties in the considered materials. In order to prevent that global performance diverges in the optimization process, limits of the material properties have to be fed into the process, be these upper limits for the Seebeck coefficient and the electrical conductivity and a lower limit for the thermal conductivity, or averages of the TE properties (respectively the figure of merit), of the power factor, or of the efficiency. The optimization strategy we are speaking about here is an optimization strategy due to the material itself. Of course there is the possibility of optimizing other design parameters like the length of an element, its cross-sectional area or the ratio of both [56–62]. Another aspect would be to include contact resistances into the considerations [54,63–69], but this is beyond the scope of this work. Clearly distinguishing from design optimization of real TE devices we will consider here an "ideal" thermogenerator or Peltier device where no thermal losses due to radiation or thermal bypasses, no thermal or electric contact resistance, and only one-dimensional flow is assumed, that is, electrical current and heat flux are parallel, see also [64]. Every nonparallel arrangement without a magnetic field, where you have an arbitrary angle between the temperature gradient and the electrical current leads to a reduction of the performance which is deduced from a generalized figure of merit shown by Gryaznov et al. [70]. For the sake of simplification, a single but representative (segmented or continuously graded) TE generator or cooler element (p-type or n-type, of element length L and constant cross-sectional area A_c) is often considered as part of a TE device or as a single-element device.* Doing this, optimization strategies are based on fixed parameters L and fixed boundary temperatures T_a and T_s within the framework of a unified 1D model for both TEG and TEC (see Figure 4.1 and [29]), where T_a is the temperature at the heat absorbing side (hot side for TEG, but cold side for TEC), and T_s denotes the heat sink temperature which is in many cases fixed not far from the room temperature. The total heat flux and its components in Figure 4.1 are indicated by the symbols $\mathbf{q} = -\kappa\nabla T + \alpha T, \mathbf{j} =: \mathbf{q}_\kappa + \mathbf{q}_\pi$ with Fourier and Peltier heat fluxes $\mathbf{q}_\kappa$ and $\mathbf{q}_\pi$, respectively (see also [30]).† Note that all flows are counted positive according to right-headed arrows, thus the magnitude of the flow vectors with left-headed arrows will adopt negative numerical values in the 1D formulae. The different arrow orientations just shall give an idea on the physical flow directions.

* In a 1D approach the segmentation or grading is clearly in the same direction as the electrical current and the heat flux are. In a quasi-1D or in a multidimensional approach you have to take care of the direction of the electrical current and the heat flow in comparison to the grading direction, see Refs. [14–18].

† One has to take care of the terms and the notation used. Normally, Q denotes the *heat* in units of 1 J, whereas with $\dot{Q} = (\partial Q/\partial t)$ the heat transfer rate, heat flow or thermal power in units of 1 W is meant. For the sake of simplicity the dot is sometimes omitted, as we do here, too. The *heat flux* is often used which is the heat transfer rate per cross-sectional area $\mathbf{q} = \frac{1}{A_c}\dot{Q} = ((1/A_c)(\partial Q/\partial t))$ in units of 1 W/m^2.

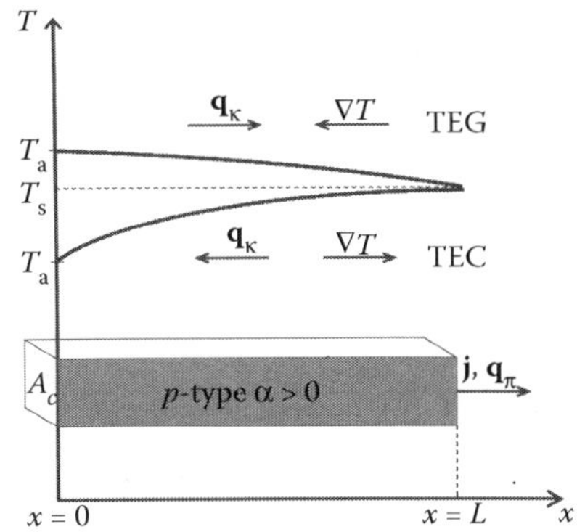

FIGURE 4.1 Unified 1D model of a TE element (single p-type pellet of length L, constant cross-sectional area A_c): lower temperature profile: Cooling operation (=TEC case with $T_a < T_s$); upper temperature profile: TE generator (=TEG case with $T_a > T_s$). Smaller bowing of the temperature profile in TEG case shall symbolize the relatively lower current in efficient operation, compared to the TEC. Note that $T(x)$ will peak in the interior of the TE element at current values only above the optimum current according to maximum efficiency and maximum coefficient of performance, respectively. This can be shown for the CPM case analytically, see footnote on page 9 of this chapter.

All calculations done here are referring to a planar arrangement of prismatic-shaped elements. Other shapes of elements need more detailed consideration, like, for example, circular shape [58,71–75], where the formulae for the power output may change while the efficiency remains independent of the shape.

On a macroscopic scale, the framework of nonequilibrium thermodynamics with Onsager's approach of a linear response theory is used [76–78]. Under isotropic conditions, the constitutive relations are

$$\mathbf{j} = \sigma\mathbf{E} - \sigma\alpha\nabla T, \quad \mathbf{q} = \alpha T\mathbf{j} - \kappa\nabla T, \tag{4.1}$$

with the electric field vector $\mathbf{E}$, electrical current density $\mathbf{j}$, temperature T, total heat flux $\mathbf{q}$, Del operator (or nabla operator) ∇, and the material properties isothermal electrical conductivity σ, resistivity $\rho = 1/\sigma$, thermal conductivity κ under zero current and Seebeck coefficient α. Note that in general all the material properties are tensors and need then a generalized description, especially if an additional magnetic field is present [70,79–86].

Assuming steady-state conditions, the governing equations result from the principles of conservation of charge and energy

$$\nabla \cdot \mathbf{j} = 0, \qquad \nabla \cdot \mathbf{q} = \mathbf{j} \cdot \mathbf{E}, \tag{4.2}$$

leading to the thermal energy balance in vector notation

$$\nabla \cdot (-\kappa\nabla T) = \frac{\mathbf{j}^2}{\sigma} - T\mathbf{j} \cdot \nabla\alpha. \tag{4.3}$$

Thereby, the electrical current density $\mathbf{j}$ is a constant due to the 1D approach. It fulfills the continuity constraint in one dimension and satisfies $\nabla \cdot \mathbf{j} = (\partial/\partial x) j_x = 0 \Rightarrow j_x = \text{const}$. Equation 4.3 makes clear that Joule's heat and a gradient in the Seebeck coefficient appear as sources of the thermal heat flux. Note that the Peltier–Thomson term $T\mathbf{j} \cdot \nabla\alpha$ contains both the usual Thomson contribution (temperature gradient effect with a temperature-dependent Seebeck coefficient) and the Peltier contribution (spatial gradient effect, see also [43]): $T\mathbf{j} \cdot \nabla\alpha = \mathbf{j} \cdot (\nabla\Pi - \alpha\nabla T)$ with the Peltier coefficient $\Pi = \alpha T$ and $\nabla\Pi = (d\Pi/dT)\nabla T = (\alpha + T(d\alpha/dT))\nabla T$.

For detailed commentary on Onsager's approach see the original articles [76–78]. Especially for application in TEs Domenicali denoted the approach of the *Onsager–de Groot–Callen* theory as a kind of "field theory" of the thermodynamics of irreversible phenomena in terms of vector functions which are dependent on temperature and position in general, see Refs. [76–78,87–99] for details.

In Section 4.2, the fundamentals of a thermodynamic optimization are introduced together with the definition of the entropy production, the TE potential and local reduced efficiencies. After that, alternative optimization variants will be discussed in Section 4.3, especially the differences taking into account local and/or temperature dependence of the properties. In Section 4.4 starting with the constant properties model (CPM), the known averaging procedures are introduced and valued. Section 4.5 is dedicated to highlight a "model-free setup" based on spatial material profiles, where you find a summary on numerical and analytical solutions of the thermal energy balance equation. The relation between the compatibility approach and optimum material grading is shown in Section 4.6. In Sections 4.7 and 4.8, a short summary and an outlook to related problems are given.

4.2 Thermodynamic Optimization and Minimum of Entropy Production

TE effects are caused by coupling between heat and charge transport of the electronic "fluid" [88]. For a deeper understanding of TE processes as processes out of thermodynamic equilibrium we highly recommend the article by Vining [100]. The description on a mesoscopic level is based on a stationary picture where all the thermodynamic potential functions are clearly defined, though the system itself produces dissipation. Since the inherent relaxation times are much smaller than the timescale of the varying potentials, this description is also a definition of a quasi-static process. As a consequence, the classical quasi-static relation $dS = \delta Q/T$ between the heat and the entropy variation is fully valid and can be expressed in terms of the entropy and heat current densities through $\mathbf{j}_S = \mathbf{q}/T$. In close connection with the general linear response theory this leads to the well-known coupled set of equations for heat flux and particle flux [87,88]

$$\begin{bmatrix} \mathbf{j}_N \\ \mathbf{q} \end{bmatrix} = \begin{bmatrix} L_{11} & L_{12} \\ L_{21} & L_{22} \end{bmatrix} \begin{bmatrix} -\frac{1}{T}\nabla\mu \\ \nabla(\frac{1}{T}) \end{bmatrix}, \tag{4.4}$$

where μ is the Gibbs electrochemical potential (as sum of the chemical potential and the electrical potential), $\mathbf{j}_N$ is the carriers flux (with $\mathbf{j} = e\mathbf{j}_N$), and $\mathbf{q}$ is the heat flux. $(1/T)\nabla\mu$ and $\nabla(1/T)$ are the forces corresponding to the electrochemical and thermal potentials, respectively. As a dissipative system, the system's evolution is driven by a minimal production of entropy where each fluctuation of any thermodynamic potential is subjected to a restoring force to equilibrium [99]. This concept had already been used by Clingman [101,102] who set out to use minimum entropy production to derive optimum device performance, for details we also refer to Ref. [103].

The symmetry of the off-diagonal terms $L_{ij} = L_{ji}$, as generally expected from Onsager's reciprocal relation, is equivalent to a minimal entropy production of the system under out-of-equilibrium conditions [99]. The kinetic coefficients L_{ij} can easily be expressed using material parameters leading to $L_{11} = (T/e^2)\sigma$, $L_{12} = -(T^2/e^2)\sigma S_N$, $L_{22} = (T^3/e^2)\sigma S_N^2 + T^2\kappa$, where e is the particle's charge, and $S_N = \alpha e$ is the so-called entropy per carrier, see, for example, [8,88,93,94]. Using the local expansion $\nabla 1/T = -1/T^2\nabla T$ and $\mathbf{E} = -\nabla\mu/e$, (4.1) is reproduced [100].

For decades, the figure of merit $z = \alpha^2\sigma/\kappa$ has provided a measure of the quality of a TE material, for details see Sections 4.4.2 and 4.4.3. A general rule is, that if a material is good (high zT) then it is good in both TEG and cooler applications.

4.2.1 Volumic Entropy Production

Since the entropy flux is $\mathbf{j}_S = \mathbf{q}/T$, it follows that the volumic entropy production ν_S is directly given by $\nabla \cdot \mathbf{j}_S = \nabla \cdot (\mathbf{q}/T)$ which gives with Equations 4.1 and 4.2

$$\nu_S = \nabla \cdot \mathbf{j}_S = \mathbf{q} \cdot \nabla\left(\frac{1}{T}\right) + \frac{\mathbf{j} \cdot \mathbf{E}}{T} = -\kappa\nabla T \cdot \nabla\left(\frac{1}{T}\right) + \frac{1}{T}\frac{j^2}{\sigma}, \tag{4.5}$$

where we identify the Joule and nonisothermal conduction contributions to the entropy. This expression can be rewritten with $\mathbf{E} = -\nabla\mu/e$ in the form $\nabla \cdot \mathbf{j}_s = \mathbf{q} \cdot \nabla(1/T) + (-\nabla\mu/(eT)) \cdot \mathbf{j}$. One can notice that the volumic entropy production is simply obtained from the summation of the force–flux products [87,88].

4.2.2 TE Potential

The relative current density u is defined* as the ratio of electrical current density $\mathbf{j}$ to the Fourier fraction of the heat flux $\mathbf{q}_\kappa$ with respect to the flow direction of $\mathbf{j} = j\,\mathbf{n}$

$$u = -\frac{\mathbf{j} \cdot \mathbf{n}}{\kappa\nabla T \cdot \mathbf{n}} \quad \text{resp.} \quad 1/u = -\frac{\kappa\nabla T \cdot \mathbf{j}}{\mathbf{j} \cdot \mathbf{j}}. \tag{4.6}$$

Note again that the fluxes are in parallel and Equation 4.6 is just the ratio of the projections onto the direction of the fluxes. Alternatively you can choose your coordinate system such that the flux direction is similar to one axis. Heat and particle fluxes can be combined giving

$$\mathbf{q} = \Phi\,\mathbf{j} = \left[\alpha T + \frac{1}{u}\right]\mathbf{j}, \tag{4.7}$$

where Φ is the TE potential as defined by Snyder [33–36,104–106]. These expressions allow a simple derivation of the volume heat production ν_q from

$$\nu_q = \nabla \cdot \mathbf{q} = \mathbf{j} \cdot \nabla\Phi = \mathbf{j} \cdot \nabla\left[\alpha T + \frac{1}{u}\right]. \tag{4.8}$$

Then we directly obtain the heat production density from the degradation of the potential Φ. Since the entropy flux is $\mathbf{j}_S = \Phi\,\mathbf{j}/\mathbf{T}$, the volumic entropy production becomes

$$\nu_S = \nabla \cdot \mathbf{j}_S = \mathbf{j} \cdot \nabla\left(\frac{\Phi}{T}\right) \quad \text{or} \quad \nu_S = \frac{\nu_q}{T} + \mathbf{q} \cdot \nabla\left(\frac{1}{T}\right). \tag{4.9}$$

This is in agreement with the classical Onsager formulation where the volumic entropy production is given by the summation of the flux–force product. Then, for a given material, the Φ potential gives a direct measure of the total volumic heat and entropy production by the respective degradation of $\nabla\Phi$

* The gradient ∇T is a vector; avoid therefore the definition $u = -j/(\kappa\nabla T)$, instead use the 1D variants $u(x) = -j/(\kappa T'(x))'$ resp. $u(T) = -(j/\kappa)x'(T)$. For a more general definition of u see first footnote on page 19 of this chapter.

and $\nabla(\Phi/T)$. From these latter expressions we see that Φ is the correct thermodynamic potential for characterizing TE processes and systems, respectively.

4.2.3 Local Reduced Efficiency of TEG and TEC

Following [44], Section 9.2.2 and [29] we can conclude that the local performance of an infinitesimal segment of a TE element of length dx with $dT = T'(x)\,dx$ can be defined as

$$\eta_{\text{loc}} = \frac{dT}{T}\eta_r \quad \text{and} \quad \varphi_{\text{loc}} = \frac{T}{dT}\varphi_r, \tag{4.10}$$

where dT/T is the infinitesimal Carnot cycle factor for TEG and T/dT the one for TEC. As the Carnot process is a reversible one, the reduced "efficiencies"* η_r and φ_r play the role of an "irreversibility factor" which at least measures the distance to reversibility for both TEG and TEC due to a nonperfect TE engine. Such considerations were published first by E. Altenkirch [107,108].

The reduced efficiency of a TEG η_r is defined as the ratio of the products of conjugated forces and fluxes [35] where we have to pay attention to the fact that the electrical power production in a volume dV is given by the production density $\pi_{el} = \mathbf{j} \cdot \mathbf{E}$, also denoted as differential electrical power. Note that the net differential power output is given by $-\pi_{el}$, see the definition of efficiency in Section 4.3 and [22,109,110]. From Equation 4.5, we find $\mathbf{j} \cdot \mathbf{E} - T\nu_S = \frac{1}{T}\mathbf{q} \cdot \nabla T = \mathbf{j}_S \cdot \nabla T$, and with $\mathbf{j} \cdot \mathbf{E} = \nabla \cdot \mathbf{q} = \mathbf{j} \cdot \nabla\Phi$ and $\mathbf{j}_S = \Phi\,\mathbf{j}/T$ we finally get

$$\eta_r = \frac{\mathbf{j} \cdot \mathbf{E}}{\mathbf{j}_S \cdot \nabla T} = \frac{\pi_{el}}{\pi_{el} - T\nu_S} = \frac{1}{1 - T\nu_S/\pi_{el}} = \frac{1}{1 + T\nu_S/|\,\pi_{el}\,|} \Rightarrow \eta_r = \frac{T\,\nabla\Phi \cdot \mathbf{j}}{\Phi\,\nabla T \cdot \mathbf{j}}, \tag{4.11}$$

which coincides with Clingman's result [101] and corresponds to the reduced variation of the TE potential $\nabla\Phi/\Phi$ when changing the other potential $\nabla T/T$ which is coherent with a general definition of the efficiency of an out-of-equilibrium thermodynamic process of coupled fluctuating parameters. The reduced efficiency expression can be rewritten from u and Φ expressions, that is, with $u = 1/(\phi - T\alpha)$ and $(\nabla\Phi \cdot \mathbf{j})/(\nabla T \cdot \mathbf{j}) = \alpha(1 - (\alpha/z)u)$ respectively $u = -(z/\alpha^2)(\nabla\Phi \cdot \mathbf{j})/(\nabla T \cdot \mathbf{j}) + z/\alpha$ [34,35]. The result is for TEG

$$\eta_r = \frac{1 - u\alpha/z}{1 + 1/(uT\alpha)} = \frac{u\alpha/z(1 - u\alpha/z)}{u\alpha/z + 1/(zT)} \tag{4.12a}$$

or

$$\eta_r = \frac{1 - \alpha/[z(\Phi - T\alpha)]}{1 + [z(\Phi - T\alpha)]/(zT\alpha)} = \frac{\alpha T}{\Phi}\left(1 - [zT\{\Phi/(\alpha T) - 1\}]^{-1}\right). \tag{4.12b}$$

An analogous approach can be found for the reduced coefficient of performance of a TEC, φ_r. As a consequence of the underlying TE effects (which are inverse to each other, and alike are the definitions of the global performance parameters, efficiency η and coefficient of performance φ),† the reduced coefficient of performance φ_r is inversely defined

* In Refs. [29] and [31] reduced efficiencies $\eta_r^{(g)} \equiv \eta_r, \eta_r^{(c)} \equiv \varphi_r$ are introduced for both TEG and TEC, respectively.

† We follow Sherman's notation here and use φ instead of *C.O.P.* in TEC formulae.

$$\varphi_r = \frac{\mathbf{j}_S \cdot \nabla T}{\mathbf{j} \cdot \mathbf{E}} = \frac{\pi_{el} - T\nu_S}{\pi_{el}} = 1 - T\nu_S/\pi_{el} = \frac{\Phi}{T}\frac{\nabla T \cdot \mathbf{j}}{\nabla \Phi \cdot \mathbf{j}}. \tag{4.13}$$

For a direct comparison of TEG and TEC we recommend a unified 1D model for both generator and cooler single elements [29]. Note that $u(T)$ differs formally only by sign if TEG and TEC are operated under reversed boundary temperatures, but otherwise in the same working conditions. For this case of directly comparing TEG and TEC we find formally $\varphi_r = 1/\eta_r$, and the reduced efficiencies present a maximum for $u = s$; where s is the compatibility factor [35] of a TEG, but $u_{opt} = s^{(g)} = \left(\sqrt{1+zT} - 1\right)/(\alpha T)$ of a TEG, but $u_{opt} = s^{(c)} = \left(-\sqrt{1+zT} - 1\right)/(\alpha T)$ of a TEC.

The reduced efficiency and local coefficient of performance are defined as functions of u in their ranges of typical use ($0 \le u \le 2s^{(g)}$ for TEG and $2s^{(c)} \le u \le 0$ for TEC). In the special situation of maximum local TEG efficiency ($u = s^{(g)} > 0$) and maximum local TEC coefficient of performance ($u = s^{(c)} < 0$) the two values are equivalent $\eta_{r,opt} = \varphi_{r,opt} = \left(\sqrt{1+zT} - 1\right)/\left(\sqrt{1+zT} + 1\right)$, as η_r and φ_r are local irreversibility factors. This again shows that zT is a thermodynamic materials quantity determining the maximum irreversibility factor that is the same for both interrelated TE effects, Seebeck and Peltier.

The equivalent optimal TE potential is given by

$$\Phi_{opt}^{(g/c)} = \alpha T + \frac{1}{s^{(g/c)}} = \alpha T \left[\frac{\sqrt{1+zT}}{\sqrt{1+zT} \mp 1}\right], \tag{4.14}$$

where the minus sign applies for TEG, but the plus sign for TEC.

Furthermore, we find

$$\eta_{r,opt} = \left(2\frac{\Phi_{opt}^{(g)}}{\alpha T} - 1\right)^{-1} \text{ and } \varphi_{r,opt} = 2\frac{\Phi_{opt}^{(c)}}{\alpha T} - 1.$$

The total efficiency η and the total coefficient of performance φ, respectively, of a finite generator and a cooler element, respectively, are obtained by summing up local contributions based on the reduced efficiency all over the TE element in an integral sense, see Equations 4.31a and 4.31b. The particular case of maximum performance of an infinitely staged* TEG and TEC has been investigated by Sherman et al. [111,112], see Section 4.6.1.

4.2.4 Thermodynamic Optimization

For each TE device it is known that the current has to be adjusted to an optimal value to reach maximum efficiency or coefficient of performance, respectively. The thermodynamic background is explained here. The local treatment leads to an additional requirement concerning TE element design. In Figure 4.2, we compare the reduced efficiency η_r of a TEG for varied values of the dimensionless figure of merit zT with the optimal reduced efficiency $\eta_{r,opt}$ (see also [113]). One can notice that, for a given element, the maximal efficiency can be obtained if the electrochemical potential and the temperature are biased in order to reach $\Phi = \Phi_{opt}$ all over the system. This "thermodynamic biasing" shows the importance of the working conditions of the TE engine, even at the local scale. As a consequence from Equation 4.14, FGM must be optimized from the material's point of view, leading to a $Tz(T)$ as large as possible, but also from the

* The device (or TE element) is broken up into an infinite number of stages. Note that the terms "perfectly infinitely staged element" and "self-compatible element" as introduced below can be used synonymously.

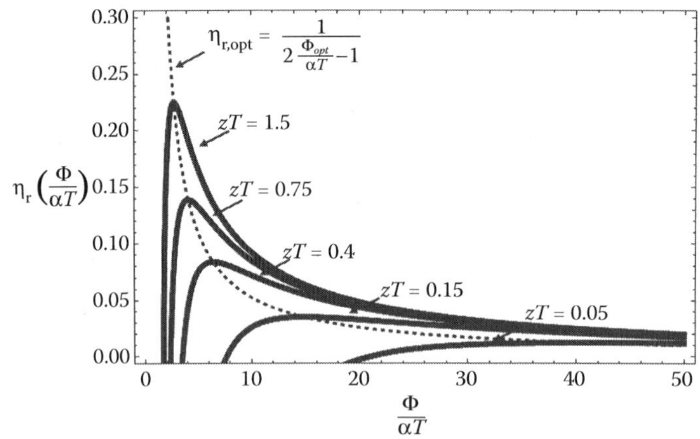

FIGURE 4.2 Reduced efficiency (irreversibility factor) for TEG as a function of the TE potential (in units of αT) $\Phi/(\alpha T)$.

working conditions point of view. This means in particular that the temperature distribution has to be adjusted to optimum shape. Otherwise the reduced efficiency of each cell (or segment) of a FGM will never reach its maximal attainable value simultaneously.

To sum up, we can draw three conclusions:

1. Simply adjusting maximum zT in each segment of graded TE elements is not sufficient to maximize TE device performance because the operation conditions at each position in the element cannot be adjusted independently from each other but are interrelated by the global condition of a common electrical current crossing each segment of the element.
2. Optimizing the temperature dependence of the material, that is, $\alpha(T)$, $\sigma(T)$, $\kappa(T)$, is applicable to achieve the maximum local efficiency expected from the zT in each segment of a chemically homogeneous element simultaneously. The compatibility approach is sufficient to achieve this, as will be described below.
3. FGM have to be optimized from a thermodynamic point of view. This should include consideration of the temperature and temperature gradient as will be demonstrated in the next section. Suitable adjustment of $\alpha(x)$, $\sigma(x)$, $\kappa(x)$ implies suitable shaping of the temperature profile.

4.3 Material Profiles and Performance Parameters in the 1D Steady State

The reduced "efficiencies" in our one-dimensional approach are given by

$$\eta_r(x) = \frac{1}{\varphi_r(x)} = \frac{\Phi'(x)/\Phi(x)}{T'(x)/T(x)} \text{ with } \Phi(x) = \alpha(x)T(x) + \frac{1}{u(x)}. \tag{4.15}$$

Equation 4.15 makes clear that the compatibility approach together with additional thermodynamic arguments opens up new opportunities for optimizing the material profiles, see also Ref. [103] and Section 4.6.2. However, before proceeding in this way, we should consider again the TE material properties α, σ, and κ which are in general temperature- and position-dependent quantities and measurable under certain constraints. Here we want to concentrate on decoupled dependencies, that is, either a temperature or spatial dependence of the material coefficients, to gain analytical results for the performance parameters of a TE element. A numerical investigation of a coupled material's local as well as temperature dependence is given by Kaliazin et al. [13]. Given the material properties as a function of temperature, Equation 4.3 reduces in the 1D steady state to

$$\kappa(T)\frac{\partial^2 T}{\partial x^2} + \frac{d\kappa}{dT}\left(\frac{\partial T}{\partial x}\right)^2 - j\,T\frac{d\alpha}{dT}\frac{\partial T}{\partial x} = -\frac{j^2}{\sigma(T)}, \tag{4.16}$$

where a constant current density j is supposed to flow in the x-direction (see, e.g., [27]).

It is important to note that if there is a "one-to-one correspondence" of temperature T and position x, the temperature profile is a continuous and strictly monotonous one (with that a bijective function). This especially applies for maximum η and maximum φ if constant or real, temperature-dependent material properties are considered.* Then, there exists an inverse function $x(T)$ corresponding to $T(x)$ and vice versa. If these conditions are fulfilled, Equation 4.16 can be transformed into

$$\frac{d}{dx}\left(-\kappa(x)\frac{dT(x)}{dx}\right) + j\,T(x)\frac{d\alpha(x)}{dx} = \frac{j^2}{\sigma(x)} \tag{4.17}$$

with $\sigma(x) = 1/\rho(x)$. Identical temperature profiles $T(x)$ are calculated from Equations 4.16 and 4.17 if spatial material profiles over the length of the TE element are given by $\alpha(x) = \alpha[T(x)]$, $\sigma(x) = \sigma[T(x)]$, and $\kappa(x) = \kappa[T(x)]$.

It is common knowledge that the material properties, especially those of semiconductors, highly depend on the carrier concentration which can be influenced by appropriate doping agents or a variation of the chemical composition [7,19,20,114–124]. Controlling the carrier concentration gives in principle the opportunity to get a spatially dependent material even if the temperature dependence is neglected [38,39]. This approach can be generalized for FGM. All material parameters may then depend not only on the temperature, but also on the local material quality which can be practically the composition of an alloy or the concentration of a dopant. Symbolically, for example, representing the concentration of a single doping species or alloying element, the local material quality can be denoted by a concentration variable c. Thus, the gradient of the Seebeck coefficient α is, for example, [85]

$$\nabla\alpha = \nabla\alpha\,|_{T=\text{const}} + \frac{\partial\alpha}{\partial T}\nabla T$$

leading to

$$\nabla\alpha(T,c) = \left.\frac{\partial\alpha}{\partial c}\right|_{T=\text{const}}\nabla c + \left.\frac{\partial\alpha}{\partial T}\right|_{c=\text{const}}\nabla T.$$

For the 1D case, and the x-coordinate used, the gradient of Seebeck can simply be written as $\nabla\alpha = d\alpha/dx\, e_x$ if the single element profiles $c(x)$ and $T(x)$ are known:

$$\alpha(x) = \alpha[c(x),\ T(x)].$$

* Within CPM we find for the slope of the temperature profile at the sink side $T'(L) = 0$ for the maximum temperature difference ($\varphi = 0$), and $T'(L) > 0$ for $\varphi > 0$. Note that the CPM is a suitable reference for moderately temperature dependent material properties.

These considerations have been the reason for establishing the spatial coordinate as the independent coordinate for an empirical approach to 1D, steady-state problems with graded materials. Naturally, Equation 4.17 can also be used as an independent differential equation when the spatial dependence of the material parameters $\alpha(x)$, $\sigma(x)$, $\kappa(x)$ is of prime interest, see, for example, [5,21,27,30,96].

We want to remark here that the differential equation (4.16) is nonlinear whereas Equation 4.17 containing spatial material profiles is linear in T which opens broader opportunities for finding analytical solutions and applying the principle of superposition.

Once having calculated $T(x)$, all performance parameters of interest can be determined as a function of the electrical current density j, for example,

- TEG: net power output density* (electrical power output per cross-sectional area A_c):

$$p_{\text{net}}(j) = -p(j) = -P(j)/A_c = -\int_0^L [\rho(x)j^2 + j\alpha(x)T'(x)]\,dx,$$

  where P is the electrical power output, p is the corresponding density.
- TEC: cooling power density (absorbed heat per time and cross-sectional area):

$$q_a(j) = Q_a(j)/A_c = [-\kappa(x)T'(x) + j\alpha(x)T(x)]_{x=0},$$

  where Q_a is the cooling power (absorbed heat per time unit).†
- TEG: efficiency $\eta(j) = p_{\text{net}}(j)/q_a(j)$
- TEC: coefficient of performance $\varphi(j) = Q_a(j)/P(j) = q_a(j)/p(j)$.

Concerning the notation we refer to Ref. [30]. For fixed material profiles (given gradients), the optimum current density can be calculated from the maximum of the device performance. For example, for arbitrary continuous monotonic gradient functions of the material profiles $\alpha(x)$, $\sigma(x)$, and $\kappa(x)$, the calculation of the temperature profile has been done numerically up to now either by a 1D finite-element method code (1D TE FEM) or the algorithm of multisegmented elements as well as other approaches, see for example, [2,3,5,6,12,13,23,26,29,125,111,112] and Section 4.5.1.

There is no doubt that global maximization of a performance parameter of the element as an integral device is a suitable guideline for an empirical approach to numerical device optimization, see for example, [7,114]. The target of 1D TE FGM research, however, is to find a set of optimal profiles $\alpha(x)$, $\sigma(x)$, $\kappa(x)$ for maximizing the performance of TE devices. Here we make use of the fact that analytical solutions of the generalized heat equation, Equation 4.17 can be found for spatially linear material profiles (see Section 4.5.2). These can actually be used as the starting point for finding more general optimal material profiles (regarding the profile shape).

Further on we will refer to the efficiency η of a TEG and to the coefficient of performance φ of a TEC, because maximization of these global device parameters can be deduced to local optimization, see Ref. [31] and Section 4.6. The power output (TEG) and the pumped heat (TEC) have to be treated separately (see Section 4.8).

* Power output is defined here according to thermodynamic rules: quantities put into the system are positive.

† Note that the dot for the time derivative is often omitted $Q_a \equiv \dot{Q}_a$.

4.4 The Maximum TE Performance

Throughout TE modeling, TE performance was initially considered assuming temperature-independent materials. The CPM is an adequate as well as historical reference for comparing both TE materials and devices.

4.4.1 Maximum Performance

The efficiency η of a TEG and the coefficient of performance φ of a TEC can be calculated from the temperature-dependent material parameters of n- and p-type elements, geometry, current, thermal, and electrical contact resistances [34] from suitably defined global averages of TE quantities.

Here we define the TE device figure of merit ZT in contrast to the materials figure of merit zT [126] using the well-known formulae for maximum efficiency and coefficient of performance for constant material properties [115,117,126–129]

$$\eta_{max} = \frac{T_a - T_s}{T_a} \frac{\sqrt{1+ZT} - 1}{\sqrt{1+ZT} + T_s/T_a} \tag{4.18a}$$

and

$$\frac{1}{\varphi_{max}} = \frac{T_s - T_a}{T_a} \frac{\sqrt{1+ZT} + 1}{\sqrt{1+ZT} - T_s/T_a}, \tag{4.18b}$$

where $T_a = T_h$ and $T_s = T_c$ for TEG and $T_a = T_c$ and $T_s = T_h$ for TEC. Given a maximum performance value one can subsequently solve for the device figure of merit ZT. For a precise definition, the values of T_s and T_a as well as whether the device is TEC or TEG must be specified—usually the T in ZT is assumed to be $T_m = 1/2(T_a + T_s)$. In general, this ZT also depends on the exact temperature-dependent material properties, exact geometry, even including nonidealities such as electrical and thermal losses (e.g., contact resistances, parasitic losses) and nonoptimized geometric parameters. However, since even the ideal CPM model requires electrical current optimization to achieve the maximum efficiency determined by ZT, the current should be adjusted (experimentally or theoretically) to achieve maximum efficiency when using this definition of ZT. For an ideal single element without parasitic losses the device ZT is equal to the material's zT within the framework of the CPM.*

4.4.2 Approximating Effective Device Figure of Merit

For an actual TE module, where the exact properties, dimensions, and interfaces of all the materials involved are not known, exact quantitative description as a device is restricted to the measurable parameters related to the TE material properties α, σ, and κ: the device effective Seebeck coefficient α_{eff}, electrical resistance R, and thermal conductance K. The exact procedures for measuring α_{eff}, K, R are not universally accepted [120,130–151] and therefore do not give exactly the same results. Nevertheless such values can be used to define an effective $Z_{eff}T$ as

$$ZT = z_{eff}T = \frac{\alpha_{eff}^2 T}{RK} \tag{4.19}$$

* Formulae (4.18) can be derived straightforwardly by using Equations 4.9 through 4.12 in Ref. [30], as well as within the symmetric CPM model where the n- and p-leg have equal properties with opposite Seebeck coefficients. In Ref. [30], you may also find the CPM formulae for the appropriate optimal current densities.

where it is common to use $T = T_m$. A good TE material is characterized by a large Seebeck coefficient to produce the TE voltage, a low thermal conductivity for limiting the dissipative Fourier heat flow throughout the device exposed to a temperature gradient and a low electrical resistivity (high electrical conductance) to minimize Joule's heating. Altogether these conditions are combined into the figure of merit [11,57,62,85,115,117,126,128,129,152–154]. By definition, see Equation 4.18, two devices with the same ZT (operated under the same conditions) have the same performance. Two devices with the same or similar $Z_{\text{eff}}\, T$ should reach similar performance to each other under similar conditions (as both TEG and TEC) but they need not be exactly equal.

Physically, it may seem most justified to use averaged materials properties to estimate the effective figure of merit and power factor f_{eff} such as

$$\frac{\alpha_{\text{eff}}^2}{RK} = \frac{\alpha_{\text{eff}}^2}{\left[\rho(x)\right]_{\text{av}}}\left[\frac{1}{\kappa(x)}\right]_{\text{av}} \quad \text{and} \quad f_{\text{eff}} = \frac{\alpha_{\text{eff}}^2}{\left[\rho(x)\right]_{\text{av}}}, \tag{4.20}$$

with

$$\begin{aligned} \alpha_{\text{eff}} &= \bar{\alpha},\ R = \frac{1}{A}\int \rho(x)\mathrm{d}x = \frac{L}{A}[\rho(x)]_{\text{av}}, \text{ and} \\ \frac{1}{K} &= \frac{1}{A}\int \frac{1}{\kappa(x)}\mathrm{d}x = \frac{L}{A}\left[\frac{1}{\kappa(x)}\right]_{\text{av}}, \end{aligned} \tag{4.21}$$

but also other formulations are used. Note that the overbar in Equation 4.21 corresponds to the average over temperature. For details of the averaging, see the next subsection.

4.4.3 Averaging the Material Profiles

At this point, we refer to a great variety of scientific work in experimentally measuring or determining the temperature dependence of the material coefficients, for an overview see Refs. [33,85,117,124,137, 152,153,155–159] and references therein.

The quantities zT and ZT are used here as single quantities in contrast to the traditional use of z or Z with units of K^{-1} (the traditional "figure of merit"). The dimensionless form is more fundamental with zT always being found together in thermodynamic equations. In CPM calculations, which are often applied for approximating the efficiency or coefficient of performance even for spatially or temperature-dependent material profiles (see, e.g., [39]), T is a reference temperature or an average of T_a and T_s. There is no doubt that zT is the "true" figure of merit from the thermodynamic point of view.* As early as in 1957 Ioffe considered that there are mainly two possibilities for averaging TE material properties in devices [117]: first, the average over space (denoted with the index "av") and second, temperature average of the Seebeck coefficient is marked with an overbar, see equation below. Both can be related to each other if the temperature profile is monotonous, for example, for Seebeck

$$\begin{aligned} \bar{\alpha} &= \frac{1}{T_s - T_a}\int_{T_a}^{T_s} \alpha(T)\mathrm{d}T = \frac{1}{L}\int_0^L \alpha(x)\frac{T'(x)}{(T_s - T_a)/L}\mathrm{d}x \\ &\approx \frac{1}{L}\int_0^L \alpha(x)\mathrm{d}x = \alpha_{\text{av}}, \end{aligned}$$

if $\alpha(x) = \alpha[T(x)]$, that is, if there is no explicit spatial dependence.

* Note that there is no apparent upper limit to zT although maximum zT has been discussed by several authors [160–165]. Already Harman and Honig derived $zT < 18$ from the viewpoint of elementary transport theory, see Ref. [85].

Note that $\int_0^L T'(x)\mathrm{d}x = \int_{T_a}^{T_s} \mathrm{d}T = T_s - T_a$; therefore the function $LT'(x)/(T_s - T_a)$ varies around unity, and both averages are close together for moderate gradients. Further note that both averages coincide in general for a linear temperature profile, but also for the CPM although it is linked to a particular non-linear profile (parabola). For temperature-dependent material Ioffe suggested

$$\rho_{av} = \ln\left(\frac{T_a}{T_s}\right) \int_{T_a}^{T_s} \rho(T)\frac{1}{T}\mathrm{d}T.$$

as an averaging formula for the resistivity. He stated there that the majority of the semiconductors have $\rho T^{-n} = \text{const.}$ as the temperature dependence of the resistivity. For that you get

$$\rho_{av} = \frac{1}{n}\ln\left(\frac{T_a}{T_s}\right)(\rho_a - \rho_s),$$

see Ref. [117]. The different averaging procedures are further discussed below in the context of the figure of merit.

For nonplanar-shaped elements, for example, ring-shaped elements, see [58,71–75,166], where the current flows are not one dimensional, the product of RK is independent of the shape. So does the effective figure of merit, see Equation 4.19. This can be shown, for example, with the relation

$$RK = \kappa\rho = \frac{\kappa}{\sigma}$$

for an ideal single and prismatic element, where the global parameters resistance $R = \rho(L/A_c)$ and thermal conductance $K = \kappa(A_c/L)$ are correlated to the local ones ($\kappa, \rho = \sigma^{-1}$), see for example, [128].

Doubtless, a method of approximating the device figure of merit ZT from the materials data is necessary for temperature-dependent or spatially dependent material to estimate the performance as given above, see Equation 4.18. Typically averaged values are used. In most cases either all material parameters are taken as temperature dependent (spatially dependent) or all are averaged. Averaging over space might seem adequate as the graded element is considered as a serial connection of segments. On the other hand, averaging over T is very practicable since it does not require knowledge of the temperature profile but can be simply deduced from temperature-dependent material properties. Formally, z_{av} can be calculated by averaging $\alpha^2(\sigma/\kappa)$ locally with respect to temperature or space or from individual averages (over temperature respectively over space) of each material profile which would lead to $z_{av} = \alpha_{av}^2(\sigma_{av}/\kappa_{av})$ for the purpose of easy analytical treatment. The latter has been done, for example, by Kaliazin et al. [13]. Alternatively, Ioffe [117], Moizhes [167], Borrego [168,169], and Efremov/Pushkars [170] defined

$$z_{eff} = \frac{(\bar{\alpha})^2}{(\overline{\rho\kappa})} = \frac{\left(\int_{T_a}^{T_s} \alpha(T)\,\mathrm{d}T\right)^2}{(T_s - T_a)\int_{T_a}^{T_s} \rho(T)\,\kappa(T)\,\mathrm{d}T}, \tag{4.22}$$

an averaging which we also recommend for purely temperature-dependent material properties. The formula is obtained starting with the calculation of the temperature distribution for the ideal case

($j = 0$), where you find from the thermal energy balance (4.16) that the heat flux is a constant and it follows that

$$qL = \int_0^L \kappa \frac{dT}{dx}\,dx = \int_{T_a}^{T_s} \kappa(T)\,dT = \bar{\kappa}\Delta T,$$

where $\Delta T = T_s - T_a$.* So you find in the absence of a current that the heat flux becomes

$$Q = \bar{\kappa} A_c \frac{\Delta T}{L} = K\Delta T \Rightarrow K = \bar{\kappa}\frac{A_c}{L}.$$

The overall resistance of the TE element can be easily obtained

$$\begin{aligned} R &= \frac{L}{A_c}\rho_{av} = \frac{1}{A_c}\int_0^L \rho(x)\,dx = \frac{1}{A_c}\int_{T_a}^{T_s} \rho(T)\frac{dx}{dT}\,dT \\ &= \frac{L}{\bar{\kappa}A_c}\frac{1}{\Delta T}\int_{T_a}^{T_s} \rho(T)\,\kappa(T)\,dT \\ \Rightarrow R &= \frac{\overline{\rho\kappa}}{\bar{\kappa}}\frac{L}{A_c}. \end{aligned}$$

Thus, we find for the locally averaged resistivity $\rho_{av} = \overline{\rho\kappa}/\bar{\kappa}$ which is nothing else than a weighted average of the (temperature dependent) resistivity by the (temperature dependent) thermal conductivity. The product RK is then the temperature average over the product of (temperature-dependent) resistivity and thermal conductivity $\overline{\rho\kappa}$. It is stated by the authors [117,167,169,170,171] that Equation 4.22 is valid only for the zero current limit but the use of these averaged properties still results in meaningful estimates of the device performance. Of course the exact method is to calculate the performance as described in Section 4.4.1.

Min et al. [172] also highlighted a method to calculate an effective $Z_{eff}T$, especially if large temperature differences are present.

The concept of effective figure of merit and power factor had been discussed in the framework of composite materials as the so-called *effective medium theory* [173,174–181]. In all of these cases a random distribution of different materials is supposed. It was found that this can enhance the power factor but not the efficiency of the TE material. This is clearly different from the situation as discussed here where we have not got a random inhomogeneity but a directed grading or segmentation.

4.5 Model-Free Setup Based on Spatial Material Profiles

In this section, a setup is used where an independent and free variability of the spatial material profiles $\alpha(x)$, $\sigma(x)$, and $\kappa(x)$ is assumed primarily, and where the temperature profile $T(x)$ can be calculated directly from Domenicali's thermal energy balance [94]. Formula (4.17) is rewritten here as

$$\kappa(x)T''(x) + \kappa'(x)T'(x) - j\,\alpha'(x)T(x) = -\rho(x)\,j^2. \tag{4.23}$$

* Note that the sign of ΔT can be both positive or negative depending on how the TE element operates (TEG/TEC).

Once knowing $T(x)$, all application-relevant device properties and performance parameters can be derived. Executing the calculation in a loop while varying the current allows selecting optimum operation parameters (optimum load resistance and current, respectively) and deducing maximum performance values.

The aim of such investigations is to identify optimum combinations of material gradients along a graded generator or cooler element. However, the practical accessibility of optimum gradient schemes is limited by the constraint of maximum zT of available materials, by the magnitude of the compositional gradient (extremely steep profiles are difficult to prepare in a controlled manner), and by interrelations between the TE properties due to the solid-state nature of the TE materials, for example, the Wiedemann–Franz law connecting electronic part of thermal conductivity and electrical conductivity.

4.5.1 Parameter Studies via Numerical Calculations

Fundamental parameter studies [23,25] have demonstrated quantitative improvement of performance achievable by material grading. Numerical calculations were based on the aforementioned model-free setup applying arbitrarily but suitably chosen continuous monotonic gradient functions for all material profiles (see e.g., [24])

$$y(x) = y_a + \frac{1 - \exp(k_y x/L)}{1 - \exp(k_y)}(y_s - y_a), \quad y = \alpha, \sigma, \kappa \tag{4.24}$$

which fix the values y_a and y_s at the absorbing and sink side, respectively, but allow varying the curvature by the numerical parameter k_y. It was found that best performance is achieved close to linear profiles in certain cases (see Figure 4.3), whereas an unfavorable combination of slopes such as strong gradients of α and σ oriented in parallel lead to a significant drop of the performance. Further it is worth mentioning that, related to the higher optimum current density, achievable effects are larger for TECs than for a TEG although the operating temperature difference was assumed much larger for generators. The strongest improvement for Peltier elements can be achieved at the maximum temperature difference if steep material gradients are selected. Concerning a detailed discussion we refer to Refs. [23,25].

4.5.2 Analytical Solution Based on Linear Material Profiles

Variation restricted to linear spatial material profiles provides hints to favorable configurations of material profiles that deliver optimum performance. The linear properties model (LPM) is—as well as the CPM—of course limited in comparison with real temperature-dependent materials, but it can give hints, for example, for optimized stacking schemes.

Investigations based on linear spatial profiles have already been done in the mid-1960s by Ybarrondo [38] and Ybarrondo and Sunderland [39]. Although they could give an analytical solution for the temperature profile (for a TE cooler with mixed boundary conditions), the performance could not be determined quantitatively by them due to the lack of suitable computation tools. Therefore, we revisit the corresponding solutions in particular to gain information about the influence of the slope of the material parameters on the performance, and we calculate the corresponding optimal values in our particular case of linear spatial profiles as an interesting example of FGM.

Assuming linear profiles, the thermal energy balance (4.23) can be written in the notation

$$[a(\kappa)\,\xi_\kappa + b(\kappa)\,x]\,T''(x) + b(\kappa)T'(x) - j\,b(\alpha)T(x) = -j^2\,[a(\sigma)\xi_\sigma + b(\sigma)x]^{-1} \tag{4.25}$$

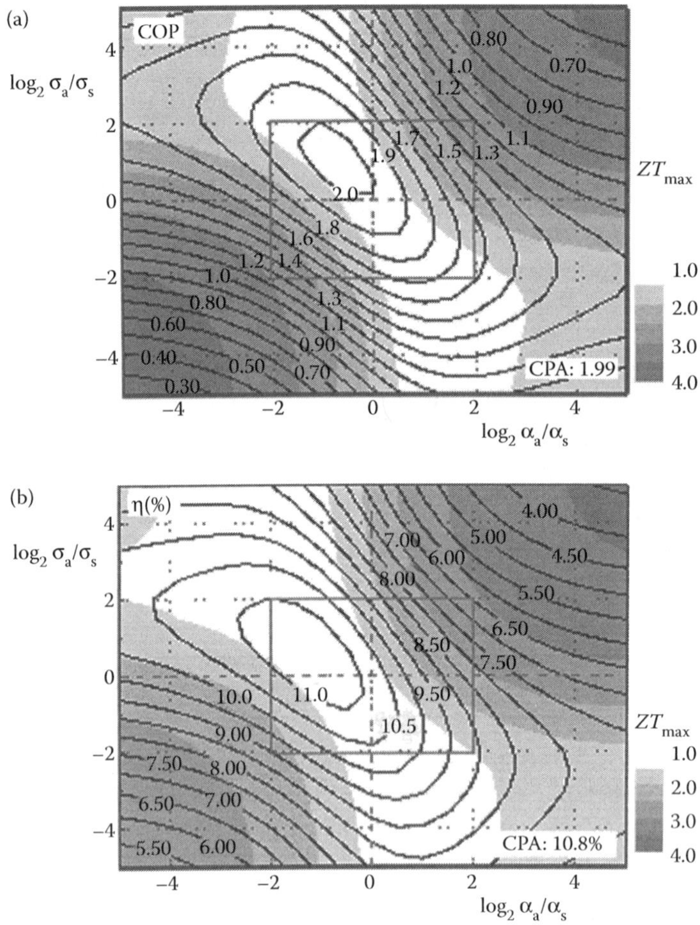

FIGURE 4.3 Variation of the maximum φ of a TEC for $T_a = 280$ K, $T_s = 300$ K (a), and maximum efficiency of a TEG for $T_a = 600$ K, $T_s = 300$ K (b) plotted versus the chosen constant gradients of the Seebeck coefficient and electrical conductivity (see also Ref. [23]); under the constraint of $(zT)_{av} = $ const.; double-logarithmic plot: $\log_2(\alpha_a/\alpha_s)$ (abscissa), $\log_2(\sigma_a/\sigma_s)$ (ordinate) in both figures; The inner frame marks the region of small to moderate gradients. The gray-tone background indicates the maximum zT which is found locally within the respective gradient configuration. For the calculation we chose $\alpha_{av} = 180$ μV/K, $\sigma_{av} = 140{,}000$ S/m and κ_{av} such that the spatial average of the figure of merit is unity, that is, $(zT)_{av} = 1$.

using the abbreviations

$$\xi_y = \frac{y_a}{y_s}, \quad a(y) = \frac{2y_{av}}{1+\xi_y}, \quad \text{and} \quad b(y) = \frac{2y_{av}}{L}\frac{(1-\xi_y)}{(1+\xi_y)} = \frac{\Delta y}{L}$$

with $y = \kappa, \sigma, \alpha$. It is obvious that b gives the appropriate slope of the material profile, for example, $\kappa'(x) = b(\kappa)$. As free parameters for the linear profiles of the material we chose the fixed spatial average y_{av} to compare with CPM and the grading parameters ξ_y. Note that there is another form of writing the material profiles with the chosen abbreviations, for example, $\kappa = a(\kappa)\xi_\kappa + b(\kappa)x$, with $\xi_\kappa = (\kappa_a/\kappa_s)$ being a primary characteristics of the linear profile. To get the Bessel equation from the homogeneous part of the thermal energy balance we further use the substitution

$$
\begin{aligned}
w(x) &= -2\sqrt{\frac{-j\, b(\alpha)}{b^2(\kappa)}}\sqrt{\kappa(x)} \\
&= -2L\sqrt{\frac{j(\alpha_a - \alpha_s)}{L(\kappa_s - \kappa_a)^2}}2\sqrt{\kappa(x)} \equiv A\sqrt{\kappa(x)}
\end{aligned} \tag{4.26}
$$

leading to $T''(w) + w^{-1}T'(w) + T(w) = 0$ for the homogeneous part of Equation 4.25 which is equivalent to a Bessel differential equation of order 0 (after multiplying with w^2). The completely transformed inhomogeneous differential equation is finally [182]

$$
T''[w(x)] + w^{-1}(x)T'[w(x)] + T[w(x)] = \frac{j}{b(\alpha)\sigma_{av}}. \tag{4.27}
$$

The temperature profile is found in terms of Bessel functions $Y_0(w)$, $J_0(w)$ with w given by Equation 4.26 for the general case $\xi_\alpha \neq 1$, $\xi_\sigma \neq 1$, $\xi_\kappa \neq 1$

$$
\begin{aligned}
T(x) &= C_1 Y_0[w(x)] + C_2 J_0[w(x)] \\
&\quad + \frac{\pi j^2 L}{\Delta\kappa}\left[J_0[w(x)]\int_0^L \frac{Y_0[w(x)]}{\sigma(x)}\mathrm{d}x - Y_0[w(x)]\int_0^L \frac{J_0[w(x)]}{\sigma(x)}\mathrm{d}x\right].
\end{aligned} \tag{4.28}
$$

The free constants C_1 and C_2 can be fixed according to the given boundary conditions. For an overview on Bessel functions we refer to standard textbooks [183,184]. The tables of all particular solutions of Equation 4.27 are given in the appendix of Ref. [182].

Although the CPM as well as the LPM cases appear physically simple, these are not at all simple from the point of view of materials' fabrication [7,19,20,114]; the question how to obtain TE materials with linear material profiles is still an open issue.

4.5.3 TEG Performance Optimization via Linear Grading

Once the temperature profile $T(x)$ throughout an element has been calculated, the performance parameter, for example, power output density p and efficiency η of a TEG depending on the electrical current density j and the grading parameters ξ_α, ξ_σ, and ξ_κ can be determined and can consequently be optimized according to the variation of these. Considering power output here, we find Equation 4.29 for a constant electrical conductivity ($\xi_\sigma = 1$), which is supposed to be in the vicinity of the maximal performance, after integration by parts [40]

$$
p \equiv p(j,\xi_\alpha,\xi_\kappa) = -\frac{j^2 L}{\sigma_{av}} - \frac{2\alpha_{av}}{1+\xi_\alpha}[T_s - T_a\xi_\alpha - (1-\xi_\alpha)T_{av}] \tag{4.29}
$$

with $T_{av} = L^{-1}\int_0^L T(x)\mathrm{d}x$. In Equation 4.29 the grading of κ enters through that spatial average of the temperature T_{av}. Here, the particular case of having only one linear profile for the Seebeck coefficient ($\xi_\alpha \neq 1$) shall be shortly discussed. For that the material parameters are chosen as $\alpha_{av} = 180$ μV/K, $\kappa_{av} = 1.35$ W/m K, and $\sigma_{av} = 140{,}000$ S/m, exemplarily. Thus zT at $T = 300$ K equals ~1. In the CPM case, a simple extremal problem to get optimum current densities j_{opt} leading to the optimal value of the performance parameter p_{max} and η_{max} has to be solved, whereas for LPM one ends up with a multidimensional extremal problem with a system of equations leading to the optimum current densities j_{opt} and grading parameters ξ_{opt}. Both TEG performance parameters are displayed in Figure 4.4 for a TE element of length $L = 5$ mm and a temperature difference of $\Delta T = 300$ K. The results for CPM and LPM are compared in Table 4.1.

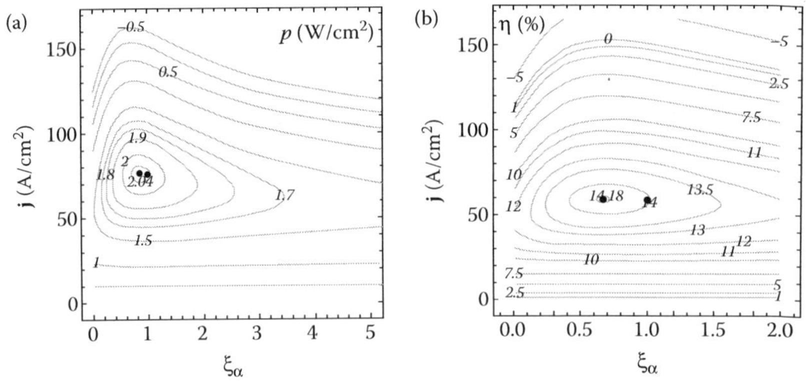

FIGURE 4.4 Performance parameter in dependence of j and ξ_α for $T_a = 600$ K and $T_s = 300$ K, $L = 5$ mm. The other two grading parameters ξ_κ and ξ_σ are set to unity, that is, σ and κ are constant; (a) electrical power output density p in W/cm², (b) efficiency η in percent. In comparison to Figure 4.3 there is no constraint for zT in this calculation. We fixed the material parameters to $\alpha_{av} = 180$ μV/K, $\sigma_{av} = 140{,}000$ S/m and $\kappa_{av} = 1.35$ W/m K.

TABLE 4.1 Values of the Optimum Performance Parameters for the Given Material

	Maximum Output Power Output Density			Maximum Efficiency		
	$j_{p,opt}$ (A/cm²)	$\xi_{\alpha,opt}$	p^{max} (W/cm²)	$j_{\eta,opt}$ (A/cm²)	$\xi_{\alpha,opt}$	η^{max} (%)
CPM	75.6	1	2.041	58.5	1	14.03
LPM	76.1	0.85	2.048	59.0	0.67	14.19

By linear grading of α both the power output and the efficiency can be slightly increased. As for CPM the optimal parameters are different, depending on what has to be optimized, p or η, but we find the same tendency in the optimization strategy. Obviously, there is only a slight increase when assuming $\kappa = $ const.; with linear profiles $\alpha(x)$ and $\kappa(x)$ the increase of power output and efficiency in the TEG case is more pronounced, see Refs. [40,182,185].

The same effect is expected for the TEC case with linear profiles $\alpha(x)$ and $\sigma(x)$. Note that all these conclusions are found supposing fixed spatial averages of α, σ, and κ as a limit to the material performance. Hence, in the light of the following has to be mentioned that the optimum configurations which exceed the CPM in performance are bound to a slightly higher z_{eff}, see Equation 4.20, than in the CPM case. This applies both to the results given in Figures 4.3 and 4.4. The observed nonsymmetry is mainly founded in the nonsymmetric contribution of α (squared) and σ (linear) to the figure of merit. This is another, numerical evidence on the conclusion of Sections 4.4.2 and 4.4.3 that fixing the spatial averages of α, σ, and κ is not a physically justified reference of "equally good" material, neither is the fixing of the spatial average of zT as was applied in the calculations for Figure 4.3.

4.6 Compatibility Approach and Optimum Material Grading

In 2002/2003, Snyder and Ursell [35,36,106] introduced the *compatibility factor* as a second characteristic besides the zT for optimizing the performance of TE devices. By introducing the relative current density u and the reduced efficiency η_r (see Section 4.2), the TE power generation could be formulated at first at the level of intensive state variables. Building on earlier investigations dating from the 1960s [39,53,85,101,111,115,186,187], this concept has been further developed in successive works [28,29,30, 46,104,105] on the basis of a one-dimensional, stationary and unifying model with material grading for the TEG and the TEC.

The compatibility approach is an alternative to Ioffe's description using global power terms which is very often used for technological applications, but is certainly not suitable for locally characterizing TE processes or even for local optimization purposes.

The advantage of using the relative current density is that the multidimensional TE problem can be reduced to a one-dimensional heat flow problem* formulated in $u(T)$ where the governing equation can be evaluated from the thermal energy balance [35]

$$\frac{\mathrm{d}}{\mathrm{d}T}\left(\frac{1}{u}\right) = -T\frac{\mathrm{d}\alpha}{\mathrm{d}T} - u\rho\kappa \quad \text{or} \quad u'(T) = \tau u^2 + \rho\kappa\, u^3, \tag{4.30}$$

alternatively, with the Thomson coefficient $\tau(T) = T\,\mathrm{d}\alpha/\mathrm{d}T$. Basically, Equation 4.30 describes a homogeneous TE element with temperature-dependent material properties, whereby (4.30) holds for both the "pump up" ($T'(x) > 0$) and "pump down" ($T'(x) < 0$) cases.† Further note that TEG and TEC cases are only distinguished by the sign of $u(T)$ if the same (temperature dependent) material properties, the same current density, but reversed boundary temperatures T_a and T_s are applied ($T_a > T_s$ for TEG, but $T_a < T_s$ as usual for TEC).

Meanwhile it has been shown that sufficient compatibility is—besides a high figure of merit z_{eff}—essential for efficient operation of a TE device, and that compatibility will facilitate rational materials selection, device design, and FGM engineering, see, for example, [26,33,34,46,104–106]. Currently, compatibility factors $s = u_{\mathrm{opt}}$ are available for performance parameters arising from the local contribution to the TE material; these are:

- TEG: compatibility factor for maximal electric power ($s^{(g,P)}$ or $s^{(P)}$) [28]: $s^{(P)} = z/(2\alpha)$
- TEG: compatibility factor for maximal η ($s^{(g)}$) [35]: $s^{(g)} = \frac{\sqrt{1+zT}-1}{\alpha T}$
- TEC: compatibility factor for maximal φ ($s^{(c)}$) [29]: $s^{(c)} = \frac{-\sqrt{1+zT}-1}{\alpha T}$

The importance of the compatibility approach has been demonstrated first for a segmented TE generator [34,35,106]. Snyder pointed out that, if the compatibility factors in segmented devices with given material differ by a factor of 2 or more, the maximum efficiency of a TEG can in fact decrease by segmentation. Compatibility is therefore of essential importance for a rational material selection in segmented devices.

Alternatively, compatibility can be considered in the same material [35,36,106]. One of the major objectives in FGM research is to find optimal graded TE elements (the so-called self-compatible elements, where $u(x) \approx s(x)$) to achieve maximum performance. Sherman et al. [111] referred to perfectly infinitely staged material which gives the maximum efficiency or coefficient of performance. Within the new concept of self-compatibility the guidelines toward a perfectly infinitely staged, that is, ideally self-compatible material are now well defined. However, because both zT and s depend on the same materials parameters, a compromise between high zT and self-compatibility will need to be reached for true optimization. Example calculations of self-compatible elements for maximum η (TEG) and maximum φ (TEC) are presented in Section 4.6.2, concerning power-related compatibility, see Section 4.8.

* For the history of local definition of efficiency see also Ref. [186]; in Ref. [70] the three-dimensional vector version is presented. A more general definition of u seems possible when writing the relative current density in terms of fluctuating currents which are indeed 3D.

† Erratum: Ref. [29], Equation 4.9, right-hand side, the ± sign has to be replaced by a minus sign valid for both the pump up and the pump down situation.

4.6.1 FGM and Self-Compatibility

The exact solution for η and φ with respect to temperature, that is, for temperature-dependent material parameters are the following integrals taken from Ref. [31] for TEG:

$$\ln(1-\eta) = \int_{T_a}^{T_s} K(u(T),T)\,dT \quad \text{with} \quad T_s \le T \le T_a, \tag{4.31a}$$

and for TEC

$$\ln\left(1+\frac{1}{\varphi}\right) = \int_{T_a}^{T_s} K(u(T),T)\,dT \quad \text{with} \quad T_a \le T \le T_s, \tag{4.31b}$$

where we have one kernel $K(u(T), T)$ for integrals of both génerator and cooler

$$K(u,T) = \frac{1}{T}\eta_r(u,T) = \frac{1}{T}\frac{1}{\varphi_r(u,T)} = \frac{1}{T}\frac{u\alpha/z(1-u\alpha/z)}{u\alpha/z+1/(zT)} = \frac{\alpha}{z}\frac{(z-u\alpha)}{(u^{-1}+\alpha T)}. \tag{4.32}$$

Note that K contains the reduced efficiencies η_r for TEG, respectively, φ_r for TEC. Alternatively, the integral kernel K can be formulated with the TE potential Φ, see Equation 4.12. For this case, Snyder [34,35] has shown that η is simply given by the relative change of the TE potential with temperature variation; an analogous relation can be found for the coefficient of performance (see [29])

$$\eta = 1-\frac{\Phi(T_s)}{\Phi(T_a)} \quad \text{and} \quad \varphi = \left(\frac{\Phi(T_s)}{\Phi(T_a)}-1\right)^{-1}. \tag{4.33}$$

This result points to the importance of the TE potential as a function of state, for details see also Ref. [103].

If we assume the feasibility to achieve complete self-compatibility (infinite staging) we can apply $u = s^{(g)}$ and $u = s^{(c)}$ to the integrals (4.31a) and (4.31b), respectively, so that they take their maximal values with the optimal reduced efficiency $\eta_{r,opt} = \varphi_{r,opt} = \left(\sqrt{1+zT}-1\right)/\left(\sqrt{1+zT}+1\right)$ for both TEG and TEC [111,117]. Then fully self-compatible performance parameters η_{sc} and φ_{sc} are given by

$$\ln(1-\eta_{sc}) = \int_{T_a}^{T_s} \frac{\eta_{r,opt}}{T}\,dT = \int_{T_a}^{T_s} \frac{1}{T}\frac{\sqrt{1+zT}-1}{\sqrt{1+zT}+1}\,dT \tag{4.34a}$$

and

$$\ln\left(1+\frac{1}{\varphi_{sc}}\right) = \int_{T_a}^{T_s} \frac{1}{T\varphi_{r,opt}}\,dT = \int_{T_a}^{T_s} \frac{1}{T}\frac{\sqrt{1+zT}+1}{\sqrt{1+zT}-1}\,dT \tag{4.34b}$$

or, using Sherman's notation* [111]

* Note that Sherman et al. used the function $\epsilon(T)$ to denote the optimal reduced efficiency.

$$\eta_{sc} = 1 - \exp\left(-\int_{T_s}^{T_a} \frac{1}{T} \frac{\sqrt{1+zT}-1}{\sqrt{1+zT}+1} \mathrm{d}T \right) \tag{4.35a}$$

and

$$\varphi_{sc} = \left[\exp\left(\int_{T_a}^{T_s} \frac{1}{T} \frac{\sqrt{1+zT}+1}{\sqrt{1+zT}-1} \mathrm{d}T \right) - 1 \right]^{-1}. \tag{4.35b}$$

We expressly emphasize, however, that the integrals (4.34) and (4.35) do not have extremal properties concerning the zT value. Analytical expressions of these integrals can be found for $z = \text{const.}$ as well as for $zT = \text{const.}$; a summary of all integral approximations is presented in the appendix of Ref. [31]. The constraint $zT = \text{const.}$ has proven to be advantageous for the construction of self-compatible elements, see next section.

By using the transformation $\int_{T_a}^{T_s} \mathrm{d}T = \int_0^L T'(x)\,\mathrm{d}x$ with a monotonous $T(x)$, we find equivalent integrals with the kernel related to the spatial coordinate

$$K^*(u(x),x) = K[u(T(x)),T(x)]T'(x). \tag{4.36}$$

Analogous results are obtained from both formulations if $\alpha(x) = \alpha[T(x)]$, $\sigma(x) = \sigma[T(x)]$ and $\kappa(x) = \kappa[T(x)]$.

The integral representations for η and φ have been discussed in Ref. [31] within the framework of variational calculus. It was found that the optimal u can only be transferred from T to x and vice versa if the optimal (and monotonous) temperature profile is known; the latter must be consistent if $u(x) = s(x)$ has to be fulfilled locally. Then, fully self-compatible performance parameters η_{sc} and φ_{sc} are given with the integral kernel K^* (see Equation 4.36) by the integrals

$$\ln(1 - \eta_{sc}) = \int_0^L K^*(s^{(g)}(x),x)\,\mathrm{d}x \tag{4.37a}$$

and

$$\ln\left(1 + \frac{1}{\varphi_{sc}}\right) = \int_0^L K^*(s^{(c)}(x),x)\,\mathrm{d}x \tag{4.37b}$$

These considerations implicate that two strategies can be established to achieve Snyder's criterion $u = s$ in a local sense, as the condition of self-compatibility, for all infinitesimal segments of a TE element (within the interval $0 \le x \le L$ or $T_a \le T < T_s$):

(A) Optimization based on Equation 4.30 and the criterion $u(T) = s(T)$, mainly used for temperature-dependent materials.
(B) Optimization based on Equation 4.23 and the criterion $u(x) = s(x)$ for FGM containing an explicit dependence of the properties on x.

Consistent optimization results are obtained if equivalent material profiles are used. However, it is important to note that constraints or a performance limit must be regarded as well, see Section 4.6.2 for an example.

A central problem is that only two governing equations are available for both optimization strategies when referring to thermoelectricity from a phenomenological point of view. They are, in general, not sufficient for calculating all three optimal material profiles. In addition, the temperature profile $T(x)$ has to be calculated in a consistent manner when $u(x) = s(x)$ is used as (thermodynamic) optimization criterion; it can be rewritten as a first-order differential equation for the optimum temperature profile based on the "coordinate" zT [30]

$$\frac{dT}{dx} = \frac{j_o}{\sigma\alpha} f(zT) \quad \text{with} \quad f(zT) = \frac{zT}{1 \pm \sqrt{1 + zT}}. \tag{4.38}$$

The positive sign applies to the TEC ($f = f^{(c)}$), but the negative one to the TEG ($f = f^{(g)}$).

An optimization strategy referring to item (B) has been proposed in [30]. It has become apparent that self-compatible elements can only be constructed based on an optimum combination of material profiles whereas there is not only a single, uniquely defined set of $\alpha(x)$, $\sigma(x)$, and $\kappa(x)$ but a manifold with two degrees of freedom. Only one profile out of the three properties can be calculated based on the optimization criterion found while two material profiles can be specified arbitrarily to fix an optimum set. The remaining degrees of freedom can be used, for example, to involve interrelations between the TE properties due to solid state nature of the TE materials. This strategy has been tested in Ref. [30] with presumed constant gradients of α and σ having opposite directions, and the thermal conductivity κ has been optimized.

From first results published in Ref. [30] it can be concluded that there is only a little reserve for TEG performance improvement when using optimized material gradients, but much more potential for the performance improvement of a TEC. However, we should emphasize here that the choice of the given profiles determines greatly the increase in performance from the effect of the self-compatibility: Optimization should be based on preset profiles $\alpha(x)$ and $\kappa(x)$ for TEG, but on profiles $\alpha(x)$ and $\sigma(x)$ for TEC. In any case, an ultimate performance limit has to be set, for example, by a $z_{max}(T)$ curve or by a constraint $z = \text{const.}$ or $zT = \text{const.}$, respectively, whereby the constant should be related to an average of the figure of merit z or zT, if direct comparison to a real material shall be made.

The number of predefined profiles can be reduced to only one, when the zT constraint is applied, while an important fact is that this constraint may not be realistic from the point of view of materials' actual availability or preparation.

4.6.2 Self-Compatible Elements for Maximum η and Maximum φ

Throughout this section, a constant dimensionless figure of merit $zT = k_0 = \text{const.}$ is assumed. Then, the optimal reduced efficiency is also a constant*

$$\eta_{r,opt} = \varphi_{r,opt} = \frac{\sqrt{1 + zT} - 1}{\sqrt{1 + zT} + 1} \Rightarrow \eta_{r0} = \frac{\sqrt{1 + k_0} - 1}{\sqrt{1 + k_0} + 1},$$

and the ideally self-compatible performance parameters η_{sc} for TEG and φ_{sc} for TEC are given by Ref. [31]

$$1 - \eta_{sc} = \left(\frac{T_s}{T_a}\right)^{\eta_{r0}} \quad \text{for TEG, and } 1 + \frac{1}{\varphi_{sc}} = \left(\frac{T_s}{T_a}\right)^{1/\eta_{r0}} \quad \text{for TEC.} \tag{4.39}$$

Further, we define here $\eta_{r0} = -\omega_{0,g}/\omega_{0,c}$ with $\omega_{0,g} = \sqrt{1 + k_0} - 1$ for TEG and $\omega_{0,c} = -\left(\sqrt{1 + k_0} + 1\right)$ for TEC. Equation 4.30 shall be evaluated now for the optimal $u_{opt} = s$. Note that the material is

* Note that we use only one constant η_{r0} for both TEG and TEC without distinction, because they are identical $\eta_{r0} = \varphi_{r0}$.

represented in Equation 4.30 by $\alpha(T)$ and the product $\rho(T)\kappa(T)$; both can be optimized based on the constraint $zT = k_0$. Applying the compatibility factor $s^{(g)}$ respectively $s^{(c)}$, we find

$$\frac{1}{u} \equiv \frac{1}{s^{(g)}} = \frac{\alpha T}{\omega_{0,g}} \quad \text{for TEG} \quad \text{and} \quad \frac{1}{u} \equiv \frac{1}{s^{(c)}} = \frac{\alpha T}{\omega_{0,c}} \quad \text{for TEC.} \tag{4.40}$$

The left-hand side of Equation 4.30 is then for TEG

$$\begin{aligned}\frac{d}{dT}\left(\frac{1}{u}\right) &\equiv \frac{d}{dT}\left(\frac{1}{s^{(g)}}\right) = \frac{d}{dT}\left(\frac{\alpha T}{\omega_{0,g}}\right)\\ &= \frac{1}{\omega_{0,g}}\frac{d}{dT}(\alpha T) = \frac{1}{\omega_{0,g}}\left[\alpha + T\frac{d\alpha}{dT}\right]\end{aligned} \tag{4.41}$$

and for TEC

$$\begin{aligned}\frac{d}{dT}\left(\frac{1}{u}\right) &\equiv \frac{d}{dT}\left(\frac{1}{s^{(c)}}\right) = \frac{d}{dT}\left(\frac{\alpha T}{\omega_{0,c}}\right)\\ &= \frac{1}{\omega_{0,c}}\frac{d}{dT}(\alpha T) = \frac{1}{\omega_{0,c}}\left[\alpha + T\frac{d\alpha}{dT}\right].\end{aligned} \tag{4.42}$$

For the right-hand side, the definition of the figure of merit gives together with the constraint

$$z = \frac{\alpha^2\sigma}{\kappa} = \frac{\alpha^2}{\rho\kappa} \Rightarrow \rho\kappa = \frac{\alpha^2}{z} = \frac{\alpha^2 T}{k_0}, \tag{4.43}$$

so that we have $s_g\rho\kappa = \omega_{0,g}\,\alpha/k_0$ for TEG, and $s_c\rho\kappa = \omega_{0,c}\,\alpha/k_0$ for TEC.

Thus, for both cases, the differential equation for the optimum Seebeck profile $\alpha(T)$ is given by

$$\frac{1}{\omega_0}\left[\alpha + T\frac{d\alpha}{dT}\right] = -T\frac{d\alpha}{dT} - \frac{\omega_0\alpha}{k_0}, \tag{4.44}$$

where $\omega_0 \equiv \omega_{0,g}$ for TEG and $\omega_0 \equiv \omega_{0,c}$ for TEC, respectively.

Equation 4.44 can be solved by separation of variables to get a power-law behavior for the optimal Seebeck coefficient where $\alpha_{ref} = \alpha(T_{ref})$:

$$\text{TEG: } \alpha(T) = \alpha_{ref}\left[\frac{T}{T_{ref}}\right]^{k_g}$$

with

$$k_g = -2\frac{\omega_{0,g}}{k_0} = \frac{2\left(1 - \sqrt{1 + k_0}\right)}{k_0} = \eta_{r0} - 1 \tag{4.45a}$$

$$\text{TEC: } \alpha(T) = \alpha_{ref}\left[\frac{T}{T_{ref}}\right]^{k_c}$$

with

$$k_c = -2\frac{\omega_{0,c}}{k_0} = \frac{2\left(1 + \sqrt{1 + k_0}\right)}{k_0} = \frac{1}{\eta_{r0}} - 1 \tag{4.45b}$$

As a result, this optimization strategy provides an analytical expression for the optimum Seebeck profile $a(T)$ for both TEG and TEC. Note that the differences in the optimum Seebeck profiles for TEG are marginal for different values k_0 (see Figure 4.5a), and that $\alpha(T)$ is only slightly curved especially for the small temperature interval ($\Delta T = 20$ K) shown for TEC in Figure 4.5b. With given Seebeck coefficient, a second expression is found for the product $\rho(T)\ \kappa(T) = T\alpha^2(T)/k_0$ (see Figure 4.6), whereas an infinite number of profiles $\kappa(T)$ and $\rho(T) = 1/\sigma(T)$ can be found which fulfill the optimal $\rho(T)\kappa(T)$ product. Note that these optimization results are valid for a homogeneous element which is described by Equation 4.30 based on temperature-dependent material parameters. For that reason, the results presented here are only suitable for FGM in limited circumstances when further local contributions to the material properties are marginal. In this case, the optimal spatial Seebeck profile $\alpha(x) = \alpha(T(x))$ is

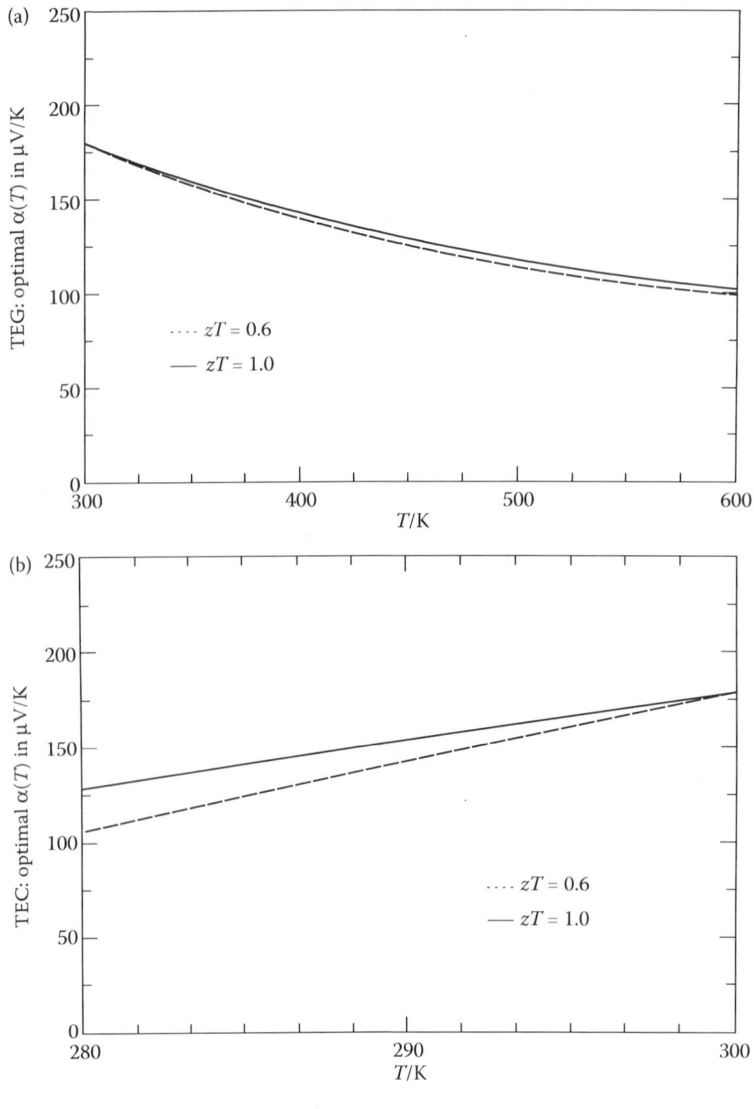

FIGURE 4.5 Optimal profile of the Seebeck coefficient for TEG ((a) $T_a = 600$ K, $T_s = 300$ K) and TEC ((b) $T_a = 280$ K, $T_s = 300$ K) plotted over temperature for the constraints $zT = 0.6$ and $zT = 1$ with reference value $\alpha_{ref} = 180$ μV/K for $T_{ref} = 300$ K.

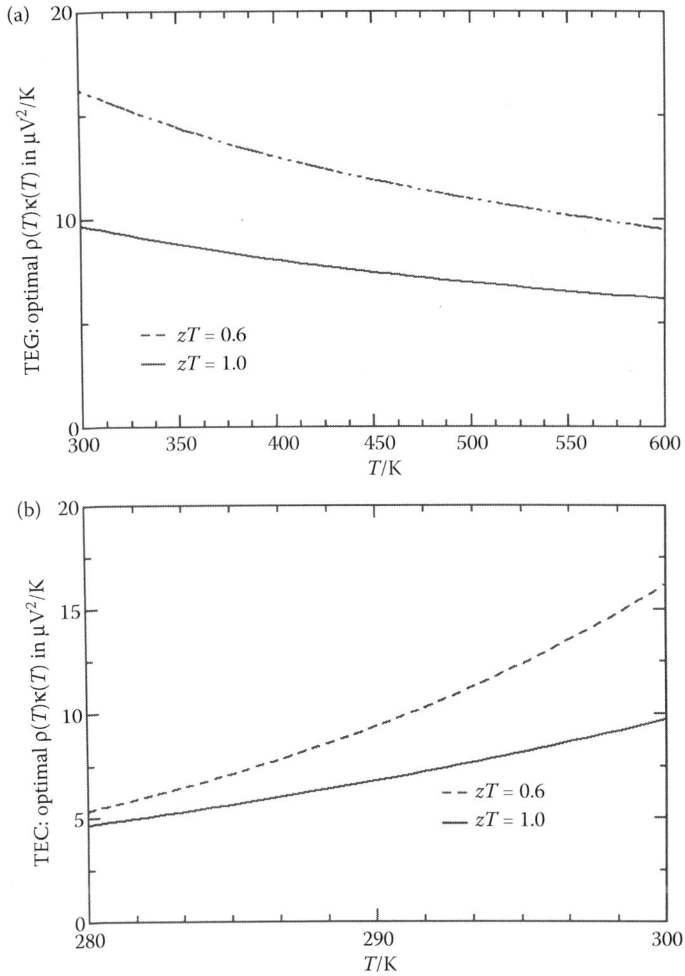

FIGURE 4.6 Optimal profile for the product $\rho(T)\kappa(T) = T\alpha^2(T)/k_0$ for TEG ((a) $T_a = 600$ K, $T_s = 300$ K); and TEC ((b) $T_a = 280$ K, $T_s = 300$ K) plotted for the constraints $zT = 0.6$ and $zT = 1$ according to the optimal Seebeck profiles shown in Figure 4.5.

increasing with increasing coordinate x for both TEC and TEG as the temperature profile $T(x)$ calculated from Equation 4.16 is increasing in the TEC but decreasing in the TEG.

Within optimization strategy B, an optimal $T(x)$ can be found from Equation 4.38 together with one optimal set of spatial profiles. Specifics are discussed here using the constraint $z(x)\ T(x) = k_0 = \text{const.}$ whereby fully self-compatible performance values are given again by the integrals (4.39).

Applying Equation 4.15 for the case of optimal reduced efficiency, we get with the optimal TE potential from Equation 4.14

$$\text{TEG:}\quad \eta_{r0} = \frac{T(x)\Phi'(x)}{T'(x)\Phi(x)} = \frac{\alpha'(x)T(x)}{T'(x)\alpha(x)} + 1, \quad \text{and} \quad \text{TEC:}\quad \frac{1}{\eta_{r0}} = \frac{\alpha'(x)T(x)}{T'(x)\alpha(x)} + 1,$$

leading to similar differential equations for both TEG and TEC

$$\text{TEG:}\quad \frac{\alpha'(x)}{\alpha(x)} = (\eta_{r0} - 1)\frac{T'(x)}{T(x)} \equiv k_g\,\frac{T'(x)}{T(x)} \tag{4.46a}$$

and

$$\text{TEC: } \frac{\alpha'(x)}{\alpha(x)} = \left(\frac{1}{\eta_{r0}} - 1\right)\frac{T'(x)}{T(x)} \equiv k_c \frac{T'(x)}{T(x)} \quad (4.46b)$$

A simple integration gives a correlation between the optimal temperature profile and the optimal, spatial Seebeck coefficient for both TEG and TEC (again with $\alpha_{ref} = \alpha(T_{ref})$) which is equivalent to Equations 4.45a and 4.45b:

$$\text{TEG: } \alpha(x) = \alpha_{ref}\left[\frac{T(x)}{T_{ref}}\right]^{k_g}, \quad \text{TEC: } \alpha(x) = \alpha_{ref}\left[\frac{T(x)}{T_{ref}}\right]^{k_c}. \quad (4.47)$$

Thus, only one material profile must be predefined when using the constraint $zT = \text{const.}$ In fact, Equation 4.47 represents a third optimization equation within variant B whereas the search for optimal, spatial profiles is then based on only one given profile, for example, $\kappa(x)$ for maximum η (TEG) and $\sigma(x)$ for maximum φ (TEC), respectively.

4.7 Summary

Fully self-compatible elements are characterized by a set of optimal material profiles whereas optimization is done here within the framework of a 1D model with fixed length of the active material and fixed boundary temperatures. An optimal set of spatial profiles can be found within a thermodynamic optimization procedure together with an optimal temperature profile $T(x)$. However, it has become clear in the last section that all profiles cannot be found simultaneously in a direct way; at least one profile has to be predefined or appropriate constraints established. Nevertheless, there is hope that a combination of thermodynamic optimization based on the appropriate constraints will give a real chance to come closer to the optimal material gradients. For a steady and monotonous optimal temperature profile, the optimal spatial profiles can be transformed into related profiles $\alpha(T)$, $\sigma(T)$, and $\kappa(T)$ in order to compare with real temperature-dependent materials data. Alternatively, stacked elements can be used to approach the optimal spatial profiles. However, it would be wrong to reduce FGM research to investigations based on the temperature dependence of given homogeneous or continuously graded materials.

The case of maximum cooling, that is, determining ΔT_{max} in the TEC case,* is obtained when considering the coefficient of performance in the limit $\varphi \to 0$. It is assumed from the previous section (see Figure 4.5) that we will get (with a monotonous $T(x)$) a monotonous, optimal Seebeck profile $\alpha(x) = \alpha[T(x)]$ also in the case of ΔT_{max}.

We expect that the use of self-compatible elements is probably the most efficient way to accomplish direct energy conversion in TEs.

4.8 Outlook and Related Problems

Further problems concern the electrical power output of a TEG and the heat pumping mode of a TEC at arbitrary ΔT. The power output is—unlike η and φ—a purely electrical quantity.

For this reason, the integral kernel does not only depend on u but also on j explicitly; from results published in Refs. [28,30] we find the integral for the net electrical power output density

* Referring to CPM, we find $T'(L) = 0$ and hence $u(L) \to -\infty$ at the heat-sink side for maximum cooling; this divergence is not found when using optimal gradients.

$$p_{\text{net}} = -\frac{P}{A_c} = \int_{T_s}^{T_a} K^P(j,u)\,\mathrm{d}T = -\int_{T_a}^{T_s} K^P(j,u)\,\mathrm{d}T \tag{4.48}$$

with

$$K^P(j,u) = j\alpha\left(1 - u\frac{\alpha}{z}\right).$$

Within the concept of power-related compatibility [28] the power compatibility factor has been found to be $s^{(P)} = z/(2\alpha) = \alpha\sigma/(2\kappa)$. Proof can be given also from Equation 4.48 if we evaluate the derivation of the kernel K^P with respect to j

$$\frac{\partial K^P}{\partial j} = \alpha\left(1 - u\frac{\alpha}{z}\right) - j\alpha\frac{\alpha}{z}\frac{\partial u}{\partial j} = \alpha - 2u\frac{\alpha^2}{z} = 0$$

$$\Rightarrow u_{\text{opt},P} = \frac{z}{2\alpha} = s^{(g,P)}, \tag{4.49}$$

where $\partial u/\partial j = u/j$ has been used which follows from the definition of u. The power compatibility factor $s^{(P)}$ has been proved to be the first-order approximation of the TEG's efficiency compatibility factor. This means that optimization strategies for power output and η will lead (apart from minor differences) to similar results especially for small zT values.

The target of the optimization procedure for the power output of a TEG of fixed length is to find not only the optimum u but also explicitly the optimum electrical current density. Results are published in a separate chapter [110], therewith continuing previous investigations on graded TEGs [12,33,39,106,187].

Considering maximum performance parameters p, η, and φ for the CPM case (averages), there is a monotonous temperature profile as well as a monotonous $u(x)$ over the length of the TE element ($u > 0$ for TEG, but $u < 0$ for φ of a TEC). The situation changes essentially if TEC modes close to maximum heat pumping shall be investigated. In this case, we find a changing sign of $T'(x)$ because $T(x)$ peaks in the interior of the TE element. Clearly, based on a suitable integral formulation, this mode could also be treated in an appropriate manner. However, it should be noted that then "pump up" ($T'(x) > 0$) and "pump down" ($T'(x) < 0$) sections have to be treated separately. It has also to be considered whether (following Sherman's intention)* $y = 1/u$ is the more practical variable [111] when performing calculations on TEC. This formulation may also be beneficial when examining maximum heat pumping and the TE heater, respectively.

References

1. L.I. Anatychuk. Current status and some prospects of thermoelectricity. In *Proceedings of the 5th European Conference on Thermoelectrics*, Odessa (Ukraine), September 2007, http://ect2007.its.org/proc-contents.html; http://ect2007.its.org/system/files/u1/pdf/04.pdf.
2. L.I. Anatychuk, O.J. Luste, and L.N. Vikhor. Optimal functions as an effective method for thermoelectric devices design. In *Fifteenth International Conference on Thermoelectrics*, Pasadena, CA, USA, pp. 223–226, March 26–29, 1996.

* Sherman used the temperature T as independent variable; thus $y(T) = -\kappa(T)/[j\,x'(T)]$ (see Equation 4.40 in the original paper [111]).

3. L.I. Anatychuk and L.N. Vikhor. Functionally graded materials and new prospects for thermoelectricity use. In *Proceedings ICT '97. XVI International Conference on Thermoelectrics*, Dresden, Saxony, Germany, pp. 588–591, August 26–29, 1997.
4. L.I. Anatychuk and L.N. Vikhor. Physics and methods of FGTM design. In *Twenty-Second International Conference on Thermoelectrics—ICT 2003*, La Grande-Motte, France, pp. 425–428, August 17–21, 2003.
5. Z. Bian and A. Shakouri. Beating the maximum cooling limit with graded thermoelectric materials. *Applied Physics Letters*, 89(21):212101, 2006.
6. Z. Bian, Hongyun Wang, Qiaoer Zhou, and Ali Shakouri. Maximum cooling temperature and uniform efficiency criterion for in-homogeneous thermoelectric materials. *Physical Review B (Condensed Matter and Materials Physics)*, 75(24):245208, 2007.
7. Z. Dashevsky, Y. Gelbstein, I. Edry, I. Drabkin, and M.P. Dariel. Optimization of thermoelectric efficiency in graded materials. In *Proceedings of the 22nd International Conference on Thermoelectrics*, La Grande-Motte, France, pp. 421–424, August 17–21, 2003.
8. C.A. Domenicali. Irreversible thermodynamics of thermoelectricity. *Reviews of Modern Physics*, 26(2):237–275, 1954.
9. N. Fuschillo. Thermoelectric phenomena. In I.B. Cadoff and E. Miller, editors, *Thermoelectric Materials and Devices*, Materials Technology Series, Chapter 1, pp. 1–17. Reinhold Publishing Cooperation, New York, 1960.
10. F. Gascoin, C. Goupil, C. de Vaulx, and A. Papavero. Non-homogeneous thermoelectric material study using the linear response approach. In *Proceedings of the 29th International Conference on Thermoelectrics*, Shanghai, China, pp. 305–309, 2010.
11. H.J. Goldsmid and J.W. Sharp. The thermal conductivity of inhomogeneous thermoelectric materials. *Physica Status Solidi (b)*, 241(11):2571–2574, 2004.
12. L. Helmers, E. Müller, J. Schilz, and W.A. Kaysser. Graded and stacked thermoelectric generators—Numerical description and maximisation of output power. *Materials Science and Engineering B*, 56(1):60–68, 1998.
13. A.E. Kaliazin, V.L. Kuznetsov, and D.M. Rowe. Rigorous calculations related to functionally graded and segmented thermoelements. In *Proceedings ICT 2001. XX International Conference on Thermoelectrics*, Beijing, People's Republic of China, pp. 286–292, June 8–11, 2001.
14. M. Korzhuev. Increase in the efficiency of thermoelectric heaters by using inhomogeneous branches. *Technical Physics*, 54:1241–1243, 2009. 10.1134/S106378420908026X.
15. M. Korzhuev. Symmetry analysis of thermoelectric energy converters with inhomogeneous legs. *Journal of Electronic Materials*, 39:1381–1385, 2010. 10.1007/s11664-010-1330-1.
16. M. Korzhuev and E. Avilov. Use of the Harman technique for figure of merit measurements of cascade thermoelectric converters. *Journal of Electronic Materials*, 39:1499–1503, 2010. 10.1007/s11664-010-1301-6.
17. M.A. Korzhuev and I.V. Katin. On the placement of thermoelectric generators in automobiles. *Journal of Electronic Materials*, 39:1390–1394, 2010. 10.1007/s11664-010-1332-z.
18. M.A. Korzhuev, L.D. Ivanova, L.I. Petrova, Yu.V. Granatkina, and T.E. Svechnikova. The multistage thermoelectric devices with inhomogeneous legs. In *Proceedings of the 5th European Conference on Thermoelectrics*, Odessa (Ukraine), September 2007, http://ect2007.its.org/proc-contents.html; http://ect2007.its.org/system/files/u1/pdf/35.pdf.
19. V.L. Kuznetsov. Functionally graded materials for thermoelectric applications. In D.M. Rowe, editor, *CRC Handbook of Thermoelectrics: Macro to Nano*, Chapter 38. Taylor & Francis, Boca Raton, FL, 38-1–38-12, December 2006.
20. V.L. Kuznetsov, L.A. Kuznetsova, A.E. Kaliazin, and D.M. Rowe. High performance functionally graded and segmented Bi_2Te_3-based materials for thermoelectric power generation. *Journal of Materials Science*, 37:2893–2897, 2002. 10.1023/A:1016092224833.
21. G.D. Mahan. Inhomogeneous thermoelectrics. *Journal of Applied Physics*, 70(8):4551–4554, 1991.

22. E. Müller, Č. Drašar, J. Schilz, and W.A. Kaysser. Functionally graded materials for sensor and energy applications. *Materials Science and Engineering A*, 362(1–2):17–39, 2003. Papers from the German Priority Programme (Functionally Graded Materials).
23. E. Müller, G. Karpinski, L.M. Wu, S. Walczak, and W. Seifert. Separated effect of 1D thermoelectric material gradients. In P. Rogl, editor, *25th International Conference on Thermoelectrics*, Vienna, Austria, pp. 204–209. IEEE, Piscataway, NJ, 08, August 6–10, 2006.
24. E. Müller, G. Karpinski, L.M. Wu, S. Walczak, and W. Seifert. Numerically based design of thermoelectric elements. In M. Rowe, editor, *4th European Conference on Thermoelectrics*, Cardiff University, Wales UK, April 9–11, 2006.
25. E. Müller, S. Walczak, and W. Seifert. Optimization strategies for segmented Peltier coolers. *Physica Status Solidi (a)*, 203(8):2128–2141, 2006.
26. E. Müller, S. Walczak, W. Seifert, C. Stiewe, and G. Karpinski. Numerical performance estimation of segmented thermoelectric elements. In T.M. Tritt, editor, *ICT 2005—24th International Conference on Thermoelectrics*, Clemson University, Clemson, SC, USA, pp. 352–357. Institute of Electrical and Electronics Engineers, Inc., Piscataway, NJ, USA, June 19–23, 2005.
27. W. Seifert, M. Ueltzen, and Eckhard Müller. One-dimensional modelling of thermoelectric cooling. *Physica Status Solidi (a)*, 1(194):277–290, 2002.
28. W. Seifert, E. Müller, G.J. Snyder, and S. Walczak. Compatibility factor for the power output of a thermogenerator. *Physica Status Solidi (RRL)*, 1(6):250–252, 2007.
29. W. Seifert, E. Müller, and S. Walczak. Generalized analytic description of one dimensional non-homogeneous TE cooler and generator elements based on the compatibility approach. In P. Rogl, editor, *25th International Conference on Thermoelectrics*, Vienna, Austria, pp. 714–719. IEEE, Piscataway, NJ, 08 August 6–10, 2006.
30. W. Seifert, E. Müller, and S. Walczak. Local optimization strategy based on first principles of thermoelectrics. *Physica Stat. Sol. (a)*, 205(12):2908–2918, December 2008.
31. W. Seifert, K. Zabrocki, G.Jeffrey Snyder, and E. Müller. The compatibility approach in the classical theory of thermoelectricity seen from the perspective of variational calculus. *Phyica Status Solidi (a)*, 207(3):760–765, 2010.
32. I. Shiota and Y. Miyamoto, editors. *Functionally Graded Material 1996*, AIST Tsukuba Research Center, Tsukuba, Japan, October 1996. *Proceedings of the 4th International Symposium on Functionally Graded Materials*, Elsevier.
33. G.J. Snyder. Application of the compatibility factor to the design of segmented and cascaded thermoelectric generators. *Applied Physics Letters*, 84(13):2436–2438, 2004.
34. G.J. Snyder. Thermoelectric power generation: Efficiency and compatibility. In D.M. Rowe, editor, *CRC Handbook of Thermoelectrics: Macro to Nano*, Chapter 9, p. 9-1–9-26. Taylor & Francis, Boca Raton, FL, 2006.
35. G.J. Snyder and T.S. Ursell. Thermoelectric efficiency and compatibility. *Physical Review Letters*, 91(14):148301, 2003.
36. G.J. Snyder. Design and optimization of compatible, segmented thermoelectric generators. In *Twenty-Second International Conference on Thermoelectrics—ICT 2003*, La Grand-Motte, France, pp. 443–446, August 17–21, 2003.
37. J. Tauc. The theory of the thermal EMF of semi-conductors. *Czechoslovak Journal of Physics*, 3:282–302, 1953.
38. L.J. Ybarrondo. *Effects of surface heat transfer and spatial property dependence on the optimum performance of a thermoelectric heat pump*. PhD thesis, Georgia Institute of Technology, 1964.
39. L.J. Ybarrondo and J.E. Sunderland. Influence of spatially dependent properties on the performance of a thermoelectric heat pump. *Advanced Energy Conversion*, 5(4):383–405, 1965.
40. K. Zabrocki, E. Müller, and W. Seifert. One-dimensional modeling of thermogenerator elements with linear material profiles. *Journal of Electronic Materials*, 39:1724–1729, 2010. 10.1007/s11664-010-1179-3.

41. C.A.B. Ball, W.A. Jesser, and J.R. Maddux. The distributed Peltier effect and its influence on cooling devices. In *Proceedings of the 14th International Conference on Thermoelectrics*, St. Petersburg, Russia, pp. 305–309, June 27–30, 1995.
42. I.M. Belov, V.P. Volkov, and O. Manyakin. Optimization of Peltier thermocouple using distributed Peltier effect. In *Eighteenth International Conference on Thermoelectrics*, Baltimore, MD, USA, pp. 316–318, August 29–September 2, 1999.
43. R.J. Buist. The extrinsic Thomson effect. In *Proceedings of the 14th International Conference on Thermoelectrics*, St. Petersburg, Russia, pp. 301–304, June 27–30, 1995.
44. M. Chen, L.A. Rosendahl, T.J. Condra, and J.K. Pedersen. Numerical modeling of thermoelectric generators with varying material properties in a circuit simulator. *IEEE Transactions on Energy Conversion*, 24(1):112–124, 2009.
45. D. Ebling, K. Bartholomé, M. Bartel, and M. Jägle. Module geometry and contact resistance of thermoelectric generators analyzed by multiphysics simulation. *Journal of Electronic Materials*, 39:1376–1380, 2010. 10.1007/s11664-010-1331-0.
46. G. Fraisse, M. Lazard, C. Goupil, and J.Y. Serrat. Study of a thermoelement's behaviour through a modelling based on electrical analogy. *International Journal of Heat and Mass Transfer*, 53(17–18):3503–3512, 2010.
47. M. Freunek, M. Müller, T. Ungan, W. Walker, and L.M. Reindl. New physical model for thermoelectric generators. *Journal of Electronic Materials*, 38(7):1214–1220, 2009.
48. A. Jacquot, M. Jägle, J. König, D.G. Ebling, and H. Böttner. Theoretical study of the Harman-method for evaluating the thermoelectric performance of materials and components at high temperature. In *Proceedings of the 5th European Conference on Thermoelectrics*, Odessa (Ukraine), September 2007, http://ect2007.its.org/proc-contents.html; http://ect2007.its.org/system/files/u1/pdf/24.pdf.
49. M. Jägle. Simulating thermoelectric effects with finite element analysis using COMSOL. In *Proceedings of the 5th European Conference on Thermoelectrics*, Odessa (Ukraine), September 2007, http://ect2007.its.org/proc-contents.html; http://ect2007.its.org/system/files/u1/pdf/56.pdf.
50. A. López, F. Villasevil, G. Noriega, and D. Platzek. Thermoelectric integrated numerical modeling process of a temperature and humidity control device apply to vehicles for fogging preventing. In *Proceedings of the 5th European Conference on Thermoelectrics*, Odessa (Ukraine), September 2007, http://ect2007.its.org/proc-contents.html; http://ect2007.its.org/system/files/u1/pdf/29.pdf.
51. A.M. Morega, M. Morega, and M.A. Panait. Structural optimization of a thermoelectric generator by numerical simulation. *Revue Roumaine des sciences techniques—Serie Électrotechnique et Énergétique*, 55(1):3–12, 2010.
52. M.H. Norwood. A comparison of theory and experiment for a thermoelectric cooler. *Journal of Applied Physics*, 32(12):2559–2563, 1961.
53. M. Power and R.A. Handelsman. Generalized calculation of thermoelectric efficiency. *Advanced Energy Conversion*, 1:45–60, 1961.
54. D.M. Rowe and G. Min. Evaluation of thermoelectric modules for power generation. *Journal of Power Sources*, 73(2):193–198, 1998.
55. E. Sandoz-Rosado and R. Stevens. Robust finite element model for the design of thermoelectric modules. *Journal of Electronic Materials*, 39:1848–1855, 2010. 10.1007/s11664-010-1077-8.
56. I.B. Cadoff and E. Miller. The thermoelectric circuit. In I.B. Cadoff and E. Miller, editors, *Thermoelectric Materials and Devices*, Materials Technology Series, Chapter 2, pp. 18–32. Reinhold Publishing Cooperation, New York, 1960.
57. D.L. Kerr. Design calculations for thermoelectric generators. In I.B. Cadoff and E. Miller, editors, *Thermoelectric Materials and Devices*, Materials Technology Series, Chapter 16, pp. 227–249. Reinhold Publishing Cooperation, New York, 1960.
58. G. Min. Thermoelectric module design theories. In D.M. Rowe, editor, *CRC Handbook of Thermoelectrics: Macro to Nano*, Chapter 11, p. 11-1–11-15. Taylor & Francis, Boca Raton, FL, 2006.

59. G. Min and D.M. Rowe. Peltier devices as generators. In D.M. Rowe, editor, *CRC Handbook of Thermoelectrics*, Chapter 38, pp. 479–488. CRC Press, Boca Raton, FL, 1995.
60. R.G. Sickert. A thermoelectric refrigerating system for submarines. *Electrical Engineering*, 79(5):364–371, 1960.
61. R.W. Ure, Jr. and R.R. Heikes. *Theoretical Calculation of Device Performance*, Chapter 15, pp. 458–517. Interscience Publishers, Inc., New York, 1961.
62. R.H. Vought. Design calculations for Peltier cooling. In I.B. Cadoff and E.Miller, editors, *Thermoelectric Materials and Devices*, Materials Technology Series, Chapter 17, pp. 250–274. Reinhold Publishing Cooperation, New York, 1960.
63. L.I. Anatychuk and O.J. Luste. Physical principles of microminiaturization in thermoelectricity. In *Fifteenth International Conference on Thermoelectrics*, Pasadena, CA, USA, pp. 279–287, March 26–29, 1996.
64. M.H. Cobble. Calculations of generator performance. In D.M. Rowe, editor, *CRC Handbook of Thermoelectrics*, Chapter 39, pp. 489–501. CRC Press, Boca Raton, FL, 1995.
65. G. Min and D.M. Rowe. "Symbiotic" application of thermoelectric conversion for fluid preheating/power generation. *Energy Conversion and Management*, 43(2):221–228, 2002.
66. G. Min and D.M. Rowe. Improved model for calculating the coefficient of performance of a Peltier module. *Energy Conversion and Management*, 41(2):163–171, 2000.
67. G. Min and D.M. Rowe. Optimisation of thermoelectric module geometry for 'waste heat' electric power generation. *Journal of Power Sources*, 38(3):253–259, 1992.
68. G. Min and M. Yatim. Variable thermal resistor based on self-powered Peltier effect. *Journal of Physics D: Applied Physics*, 41(22):222001, 2008.
69. D.M. Rowe and G. Min. Design theory of thermoelectric modules for electrical power generation. *IEEE Proceedings—Science, Measurement and Technology*, 143(6):351–356, 1996.
70. O.S. Gryaznov, B.Ya. Moizhes, and V.A. Nemchinskii. Generalized thermoelectric efficiency. *Soviet Physics-Technical Physics*, 23(8):975–980, 1978. Translated from *Zhurnal tekhnicheskoi fiziki* **48**, pp. 1720–1728.
71. K. Landecker. Some aspects of the performance of refrigerating thermojunctions with radial flow of current. *Journal of Applied Physics*, 47(5):1846–1851, 1976.
72. K. Landecker. On power-generating thermojunctions with radial flow of current. *Solar Energy*, 19(5):439–443, 1977.
73. K. Landecker. Heat transport in coaxial thermoelectric disks with radial flow of current. *Journal of Applied Physics*, 49(9):4939–4941, 1978.
74. G. Min and D.M. Rowe. Ring-structured thermoelectric module. *Semiconductor Science and Technology*, 22(8):880, 2007.
75. S.B. Riffat and G.Q. Qiu. Design and characterization of a cylindrical, water-cooled heat sink for thermoelectric air-conditioners. *International Journal of Energy Research*, 30:67–80, 2006.
76. L. Onsager. Reciprocal relations in irreversible processes. i. *Physical Review*, 37(4):405–426, 1931.
77. L. Onsager. Reciprocal relations in irreversible processes. ii. *Physical Review*, 38(12):2265–2279, 1931.
78. L. Onsager. Theories and problems of liquid diffusion. *Annals of the New York Academy of Sciences*, 46:241–265, 1945.
79. L. Anatychuk. The law of thermoelectric induction and extending the capabilities of its application. *Journal of Electronic Materials*, 39:1869–1874, 2010. 10.1007/s11664-009-0983-0.
80. L. Anatychuk and R. Kobylyansky. Power properties of the transverse short-circuited thermoelement. *Journal of Electronic Materials*, 39:1704–1707, 2010. 10.1007/s11664-010-1205-5.
81. L.I. Anatychuk. Original thermoelectric development of physical concepts, materials and related technologies and applications. In *Proceedings of XX International Conference on Thermoelectrics (ICT 2001)*, Beijing, People's Republic of China, pp. 35–41, June 8–11, 2001.

82. L.I. Anatychuk and O.J. Luste. Thermoelectric eddy currents. Calculation and control methods. In *Proceedings ICT '97. XVI International Conference on Thermoelectrics*, Dresden, Saxony, Germany, pp. 595–598, August 26–29, 1997.
83. I.S. Buda and V.S. Lutsyak. The theory of an anisotropic thermoelectric cooler in a magnetic field. *Physica Status Solidi (a)*, 147(2):491–496, 1995.
84. I.S. Buda, V.S. Lutsyak, U.M. Khamets, and L.A. Shcherbina. Thermodynamic definition of the thermoelectric figure of merit of an anisotropic medium. *Physica Status Solidi (a)*, 123(2):K139–K143, 1991.
85. T.C. Harman and J.M. Honig. *Thermoelectric and Thermomagnetic Effects and Applications*. McGraw-Hill Book Company, New York, 1967.
86. A.G. Samoilowitsch and L.L. Korenblit. Gegenwärtiger Stand der Theorie der thermoelektrischen und thermomagnetischen Erscheinungen in Halbleitern. *Fortschritte der Physik*, 1:486–554, 1953. Unshortened translation of the article published in *Uspechi Fiz. Nauk* **49**, p. 243 and p. 337, 1953.
87. H.B. Callen. *On the Theory of Irreversible Processes*. PhD thesis, MIT, 1947.
88. H.B. Callen. The application of Onsager's reciprocal relations to thermoelectric, thermomagnetic, and galvanomagnetic effects. *Physical Review*, 73(11):1349–1358, 1948.
89. H.B. Callen. *Thermodynamics and an Introduction to Thermostatistics*. 2nd edition, John Wiley & Sons, Inc., New York, USA, 1985.
90. S. de Groot and P. Mazur. *Non-equilibrium Thermodynamics*. Dover, London, 1984 (reprinted).
91. S.R. de Groot. *Thermodynamics of Irreversible Processes*. North-Holland Publishing Company, Amsterdam, 1963.
92. K.G. Denbigh. *The Thermodynamics of the Steady State*. John Wiley & Sons, Inc., New York, 1951.
93. C.A. Domenicali. Irreversible thermodynamics of thermoelectric effects in inhomogeneous, anisotropic media. *Physical Review*, 92(4):877–881, 1953.
94. C.A. Domenicali. Stationary temperature distribution in an electrically heated conductor. *Journal of Applied Physics*, 25(10):1310–1311, 1954.
95. D.D. Fitts. *Nonequilibrium Thermodynamics: A Phenomenological Theory of Irreversible Processes in Fluid Systems*. McGraw-Hill, New York, 1962.
96. G.D. Mahan. Density variations in thermoelectrics. *Journal of Applied Physics*, 87(10):7326–7332, 2000.
97. N. Pottier. *Physique statisitique hors équilibre, processus irréversibles linéaires*. Savoirs Actuels EDP Sciences/CNRS Editions, France, 2007.
98. I. Prigogine. *Introduction to Thermodynamics of Irreversible Processes*. 3rd edition, John Wiley & Sons, Inc., New York, July 1968.
99. Y. Rocard. *Thermodynamique*. 2nd edition, Masson, Paris, 1967.
100. C.B. Vining. The thermoelectric process. In T.M. Tritt, M.G. Kanatzidis, Jr. H.B. Lyon, and G.D. Mahan, editors, *Materials Research Society Symposium Proceedings: Thermoelectric Materials-New Directions and Approaches*, San Francisco, CA, USA, pp. 3–13. Materials Research Society, Warrendale, PA, USA, March 31–April 3, 1997.
101. W.H. Clingman. Entropy production and optimum device design. *Advanced Energy Conversion*, 1:61–79, 1961.
102. W.H. Clingman. New concepts in thermoelectric device design. *Proceedings of the IRE*, 49(7):1155–1160, 1961.
103. C. Goupil, W. Seifert, K. Zabrocki, E. Müller, and G. Jeffrey Snyder. Thermodynamics of thermoelectric phenomena and applications. *Entropy*, 13(8), 1481–1517, 2011.
104. M. Lazard, E. Rapp, and H. Scherrer. Some considerations towards design and optimization of segmented thermoelectric generators. In *Proceedings of the 5th European Conference on Thermoelectrics*, Odessa (Ukraine), September 10–12, 2007, http://ect2007.its.org/system/files/u1/pdf/46.pdf
105. M. Lazard. Heat transfer in thermoelectricity: Modelling, optimization and design. In *Proceedings of the 7th IASME/WSEAS International Conference on Heat Transfer, Thermal Engineering and Environment*, Moscow, Russia, pp. 129–134, August 20–22, 2009.

106. T.S. Ursell and G.J. Snyder. Compatibility of segmented thermoelectric generators. In *Twenty-First International Conference on Thermoelectrics*, Long Beach, CA, USA, pp. 412–417, August 25–29, 2002.
107. E. Altenkirch. Über den Nutzeffekt der Thermosäulen. *Physikalische Zeitschrift*, 10:560–580, 1909.
108. E. Altenkirch. Elektrothermische Kälteerzeugung und reversible elektrische Heizung. *Physikalische Zeitschrift*, 12:920–924, 1911.
109. J. Schilz, E. Müller, L. Helmers, Y.S. Kang, Y. Noda, and M. Niino. On the composition function of graded thermoelectric materials. *Materials Science Forum*, 308–311:647–652, 1999.
110. W. Seifert, K. Zabrocki, E. Müller, and G.J. Snyder. Power-related compatibility and maximum electrical power output of a thermogenerator. *Physica Status Solidi (a)*, 207(10):2399–2406, 2010.
111. B. Sherman, R.R. Heikes, and R.W. Ure Jr. Calculation of efficiency of thermoelectric devices. *Journal of Applied Physics*, 31(1):1–16, 1960.
112. B. Sherman, R.R. Heikes, and R.W. Ure Jr. Calculation of efficiency of thermoelectric devices. In I.B. Cadoff and E. Miller, editors, *Thermoelectric Materials and Devices*, Materials Technology Series, Chapter 15, pp. 199–226. Reinhold Publishing Cooperation, New York, 1960. Reprint of [111].
113. C. Goupil. Thermodynamics of the thermoelectric potential. *Journal of Applied Physics*, 106:104907, 2009.
114. Z. Dashevsky, S. Shusterman, M.P. Dariel, and I. Drabkin. Thermoelectric efficiency in graded indium-doped PbTe crystals. *Journal of Applied Physics*, 92(3):1425–1430, 2002.
115. P.H. Egli. *Thermoelectricity*. John Wiley & Sons, Inc., New York, 1960.
116. Y. Gelbstein, Z. Dashevsky, and M.P. Dariel. High performance n-type PbTe-based materials for thermoelectric applications. *Physica B: Condensed Matter*, 363(1–4):196–205, 2005.
117. A.F. Ioffe. *Semiconductor Thermoelements and Thermoelectric Cooling*. Infosearch, Ltd., London, 1957.
118. A.F. Joffe and L.S. Stil'bans. Physical problems of thermoelectricity. *Reports on Progress in Physics*, 22(1):167, 1959.
119. V. Jovovic and J. Heremans. Doping effects on the thermoelectric properties of $AgSbTe_2$. *Journal of Electronic Materials*, 38:1504–1509, 2009. 10.1007/s11664-009-0669-7.
120. P.H. Klein. Properties affecting the utility of thermoelectric materials. In I.B. Cadoff and E. Miller, editors, *Thermoelectric Materials and Devices*, Materials Technology Series, Chapter 5, pp. 55–83. Reinhold Publishing Cooperation, New York, 1960.
121. D.K.C. MacDonald. *Thermoelectricity: An Introduction to the Principles*. Dover Publications, Inc., New York, 2006.
122. F.D. Rosi, B. Abeles, and R.V. Jensen. Materials for thermoelectric refrigeration. *Journal of Physics and Chemistry of Solids*, 10(2–3):191–200, 1959.
123. F.D. Rosi, E.F. Hockings, and N.E. Lindenblad. Semiconducting materials for thermoelectric power generation. *RCA (Radio Cooperation of America) Review*, 22:82–121, 1961.
124. G.J. Snyder and E.S. Toberer. Complex thermoelectric materials. *Nature Materials*, 7:105–114, 2008.
125. T.P. Hogan and T. Shih. Modeling and characterization of power generation modules based on bulk materials. In D.M. Rowe, editor, *CRC Handbook of Thermoelectrics: Macro to Nano*, Chapter 12, pp. 12-1–12-23. Taylor and Francis, Boca Raton, FL, 2006.
126. H.J. Goldsmid. *Electronic Refrigeration*. Pion, London, 1986.
127. I.B. Cadoff and E. Miller, editors. *Thermoelectric Materials and Devices*, Materials Technology Series, New York, 1960. Reinhold Publishing Cooperation. Lectures presented during the course on Thermoelectric Materials and Devices sponsored by the Department of Metallurgical Engineering in cooperation with the Office of Special Services to Business and Industry, New York, NY, June 1959 and 1960.
128. S.I. Freedman. Thermoelectric power generation. In G.W. Sutton, editor, *Direct Energy Conversion, Inter-University Electronic Series*, Vol. 3, Chapter 3, pp. 105–180. McGraw-Hill Book Company, New York, St. Louis, San Francisco, 1966.

129. R.R. Heikes and R.W. Ure Jr. *Thermoelectricity: Science and Engineering.* Interscience Publishers, Inc., New York, 1961.
130. U. Birkholz, E. Groß, M. Riffel, H. Roth, U. Stöhrer, and W. Wittmer. Measurement of the efficiency of a HMS-$FeSi_2$ thermoelectric generator. In *Proceedings of the 11th International Conference on Thermoelectrics*, pp. 51–54, Arlington, TX, USA, October 7–9, 1992.
131. U. Birkholz. Zur experimentellen Bestimmung der thermoelektrischen Effektivität von Halbleitern. *Solid-State Electronics*, 1(1):34–38, 1960.
132. A.E. Bowley, L.E.J. Cowles, G.J. Williams, and H.J. Goldsmid. Measurement of the figure of merit of a thermoelectric material. *Journal of Scientific Instruments*, 38(11):433, 1961.
133. A.D. Downey and T.P. Hogan. Circuit model of a thermoelectric module for ac electrical measurements. In *24th International Conference on Thermoelectrics*, Clemson University, Clemson, SC, USA, June 19–23, 2005, *ICT 2005*, pp. 79–82, Piscataway, NJ, USA, June 2005.
134. A.D. Downey, T.P. Hogan, and B. Cook. Characterization of thermoelectric elements and devices by impedance spectroscopy. *Review of Scientific Instruments*, 78(9):093904, 2007.
135. T.C. Harman. Special techniques for measurement of thermoelectric properties. *Journal of Applied Physics*, 29(9):1373–1374, 1958.
136. T.C. Harman. Measurment of thermoelectric properties by using Peltier heat. In P.H. Egli, editor, *Thermoelectricity*, Chapter 14, pp. 235–255. John Wiley & Sons, Inc., New York, 1960.
137. T.C. Harman, J.H. Cahn, and M.J. Logan. Measurement of thermal conductivity by utilization of the Peltier effect. *Journal of Applied Physics*, 30(9):1351–1359, 1959.
138. T.C. Harman. Measurement of pertinent thermoelectric properties. In I.B. Cadoff and E. Miller, editors, *Thermoelectric Materials and Devices*, Materials Technology Series, Chapter 6, pp. 84–97. Reinhold Publishing Cooperation, New York, 1960.
139. H. Iwasaki and H. Hori. Thermoelectric property measurements by the improved Harman method. In *24th International Conference on Thermoelectrics*, Clemson University, Clemson, SC, USA, June 19–23, 2005, *ICT 2005*, pp. 513–516, Piscataway, NJ, USA, 2005.
140. H. Iwasaki, M. Koyano, and H. Hori. Evaluation of the figure of merit on thermoelectric materials by Harman method. *Japanese Journal of Applied Physics*, 41(Part 1, No. 11A):6606–6609, 2002.
141. K. Katayama and H. Umemiya. Measurement of thermal properties of semiconductor by utilization of the Peltier effect. *Bulletin of JSME*, 12(50):265–272, 1969.
142. P.H. Klein. The *S*–ρ meter: A device for rapid measurement of Seebeck coefficients and electrical resistivities at temperatures up to 1000 C. In P.H. Egli, editor, *Thermoelectricity*, Chapter 15, pp. 256–269. John Wiley & Sons, Inc., New York, 1960.
143. W. Kobayashi, W. Tamura, and I. Terasaki. Thermal conductivity and dimensionless figure of merit of thermoelectric rhodium oxides measured by a modified Harman method. *Journal of Electronic Materials*, 38:964–967, 2009. 10.1007/s11664-009-0788-1.
144. E. Müller, C. Stiewe, D.M. Rowe, and S.G.K. Williams. Approaches to thermoelectric standardization. In D.M. Rowe, editor, *CRC Handbook of Thermoelectrics: Macro to Nano*, Chapter 26, pp. 26-1–26-30. Taylor and Francis, Boca Raton, FL, 2006.
145. A. Muto, D. Kraemer, Q. Hao, Z.F. Ren, and G. Chen. Thermoelectric properties and efficiency measurements under large temperature differences. *Review of Scientific Instruments*, 80(9):093901, 2009.
146. A.W. Penn. The corrections used in the adiabatic measurement of thermal conductivity using the Peltier effect. *Journal of Scientific Instruments*, 41(10):626, 1964.
147. L. Rauscher, S. Fujimoto, H.T. Kaibe, and S. Sano. Efficiency determination and general characterization of thermoelectric generators using an absolute measurement of the heat flow. *Measurement Science and Technology*, 16(5):1054, 2005.
148. L. Rauscher, H.T. Kaibe, H. Ishimabushi, S. Sano, E. Müller, and D. Platzek. New approach for highly accurate efficiency determination of thermoelectric generator modules. In *ICT 2003, Twenty-Second International Conference on Thermoelectrics*, pp. 508–511, La Grande-Motte, France, August 17–21, 2003.

149. A.J. Reddy, E. Siivola, P. Thomas, G. Krueger, and R. Venkatasubramanian. Measurement and analysis of power conversion efficiency in thin-film and segmented thermoelectric devices. In *24th International Conference on Thermoelectrics*, Clemson University, Clemson, SC, USA, June 19–23, 2005, *ICT 2005*, pp. 72–75, Piscataway, NJ, USA, 2005.
150. E. Sandoz-Rosado and R. Stevens. Experimental characterization of thermoelectric modules and comparison with theoretical models for power generation. *Journal of Electronic Materials*, 38:1239–1244, 2009. 10.1007/s11664-009-0744-0.
151. U. Stöhrer. Measurement of the transport properties of $FeSi_2$ and HMS by utilization of the Peltier effect in the temperature range 50–800 degrees C. *Measurement Science and Technology*, 5(4):440, 1994.
152. D.M. Rowe, editor. *CRC Handbook of Thermoelectrics*. RC, Boca Raton, FL, 1995.
153. D.M. Rowe, editor. *CRC Handbook of Thermoelectrics: Macro to Nano*. CRC Press, Boca Raton, FL, 2006.
154. D.M. Rowe and C.M. Bhandari. *Modern Thermoelectrics*. Reston Publishing Company, Inc., 11480 Sunset Mills Roads, Reston, Virginia 22090, 1983.
155. S.R. Brown, S.M. Kauzlarich, F. Gascoin, and G.J. Snyder. Yb_{14} Mn Sb_{11}: New high efficiency thermoelectric material for power generation. *Chemistry of Materials*, 18(7):1873–1877, 2006.
156. T.C. Harman, P.J. Taylor, M.P. Walsh, and B.E. LaForge. Quantum dot superlattice thermoelectric materials and devices. *Science*, 297(5590):2229–2232, 2002.
157. A.J. Minnich, M.S. Dresselhaus, Z.F. Ren, and G. Chen. Bulk nanostructured thermoelectric materials: Current research and future prospects. *Energy & Environmental Science*, 2(5):466–479, 2009.
158. H. Süßmann and E. Müller. Verification of a transport model for p-type $(Bi_{0.5}\,Sb_{0.5})\,Te_3$ and $(Bi_{0.25}\,Sb_{0.75})\,Te_3$ mixed crystals by means of temperature dependent thermoelectric properties below room temperature. In M.V. Vedernikov, M.I. Federov, and A.E. Kaliazin, editors, *Fourteenth International Conference on Thermoelectrics*, pp. 1–7, St. Petersburg, Russia, June 27–30, 1995. International Thermoelectric Society.
159. R. Venkatasubramanian, E. Siivola, T. Colpitts, and B. O'Quinn. Thin-film thermoelectric devices with high room-temperature figures of merit. *Nature*, 413:597–602, 2001.
160. M.R. El-Saden. Irreversible thermodynamics and the theoretical bound on the thermomagnetic figure of merit. *Journal of Applied Physics*, 33(10):3145–3146, 1962.
161. H.J. Goldsmid. A new upper limit to the thermoelectric figure-of-merit. In D.M. Rowe, editor, *CRC Handbook of Thermoelectrics: Macro to Nano*, Chapter 10, pp. 10-1–10-10. Taylor and Francis, Boca Raton, FL, 2006.
162. H. Littman. A clarification of the theoretical upper bound on the thermoelectric "figure of merit" derived from irreversible thermodynamics. *Journal of Applied Physics*, 33(8):2655–2656, 1962.
163. H. Littman and B. Davidson. Theoretical bound on the thermoelectric figure of merit from irreversible thermodynamics. *Journal of Applied Physics*, 32(2):217–219, 1961.
164. E.S. Rittner. Comment on Theoretical bound on the thermoelectric figure of merit from irreversible thermodynamics. *Journal of Applied Physics*, 33(8):2654–2655, 1962.
165. C.B. Vining. The thermoelectric limit $ZT \sim 1$: Fact or artifact. In K.R. Rao, editor, *The XIth International Conference on Thermoelectrics*, pp. 223–231, University of Texas, Arlington, TX, USA, October 7–9, 1992.
166. J.H. Lienhard IV and J.H. Lienhard V. *A Heat Transfer Textbook*. 3rd edition, Phlogiston Press, Cambridge, MA, USA, 2008.
167. B. Ya. Moizhes. The influence of the temperature dependence of physical parameters on the efficiency of thermoelectric generators and refrigerators. *Soviet Physics-Solid State*, 2:728–737, 1960. Translated from *Fizika Tverdogo Tela*.
168. J.M. Borrego. Carrier concentration optimization in semiconductor thermoelements. *IEEE Transactions on Electron Devices*, 10(6):364–370, 1963.
169. J.M. Borrego. Approximate analysis of the operation of thermoelectric generators with temperature dependent parameters. *IEEE Transactions on Aerospace*, 2:4–9, 1964.

170. A.A. Efremov and A.S. Pushkars. Energy calculation of thermoelements with arbitrary temperature dependence of thermoelectric properties of materials by heat balance technique. *Energy conversion*, 11(3):101–104, 1971.
171. W. Yi-Dong, W. Zhi-Min, and H. Yuan-Jin. A new method for calculating the thermoelectric efficiency. *Chinese Physics Letters*, 21(9):1848, 2004.
172. G. Min, D.M. Rowe, and K. Kontostavlakis. Thermoelectric figure-of-merit under large temperature differences. *Journal of Physics D: Applied Physics*, 37(8):1301–1304, 2004.
173. S. Kirkpatrick. Percolation and conduction. *Reviews of Modern Physics*, 45(4):574–588, 1973.
174. D.J. Bergman and L.G. Fel. Enhancement of thermoelectric power factor in composite thermoelectrics. *Journal of Applied Physics*, 85(12):8205–8216, 1999.
175. D.J. Bergman and O. Levy. Composite thermoelectrics—Exact results and calculational methods. In G.A. Slack, D.D. Allred, and C.B. Vining, editors, *Modern Perspectives On Thermoelectrics and Related Materials*, Vol. 234, pp. 39–45. Pittsburgh, PA, 1991.
176. D.J. Bergman and L.G. Fel. Enhancement of thermoelectric power factor in composite thermoelectrics. In *Eighteenth International Conference on Thermoelectrics*, pp. 76–79, 1999.
177. C. Herring. Effect of random inhomogeneities on electrical and galvanomagnetic measurements. *Journal of Applied Physics*, 31(11):1939–1953, 1960.
178. R. Landauer. Electrical transport and optical properties of inhomogeneous media. In J.C. Garland and D.B. Tanner, editors, *Proceedings of the First Conference on the Electrical Transport and Optical Properties of Inhomogeneous Media, AIP Conference Proceedings*, Vol. 40, pp. 2–45, Ohio State University, New York, 1978.
179. A.A. Snarskii, G.V. Adzhigai, and I.V. Bezsudnov. On the inherent figure of merit of thermoelectric composites. *Journal of Thermoelectricity*, 4:74–81, 2005.
180. J.P. Straley. Thermoelectric properties of inhomogeneous materials. *Journal of Physics D: Applied Physics*, 14(11):2101, 1981.
181. I. Webman, J. Jortner, and M.H. Cohen. Thermoelectric power in inhomogeneous materials. *Physical Review B*, 16(6):2959–2964, 1977.
182. K. Zabrocki, E. Müller, W. Seifert, and S. Trimper. Performance optimization of a thermoelectric generator element with linear material profiles in a 1D setup. *Journal of Material Research, Focus Issue: Advances in Thermoelectric Materials*, 26(15):1963–1974, 2011, doi: 10.1557/jmr.2011.91.
183. M. Abramowitz and I.A. Stegun, editors. *Handbook of Mathematical Functions with Formulas, Graphs, and Mathematical Tables*. Dover Publications, Inc., 180 Varick Street, New York, NY, 10014, 10 edition, 1972.
184. G.B. Mathews and E. Meissel. *A Treatise on Bessel Functions and their Applications to Physics*. BiblioBazaar, LLC, 2008.
186. B.Ya. Moizhes, A.V. Petrov, Yu.P. Shishkin, and L.A. Kolomoets. On the choice of the optimal mode of operation of a cascade thermoelectric element. *Soviet Physics-Technical Physics*, 7:336, 1962.
185. K. Zabrocki, W. Seifert, and E. Müller. Performance optimization of a thermoelectric generator element with linear material profiles in a 1D scheme. In H.L. Tuller, J.D. Baniecki, G.J. Snyder, and J.A. Malen, editors, *Thermoelectric Materials—Growth, Properties, Novel Characterization Methods, and Applications, Materials Research Society Symposia Proceedings*, Vol. 1267, pp. 27–32, Warrendale, PA, 2010.
187. R. Ure and R. Heikes. The figure of merit of a thermoelectric generator. *Advanced Energy Conversion*, 2:177, 1962.

5

Thermodynamics and Phase Transformations in Thermoelectric Materials: Applications to the Development of New Materials

Jean-Claude Tedenac
Montpellier 2 University

5.1 Introduction

This chapter evolves out of many years of research in materials for thermoelectrics. The study of phase diagrams is necessary to understand the synthesis of established thermoelectric materials, enhance the material properties, and contribute to the development of new and better materials. During the last century, the development of such materials was made with the help of experimental phase diagrams. This approach was not so bad; in fact, the materials were based on few well-known systems. Now, with the development of new materials such as the intermetallics, this thermodynamic approach is necessary for tailoring such materials; also, they are multicomponent (e.g., some skutterudites contain more then five elements).

Consequently, information on phase transformation, thermodynamic stabilities, and process modeling of the new materials presently studied for thermoelectricity are crucial.

Moreover, the processes involved in the fabrication of a thermoelectric component lead to other problems:

1. Solidification processes cause segregations in single phases and, depending on the composition of the melt, eutectic precipitations.
2. Temperature dependence of the material composition causes point defects leading to imperfect electronic properties.
3. Hot-pressing processes cause element diffusion in the grains and at the boundaries leading to nonequilibrium material with properties changing with time and temperature.

In the scope of the present development of materials for recovering waste energy and nanomaterials, it is really necessary to study the thermal stabilities of phases in the light of a thermodynamic approach.

Thermodynamics is a powerful science for studying the materials. The principles of thermodynamics of heterogeneous equilibria were developed by J.W. Gibbs [1]. He fixed the rules applicable to the phase relationships in a system. More recently, Prigogine and Defay [2] explained that beyond the classical thermodynamics, the affinity function or driving force of an internal process can be evaluated as an ordinary thermodynamic quantity.

These two approaches founded the modern chemical thermodynamics of the heterogeneous systems.

In this view it is very important to make a distinction between external and internal variables which refer to the production of entropy in an internal process.

The first concept to be introduced is the equation of state which is defined for a unary system as follows:

$$F(T,p,V) = 0 \tag{5.1}$$

It means that one of the three variables is defined by the other two.

At this stage, considering the materials, it is necessary to introduce the concept of system. Then we have to distinguish between unary and multinary systems. A unary system consists of a single component, mainly an element of the periodic table. By adding more components, we define a multinary system and in such cases, we have to add additional variables as amount of components.

For the definition of all terms useful in phase equilibria study, any general book can be consulted such as Saunders and Miodownick [3], Hillert [4], or Hack [5].

5.2 Basic Concepts

5.2.1 Definition of Terms and Symbols

Thermodynamic temperature	T
Gas constant	8.31451 J mol^{-1} K^{-1}
Pressure	P
Volume	V
Heat quantity	Q
Number of moles of component i	N_i
Total number of moles	$N, (N = \Sigma_{i=1}^{j} N_i)$
Avogadro number	N_A (6.023 × 10^{23} atoms per mole)
Molar fraction of component i	$x_i = N_i/N$
Fraction of constituent i*	y_{I}
Fractions of the constituent in a phase	Y
Total Gibbs energy of the system	G
Gibbs energy per mole of components of the system	G_{m}
Gibbs energy per mole of components of the phase α	G_{m}^{α}
Gibbs energy per mole of component i of the phase α	G_i^{α}

Gibbs energy per mole of the pure component *i* of the phase α	$^0G_i^\alpha$
Chemical potential of component *i*	m_i
Activity of component *i*	$a_i = \exp[\mu_i/RT]$

5.2.2 Important Thermodynamic Relations

Gibbs energy function	$G = G(T, p, N_i)$
Entropy	$S = -T(\partial G_m/\partial T)_{p,N_i}$
Enthalpy	$H = G + TS = G - T(\partial Gm/\partial T)_{p,N_i}$
Volume	$V = -T\ (\partial G/\partial p)_{T,N_i}$
Chemical potential of component	$\mu_i = -T(\partial G/\partial N_i)_{T,N_j \neq N_i}$
Heat capacity	$C_p = -T(\partial^2 G/\partial p \partial T)_{p,N_i}$
Thermal expansion	$\alpha = \frac{1}{V}(\partial^2 G/\partial p \partial T)_{N_i}$
Isothermal compressibility	$\kappa = -\frac{1}{V}(\partial^2 G/\partial p^2)_{T,N_i}$
Bulk modulus	$B = 1/K$

5.2.3 Gibbs Phase Rule

The well-known phase rule of J.W. Gibbs is the core of this part of thermodynamics. It stands that the maximum number of phases stable simultaneously is determined by the number of independent variables of state according to the equation

$$V = c + 2 - f$$

where V is the degree of freedom, c is the number of independent components, and f is the number of stable phases taking place in equilibrium.

5.2.4 Crystallography

The thermodynamic modeling of the solid phases of a system must be done in two ways. In the absence of any crystallographic information, the modeling can be assumed as a classical Redlich–Kister equation. When the phase is correctly described from a structural point of view, the modeling must be done in a sublattice description taking into account the Wyckoff positions and the coordination surrounding of elements. An accurate description of multicomponent systems requires physical phase descriptions related to the crystal structure. Ordered intermetallic compounds are commonly described by sublattice models based on Wyckoff site occupancy. In this description each sublattice is occupied by one kind of atoms. But in order to have a good description, a combination of similar sublattices is necessary to reduce the number of adjustable parameters.

5.2.5 Phase Diagrams and Relevant Parameters

In some cases of materials containing volatile elements, the problem of the pressure of component is a problem for understanding the chemical behavior of the material. As an example, Figure 5.1 presents the phase diagram of the system indium–selenium studied in former time from the thermoelectric point of view. The calculation is made without the gas phase in the number of phases taking part at equilibrium. In Figure 5.1, the gas phase is present, and in the region of selenium-rich side, it is easy to understand that near the melting temperature of compounds, the vapor pressure is not negligible. In such cases, this thermodynamic approach can help in the synthesis processes.

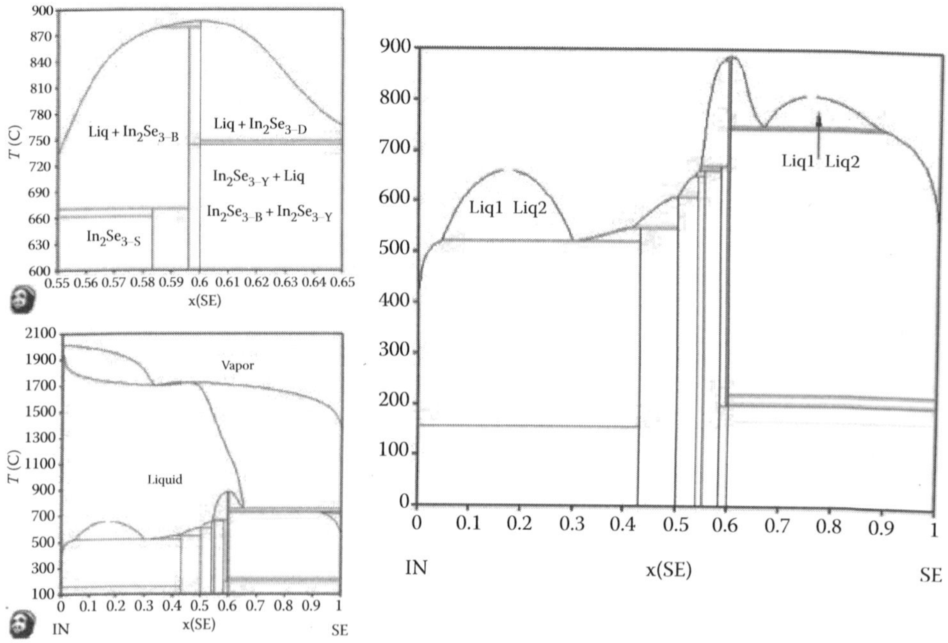

FIGURE 5.1 Calculated phase diagram of the indium–selenium system. The two small figures show the region around In_2Se_3 and the calculation taking into account the gas phase participating at equilibria.

5.3 Application of Phase Diagrams in Lead Telluride-Based Materials

5.3.1 Lead Telluride Material

The crystal structure of lead telluride is rocksalt at normal temperature and normal pressure with a four-formula unit (Figure 5.2). The lattice constant is 0.6461 nm [6] and the density is 8250 kg m^{-3}.

Lead telluride belongs to the IV–VI semiconductor group. This compound has a slight departure from the stoichiometric composition, which creates electrically active defects. The carrier concentration in PbTe materials changes with the chemical composition (p-type in the metal-rich side and n-type in the tellurium-rich side). It was well known that the defects mainly responsible for that

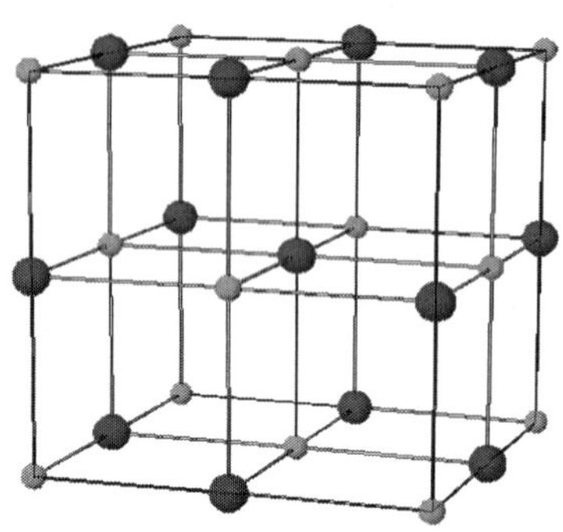

FIGURE 5.2 Crystal structure of lead telluride.

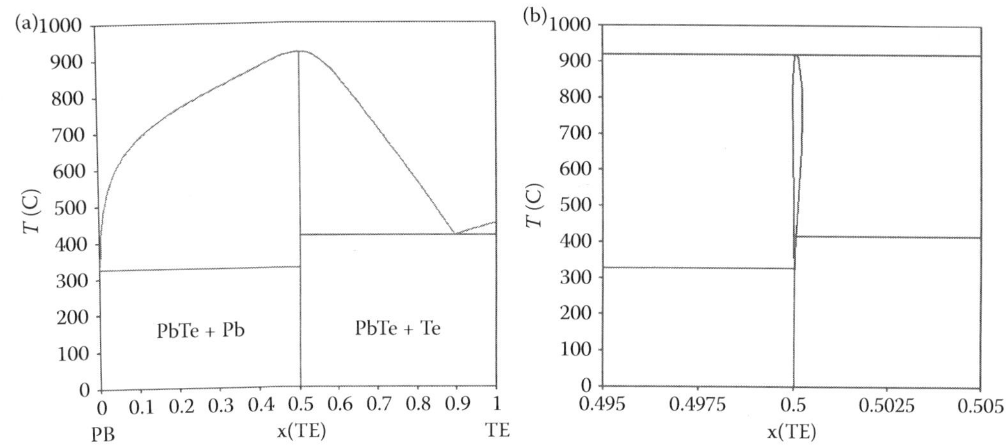

FIGURE 5.3 (a) Phase diagram of the Pb–Te system. (b) Extension in the central part of the diagram showing the nonstoichiometric region around Pb–Te at high temperature.

situation are doubly ionized vacancies on the tellurium side and on the lead-rich side, they are tellurium vacancies.

The congruent melting point is on the side rich in tellurium and materials obtained by solidification from the liquid state are generally *p*-type with a carrier concentration of around 10^{24} m^{-3}. By annealing in an atmosphere saturated with lead or tellurium, one can change the conduction type of the material and get the number of carriers up to 1.5×10^{24} m^{-3} for the *n*-type and 8×10^{24} m^{-3} for the *p*-type [6].

5.3.2 Binary and Related Ternary Phase Diagrams

Phase diagram description is very important for tailoring such materials. The two compounds PbTe and SnTe (narrow band-gap semiconductors) display a complete solubility in the whole composition range. The synthesis of ternary alloys makes it suitable for thermoelectrics. A thermodynamic database has been constructed by H. L. Lukas ([7] and cited in Ref. [8]). The free energy function description of the system has been modeled in order to describe the departure from stoichiometry at all temperatures. This problem is well described in Figure 5.3a and b, where the high-temperature range homogeneity appears.

Usually, ternary alloys based on 20–30 mol% of tin are studied for thermogenerator materials [8]. Some important problems in the constitution of materials can be solved using the corresponding phase diagrams. The definition of the homogeneity range of (PbSn)Te materials in the whole temperature range can easily be made by a thermodynamic description of the section. The thermoelectric material of *p*-type presenting the higher *ZT* is close to the composition $Pb_{0.8}Sn_{0.2}Te$. It can be described by a cut (isopleth section) of the ternary lead–tin–tellurium at 20 at% of tin.

The section is presented in Figure 5.4a and b. In the central part of the phase diagram, the homogeneity zone of the thermoelectric material is shown. This description is very useful for determining precisely the conditions of quenching and the synthesis of the single-phased material, because in the ternary system, the ternary phase region entails some problems in synthesis.

5.4 Description of Systems by the Calphad Method

5.4.1 General Calphad Technique

The Calphad method has been widely developed for metallic systems, oxide systems, and semiconductor systems as, for example, gallium arsenide [9], cadmium telluride [10], and lead telluride systems [8].

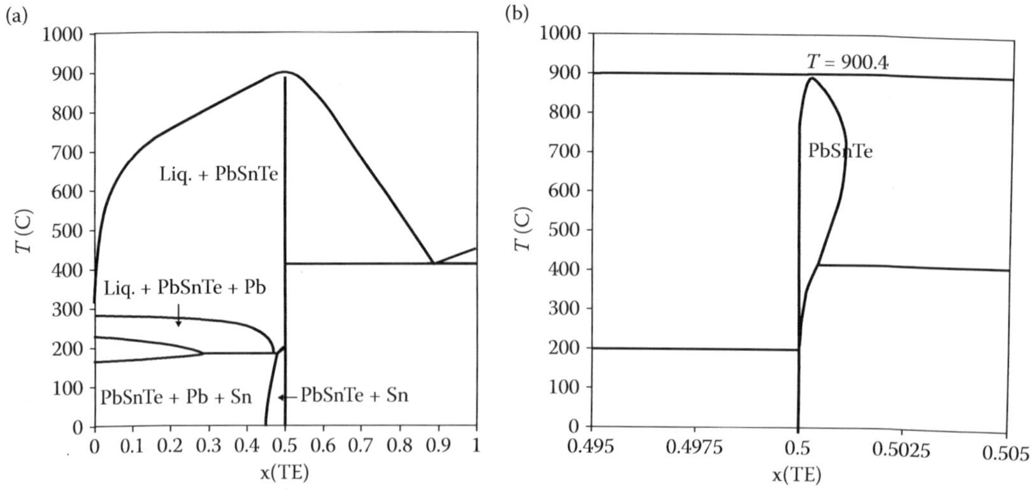

FIGURE 5.4 (a) Phase diagram of the Pb–Sn–Te, section between $Pb_{0.8}$–$Sn_{0.2}$–Te system. (b) Extension in the central part of the diagram showing the nonstoichiometric region.

In the case of semiconductor materials and therefore of thermoelectrics, it is necessary to introduce the point defects in the modeling of the phases in order to map the defects as a function of the chemical composition as in GaN [11]. This characteristic, important for physical applications, was not taken into account in most of the previous assessment of semiconductor systems. In the last decade, the defect modeling in semiconductors has been introduced and used by some authors (e.g., in the Cd–Te). Nowadays, the defects are obtained by the first principle calculations and put into Calphad in the same way as the other input.

Calculation of phase equilibria in a multicomponent system is obtained by the minimization of the total Gibbs energy of a system, where G is a summation of the Gibbs energy of all phases that take part in each equilibrium as is expressed below:

$$G = \sum_{i=1}^{j} n_i G_i^0 = \min \tag{5.2}$$

The thermodynamic description of the whole system requires the assignment of thermodynamic functions to each phase. The main interest of the Calphad method is the use of a variety of models to describe the free-energy functions of the various phases of the system as a function of temperature, pressure, and composition. Then, the Gibbs energy function of a i phase can be written as

$$G_m = \sum_i x_i \left({}^iG_0 - {}^iG_0^{\text{phys.}} \right) + G_{\text{m}}^{\text{phys.}} - TS_{\text{m}}^{\text{ideal.}} + {}^EG_i^{\text{m}} \tag{5.3}$$

The effects of particular physical phenomena (as magnetism) are taken into account by subtracting them from the description of the end-members (reference state) and it is introduced for the solution through the contribution by G^{phys}. The last term on the right-hand side is the excess term.

Usually the temperature dependence of the Gibbs energy is expressed as power series of T where a, b, c, and d_n are coefficients and n are integers (Equation 5.4). The pure elements are represented by the same function according to the database of A. Dinsdale [12].

$$G = a + b.T + c.T.\ln(T) + \Sigma d_n.T^n \tag{5.4}$$

By taking into account all the parameters, the Gibbs energy of phases are then represented by three contributions as expressed in Equation 5.5).

$$G^0 = \text{ref } G^0 + G^{\text{ideal}} + {}^{E}G(4) \tag{5.5}$$

The first term of the right-hand side corresponds to the Gibbs energy of a mechanical mixture of the components, the second one corresponds to the entropy of mixing for an ideal solution, and the third term, the so-called excess term, represents all the deviations from ideality.

Thermodynamic modeling of phases is the core of the Calphad approach. Some systems have few strictly stoichiometric compounds, so it is important to model phases deviating from the ideal stoichiometry. The most used model for solution phases are random substitutional solution, or ordered sublattices. In ordered solid phases, the Wagner–Shottky model is used for describing small deviations from stoichiometry (which are noninteracting defects). Many intermetallic binary compounds are generally nonstoichiometric and can exist in a large range of composition. In this case the most used model is the sublattice model (Equation 5.4). Moreover, additional information from physical properties (chemical potentials of electrons and holes measured by the transport experiment) can be put inside this model and can be implemented in the database. The general Calphad method is described in Figure 5.5.

All the models can be used for ternary and multicomponent systems by adding high-order interaction terms in the expression of the excess free energy.

In such high-ordered systems, the Gibbs energy must be calculated from extrapolation of the excess quantities of constituent subsystems. Several methods can be used and among them the geometrical Muggianu method is the most used (Equation 5.4). Consequently, the Gibbs energy of an n-component solution phase must be determined by the $n-1$ energies using the method given in Equation 5.6.

In fact, the relationship between chemical potentials of the phases and the system gives nonlinear equations that can be solved by numerical methods such as Newton–Raphson methods. The Calphad-type packages use mathematical methods to minimize the Gibbs functions as it is shown below.

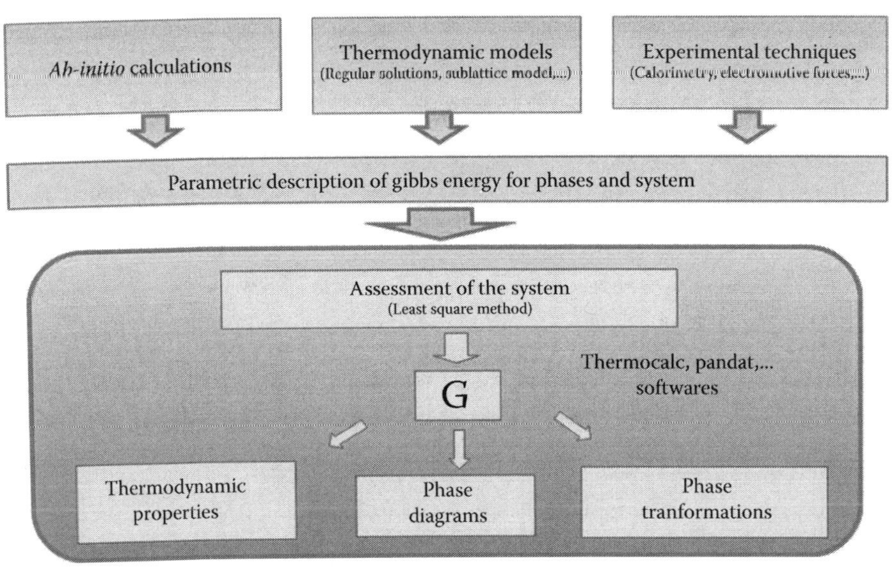

FIGURE 5.5 The general Calphad method.

$$G = \sum_i x_i G_i^0 + RTx_i \ \text{In}\ x_i + G^{ex}$$

$$\begin{aligned} &\text{Stage 1 binaries. Assessment of binaries : } G^{ex}_{bin.} \\ &\text{Stage 2 ternaries. Extrapolation } G^{ex}_{bin.} + G^{ex}_{ter.} \\ &\text{Stage 3 quaternaries. Extrapolation } G^{ex}_{bin.} + G^{ex}_{qua.} \\ &\text{Stage } N+1 \text{ high-order systems} \end{aligned} \tag{5.6}$$

The experimental data used for these calculations are those concerning the phase diagram measurements, enthalpy of formation of the compounds, temperature dependence of C_p and chemical potentials measured by electrochemistry. According to the description adopted in the Scientific Group Thermodata Europe (SGTE) database, the Gibbs energy of pure elements is taken with reference to the enthalpy of the elements in the SER state (standard element reference), the elements are in their stable state at $p = 10^5$ Pa and $T = 298.15$ K.

5.4.2 Description of Antimonide Systems

In the next generation of thermoelectric systems, materials will be built up with intermetallic phases. Many of these materials are based on antimony systems such as Sb–Zn or Co–Sb and they will be multicomponent in order to optimize thermoelectric properties.

The optimized skutterudite thermoelectric material belongs to the multicomponent system: Ce–Co–Fe–Ni–Sb. The purpose of this research is to build a database with five elements in order to know the material's behavior. As it was reported in Section 5.3.1, in a general Calphad procedure, the database is built step by step and at present we have a good knowledge of the ternaries Co–Fe–Sb and Co–Ni–Sb. They can be considered as the key systems for studying the skutterudite materials. The next section presents these results.

5.4.2.1 Zinc–Antimony Alloy Phases

The zinc antimonide materials have been demonstrated to have high ZT values (the higher is for the Zn_4Sb_3 phase, then ZnSb) [13]. Moreover, they are suitable for working at lower temperatures than skutterudite. Having a high figure of merit initially, those thermoelectric materials could have good properties with some achievement in the metallurgical processes [14]. The problem occurring in zinc antimony materials is due to difficulties in the crystal growth of materials. These difficulties are easily understood by a complicated phase diagram. From a recent paper [15] the different phase equilibria determinations lead to six different versions of this phase diagram. These differences were due to real difficulties in the understanding of the phase stabilities and consequently to the phase equilibria determination.

Using a Calphad analysis we have determined the phase relationship in this binary system and explained the behavior of the material. Figure 5.6a and b presents the calculated phase diagram obtained in this way. All the important information for crystal growth (temperatures, compositions, phase stabilities, liquidus shape, etc.) are indicated in this diagram [16].

Moreover, by a crystal structure analysis of the compounds, information about the sublattice modeling can be used. Associating phase diagram and crystal chemistry, the two intermetallic compounds can be modeled by a four-sublattice model and according to Ref. [3], the model for the Gibbs energy calculations should be a four-sublattice model with two antimony sublattices (Sb1, Sb2), one zinc and one vacancy sublattice including zinc as interstitial.

The problem existing in zinc antimony materials is due to difficulties in the crystal growth of materials. Having initially a high figure of merit, those thermoelectric materials could have better properties with some achievement in the metallurgical processes [13]. These difficulties are easily understood by studying a complicated phase diagram.

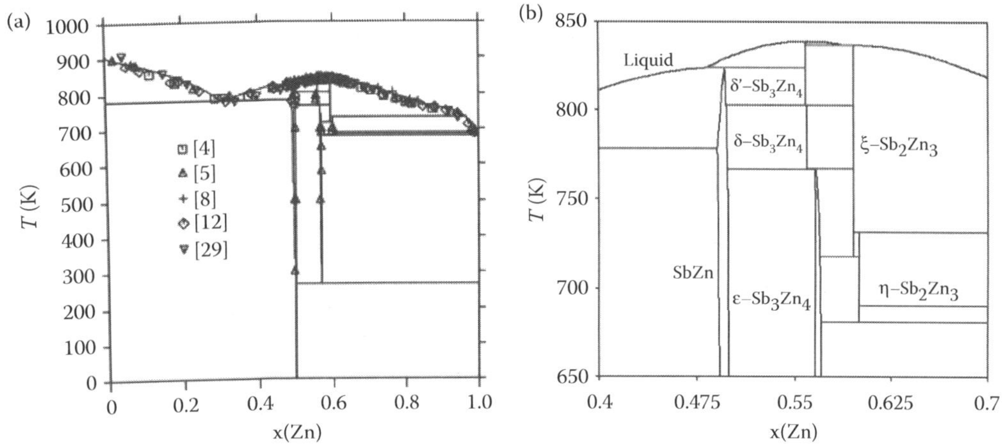

FIGURE 5.6 (a) The calculated Sb–Zn phase diagram after Ref. [14] compared with the experimental determinations. (b) The central part of the phase diagram showing the phase transformations before melting.

5.4.2.2 Skutterudite Phases

The ternary Co–Ni–Sb will be presented in this section.

One key system for studying these materials where the skutterudite phase should exist with a certain homogeneity range is the Co–Ni–Sb ternary. First, it is necessary to assess the border binaries (Co–Sb, Ni–Sb).

Most of the binary systems having a skutterudite-type compound were experimentally studied but no thermodynamic assessment was available until now. The skutterudite phase (cobalt triantimonide) is a stable compound and the nickel materials appear unstable; the extension of solid solution has to be studied in the ternary.

First, the Co–Sb system has been modeled [15–17]. In this system, three intermetallic compounds are formed: CoSb, $CoSb_2$, and $CoSb_3$. In nickel–antimony, only two compounds are formed: CoSb and $CoSb_2$; consequently, in this assessment the regular solution model was adopted for the liquid and terminal solid solution. Concerning the intermetallic compounds, there is no experimental information on homogeneity range, so the Gibbs energy of $CoSb_2$ and $CoSb_3$ are stoichiometrically modeled and described by a subregular model. The 1:1 compound was described with a three-sublattice model in order to take into account the departure from stoichiometry on both sides of the solid solution. The modeled phase diagrams are presented in Figure 5.7a. Figure 5.7b shows the shape of the Gibbs energy functions at 500°C for the binary Co–Sb; they are significant of the relevance of the models.

Table 5.1 lists the enthalpy of formation of the two skutterudite phases based on cobalt (stable) and nickel (metastable). One can see that the enthalpy of formation of $NiSb_3$ is the lowest among all compounds. It indicates that the range homogeneity of the ternary solid solution should be restricted to low nickel content. It has been calculated at different temperatures.

From the models describing phase in binary systems, modeling ternaries [18] has been undertaken, especially as we had some experimental data. Figures 5.8 and 5.9 present the results of calculations for the isothermal sections at 450°C and 540°C, temperatures that are typical operating thermogenerators for motor vehicles. It is observed that in this temperature range the chemical compositon of the phase changes significantly between pure nickel phase and 5 and 8 mol% which leads to the precipitation of phases are $NiSb_2$ and/or NiSb depending on conditions, which is not very good for the thermoelectric properties. Thus, the synthesis of single-phase materials must be limited in composition to $(Co_{0.95}Ni_{0.05})_3Sb$.

Therefore, the liquidus curve and the invariant temperatures are known with good accuracy. This result will be an aid for understanding the thermodynamic behavior of the ternary and quaternary

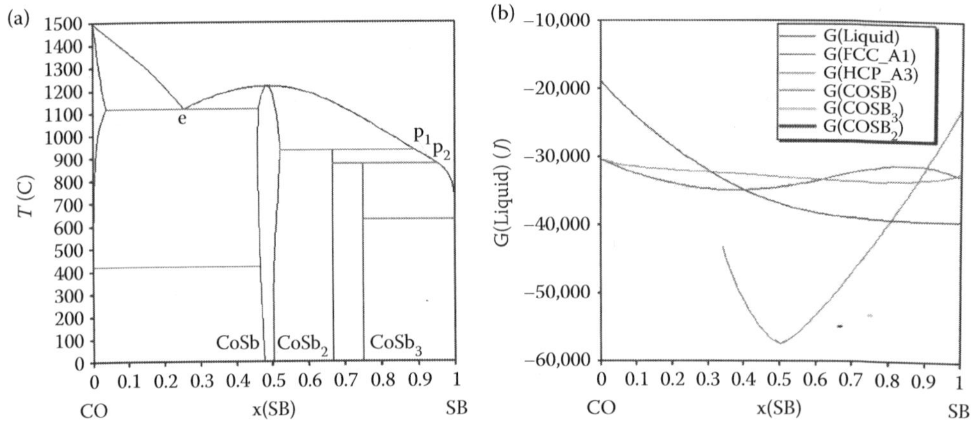

FIGURE 5.7 (a) Phase diagram of cobalt–antimony showing the three intermetallic compounds. The $CoSb_3$ corresponds to the "skuterrudite" phase. (b) Gibbs energy diagram of the cobalt–antimony system showing the relative stabilities of the solid phase at 500°C.

TABLE 5.1 Enthalpies of Formation of Compounds in the Co–Sb and Ni–Sb Systems

System Ni–Sb	System Co–Sb
$G^{NiSb} = -69{,}039$ J mol^{-1}	$G^{CoSb} = -57{,}327$ J mol^{-1}
$G^{NiSb2} = -60{,}804$ J mol^{-1}	$G^{CoSb2} = -54{,}869$ J mol^{-1}
$G^{NiSb3} = -39{,}165$ J mol^{-1}	$G^{CoSb3} = -53{,}163$ J mol^{-1}

phases. It has been possible to calculate the temperature of the metastable melting point of $CoSb_3$ (1285 K instead of 1210 K for the peritectic decomposition of the 1:2 phase). One can observe that these temperatures are not so far and it could explain the problems occurring in the fabrication of the materials.

As it was explained in Section 5.1, modeling a ternary system requires knowing the binaries. When the binary phases exist in the form of ternary solid solution, the models used to describe the

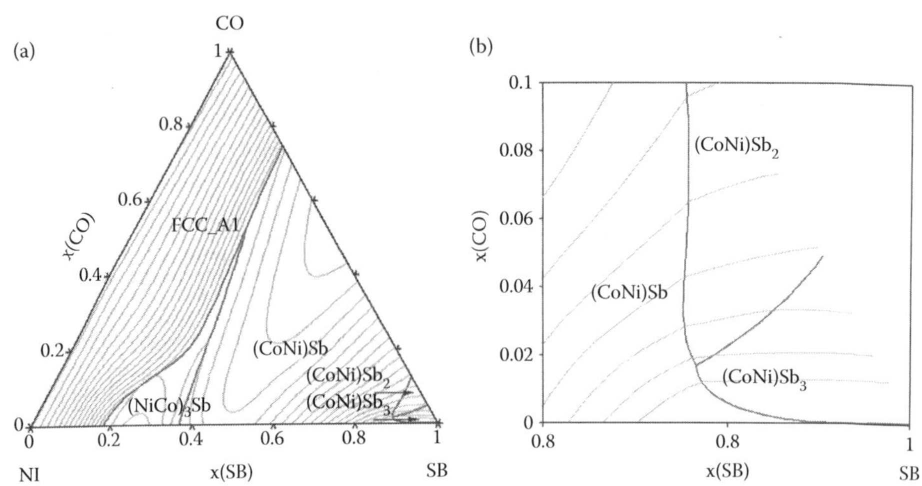

FIGURE 5.8 Phase diagram of cobalt–nickel–antimony. (a) Projection of the liquidus surface showing the crystallization fields of the solid phases. (b) The antimony-rich region showing the fields of $CoSb_2$ and $CoSb_3$ compounds.

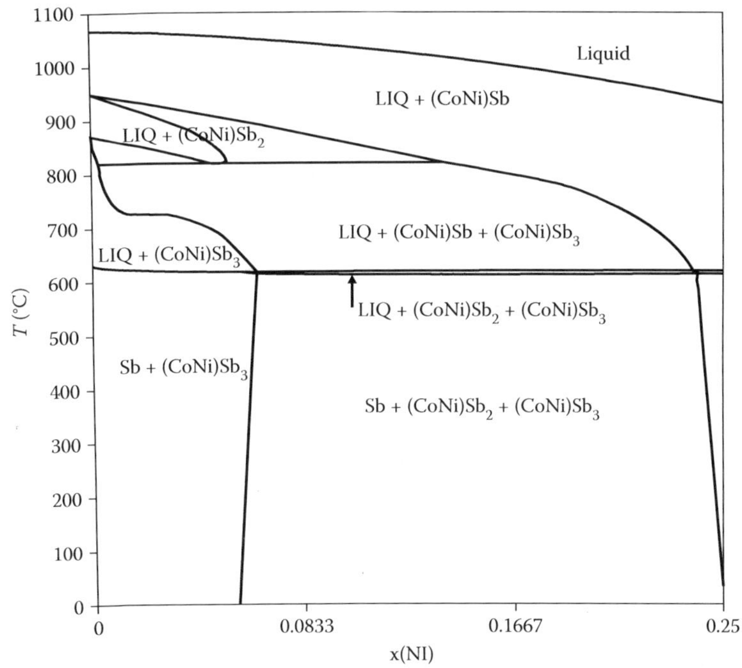

FIGURE 5.9 The isopleth section of the Co–Ni–Sb ternary between the $NiSb_3$ and $CoSb_3$ compositions.

phases must be mutually compatible, it is usually the biggest challenge of scaling up in phase diagrams. In that case one can observe that a phase is dominant (the 1:1 nickel arsenide phase) and a good modeling of the ternary and related binaries depends on the model used in this phase. In the present system, results in Figure 5.8a and b shows the liquidus projections, where one can see the phase crystallization fields.

In Figure 5.9, we present the isopleth section between $CoSb_3$ and metastable $NiSb_3$ where one can see the field of the ternary solid solution $(Co,Ni)Sb_3$. It shows the difficulty to crystallize that phase as a pure material. On the cobalt-rich side of the section we can see the presence of antimony in two or three phase fields. This is a problem for further applications. This is due to the fact that all reactions in ternary are peritectic or transitory and the pure phases cannot crystallize without accurate thermal treatment. The published results on the materials synthesis in different laboratories show many phase precipitations in materials and they can be explained at the light of such thermodynamic analysis.

5.5 Conclusions

By using the Calphad method we described phase diagrams of some thermoelectric materials and show that the possibilities of this kind of research in the description of systems Sb–Zn and Co–Ni–Sb have been obtained from a self-consistent set of thermodynamic parameters describing the Gibbs energy of each phase depending on the composition and temperature. Building a database allows scientists to calculate thermodynamic properties and represent all types of functions. Such liquidus surface projections, isothermal sections, and values of thermodynamic functions can be derived from the calculations. With these thermodynamic descriptions, one can then have various calculations of practical interest for the development of materials, using the principles established by Scheil and Gulliver [18,19].

References

1. J.W. Gibbs, On the equilibrium of heterogeneous substances, *Trans. Conn. Acad. III*, 108–248, 1778, 343–524.
2 I. Prigogine, R. Defay, Traité de Thermodynamique, *Desoer (Liege)*, 1, 1950, 38.
3. N. Saunders, P. Miodownick, *CALPHAD, A Comprehensive Guide*, Pergamon, Exeter, 1998, p. 261.
4. M. Hillert, *Phase Equilibria, Phase Diagrams and Phase Transformations*, 2nd (edn.), Cambridge University Press, 2008.
5. K. Hack, *The SGTE Casebook: Thermodynamics at Work*, 2008, 11–42.
6. R.-F. Brebick, R.S. Allgaier, Deviation from stoichiometry and lattice defects in IV–VI compounds and alloys, *J. Chem. Phys.* 36(5), 1962, 1283–1289.
7. U. Kattner, Thermodynamic modeling of multicomponent systems, *J. Min. Met. Mat.*, 49(12), 1997, 14–19.
8. N. Bouad, PhD thesis, Montpellier (France), 2001.
9. I. Ansara, C. Chatillon, H.L. Lukas, T. Nishizawa, H. Ohtani, K. Ishida, M. Hillert et al., A binary database for III–V semiconductor systems, *Calphad*, 18(2), 1994, 177–222.
10. U. Kattner, H.L. Lukas, G. Petsow, B. Gather, E. Irle, R. Blachnik, Excess enthalpy measurementsand évaluation of the Sn–Pb system, *Z. fur Metallk.*, 79, 1988, 32–40.
11. J.-B. Li, J.C. Tedenac, Thermodynamic modeling of native point defects and GaN semiconductors dopants, *J. Elec. Mat.*, 31(4), 2002, 321–326.
12. A. Dinsdale, SGTE data for pure éléments, *Calphad*, 15, 1991, 317–425.
13. J.P. Fleurial, T. Caillat, A. Borshevsky, D.T. Morelli, G.P. Meisner, High figure of merit in Ce-filled skutterudites, *Proc. 15th Int. Conf. Thermoelectrics*, Piscataway, NJ, 1996, p. 91.
14. L. Ya, J.C. Tedenac, Thermodynamic modeling of the Cd–Sb–Zn ternary system, *Calphad*, 33(4), 2009, 684–694.
15. Y. Zhang, L. Changrong, Z. Du, T. Geng, The thermodynamic assessment of the Co–Sb system, *Calphad*, 32(1)(March), 2008, 56–63.
16. Y. Zhang, L. Changrong, Z. Du, C. Guo, A thermodynamic assessment of Ni–Sb system, *Calphad*, 32(2), 2008, 378–388.
17. Y. Zhang, L. Changrong, Z. Du, C. Guo, J.C. Tedenac, The thermodynamic assessment of the ternary Co–Ni–Sb system, *Calphad*, 33(2), 2009, 405–414.
18. V. Scheil, *Z. fur Metallk.*, 34, 1942, 70.
19. G.H. Gulliver, *Metal and Alloys*, Verlag-Griffin, Vol. 25, 1922, pp. 120–157.

6

First Principles Calculations of Electron Transport Properties in Disordered Thermoelectrics

Janusz Tobola
AGH University of Science and Technology

Laurent Chaput
University of Nancy

6.1 Introduction

Electron transport properties of crystalline materials are associated with the most fascinating phenomena investigated in condensed matter physics. Electrical conductivity σ, electron thermal conductivity κ_e, thermopower α, as well as Hall coefficient R_H are quantities commonly used to probe electron transport in solids and to roughly differentiate metals, semimetals, semiconductors, and insulators. The quantum equations of electron motion in crystals provide an efficient junction between microscopic understanding of charge and heat transport in terms of Fermi surface (FS) features on the one hand, and measured physical quantities (σ, κ_e, α, and R_H) on the other.[1–3] Density functional theory (DFT) techniques are well-established theoretical methods to calculate electronic structure in materials science. Interestingly, in the last two decades, the first principles band structure computations have started to play an important role not only in the study of electronic and magnetic properties in solids but also to enable more realistic interpretation of transport properties, including complex multiatom thermoelectric materials. Since electron transport behaviors are directly related to electronic states near the Fermi energy (ϵ_F), searching for accurate information on **k**-space electron features appears particularly valuable to model or even to predict thermoelectric behaviors.

Actually, first principles methods attempt to determine, in the framework of employed approximations, ground-state properties of solid systems. Hence, finite temperature effects, especially interesting in investigations of thermopower $\alpha(T)$ or electrical conductivity $\sigma(T)$, are not explicitly taken into account in such calculations. Modifications of ground-state electronic properties with temperature are simply approximated using the Fermi–Dirac distribution function $f(E,T)$. Note, that this function changes markedly only in the energy range of few k_BT around E_F and this very narrow energy window must be inspected in detail to reliably calculate kinetic parameters for electrons. The temperature effects can also be included by varying lattice parameters and interatomic distances of a system in electronic structure calculations.

In this chapter, we briefly present two ways to calculate electron transport coefficients on the basis of DFT electronic structure calculations. First, linear equations of nonequilibrium thermodynamical systems, in terms of external fields (temperature gradient ∇T and electric field $\vec{\mathcal{E}}$) yielding charge and heat currents ($\vec{J}$ and $\vec{J}_Q$, respectively), are discussed. This theory combined with the Boltzmann equation and the relaxation time approach allow us to determine the Onsager coefficients that directly enter temperature-dependent expressions for electrical conductivity $\sigma(T)$ and Lorenz factor $L(T)$ as well as thermopower $\alpha(T)$ as a function of carrier concentration n, p. We show the illustrative results of the full-potential linearized augmented plane waves (FLAPW)[4] calculations and the Boltzmann approach, as applied to real TE materials (e.g., doped $CoSb_3$ skutterudite), where the effect of doping has been simulated by a rigid shift of E_F on density of states (DOS) of the ordered host system.

Second, we present the results of residual electrical conductivity σ_0 and Seebeck coefficient calculations as determined from direct FS integration of k-dependent velocities, and lifetimes of electrons in chemically disordered materials. It is worth noting that in alloys, electronic bands $\epsilon(\vec{k})$ and then FS have complex energy values with its nonvanishing imaginary part Im ϵ related to chemical disorder. The results of illustrative calculations with the use of the Korringa–Kohn–Rostoker technique and the coherent potential approximation (KKR–CPA),[5] are shown for disordered half-Heusler phases to determine the thermopower slope a/T using the Mott's formula.[1] Moreover, simplified estimation of Seebeck coefficients, based only on accurate analysis of DOS near ϵ_F is also presented as applied to some TE materials. Finally, the role of some crystal point defects (antisite imperfections or vacancies) in more reliable determination of electronic structure and electron transport properties is discussed in view of KKR–CPA calculations.

6.2 Linear Nonequilibrium Thermodynamics

6.2.1 General Equations

Thermodynamics describes the exchange of energy between equilibrium states using the first law, which says that energy is conserved during these processes, and with the second law it indicates their direction, since entropy must be created. When matter or energy is exchanged with a system initially at equilibrium, applying, for example, an electric field $\vec{\mathcal{E}}$ or a temperature gradient ∇T, it will then respond, trying to establish again an equilibrium state by displacing energy and matter and therefore producing fluxes and entropy. Now, one can ask about the internal forces that create these fluxes. In a simple view, one could simply reply that the electric field and the temperature gradient generate the fluxes of charged particles and heat. However, such an approach does not fully reveal symmetries of the system. The other possibility is to focus on possible ways of entropy creation.[6] If the entropy of the equilibrium state can be written as $dS = \sum_i f_i dX_i$, with X_i the extensive quantities describing the system and f_i their intensive conjugate variables, then it can be shown that the entropy produced per unit volume and unit time is

$$\sigma_S = \sum_i \vec{J}_{X_i} \cdot \nabla f_i(\vec{r},t) \tag{6.1}$$

The argument just comes from the balance equation for the entropy density $s(\vec{r},t)$,

$$\frac{\partial s}{\partial t} + \nabla \cdot \vec{J}_S = \sigma_S \tag{6.2}$$

together with $ds = \Sigma_i f_i dx_i$, or $\vec{J}_S = \Sigma_i f_i \vec{J}_{X_i}$ where J_{X_i} is the flux of the quantity X_i and its volume density x_i. In full details, this gives

$$\sigma_S = \sum_i f_i \frac{\partial x_i}{\partial t} + \sum_i f_i \nabla \cdot \vec{J}_{X_i} + \sum_i \nabla f_i \cdot \vec{J}_{X_i} = \sum_i \nabla f_i \cdot \vec{J}_{X_i} \tag{6.3}$$

The central assumption in the previous derivation is to write $ds(\vec{r},t) = \Sigma_i f_i dx_i(\vec{r},t)$ which means that one supposes the equilibrium relationship between the entropy and the extensive variables to be valid locally, within a volume element d^3r at any time t. This is called the local equilibrium hypothesis.

Under such an assumption, Equation 6.1 allows us to identify the ∇f_i as the forces creating the fluxes

$$\vec{J}_{X_i} = g(\nabla f_1, \nabla f_2, \ldots) \simeq \sum_k L_{ik} \nabla f_k \tag{6.4}$$

In the second equation, we have assumed the forces to be small, and the first term has been omitted in the Taylor expansion because there is no net current without any forces. The advantage to identify the forces in this way comes from the symmetry relations they fulfill

$$L_{ij} = L_{ji} \tag{6.5}$$

In transport experiments, this means that there is no need to measure all the coupling coefficients. But beyond this practical interest, these equations are a consequence of the invariance of microscopic equations with time inversion.[7,8]

6.2.2 Thermoelectricity

In the case of thermoelectric phenomena, one considers effect on the sample of temperature gradient ∇T and electric field $\vec{\mathcal{E}}$ that can be expressed in terms of the electric potential $\vec{\mathcal{E}} = -\nabla\phi$. The resulting entropy change of the system is given by

$$dS = \frac{1}{T} dE - \frac{\mu + q\phi}{T} dN = \frac{1}{T} dE - \frac{\bar{\mu}}{T} dN \tag{6.6}$$

where E denotes energy, T temperature, N number of particles with charge q, μ the chemical potential, and $\bar{\mu} = \mu + q\phi$ the electrochemical potential. Therefore, one obtains

$$\sigma_S = \nabla\left(\frac{1}{T}\right) \cdot \vec{J}_E - \nabla\left(\frac{\bar{\mu}}{T}\right) \vec{J}_N \tag{6.7}$$

However, the fluxes of particles, $\vec{J}_N$, and energy, $\vec{J}_E$, cannot be easily measured, and it is better to deal with the fluxes of charges, $\vec{J} = q\vec{J}_N$, and heat, $\vec{J}_Q = \vec{J}_E - \bar{\mu}\vec{J}_N$. The production of entropy can then be rewritten as

$$\sigma_S = \nabla\left(\frac{1}{T}\right)\cdot \vec{J}_Q - \frac{1}{qT}\cdot \vec{J} \tag{6.8}$$

which gives

$$\vec{J} = \mathcal{L}_{11}\left(\frac{-\nabla\bar{\mu}}{q}\right) + \mathcal{L}_{12}\left(\frac{-\nabla T}{T}\right) \tag{6.9}$$

$$\vec{J}_Q = \mathcal{L}_{21}\left(\frac{-\nabla\bar{\mu}}{q}\right) + \mathcal{L}_{22}\left(\frac{-\nabla T}{T}\right) \tag{6.10}$$

with the property $\mathcal{L}_{12} = \mathcal{L}_{21}$. The relations of the $\mathcal{L}_{ij}$ coefficients with the L_{EE}, L_{EN}, and L_{NN} ones can be easily established when replacing $\vec{J}$ and $\vec{J}_Q$ by their expressions in terms of $\vec{J}_E$ and $\vec{J}_N$. It yields

$$\vec{J} = q\vec{J}_N = \frac{q^2}{T}L_{NN}\left(\frac{-\nabla\bar{\mu}}{q}\right) + \frac{q}{T}(L_{NE} - \bar{\mu}L_{NN})\left(\frac{-\nabla T}{T}\right) \tag{6.11}$$

$$\vec{J}_Q = \vec{J}_E - \bar{\mu}\vec{J}_N = \frac{q}{T}(L_{EN} - \bar{\mu}L_{NN})\left(\frac{-\nabla\bar{\mu}}{q}\right) + \frac{1}{T}(L_{EE} - 2\bar{\mu}L_{EN} + \bar{\mu}^2L_{NN})\left(\frac{-\nabla T}{T}\right) \tag{6.12}$$

and finally,

$$\mathcal{L}_{11} = \frac{q^2}{T}L_{NN} \tag{6.13}$$

$$\mathcal{L}_{12} = \frac{q}{T}(L_{NE} - \bar{\mu}L_{NN}) \tag{6.14}$$

$$\mathcal{L}_{22} = \frac{1}{T}(L_{EE} - 2\bar{\mu}L_{EN} + \bar{\mu}^2L_{NN}) \tag{6.15}$$

In transport experiments, the measured quantities are the electrical conductivity σ, the thermopower α, and the thermal conductivity κ_e. They can directly be expressed in terms of the $\mathcal{L}_{ij}$, previously defined using the correct boundary conditions in Equations 6.9 and 6.10.

In a material with uniform density, $\bar{\mu}(\vec{r}) = \mu(T(\vec{r})) + q\phi(\vec{r})$, we obtain

- For the electrical conductivity,

$$(\nabla T = 0) \Rightarrow \vec{J} = \mathcal{L}_{11}(-\nabla\phi) \Rightarrow \sigma = \mathcal{L}_{11} = (q^2/T)L_{NN} \tag{6.16}$$

- For the electronic part of thermal conductivity,

$$(\vec{J} = 0) \Rightarrow -\nabla\bar{\mu}/q = \mathcal{L}_{11}^{-1}\mathcal{L}_{12}\nabla T/T \Rightarrow \vec{J}_Q = -1/T(\mathcal{L}_{22} - \mathcal{L}_{21}\mathcal{L}_{11}^{-1}\mathcal{L}_{12})\nabla T \tag{6.17}$$

$$\kappa_e = \frac{1}{T}(\mathcal{L}_{22} - \mathcal{L}_{21}\mathcal{L}_{11}^{-1}\mathcal{L}_{12}) \tag{6.18}$$

$$\kappa_e = \frac{1}{T^2}(L_{EE} - L_{EN}L_{NN}^{-1}L_{NE}) \tag{6.19}$$

- For the thermopower,

$$(\vec{J} = 0) \Rightarrow -\nabla\bar{\mu}/q = \mathcal{L}_{11}^{-1}\mathcal{L}_{12}\nabla T/T \tag{6.20}$$

$$\alpha = \frac{1}{T}\mathcal{L}_{11}^{-1}\mathcal{L}_{12} \tag{6.21}$$

$$\alpha = \frac{1}{qT}(L_{NN}^{-1}L_{NE} - \bar{\mu}) \tag{6.22}$$

From the above equations, we now have a clear view of the heat current. Using Equation 6.10, we obtain

$$\vec{J}_Q = -\kappa_e\nabla T + T\alpha\vec{J} \tag{6.23}$$

The first term is the electronic thermal conduction, whereas the second one comes from the transport of electronic charge and can be understood[9] in terms of charged particles carrying entropy $s_0 = q\alpha$.

6.2.3 The Boltzmann Equation

In the previous paragraph, we have expressed the transport coefficients σ, α, and κ_e in terms of the Onsager coefficients $\mathcal{L}_{ij}$. Now, a microscopic theory is required to calculate them. As far as electrons are concerned, there are two main components of such a model that are accounted for, that is, the energy spectrum defined by the band structure $\epsilon_n(\vec{k})$ within the approximation of independent electrons, as well as the distribution function, which determines the probability of electron to be found in a given energy state. At equilibrium, such probability can be directly determined from the Fermi–Dirac distribution function, $f_0(\epsilon) = 1/(e^{(\epsilon-\mu)/(k_BT)} + 1)$. However, upon applying external fields (∇T and $\vec{\mathcal{E}}$) on the system, the form of the distribution function becomes unknown and should be calculated. In a bulk material, one usually uses semiclassical theory to obtain the average occupation of the state of wavevector $\vec{k}$ and band index n, the so-called one-particle occupation function, $f_n(\vec{r},t)$. It is done using the Boltzmann equation, assuming that the number of particles leaving the phase space element d^3rd^3k per unit time is proportional to the deviation from f_0

$$\frac{df}{dt} = \frac{\partial f}{\partial t} + \frac{d\hbar\vec{k}}{dt}\frac{\partial f}{\hbar\vec{k}} + \frac{d\vec{r}}{dt}\frac{\partial f}{\partial\vec{r}} = \frac{f_0 - f}{\tau(\vec{k})} \tag{6.24}$$

where $d(\hbar\vec{k})/dt$ represents the external force $-e\vec{\mathcal{E}}$ and $d\vec{r}/dt$ is identified with the group velocity of the electrons, $\vec{\upsilon}_n(\vec{k}) = \frac{1}{\hbar}((\partial\epsilon_n(\vec{k}))/\partial\vec{k})$

$$\frac{\partial f}{\partial t} - \frac{e}{\hbar}\frac{\partial f}{\vec{k}}\cdot\vec{\mathcal{E}} + \vec{\upsilon}_n(\vec{k})\cdot\frac{\partial f}{\partial\vec{r}} = \frac{f_0 - f}{\tau(\vec{k})} \tag{6.25}$$

$\tau(\vec{k})$ is the relaxation time, that is, the time necessary to allow the system relaxing to equilibrium due to electron collisions.

If we are able to solve Equation 6.25, it is then straightforward to calculate the statistical average of the charge and heat currents as follows:

$$\vec{J} = (-\mathrm{e})\vec{J}_{\mathrm{N}} = \frac{1}{V}\sum_{nk}(-\mathrm{e})\vec{\upsilon}_n(\vec{k})f_n(\vec{k}) \tag{6.26}$$

$$\vec{J}_Q = \vec{J}_{\mathrm{E}} - \mu\vec{J}_{\mathrm{N}} = \frac{1}{V}\sum_{nk}(\epsilon_n(\vec{k}) - \mu)\vec{\upsilon}_n(\vec{k})f_n(\vec{k}). \tag{6.27}$$

However, there is no closed solution known for this equation. But considering stationary conditions for the system, that is, $\partial f/\partial t = 0$, as well as assuming the temperature gradient ∇T and electric field $\vec{\mathcal{E}}$ to be small, the equation can be linearized, $\nabla T(\partial f/\partial T) \simeq \nabla T(\partial f_0/\partial T)$ and $\vec{\mathcal{E}}\cdot(\partial f/\partial\vec{k}) \simeq \vec{\mathcal{E}}\cdot(\partial f_0/\partial\vec{k})$, to get the solution

$$f = f_0 - \tau(\vec{k})\vec{\upsilon}_n(\vec{k})\cdot\left(-e\frac{\partial f_0}{\partial\epsilon_n(\vec{k})}\vec{\mathcal{E}} + \frac{\partial f_0}{\partial T}\nabla T\right) \tag{6.28}$$

Using Equations 6.26 and 6.27, the charge and heat currents can therefore be written as

$$\begin{aligned}\vec{J} &= \frac{1}{V}\sum_{nk}\mathrm{e}^2\frac{\partial f_0}{\partial\mu}(\tau(\vec{k})\vec{\upsilon}_n(\vec{k})\otimes\vec{\upsilon}_n(\vec{k}))\cdot\vec{\mathcal{E}}\\ &+\frac{1}{V}\sum_{nk}(-\mathrm{e})(\epsilon_n(\vec{k})-\mu)\frac{\partial f_0}{\partial\mu}(\tau(\vec{k})\vec{\upsilon}_n(\vec{k})\otimes\vec{\upsilon}_n(\vec{k}))\cdot\left(\frac{-\nabla T}{T}\right)\end{aligned} \tag{6.29}$$

$$\begin{aligned}\vec{J}_Q &= \frac{1}{V}\sum_{nk}(-\mathrm{e})(\epsilon_n(\vec{k})-\mu)\frac{\partial f_0}{\partial\mu}(\tau(\vec{k})\vec{\upsilon}_n(\vec{k})\otimes\vec{\upsilon}_n(\vec{k}))\cdot\vec{\mathcal{E}}\\ &+\frac{1}{V}\sum_{nk}(\epsilon_n(\vec{k})-\mu)^2\frac{\partial f_0}{\partial\mu}(\tau(\vec{k})\vec{\upsilon}_n(\vec{k})\otimes\vec{\upsilon}_n(\vec{k}))\cdot\left(\frac{-\nabla T}{T}\right)\end{aligned} \tag{6.30}$$

A comparison with Equations 6.9 and 6.10 allows finally to obtain the Onsager coefficients

$$\mathcal{L}_{ij} = \int d\epsilon\left(\frac{\epsilon-\mu}{-e}\right)^{i+j-2}\sigma(\epsilon) \tag{6.31}$$

where $\sigma(\epsilon)$ represents the transport function defined as[10]

$$\sigma(\epsilon) = \frac{\mathrm{e}^2}{V}\sum_{nk}\tau(\vec{k})\vec{\upsilon}_n(\vec{k})\otimes\vec{\upsilon}_n(\vec{k})\delta(\epsilon-\epsilon_n(\vec{k})) \tag{6.32}$$

It is an energy-dependent conductivity and, as shown by Equation 6.31, its average value k_BT around the Fermi energy (ϵ_F) gives the ordinary electrical conductivity. The above equations also show that the higher moments of $\sigma(\epsilon)$ describe the other transport coefficients as thermopower α and electronic thermal conductivity κ_e. The transport function appears therefore as the key quantity defined in this paragraph and it needs to be accurately computed to study realistic systems. However, before analyzing illustrative examples of the aforementioned calculations (next paragraph) it is worth mentioning the approximations that have been made to obtain these results. The first of them is the local equilibrium hypothesis. In order to satisfy this condition, the variation of investigated quantity due to the applied fields should be smaller than the thermodynamic fluctuations over the volume element d^3r. Furthermore, the Bloch–Boltzmann equation 6.25 is a semiclassical approximation and therefore the classical limit criteria of quantum mechanics should be fulfilled.[11] It states that the variation of the de Broglie wavelength ($\lambda = h/p$) needs to be small on a distance of the order of λ itself, $|d\lambda/dx| \ll 1$. If the force F is applied, this criterion can be rewritten as $m\hbar F/p^3 \ll 1$ or $\lambda F \ll \bar{\epsilon}$, where $\bar{\epsilon}$ is the average kinetic energy of electrons. One has $\bar{\epsilon} \sim k_B T$ for semiconductors, and $\bar{\epsilon} \sim \epsilon_F$ for metals.

In our case, the forces are $|\nabla\mu| = |\partial\mu/\partial T||\nabla T|$ and $e\mathcal{E}$. The validity conditions for the Boltzmann equation are therefore

$$\lambda\left|\frac{\partial\mu}{\partial T}\nabla T\right| \ll k_B T \text{ or } \epsilon_F \tag{6.33}$$

$$\lambda e\mathcal{E} \ll k_B T \text{ or } \epsilon_F \tag{6.34}$$

as well as $\lambda \ll l$ (l—the sample size). The other approximation we made is the relaxation time. To derive $\tau(k)$ one usually uses the Fermi golden rule, which shows that we consider the scattering processes by impurities or phonons "one by one." Therefore, to be consistent, electron propagation should not be affected by two scattering centers. It appears true when the duration of the scattering Δt is short with respect to the time interval between two scattering events

$$\tau \gg \Delta t \Rightarrow \tau \gg \hbar/\Delta E \Rightarrow \tau \gg \hbar/k_B T$$

If $\bar{\upsilon}$ is the average velocity and $\Lambda = \tau\bar{\upsilon}$ the mean free path, the above equation can also be written as

$$\Lambda \gg \lambda$$

6.2.4 Application to TE Materials

In principle, Equations 6.31 and 6.32 allow calculating the transport coefficients σ, α, and κ_e from the band structure $\epsilon_n(\vec{k})$ once we know the relaxation time $\tau(\vec{k})$. To obtain $\tau(\vec{k})$, it is needed to calculate the diffusion of electrons by impurities, phonons, other electrons, and so on, using as mentioned above, for example, the Fermi golden rule. However, such advanced calculations have not been performed yet. In the next section, we attempt to explain how to account for electron scattering by impurities and defects in the first principles calculations. But for now we first review what is possible to do using simple models for $\tau(\vec{k})$, and therefore requiring no more than the knowledge of the band structure of the crystalline system.

At first, we ignore the details of the scattering processes, and take $\tau(\vec{k})$ to be a constant τ, but this value is unknown. It is equivalent to say the system relaxes to equilibrium following a statistical Poisson's process. We have no information about what causes the transitions allowing the system to relax, except for the assumption that they are isotropic and unrelated to the distant past. This approximation is of course of no use for the electrical conductivity, which is proportional to τ. However, the situation is different for the thermopower α and the Lorenz number $L = \kappa_e/(T\sigma)$ which are the key quantities for the thermoelectric figure of merit, $ZT \propto \alpha^2/L$. This comes from the fact that they can be expressed as a ratio of the Onsager coefficients, and therefore being *independent* of a constant relaxation time. In particular,

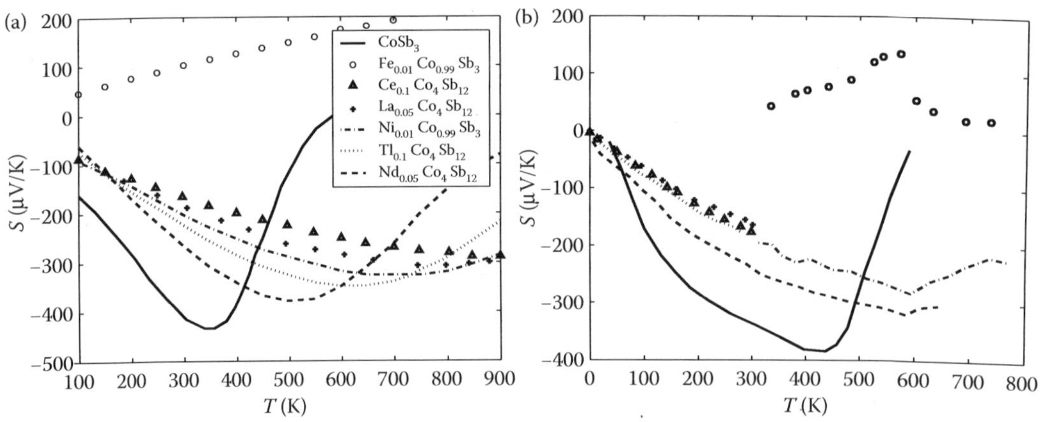

FIGURE 6.1 (a) Calculated thermopower. (b) Experimental thermopower from Refs. 14–20.

$$\alpha = \frac{1}{T}\frac{\mathcal{L}_{12}}{\mathcal{L}_{11}} = \frac{1}{-eT}\frac{\int d\epsilon(\epsilon-\mu)\sum_{nk}\vec{\upsilon}_n(\vec{k})\vec{\upsilon}_n(\vec{k})\delta(\epsilon-\epsilon_n(\vec{k}))}{\int d\epsilon\sum_{nk}\vec{\upsilon}_n(\vec{k})\vec{\upsilon}_n(\vec{k})\delta(\epsilon-\epsilon_n(\vec{k}))} \tag{6.35}$$

Several research groups have already used this methodology to calculate thermoelectric parameters in different materials.[10,12,13] As an illustrative example, let us consider doped compounds based on skutterudite $CoSb_3$.[14] The current strategy in these TE materials is to insert atoms to the voids of the structure, expecting they behave as isolated Einstein-like oscillators and therefore reduce the thermal conductivity, since they should scatter the crystal phonons. Certainly, insertion changes the number of electrons in the system and one is mainly interested to know how the thermopower and Lorenz number will be altered. These calculations have been performed using the method explained above, and in Figure 6.1 we compare the thermopower with the experimental measurements. The agreement is clearly good and we obtain that the absolute value of the thermopower decreases as the doping electron concentration increases.

Figure 6.2 shows the Lorenz number as a function of the electron concentration. Negative values refer to p doping. One can see that for some carrier concentrations it passes through a minimum. This can indeed be shown analytically[10] and is of particular importance to obtain a large value of ZT.

It is possible to use other models for the relaxation time[21] but L and α is no longer independent of τ. However, there still appears a cancellation between numerator and denominator of Equation 6.35, indicating that L and α is only weakly dependent on $\tau(\vec{k})$ as long as it is a smooth function around ϵ_F. It is, for example, not the case for the Kondo compounds,[22,23] but in many others cases this approach allows to use rudimentary approximations for $\tau(\vec{k})$ and the constant relaxation time being the simplest one.

Noteworthy, the aforementioned strategy to calculate electronic transport coefficients has also been applied to other family of thermoelectric materials, namely Ni(Ti,Zr,Hf)Sn half-Heusler systems.[24] Experimentally, it was found that alloying isoelectronic semiconductors as, for example, NiTiSn and NiHfSn gave rise to substantial increase of thermopower.[25] It was interesting to know whether this unusual behavior is connected with some FS features, appearing upon chemical disorder or there are other reasons for such a unique behavior. Actually, our theoretical study[24] came to the conclusion that the electronic structure only slightly changes with Ti/Hf or Ti/Zr content, and it cannot explain the thermopower increase. Conversely, the apparently different values of Seebeck coefficient measured in nominally isoelectronic Ni(Ti,Hf,Zr)Sn systems in different experiments are mainly related to the different carrier concentrations, as available from Hall measurements (the phenomenon likely

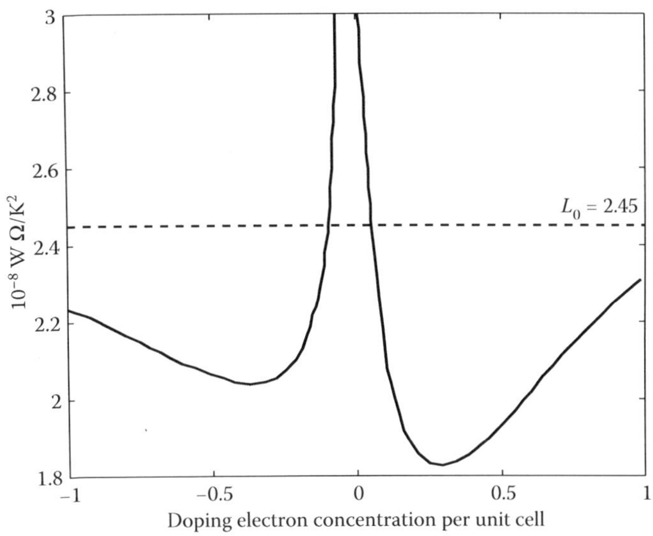

FIGURE 6.2 Lorenz factor at 300 K. The dashed lines show the constant metallic value.

corresponding to the variable magnitude of defects occurring in various samples). Hence, a marked enhancement of thermopower in $NiTi_{0.5}Hf_{0.5}Sn$ with respect to the values observed in parent compounds NiTiSn and NiHfSn was presumably interpreted in terms of much lower carrier concentration, detected in the former.

An important and rather general observation from the aforementioned studies is that the experimental thermopower changes strongly with the carrier concentration n and clearly shows that this quantity should be regarded as a significant factor, when optimizing the thermopower of alloys, since the $\alpha(n)$ variation alone is sufficient to reproduce quite well the experimental results, when both thermopower and carrier concentration data are available.

Finally, in typical TE systems, the thermopower is expected to tend linearly to zero at very low temperature ($\alpha \sim T$), and it can be reasonably extrapolated from $\alpha(T)$ points computed at a higher temperature, if the numerics become difficult. On the other hand, theoretical predictions of the Seebeck coefficient behavior seems to be physically reliable and interesting for applications in a higher temperature range (100–1200 K).

6.3 Multiple Scattering Theory Approach

The Green's function (GF) formalism appears to be naturally adapted to resolve equations of motion for electrons in solids, since it is directly related to fundamental electronic property as DOS, $n(\epsilon)$:

$$n(\epsilon) = -\frac{1}{\pi}\int_{WS} \text{Im}\, G(\vec{r},\vec{r};\epsilon)\, d\vec{r} \tag{6.36}$$

In the presence of external potential V, GF of the Hamiltonian H, $G = (z - H)^{-1}$ (z is a complex energy) defined using the Dyson's equation

$$G = G_0 + G_0 V G \tag{6.37}$$

and can be expanded in infinite series

$$G = G_0 + G_0VG_0 + G_0VG_0VG_0 + \cdots \tag{6.38}$$

The first term represents a free propagation and the higher terms describe more and more complex scattering processes, when other centers (sites) are involved. In order to simplify Equation 6.37, the scattering operator $T = V + VG_0T$ is introduced, yielding

$$G = G_0 + G_0TG_0 \tag{6.39}$$

Equation 6.39 is more advantageous than Equation 6.37, since one obtains a direct expression for all electron GF, but analytical properties of the operators T and G are quite similar as they both possess first-order poles at $z = \epsilon$ energies. Hence, investigating T operator is easier but derived information is almost equivalent to those obtained when studying GF. In crystals, atomic potentials are periodically displayed in the space and the effective one-electron potential can be written as $V = \Sigma_i V_i$ (for the sake of simplicity, one type of atom constituting lattice is considered), where i determines atom positions with respect to the ith Wigner–Seitz cell. As the first approach, one can imagine the crystal as a set of non-overlapping spheres with the potential spherically symmetric inside and constant outside these spheres, known as the muffin-tin (MT) model of crystal potential. In the case of full potential, the MT spheres are replaced with Voronoi polyhedrons and crystal potential can be expressed, that is, using the multipole expansion. It was shown that the multiple scattering equations,[26] and especially scattering operator can be reduced, within the MT model, to the form containing only two operators, that is, the single-site scattering function t_i, which is defined as the scattering operator that can be obtained if only site i is present in the crystal, and GF of free particle G_0.[27]

$$t_i = V_i + V_iG_0t_i \tag{6.40}$$

$$T(\epsilon) = \sum_i V_i + V_iG_0T \tag{6.41}$$

$$= \sum_i t_i(\epsilon) + \sum_{j\neq i} t_i(\epsilon)G_0(\epsilon)t_j(\epsilon) + \sum_{j\neq i}\sum_{m\neq j} t_i(\epsilon)G_0(\epsilon)t_j(\epsilon)G_0(\epsilon)t_m(\epsilon) + \cdots \tag{6.42}$$

From Equation 6.42 one learns that the full description of electron propagation includes single, double, triple, and higher-order scattering processes. To simplify Equation 6.42, one can introduce the *path scattering operator* τ^{ij} that physically describes scattering between i and j sites (but other scattering centers are also involved), yielding the following expression:

$$T = \sum_{ij} \tau^{ij} \quad \text{with} \tag{6.43}$$

$$\tau^{ij}(\epsilon) = t_i(\epsilon)\delta_{ij} + \sum_{m\neq i} t_i(\epsilon)G_0(\epsilon)\tau^{mj}(\epsilon) \tag{6.44}$$

It is interesting to note that operators $\tau^{ij}(\epsilon)$ (similar to property of the MT potentials) are selectively related to ith and jth MT spheres, that is, $<\vec{r}\left|\tau^{ij}(\epsilon)\right|\vec{r}'> = \tau^{ij}(\vec{r}_i,\vec{r}_j,\epsilon)$ does not vanish when $r_i \le R_i^{\mathrm{MT}} \wedge r_j \le R_j^{\mathrm{MT}}$.

The KKR method is mostly based on the multiply scattering equations, and similar to other band theory techniques, it tends to determine electron energy eigenvalues, being equivalent to searching for poles of GF (coinciding with poles of the scattering operator T). Bearing in mind the local character of $\tau^{ij}(\epsilon)$ and spherical symmetry of crystal potential, the most convenient basis of functions are spherical harmonics $Y_{lm}(\hat{r})$ for angular part combined with spherical Bessel $j_l(r)$ and Neumann $n_l(r)$ functions for radial part of Schrödinger equation. Performing Fourier transformation of matrix elements $\tau^{ij}(\epsilon)$

$$\tau^{ij}(\vec{k},\vec{k}';\epsilon) = \int e^{-ikr_i}\tau^{ij}(\vec{r}_i,\vec{r}_j,\epsilon)e^{ik'r_j}d^3r_id^3r_j \tag{6.45}$$

and expanding them in spherical harmonics Y_L $(L \equiv lm)$, we obtain

$$\tau^{ij}(\vec{k},\vec{k}';\epsilon) = (4\pi)^2\sum_{LL'} Y_L(\breve{k})\tau^{ij}_{LL'}(k,k';\epsilon)Y_{L'}(\breve{k}') \tag{6.46}$$

Similar transformations done for Equation 6.44 and limiting our considerations to the so-called *on the energy shell* elements $\left(k = k' \equiv \kappa = \sqrt{\epsilon}\right)$ yield

$$\tau^{ij}_{LL'}(\kappa) = \delta_{ij}t^i_l(\epsilon) + \sum_{i\neq m}\sum_{L''L'''} t^i_{LL''}B^{im}_{L''L'''}(\kappa)\tau^{mj}_{L'''L'}(\kappa) \tag{6.47}$$

The introduced matrix $B^{im}_{LL'}(\kappa)$ depends only on the symmetry of lattice, since it is related to GF of free particle, which can be expanded in the spherical harmonics as follows:

$$G_0(\vec{r}_i - \vec{r}_j';\epsilon) = \sum_{LL'} Y_L(\hat{r}_i)j_l(\kappa r_i)B^{ij}_{LL'}(\kappa)j_{l'}(\kappa r_j')Y_{L'}(\hat{r}_j') \tag{6.48}$$

It is interesting to note that scattering properties of crystal potential are fully included in the matrix t^i_l, which is commonly known to be related to the phase shift δ^i_l of l-partial wave-function

$$t^i_l(\kappa) - -\kappa^{-1}e^{i\delta^i_l}\sin\delta^i_l(\kappa) \tag{6.49}$$

All the aforementioned equations lead to the KKR determinant, which allows establishing electron dispersion relations $\epsilon(\mathbf{k})$ from

$$\det[t^{-1}(\epsilon) - B(\mathbf{k},\epsilon)] = 0 \tag{6.50}$$

Equation 6.50 contains the matrix $B(\mathbf{k}, \epsilon)$ which is the lattice Fourier transform of $B^{im}_{LL'}(\kappa)$ and is often called the KKR *matrix of constants structure* that can also be expressed in a slightly different form,[28] being very convenient for numerical determination of electronic bands $\epsilon(\vec{k})$ in multiatom complex systems. It only depends on the symmetry of the investigated system, while scattering properties of crystal potential are included in the matrix $t^{-1}(\epsilon)$. Actually, we do not intend to go into further details of the GF formalism (see, e.g., Refs. 5 and 26 and references therein), but it is worth underlining that based on this methodology one can straightforwardly define an efficient medium that permits calculating the electronic structure of chemically disordered systems, which is especially relevant to doped TE materials.

In the case of binary $A_{1-x}B_x$ alloy, one replaces a disordered sublattice by an effective medium, which recovers translational symmetry of crystal potential, being lost in chemically disordered systems. One way is based on construction of super-cell systems by multiplying primitive unit cells to approach the real stoichiometry of the system. This approach generally results in lowering of symmetry of crystal unit cell and sometimes meets some barriers related to numerical stability of self-consistent results (problem of convergence) and/or very time-consuming calculations due to a large number of atoms in the unit cell. Attempting to investigate doping (low concentration of substituted elements), a typical situation in thermoelectric systems, one needs to build huge unit cells (1% of the substituted element requires 100 times larger unit cell).

Other efficient approach, which allows retrieving translational crystal symmetry lost in a disordered system, is the coherent potential approximation (CPA) (see, e.g., Refs. 27 and 29). As we have shown above, electron propagation can be described in terms of all electron GF (Equation 6.38) or in terms of scattering function (Equation 6.42). In the case of disordered crystal site, that is, occupied randomly by A and B atoms, the CPA model replaces this site by an effective medium characterized by the scattering function T_{CP} and is defined as follows:

$$T_{\mathrm{CP}} = c_{\mathrm{A}}T_{\mathrm{A}} + c_{\mathrm{B}}T_{\mathrm{B}} \tag{6.51}$$

or equivalently with the use of GF

$$G_{\mathrm{CP}} = c_{\mathrm{A}}G_{\mathrm{A}} + c_{\mathrm{B}}G_{\mathrm{B}} \tag{6.52}$$

with $c_{\mathrm{A}} + c_{\mathrm{B}} = 1$. G_{A} and G_{B} (or scattering operator functions T_{A} and T_{B}) characterize ordered systems, when the "disordered" site is occupied exclusively by atom A or B, respectively. The CPA condition can also be extended to cases of more complex disorder, when, for example, the same site is filled with more than two atoms:

$$G_{\mathrm{CP}} = \sum_i G_{\mathrm{A}_i} c_{\mathrm{A}_i} \tag{6.53}$$

where G_{A_i} denotes GF of A_i atom occupying the site with concentration c_{A_i} ($\sum_i c_{\mathrm{A}_i} = 1$).

In practice, electronic structure calculations using the KKR–CPA method,[5,28] the CPA condition is solved in an iteration way and its convergence is guaranteed by the analyticity of the GF in the complex energy plane, called Herglotz property ($ImG_{\mathrm{CP}} < 0$). The main advantage of the CPA approach is related to the fact that introduced chemical disorder (treated as random distribution of atoms) does not change the symmetry of the unit cell, compared to other approaches based on super-cell computations.

Noteworthy, the KKR–CPA technique also allows treating extra crystal imperfections, that is, disordered systems containing impurities, vacancies, or antisite defects. The role of vacancy defects is particularly important in semiconducting materials due to its strong influence on the position of ϵ_{F} and consequently on electron transport characteristics. In such computations, a vacancy is treated in the same way as other atoms, having self-consistent crystal potential with, however, null nuclear charge ($Z = 0$). Another beneficial aspect of electronic structure calculations of doped systems may concern the use of total energy KKR–CPA calculations to study preferential occupation of impurities entered to the host. It appears particularly helpful to investigate the site selectivity in some ambiguous cases (undetectable by standard diffraction measurements), having in general tremendous effect on electron transport behaviors.

It is worth noting that in electronic structure calculations of disordered materials, the Fermi level ϵ_{F} is determined (like in ordered systems) from the generalized Lloyd formula,[30] which does not require integration over the occupied DOS. The accurate positioning of ϵ_{F} appears particularly important, when studying doped systems (very low impurity content), being close to the metal–semiconductor cross-overs. Finally, total, site-decomposed as well as l-decomposed DOS are calculated in the reciprocal space using a standard technique of small tetrahedrons.

6.3.1 Thermoelectric Properties of Disordered Systems

The above-mentioned complex energy bands of electrons in chemically disordered samples may provide a complementary attitude to calculate electron transport properties of thermoelectric systems. Initially, this concept was applied to study evolution of electron transport properties in metallic binary alloys.[31,32] This idea has also been successfully applied to more complex multiatom systems as, for example, half-Heusler alloys that exhibit metal–semiconductor transition.[33,34] The residual electrical conductivity σ_0 and linear term of thermopower α/T (both extrapolated to $T = 0$) are deduced from the ground-state kinetic parameters of electrons, which are actually determined without any adjustable parameter (e.g., the relaxation time). At low temperature, the Seebeck coefficient is calculated from well-known Mott's formula[1] which is the $T \to 0$ limit of Equations 6.31 and 6.22:

$$\frac{\alpha}{T} = -\frac{\pi^2 k_B^2}{3e} \frac{\partial \ln \sigma(\epsilon)}{\partial \epsilon}\bigg|_{\epsilon_F} \tag{6.54}$$

where the transport function $\sigma(\epsilon)$ can be found from integration over the constant energy surface $\Sigma(\epsilon)$:

$$\sigma(\epsilon) = -\frac{2e^2}{3(2\pi)^2 \hbar} \int_{\Sigma(\epsilon)} dS_k \upsilon_k \tau_k \tag{6.55}$$

υ_k and τ_k correspond to the group velocity and lifetime (not fully equivalent to relaxation time, since it essentially describes scattering due to chemical disorder) of an electron with k-vector, respectively. In alloys, the energy bands $\epsilon_n(k)$ having complex values are derived from electronic structure computations (e.g., the KKR–CPA method), accounting for chemical disorder effects. In practice, one extracts the complex energy dispersion curves for accurately converged effective potentials and electron charges. The group velocity of the electron is then determined by numerical differentiation of the real part of $\epsilon(k)$:

$$\upsilon_{k,n} = \frac{1}{\hbar} \nabla_k [\mathrm{Re}\,\epsilon_n(k)] \tag{6.56}$$

and, from imaginary part of $\epsilon(k)$, the electron lifetime is computed:

$$\tau_{k,n} = \frac{h}{\mathrm{Im}\,\epsilon_n(k)} \tag{6.57}$$

where index n numbers bands crossing the Fermi energy. Velocities and lifetimes of electrons are determined for different FS sheets and next averaged over this surface in reciprocal space. We can note, from Equations 6.54 and 6.55, that the physically interesting energy range for the evaluation of transport coefficients corresponds to complex energy bands near ϵ_F. So, one has to investigate accurately constant energy surface shapes in the vicinity of the Fermi energy, in order to obtain the energy derivative in Equation 6.56. From the numerical point of view, proper extraction of electron transport coefficients is not always an easy task, since such computations strongly depend on the FS topology. Unfortunately, computational results become unstable when either the Fermi level tends to the band gap edge (vanishing DOS) or the concentration of impurity atom tends to zero ($x \to 0$). In the first case, FS is so small that better and better numerical accuracy is required. In the second case, the decreasing chemical disorder implies a smaller and smaller value of the imaginary part of energy band, which next results in huge values of electron lifetimes (not physically admitted). Consequently, one has to avoid these two cases when calculating the conductivity and thermopower from Equations 6.54 and 6.55, respectively.

6.3.2 Illustrative Examples of KKR–CPA Calculations

Half-Heusler phases still belong to attracting thermoelectric systems,[35–39] since there are many examples of XYZ (X and Y are transition metals elements, while Z is an *sp* element) compounds that exhibit semiconducting behaviors. It was established from many first principles studies (see, e.g., Refs. 40–46) that half-Heusler systems satisfying the phenomenological rule of 18-valence electron count (V EC) per formula exhibit an energy gap at the Fermi level.[42,43] The fact that this crystal structure offers wide opportunities of doping/substituting on three crystallographic sites make them interesting, in spite of the fact that the main problem of these materials is related to their high thermal conductivity, likely due to relatively dense structure. Interestingly, it was also revealed[47,48] that alloying two ordered compounds exhibiting metallic behaviors may lead to metal–semiconductor–metal crossovers accompanied by the change of the Seebeck coefficient sign, depending on alloy composition. Such an uncommon electronic structure may open prospects to tune thermopower value and sign and then TE properties within the same structure with slightly different V EC. Figure 6.3 presents the evolution of the DOS at ϵ_F in $Fe_{1-x}Ni_xTiSb$ versus V EC (V EC = 17 and V EC = 19 corresponds to FeTiSb- and NiTiSb-ordered compounds, respectively), where the crossover from hole-like metal through semiconductor to electron-like metal is observed. This behavior remains in full agreement with electrical conductivity and thermopower measurements that showed a change in sign of the Seebeck coefficient when V EC ~ 18 (Figure 6.4). The thermopower calculated[33] using calculated velocities and lifetimes of electrons at the FS and the Mott's formula (Equation 6.56) are in reasonable agreement with experimental data (Figure 6.4). It is interesting to note that the computed *k*-dependent velocities and lifetimes vary strongly with alloy composition as well as with the curvature of FS sheets.

Similar procedure was also applied to other half-Heusler $Ti_{1-x}Sc_xNiSn$, exhibiting semiconductor–metal transition.[49] From the complex energy bands calculated from the KKR–CPA method, the linear slope of thermopower α/T was determined, that remains even in better agreement with experimental points at room temperature (Figure 6.5) than it was observed in $Fe_{1-x}Ni_xTiSb$. Figure 6.6 presents in more detail the variations of electron velocities and lifetimes in $Ti_{1-x}Sc_xNiSn$ for $0.1 < x < 0.6$ as deduced from averaging of υ_k and t_k over the FS as well as the corresponding values obtained from averaging over three independent hole-like FS sheets. On the whole, these values are of the order of 10^6 m/s and 10^{-9} s, respectively. Note, that the electron lifetime is related to chemical disorder only, and other scattering processes that may affect relaxation time are not accounted for. A relatively good agreement between theory and experiment obtained in the aforementioned systems rather indicates that electron scattering is dominated by chemical disorder. It seems interesting to verify to what extent the kinetic parameters

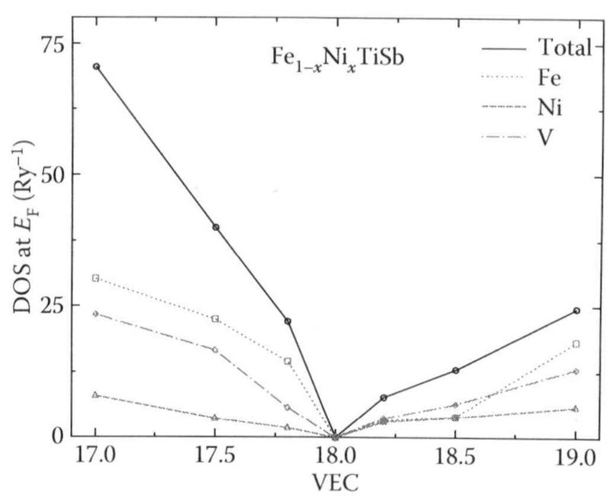

FIGURE 6.3 Metal–semiconductor–metal crossover from DOS (per f.u. and per spin) variation at ϵ_F in half-Heusler $Fe_{1-x}Ni_xTiSb$ versus valence electron count (V EC). Ordered compounds FeTiSb and NiTiSb correspond to V EC = 17 and V EC = 19, respectively.

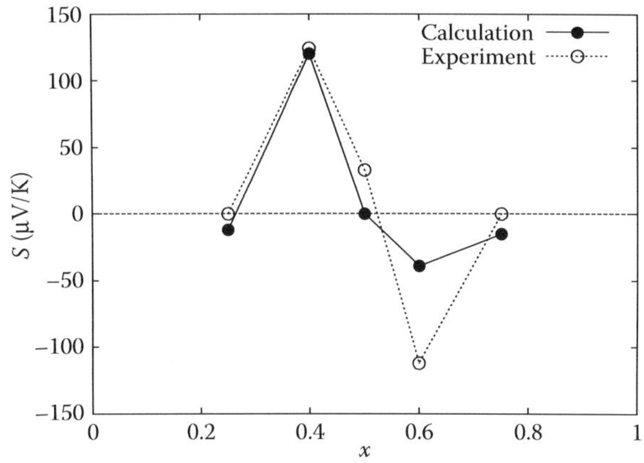

FIGURE 6.4 Room temperature thermopower in half-Heusler $Fe_{1-x}Ni_xTiSb$.

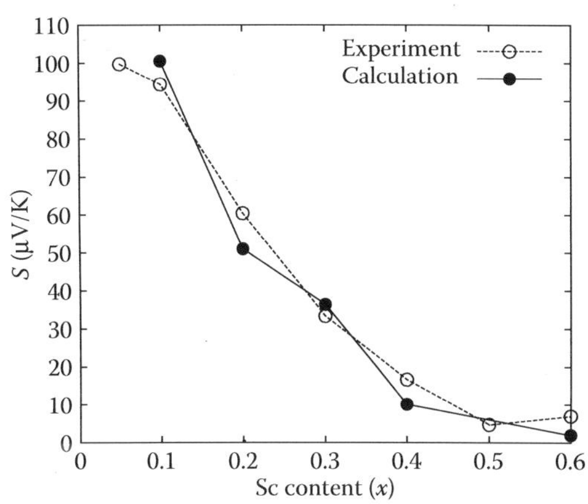

FIGURE 6.5 RT thermopower in half-Heusler $Ti_{1-x}Sc_xNiSn$.

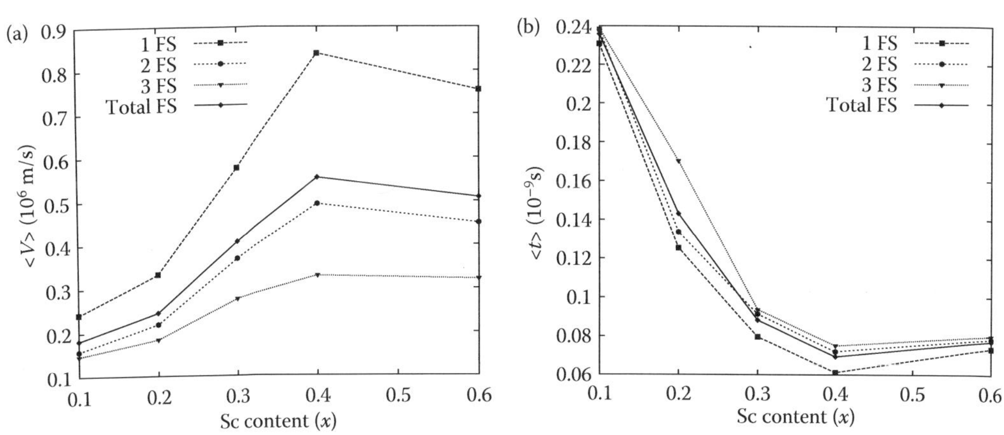

FIGURE 6.6 Variation of electron velocities (a) and lifetimes (b) averaged over FS in half-Heusler $Ti_{1-x}Sc_xNiSn$ derived from complex energy bands KKR–CPA calculations. The contributions from three FS sheets are also plotted.

of electrons near FS influence the calculated thermopower α/T values. One can roughly assume that electrical conductivity is mostly related to DOS characteristics $n(E)$ near ϵ_F. It means that $\sigma(E) \sim e\mu(E)$ $n(E)$ and when electron mobility weakly varies with energy, that is, $\mu(E) \sim \mu_0$. With the use of Mott's formula the thermopower slope can be tentatively related to DOS as follows:

$$\frac{\alpha}{T} \sim -\frac{\partial \ln n(\epsilon)}{\partial \epsilon}\bigg|_{\epsilon_F} \tag{6.58}$$

Such treatment permits roughly estimating thermopower, if at least DOS is calculated in a realistic way, that is, accounting for alloy effects. In the case of the aforementioned $Ti_{1-x}Sc_xNiSn$, the KKR–CPA calculations yielded α/T (in $\mu V/K^2$): 0.356, 0.278, 0.122, and 0.088 for $x = 0.1$, 0.2, 0.4, and 0.6, respectively. At selected temperature $T = 90$ K, these coefficients resulted in the thermopower values 32, 25, 11, and 8 μV/K, which can be compared to experimental points 30, 13, 5, and 4 μV/K, respectively. Hence, it seems that such simplified calculations tend to overestimate the calculated thermopower in the investigated alloys,[34] but it may be really helpful in the tentative analysis of TE properties.

Similar thermopower estimation from KKR–CPA DOS calculations were also performed in other TE materials, as Zintl phases $Mo_{3-x}T_xSb_7$ with $T =$ Fe, Ru[50,51] or $Mo_3Sb_{7-x}Te_x$.[52–55] The idea of searching for attracting TE properties in Mo_3Sb_7-based compound has arrived from the fact that when filling valence bands in Mo_3Sb_7 by substitution of Mo or Sb with elements richer in electrons, metal–semiconductor crossovers were expected. This is due to the presence of the energy gap on DOS,[53,56] which leaves two holes per Mo_3Sb_7 unit just above the Fermi level. Hence, Mo_2RuSb_7 or $Mo_3Sb_5Te_2$ were believed to be semiconductors, but experimentally neither Mo-site in $Mo_{3-x}Ru_xSb_7$ nor Sb-sites in $Mo_3Sb_{7-x}Te_x$ were filled more than $x = 0.8$[51] and $x = 1.8$,[55] respectively. However, theoretically predicted effect of strong decreasing DOS at E_F was indirectly observed on marked change of temperature-dependent electrical resistivity and thermopower characteristics, and directly on specific heat measurements.[51] Analyzing the DOS in $Mo_{3-x}Ru_xSb_7$, the room temperature thermopower was estimated from Equation 6.58 to be as large as 23, 40, and 58 μV/K, being in fair agreement with the experimental data 19, 33, and 62 μV/K[51] for $x = 0.25$, 0.5, and 0.8, respectively. In the case of $Mo_3Sb_{7-x}Te_x$,[57] such comparison of thermopower computed for $T = 100$ K (11, 16, and 35 μV/K) with experimental values (8, 12, and 22 μV/K) is more or less satisfying, but this simplified approach tends to overestimate the measured Seebeck coefficients. Even more spectacular comparison has been recently obtained in $Mo_{3-x}Fe_xSb_7$ with $x \sim 0.5$, where metallic-like behaviors of resistivity curves were accompanied by almost linear increase of thermopower in the wide range of temperature from 4 to 1000 K, with the experimental $\alpha/T \sim 0.10$ $\mu V/K^2$ slope[50] being excellently interpreted from variation of DOS in the vicinity of ϵ_F resulted in $\alpha/T \sim 0.11$ $\mu V/K^2$.

6.3.3 Defects Influence on Thermoelectric Properties

Another possibility of application of KKR–CPA methodology concerns studying of point defects in TE materials. It is well known that impurities, antisite imperfections, vacancy defects, and atoms filling interstitial voids in semiconductors are a particular type of chemical disorder that may affect electronic states in near ϵ_F and actually govern electron transport properties. Hence, theoretical insight into such behaviors appears particularly interesting and provides a better understanding of observed phenomena as well as it may support experimental searching for more efficient TE systems (even as obtained results have more qualitative than quantitative character). Let us present a few examples of the use of electronic structure calculations in disordered thermoelectrics containing defects.

FeVSb crystallizing in half-Heusler structure is an interesting example of a "dirty" metal that has been considered for TE application.[58,59] This compound exhibits ground-state properties, which appear to be strongly covered by antisite defects found in real samples. The perfectly ordered compound shows

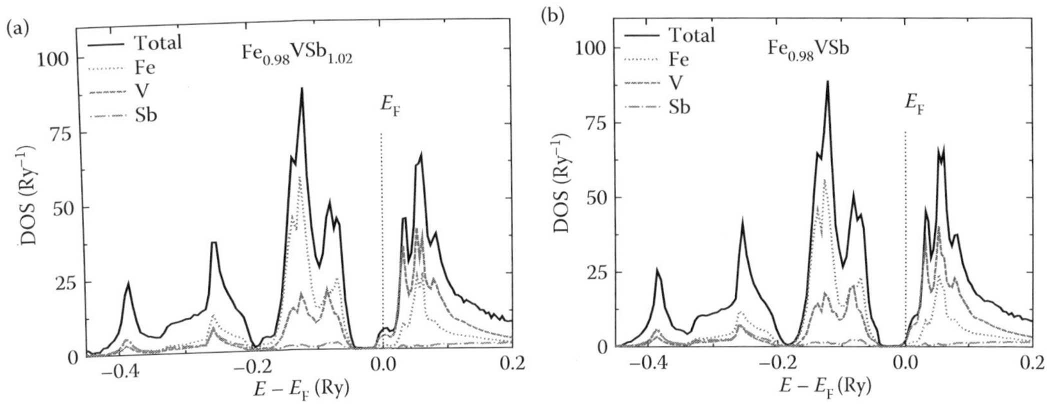

FIGURE 6.7 Total and site-decomposed DOS in FeVSb containing antisite or vacancy defects: an excess of Sb atoms on Fe-site in $Fe_{1-x}VSb_{1+x}$ (a) and an atom deficiency on Fe-site in $Fe_{1-x}VSb$ (b).

an energy gap of ~0.4 eV at the Fermi level, but as detected from electron microprobe analysis (EMPA) and ^{57}Fe Mossbauer spectroscopy analysis, there is an excess of antimony presumably on Fe-site or an atom deficiency on Fe-site. Figure 6.7 clearly presents that both defects markedly modify conduction band edge, mainly due to the appearance of an additional cusp on DOS attributed essentially to the V-site. But the most striking electronic structure feature corresponds to the ϵ_F jump across the energy gap, which falls into finite DOS. This is due to the fact that *d*-states energy for Sb impurities (*s*–*p* element) or Fe vacancies are much higher than for an Fe atom (transition element). So, there is a strong repulsive perturbation in the *d* channel on an impurity site, which expels the *d*-states of this site from the valence band up into the lower part of the conduction band. A total of 10 filled *d*-states is thus transferred from the valence band to the conduction band. Since five valence electrons are gained with an extra Sb atom and eight valence electrons lost with the removed Fe atom, this gives a total of seven electrons in the conduction band for each Sb impurity on the Fe-site (or two electrons for Fe vacancies). Such behavior corresponds well with the metallic-like resistivity at low temperature accompanied by large and negative thermopower, due to the Fermi level lying on increasing DOS.[59] Interestingly, bearing in mind the remarkable effect on electronic states near E_F upon inclusion of defects shown in FeVSb, in Zr-doped FeVSb also, unusual electron transport behaviors (change of thermopower sign as well as radical changes of resistivity curves with Zr contents) were successfully interpreted in terms of electron/hole compensation, that have arisen from native defects, as also detected from EMPA analysis in Zr-doped samples.

Mg_2X (X = Si, Ge, and Sn) semiconductors with antifluorite-type structure belong to very promising TE materials due to low-density composition containing nontoxic and relatively cheap elements. In particular, their solid solutions, for example, Mg_2(Si–Sn) tuned by intentional doping may exhibit quite high *ZT*.[60,61] However, substitution may also lead to an occurrence of extra defects, as a vacancy defect on Mg-site observed in Sb-substituted $Mg_2Si_{1-x}Sb_x$.[62] It was revealed that at low content, antimony on Si-site behaved as electron donor leading to semiconductor–metal transition and smaller thermopower, but with higher Sb doping, the samples again exhibited semiconducting-like resistivity accompanied by increased Seebeck coefficient. It was found that with increasing Sb content in $Mg_2Si_{1-x}Sb_x$, an important amount of Mg-site vacancy defects appeared, and the aforementioned TE properties were tentatively interpreted in terms of charged vacancy on Mg-site, playing a role of a double-hole donor, which together with Sb substitution delivering electrons, leads to electron–hole compensation. Figure 6.8 presents DOS in $Mg_2Si_{0.8}Sb_{0.2}$ and $Mg_{1.9}Si_{0.8}Sb_{0.2}$,[63] which confirms the mechanism of $n-p$ carriers compensation well.

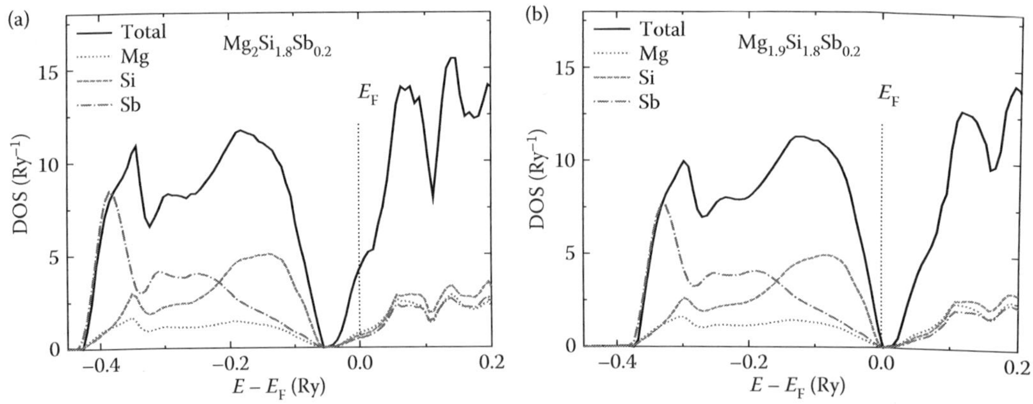

FIGURE 6.8 Total and site-decomposed DOS in $Mg_{2-\delta}Si_{1-x}Sb_x$ illustrating effects of Si/Sb alloying and vacancy defects on Mg-site leading to the metallic $Mg_2Si_{0.8}Sb_{0.2}$ (a) and the semiconducting $Mg_{1.9}Si_{0.8}Sb_{0.2}$ (b).

6.4 Conclusions

In Section 6.1 of this chapter, we introduced the transport coefficients σ, α, and κ_e in a phenomenological way using the Onsager coefficients. The Boltzmann equation was then used to link these coefficients with the quantum theory of the electronic structure. As a result, the Onsager coefficients have been expressed in terms of the transport function $\sigma(\epsilon)$. The transport function contains two important terms, that is, υ^2 which is mainly the kinetic energy of electrons, and $\tau(\vec{k})$ which denotes the relaxation time, including different types of electrons scattering.

In Sections 6.2 and 6.3, the aforementioned theory has been used to investigate the transport properties of real TE compounds within two different approximations for $\tau(\vec{k})$.

In Section 6.2, we have shown without direct calculation of $\tau(\vec{k})$ that such approach allows obtaining the thermopower and the Lorenz factor values in agreement with experiments, as long as $\tau(\vec{k})$ is independent of the wave vector $\vec{k}$. In particular, we have presented that this methodology leads to a better understanding of thermopower variation with temperature and carrier concentration, in $CoSb_3$-based skutterudites and half-Heusler Ni(Ti,Zr,Hf)Sn system, where the variation of the number of charge carriers is really the important parameter to model the effect of doping.

In Section 6.3, $\tau(\vec{k})$ has been calculated within the coherent potential approximation. In principle, this approximation includes only electron scattering by impurities, but it appears as a relevant limit to consider for electrons transport at a low temperature, allowing consequently to calculate the residual resistivity by integrating electron kinetic parameters over the FS. KKR–CPA calculations on several alloys like half-Heusler $Ti_{1-x}Sc_xNiSn$ have shown that this approach also reproduces well the value of thermopower at low temperature, and this is probably more general. We have also illustrated in some cases of TE materials that the electronic structure calculations accounting for point defects provide additional tool to better explain electron transport properties of disordered thermoelectrics.

On the whole, it appears clear from the discussion in the chapter that the methods explained in Sections 6.2 and 6.3 both neglect the scattering of electrons by phonons in the relaxation time $\tau(\vec{k})$. We should bear in mind that lattice vibrations are largely responsible for the behavior of the transport coefficients at high temperature. Currently, the *ab initio* calculations of the electron–phonon interactions for this purpose are becoming the object of intensive investigations, as well as the calculations of the lattice thermal conductivity, being of prime interest for thermoelectricity.

Acknowledgment

One author (J.T.) acknowledges the partial support of the Polish National Science Centre (Grant No. 2011/02/A/ST3/00124).

References

1. N. F. Mott and H. Jones, *The Theory of the Properties of Metals and Alloys*, Dover, New York, 1958.
2. F. J. Ziman, *Electrons and Phonons,* Oxford University Press, London, 1960.
3. F. J. Blatt, *Physics of Electronic Conduction in Solids,* McGraw-Hill Book Co., New York, 1968.
4. P. Blaha, K. Schwarz, G. K. H. Madsen, D. Kvasnicka, and J. Luitz, *WIEN2K: An Augmented Plane Wave Local Orbitals Program for Calculating Crystal Properties,* Karlheinz Schwarz, Techn. Universitat Wien, Austria, ISBN 3-9501031-1-2., 2001.
5. A. Bansil, S. Kaprzyk, P. E. Mijnarends, and J. Tobola, Electronic structure and magnetism of $Fe_{3-x}V_xX$ (X = Si, Ga, and Al) alloys by the KKR-CPA method, *Phys. Rev. B* **60**, 13396, 1999.
6. I. Prigogine and D. Kondepudi, *Thermodynamique,* Odile Jacobs, Paris, France, 1999.
7. L. Onsager, Reciprocal relations in irreversible processes. I, *Phys. Rev.* **37**, 405, 1931.
8. L. Onsager, Reciprocal relations in irreversible processes. II, *Phys. Rev.* **38**, 2265, 1931.
9. N. Pottier, *Physique Statistique Hors D'équilibre: Processus Irrversibles Linaires,* EDP SCIENCES, 2007.
10. L. Chaput, P. Pécheur, J. Tobola, and H. Scherrer, Transport in doped skutterudites: *Ab initio* electronic structure calculations, *Phys. Rev. B* **72**, 085126, 2005.
11. L. D. Landau and L. M. Lifshitz, *Quantum Mechanics Non-Relativistic Theory*, 3rd Edition, Volume 3, Butterworth-Heinemann, Elsevier Science Ltd, 1981.
12. L. Lykke, B. B. Iversen, and G. K. H. Madsen, Electronic structure and transport in the low-temperature thermoelectric $CsBi_4Te_6$: Semiclassical transport equations, *Phys. Rev. B* **73**, 195121, 2005.
13. D. J. Singh and I. I. Mazin, Calculated thermoelectric properties of La-filled skutterudites, *Phys. Rev. B* **56**, R1650, 1997.
14. A. Caillat, T. Borshchevsky, and J. P. Fleurial, Properties of single crystalline semiconducting $CoSb_3$, *J. Appl. Phys* **80**, 4442, 1996.
15. C. Uher, Skutterudites: Prospective novel thermoelectrics, *Semicond. Semimetals* **69**, 139, 2001.
16. D. T. Morelli, G. P. Meisner, B. Chen, S. Hu, and C. Uher, Cerium filling and doping of cobalt triantimonide, *Phys. Rev. B* **56**, 7376, 1997.
17. B. C. Sales, B. C. Chakoumakos, and D. Mandrus, Thermoelectric properties of thallium-filled skutterudites, *Phys. Rev. B* **61**, 2475, 2000.
18. G. S. Nolas, J. L. Cohn, and G. A. Slack, Effect of partial void filling on the lattice thermal conductivity of skutterudites, *Phys. Rev. B* **58**, 164, 1998.
19. V. L. Kuznetsov, L. A. Kuznetsova, and D. M. Rowe, Effect of partial void filling on the transport properties of $Nd_xCo_4Sb_{12}$ skutterudites, *J. Phys.: Condens. Matter* **15**, 5035, 2003.
20. S. Katsuyama, Y. Shichijo, M. Ito, K. Majima, and H. Nagai, Thermoelectric properties of the skutterudite $Co_{1-x}Fe_xSb_3$ system, *J. Appl. Phys.* **84**, 6708, 1998.
21. L. Chaput, G. Hug, P. Pécheur, and H. Scherrer, Anisotropy and thermopower in Ti_3SiC_2, *Phys. Rev. B* **71**, 121104, 2005.
22. G. D. Mahan, Good Thermoelectrics, *Solid State Physics* **51**, 81, 1997.
23. V. S. Oudovenko and G. Kotliar, Thermoelectric properties of the degenerate Hubbard model, *Phys. Rev. B* **65**, 075102, 2002.
24. L. Chaput, J. Tobola, P. Pecheur, H. Scherrer, and S. Kaprzyk, Electronic structure and thermopower of $Ni(Ti_{0.5}Hf_{0.5})Sn$ and related half-Heusler phases, *Phys. Rev. B* **73**, 045121, 2006.
25. S. Sakurada and N. Shutoh, Effect of Ti substitution on the thermoelectric properties of (Zr,Hf)NiSn half-Heusler compounds, *App. Phys. Lett.* **86**, 082105, 2005.

26. W. H. Butler, P. H. Dederichs, A. Gonis, and R. Weaver, *Applications of Multiply Scattering Theory in Material Science, MRS Symposia Proceedings, No. 253*, Materials Research Society, Pittsburgh, 1992.
27. B. L. Gyorffy, Coherent-potential approximation for a nonoverlapping-muffin-tin-potential model of random substitutional alloys, *Phys. Rev. B* **5**, 2382, 1972.
28. T. Stopa, S. Kaprzyk, and J. Tobola, Linear aspects of the Korringa–Kohn–Rostoker formalism, *J. Phys.: Condens. Matter* **16**, 4921, 2004.
29. P. Soven, Coherent-potential model of substitutional disordered alloys, *Phys. Rev.* **156**, 809, 1967.
30. S. Kaprzyk and A. Bansil, Green's function and a generalized Lloyd formula for the density of states in disordered muffin-tin alloys, *Phys. Rev. B* **42**, 7358, 1990.
31. W. H. Butler and G. Stocks, Calculated electrical conductivity and thermopower of silver-palladium alloys, *Phys. Rev. B* **29**, 4217, 1984.
32. J. Banhart and H. Ebert, First-principles calculation of the thermoelectric power of disordered alloys, *Solid State Comm.* **94**, 445, 1995.
33. T. Stopa, J. Tobola, and S. Kaprzyk, Effect of Fermi surface shape on transport properties in $Fe_{1-x}Ni_xTiSb$ half-Heusler alloys, in *Proceedings of the 2nd European Conference on Thermoelectrics* (Krakow, Poland, 2004).
34. T. Stopa, J. Tobola, S. Kaprzyk, E. K. Hlil, and D. Fruchart, Resistivity and thermopower calculations in half-Heusler $Ti_{1-x}Sc_xNiSn$ alloys from the KKR-CPA method, *J. Phys.: Condens. Matter* **18**, 6379, 2006.
35. F. G. Aliev, F. G. Brandt, V. Moshchalkov, R. V. Skolozdra, and A. I. Belogorov, Gap at the Fermi level in the intermetallic vacancy system RNiSn (R = Ti, Zr, Hf), *Z. Phys. B: Condens. Matter* **75**, 167, 1989.
36. J. Pierre, R. Skolozdra, J. Tobola, S. Kaprzyk, C. Hordequin, M. A. Kouacou, I. Karla, R. Currat, and E. Lelievre-Berna, Properties on request in semi-Heusler phases, *J. Alloys Compd.* **262–263**, 101, 1997.
37. C. C. Uher, J. Yang, S. Hu, D. T. Morelli, and G. P. Meisner, Transport properties of pure and doped MNiSn (M = Zr, Hf), *Phys. Rev. B* **59**, 8615, 1999.
38. H. Hohl, A. P. Ramirez, C. Goldmann, G. Ernst, B. Wolfing, and E. Bucher, Efficient dopants for ZrNiSn-based thermoelectric materials, *J. Phys.: Condens. Matter* **11**, 1697, 1999.
39. S. J. Poon, Electronic and thermoelectric properties of Half-Heusler alloys, *Semicond. Semimetals* **70**, 37, 2001.
40. S. Ogut and K. Rabe, Band gap and stability in the ternary intermetallic compounds NiSnM (M = Ti, Zr, Hf): A first-principles study, *Phys. Rev. B* **51**, 10443, 1995.
41. A. Slebarski, A. Jezierski, A. Zygmunt, S. Mahl, and M. Neumann, Suppression of the gap energy in Zr-Ni-Sn and Ti-Ni-Sn by partial substitution of Zr and Ti by Ce, *Phys. Rev. B* **57**, 9544, 1998.
42. J. Tobola, J. Pierre, S. Kaprzyk, R. V. Skolozdra, and M. A. Kouacou, Crossover from semiconductor to magnetic metal in semi-Heusler phases as a function of valence electron concentration, *J. Phys.: Condens. Matter* **10**, 1013, 1998.
43. J. Tobola and J. Pierre, Electronic phase diagram of the XTZ (X = Fe, Co, Ni; T = Ti, V, Zr, Nb, Mn; Z = Sn, Sb) semi-Heusler compounds, *J. Alloys Compd.* **296**, 243, 2000.
44. P. Larson, S. D. Mahanti, and M. G. Kanatzidis, Structural stability of Ni-containing half-Heusler compounds, *Phys. Rev. B* **62**, 12754, 2000.
45. I. Galanakis, P. H. Dederichs, and N. Papanikolaou, Origin and properties of the gap in the half-ferromagnetic Heusler alloys, *Phys. Rev. B* **66**, 134428, 2002.
46. B. R. K. Nanda and I. Dasgupta, Electronic structure and magnetism in half-Heusler compounds, *J. Phys.: Condens. Matter* **15**, 7307, 2003.
47. J. Tobola, L. Jodin, P. Pecheur, H. Scherrer, and S. Kaprzyk, Composition-induced metal-semiconductor-metal crossover in half-Heusler $Fe_{1-x}Ni_xTiSb$, *Phys. Rev. B* **64**, 155103, 2001.
48. J. Tobola, L. Jodin, P. Pecheur, and G. Venturini, Unusual electron structure and electron transport properties of some disordered half-Heusler phases, *J. Alloys Compd.* **383**, 328, 2004.

49. A. Horyn, A. Bodak, L. Romaka, Y. Gorelenko, A. Tkachuk, V. Davydov, and Y. Stadnyk, Crystal structure and physical properties of (Ti,Sc)NiSn and (Zr,Sc)NiSn solid solutions, *J. Alloys Compd.* **363**, 10, 2004.
50. C. Candolfi, B. Lenoir, A. Dauscher, B. Malaman, E. Guilmeau, J. Hejtmanek, and J. Tobola, High thermoelectric power factor in Fe-substituted Mo_3Sb_7, *App. Phys. Lett.* **96**, 262103, 2010.
51. C. Candolfi, B. Lenoir, A. Dauscher, J. Hejtmanek, and J. Tobola, Electronic band structure, transport, and magnetic properties of $Mo_{3-x}Ru_xSb_7$, *Phys, Rev. B* **80**, 155127, 2009.
52. F. Gascoin, J. Rasmussen, and G. J. Snyder, High temperature thermoelectric properties of $Mo_3Sb_{7-x}Te_x$ for $x = 1.6$ and 1.5, *J. Alloys Compd.* **427**, 324, 2007.
53. E. E. Dashjav, A. Szczepenowska, and H. Kleinke, Optimization of the thermopower of the antimonide Mo_3Sb_7 by a partial Sb/Te substitution, *J. Mater. Chem.* **12**, 345, 2002.
54. N. Soheilnia, H. Xu, H. Zhang, T. M. Tritt, I. Swainson, and H. Kleinke, Thermoelectric properties of $Re_3Ge_{0.6}As_{6.4}$ and Re_3GeAs_6 in comparison to $Mo_3Sb_{5.4}Te_{1.6}$, *Chem. Mater.* **19**, 4063, 2007.
55. C. Candolfi, B. Lenoir, A. Dauscher, J. Tobola, S. J. Clarke, and R. I. Smith, Neutron diffraction and *Ab initio* studies of Te site preference in $Mo_3Sb_{7-x}Te_x$, *Chem. Mater.* **20**, 6556, 2008.
56. B. Wiendlocha, J. Tobola, M. Sternik, S. Kaprzyk, K. Parlinski, and A. M. Oles, Superconductivity of Mo_3Sb_7 from first principles, *Phys. Rev. B* **78**, 060507R, 2008.
57. C. Candolfi, B. Lenoir, A. Dauscher, J. Hejtmanek, and J. Tobola, Low-temperature galvanomagnetic, magnetic, and thermoelectric properties of $Mo_3Sb_{7-x}Te_x$ ($x = 0.0$, 0.3, 1.0, 1.6, and 1.8), *Phys. Rev. B* **79**, 235108, 2009.
58. D. Young, P. Khalifah, R. J. Cava, and A. P. Ramirez, Thermoelectric properties of pure and doped FeMSb (M = V, Nb), *J. Appl. Phys.* **87**, 317, 2000.
59. L. Jodin, J. Tobola, P. Pecheur, H. Scherrer, and S. Kaprzyk, Effect of substitutions and defects in half-Heusler FeVSb studied by electron transport measurements and KKR-CPA electronic structure calculations, *Phys. Rev. B* **70**, 184207, 2004.
60. Y. Noda, H. Kon, Y. Furukawa, N. Otsuka, I. A. Nishida, and K. Masumoto, Preparation and thermoelectric properties of $Mg_2Si_{1-x}Ge_x$ ($x = 0.0$–0.4) solid solution semiconductors, *Mater. Trans. JIM* **33**, 845, 1992.
61. V. K. Zaitsev, M. I. Fedorov, E. A. Guriewa, I. S. Eremin, P. P. Konstantinov, A. Y. Samunin, and M. V. Vedernikov, Highly effective $Mg_2Si_{1-x}Sn_x$ thermoelectrics, *Phys. Rev. B* **74**, 0452071, 2006.
62. G. Nolas, D. Wang, and M. Beekman, Transport properties of polycrystalline $Mg_2Si_{1-y}Sb_y$ ($0 \leq y < 0.4$), *Phys. Rev. B* **76**, 235204, 2007.
63. J. Tobola, S. Kaprzyk, and H. Scherrer, Mg-vacancy-induced semiconducting properties in $Mg_2Si_{1-x}Sb_x$ from electronic structure calculations, *J. Electron. Mater.* **39**, 2064, 2010.

7

New Thermoelectric Materials with Precisely Determined Electronic Structure and Phonon Dispersion

Tsunehiro Takeuchi
Nagoya University

7.1 Introduction

Thermoelectric materials have attracted considerable interest because the waste heat emitted from a variety of heat sources, such as industries, homes, automobiles, and so on, is effectively recovered as useful electricity by the use of the thermoelectric generator in which thermoelectric materials are used as one of the main components. The efficiency of energy conversion in thermoelectric generators increases with increasing magnitude of dimensionless figure of merit, $ZT = S^2\sigma T/\kappa$, of the thermoelectric materials used in the generator. Here, S, σ, and κ indicate the Seebeck coefficient, electrical conductivity, and thermal conductivity, respectively. The materials capable of possessing a large magnitude of ZT are highly required to construct a thermoelectric generator of high performance. A large number of investigations have been conducted to obtain such materials for more than 50 years. Unfortunately, however, the magnitude of ZT of bulk materials was limited to small values <2, and the resulting efficiency of energy conversion η is not high enough to cause a wide use of the thermoelectric generators.

It is often argued that one of the main reasons for the difficulty in increasing ZT is the complicated relation among the three thermoelectric properties, S, σ, and κ. Since one of them cannot be controlled without affecting the others, it is widely believed that the balanced condition for the maximum ZT is hardly obtained. The difference of the theoretically predicted behaviors of thermoelectric properties from the experimentally observed ones also makes it difficult to develop a material possessing a very high ZT value. In order to develop thermoelectric generators of high performance, therefore, we should know the factors for the three thermoelectric properties from a microscopic point of view and construct the guiding principle to optimize the balance of three thermoelectric properties.

In this chapter, a new strategy to develop a thermoelectric material is proposed on the basis of the fact that the unusual behaviors of thermoelectric properties can be quantitatively accounted for by using the information about the fine electronic structure near the chemical potential and the phonon dispersions. In Section 7.2, the theories capable of quantitatively predicting the thermoelectric properties are briefly explained together with the problems in the approximations usually used to estimate the properties of thermoelectric materials. In Section 7.3, the experimental evidences to probe the validity of the theories are described. The relation between phonon dispersion and lattice thermal conductivity is briefly described in Section 7.4. The new strategy to develop thermoelectric materials is introduced in Section 7.5, and the development of materials using the newly proposed strategy is shown in Section 7.6. After pointing out the remaining problems in Section 7.7, we shall summarize this chapter.

7.2 Linear Response Theory

7.2.1 The Energy Range of Electrons Dominating the Thermoelectric Properties

The electron transport properties in solids are usually calculated in the context of linear response theory, regardless of the quantum state of the system: metal, semiconductor, band insulator, or the localized insulator. In the linear response theory, the variation of electron distribution caused by the external field, such as temperature gradient, electrical field, magnetic field, and so on, is converted into the energy function by assuming the linear relations. In the case of describing thermoelectric properties, we have to consider the effect of electrical field **E** and temperature gradient ∇T on the distribution of electrons. The electrical current **J**, therefore, is composed of two terms as it is shown in the following equation [1–3]:

$$\mathbf{J} = K_0\mathbf{E} - \frac{1}{|e|T}K_1(-\nabla T) \tag{7.1}$$

The function K_n is known as the transport integral shown below.

$$K_n = \int_{-\infty}^{\infty} \sigma(\varepsilon,T)(\varepsilon-\mu)^n\left(-\frac{\partial f_{\mathrm{FD}}(\varepsilon,T)}{\partial\varepsilon}\right)d\varepsilon \tag{7.2}$$

Here, $\sigma(\varepsilon,T)$, e, $f_{\mathrm{FD}}(\varepsilon,T)$, and μ represent spectral conductivity, the unit charge of electron, the Fermi–Dirac distribution function, and chemical potential, respectively. The first and second terms of Equation 7.1 represent the contribution of the electrical field and that of the temperature gradient, respectively. It would be worth noting here that Equation 7.1 is sometimes called the *Onsager reciprocal relation*. The chemical potential of conduction electrons μ is a function of temperature and is defined as the energy difference between $N+1$ electron system and N electron system kept at a given temperature. The chemical potential is sometimes called the Fermi level. It becomes equal to the Fermi energy at absolute zero, but gradually deviates from the Fermi energy with increasing temperature due to the energy dependence of electron density of states.

Since each electron carries the energy of $(\varepsilon-\mu)$, we obtain the following formula that represents the energy flow $\mathbf{J}_{\mathrm{Qel}}$ by multiplying $\sigma(\varepsilon,T)$ persisting in the right side of Equation 7.1 by $-((\varepsilon-\mu)/|e|)$

$$\mathbf{J}_{\text{Qel}} = -\frac{1}{|e|}K_1\mathbf{E} + \frac{1}{e^2T}K_2(-\nabla T) \tag{7.3}$$

Here, we employ the following formulae for defining electrical conductivity σ, Seebeck coefficient S, and electron thermal conductivity κ_{el}.

$$\mathbf{J} = \sigma\mathbf{E} \tag{7.4}$$

$$\mathbf{E} = S\nabla T \tag{7.5}$$

$$\mathbf{J}_{\text{Qel}} = \kappa_{\text{el}}(-\nabla T) \tag{7.6}$$

The equation representing electrical conductivity $\sigma(T)$ can be deduced from Equations 7.4 and 7.1 in absence of temperature gradient. The equation of Seebeck coefficient $S(T)$ is also obtained by using the relation between **E** and ∇T deduced from Equation 7.1 under the assumption of no electrical current. By substituting the product of the resulting $S(T)$ and ∇T for **E** in Equation 7.3, we can also obtain the formula of electron thermal conductivity $\kappa_{\text{el}}(T)$. The resulting equations for $\sigma(T)$, $S(T)$, and $\kappa_{\text{el}}(T)$ are summarized in Equations 7.7 through 7.9.

$$\sigma(T) = \int \sigma(\varepsilon,T)\left\{-\frac{\partial f_{\text{FD}}(\varepsilon,T)}{\partial\varepsilon}\right\}d\varepsilon \tag{7.7}$$

$$S(T) = -\frac{1}{|e|T}\frac{\int \sigma(\varepsilon,T)(\varepsilon-\mu)\left\{-\partial f_{\text{FD}}(\varepsilon,T)/\partial\varepsilon\right\}d\varepsilon}{\int \sigma(\varepsilon,T)\left\{-\partial f_{\text{FD}}(\varepsilon,T)/\partial\varepsilon\right\}d\varepsilon} \tag{7.8}$$

$$\kappa_{\text{el}}(T) = \frac{1}{e^2T}\int \sigma(\varepsilon,T)(\varepsilon-\mu)^2\left\{-\frac{\partial f_{\text{FD}}(\varepsilon,T)}{\partial\varepsilon}\right\}d\varepsilon - \frac{1}{e^2T}\frac{\left\{\int\left[\sigma(\varepsilon,T)(\varepsilon-\mu)\left\{\partial f_{\text{FD}}(\varepsilon,T)/\partial\varepsilon\right\}d\varepsilon\right]^2\right\}}{\int \sigma(\varepsilon,T)\left\{-\partial f_{\text{FD}}(\varepsilon,T)/\partial\varepsilon\right\}d\varepsilon} \tag{7.9}$$

These equations are hereafter called *the original equations* for describing thermoelectric properties.

Before showing the usually used mathematical transformation, it would be better to briefly explain the characteristics of these original equations because the energy range of electrons to determine the behaviors of thermoelectric properties, which play a key role in constructing a guiding principle for developing new thermoelectric materials, is easily understood from the original equations without using the generally used, complicated mathematical transformations.

It is of great importance to know that all the integrands in Equations 7.7 through 7.9 consist of the spectral conductivity $\sigma(\varepsilon,T)$ and one of the following functions: $F_{\text{W0}}(\varepsilon,T) = (-\partial f_{\text{FD}}/\partial\varepsilon)$, $F_{\text{W1}}(\varepsilon,T) = (\varepsilon-\mu)(-\partial f_{\text{FD}}/\partial\varepsilon)$, or $F_{\text{W2}}(\varepsilon,T) = (\varepsilon-\mu)^2(-\partial f_{\text{FD}}/\partial\varepsilon)$. These functions plotted in Figure 7.1 are tentatively called *the window functions* because all the functions possess nontrivial value only in the very narrow energy range of a few k_BT centered at the chemical potential μ to limit the energy range of the integrations. It should also be very important to mention that the window functions can be calculated without any difficulty at a given temperature because those functions contain only the energy derivative of Fermi–Dirac distribution function. The thermoelectric properties, $\sigma(T)$, $S(T)$, and $\kappa_{\text{el}}(T)$, are capable of being estimated from the calculated window functions by means of numerical integration, provided that the magnitude of spectral conductivity $\sigma(\varepsilon,T)$ is precisely determined as a function

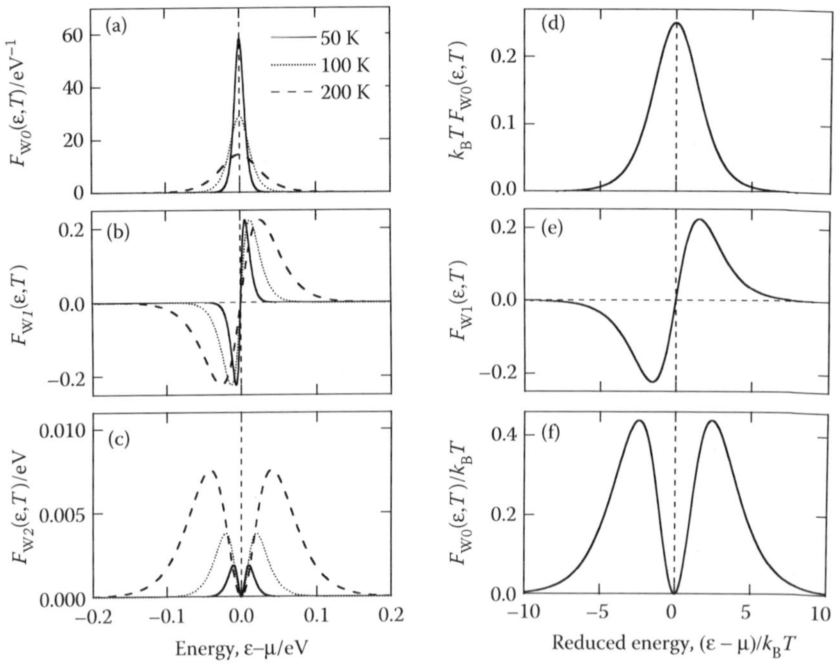

FIGURE 7.1 Energy dependence of window functions. Three window functions at 50, 100, and 200 K are plotted as a function of energy in (a)–(c). The window functions are also plotted as a function of reduced energy in (d)–(f).

of energy. It is, therefore, strongly argued that the quantitative evaluation of spectral conductivity is of crucial importance for estimating the behavior of thermoelectric properties.

The electrical conductivity $\sigma(T)$ is determined in association solely with the window function $F_{W0}(\varepsilon,T)$, which possesses a single peak exactly at the chemical potential. The full-width-half-maximum of the peak is about 3.5 k_BT, and it increases with increasing temperature due to the involved factor T. The integrated intensity of $F_{W0}(\varepsilon,T)$ over the whole energy range always represents unity regardless of temperature. This fact indicates that the function $F_{W0}(\varepsilon,T)$ provides us with the probability of spectral conductivity $\sigma(\varepsilon,T)$ at a given energy to contribute toward the electron conductivity $\sigma(T)$. It is expressed, in other words, that the electrical conductivity $\sigma(T)$ at a given temperature corresponds to the weighted average value of spectral conductivity $\sigma(\varepsilon,T)$ calculated so that $F_{W0}(\varepsilon,T)$ to plays the role of distribution function.

In Equation 7.8 for describing Seebeck coefficient, the two window functions, $F_{W1}(\varepsilon,T)$ and $F_{W0}(\varepsilon,T)$, persist in the numerator and denominator, respectively. It would be easily realized that the denominator is simply transformed into the electrical conductivity $\sigma(T)$. The window function $F_{W1}(\varepsilon,T)$ in the numerator is an odd function of $(\varepsilon - \mu)$, and possesses a local maximum of positive sign and a local minimum of negative sign at $\varepsilon - \mu \sim 1.6\ k_BT$ and $\sim -1.6\ k_BT$, respectively. It means that the Seebeck coefficient becomes large when the spectral conductivity around $\varepsilon - \mu = 1.6\ k_BT$ is much larger or much smaller than that around $\varepsilon - \mu = -1.6\ k_BT$.

Let us now consider the electron thermal conductivity $\kappa_e(T)$. The first term of Equation 7.9 involves the window function $F_{W2}(\varepsilon,T)$ that possesses two peaks of positive sign at $(\varepsilon - \mu) \sim 2.4\ k_BT$ and $(\varepsilon - \mu) \sim -2.4\ k_BT$ but almost no intensity near the chemical potential. This fact indicates that the electrons contributing to the electrical conductivity are definitely different from those contributing to the electron thermal conductivity: the former persists in the vicinity of chemical potential while the latter persists at around $(\varepsilon - \mu) \sim \pm 2.4\ k_BT$. Some of the readers may think this fact strange because the Wiedemann–Franz law that represents a direct relation between electrical conductivity and electron thermal conductivity as $\kappa_{el} = L_0\sigma T$, where $L_0 = k_B^2\pi^2/3e^2 \sim 2.44 \times 10^{-8}\ \mathrm{W\Omega K^2}$, indicates a constant known as the Lorenz number, is applicable for a large number of metallic materials. The validity and limitation of the Wiedemann–Franz law will be explained in Section 7.2.3 in detail.

The second term of Equation 7.9 always possesses a negative value to reduce the electron thermal conductivity. By rewriting it into $\kappa_{el_{2nd}}(T) = -S^2\sigma T$, we realized that the second term of Equation 7.9 generally provides us with a trivial magnitude to be safely ignored. For example, we obtain a very small magnitude of -3×10^{-2} W/m/K from the condition of typical metals: $S = 1$ μV/K and $\sigma = 1 \times 10^8\ \Omega^{-1}\ \text{m}^{-1}$ ($\rho = 1\ \mu\ \Omega$ cm) at 300 K. In some cases, however, $\kappa_{el_{2nd}}(T)$ possesses a nontrivial value especially when materials possess a large magnitude of power factor PF $= S^2\sigma$. For example, $\kappa_{el_{2nd}} = -1.2$ W/m/K is calculated from $S = 200$ μV/K and $\sigma = 1 \times 10^5\ \Omega^{-1}\ \text{m}^{-1}$ ($\rho = 1$ mΩ cm) at 300 K.

By considering the energy dependence of three window functions described above, we now know the characteristic energy ranges in which electrons dominantly contribute to each thermoelectric property: Electrical conductivity is determined from the spectral conductivity in the very narrow energy range of a few k_BT centered at the chemical potential. Seebeck coefficient is determined from the difference in the averaged spectral conductivity at around $(\varepsilon - \mu) = 1.6\ k_BT$ and $(\varepsilon - \mu) = -1.6\ k_BT$. The electron thermal conductivity is determined from the averaged spectral conductivity at around $(\varepsilon - \mu) = \pm 2.4\ k_BT$. It would be naturally understood, therefore, that we should understand the energy dependence of spectral conductivity to investigate the behaviors of thermoelectric properties.

7.2.2 Spectral Conductivity

If the semiclassical Bloch–Boltzmann theory is employed, the spectral conductivity becomes a function of electronic density of states, group velocity, and relaxation time. As a typical example, the spectral conductivity of isotropic system deduced using the Bloch–Boltzmann theory is shown below.

$$\sigma(\varepsilon, T) = \frac{e^2}{3} N(\varepsilon) \upsilon^2(\varepsilon) \tau(T) \tag{7.10}$$

Since the quantum states of electrons are more rigorously taken into account in the Kubo–Greenwood formula, one may think that the Kubo–Greenwood formula should be used rather than the semiclassical Boltzmann equation. However we intentionally employ the semiclassical Boltzmann equation in this section because the semiclassical Bloch–Boltzmann theory is much easier to handle and is capable of explaining well the behavior of electrical properties of typical metals and semiconductor, in which both the Bloch states and its mean free path are well defined. Nearly all the thermoelectric materials belong to the class of such materials.

The relaxation time of electrons generally has less obvious energy dependence in the very narrow energy range where the three window functions possess nontrivial values. A large number of high-resolution angle-resolved photoemission spectroscopy data, from which the lifetime of the Bloch states is quantitatively estimated, makes this argument convincing [4–7]. Besides, the group velocity for many metallic materials is kept almost constant in the narrow energy range that we have to consider. It is safely argued, therefore, that the electron density of states generally determines the spectral conductivity as the most dominant factor. We have to keep in mind, however, that this argument is not valid for all metallic materials. For instance, we have to seriously consider the energy and momentum dependence of group velocity to estimate the spectral conductivity especially when several bands persist in the vicinity of the chemical potential. The relaxation time also possesses nontrivial energy and momentum dependences in some materials due to the effect of collective excitations.

For developing thermoelectric materials, we should find the materials simultaneously possessing a large magnitude of both electrical conductivity and Seebeck coefficient. The relation between the window functions and spectral conductivity described above suggests that this condition is realized for the degenerate semiconductors, in which the chemical potential is located inside a band near the band edge. Indeed, many thermoelectric materials have been developed using the semiconducting materials. It is also clearly understood that the fine electronic structure in the narrow energy range of a few k_BT centered at the chemical potential is of great importance to increase the magnitude of dimensionless figure of merit. This fact lets us realize that new thermoelectric materials of high performance could be developed by considering the fine electronic structure that has not been taken into account for the degenerate semiconductors.

7.2.3 Generally Used Approximations

In this section, the generally used theoretical formulae for describing thermoelectric properties are deduced from the original equations 7.7 through 7.9, clearly denoting the approximations used in the mathematical transformations. It is also explained that serious problems arise from the inappropriate approximations used to deduce the well-known formulae.

7.2.3.1 Metals

For many metallic materials, the spectral conductivity is well approximated with a linear function of energy in the energy range where the three window functions possess a meaningful value. The validity of this argument is well understood by taking a look, as a typical example, at the density of states and energy momentum dispersion of fcc copper shown in Figure 7.2. Obviously the density of states of fcc copper is safely assumed to be a linear function of energy in the vicinity of chemical potential. Besides, the group velocity, which is determined by the gradient of energy eigenvalue in the momentum space, is kept constant in the energy range of our interest. Although the estimation of relaxation time is not straightforward but rather difficult, we tentatively used the energy-independent hypothesis for relaxation time because it is often used theoretically and confirmed for some materials experimentally [4–6]. As a result, the spectral conductivity of typical metals is safely approximated with a linear function of energy as $\sigma(\varepsilon,T) \approx \sigma(\mu,T) + (\varepsilon - \mu)[\partial\sigma(\varepsilon,T)/\partial\varepsilon]_{\varepsilon=\mu}$.

Notably, in the case of the spectral conductivity linearly varying with energy, Equation 7.7 for the electrical conductivity is mathematically transformed into a very simple form. Since the window function $F_{W0}(\varepsilon,T)$ is an even function of $(\varepsilon - \mu)$, the second term of spectral conductivity, which is an odd function of $(\varepsilon - \mu)$, disappears during the integration in Equation 7.7. The integrated intensity of $F_{W0}(\varepsilon,T)$ over the whole energy range becomes unity regardless of temperature, and therefore the remaining first term provides us with the following equation:

$$\sigma(T) = \sigma(\mu,T) \tag{7.11}$$

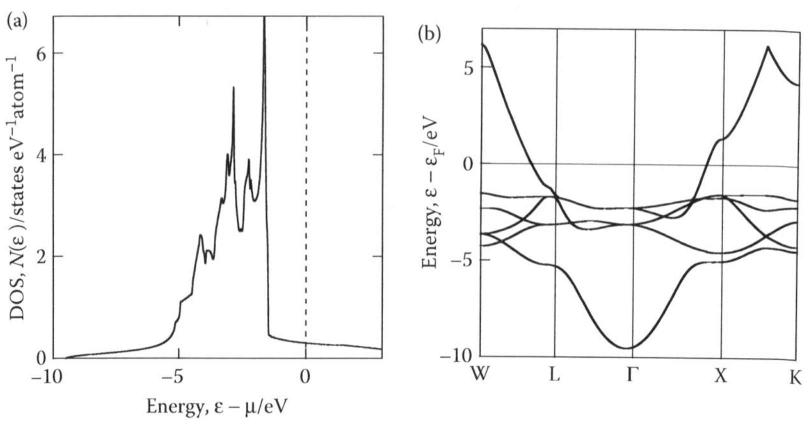

FIGURE 7.2 (a) Electronic density of states and (b) energy-momentum dispersion of fcc Cu.

Equation 7.11 suggests that the electrical conductivity of metals can be estimated solely from the spectral conductivity at the chemical potential, despite the fact that not only the electrons exactly at the chemical potential but also those near the chemical potential contribute to the thermoelectric properties.

Let us now deduce the well-known Mott formula for the Seebeck coefficient. Under the assumption of spectral conductivity linearly varying with energy, $\sigma(\varepsilon,T) \approx \sigma(\mu,T) + (\varepsilon-\mu)[\partial\sigma(\varepsilon,T)/\partial\varepsilon]_{\varepsilon=\mu}$, the first term of spectral conductivity, which is kept constant over the whole energy range, disappears from the integral persisting in the numerator of Equation 7.8, because the window function $F_{W1}(\varepsilon,T)$ is an odd function of energy $(\varepsilon-\mu)$. By using the following integration,

$$\int(\varepsilon-\mu)F_{W1}(\varepsilon,T)d\varepsilon = \int F_{W2}(\varepsilon,T)d\varepsilon = \frac{k_B^2T^2\pi^2}{3}, \tag{7.12}$$

one can easily obtain the famous Mott formula [1–3], which represents the behavior of Seebeck coefficient of typical metallic samples.

$$S(T) = -\frac{\pi^2k_B^2T}{3|e|}\left(\frac{\partial\ln\sigma(\varepsilon,T)}{\partial\varepsilon}\right)_{\varepsilon=\mu} \tag{7.13}$$

According to the Mott formula, it is understood that $S(T)$ of metals is directly proportional to absolute temperature.

The electron thermal conductivity of metallic samples is generally described by using the Wiedemann–Franz law, $\kappa_{el} = L_0\sigma T$. Notably, the Wiedemann–Franz law is easily deduced from Equation 7.9 under the assumption of spectral conductivity linearly varying with energy, $\sigma(\varepsilon, T) \approx \sigma(\mu, T) + (\varepsilon-\mu)[\partial\sigma(\varepsilon, T)/\partial\varepsilon]_{\varepsilon=\mu}$. Let us now ignore the second term of Equation 7.9 because it generally provides us with a negligibly small magnitude for many metallic materials. Equation 7.9 is then transformed into the following formula:

$$\kappa_{el}(T) = \frac{1}{e^2T}\int\sigma(\varepsilon,T)F_{W2}(\varepsilon,T)\,d\varepsilon = \frac{1}{e^2T}\int\sigma(\mu,T)F_{W2}(\varepsilon,T)\,d\varepsilon = \frac{\pi^2k_B^2T}{3e^2}\sigma(\mu,T) = L_0\sigma(\mu,T)T \tag{7.14}$$

The second term of spectral conductivity disappeared because it is an odd function of $(\varepsilon-\mu)$ and the $F_{W2}(\varepsilon,T)$, which persists in the integral with $\sigma(\varepsilon,T)$, is an even function of $(\varepsilon-\mu)$. It is easily understood that the Wiedemann–Franz law can be deduced from Equations 7.11 and 7.14.

It would be very useful to mention that the integration shown in Equation 7.12 can be transformed into the following equation:

$$\int F_{W2}(\varepsilon,T)\,d\varepsilon = \frac{k_B^2T^2\pi^2}{3} = L_0e^2T^2 \tag{7.15}$$

The consequently obtained equation unambiguously indicates that the Lorenz number L_0 is obtained simply from the weight of window function $F_{W2}(\varepsilon,T)$. Although the Lorenz number has been often argued to possess temperature dependence for describing the deviation of electron thermal conductivity from the Wiedemann–Franz law, Equation 7.15 clearly indicates that the unusual thermal conductivity should not be responsible for the temperature dependence of Lorenz number but for the energy dependence of spectral conductivity.

The electrical resistivity is determined from the spectral conductivity averaged over the very narrow energy range of a few k_BT in width centered at the chemical potential, while the electron thermal conductivity is determined from the spectral conductivity averaged around $(\varepsilon-\mu) = \pm2.4\ k_BT$. If the spectral conductivity is well approximated with a linear function of energy, the averaged spectral conductivity for both the electrical conductivity and electron thermal conductivity becomes equal to $\sigma(\mu,T)$

regardless of temperatures because the window functions $F_{W0}(\varepsilon,T)$ and $F_{W2}(\varepsilon,T)$ used to calculate these properties are the even function of $(\varepsilon - \mu)$. Therefore both properties are directly related with each other via $\sigma(\mu,T)$, and the Wiedemann–Franz law consequently becomes usable. The consideration described above clearly indicates that the Wiedemann–Franz law is valid only when the spectral conductivity is approximated with a linear function of energy near the chemical potential and $S^2\sigma T$ provides us with a negligibly small magnitude.

7.2.3.2 Semiconductors

Let us now explain the formulae of thermoelectric properties generally used for semiconductors. The following formula, which was originally deduced in the classical Drude theory [3], is frequently employed to describe the electrical conductivity of semiconducting materials, including the semiconducting thermoelectric materials.

$$\sigma = \frac{ne^2\tau}{m^*} = ne\mu_{\mathrm{m}} \tag{7.16}$$

Here, m^* and μ_{m} represent the effective mass of electrons and the carrier mobility, respectively. Equation 7.16 is very useful to discuss the electrical conduction in intrinsic semiconductors and the impurity semiconductors of dilute carriers within the classical dynamics.

Equation 7.16 is obtainable not only from the classical Drude theory but also from the semiclassical Boltzmann transport equation of isotropic systems, provided that (a) the spectral conductivity is a linear function of energy and (b) the energy-momentum dispersion of the band near the chemical potential is well approximated with a parabolic shape as $\varepsilon = \hbar^2 k^2/(2m^*)$. Condition (b) is definitely acceptable for both the intrinsic semiconductors and the semiconductors possessing dilute carriers because the energy-momentum dispersion of materials possessing a periodic structure can be generally approximated with a parabolic shape at the band edge. Condition (a) is also acceptable at a low temperature for any material.

We should realize, however, that both conditions (a) and (b) become invalid at high temperature because the energy range for electrical conduction becomes considerably wider to cover the energy range where nontrivial energy dependences persist in the electronic structure. It should also be mentioned that many thermoelectric semiconductors are classified into the group of degenerate semiconductors for which the band(s) crosses the chemical potential, and therefore the energy-momentum dispersion near the chemical potential is no longer approximated with a parabolic shape. Thus, Equation 7.16 is not generally usable for the thermoelectric semiconductors, especially at a high temperature.

Let us consider the inappropriate approximations of both the Fermi–Dirac distribution function and spectral conductivity used for the semiconductors.

$$f_{\mathrm{FD}}(\varepsilon,T) = \frac{1}{1+\exp\left(\dfrac{\varepsilon-\mu}{k_{\mathrm{B}}T}\right)} \cong \exp\left(-\frac{\varepsilon-\varepsilon_{\mathrm{edge}}}{k_{\mathrm{B}}T}\right) \tag{7.17}$$

$$\sigma(\varepsilon,T) \propto (\varepsilon-\varepsilon_{\mathrm{edge}})^{\frac{3}{2}+\gamma} \tag{7.18}$$

These approximations are frequently employed to obtain the equations for Seebeck coefficient of semiconductors [2,8]. Equation (7.17) is usable only when the energy range we consider is far apart from the chemical potential, and the chemical potential is pinned at the edge of the band $\varepsilon_{\mathrm{edge}}$. This condition is satisfied only when the very small amount of carriers persisting in the band at absolute zero and the dominant energy range of electron transport is far apart from the band edge. The latter condition is hardly justified unless the temperature is extremely high. Equation 7.17, on the other hand, is obtained from Equation 7.10 and the relations of $N(\varepsilon) \propto (\varepsilon - \varepsilon_{\text{edge}})^{1/2}$, $\upsilon \propto (\varepsilon - \varepsilon_{\text{edge}})^{1/2}$, and $\tau \propto (\varepsilon - \varepsilon_{\text{edge}})^{\gamma}$. The relations about the electronic density of states and group velocity, $N(\varepsilon) \propto (\varepsilon - \varepsilon_{\text{edge}})^{1/2}$ and $\upsilon \propto (\varepsilon - \varepsilon_{\text{edge}})^{1/2}$, are deduced from the single parabolic band model of $\varepsilon = \hbar^2 k^2/(2m^*)$. The approximation of relaxation time $\tau \propto (\varepsilon - \varepsilon_{\text{edge}})^{\gamma}$ has sometimes been used to understand the electron transport properties of semiconducting materials, although its validity has not been established well for many materials. By using these rough approximations, one can deduce the following formula for the Seebeck coefficient of semiconducting materials from the original Equation 7.8.

$$
\begin{aligned}
S(T) &= -\frac{1}{|e|T} \frac{\int \sigma(\varepsilon,T)(\varepsilon - \varepsilon_{\text{edge}}) \left\{ -\frac{\partial f_{\text{FD}}(\varepsilon,T)}{\partial \varepsilon} \right\} d\varepsilon}{\int \sigma(\varepsilon,T) \left\{ -\frac{\partial f_{\text{FD}}(\varepsilon,T)}{\partial \varepsilon} \right\} d\varepsilon} + \frac{\mu - \varepsilon_{\text{edge}}}{|e|T} \\
&= -\frac{1}{|e|T} \frac{\int (\varepsilon - \varepsilon_{\text{edge}})^{\frac{5}{2}+\gamma} \exp\left\{ -\frac{\varepsilon - \varepsilon_{\text{edge}}}{k_B T} \right\} d\varepsilon}{\int \left(\varepsilon - \varepsilon_{\text{edge}}\right)^{\frac{3}{2}+\gamma} \exp\left\{ -\frac{\varepsilon - \varepsilon_{\text{edge}}}{k_B T} \right\} d\varepsilon} + \frac{\mu - \varepsilon_{\text{edge}}}{|e|T} \\
&= -\frac{k_B}{|e|} \frac{\int x^{\frac{5}{2}+\gamma} \exp(-x)\, dx}{\int x^{\frac{3}{2}+\gamma} \exp(-x)\, dx} + \frac{\mu - \varepsilon_{\text{edge}}}{|e|T} \\
&\approx -\frac{k_B}{|e|} \frac{\Gamma\left(\frac{7}{2} + \gamma\right)}{\Gamma\left(\frac{5}{2} + \gamma\right)} + \frac{\mu - \varepsilon_{\text{edge}}}{|e|T} = -\frac{k_B}{(|e|)\left(\frac{5}{2} + \gamma + \frac{\mu - \varepsilon_{\text{edge}}}{k_B T}\right)}
\end{aligned}
\tag{7.19}
$$

Here, $\Gamma(z) = \int_0^\infty t^{z-1} e^t dt$ represents the Gamma function.

It is stressed that the assumptions used to deduce Equation 7.19 are definitely inappropriate for the thermoelectric semiconductors. Therefore we should avoid using Equation 7.19 for discussing the Seebeck coefficient of thermoelectric materials.

The electron thermal conductivity of the semiconducting materials is often discussed using the Wiedemann–Franz law in the same way as that of metallic solids. The use of Wiedemann–Franz law is not seriously wrong but provides us with reasonable results in many cases. However, the assumption of spectral conductivity linearly varying with energy near the chemical potential, which is used to deduce the Wiedemann–Franz law, is generally invalid for the thermoelectric semiconductors, and the electron thermal conductivity begins to deviate from that estimated from the Wiedemann–Franz law, especially at high temperatures.

Invalidity of the Wiedemann–Franz law for the electron thermal conductivity of semiconductors is most emphasized when electrons and holes are simultaneously excited in the conduction band and valence band across the band gap, respectively. This condition frequently occurs at high temperatures, and the consequently observed unusual increase of electron thermal conductivity has been discussed in terms of *bipolar diffusion effect* [9].

It has been frequently argued that the bipolar diffusion effect is brought about by the diffusion of electron–hole pairs moving toward the same direction due to the temperature gradient in the sample. In this situation, the charge of electrons and holes is compensated with each other, and the heat current occurs without causing electrical current to make the Wiedemann–Franz law invalid. This argument is seemingly convincing within the classical statistics. We should note, however, that the bipolar diffusion effect is more rigorously accounted for with Equation 7.10 using the information about electronic structure near the chemical potential without considering the old-fashioned electron–hole pairs [10,11]. The bipolar diffusion effect of electron thermal conductivity is caused mainly because the energy range of electrons determining the electron thermal conductivity is different from that of electrical conductivity. The detailed formulation for rigorously explaining the bipolar diffusion effect was shown in the literature, in which a typical example showing this effect on thermal conductivity is accounted for together with the fact that the bipolar diffusion effect is quantitatively reproduced by Equation 7.9 using information about electronic structure.

Before finishing this section, it is emphasized again that very rough approximations are used mainly for electronic structure to deduce the famous equations for the thermoelectric properties of metallic materials and semiconducting materials. The spectral conductivity is assumed to linearly vary with energy for the metallic materials, while the simple parabolic band is used in the theory of the semiconducting materials. Therefore, the consequently obtained equations, such as the Mott formula and Wiedemann–Franz law, are applicable only for limited materials. The electronic structure of the typical thermoelectric materials shown in the literature [12–16] definitely proves that these approximations are unacceptable for these thermoelectric materials. The invalid approximations of electronic structure must be responsible for the difficulty in predicting the temperature dependence of thermoelectric properties of many thermoelectric materials.

7.3 Evidences Proving the Validity of Linear Response Theory

Now, it is clearly understood that the generally used equations for the thermoelectric properties, such as the Mott formula, cannot be used for typical thermoelectric materials because of their nontrivial, fine electronic structure near the chemical potential. The problems in quantitatively predicting that the thermoelectric properties are caused by the unacceptable approximations about the electronic structure introduced during the mathematical transformations from the original equations 7.7 through 7.9. Therefore, we avoided using such inappropriate approximations by directly employing the original equations together with the precisely determined electronic structure. Since the thermoelectric properties (S, σ, and κ_{el}) can be easily calculated if the spectral conductivity and chemical potential are obtained, it is of great importance to precisely deduce the spectral conductivity for evaluating the thermoelectric properties. For obtaining spectral conductivity, we needed to know the information about electronic structure and electron scatterings. The first-principles calculations and the high-resolution photoemission spectroscopy are well developed for this purpose and used in our recently performed studies [16]. Information about the electron scattering, on the other hand, is not very easy to estimate generally, but can be obtained from the spectral shape of high-resolution angle-resolved photoemission spectroscopy although it is obtainable only for the occupied states of single crystals [17–21].

Seebeck coefficient is selected as the first example of the three thermoelectric properties because we do not have to seriously be concerned about the unknown electron scattering in calculating the Seebeck coefficient. This situation is easily understood by knowing that the relaxation time, which generally has very weak energy dependence near the chemical potential, comes out from the integrals in Equation 7.8 and disappears because it persists in both the denominator and the numerator. Figure 7.3 shows, the measured and the calculated Seebeck coefficient with the electronic structure used for the calculations. As typical examples, we selected (a) fcc copper, (b) a structure complex alloy $Al_{82.6-x}Re_{17.4}Si_x$ ($7 < x < 12$) consisting of 138 atoms in the unit cell [16], and (c) a layered copper oxide $(Bi,Pb)_2SrCuO_{6+d}$ [18]. For obtaining the electronic structure of fcc copper and $Al_{82.6-x}Re_{17.4}Si_x$ ($7 < x < 12$), we used the first-principles calculations. The validity of calculated electronic structure of $Al_{82.6-x}Re_{17.4}Si_x$ was conformed by the experimentally determined electron specific heat coefficient [22]. For investigating the electronic structure of $(Bi,Pb)_2SrCuO_{6+d}$, we employed the high-resolution angle-resolved photoemission spectroscopy.

The electronic structure of fcc copper near the chemical potential is characterized by the energy-independent group velocity and the density of states linearly varying with energy. These characteristics lead to the spectral conductivity linearly varying with energy as it is assumed for deducing the Mott formula. The fcc copper possesses the Seebeck coefficient linearly increasing with temperature as it is suggested by the Mott formula, and this behavior is reproduced well by the calculation. The shallow

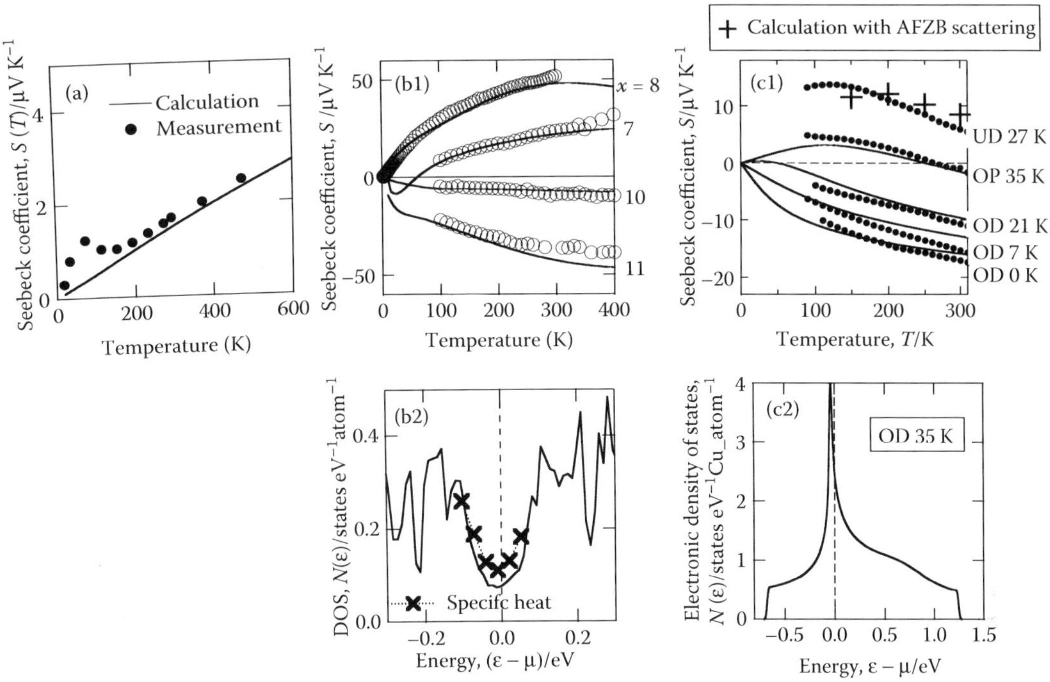

FIGURE 7.3 Calculated (solid lines) and measured (markers) Seebeck coefficient of (a) fcc Cu, (b) $Al_{82.6-x}Re_{17.4}Si_x$ structure complex alloys, and (c) $(Bi,Pb)_2Sr_2CuO_{6+d}$ cuprate superconductors. The electronic density of states used for the calculation is shown in (b2) and (c2) for $Al_{82.6-x}Re_{17.4}Si_x$ and $(Bi,Pb)_2Sr_2CuO_{6+d}$, respectively. OD (overdoped), OP (optimally doped), and UD (underdoped) indicate the doping conditions of cuprate superconductors, and the following number is the critical temperature of superconducting phase. The calculated Seebeck coefficient of UD 27K $(Bi,Pb)_2Sr_2CuO_{6+d}$ is shown with crosses because of the difficulty of calculation under the anisotropic scattering in association with the fluctuating antiferromagnetic ordering. (From Takeuchi, T. et al. 2004. *Physical Review B*, **70**: 146402 1–4; Kondo, T. et al. 2006. *Physical Review B*, **74**: 225411 1–9.)

peak of Seebeck coefficient observed only for the measured data at low temperatures is attributed to the phonon drag effect that was not taken into account in the calculation.

The Seebeck coefficient of $Al_{82.6-x}Re_{17.4}Si_x$ ($7 < x < 12$) shows the nonlinear temperature dependence regardless of Si concentration x [17]. This unusual temperature dependence, which cannot be accounted for with the Mott formula, was quantitatively reproduced by the calculations using Equation 7.8 and the precisely determined spectral conductivity. This result let us realize that the unusual, nonlinear temperature dependence of Seebeck coefficient observed for $Al_{82.6-x}Re_{17.4}Si_x$ ($7 < x < 12$) was caused by the narrow pseudogap persisting near the chemical potential. The presence of such a narrow pseudogap prevents us from using the spectra conductivity linearly varying with energy, and consequently the Mott formula becomes invalid.

The Seebeck coefficient of $(Bi,Pb)_2SrCuO_{6+d}$ possessing a relatively high hole concentration shows that the negative value and linearly varies with increasing temperature [18]. However, the samples of a smaller hole concentration show unusual temperature dependence: it possesses a positive value at low temperatures and increases with increasing temperature. It turns out to decrease after becoming maximal at a characteristic temperature T_{peak}. The sign of the Seebeck coefficient eventually becomes negative at a high temperature above T_0. Both T_{peak} and T_0 increase with decreasing hole-concentration. Surprisingly, these unusual behaviors of Seebeck coefficient were quantitatively reproduced by calculation using Equation 7.8 and experimentally determined electronic structure as shown in Figure 7.3c.

It is consequently understood that these behaviors are caused by the fine structure in spectral conductivity in association with both the van Hove singularity and anisotropic electron scatterings.

The unusual behaviors of Seebeck coefficient observed for other materials were also quantitatively accounted for by using the same analyzing technique [20,21,23]. The agreement between the calculated and measured Seebeck coefficient definitely proves the validity of Equation 7.8 that was deduced under the linear response theory.

Let us now consider electrical conductivity. The quantitative estimation of electrical conductivity is slightly difficult because of the difficulty in obtaining the information about the magnitude of relaxation time. Nevertheless, if the angle-resolved photoemission spectroscopy is employed, one can obtain it from the spectral shape, provided that single-grained samples and their fresh surface are prepared in the ultrahigh vacuum chamber. We did this experiment for some materials, and show here the results obtained for $(Bi,Pb)_2SrCuO_{6+d}$ as one of the typical examples.

The layered cuprate $(Bi,Pb)_2SrCuO_{6+d}$ possesses anisotropic electron scattering due to the fluctuating antiferromagnetic ordering: the electrons persisting near the antiferromagnetic Brillouin zone boundary are more frequently scattered than the others [19]. By considering this tendency using the information obtained from the angle-resolved photoemission spectroscopy data, the experimentally observed electrical resistivity was quantitatively reproduced by the calculation using the spectral conductivity obtained from the Boltzmann equation. The calculated electrical resistivity is shown with the measured one in Figure 7.4e and f.

The electron thermal conductivity of structure complex alloy $Al_{82.6-x}Re_{17.4}Si_x$ $(7 < x < 12)$ shown in Figure 7.5 is picked up here as one of the typical examples indicating the importance of fine electronic

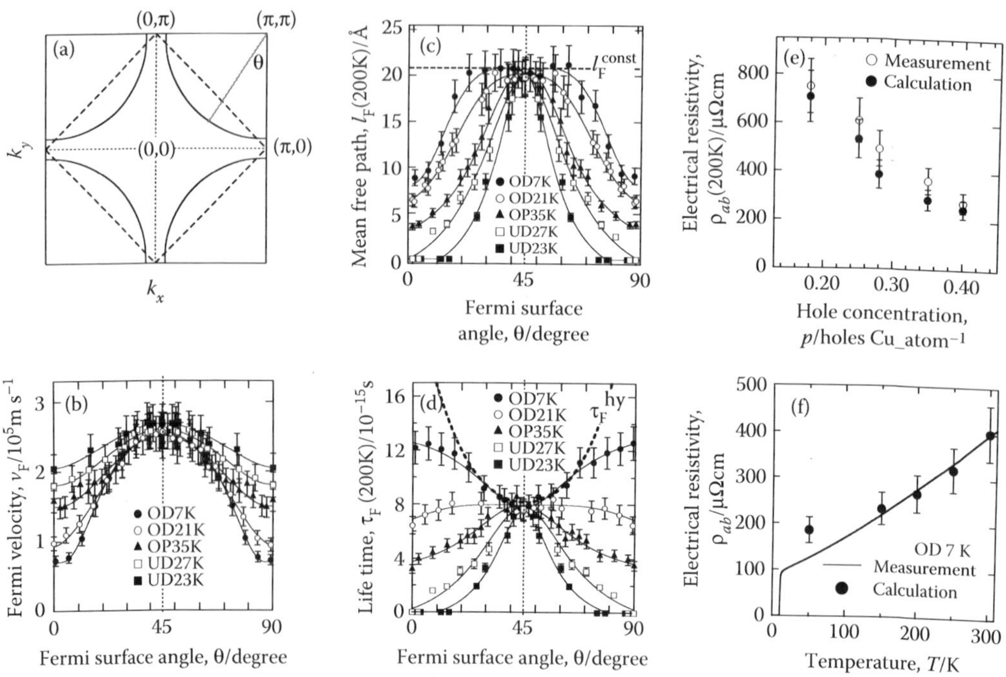

FIGURE 7.4 (a) Fermi surface in the first Brillouin zone, (b) Fermi velocity, (c) mean free path, (d) lifetime of Bloch states of $(Bi,Pb)_2Sr_2CuO_{6+d}$ are plotted together with the calculated and the measured electrical resistivity in (e) and (f). It is certainly understood that the electrical resistivity is determined by the combinational effect of anisotropic group velocity, mean free path, and lifetime. The measured electrical resistivity was quantitatively reproduced by the calculation. This fact indicates that the electrical resistivity can be calculated within the context of linear response theory, provided that lifetime of Bloch states are experimentally obtained together with the electronic structure. (From Kondo, T. et al. 2006. *Physical Review B*, **74**: 225411 1–9.)

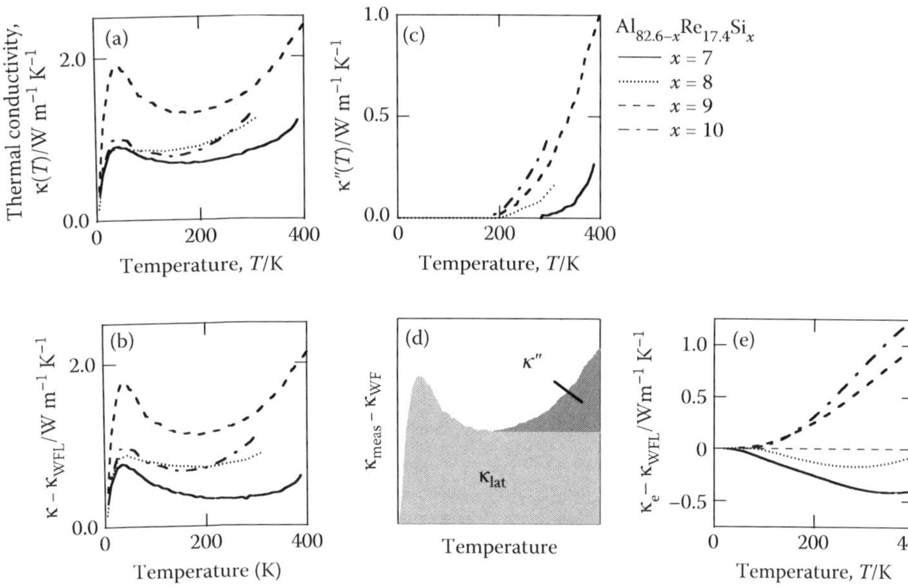

FIGURE 7.5 (a) Thermal conductivity of $Al_{82.6-x}Re_{17.4}Si_x$ alloys possessing the bipolar diffusion effect. (b) The thermal conductivity increases with increasing temperature even if the electron thermal conductivity is roughly removed by using the Wiedemann–Franz law. (c) The unusual increase in thermal conductivity was extracted by assuming that (d) the lattice thermal conductivity becomes constant at high temperature. (e) The calculation using the electron density of states reproduced well the measured unusual thermal conductivity. The calculated value qualitatively reproduced the measured one. (From Takeuchi, T. 2009. *Zeitschrift für Kristallographie*, **224**: 35–38; Takeuchi, T. 2009. *Journal of Electronic Materials*, **8**: 1354–1359.)

structure near the chemical potential. The electronic density of states near the chemical potential of this material is characterized by the narrow pseudogap of a few hundred meV in width. This fact unambiguously indicates that the Wiedemann–Franz law is not applicable for this material. Indeed, the $Al_{82.6-x}Re_{17.4}Si_x$ ($7 < x < 12$) alloys show unusual increase in thermal conductivity at high temperatures as that predicted by the bipolar diffusion effect [10,11], despite the fact that the electronic structure cannot be described by the two-band model that was used to formulate the bipolar diffusion effect. The unusual increase in thermal conductivity remained observable after removing the electron thermal conductivity that was estimated from the Wiedemann–Franz law.

The electron thermal conductivity of $Al_{82.6-x}Re_{17.4}Si_x$ ($7 < x < 12$) was calculated by using Equation 7.9 and the spectral conductivity used to estimate the Seebeck coefficient and the resulting value represented well the measured ones for both the temperature dependence and the carrier concentration dependence [10,11]. As a result of calculation, we realized that the averaged spectral conductivity at around $(\varepsilon - \mu) = \pm 2.4k_BT$ is always larger than that around $(\varepsilon - \mu) = 0$ due to the presence of pseudogap at the Fermi energy, and the difference becomes more evident with increasing temperature. This larger averaged weight of spectral conductivity at around $(\varepsilon - \mu) = \pm 2.4k_BT$ than that around $(\varepsilon - \mu) = 0$ makes the electron thermal conductivity of this material much larger than that expected from the Wiedemann–Franz law.

The calculations for the unusual thermoelectric properties definitely proved that the original equations 7.7 through 7.9 are useful to quantitatively estimate the electronic properties of thermoelectric materials, provided that the reliable spectral conductivity is obtained on the basis of the information about the fine electronic structure near the chemical potential. It is emphasized, therefore, that the investigation into the fine electronic structure near the chemical potential is of great importance to develop new, practical, thermoelectric materials.

7.4 Lattice Thermal Conductivity and Phonon Dispersion

Before constructing the guiding principle of developing practical thermoelectric materials, we should know the lattice thermal conductivity from the microscopic point of view. The lattice thermal conductivity is determined by the phonon dispersion and relaxation time of phonon scattering, provided that the wave packets of phonon are well defined in the materials. If the wave packets are well defined in the material, one can safely employ the classical theory that uses the relaxation time approximation and the Boltzmann factor at high temperatures. The lattice thermal conductivity is estimated from three factors: specific heat at constant volume C_v, mean group velocity v_G, and relaxation time τ.

$$\kappa_{lat} = (1/3)C_v v_G^2 \tau = (1/3)C_v v_G \lambda \tag{7.20}$$

Here, $\lambda = v_G\tau$ indicates the mean free path of phonons. The Dulong–Petit law suggests that the molar specific heat of any thermoelectric material becomes $3R$ at high temperatures above the Debye temperature, and therefore the specific heat at constant volume C_v would not possess any significant variation at temperatures where thermoelectric devices are used. This consideration lets us naturally realize that the significant reduction of both group velocity and relaxation time would provide us with the small magnitude of lattice thermal conductivity.

The materials consisting of heavy elements generally possess a small lattice thermal conductivity because the heavy mean atomic weight leads to the reduction of the energy of phonons and consequently to the reduction of sound velocity and increase of umklapp scattering provability at a given temperature. In the same manner as the materials of heavy constituent elements, the materials characterized by rather weak bonds between the atoms in the unit cell should possess a small lattice thermal conductivity.

The disordered materials, such as amorphous solids, also have a small magnitude of lattice thermal conductivity due to the small relaxation time in association with the intense phonon scattering by the disordered structure. Unfortunately, however, this characteristic always coincides with the small magnitude of electrical conductivity because of the small relaxation time of conducting carriers caused by the disordered structure. Since the ratio of lattice thermal conductivity to the electron thermal conductivity, κ_{lat}/κ_{el}, plays a more important role than that of bare κ_{lat} for obtaining a large magnitude of ZT, the use of disordered materials would not be best suited for developing a practical thermoelectric material.

The materials showing the optical phonon branches at low energies are capable of possessing a very small thermal conductivity. The materials with a rattling mode are a typical example [24] because the rattling mode provides us with a less dispersive optical phonon blanch at very low energy. These optical phonon branches staying at very low energies are intensively excited even at low temperatures far below 300 K, and consequently leads to both the intensified umklapp process of phonon scattering and the reduction of mean group velocity of phonons. These characteristics unambiguously cause a small magnitude of lattice thermal conductivity.

The same mechanism works in the complex structure alloys [25]. Since the unit cell of these materials contains a large number of atoms, the first Brillouin zone is multiply folded. The three acoustic phonon branches, which mainly conduct heat, stay below a very small energy and a large number of optical phonon branches are densely distributed over a wide energy range. Some of the optical phonon branches stay in low energies, and hence the lattice thermal conductivity becomes very small due to the optical branches in the same manner as that in the ratting materials. One may also realize that many of the famous thermoelectric materials are characterized by the large size of unit cell or large number of atoms in it. The lattice thermal conductivity of the samples with an artificially prepared, well-ordered superlattice should have the small thermal conductivity because of its small first Brillouin zone. This fact is easily confirmed in the previously reported papers, although the role of phonon dispersion was not clearly mentioned [26,27].

We should comment on the frequently used models of phonon dispersion. The Debye model has been sometimes employed to understand the lattice thermal conductivity. It should be remembered, however, that the Debye model is usable only at very low temperatures far below the Debye temperature, which is much lower than the temperature range where thermoelectric materials are supposed to be used. Thus, we should abandon the Debye model in which phonon dispersion is oversimplified. The use of the Einstein mode cannot be recommended either because the acoustic phonon branches that mainly carry the heat are not taken into account in this model. Therefore, instead of using the Debye model or the Einstein model, we should use the more detailed information about phonon dispersion to investigate the behavior of lattice thermal conductivity.

7.5 Guiding Principle to Produce New Thermoelectric Materials

7.5.1 Two Factors for Increasing *ZT*

In this section, we construct a guiding principle to produce a new, practical, thermoelectric material in terms of fine electronic structure and phonon dispersion.

Recently, the developments in the first-principles band calculation and the photoemission spectroscopy have allowed us to obtain reliable information about the electronic structure. By using the original equations 7.7 through 7.9 together with the spectral conductivity calculated from the electronic structure, we succeeded in interpreting the unusual behaviors in electron transport properties, $\sigma(T)$, $S(T)$, and $\kappa_{el}(T)$, observed for some different groups of material [16–23,26]. Keeping this fact in mind, we mathematically transformed the *ZT* formula into a slightly modified form.

$$ZT = \frac{S^2\sigma T}{\kappa_{el} + \kappa_{lat}} = \frac{S^2\sigma T}{\kappa_{el}}\left(\frac{1}{1+\kappa_{lat}/\kappa_{el}}\right) \tag{7.21}$$

Two factors can be extracted from Equation 7.21. The first factor,

$$A = \frac{S^2\sigma T}{\kappa_{el}}, \tag{7.22}$$

is determined by the spectral conductivity $\sigma(\varepsilon,T)$ and chemical potential μ. The second factor

$$B = \frac{1}{1+\kappa_{lat}/\kappa_{el}}, \tag{7.23}$$

on the other hand, is determined by the ratio of lattice thermal conductivity to that of electrons, $R = \kappa_{lat}/\kappa_{el}$. The magnitude of factor B is unity at $R = 0$ and gradually decreases toward zero with increasing R. This fact unambiguously indicates that the magnitude of factor A represents the maximum value of *ZT* obtainable at a given spectral conductivity and a given chemical potential, while the factor B always makes *ZT* value smaller than A.

7.5.2 Conditions to Increase the Magnitude of *A*

The factor A is solely determined from electronic structure: spectral conductivity and chemical potential. It would also be very important to note here that the factor A is not affected by the magnitude of the spectral conductivity but only by its energy dependence. It is explained, in other words, that even if the normalized spectral conductivity $\sigma_{norm}(\varepsilon,T) = \sigma(\varepsilon,T)/C$ was substituted for the original spectral conductivity $\sigma(\varepsilon,T)$ in the calculation of A, the same result is obtained because the energy-independent factor C comes out from the integrals in Equations 7.7 through 7.9 and disappears from Equation 7.22 due to its presence in both the denominator and numerator. The same is true even if the factor C varies with varying temperature, provided that C has no energy dependence or negligibly small energy dependence. Considering this fact, the magnitude of factor A obtainable for real materials is calculated in this subsection.

For the calculation of factor A, it would be better for readers to start with the simplest electronic structure, the single parabolic band model in the one- to three-dimensional space, partly because the electronic structure near the band edge is generally well accounted for with this model, and partly because a large ZT value is obtainable when the chemical potential is located near the band edge such as that in degenerate semiconductors. The energy-momentum dispersion in this model is expressed as $\varepsilon = \hbar^2\mathbf{k}^2/(2m^*)$. For the single parabolic band model, Equation 7.10 would be applicable for describing the spectral conductivity.

The temperature dependence of electrical resistivity in metals is generally caused by the temperature dependence of relaxation time, $\tau(\varepsilon,T)$. This fact seemingly suggests that we have to seriously concern about the effect of the complicated temperature dependence of relaxation time when calculating A. However, if the mean free path or relaxation time has negligibly small energy dependence in the vicinity of chemical potential, those terms disappear from Equation 7.21 even if those terms possess a strong temperature dependence. Note here that as far as we observed in our photoemission experiments of various metallic materials, including the materials of strong electron correlation, the energy dependence of the relaxation time is negligibly small in the energy range where the electrons determine the transport properties [21]. Thus, we do not have to be seriously concerned about the temperature-dependent spectral conductivity but simply use the temperature-independent normalized spectral conductivity as $\sigma_{norm}(\varepsilon) = \sigma(\varepsilon,T)/\tau(T)$ or $\sigma_{norm}(\varepsilon) = \sigma(\varepsilon,T)/\ell(T)$.

Under the assumption of energy-independent mean free path, we calculated the normalized spectral conductivity of the single parabolic band model $\sigma_{norm}(\varepsilon)$, which is directly proportional to $(\varepsilon - \varepsilon_0)^0$, $(\varepsilon - \varepsilon_0)^{1/2}$, and $(\varepsilon - \varepsilon_0)$ for the one-, two-, and three-dimensional model, respectively. Here, ε_o represents the energy of the band edge. In the case of the energy-independent relaxation time, on the other hand, we obtain $\sigma_{norm}(\varepsilon)$ which is directly proportional to $(\varepsilon - \varepsilon_0)^{1/2}$, $(\varepsilon - \varepsilon_0)$, and $(\varepsilon - \varepsilon_0)^{3/2}$ for the one-, two-, and three-dimensional model, respectively. The calculated value of A is shown in Figure 7.6b as a function of chemical potential together with $\sigma(\varepsilon,T)$ used in the calculation in (s). Obviously, factor A possesses a larger value with a larger exponent n in $\sigma_{norm}(\varepsilon,T) \propto (\varepsilon - \varepsilon_0)^n$. This fact clearly indicates that the three-dimensional model always possesses a larger A value than that of the one- or two-dimensional model at a given value of $(\mu - \varepsilon_0)$. It is also easily realized that the value of A increases with decreasing chemical potential, and that the large magnitude of A exceeding 2 is obtainable at a given chemical potential near the band edge.

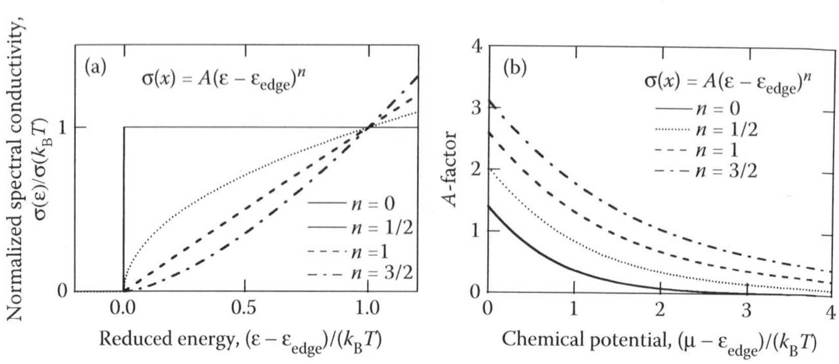

FIGURE 7.6 (a) The models of spectral conductivity and (b) the magnitude of A obtained at a given chemical potential. (From Takeuchi, T. 2009. *Materials Transactions*, **50**: 2359–2365. With permission.)

One has to remember, however, that the magnitude of factor B becomes much smaller than unity at this particular condition because the chemical potential near the band edge limits the number of electrons in the band to a small value. The small number of carriers definitely makes κ_{el} much smaller than κ_{lat}, and therefore a large R value exceeding two, a small B value less than 1/3, and consequently a small ZT value much less than unity would be obtained. It is strongly argued, therefore, that as long as the degenerate semiconductors possessing a very simple parabolic band near the band edge are used, we can never obtain a large ZT value, much larger than unity.

By knowing that the large exponent n in $\sigma_{norm}(\varepsilon) \propto (\varepsilon - \mu)^n$ indicates more significant variation of $\sigma_{norm}(\varepsilon)$ with varying energy near the chemical potential, one may realize some other electronic structures capable of leading to the large magnitude of A. One typical example is $\sigma_{norm}(\varepsilon)$ consisting of a step function as that used in the low-dimensional subbands model [28]. Thus, the chemical potential dependence of A was calculated for a hypothetical spectral conductivity consisting of a step function, in which the magnitude of spectral conductivity increases from Int2 to Int1 at a given energy ε_{step}. This model of spectral conductivity is shown in Figure 7.7 together with the consequently obtained A value. Obviously, the magnitude of A increases with increasing Int1/Int2, and a very large magnitude of A exceeding 3 is easily obtained in this hypothetical model. In sharp contrast to the free-electron model, this model provides us with a finite magnitude of spectral conductivity at the energy range below ε_{step}. This finite magnitude of Int2 leads to a finite electrical conductivity even with the condition of $\varepsilon < \varepsilon_{step}$. The finite value of κ_{el} caused by the finite magnitude of $\sigma_{norm}(\varepsilon)$ near the chemical potential must lead to a relatively large magnitude of the factor B. The consequently obtained large magnitudes both in A and B would provide us with a large magnitude of ZT exceeding unity.

Unfortunately, however, it is not very easy to obtain a step function with a large Int1/Int2 ratio in the real $\sigma_{norm}(\varepsilon)$ of bulk materials. Here, a model consisting of two bands possessing a fairly different energy width is proposed as a possible electronic structure showing a step-like behavior in its spectral conductivity [29]. By placing the narrow band at the energy near the band edge of wider band, we can obtain a step from the small magnitude of wider band to the large magnitude of narrower band, such as the step from Int2 to Int1 in the step function model. The result obtained from the hypothetical spectral function consisting of single step function suggests that the magnitude of A becomes larger with larger Int1/Int2 ratio with the chemical potential located at the energy range where the smaller intensity Int2 persists. Such a smaller magnitude is obtainable near the band edge rather than the center of the band. It is, therefore, naturally realized that the chemical potential must be located at the energy range between the edge ε_0 of the wider band and the step ε_{step}.

Figure 7.8a and b shows the spectral conductivity calculated using two tight-binding bands of a three dimensional simple cubic lattice under the assumption of energy-independent mean free path and that

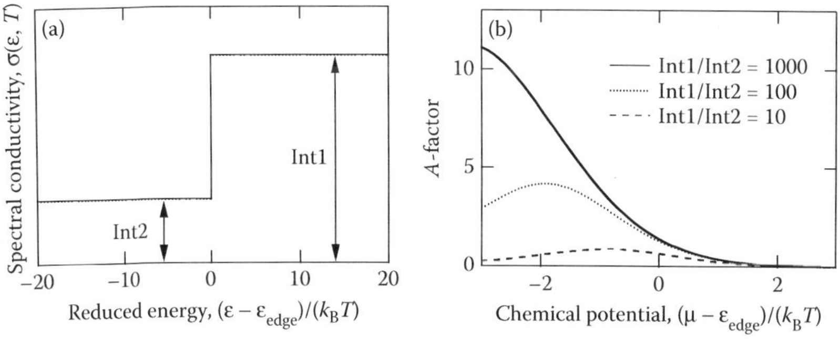

FIGURE 7.7 (a) The models of spectral conductivity possessing a step structure and (b) the magnitude of A obtained at a given chemical potential. (From Hicks, L. D., Dresselhaus, M. S. 1993. *Physical Review B*, **47**: 16631–16634. With permission.)

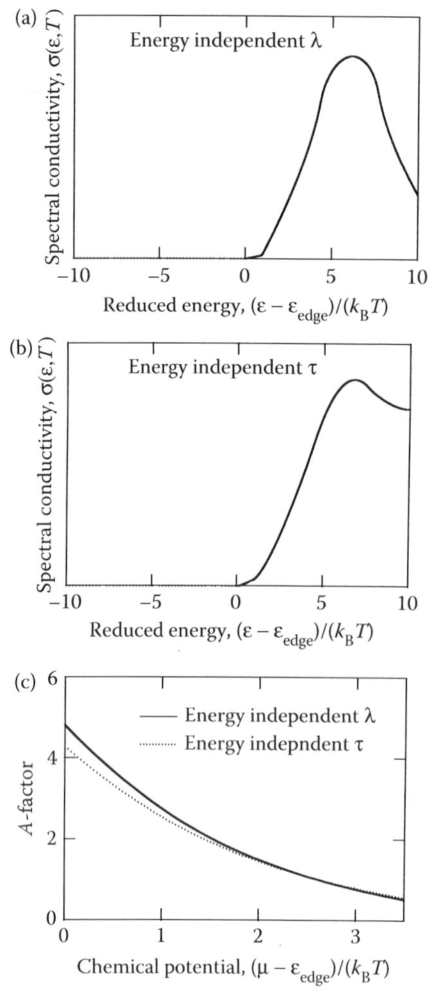

FIGURE 7.8 Spectral conductivity of two-overlapping-band model. The mean free path is assumed to be energy independent in (a), while the relaxation time is energy independent in (b). The resulting magnitude of *A* factor is shown in (c). Obviously, a larger magnitude of *A* factor than that of single parabolic band model is obtainable for the two-overlapping-band models. (From Takeuchi, T. 2009. *Materials Transactions*, **50**: 2359–2365. With permission.)

of energy-independent relaxation time. The wider band is tentatively assumed to have 100 k_BT energy width. The narrower band of 10 k_BT width is located near the band edge of the other band with its center at 6 k_BT from the edge. This is also shown in Figure 7.8a and b.

The magnitude of *A* was calculated as a function of chemical potential. The resulting *A* is shown in Figure 7.8c as a function of chemical potential. The structure of the tight-binding band is well approximated with the single parabolic band model at its band edge, and therefore the chemical potential dependence of *A* value solely with the wide band shows almost the same behavior with that of the three-dimensional free-electron model shown in Figure 7.6. Notably, the magnitude of *A* is enhanced when the narrow band is additionally located near the band edge of the wide band, despite the fact that the carrier concentration is not modified by the additional narrow band.

The semiconductors or semimetals consisting of simple metal elements and transition metal elements are capable of possessing the electronic structure showing the characteristics of two-overlapping-band model. In such alloys, the *s*,*p* bands would play the role of the wide band, while the *d* band of transition metal elements plays the role of the narrow band. Kondo semiconductors [30] are also considered as a plausible candidate because the resonance band works as the narrow band while the *s,p* band as the wider one. The recently predicted topological insulators [31], into which Bi_2Te_3 and Bi_3Se_3 are classified, also possess the same characteristics in their electronic structure as in the care of the surface band and bulk bands to play the role of wide band and narrow band, respectively.

7.5.3 Conditions to Increase the Magnitude of *B*

The factor *A* is transformed into the more simple form of $A \sim S^2/L_0$ by the use of the Wiedemann–Franz law, despite the fact that the Wiedemann–Franz law is not rigorously valid for all solid materials. Since *ZT* is obtained as the product of $A \sim S^2/L_0$ and $B = 1/(1+(\kappa_{lat}/\kappa_{el}))$, the electron thermal conductivity should be as large as possible for increasing the large magnitude of both *B* and *ZT*, provided that the magnitude of Seebeck coefficient is kept large. This is a surprising fact because the smaller thermal conductivity has been considered as one of the main strategy for increasing magnitude of *ZT*.

The magnitude of conductivity ratio of $R = \kappa_{lat}/\kappa_{el}$ is more important than that of lattice thermal conductivity κ_{lat} because the large magnitude of *B* is obtainable when the conductivity ratio is large. Thus, we should find the methods to reduce the conductivity ratio $R = \kappa_{lat}/\kappa_{el}$ rather than the methods to reduce both κ_{lat} and κ_{el}. This condition for a large magnitude of *B* would be accounted for with Slack's proposed idea of '*Electron Crystal & Phonon Glass*' [32]. The reduction of packing density of grains in ceramics, the introduction of scatterers, and the reduction of grain size would not generally provide us with a drastic enhance in the magnitude of either *B* or *ZT* because these techniques reduce not only κ_{lat} but also κ_{el}.

The heavily doped degenerate semiconducting materials, in which many of the thermoelectric materials are classified, generally possess an electrical conductivity less than $1 \times 10^{-5}\Omega^{-1}m^{-1}$, with which the electron thermal conductivity is roughly estimated to stay below 0.7 W/m K. By considering that the minimum lattice thermal conductivity of solids would be limited above ~0.5 W/m K, one may realize that the lattice thermal conductivity should be reduced to be comparable with that of the minimum lattice thermal conductivity for obtaining the large magnitude of *B* factor exceeding 0.2. Since it is not very easy but rather difficult to obtain the large *A* factor exceeding 5 under the carrier concentration leading to $\sigma > 1 \times 10^{-4}\Omega^{-1}m^{-1}$, the reduction of lattice thermal conductivity into the value comparable with the minimum thermal conductivity is of great importance.

In order to reduce lattice thermal conductivity, we should consider the phonon dispersion. As it was discussed in Section 7.4, the presence of optical phonon branches in the low energy range cause very small lattice thermal conductivity. This condition is realized for the materials with rattling atoms or those with large lattice constant. Thus, if we find the material possessing such a crystal structure and an appropriate electronic structure leading to the large *A* factor, that material must possess a large magnitude of a large *ZT* value to become a practical thermoelectric material.

Unfortunately, it is not very easy to find such an ideal material. Since it would be possible to find the material to possess the appropriate electronic structure, we should consider some special methods to reduce the lattice thermal conductivity. The use of well-ordered, artificially introduced superlattice, which must produce optical phonon branches at the low energy range, is one of the most efficient methods to reduce the lattice thermal conductivity without greatly altering the electronic structure. The large magnitude of *A* exceeding 3 together with relatively large magnitude of *B*, which is obtained by the use of superlattice, naturally leads to a large magnitude of *ZT* exceeding unity.

7.6 Development of New Thermoelectric Materials

Two typical examples of material development using the information about the electronic structure and phonon dispersion are shown in this section.

Recently, a new, environmentally friendly, cheap, thermoelectric material was developed on the basis of two-overlapping-band model [30]. The Si_2Ti-type Al–Mn–Si (C54-phase, Pearson Symbol: *oF*24,

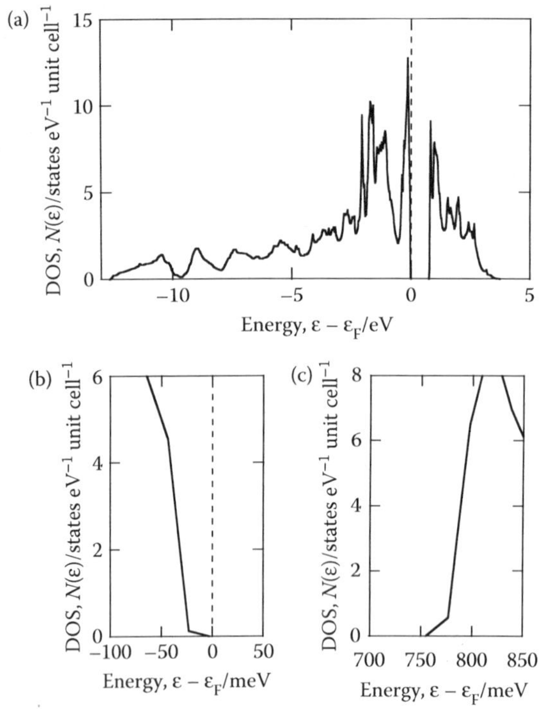

FIGURE 7.9 (a) Electronic density of states calculated for Si_2Ti-type Al–Mn–Si alloy. The electronic density of states near the band edge is enlarged and shown in (b) and (c). The density of states shows a step structure at the lower edge of the conduction band and the higher edge of valence band. (From Takeuchi, T. et al. 2010. *Materials Transactions*, **51**: 1127–1135. With permission.)

Space group: *Fddd*) is the material possessing a large magnitude of Seebeck coefficient exceeding 300 μV/K [33,34]. The electronic structure obtained by the first-principles band calculation using the unit cell consisting of 24 atoms ($Al_8Mn_8Si_8$) is shown in Figure 7.9. Although the step structure near the band edge is not obviously emphasized, the large density of states just below/above the band gap of the $Al_8Mn_8Si_8$ C54 phase could result in a large Seebeck coefficient with positive or negative sign at slightly different electron concentrations.

Figure 7.10a through d shows the Seebeck coefficient, electrical resistivity, thermal conductivity, and resulting *ZT* of $Al_{32}Mn_{34}Si_{34}$ and $Al_{33}Mn_{34}Si_{33}$ C54 phases. The magnitude of Seebeck coefficient $|S|$ of both samples linearly increases with increasing temperature, and it eventually exceeds 300 μV/K at about 400 ~ 500 K. The maximum magnitude of $|S|$ observed for $Al_{32}Mn_{34}Si_{34}$ and $Al_{33}Mn_{34}Si_{33}$ is 330 μV/K at $T_{peak} = 490$ K and 320 μV/K at $T_{peak} = 400$ K, respectively. Notably, the Seebeck coefficient of $Al_{32}Mn_{34}Si_{34}$ is kept negative over the whole temperature range of the measurement, while $Al_{33}Mn_{34}Si_{33}$ always possesses the positive Seebeck coefficient. After becoming maximal at T_{peak}, $|S|$ of both compounds turns out to decrease with increasing temperature. This fact indicates that the excitation of electrons across the gap becomes nontrivial at high temperatures above T_{peak}.

The less obvious temperature dependence of electrical resistivity suggests that the Al–Mn–Si C54 phase is characterized by the disordering in the structure. This fact indicates that the electron mean free path is already limited to the smallest one, and decrease of the grain size does not significantly affect the magnitude of electrical resistivity while it would definitely reduce the lattice thermal conductivity. This consideration can be confirmed in the difference of the ratio κ_{lat}/κ_{el} for samples possessing different grain size. The sample of small grain size, which is prepared by mechanical grinding and pulsed current

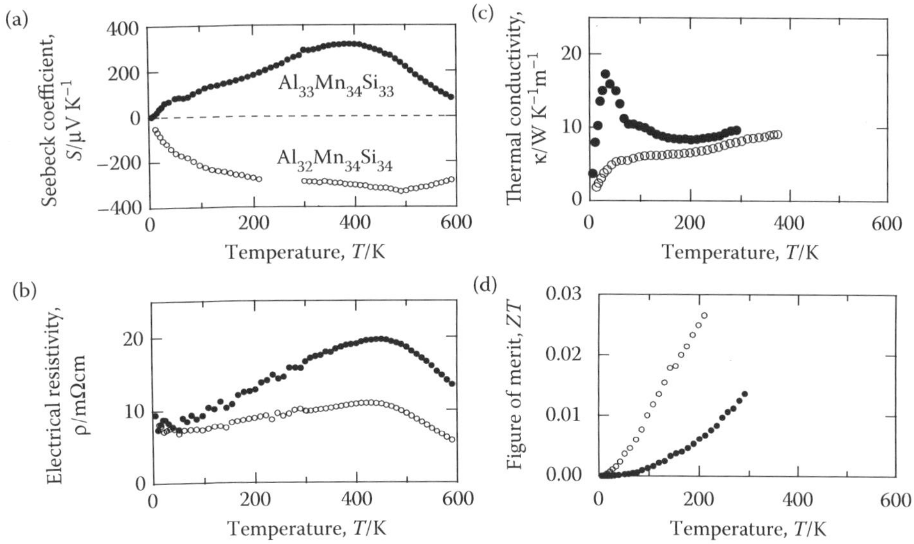

FIGURE 7.10 Thermoelectric properties of Si_2Ti-type Al–Mn–Si alloys developed on the basis of electronic structure analysis. Very large magnitude of Seebeck coefficient was observable both for *n*-type and *p*-type materials. (a) Seebeck coefficient, (b) electrical resistivity, (c) thermal conductivity, and (d) dimensionless figure of merit. (From Takeuchi, T. et al. 2010. *Materials Transactions*, **51**: 1127–1135. With permission.)

sintering, possesses 40% smaller magnitude of κ_{lat}/κ_{el} than that of the large grain size sample prepared solely by induction melting.

The lattice thermal conductivity of the Al–Mn–Si *C*54 phase was reduced by partially substituting heavy elements for the constituent elements. When 15% the manganese in the Al–Mn–Si *C*54 phase is replaced by ruthenium, the lattice thermal conductivity is decreased more than 25%. The reduction in the thermal conductivity is caused mainly by the intensified phonon scatterings and partly by the slightly modified phonon dispersion.

As a result of the sample preparation techniques using the information about electronic structure and phonon dispersion, the Al–Mn–Ru–Si thermoelectric material possessing the largest *ZT* value of 0.3 at 400 K was successfully developed. Although the value of *ZT* is not sufficiently large as the practical material, the strategy and techniques used to develop the Al–Mn–Ru–Si thermoelectric material is usable for many other thermoelectric materials to effectively increase the *ZT* value.

As the second example, the improvement of thermoelectric properties of Fe_2VAl is introduced briefly [35]. The Heusler-type Fe_2VAl, which possesses a large power factor $PF = S^2\sigma$ exceeding 3×10^{-3} W m^{-1} K^2, is one of the most plausible materials obtainable in the environment-friendly system [36,37]. It has been argued that the large poser factor of Fe_2VAl is caused by the presence of pseudogap staying across the Fermi energy. This argument is presumably true because the pseudogap observed in the electrical density of states of Fe_2VAl produces the step structure both at the low-energy side and the high-energy side of the minimum density of states. The mechanical strength, corrosion resistance, and cheap constituent elements let us believe that Fe_2VAl is one of the most plausible candidates for practical thermoelectric material, but its large magnitude of lattice thermal conductivity κ_{lat} prevents us from obtaining a large *ZT* value. Thus, we tried to reduce the lattice thermal conductivity of Fe_2VAl from a microscopic point of view.

The electrical resistivity of Fe_2VAl does not possess significant variation at high temperatures above 300 K, and therefore the mean free path of electrons in this material is supposed to be in the lowest limit, such as that in the Ioffe–Regel limit. For the materials staying in the Ioffe–Regel limit, one of the most efficient ways to reduce the lattice thermal conductivity is the reduction of grain size and the partial

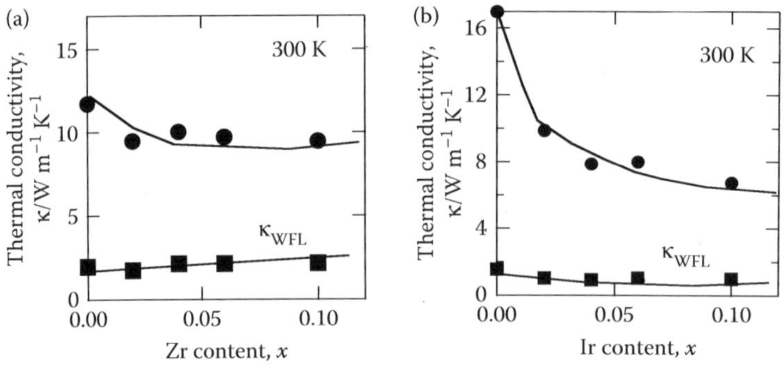

FIGURE 7.11 Thermal conductivity reduction with partial substitution of constituent elements. The thermal conductivity of $Fe_2V_{1-x}Zr_xAl_{0.9-x}Si_{0.1+x}$ ($x = 0$, 0.02, 0.04, 0.06, and 0.1) and $Fe_{2-y}Ir_yV_{0.9-y}Ti_{0.1+y}Al$ ($y = 0$, 0.02, 0.04, 0.06, and 0.1) is shown in (a) and (b). The heavier Ir atom possesses a more strong effect in reducing thermal conductivity. (From Goto, H. et al. 2010. *Materials Transactions*, **51**: 2139–2144. With permission.)

substitution of heavy atoms for the constituent elements, both of which effectively reduce the mean free path of phonons. These two methods had already been confirmed to work in the previously denoted Al–Mn–Si *C*54 phase possessing weak temperature dependence of electrical resistivity.

Figure 7.11 shows the thermal conductivity of $Fe_2V_{1-x}Zr_xAl_{0.9-x}Si_{0.1+x}$ ($x = 0$, 0.02, 0.04, 0.06, and 0.1) and $Fe_{2-y}Ir_yV_{0.9-y}Ti_{0.1+y}Al$ ($y = 0$, 0.02, 0.04, 0.06, and 0.1). In the series of samples, 3*d* transition metal elements V and Fe are partially replaced with 4*d* element Zr and Ir, respectively. For both series of compounds, the thermal conductivity drastically decreases with increasing Zr or Ir concentration, and reaches 10 W/m K and 7 W/m K at $x = 0.1$ and $y = 0.1$, respectively. It should be stressed that only 2.5 at. % substitution of 4*d* or 5*d* transition metal elements for the constituent 3*d* transition metal elements leads to more than 50% reduction of thermal conductivity. It is also easily realized that the substitution of Ir for Fe definitely provided us with more significant reduction of thermal conductivity than that of Zr for V. This difference in the reduction of lattice thermal conductivity must be brought about by the difference in atomic weight between Ir and Zr. Note here that Ir is one of the 5*d* transition metal elements and its atomic mass is 192.22 which is much larger than 91.224 of the 4*d* transition metal Zr.

These experimental results indicate that the partial substitution of heavy elements for the constituent elements of thermoelectric materials definitely reduces thermal conductivity. We have to know, however, that if the impurity states caused by the partial element substitution are located in the pseudogap, the absolute value of Seebeck coefficient is supposed to be reduced. Therefore, the impurity states near the Fermi energy was investigated by means of first-principles calculations. The band calculation of Fe_2VAl suggests that the pseudogap is located at $-0.2 \leq \varepsilon - \varepsilon_F \leq 0.3$ eV. It is of great importance to identify the elements that do not cause any significant variations of electronic structure in this particular energy range. For the investigation into the effect of element substitution, the band calculations could not be very useful because the periodic arrangement of the unit cell required for the band calculations is destroyed by the partial substitution of the other elements. Cluster calculations, on the other hand, are appropriate for investigating the energy eigenvalue of the localized states produced by the impurity elements.

From the calculated partial density of states of the impurity atoms shown in Figure 7.12, it is realized that Ru, Rh, and Os are usable as the substitute for Fe, while Nb, Mo, Ta, and W are appropriate for substituting V. Ir is also considered as one of the candidate substitutes for Fe, but it is supposed to work only in *p*-type material. Two elements Ta and W were selected as the best substitutes because those elements are relatively cheap, environmentally friendly, and appropriate for both *n*-type and *p*-type materials. Figure 7.13 shows Seebeck coefficient of Fe_2VAl-based alloys, in which the V atoms are partially substituted with Ta or W. The Si and Ti were used to control the carrier concentration in the samples. Although

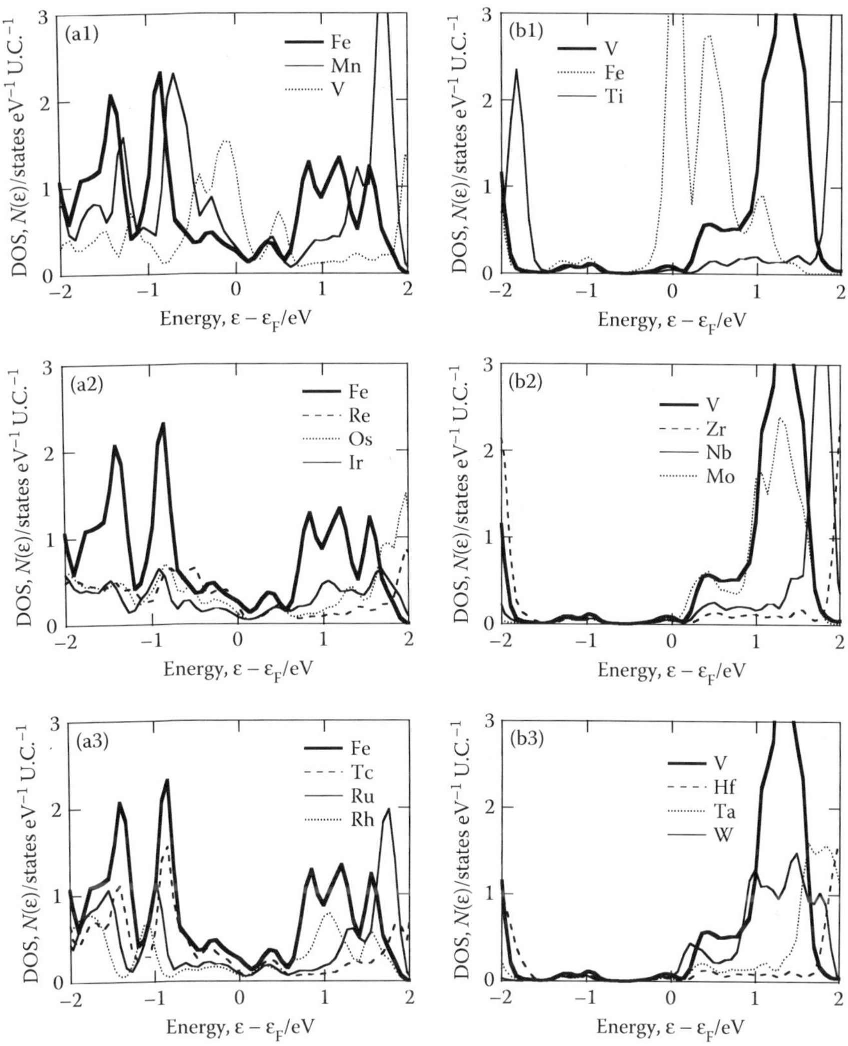

FIGURE 7.12 Partial density of states in Fe_2VAl-based alloys. One of the iron atoms in the unit cell is replaced with 3*d*, 4*d*, and 5*d* transition metal atom, and the impurity density of states are shown in (a1), (a2), and (a3), respectively. Impurity states formed by the 3*d*, 4*d*, and 5*d* transition metal atom for V are shown in (b1), (b2), and (b3), respectively. Since the pseudogap is supposed to cause the large Seebeck coefficient, we identified the impurity elements that would not fill the pseudogap at $-0.5\text{eV} < \varepsilon - \varepsilon_F < 0.3$ eV.

Seebeck coefficient is sensitive to the electronic structure near the chemical potential, it is unambiguously confirmed that the large amount of heavy element substitution, which was identified by the first-principles calculations, obviously cause no significant rediction on the magnitude of Seebeck coefficient or no significant variation in its temperature dependence.

As a result of the partial element substitution, the maximum magnitude of *ZT* observable for the Fe_2VAl-based alloys increased from 0.1 to 0.3. If the superlattice structure were introduced on these Fe_2VAl-based alloys, the magnitude of *ZT* would be more significantly increased due to the reduction of lattice thermal conductivity.

The results obtained for the Al–Mn–Si C54-phase and Fe_2VAl-based materials indicate that the thermoelectric properties are effectively improved by the material design constructed using the information

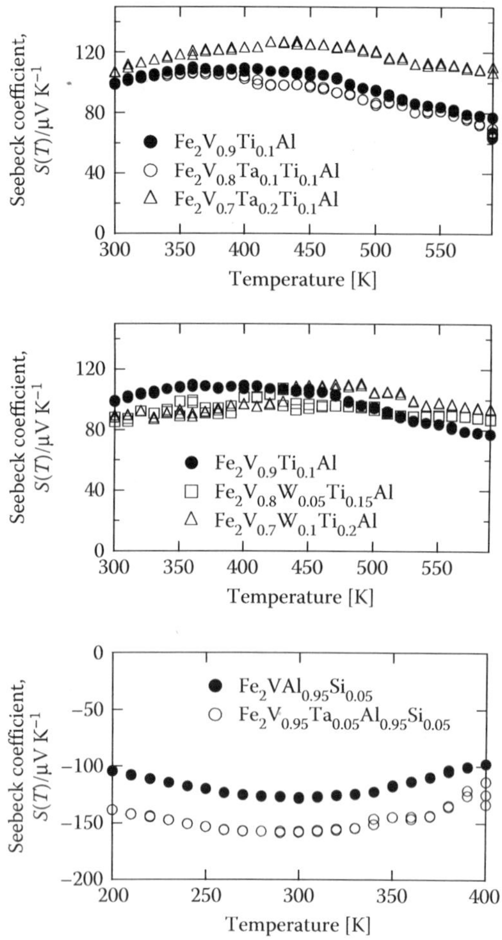

FIGURE 7.13 Seebeck coefficient of the Fe_2VAl-based alloy in which V is partially substituted by the 5*d* transition metal elements which are determined by the first-principles calculations. The magnitude of Seebeck coefficient is kept large even with the 5*d* elements substitution.

mainly about electronic structure and partly about phonon dispersion. We believe that the investigation into the electronic structure and phonon dispersion would become one of the common techniques to develop new, practical, thermoelectric materials.

7.7 Remaining Problems

Before summarizing this chapter, some remaining problems about "the strategy to develop thermoelectric materials in terms of electronic structure and phonons dispersion" are briefly mentioned here.

Although the use of first-principles calculations, high-resolution angle-resolved photoemission spectroscopy, x-ray inelastic scattering, neutron inelastic scattering, and so on all of which are capable of revealing the electronic structure or phonon dispersion of thermoelectric materials, is becoming common, especially in the field of physics, these special techniques have not been commonly used to develop functional materials, including thermoelectric materials. Wider spread of these special techniques should be required for the wider use of the new strategy for developing thermoelectric materials.

It should be mentioned, in addition, that the deep understanding of theories describing the thermoelectric properties is also very important for the proper use of the new strategy. All the theories described in this section are not new but rather old, established in the early twentieth century. In spite of this fact, the development of the thermoelectric materials has been conducted using the classical Drude theory. This method had led to success but it was limited because the Drude theory provides us with the same results as that of the semiclassical Boltzmann theory only if the electronic structure is well described with the single parabolic band model. Drastic improvements of material performance have not been obtained over the past few decades before the 1990s. The performance of thermoelectric materials has slightly improved after the 1990s due to some new ideas, such as the phonon glass electron crystal with the rattling effect, the quantum effect of low-dimensional structure, the Kondo peak in heavy fermion system, and so on. Many readers may realize that these newly proposed ideas are well accounted for with the same characteristics in electronic structure and phonon dispersions for better thermoelectric properties described in this chapter. Thus, it is strongly argued again that the proper understanding of theories, especially on its relation with electronic structure and phonon dispersion, must lead us to obtaining new, practical, cheap, environmentally friendly, thermoelectric materials of high performance.

7.8 Summary

In this chapter, the linear response theory that can be used to predict the thermoelectric properties of solids was explained together with its relation to the fine electronic structure near the chemical potential. Since the calculation of thermoelectric properties proved the validity of this theory by quantitatively reproducing the experimentally observed data, a guiding principle to develop new, practical, thermoelectric materials was constructed on the basis of the linear response theory in terms of electronic structure and phonon for their characteristics to lead to the plausible thermoelectric properties. A new thermoelectric material Al–Mn–Si *C*54 phase was successfully developed using this strategy, and the thermoelectric properties of Fe_2VAl-based thermoelectric materials were also successfully improved by considering the characteristics of electronic structure and phonon dispersion. Thus, it is stressed, in conclusion, that the development of thermoelectric materials of high performance would be accelerated if the validity of the newly proposed idea for developing thermoelectric materials in terms of electronic structure and phonon dispersion becomes widely recognized.

References

1. Ziman, J. M. 2001. *Electrons and Phonons: The Theory of Transport Phenomena in Solids*, Oxford, Oxford University Press.
2. Mott, N. F., Jones, H. 1958. *The Theory of the Properties of Metals and Alloys*, New York, Dover Publications.
3. Ashcroft, N. W., Mermin, N. D. 1979. *Solid State Physics*, Toronto, Thomson Learning Publishers.
4. Ingle, N. J. C., Shen, K. M., Baumberger, F., Meevasana, W., Lu, D. H., Shen Z.-X., Damascelli, A. et al. 2005. Quantitative analysis of Sr_2RuO_4 angle-resolved photoemission spectra: Many-body interactions. *Physical Review B*, **72**: 205114 1–9.
5. Cui, X. Y., Shimada, K., Sakisaka, Y., Kato, H., Hoesch, M., Oguchi, T., Aiura, Y., Namatame, H. Taniguchi, M. 2010. Evaluation of the coupling parameters of many-body interactions in Fe(110). *Physical Review B*, **82**: 195132 1–9.
6. Shimada, K. 2007. High-resolution photoemission spectroscopy of solids using synchrotron radiation. *Lecture Notes in Physics*, **715**: 85–112.
7. Takeuchi, T., Kondo, T., Kitao, T., Soda, K., Shikano, M., Funahashi, R., Mikami, M., Mizutani, U. 2005. Electronic structure near the Fermi level in the $Ca_3Co_4O_9$ layered cobalt oxide. *Journal of Electron Spectroscopy and Related Phenomena*, **144–147**: 849–852.
8. Ioffe, A. F. 1957. *Semiconductor Thermoelements and Thermoelectric Cooling*. London, Infosearch Ltd.
9. Drabble, J. R., Goldsmid, H. J. 1961. *Thermal Conduction in SemiConductors*. New York Pergamon Press, pp. 117–125.

10. Takeuchi, T. 2009. Thermal conductivity of the Al-based quasicrystals and approximants. *Zeitschrift für Kristallographie*, **224**: 35–38.
11. Takeuchi, T. 2009. Unusual increase of electron thermal conductivity caused by a pseudogap at the fermi level. *Journal of Electronic Materials*, **8**: 1354–1359.
12. Mishra, S. K., Satpathy, S., Jepsen, O. 1997. Electronic structure and thermoelectric properties of bismuth telluride and bismuth selenide. *Journal of Physics: Condensed Matter*, **9**: 461–470.
13. Fornari, M., Singh, D. J. 1999 Electronic structure and thermoelectric prospects of phosphide skutterudites. *Physical Review B*, **59** 9722–9724.
14. Madsen, G. K. H., Schwarz, K., Blaha, P., Singh, D. J. 2003. Electronic structure and transport in type-I and type-VIII clathrates containing strontium, barium, and europium. *Physical Review B*, **68**: 125212 1–7.
15. Yang, H.-B., Wang, S.-C., Sekharan, A. K. P., Matsui, H., Souma, S., Sato, T., Takahashi, T. et al. 2004. ARPES on $Na_{0.6}CoO_2$: Fermi surface and unusual band dispersion. *Physical Review Letters*, **92**: 246402 1–4.
16. Takeuchi, T., Otagiri, T., Sakagami, H., Kondo, T., Mizutani, U., Sato, H. 2004. Thermoelectric properties of $Al_{82.6-x}Re_{17.4}Si_x$ 1/1-cubic approximants. *Physical Review B*, **70**: 146402 1–4.
17. Takeuchi, T., Kondo, T., Takami, T., Takahashi, H., Ikuta, H., Mizutani, U., Soda, K. et al. 2004. Contribution of electronic structure to the large thermoelectric power in layered cobalt oxides. *Physical Review B*, **69**: 125410 1–9.
18. Kondo, T., Takeuchi, T., Yokoya, T., Tsuda, T., Shin, S., Mizutani, U. 2005. Hole concentration dependence of the thermoelectric power in $(Bi,Pb)_2(Sr,La)_2CuO_{6+d}$. *Physical Review B*, **72** : 024533 1–9.
19. Kondo, T., Takeuchi, T., Tsuda, T., Shin, S. 2006. Electrical resistivity and scattering processes in the $(Bi,Pb)_2(Sr,La)_2CuO_{6+d}$ studied by angle-resolved photoemission spectroscopy. *Physical Review B*, **74**: 225411 1–9.
20. Komoto, H., Takeuchi, T. 2009. Large thermoelectric power of $La_{2-x}Sr_xCuO_4$ possessing two-dimensional electronic structure. *Journal of Electronic Materials*, **38**: 1365–1370.
21. Kitao, T., Takeuchi, T., Kondo, T., Ikuta, H., Mikami, M., Shikano, M., Funahashi. R. 2006. Electronic structure and the thermoelectric power of Na_xCoO_2 and $Ca_3Co_4O_9$ and layered cobalt oxides. *Transactions of the Materials Research Society of Japan*, **31**: 367–370.
22. Takeuchi, T., Onogi, T., Sato, H., Otagiri, T., Mizutani, U., Kato, K., Kamiyama, T. 2003. Contribution of local atomic arrangements and electronic structure to high electrical resistivity in $Al_{82.6-x}Re_{17.4}Si_x$ $(7 \leqq x \leqq 12)$ 1/1–1/1–1/1 approximant. *Physical Review B*, **68**: 184203 1–10.
23. Takeuchi, T., Kaga, H., Okada, Y., Ikuta, H., Baba, T., Tsuda, S., Shin, S. 2007. Quantative evaluation of the thermoelectric power of $(Bi,Pb)_2Sr_2CaCu_2O_{8+d}$ from the experimentally determined electronic structure. *Journal of Electron Spectroscopy and Related Phenomena*, **156–158**: 452–456.
24. Lee, C.H., Hase, Sugawara, H., Yoshizawa, H., Sato, H. 2006. Low-lying optical phonon modes in the filled skutterudite $CeRu_4Sb_{12}$. *Journal of the Physical Society of Japan*, **75**: 123602 1–6.
25. Takeuchi, T., Nagasako, N., Asahi, R., Mizutani, U. 2006. Extremely small thermal conductivity of the Al-based Mackay-type 1/1-cubic approximants. *Physical Review B*, **74**: 054206 1–12.
26. Venkatasubramanian, R., Siivola, E., Colpitts, T., O'Quinn, B. 2001. Thin-film thermoelectric devices with high room-temperature figures of merit. *Nature*, **413**: 597–602.
27. Touzelbaev, M. N., Zhou, P., Venkatasubramanian, R., Goodson, K. E. 2001. Thermal characterization of Bi_2Te_3/Sb_2Te_3 superlattices. *Journal of Applied Physics*, **90**: 763–767.
28. Hicks, L. D., Dresselhaus, M. S. 1993. Thermoelectric figure of merit of a one-dimensional conductor, *Physical Review B*, **47**: 16631–16634.
29. Takeuchi, T. 2009. Conditions of spectral conductivity to optimize thermoelectric properties for developing practical thermoelectric materials. *Materials Transactions*, **50**: 2359–2365.
30. Adroja, D. T., Padalia, B. D., Bhatia, S. N., Malik, S. K. 1992. Thermoelectric power of the Kondo-lattice system: CePdSn. *Physical Review B*, **45**: 477–479.
31. Kane, C. L., Moore, J. E. 2011. Topological insulators. *Physics World*, **24**: 32–36.

32. Slack, G. A., Tsoukala, V. 1994. Some properties of semiconducting $IrSb_3$. *Journal of Applied Physics*, **76**: 1665–1671.
33. Takeuchi, T., Toyama, Y., Yamamoto, A., Hazama, H., Asahi, R. 2010. Thermoelectric properties of Si_2Ti-type Al-Mn-Si alloys. *Materials Transactions*, **51**: 1127–1135.
34. Toyama, A., Hazama, H., Asahi, R., Takeuchi, T. 2011. Thermoelectric properties of Si_2Ti-type Al-(Mn, Cr, Fe)-Si alloys. *Journal of Electronic Materials*, **40**: 1052–1056.
35. Goto, H., Terazawa, Y., Mikami, M., Takeuchi, T. 2010. The effects of element substitution on electronic structure, electron transport properties, and lattice thermal conductivity of Fe_2VAl thermoelectric material. *Materials Transactions*, **51**: 2139–2144.
36. Nishino, Y., Kato, H., Mizutani, U. 2001. Effect of off-stoichiometry on the transport properties of the Heusler-type Fe_2VAl compound. *Physical Review B*, **63**: 233303 1–4.
37. Nishino, Y., Deguchi, S., Mizutani, U. 2006. Thermal and transport properties of the Heusler-type $Fe_2VAl_{1-x}Ge_x$ ($0 \leq x \leq 0.20$) alloys: Effect of doping on lattice thermal conductivity, electrical resistivity, and Seebeck coefficient. *Physical Review B*, **74**: 115115 1–6.

8

Entropy Flow in Interactive Semiconductor/Metal Nanoensembles

Dieter M. Gruen
Argonne National Laboratory

8.1 Introduction

It is probably not an overstatement to say that multitudes of people have been provided an opportunity to live on earth in relative comfort because of the availability of plentiful electricity. Continued availability of this convenient form of energy is threatened today by ever-increasing demand due to increases in population and per capita consumption as well as the twin problems of decreasing availability of fossil fuels accompanied by problems of climate change associated with their use. We take electricity so much for granted that it is tempting to forget that commercial production did not begin until 1881 after Edison's determined work on power stations had borne fruit. It is good to remind ourselves that this occurred in fact 50 years after Michael Faraday's discovery of the principles of electromagnetic induction in 1831, the basic foundation for all subsequent developments involving the large-scale production of electricity. If history is a guide, then one might expect that major changes in large-scale electricity production will take a considerable period of time for implementation after the basic developmental work on which such a new technology would be based has been accomplished.

One of the manifestations of today's "energy crisis" is a deep concern about the future of electricity availability. We are in a period of profound reexamination of all aspects of the conversion of energy to electricity because of the necessity to supplant current energy sources with alternative and sustainable nuclear and solar sources. We are fortunate that such a reexamination can be expected ultimately to result in new technologies based on the enormous advances in scientific understanding that have occurred in the past 100 years. Since the discovery of electromagnetic induction, we have seen startling experimental advances in the fields of solid-state physics and chemistry that are in part due to insights gained from the quantum theory whose origins can be dated to the beginning of the twentieth century. These revolutionary scientific discoveries have enabled us to consider, for example, all solid-state technologies that do not require rotating machinery for electricity production. Of these, photovoltaic electricity generation is today the most advanced from a practical point of view. However, there are other solid-state technologies for power generation, such as thermionic and thermoelectric conversion that may offer advantages for example in conversion efficiency and insofar as resource limitations and environmental impacts are concerned [1]. These less well known but nonetheless promising approaches are also being studied worldwide and it is electricity generated by thermoelectric materials that is the subject of the present discussion.

State-of-the-art thermoelectric materials have heat-to-electricity conversion efficiencies of about 10%. One has to be able to reach efficiencies of 30–40% in order to compete with engines using electromagnetic induction technology. Clearly, to meet this enormous challenge, breakthrough new materials are required that can reach figure of merit values for ZT of four or above compared to present-day values near unity. Although, thermodynamics does not place an upper limit on ZT, other considerations, such as the Wiedemann–Franz law for example show that achieving $ZT = 4$ presents major materials hurdles.

This chapter presents and elucidates one such approach from both experimental as well as theoretical points of view. We examine the role entropy plays as a factor determining thermoelectric conversion efficiency. Synthetic procedures to enhance entropic flows will be explored and theoretical considerations discussed that inform the experiments. The starting point is the very definition of thermopower in terms of the entropy transported along a temperature gradient per added charge carrier. Since thermopower enters the figure of merit as the square of that quantity, its optimization should obviously be of the highest priority in the quest for high ZT. Of course, extremely important also is a low value of the thermal conductivity as well as a high value of the electrical conductivity.

Many years ago, Jonker [2,3] showed the existence of a strong coupling between thermopower and carrier density. At constant density of states, the former decreases linearly as a logarithmically increasing function of the latter. The existence of this coupling is one of the major factors that has frustrated the achievement of a high power factor (PFs), the product of the square of the thermopower and the electrical conductivity. The supposition underlying the approach to thermoelectricity presented in this chapter is that the achievement of high ZT requires at least the efficient conversion of configurational entropy to electronic entropy. It follows from that supposition that one must be able to optimize those two quantities simultaneously. The theoretical foundation for these statements will be presented in the body of this chapter. For now, it will suffice to say that the nanoensemble approach with which this chapter is concerned was adopted by us in order to create possibilities from a materials point of view for the simultaneous optimization of these two very important quantities. If successful, this approach should enable one to treat thermopower and electrical conductivity as independent experimental variables. In principle, such an approach has the potential to achieve higher PFs than possible up to now. This chapter focuses on the problems that must be overcome to make entropy transport across interfaces a viable alternative way to achieve high-efficiency thermoelectrics.

8.2 Entropy of Mixing of Charge Carriers with Quantum States

A crucial unsolved problem in thermoelectricity is the enhancement of the conversion efficiency of heat to electricity. This can only be done by discovering and developing ever more efficient thermoelectric materials. Since thermoelectric devices are heat engines, the Carnot efficiency is the highest achievable efficiency. For power generation, thermoelectrics operating in the temperature range of about 300–1200 K, the Carnot efficiency is 75%. Unfortunately, the performance of state-of-the-art materials is only about 10% leaving much room for improvement. Although, reaching 30–40% efficiencies would constitute a very important breakthrough in energy conversion technology and is not theoretically excluded, the achievement of this goal is daunting indeed.

It would clearly be advantageous to base an effort in thermoelectric materials development on a complete theoretical treatment of the thermoelectric effect. However, such a treatment grounded on a detailed transport theory involving aspects of Onsager's irreversible thermodynamics [4] as well as Feynman's variational approach to electron–phonon interactions [5] is not available at present. It appears to us that any such theory, should it be developed, will have to involve entropy as a central concept underlying thermoelectric performance. Thus, we have chosen to concentrate our attention on a search for a new class of thermoelectric materials with properties designed simultaneously to optimize both the transformation of lattice to carrier entropy and the transport of carrier entropy in a thermal gradient.

The thermoelectric power, α, according to a thoroughgoing analysis by Callen [6] based on irreversible thermodynamics, can be considered as the entropy transported per coulomb of electron flow. According to this view, the fundamental quantities associated with variations in the magnitude of the thermopower then are due to changes in the amounts of carrier entropy. It follows that thermoelectricity is due to the transformation of heat, lattice vibrational entropy, into electronic entropy. Clausius coined the term entropy which means transformation in the 1850s when he and Lord Kelvin found entropy to be a necessary concept in the formulation of the second law of thermodynamics. Only later was entropy associated with probability as expressed in the famous equation:

$$S = k \ln W \tag{8.1}$$

where S equals entropy and W the number of states of a system. Although, the Boltzmann formalism codified in 1882 in Equation 8.1 has many interpretations, it was Planck who recognized its importance as the link between classical thermodynamics and quantum theory. In the intervening years, much has been done to develop an understanding of the relationship between entropy and thermoelectricity based on quantum mechanical considerations [7]. Here, use is made of some of the salient results of this extensive body of work. In the following, a heuristic attitude is taken both as regards these theoretical developments and particularly their potential applications to the development of new materials. The approach taken here is clearly only one of a number of different ways that insights into carrier entropy transport might be translated into the development of new thermoelectric materials.

If one restricts oneself to noninteracting charge carriers, then the contribution to the Seebeck coefficient is the entropy of mixing associated with adding a single-charge carrier. Considering an idealized situation in which n fermion charge carriers are distributed among N states, one can derive the following expression, after applying the Stirling approximation [8]:

$$\mathrm{d}S/\mathrm{d}n = k \ln(1 - c/c) \tag{8.2}$$

where $c = n/N$. The validity of the underlying assumption of noninteracting carriers depends very much on the details of the structure and composition of the particular material under investigation. In any event, one of the aims of the current development effort is to synthesize materials where the degree of carrier interaction with phonons is one of the variables over which one can exercise a certain amount of control.

It is instructive to examine Equation 8.2 in more detail in order better to understand some of the basic factors influencing the magnitude of S. One would like to be able to achieve as large a value of S as possible since it enters the expression for the figure of merit as the square. Clearly, increasing S can be achieved by decreasing c. At constant N, a decrease in c means a corresponding decrease in n. Since the product of n and the carrier mobility equals the conductivity, σ, the reduction in n has a negative effect on the magnitude both of the PF and the figure of merit, Z. Alternatively, one can decrease c by increasing N while keeping n fixed. As can be seen, Equation 8.2 makes clear the underlying reason for the universally observed close coupling of thermopower and electrical conductivity. Experience shows that optimizing the PF always involves a compromise between the magnitudes of these two quantities if one restricts oneself, as customary, to monophasic systems. A reasonable conclusion considering this state of affairs might be that uncoupling thermopower from electrical transport can have only limited success if one restricts oneself to monophasic systems.

In principle, the limitation described above could perhaps be ameliorated if it were possible simultaneously to increase both n and N enabling their ratio to be kept at a low value. This approach would have the effect of increasing the conductivity without decreasing S. One way of accomplishing such a desirable outcome would be the introduction of a biphasic system one of whose components primarily controls the carrier density and mobility while the other primarily provides a high density of states within thermal energies of the Fermi level. Implementation of such an approach will require appropriate "carrier" and "state" materials that are strongly interactive in the sense of promoting the creation of interfacial regions conducive to enhancing the thermoelectric phenomena resulting from the entropy-of-mixing process.

Considering the thermoelectric effect to be composed of two linked processes, entropy transfer and entropy transport, suggests the use of a biphasic material composed of a semiconductor in contact with a metal. The nature of the junction between these two phases then comes to play a crucial role in that carriers must be able to move relatively freely from one phase to the other. To accomplish the transformation of the semiconductor's configurational entropy into carrier entropy appears to require slow carriers self-trapped in the "polaron" field of the semiconductor. On the other hand, efficient transport of that entropy requires a high carrier mobility. When carriers cross the interface and reenter the metallic phase, the resultant increase in electrical conductivity can be expected to be that characteristic of electrons in metals. Clearly, establishing an appropriate balance between carrier–lattice interactions in the two components constituting the biphasic nanoensembles will determine to a considerable extent the difference in electron mobility in the two phases. This requirement, although challenging, may be met through advances in areas such as the design of compatible semiconductor/metal orbital energy-level structures leading to ohmic junctions with barriers whose heights can be adjusted to a certain degree by a variety of different methodologies [9].

The interfacial regions in which the thermoelectric phenomena described above are active will almost surely be limited by "proximity" effects because the excess energy in the form of entropy that enables carriers to cross the interfacial barrier will be attenuated at a certain rate in the metallic phase. A relatively thin coating of the metallic phase surrounding nanostructured semiconductors is therefore a preferred embodiment. The synthesis of materials with substantial volumetric ratios of interface to bulk is desirable also from the point of view of maximizing the impact on the intrinsic thermoelectric properties of a material. These requirement could be fulfilled by bulk structures composed of nanoparticles each in turn constituted of semiconducting cores surrounded by metallic shells. Reflecting the interactive nature of the nanoparticles, we call the resulting substances bulk nanoensembles. In order to gain an understanding of the problems that have been encountered in our systematic attempt to create materials that would ultimately allow optimization of entropy transformation and entropy transport processes, we describe the sequential steps that have been taken in this quest up to now.

8.3 Thermoelectric Properties of Carbon Nanoensembles

To achieve high Carnot efficiency requires refractory nanomaterials that resist recrystallization at high temperatures. To have an impact on global electricity generation, these materials must not be resource limited, be relatively inexpensive and environmentally benign. The list of candidate materials fulfilling these demanding requirements is very short. The choice of materials is even further restricted by the need to use two materials, one to function in the role of an entropy transfer agent, the other to enhance entropy transport.

The first stage in a still-continuing materials development program was focused on the synthesis and characterization of nanocarbon ensembles composed of nanodiamond cores and nanographite (NG) shells. The experimental results as well as their theoretical treatment will now be briefly described.

Doped diamond is known to have excellent semiconducting properties and doped graphite is a well-studied semimetal. The thermoelectric properties resulting from a combination of these two materials provides considerable insight into the validity of the current approach in part because density functional theory can be applied in a relatively straightforward and rigorous way to this low atomic number monoelemental but nonetheless biphasic nanoensemble.

It was shown that reaction of mixtures of disperse ultrananocrystalline diamond (UNCD) having particle sizes of about 5 nm and 5 wt.% B_4C with methane gas at temperatures near 1200 K results in mechanically rigid compacts resulting in boron-doped nanocarbon ensembles [10,11]. The amount of boron that enters into substitutional solid solution in the nanocarbon ensembles as a dopant is only a small fraction of the quantity added initially [12]. Strongly temperature-dependent PFs ($PF = \alpha^2\sigma$, where α is the Seebeck coefficient and σ is the electrical conductivity) that increase 30–40-fold between ambient and 1000 K were found for boron-doped versus undoped material. Measurements of PFs were made up to 1200 K on samples annealed at temperatures up to 2500 K. The nanoensembles which gained between 70% and 100% in weight, with the greatest change occurring between 950 and 1050 K during reaction with methane, had densities of 1.5–1.7 g/cm^3. Ensembles heated to 1200 K are called "as-fabricated" samples to distinguish from samples that have been annealed in a mixture of Ar-4%H_2 in a furnace with SiC heating elements to 1700 K and in a carbon furnace in argon to 2500 K for varying lengths of times.

Scanning electron microscopy (SEM) micrographs were obtained using a HITACHI S-4700 field emission gun scanning electron microscope. Raman spectra at an excitation wavelength of 632.8 nm were collected at room temperature, using a Renishaw micro-Raman RM2000 on a 50× objective over the range of 50–4000 cm^{-1} using an ~1 μm spot size. X-ray diffraction data [Rigaku miniflex + diffractometer with Cu $K\alpha$ radiation ($\lambda = 0$, 154 nm)] were collected from 3 to 90° 2θ with a scan speed of 2 s/step and a step size of 0.02? Diffraction patters were taken of as fabricated and as well as of annealed samples. Phase identification was based on the International Center for Diffraction Data Base.

High-resolution transmission electron microscopy (HRTEM) was performed using a field emission FEI Tecnai F20 operating at 200 keV. The transmission electron microscopy (TEM) samples were prepared by scraping the compact bulk samples and dispersing the resulting powders in ethanol. The dispersions were ultrasonicated and subsequently deposited on a holey carbon 3 mm copper grid.

Seebeck coefficients and dc electrical conductivities were measured using an apparatus based on a well-established design [13]. It consists of a 1800 K, single-zone tube (mullite, 4.4 cm inside diameter) furnace and controller, a Keithly 2700 mm/data acquisition system, and a Keithly 6220 current source, all controlled by a computer using LABVIEW. The ends of the tube were water cooled and sealed with end caps that allowed fittings for thermocouples and gas inlet/outlet valves.

SEM measurements demonstrated that the original UNCD particles retained their shape and size though they agglomerate and are surrounded by carbon deposits that grew as a result of the reaction with methane.

The Raman scattering spectra, typical of high temperature carbon nanostructured materials under red laser excitation, reveal peaks at around 1330 and 1595 cm^{-1}, which are assigned to the disorder-induced and E_{2g} modes of graphite respectively (Figure 8.1). The *D/G* intensity peak ratio, extrapolated from the spectra after fitting of the data with a Gaussian function and subtraction of the baseline for each spectrum, is related to the cluster size or in-plane correlation length ($L\alpha$) [14,15] which turned out to be about 26 nm consistent with TEM measurements. Graphite peak intensity increased and peak width decreased with annealing temperatures above 1500 K as expected. Diamond content decreased as a function of temperature and graphitization was complete above about 1700 K.

TEM shows that reaction with methane surrounds the original UNCD particles with several kinds of sp^2 bonded carbons. Annealing of the as-fabricated ensembles at 1700 K for 12 h induces graphitization of the UNCD particles. This process becomes more pronounced as the annealing time is increased from 12 to 16 h. After the longer annealing time, the UNCD content is reduced and is replaced by a complex mixture of nanocarbon filaments, needles, and cones. Annealing at 2500 K for 4 h completely transforms the original ensembles. The amount of sp^3 bonded carbon is significantly reduced and replaced by agglomerates of nonporous structures, showing graphitic curved surfaces (Figure 8.2).

Seebeck coefficient (α) and dc electrical conductivity (σ) measurements showed that increasing annealing temperature and times strongly enhanced the Seebeck coefficients and increases the PF by as much as an order of magnitude. The 2500 K anneals produced the highest PF in part because of the

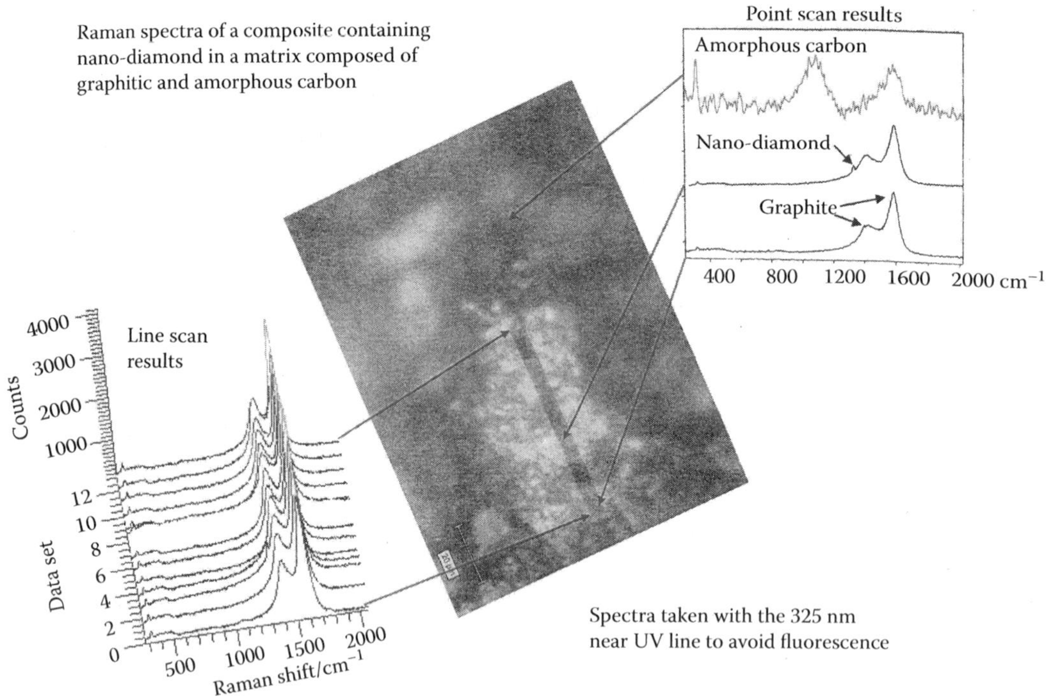

FIGURE 8.1 **(See color insert.)** Raman spectra of nanocarbon ensembles containing UNCD cores surrounded by graphitic and amorphous carbon shells.

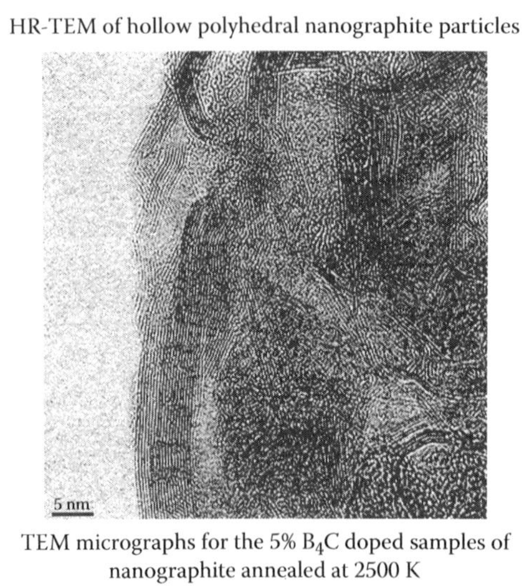

FIGURE 8.2 **(See color insert.)** High-resolution transmission electron micrograph of highly electrically conducting hollow polyhedral NG structures formed at annealing temperatures of 2500 K.

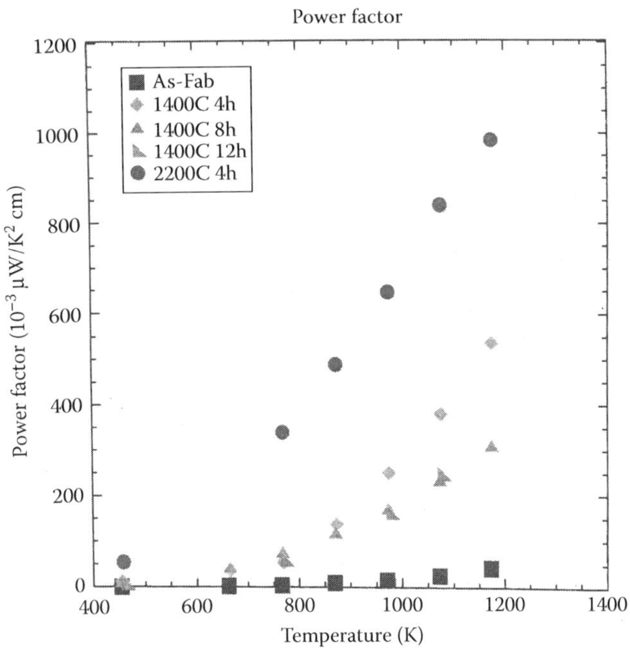

FIGURE 8.3 PFs of nanocarbon ensembles as a function of temperature after various temperature anneals.

dramatic increase in electrical conductivity (Figure 8.3). Thermal conductivities of the nanoensembles are in the range of 0.03–0.07 W/cm K [16]. Figures of merit calculated for the nanocarbon ensembles are too low to be of practical interest.

It has been known for a long time, based on band structure calculations of graphite, that boron doping induces an acceptor-like feature near the top of the valence band. Subtle details of changes in the electronic structure of graphitic materials resulting from boron substitutions, however, cannot be deduced from such calculations. On close consideration of this and other thermoelectric systems it became apparent that the details of dopant substitution in the graphite as well as in other lattices likely plays a crucial role in determining thermoelectric properties. A plausible postulate was therefore advanced [10] that a multitude of boron substitutional configurations each with a distinctive energy leading to a plethora of electronic states are intimately involved in determining the thermopower of a material. A corollary of that postulate is that various boron dopant configurations provide a mechanism by means of which large configurational entropies can be created [10].

These key insights stimulated collaboration with L. Curtiss and P. Redfern that led to a reexamination of the nature of the acceptor-like feature using density functional calculations. In the calculational work, boron substituted polyaromatics served as molecular models with results also on stacked graphene sheets [17]. Quantitative relative stabilities of Kekule versus non-Kekule structures and orbital energetics of various boron substitutional configurations were obtained which will be detailed below. Anticipating the more complete discussion to follow, one may conclude that the molecular analog calculations confirm in exquisite detail the main features of the arguments set out in Ref. [10].

The magnitude of the electrical conductivity, σ, suggests that one is dealing with hopping conductivity [18] characteristic of disordered materials with a local disordering length of a few interatomic distances. A qualitative description of the conductivity changes observed in the nanocarbon ensembles can be given on the basis of Mott's concept of hopping conductivity [19]. At annealing temperatures up to 1700 K conductivity occurs primarily along carbon chains and defected planes. At higher temperatures, graphene sheets, fullerenic structures, and other curved surfaces provide better conduction paths. Finally, at the highest annealing temperatures, 2500 K, the formation of hollow polyhedral NG structures with reduced numbers of defects provides excellent conduction paths leading to σ values of 300 S/cm and above. Clearly, a more detailed understanding of conductivity in these materials will require both extensive experimental measurements, for example, of the temperature dependence of the conductivity and of the Hall coefficients. Thoroughgoing theoretical treatment of electronic transport in these nanoporous materials will also be necessary. In any event the dramatic effects of annealing on the magnitude of α and σ can be qualitatively understood on the basis of an increased solubility of boron in graphite with temperature on the one hand and more favorable conduction paths arising from highly ordered and less defective graphite layers on the other.

The results on the nanocarbon ensembles are interesting particularly from the point of view of the light they shed on the transformation of UNCD to NG as a function of temperature. Of particular relevance is the work of Kuznetsov and Butenko [18] and Kuznetsov et al. [20] as well as that of Andersson et al. [21] This, by now, quite well-developed subject will be discussed here only to the extent that it relates to thermoelectric properties. Annealing affects the magnitudes of α and σ *occur* not only because of changes in boron solubility and ordering of graphene structures but also because of the transformation of UNCD to NG. In the as-fabricated ensembles, UNCD is the primary constituent prior to reaction with methane. The UNCD to NG transformation has been followed using XRD, HRTEM, and Raman characterization techniques. The diamond (111) reflection is at a somewhat smaller 2θ angle than the graphite (100) and (101) reflections and overlaps the latter two reflections, thus making it difficult to use this Bragg region to follow the transformation quantitatively. Data analysis is therefore based on a determination of the ratios of intensities of the diamond (002) near 75° relative to the graphite reflections near 80°. In this Bragg region, the diamond and graphite reflections are reasonably well resolved. The detailed work of Andersson et al. [21] provides baseline information regarding the temperature-dependent transformation of disperse UNCD to graphite. Their data and those described here are in excellent agreement.

The HRTEM micrographs taken after various annealing temperature treatments substantiate the conclusions reached on the basis of the XRD measurements. As an example, nano diamond ensembles (NDEs) annealed at 2100 K show a greatly diminished number of UNCD crystallites compared to the as-fabricated compacts with only a few 2–5 nm UNCD particles remaining after 2100 K treatment. Structures displaying graphitic and curved surfaces abound with equidistant sheets forming essentially defect-free layers.

Density functional theory was used to explore the concept of configurational entropy mentioned above which is an important aspect of the approach to thermoelectricity that is examined here. The basic idea is that dopant configurations allow one, under certain circumstances, to introduce a plethora of electronic states that lie within thermal energies of the Fermi level allowing a semiconductor having these characteristics efficiently to transform configurational to electronic entropy. The molecular analog calculations employed here examine in great detail the wave functions associated with each individual dopant configuration and the relevant energetics and orbital hybridizations. Acquiring a detailed understanding of the electronic structure of the solid involved in the transformation of heat to electricity represents an important step in gaining a fundamental understanding of this process.

The calculations were done with the B3LYP functional [22] along with the 6-31G* basis set. The GAUSSIAN03 set of computer codes [23] was used for all calculations. Spin-unrestricted methods were used for open-shell species. All calculations were done as molecular systems, that is, nonperiodic. Unless noted, the results are based on complete geometry optimizations, including the boron-substituted nanographene which were investigated as models both for Kekule and non-Kekule systems. In previous studies, the B3LYP density functional methods have been found to give reliable HOMO–LUMO gaps for related species [24] as well as singlet–triplet energy differences [25,26]. Stacks of nanographene sheets were also investigated and their energy levels were determined with density functional-based tight binding (DFTB) [27–29], which has proven reliable for studies on a variety of carbon-based materials [30,31].

Non-Kekule structures have an even number of π-electrons, but the topology of the sheet makes it impossible to write a resonance structure in which each π-electron is paired with another on a neighboring carbon [32,33]. Importantly, these calculations determined the stabilities of non-Kekule graphene

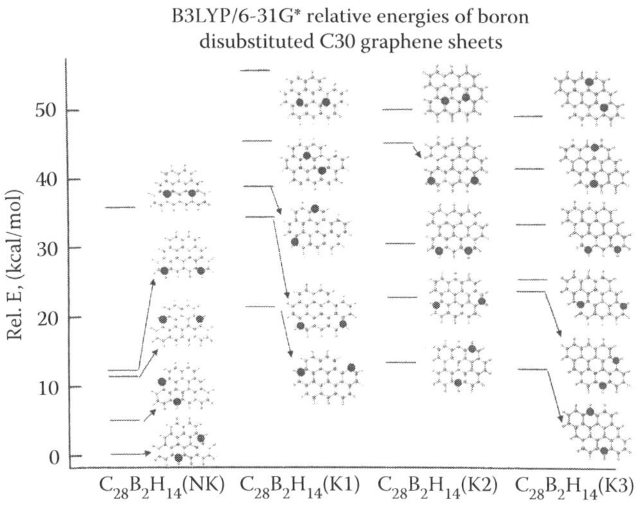

FIGURE 8.4 Relative energies between four different boron configurations comparing one non-Kekule (NK) with three different Kekule (K1, K2, K3) graphene structures each composed of $C_{28}B_2H_{14}$. Solid circles indicate where boron substitutions are located.

structures, which may very likely be present in the nanocarbon ensembles [10,11] relative to the stabilities of Kekule-type structures. For example, the B3LYP/6-31G* relative energies of different spin multiplicities of the non-Kekule $C_{22}H_{12}$ show that the triplet is 23.4 kcal/mol more stable while in the case of the $C_{30}H_{14}$ structure the triplet is 19.4 kcal/mol more stable.

For the $C_{30}H_{14}$ graphene sheet, a more extensive investigation of various substitutional sites for borons was carried out with full geometry optimization so that they would be consistent with unsubstituted all carbon sheets. Figure 8.4 illustrates the relative energies of different double boron substitutions in four different $C_{28}B_2H_{14}$ graphene structures. The results indicate that the most stable doubly substituted grapheme sheet is the one with the non-Kekule form for the pristine sheet. In addition, the results in Figure 8.4 indicate that boron substitution in the non-Kekule structure generally results in a more stable structure than for substitution in a Kekule structure, if one considers similar boron siting. The boron substitutions in non-Kekule graphene structures tend to be more stable than in the Kekule graphene structures because the high-spin non-Kekule graphene ground states have three coordinated carbons with an unpaired electron that are favorable for boron substitution.

The reaction energy for double boron substitution in $C_{30}H_{14}$ is endothermic by 55–110 kcal/mol relative to removal of two carbon atoms for different boron substituted structures considered for $C_{30}H_{14}$. The large endothermic reaction energies for boron substitution is consistent with the low solubility of boron in graphite of about 2% at high temperatures [12].

The results of the density functional calculations on the pristine and boron-doped graphene systems provide a framework for considering certain aspects of the behavior of the Seebeck (S) coefficient found for boron-doped NG [10,11]. First, the calculations indicate that in the pristine systems the electronic states are high spin when the system has a large enough number of carbon atoms. This occurs at around 72 carbon atoms for Kekule structures and at the smallest number of carbons necessary to form a non-Kekule structure. The high spin state of all carbon systems implies the existence of both p- and n-type carriers leading to small S values [34]. This conclusion is consistent with the small S values that was experimentally found for all the carbon NG ensembles without boron doping where the S values are found to be close the zero with little temperature dependence [10].

The calculations also show that boron substitution leads to low-spin nanographene sheets, unlike the pristine unsubstituted all carbon sheets. This implies that boron lowers the energy of the highest occupied orbital resulting in holes in the valence orbital energy levels. These results are consistent with a number of experimental as well as theoretical studies of boron substitution in bulk graphite, nanotubes, and grapheme nanoribbons that show boron creating holes at the top of the valence band and a lowering of the Fermi level [35,36]. These previous theoretical results described in the literature have been based on periodic density functional calculations. A similar situation occurs in the finite or molecular analog nanographene systems as well but here it is possible to show in detail that each boron "configuration" gives rise to a unique electronic state signature reflecting subtle changes in orbital energies.

In addition to the calculations on small individual graphene sheets, DFTB calculations were also done on a model of four stacked graphene sheets each consisting of about 100 carbon atoms. Each of these sheets can have almost $5 \times 10 + 3$ different configurations of doubly substituted borons of which five distinct configurations are shown in Figure 8.5 while the stack of four graphene sheets each one with

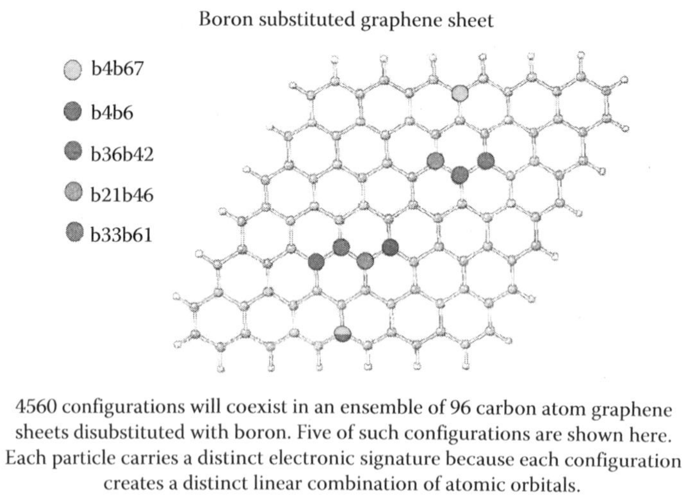

FIGURE 8.5 **(See color insert.)** Five distinct configurations of doubly substituted borons in a graphene sheet of about 100 carbon atoms.

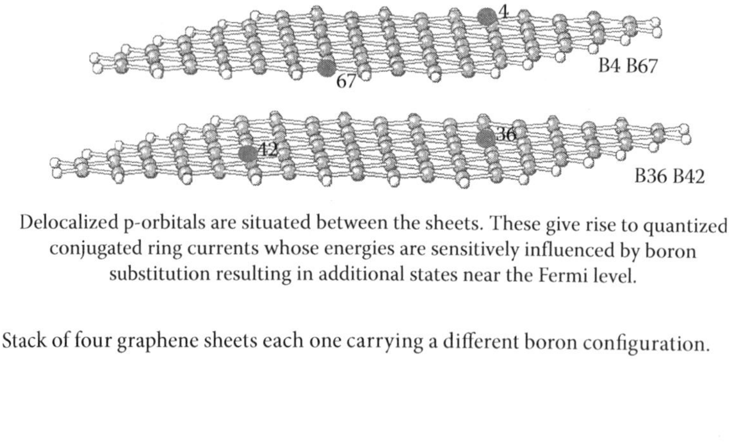

FIGURE 8.6 Stack of four graphene sheets each one carrying a different boron configuration.

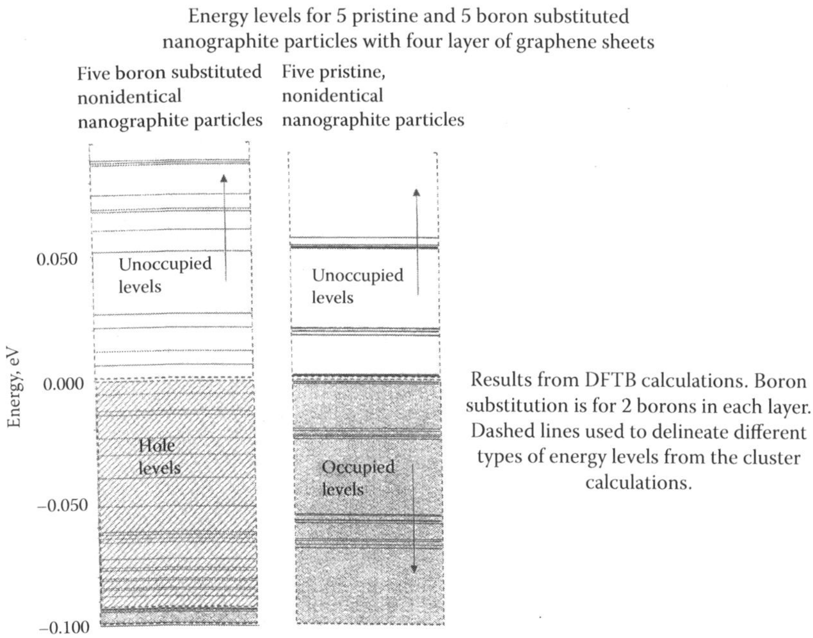

FIGURE 8.7 Calculated energy levels near the HOMO–LUMO gap for five pristine (unsubstituted) particles compared to five substituted particles each consisting of a stack such as shown in Figure 8.6.

a different configuration of borons is illustrated in Figure 8.6. The calculated energy levels near the HOMO–LUMO gap for the pristine and di-boron-substituted system of five NG particles each consisting of a stack of four 100 carbon atom sheets are illustrated in Figure 8.7. The figure shows the levels corresponding to the creation of holes due to boron substitution and a concomitant lowering of the highest-occupied energy level by 0.1 eV.

In the carbon ensembles studied experimentally [10,11], where the number of carbon atoms is 10 + 5 – 10 + 7 instead of 10 + 2, the number of possible boron geometric configurations in each nanoparticle becomes quite large. Each configuration, as we have seen, is associated with a distinct NG structure and because of the subtleties of orbital energy levels each structure gives rise to a distinct electronic state. In principle then, the creation of "holes" at the top of the valence band by the electron-deficient borons creates many empty, perhaps nearly degenerate electronic states. The model presented here suggests that boron substitution lowers the Fermi level of grapheme sheets by about 0.1 eV. The large number of empty states that owing to this circumstance can be populated thermally thus provides a mechanism for the efficient transformation of configurational to electronic entropy and thus for the conversion of heat to electricity.

8.4 Thermoelectric Properties of Silicon Carbide/Nickel Silicide Nanoensembles

Several conclusions drawn from the work on carbon nanoensembles have determined the directions of research aimed at improving thermoelectric performance. It was recognized at the outset of our work that the refractory nature of carbon materials required the development of special synthetic procedures to produce mechanically robust compacts. This was successfully accomplished using methane gas in the high-temperature chemical vapor deposition (CVD) process described above to generate diamond core/graphitic shell structures. Although, the thermopower of boron-doped diamond exceeds 300 μV/K according to published reports [37], the measured thermopower of the carbon nanoensembles did not exceed about 50 μV/K. There are at least three reasons for this observation. The first is that the content of the NG phase, which is very difficult to control in the CVD process, is so large as to obscure the expected entropy enhancement effect due to the presence of nanodiamond. Second, the sizable orbital mismatch between the sp^3 bonded diamond and the sp^2 bonded graphite precludes the existence of ohmic contacts between these two forms of carbon. Third, the annealing temperature required to achieve optimal electrical conductivity exceeds the graphitization temperature of diamond.

To overcome the disadvantages inherent in carbon nanoensembles, a different class of nanoensembles based on silicon carbide core nickel silicide shell structures is being developed. The considerably higher thermopower of doped SiC compared to graphene is expected to help achieve figures of merit of interest for energy conversion. Reaching this goal presents a considerable challenge in part because the electrical conductivity of SiC is low. Use of Ni_2Si as an electrically conducting shell will be explored as the entropy transport medium under the influence of a thermal gradient.

The synthesis of Al- and B-doped SiC was carried out using mixtures of fine-grained powders of? SiC, Al_4C_3, and B_4C. The mixtures containing 3 wt.% each of Al_4C_3 and B_4C were spark plasma sintered in a graphite die with holding times of 10 min at 2300 K and 40 MPa [38]. The resulting discs were used to fabricate 5 × 5 × 17 mm rectangular shapes on which Seebeck coefficient, electrical conductivity, XRD, and Raman measurements were carried out in a manner described in detail in Ref. [11]. The XRD and Raman spectra show that the ensembles consist primarily of 6H SiC [39]. A complete elucidation of those data is beyond the scope of this chapter since the phase equilibria even in the Al–Si/C ternary system are highly complex [40] and would be even more so in the presence of B. One cannot quantitatively determine the concentrations of substitutional Al and B from current data and one assumes that the thermoelectric behavior is due to the presence of both dopants at levels of 10 + 18 to 10 + 19/cm^3 based on known solubilities of Al and B in α SiC at comparable temperatures [41,42].

Ensembles containing 30 wt.% (11 mol%) Ni_2Si powder in addition to the components given above were used to fabricate rectangular samples of similar dimensions from compacts prepared by spark plasma sintering (SPS) processing with holding times of 10 min at temperatures up to 1900 K. Ingots of pure Ni_2Si were prepared by melting Ni_2Si powder in Al_2O_3 crucibles from which rectangular samples were also fabricated. The SPS processed ensembles as well as the pure Ni_2Si samples were characterized using all the techniques listed above [43].

As already stated, XRD shows that the ensembles processed in the absence of Ni_2Si consist primarily of 6H SiC (Figure 8.8a). However, SPS-processed powder mixtures containing 11 mol% Ni_2Si in addition to SiC (Figure 8.8b) display several peaks assigned to that compound. The absence of peaks due to other nickel silicides shows that under the conditions of SPS processing employed here, reaction between SiC and Ni_2Si does not occur except perhaps at surfaces. There is evidence from the XRD for the existence of small amounts of unreacted Al_4C_3 and B_4C in the spark plasma treated samples. The pure Ni_2Si ingot consists entirely of that phase as determined by XRD.

The thermopowers (Seebeck coefficients) and electrical conductivities as a function of temperature of SiC is displayed in Figure 8.9 and those of Ni_2Si in Figure 8.10. The magnitude of the thermopower of pure SiC increases strongly in the range of 400–800 K, and then reaches a plateau at about 350 μV/K. The electrical conductivity increases over the entire temperature regime investigated and continues to do so even at the highest measuring temperature of 1100 K reaching a value near 2S/cm. These results are consistent with those reported by Koumoto et al. [44] and Pai et al. [45,46] on boron-doped SiC. The thermopower of Ni_2Si has a value of −12 μV/K at 400 K and increases almost linearly to reach −37 μV/K near 1100 K while the electrical conductivity decreases from 3.1 × 10 + 4 to 1.2 × 10 + 4 S/cm in the same temperature interval.

The thermopower as a function of temperature of pure Ni_2Si is compared to that of the SiC/Ni_2Si ensemble in Figure 8.11. Also shown is a simulated "Expected" curve for the thermopower of Ni_2Si based on the assumption of a somewhat reduced electrical conductivity compared to pure Ni_2Si leading to an expected increase in the thermopower of Ni_2Si. Comparison of the thermopower data of the ensemble

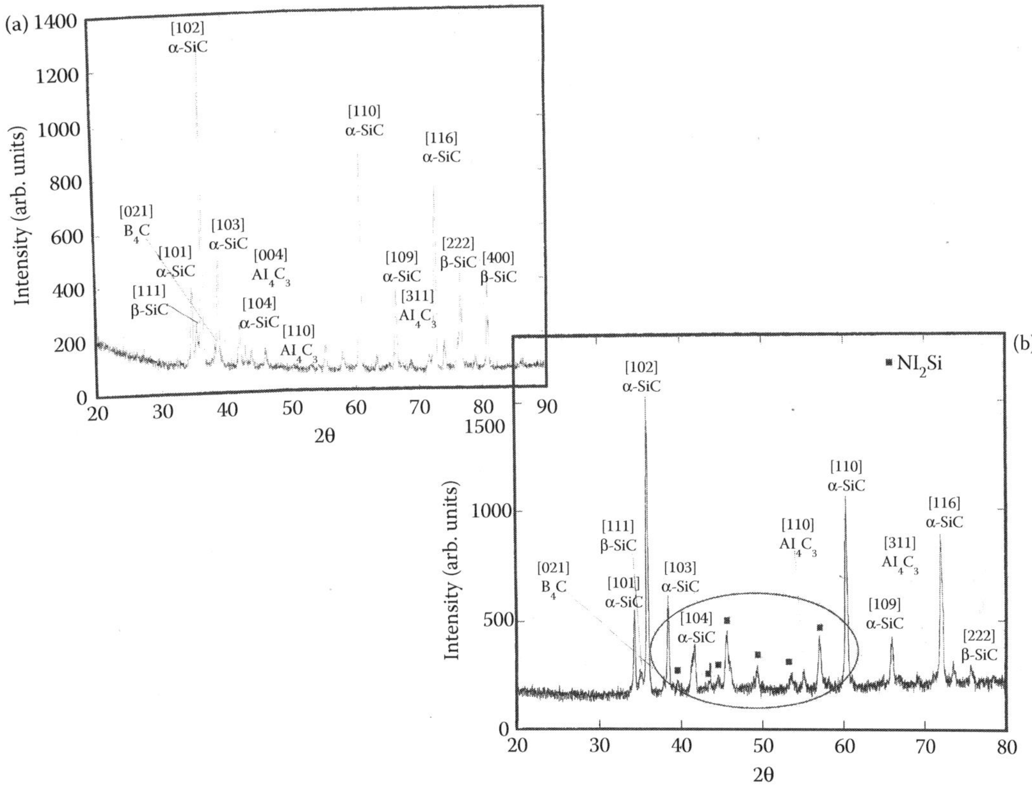

FIGURE 8.8 XRD spectra of SPS processed (a) α SiC; (b) α SiC/Ni_2Si mixture. Both (a) and (b) contain small amounts of Al_4C_3 and B_4C.

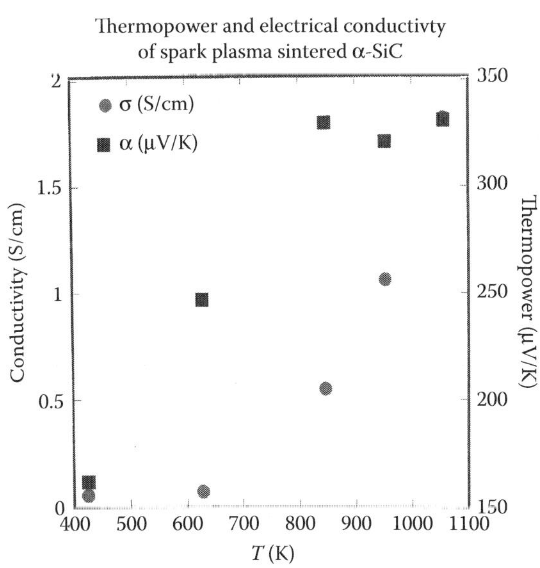

FIGURE 8.9 Thermopower and electrical conductivity of SPS processed α SiC.

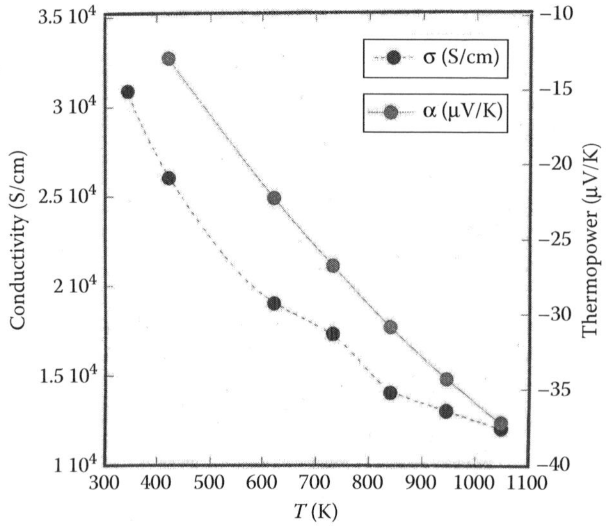

FIGURE 8.10 Thermopower and electrical conductivity of Ni_2Si.

near 1100 K with those of the "Expected" Ni_2Si data shows that a change from −40 to −20 μV/K has occurred. One way to interpret this observation is to assume that the strong decrease in (negative) thermopower of the ensemble at 1100 K is due to a "proximity effect" exerted by the presence of SiC with its large positive thermopower. Another possible interpretation might be that the observed changes are due to the direct contribution of the thermopower of SiC because the electrical conductivity of SiC itself constitutes about 2% of the total conductivity of 100 S/cm which is due overwhelmingly to the presence of highly conductive Ni_2Si. Since a contribution due to SiC conductivity would account for at most one-third of the experimentally determined change in thermopower, the "proximity" effect is considered to be the most likely explanation of the observations shown in Figure 8.11.

The measured electrical conductivity of the ensemble is close to 100 S/cm rather than about 1000 S/cm calculated on the basis of dilution of Ni_2Si by SiC. This result is thought to be due to two circumstances:

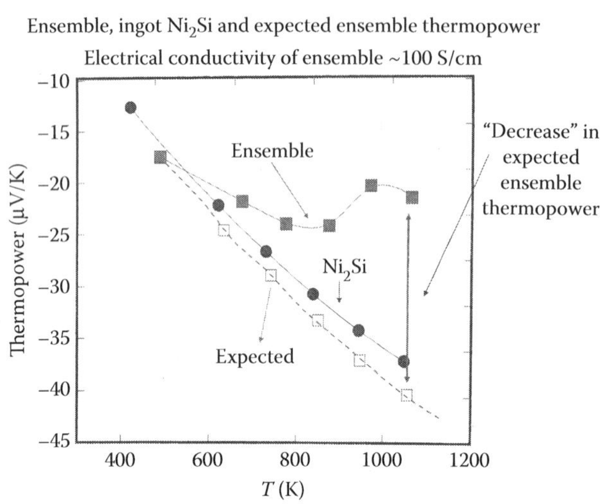

FIGURE 8.11 SiC/Ni_2Si SPS-processed ensemble, ingot Ni_2Si and "expected" ensemble thermopower.

α-SiC/Ni_2Si ensemble (5wt% Al_4C_3; 5wt% B_4C, 20wt% Ni_2Si)
EDAX analysis

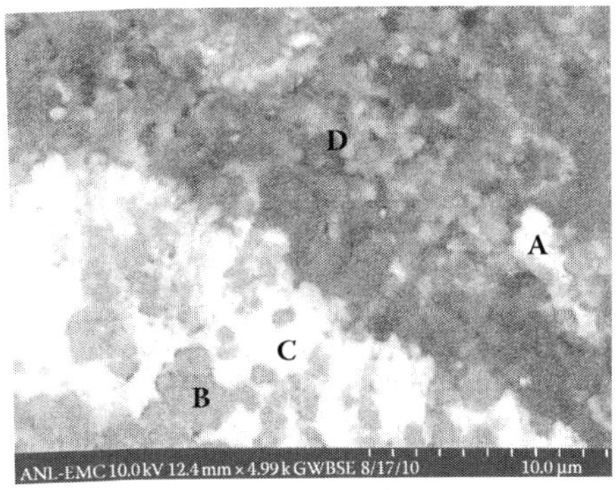

EDAX quantification
(ZAF method)

A: 52% NiI; 40% SiK; 7.5% AIK

B: 63% NiI; 32% SiK; 5% AIK

C: 31.5% NiI; 62% SiK; 6.5% AIK

D: 99% SiC

FIGURE 8.12 (**See color insert.**) SEM and EDAX analysis of SiC/Ni_2Si SPS-processed ensemble.

(1) at 11 mol%, the Ni_2Si concentration is below the percolation limit and (2) the SPS processed ensemble is very inhomogeneous. The extreme compositional inhomogeneity of the SPS-processed ensemble is made clear from an examination of Figure 8.12. An SEM backscattering and energy dispersive x-ray spectroscopy (EDAX) study of the ensemble shows regions whose elemental compositions point to various thicknesses of Ni_2Si coating SiC (A,C); regions of pure Ni_2Si (B) and regions of pure SiC (D). It is therefore not possible rigorously to interpret the thermopower and conductivity data obtained with the SPS-synthesized ensemble. To obtain definitive evidence for the existence and to make precise measurements of the magnitude of "proximity" effects will require the development of a model system that allows one to coat SiC homogeneously with Ni_2Si. Such efforts are currently underway in our laboratory. Clearly, if the interpretation of the data on SPS processed powders based on a qualitative analysis can ultimately be quantitatively verified, a basis for optimization of the "proximity" effect becomes possible. Unequivocal proof of effective entropy transport across semiconductor/metal interfaces would have profound consequences on the future development of high-performance thermoelectrics.

Density functional cluster calculations were employed to explore the concept of configurational entropy in the SiC/Ni_2Si system in a way similar to that used for the nanocarbon system. The intent was to study, in the first instance, the effect of Al substitution on the orbital energy levels of SiC clusters. A dramatic demonstration of the influence of various dopant substitutions on the orbital wave functions is graphically illustrated in Figure 8.13. The creation of individual electronic states resulting from different dopant configurations is made apparent in these panels.

Only three of the panoply of configurations that can be produced by substituting two Si atoms by Al atoms are shown in Figure 8.13. Full optimizations were done at the B3LYP/6-31 G* level. Energy levels were calculated at the PW91/6-31 G* level of theory. Importantly, the HOMO–LUMO energy differences decrease substantially with substitution from the value 4.34 eV without substitution to 0.67, 3.75, and 0.44 eV for the three Al configurations. A large gap of 3.75 eV occurs for one of the Al structures which, it turns out, is not a typical Al dopant configuration because of substantial sp^2 hybridization. These gaps would presumably decrease even more if calculations were carried out on clusters larger than the about 1 nm size used here. This conclusion is based by analogy on results for larger undoped SiC clusters reported by Nayak and coworkers [47] where the gap was found to decrease from 4.25 to 2.5 eV in going from 1 to 2 nm size clusters. The gap decrease upon substitution by the electron deficient Al concomitantly leads to the formation of holes in the valence energy levels similar to our earlier finding from calculations on B substitution in NG clusters. The range of values found for different substitutional sites reflects the different electronic structures of the configurations. Importantly, one concludes that Al substitution in clusters >1 nm would result in HOMO–LUMO gaps small enough so that hole states within thermal energies of the Fermi level are created.

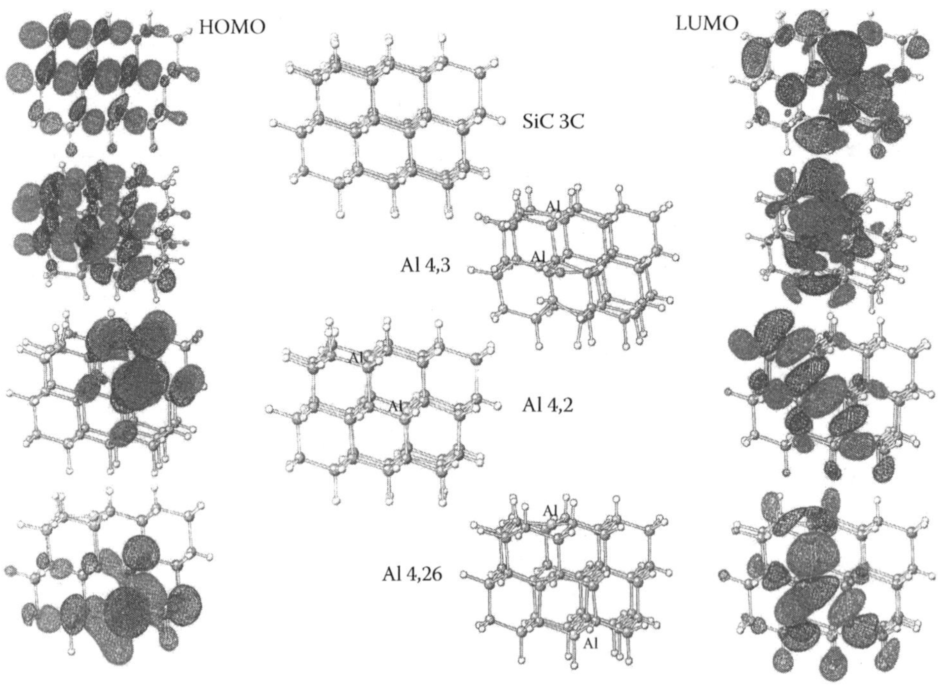

FIGURE 8.13 Density functional calculations of *unsubstituted and three* doubly Al-substituted configurations of SiC clusters showing the wavefunctions of the HOMO (left) and the LUMO (right).

The extremely sensitive dependence of the shapes of orbital wavefunctions shown in Figure 8.13 arise from very small differences in hybridized orbital composition which in turn reflect small differences in orbital energies. This of course is the underlying reason for the large number of distinct quantum states leading to sizeable configurational entropies in doped semiconductors. Many of these states, created by substitutional doping, lie within thermal energies of the Fermi level. Thermally excited carriers can therefore occupy these states and transport their excitation energy encoded as carrier entropy in a thermal gradient. Crossing the interfacial region and entering the metallic phase strongly enhances the kinetics of transport. One supposes that this sequence of events is closely associated with the basic mechanism giving rise to the thermoelectric effect. The change in entropy per added carrier is clearly an important aspect determining the efficiency of the solid-state conversion of heat to electricity.

8.5 Concluding Remarks

A novel approach to the development of high-performance thermoelectric materials is discussed focusing on the transformation of configurational to carrier entropy in a doped semiconductor core with subsequent transport of carrier entropy in a metallic shell. The concept involves the creation of bulk nanoensembles to minimize thermal and maximize electrical conductivity.

Two systems meeting the requirements of nonresource limited, environmentally benign, economically viable and thermally stable materials have been investigated: UNCD cores surrounded by nanographitic shells and silicon carbide cores surrounded by nickel silicide shells. It is believed that "proximity" effects have been observed in the latter system showing that highly entropic carriers due to the presence of doped silicon carbide are transported in metallic Ni_2Si shells. This phenomenon is currently being studied in a new model system to obtain confirmation of its existence and sufficient understanding to lead to procedures for optimization of this potentially important finding.

The experimental work has been supplemented by detailed density functional calculations. These have confirmed, in the two cases mentioned above, the supposition that a multitude of doping configurations in lattices lead to a plethora of unique electronic states frequently situated within thermal energies of the Fermi level [10]. It appears that large configurational entropies generated by the doping of semiconductors can usefully be explored by molecular analog calculations. These are currently being extended to include the coupling of plasmons to longitudinal optical phonons [48]. The elucidation of the details of this coupling, both theoretically and experimentally, as a function of the nature of the carriers as well as carrier doping concentrations, configurations, and temperature, is anticipated to result in a detailed elaboration of the understanding necessary for the development of high-performance thermoelectrics.

Acknowledgments

This work was performed under the auspices of the U.S. Department of Energy, Office of Basic Energy Sciences, under Contract No. DE-AC02-06CH11357.

References

1. M. Xie and D. M. Gruen, Potential impact of $ZT = 4$ thermoelectric materials on solar thermal energy conversion technologies, *J. Phys. Chem. B*, 114, 14339–14342, 2010.
2. G. H. Jonker, The application of combined conductivity and Seebeck-effect plots for the analysis of semiconductor properties, *Philips Res. Rep.*, 23, 131, 1968.
3. N. Mansourian-Hadavi, D. Ko, T. O. Manson, and K. R. Poeppelmeier, Quadruple and quintuple perovskite-layered cuprates ($NdDyBa_{2-x}Sr_xCu_{2+y}Ti_{2-y}O_{11-\delta}$ and $NdDyCaBa_{2-x}Sr_xCu_{2+y}Ti_{3-y}O_{14-\delta}$): Their defect chemistry and electrical properties, *J. Solid State Chem.*, 155, 216, 2000.
4. L. Onsager, Reciprocal relations in irreversible processes. I, *Phys. Rev.*, 37, 405, 1931; Reciprocal relations in irreversible processes. II, 38, 2265, 1931.
5. R. Feynman, Slow electrons in a polar crystal, *Phys. Rev.*, 97, 660, 1955.
6. H. Callen, *Phys. Rev.*, 73, 1349, 1948; *Thermodynamics*, 2nd edition, p. 316; John Wiley and Sons, New York, 1985.
7. P. M. Chaikin and G. Beni, Thermopower in the correlated hopping regime, *Phys. Rev. B*, 13, 647, 1976.
8. D. Emin, Thermoelectric power due to electronic hopping motion, *Phys. Rev. Lett.*, 35, 882, 1975; Effects of charge carriers' interactions on Seebeck coefficients, Chapter 5, in *Thermoelectrics Handbook*, D. M. Rowe (ed.), Taylor & Francis, Boca Raton, FL, 2006.
9. E. Alptekin and M. C. Ozturk, Tuning of nickel silicide Schottky barrier height on p-Type silicon by indium implantation, *IEEE Electron Device Lett.*, 30, 1272, 2009.
10. D. M. Gruen, P. Bruno, and M. Xie, Configurational, electronic entropies and the thermoelectric properties of nanocarbon ensembles, *Appl. Phys. Lett.*, 92, 143118, 2008.
11. D. M. Gruen, P. Bruno, R. Arenal, J. Routbort, D. Singh, and M. Xie, Thermoelectric power factors of nanocarbon ensembles as a function of temperature, *J. Appl. Phys.*, 105, 073710, 2009.
12. C. E. Lowell, Solid solution of boron in graphite, *J. Am. Ceram. Soc.*, 50, 142, 1967.
13. B.-S. Hong, S. J. Ford, and T. O. Mason, Equilibrium electrical property measurements in electroceramics, *Key Eng. Mater.*, 125–126, 163, 1997.
14. A. C. Ferrari and J. Robertson, Resonant Raman spectroscopy of disordered, amorphous, and diamondlike carbon, *Phys. Rev. B*, 64, 075414, 2001.
15. R. Arenal, G. Montagnac, P. Bruno, and D. M. Gruen, Multiwavelength Raman spectroscopy of diamond nanowires present in n-type ultrananocrystalline films, *Phys. Rev. B*, 76, 245316, 2007.
16. Measurements of thermal conductivity and confirmation of previously measured thermopower were made at the High Temperature Laboratory, Oak Ridge, TN (Private communication).

17. P. C. Redfern, D. M. Gruen, and L. A. Curtiss, Effect of boron substitution on the electronic structure of nanographene and its relevance to the thermoelectric transport properties of nanocarbon ensembles, *Chem. Phys. Lett.*, 471, 264, 2009.
18. V. L. Kuznetsov and Y. V. Butenko, Diamond phase transition at nanoscale, in *Ultrananocrystalline Diamond*, O. A. Shenderova and D. M. Gruen (eds), William Andrew Publishing, Norwich, NY. p. 405, 2006.
19. N. Mott, *Conduction in Non-Crystalline Materials*, 2nd edition, Oxford Science, New York, 1993.
20. V. L. Kuznetsov, Y. V. Butenko, A. L. Chuvilin, A. I. Romanenko, and A. V. Okotrub, Electrical resistivity of graphitized ultra-dispersed diamond and onion-like carbon, *Chem. Phys. Lett.*, 336, 397, 2001.
21. O. E. Andersson, B. L. V. Prasad, H. Sato, E. Enoki, Y. Hishiyama, Y. Kaburagi, M. Yoshikawa, and S. Bandow, Structure and electronic properties of graphite nanoparticles, *Phys. Rev. B*, 58, 16387, 1998.
22. A. D. Becke, Density-functional thermochemistry. III. The role of exact exchange, *J. Chem. Phys.*, 98, 5648, 1993.
23. M. J. Frisch et al., *GAUSSIAN03*, 2003.
24. D. Moran, F. Stahl, H. F. Bettinger, H. F. Schaefer, and P. V. Schleyer, Towards graphite: Magnetic properties of large polybenzenoid hydrocarbons, *J. Am. Chem. Soc.*, 125, 6746, 2003.
25. L. A. Curtiss, K. Ragavachari, P. C. Redfern, and J. A. Pople, Assessment of Gaussian-2 and density functional theories for the computation of enthalpies of formation, *J. Chem. Phys.*, 106, 1063, 1997.
26. P. C. Redfern, P. Zapol, M. Sternberg, S. P. Adiga, S. A. Zygmunt, and L. A. Curtiss, Quantum chemical study of mechanisms for oxidative dehydrogenation of propane on vanadium oxide, *J. Phys. Chem. B*, 110, 3015, 2006.
27. T. Frauenheim, G. Seifert, M. Elstner, T. Niehaus, C. Kohler, M. Amkreutz, M. Sternberg, Z. Hajnal, A. Di Carlo, and S. Suhai, Atomistic simulations of complex materials: Ground-state and excited-state properties, *J. Phys. Condens. Matter*, 14, 3015, 2002.
28. T. Frauenheim, G. Seifert, M. Elstner, Z. Hajnal, G. Jungnickel, D. Porezag, S. Suhai, and R. Scholz, A self-consistent charge density-functional based tight-binding method for predictive materials simulations in physics, chemistry and biology, *Phys. Stat. Sol. B*, 217, 41, 2000.
29. D. Porezag, T. Frauenheim, T. Kohler, G. Seifert, and R. Kaschner, Construction of tight-binding-like potentials on the basis of density functional theory: Application to carbon, *Phys. Rev. B*, 51, 12947, 1995.
30. M. Sternberg, P. Zapol and L. A. Curtiss, Carbon dimmers on the diamond (100) surface: Growth and nucleation, *Phys. Rev. B*, 68, 205330, 2003.
31. M. Sternberg, L. A. Curtiss, D. M. Gruen, K. Kedziora, D. A. Horner, P. C. Redfern, and P. Zapol, Carbon ad-dimer defects in carbon nanotubes, *Phys. Rev. Lett.*, 96, 075506, 2006.
32. G. Allinson, R. J. Bushby, J. L. Paillaud, and M. Thorntonpett, Synthesis of a derivative of triangulene; the first non-Kekule polynuclear aromatic, *J. Chem Soc.-Perkin Trans.*, 1, 385, 1995.
33. S. J. Cyvin, J. Brunvoll, and B. N. Cyvin, A periodic table for all-benzenoid hydrocarbons and enumerations of some polyhex isomers, *J. Math. Chem.*, 8, 63, 1991.
34. H. I. Zhang, J. F. Li, B. P. Zhang, K. F. Yao, W. S. Liu, and H. Wang, Electrical and thermal properties of carbon nanotube bulk materials: Experimental studies for the 328–958 K temperature range, *Phys. Rev. B*, 75, 9, 2007.
35. K. McGuire, N. Gothard, P. L. Gai, M. S. Dresselhaus, G. Sumanasekera, and A. M. Rao, Synthesis and Raman characterization of boron-doped single-walled carbon nanotubes, *Carbon*, 43, 219, 2005.
36. D. L. Carroll, P. Redlich, X. Blase, J.-C. Charlier, S. Curran, P. M. Ajayan, S. Roth, and M. Ruhle, Effects of nanodomain formation on the electronic structure of doped carbon nanotubes, *Phys Rev. Lett.*, 81, 2332, 1998.

37. A. Balducci, M. Marinelli, M. E. Morgada, G. Pucella, G. Rodriguez, M. Soccia, and G. Verona-Rinati, CVD-diamond-based thermocouple for high sensitive temperature measurements, *Microsyst. Technol. Micro Nanosyst. Inform. Storage Process Syst.*, 12, 365, 2006.
38. P. C. Redfern, D. M. Gruen, L. Curtiss, P. Bruno, J. Routbort, and D. Singh, Effect of Al and B substitution on the electronic structure and thermoelectric properties of silicon carbide nanoparticles, *Nanosci. Nanotechnol. Lett.*, 3(1), 114–118, 2011.
39. L. K. Frevel, D. R. Petersen, and C. K. Saha, Polytype distribution in silicon carbide, *J. Mater. Sci.*, 27, 1913, 1992.
40. S. I. Nakashima, M. Higashihira, K. Maeda, and H. Tanaka, Raman scattering characterization of polytype in silicon carbide ceramics: Comparison with x-ray diffraction, *J. Am. Ceram. Soc.*, 86, 823, 2003.
41. J. C. Viala, P. Fortier, and J. Bouix, Stable and metastable phase equilibria in the chemical interaction between aluminum and silicon carbide, *J. Mater. Sci.*, 25, 1842, 1990.
42. Y. Tajima and W. D. Kingery, Solid solubility of aluminum and boron in silicon carbide, *Commun. Am. Ceram. Soc.*, 65, C-27, 1982.
43. D. M. Gruen, P. Bruno, M. Mendelsohn, J. Routbort, D. Singh, P. Redfern, and L. Curtiss, Core-shell silicon carbide-transition metal silicide thermoelectric ensembles, *Symposium on Thermoelectric Materials for Solid-State Power Generation and Refrigeration*, Materials Research Society Fall Meeting, Boston, MA, p. 352, 2010.
44. K. Koumoto, M. Shimohigoshi, S. Takeda, and H. Yanagida, *Ceram. Trans.*, 2, 501, 1989.
45. H. Pai, K. Koumoto, S. Takeda, and H. Yanagida, Preparation of SiC hollow particles by gas-phase reaction and thermoelectric properties of sintered bodies, *Solid State Ion.*, 32/33, 669, 1989.
46. C. H. Pai, Thermoelectric properties of boron compound-doped alpha-SiC ceramics, *J. Ceram. Soc. Jpn.*, 112, 88, 2004.
47. X.-H. Peng, S. K. Nayak, A. Alizadeh, K. K. Varanasi, N. H. Bhate, L. B. Rowland, and S. J. K. Kumar, First-principles study of the effects of polytype and size on energy gaps in SiC nanoclusters, *J. Appl. Phys.*, 102, 024304, 2007.
48. B. B. Varga, Coupling of plasmons to polar phonons in degenerate semiconductors, *Phys. Rev.*, 137, A1896, 1965.

9

Ab Initio-Based Band Engineering and Rational Design of Thermoelectric Materials

Jiong Yang
Chinese Academy of Sciences

Xun Shi
Chinese Academy of Sciences

Wenqing Zhang
Chinese Academy of Sciences

Lidong Chen
Chinese Academy of Sciences

Jihui Yang
General Motors R&D Center

9.1 Introduction

The key issue on thermoelectric (TE) research is to search for high-efficiency TE materials, which should have high Seebeck coefficient (S), high electrical conductivity (σ), low thermal conductivity (κ), summation of the electronic part κ_e, and lattice part κ_L, and therefore a reasonably high dimensionless TE figure of merit ZT ($= S^2\sigma T/\kappa$). Identifying materials with a high ZT has proven to be an extremely challenging task since these transport properties correlate with each other: Increasing Seebeck coefficient usually means lowering electrical conductivity, and vice versa; electronic thermal conductivity also relates with electrical conductivity via Wiedemann–Franz law. TE research via experimental efforts alone is usually time consuming, and the results are not predictive due to the lack of understanding on the transport mechanism of energy carriers in materials. *Ab initio* methods can build knowledge from the study of electrons and atomic nuclei, and investigate the electronic structures of many-body solid systems. They have been firmly established over the last decades, and are considered to be an essential and useful way in helping the search for advanced materials. The approach has become more and more popular in TE materials, as well as other materials research.

In this book chapter, we summarize some of our recent work by using density functional *ab initio* method to understand the TE transport mechanism and help the design of novel TE materials. Since its publication in the 1960s,[1,2] *ab initio* method based on density functional theory (DFT) has become a powerful tool to study the electronic structures of materials, especially for their ground states. Numerous papers and textbooks can be referred to obtain basic knowledge of DFT, which will be shortly overviewed in the next section. DFT offers the possibility to explore the materials in variable aspects. Since total energies can be calculated very accurately within DFT, the structural properties, such as lattice parameters, internal coordinates, atomic configurations, and compositions can be predicted. This is important for material design since precise structural information is the basis of understanding other microscopic and macroscopic properties. Based on a reliable crystal structure, electronic structure is also able to be calculated for understanding much functionality of materials. Although the DFT nowadays has some difficulties in dealing with some systems with strongly correlated electrons, band structures, in most cases, are accurate enough to give rational interpretation and prediction of physical properties. In TE, the electrical transport properties of materials can be directly determined by their band structures, including Seebeck coefficients, electrical conductivities, and electronic thermal conductivities, with a few reasonable physical approximations introduced. Because electrical transport properties of a material strongly correlate with its electronic structures, such calculations can also provide useful information about optimization and design of TE materials.

The chapter is organized as follows. First, we briefly review the methodologies and computational details for electrical transport calculations in Section 9.2. Two good overviews of investigating TE materials by *ab initio* approach, one in *Semiconductors and Semimetals* (2001) by D. Singh[3] and another one in the *CRC Handbook* (2006) by J. Tse,[4] were published before. People may read those earlier publications if they have interest. Instead, we focus on how *ab initio* methods, when combined with transport theory, can be applied to solve the real optimization and design issues in three different typical TE material systems, covering their structural, electronic, and electrical transport properties. In-depth physical analysis has been published in separate papers, and this chapter is to primarily summarize how the DFT-based calculations help to rationalize the optimization, design, and selection of novel materials. We conclude the review in Section 9.6, and present an outlook for future work on the theoretical prediction of transport properties and design of novel TE materials.

9.2 Theoretical Methods

9.2.1 Simple Overview of DFT-Based *Ab Initio* Approach

The density functional theory originated from the work of Hohenberg, Kohn, and Sham in 1964 and 1965.[1,2] The foundations of DFT are the two Hohenberg–Kohn (HK) theorems[1]:

Theorem I: For any system of interacting particles in an external potential $V_{ext}(\mathbf{r})$, the potential $V_{ext}(\mathbf{r})$ is determined uniquely, except for a constant, by the ground state particle density $\rho_0(\mathrm{r})$.

Theorem II: A universal functional for the energy $E[\rho]$ in terms of the density $\rho(\mathrm{r})$ can be defined, valid for any external potential $V_{ext}(\mathrm{r})$. For any particular $V_{ext}(\mathrm{r})$, the exact ground state energy of the system is the global minimum value of this functional, and the density that minimizes the functional is the exact ground state density $\rho_0(\mathrm{r})$.

Under HK Theorem I, the total energy functional of a many-electron system is

$$E[\rho] = T[\rho] + E_{e-e}[\rho] + \int d^3\mathbf{r} V_{ext}(\mathbf{r})\rho(\mathbf{r}), \tag{9.1}$$

where $T[\rho]$ is the kinetic energy and $E_{e-e}[\rho]$ is the interaction energy of electrons. But the HK theorems do not provide a way to define these functionals. Kohn and Sham tried to solve the problem in 1965.[2] They introduced an auxiliary independent-particle system composed of Kohn–Sham orbitals $\psi_i(\mathrm{r})$. The sum of these orbitals equals to the particle density of the real systems

$$\rho(\mathbf{r}) = \sum_{i=1}^{N} |\psi_i(\mathbf{r})|^2, \tag{9.2}$$

where N is the number of particles. Let T_s be the independent-particle kinetic energy, and then the Kohn–Sham version of Equation 9.1 can be rewritten as

$$E_{KS}[\rho] = T_s[\rho] + E_{Hatree}[\rho] + \int d^3\mathbf{r}V_{ext}(\mathbf{r})\rho + E_{xc}[\rho]. \tag{9.3}$$

The term $E_{xc}[\rho]$ includes not only the exchange and correlation energy of interacting electrons, but also the difference between T and T_s. The exact form of $E_{xc}[\rho]$ is still unknown. Exploiting the variational principle under the HK Theorem II, and introducing Lagrange multiplier method for handling the conservation of particle number constraint, Schrödinger-like single-particle equations can be obtained:

$$H_{KS}\psi_i = \varepsilon_i\psi(\mathbf{r}), \tag{9.4}$$

with

$$\begin{aligned} H_{KS}(\mathbf{r}) &= -\frac{\hbar^2}{2m_e}\nabla^2 + V_{eff}(\mathbf{r}) \\ V_{eff}(\mathbf{r}) &= V_{Hatree}(\mathbf{r}) + V_{ext}(\mathbf{r}) + V_{xc}(\mathbf{r}). \end{aligned} \tag{9.5}$$

Here

$$V_{Hatree}(\mathbf{r}) = \frac{\delta E_{Hatree}}{\delta\rho(\mathbf{r})} \quad \text{and} \quad V_{xc}(\mathbf{r}) = \frac{\delta E_{xc}}{\delta\rho(\mathbf{r})}. \tag{9.6}$$

Equations 9.4 through 9.6 are the well-known Kohn–Sham equations. Each independent auxiliary particle "feels" the effective potential composed of other N–1 particles. The essence of Kohn and Sham's work was to separate terms that can be expressed explicitly, and leave the unknown parts to the $E_{xc}[\rho]$ term. If the exact form of $E_{xc}[\rho]$ is known, the exact ground state density and energy for the interacting system can thus be obtained. One type of commonly used $E_{xc}[\rho]$ is the local density approximation (LDA),[5] in which the exchange correlation energy is an integral over all space with each point is affected only by the local density,

$$E_{xc}[\rho] = \int d^3\mathbf{r}\rho(\mathbf{r})\varepsilon_{xc}(\rho). \tag{9.7}$$

If the density gradient is also considered, that is, the $\varepsilon_{xc}(\rho)$ in Equation 9.7 is replaced by $\varepsilon_{xc}(\rho, \nabla\rho)$, it leads to the so-called generalized gradient approximation(GGA)-type exchange correlation energy.[5] Efforts to obtain more accurate and efficient $E_{xc}[\rho]$ never stop, including modifications on LDAs or GGAs, orbital-dependent functionals, and hybrid functionals.[5] But in many cases, DFT results from LDA or GGA are good enough for the study of TE materials.

The basic quantities DFT calculations can provide are the total energy differences, as well as their derivative—forces, and the band structures. Total energies are the ground state properties of materials, so the values predicted by LDAs or GGAs will be very accurate, with an error bar within 5%.[5] This ensures DFT to make rational predictions on structural properties, as well as phonon calculations. Band structures, the eigenvalues of Equation 9.4 in the reciprocal space, are in principle related with the excitation energies of particles, that is, not the ground state properties. Figure 9.1 shows a typical band structure, which is the one of Co_4Sb_{12}, with a band gap at Γ point of 0.07 eV. The most well-known shortcoming of DFT calculations is the underestimation of semiconductors' band gaps, and the

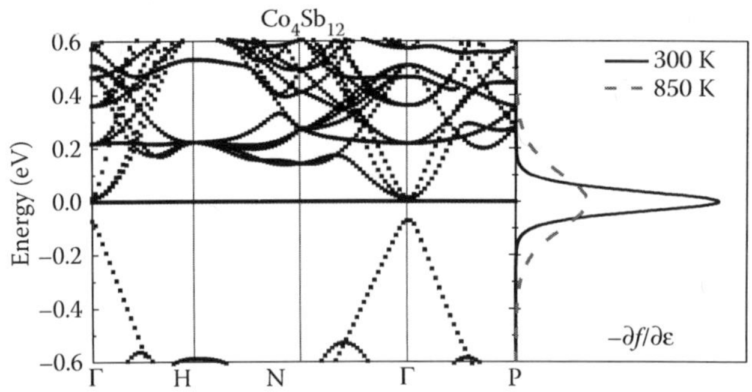

FIGURE 9.1 Band structure of Co_4Sb_{12}. The Fermi level is at the conduction band bottom of Co_4Sb_{12}. The illustrations of $(-\partial f/\partial \varepsilon)$ at 300 K and 850 K are also given.

uncertainties vary among materials.[3] But for TEs, good description of band shape near Fermi levels is preferable over the band gaps themselves, and it can be quite reliable in the DFT calculations. Most of the modern TE materials are doped semiconductors, that is, the Fermi levels are within the valence or conduction bands. So from moderate to working temperature, the contribution from opposite carriers is negligible. That is why the exact values of band gaps are not so important since they merely have influence on the turning temperature in bipolar effect. But band shape, on the other hand, determines the electronic group velocities appeared in the formula of electrical transport properties, which is a crucial parameter in the DFT-based transport simulations. The bridge combining the DFT results and macro electrical transport properties is the Boltzmann transport theory, which works for semiclassical systems with sufficient localized excitations.[6]

9.2.2 Boltzmann Transport Theory

The fundamentals of Boltzmann theory describing transport has been built for decades. In principle, it is based on a classical picture of the transport of energy carriers. But unfortunately, due to its requirement of computational resources, this method was not practicable even in the 1990s when combining with quantum mechanical calculations of electronic structures, unless simplified and analytical models for bands were used.[6,7] In recent years, by the advancement of computer technology, electrical transport calculations composed of Boltzmann transport theory and DFT band structures have been developed, and become more and more useful in evaluating transport properties under certain approximations of energy carriers. A brief summary of Boltzmann transport theory is to be given in this section, and details can be referred to the classical book by Ziman.[6]

The Boltzmann equation describes the change of energy carrier (electrons and holes) distribution function induced by external fields, such as lattice phonon scattering or defect scattering. Under steady-state conditions, the time evolution of distribution function $f(\mathbf{k}, \mathbf{r}, \mathbf{t})$ under collision should satisfy:

$$\dot{\mathbf{k}} \cdot \nabla_{\mathbf{k}} f + \mathbf{v}_{\mathbf{k}} \cdot \nabla_{\mathbf{r}} f = \left(\frac{\partial f}{\partial t} \right)_{\text{collision}} . \tag{9.8}$$

Here **k**, **v**, and **r** are the wave vector, velocity, and spatial distribution of the carriers in a classical picture, respectively.

Owing to the complexity of various carrier scattering mechanisms, it is almost impossible to get an exact solution of the Boltzmann transport equation. For simplicity, relaxation time approximation is usually introduced when the elastic processes dominate the electrical transport. In this approximation, the collision term is assumed to follow

$$\left(\frac{\partial f}{\partial t}\right)_{\text{collision}} = -\frac{f - f_0}{\tau}, \tag{9.9}$$

where f_0 is the equilibrium, that is, Fermi–Dirac distribution function, and τ is the relaxation time. Equation 9.9 describes the process of restoring f_0 from f via various scattering processes. The net effect is to restore f to f_0 exponentially with a relaxation time τ.

Based on the relaxation time approximation, electrical conductivity and Seebeck coefficient tensors of a material can be written as[8]

$$\sigma_{\alpha\beta}(T,\mu) = \frac{1}{\Omega}\int \bar{\sigma}_{\alpha\beta}(\varepsilon)\left[-\frac{\partial f_0(T,\varepsilon,\mu)}{\partial \varepsilon}\right]d\varepsilon, \tag{9.10}$$

$$S_{\alpha\beta}(T,\mu) = \frac{1}{eT\Omega\sigma_{\alpha\beta}(T,\mu)}\int \bar{\sigma}_{\alpha\beta}(\varepsilon)(\varepsilon-\mu)\left[-\frac{\partial f_0(T,\varepsilon,\mu)}{\partial \varepsilon}\right]d\varepsilon, \tag{9.11}$$

where α and β are tensor indices, Ω, μ, and e are the volume of unit cell, Fermi level of carriers, and the electron charge, respectively. The essential part of σ and S is the transport distribution (TD) function tensor defined as

$$\bar{\sigma}_{\alpha\beta}(\varepsilon) = \frac{e^2}{N}\sum_{i,\mathbf{k}} \tau_{i,\mathbf{k}} \cdot v_\alpha \cdot v_\beta \cdot \frac{\delta(\varepsilon - \varepsilon_{i,\mathbf{k}})}{d\varepsilon}, \tag{9.12}$$

where N and i are the number of **k** points sampling and band index, respectively. In TD, v_α is the αth component of the group velocity $\mathbf{v}_{i,\mathbf{k}}$ of carriers, which can be derived directly from the derivative of band structures:

$$\mathbf{v}_{i,\mathbf{k}} = \frac{1}{\hbar}\nabla_{\mathbf{k}}\varepsilon_{i,\mathbf{k}}. \tag{9.13}$$

In practice, the so-called "band-crossing" always exists and may cause problem in calculating group velocity,[9] especially for complex systems with over 50 atoms in our test. In this regard, momentum matrix method can be applied to get accurate group velocity in a large cell:

$$\mathbf{v}_{i,\mathbf{k}} = \frac{1}{m_e}\mathbf{p}_{i,\mathbf{k}} = \frac{1}{m_e}\left\langle \psi_{i,\mathbf{k}} \middle| \breve{p} \middle| \psi_{i,\mathbf{k}} \right\rangle, \tag{9.14}$$

where m_e is the mass of electron. In either way of group velocity evaluation, calculations on large **k** mesh are necessary to get converged transport properties.

In the equation of TD (Equation 9.12), when combining with quantum mechanical band structures of materials, there is only one undetermined parameter τ for transport property investigation. Generally speaking, the electronic relaxation time is related to many different types of scattering processes experienced by carriers in materials. Let $P(\mathbf{k}, \mathbf{k}')$ define the scattering probability from an occupied state **k** to an unoccupied state **k**′, and a nearly elastic collision is ensured, then the relaxation time τ under external electric field **F** can be expressed as[10,11]

$$\frac{1}{\tau_{\mathbf{k}}} = \frac{V}{(2\pi)^3}\int P(\mathbf{k},\mathbf{k}')\left[1 - \frac{\mathbf{v}_{\mathbf{k}'} \cdot \mathbf{F}}{\mathbf{v}_{\mathbf{k}} \cdot \mathbf{F}}\right]d^3\mathbf{k}'. \tag{9.15}$$

There are many forms of carrier scattering mechanisms existing in solids, including acoustic phonon scattering ($P_{acoustic}$), piezoelectric scattering ($P_{piezoelectric}$), nonpolar and polar optical phonon scattering ($P_{nonpolar\text{-}optical}$ and $P_{polar\text{-}optical}$), impurity scattering ($P_{impurity}$), small and large defect scattering, and so on. The total scattering probability contains the contribution from scattering mechanisms, that is,

$$P(\mathbf{k},\mathbf{k}') = P_{acoustic} + P_{piezoelectric} + P_{nonpolar\text{-}optical} + P_{polar\text{-}optical} + P_{impurity} + \cdots \tag{9.16}$$

Consequently, the scattering rate, or the inverse total carrier relaxation time is

$$\frac{1}{\tau} = \frac{1}{\tau_{acoustic}} + \frac{1}{\tau_{piezoelectric}} + \frac{1}{\tau_{nonpolar\text{-}optical}} + \frac{1}{\tau_{polar\text{-}optical}} + \frac{1}{\tau_{impurity}} + \cdots \tag{9.17}$$

Qualitative discussions about the fundamentals of the aforementioned scattering processes have been given for decades, including formulation of them with physically reasonable but empirical parameters. However, to evaluate τ in a quantitative way for systems containing many scattering processes is a very challenging work.

In principle, the relaxation time τ varies with $\varepsilon_{i,\mathbf{k}}$. For many cases, however, it is weakly dependent on band energy, and can be simplified as a constant, which describes the net scattering in an effective way.[12] Thus, the relaxation time is eliminated in Equations 9.10 and 9.11, and Seebeck coefficient, as well as Lorentz coefficient, are independent with the relaxation time. Although the constant relaxation time approximation (CRTA) seems simple, it gives surprisingly good results. For TE materials without extremely sharp band structure features,[3] CRTA can provide Seebeck coefficients very close to experimental values; and for the study of series of TE compounds with the same crystal structures and similar chemical compositions, the relative magnitude of σ/τ and $S^2\sigma/\tau$ and their temperature dependence are also useful for experiments. Discussions on the limitation and expansion of the approach will also be outlined in the later parts of this chapter.

For electrical transport property evaluation from theory, it is a great step to get the exact or relative values of Seebeck coefficients and electrical conductivities, but not all. Although electrical transport calculations can provide direct results like experiments, they are, in some manner, another version of band structures. Any variation on band structures will shed light on electrical transport parameters. Meanwhile, electronic structures are also necessary for deep understanding of TE materials themselves. These include analysis on chemical bonding, band characters, and even connections with the basic properties of elements or element clusters. One will have trouble to obtain systematic and predictable knowledge of TE materials with electrical transport calculations only. Furthermore, as for the good of experiments, we cannot get full predictions of electrical transport since the uncertainty of relaxation time. But on the other hand, the relations of Seebeck coefficients with compositions, chemical potentials or carrier concentrations, and temperature are obtainable. To get such information, analyses on both band structures and the electrical transport calculations are required.

9.2.3 Relations between Band Structures and Electrical Transport Properties

What kind of electronic structures can lead to good electrical performance is the key concern from theoretical point of view, and has been extensively discussed. General discussions have been given early, even without a universal conclusion.[3,13,14] Several specific aspects related to electrical transport property optimization will be discussed in this chapter, based on our recent *ab initio* transport calculations for a few specific systems. Before that, a few critical quantities are described for further understanding.

9.2.3.1 Optimal Carrier Concentrations Corresponding to Highest Power Factors

Most of the TE materials especially those traditional compounds, are narrow-band semiconductors with some doping. As seen from Equation 9.11, electrical conductivity appears in the denominator of

Seebeck coefficient when the concentration of conducting electrons (or holes) increases, electrical conductivity increases, while Seebeck coefficient decreases. The compromise of the two opposite dependences determines the peaks of power factors, which often locates close to the band edges of semiconductors. Earlier work in simple effective mass approximation for model semiconductors claimed that the optimal carrier concentrations should be around 10^{18}–10^{19} cm^{-3}.[13] But for TE materials used in high temperature, the values could reach 10^{20} cm^{-3} or even higher. These values can be easily obtained in *ab initio* calculations under a reasonable definition of carrier concentrations, that is, "electrons in conduction bands, or holes in valence bands" used in Section 9.4.

$$p = \frac{2}{\Omega} \iint_{BZ,VB} [1 - f_0(T,\varepsilon,\mu)]d\varepsilon$$

or (9.18)

$$n = \frac{2}{\Omega} \iint_{BZ,CB} f_0(T,\varepsilon,\mu)d\varepsilon$$

The estimated optimal carrier concentrations (or doping levels) are comparable to experimental results, and they can offer useful guidance for experimental optimizations of power factors. One should also be cautious that optimal doping levels are temperature dependent. If precise values are required, the electrical transport calculations should be carried out under the working temperature. Otherwise, the doping levels will be misestimated.

9.2.3.2 Density of States and TD

As presented in many references, rapidly increasing density of states (DOS) near the Fermi level, or large effective mass in a single band model, is important to get high Seebeck coefficients. From this point of view, the so-called resonant states, that is, sharply enhanced electronic states, are favorable to enhance electrical transport performances because they increase the asymmetry of electronic states around Fermi levels. However, Equations 9.10 through 9.12 show that DOS is not the only factor that affects the electrical transport properties. The TD, composed of DOS and square of group velocities, is the centerpiece of all transport coefficients. Group velocity, or the band dispersion, is also critical, especially in drawing comparison among materials with different electronic states. Generally speaking, the group velocities of *s*, *p* states are larger than those of *d* states, and localized *f* states usually possess extremely flat bands, and thus low group velocities. This implies that TE transport properties may not benefit from the very high DOS of *f* states as believed before.[14] Moreover, to get same TD slope for good electrical properties, larger group velocities mean smaller band edge DOS, which results in smaller optimal carrier concentrations. Section 9.4 will give an example of this.

9.2.3.3 Band Gaps

As might be known to all, the band gap of a TE semiconductor should be large enough to prevent bipolar effect, but at the same time it is also small enough to allow thermal excitation of carriers, especially at an elevated operating temperature T_{OP}. The effect of band gaps on transport strongly relates with the smearing range of the function $(-\partial f/\partial\varepsilon)$, which appears in the formula of electrical conductivity (Equation 9.10) and Seebeck coefficient (Equation 9.11). Assume the Fermi level lays at band edge, such as the conduction band minimum (CBM) of $CoSb_3$ in Figure 9.1, the value of $(-\partial f/\partial\varepsilon)$ fades to less than 0.1% when $|\varepsilon - \mu| > 0.21$ eV at 300 K. That means if we take 0.1% of the peak value as the criterion for minor carrier not affecting the electrical transport, the band gap should be larger than 0.21 eV when T_{OP} is at room temperature. At elevated temperature such as 850 K, the value becomes 0.61 eV. The simply estimated band gap values are not far from Mahan's "$10k_BT_{OP}$" value,[14] where k_B is the Boltzmann constant, and can be considered as one of the band gap requirements for good TE materials. Nevertheless, if TDs are quite different in the energy windows close to the CBM and valence band maximum (VBM), the band gap values which can prevent bipolar effect could be different from the above estimation. In $CoSb_3$, since TD close to CBM is orders of magnitude larger than that close to VBM, bipolar effect is hardly observed in *n*-type filled skutterudites at moderate filling fractions, although the calculated band gap of $CoSb_3$ is not very large (see Figure 9.1). However, it may be different for *p*-type skutterudites.

In the following sections, we discuss the applications of DFT-based methods to the rational design and performance optimization of three categories of TE materials: transition-metal-substituted type-I clathrates (Section 9.3), half-Heusler (HH) compounds (Section 9.4), and *n*-type filled skutterudites (Section 9.5). Section 9.3 demonstrates the usage of the approach to determine the structural properties of novel clathrates, and help the design of novel materials along with experiments. Section 9.4 is a typical theoretical search of materials with high electrical performance among series of semiconductors. For the work on the *n*-type filled skutterudites in Section 9.5, our results demonstrate the power of DFT calculations plus Boltzmann transport theory in understanding the transport mechanism of heavily doped semiconductors, in which our results predict the optimal range of carrier concentration corresponding to the best electrical properties and has been considered to be the critical step to realize *ZT* enhancement in $CoSb_3$-based materials.

Electronic structures are calculated primarily by using the projected augmented wave (PAW) method[15] implemented in Vienna *ab initio* simulation package (VASP).[16] Perdew–Burke–Ernzerhof (PBE) form GGA is adopted for the evaluation of exchange-correlation functional. For systems of small unit cell, such as HH compounds in Section 9.4, group velocity in band derivative method (Equation 9.13) is accurate enough.[8] As for complex compounds with large supercells in Sections 9.3 and 9.5, group velocity in the momentum matrix method (Equation 9.14) is adopted.

9.3 Type-I $Ba_8M_xGe_{46-x}$ Clathrates with Transition Metal Cross Substitution

Type-I clathrates have recently been identified as prospective TE materials partly due to their low thermal conductivities.[17,18] These compounds crystallize in the cubic space group $Pm\bar{3}n$ (No. 221). There are two different types of polyhedra in one unit cell, dodecahedral and tetrakaidecahedral cages. Forty-six framework sites can be categorized into three symmetrical positions, 6c, 16i, and 24k. The cages are filled by eight foreign metals. The most extensively studied type-I clathrates are $A_8M_{16}X_{30}$ (A = Sr, Ba, Eu; M = Al, Ga, In; X = Si, Ge, Sn). Among the series of $A_8M_{16}X_{30}$, $Ba(Sr)_8Ga_{16}Ge_{30}$ has the highest TE properties with *ZT* values being around 1.0 at a temperature higher than 1000 K.[19,20] Transition metal (TM) substitution for the framework atoms may be one of the solutions for further improvement of TE performance, since the *d* orbitals of TMs provide differences on the electronic structures near Fermi levels. In this section, the compositions, electronic structures, and electrical transport properties of the type-I clathrates with TM cross substitutions are studied theoretically. Making clear TMs' influence on clathrates, and the way toward higher *ZT*s are the main purposes of this section.

It has been known that the so-called Zintl–Klemm (ZK) or charge balance rule[17,21] is a very important concept when studying the compositions and structures of type-I clathrates. Therefore, we started from clarifying if the ZK rule is still effective for the clathrate structures with different TMs. The ZK concept is in principle a charge balance rule and implies that each framework atom should form a close-shell valence electron orbital, and the whole compound becomes semiconducting. Group IVA element Si, Ge, or Sn, with their four *s*, *p* valence electrons, can form semiconducting Si_{46}, Ge_{46}, and Sn_{46}, in which each constituent atom has four tetrahedral sp^3 bonds. When Ba atoms are filled into the crystal voids and donate their outmost 6*s* electrons to the conduction bands of the clathrates, group IVA framework atoms have to be partially substituted by atoms with less valence electrons in order to balance the extra electrons of Ba. We define Δn as the number of electron deficiency of the substituting atoms such as TMs comparing to group IVA host elements. For group IIIA Ga, the Δn is apparently −1. But for TMs, the inner *d* states need to be fully filled first, so Δn for Zn, Cu, and Ni (Pt) are −2, −3, and −4, respectively. Specially, Δn for a defect γ (vacancy of the framework atoms) is also −4. $A_8M_{16}X_{30}$ compounds mentioned above obey the ZK rule, and so does the reported $Ba_8Zn_8Ge_{38}$.[22] However, a few references[23,24] also showed that the compositions of many TM-containing clathrates are under debate. This is an interesting as well as important issue to be clarified since the composition is the foundation of other properties. Another challenge in type-I clathrate design is the strategy to tune band gaps, which makes the *ZT* peaks fall in the desired operating temperature range.

Site preferences of TM atoms need to be discussed specifically before a theoretical study of structure, composition, and chemical bonding. Blake et al. found that Ga atoms in $Ba_8Ga_{16}Ge_{30}$ were not randomly distributed, and the stable configurations usually satisfy the following rules: (1) Ga prefers 6c and 24k sites; (2) no Ga–Ga direct bond.[25] These two rules were roughly consistent with experiments.[25–27] For each type-I clathrates of TMs in our study, after comparing the total energies of different configurations, we find that the two rules from Blake are also valid for TM substitution, implying that TM atoms prefer 6c sites, and that no TM–TM direct bond is found in the most stable configuration. The most stable configuration is adopted in the composition research below.

Comparison between the energetics of $Ba_8Ni_4Ge_{42}$ and $Ba_8Ni_3Ge_{43}$ is taken as an example to discuss the methodology of determining the stable composition. Considering the following reactions with the same reactants:

$$\begin{aligned} &8Ba + 3Ni + 43Ge \longrightarrow Ba_8Ni_3Ge_{43} \\ &8Ba + 3Ni + 43Ge \longrightarrow \frac{3}{4}Ba_8Ni_4Ge_{42} + 2BaGe_2 + \frac{15}{2}Ge. \end{aligned} \tag{9.19}$$

The reaction formula of the second equation with secondary phase $BaGe_2$ is confirmed by experiments.[23,28] The structure of $BaGe_2$ is from the Inorganic Crystal Structural Database (ICSD),[29] with fully relaxed lattice parameters and internal atomic coordinates. In Equation 9.19, the enthalpy change of second reaction is 0.48 eV lower than the first one. Since the reaction corresponding to larger enthalpy decrease is easier to occur, our result indicates that $Ba_8Ni_4Ge_{42}$ is always the final product in experiments. Even by using different secondary phases to balance the reaction between clathrates, such as NiGe + $BaGe_2$, the relative stabilities among the two clathrates with Ni_3 and Ni_4 remain unchanged. For each $Ba_8M_xGe_{46-x}$ (M = γ, Ni, Pt, Cu, Zn, and Ga), we compare the stabilities of the clathrates with different x values to find the most stable systems, using the same approach.

Figure 9.2 shows the most stable compositions for $Ba_8M_xGe_{46-x}$(M = γ, Ni, Pt, Cu, Zn, and Ga). Our results prove that, in Ge-based type-I clathrates, ZK rule should still be followed, nearly perfectly, for

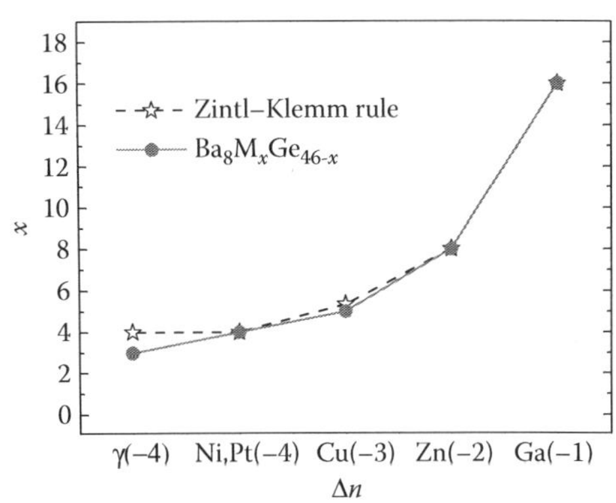

FIGURE 9.2 Calculated stable compositions for $Ba_8M_xGe_{46-x}$ (M = γ, Ni, Pt, Cu, Zn, and Ga). Zintl–Klemm curve is also shown for comparison.

ternary compounds. For Cu-substituted clathrate, the exact Zintl composition is $Ba_8Cu_{5.33}Ge_{40.67}$. Due to the limitation of supercell size in our simulations, we can only get $Ba_8Cu_5Ge_{41}$ for the composition of $Ba_8Cu_xGe_{46-x}$, very close to the ZK rule. Compositions of other ternary Ge-based type-I clathrates such as $Ba_8Ga_{16}Ge_{30}$, $Ba_8Zn_8Ge_{38}$, have been studied over the years. Most of them follow or nearly follow the ZK rule, which is consistent with our results. Ternary type-I clathrates following the ZK rule indicate that they could be semiconductors and used in TE applications. But for binary $Ba_8\gamma_xGe_{46-x}$, quite a few experimental works showed that the most stable composition is $Ba_8\gamma_3Ge_{43}$, deviating from ZK rule.[17,30] Unlike ternary type-I clathrates, γ in framework means bond breaking, thus causes extra energy increase, which competes with the ZK-required charge balance rule. It should also be noted that the compositions in Figure 9.2 are for the most stable configurations in thermodynamics. Since the sp^3 covalent bonds in type-I clathrates are hard to be broken if formed, it is also likely to get metastable phases which may violate the ZK rule.

It has been speculated that transition-metal substitution for framework atoms in the clathrates, by introducing the relatively localized *d* orbitals, could provide an approach of modifying the framework bonds, tuning band gaps, enhancing band edge DOS and thus Seebeck coefficients. We thus carefully investigated the electronic structures of these ternary type-I clathrates. The DOS of CBs of $Ba_8Ni_4Ge_{42}$ and $Ba_8Ga_{16}Ge_{30}$ are plotted in Figure 9.3a. Zero point is set to be the bottom of CB. The relatively more localized *d* states from Ni than *s*, *p* states from Ga or Ge enhance the band edge DOS in $BaNi_4Ge_{42}$, as plotted in Figure 9.3a. This is indeed favorable for the enhancement of Seebeck coefficients. Although the TM-substituted type-I clathrates follow ZK rule, the *d* states of TMs are close to the Fermi levels, especially for Ni and Cu, and make the valence bands and even conduction bands bear *d* state characters. The band gap variation is also affected by these *d* states of TMs. Figure 9.3b shows the band gaps of all the ternary $Ba_8M_xGe_{46-x}$. Generally, band gaps of the TM-substituted type-I clathrates are lower than the gap 0.54 eV of $Ba_8Ga_{16}Ge_{30}$. The values are 0.3 eV for Zn-, 0.1 eV for Cu-, and close to zero for Ni- (and Pd-, Pt-) substitution. TM substitutions shrink the band gaps, and this feature fades from Ni to

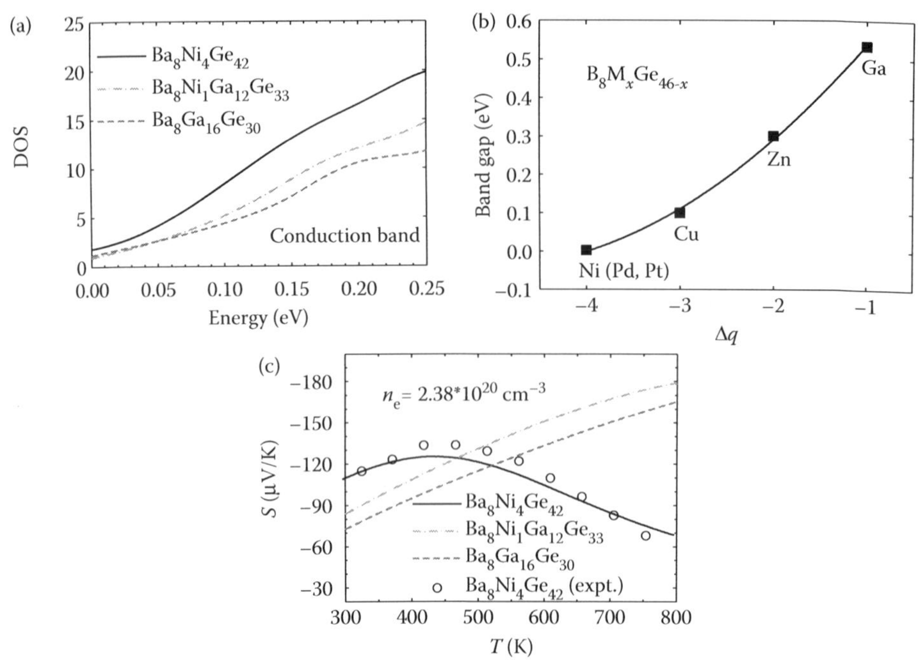

FIGURE 9.3 (a) DOS of conduction bands; (b) band gaps in $Ba_8M_xGe_{46-x}$; (c) temperature dependence of Seebeck coefficients at a given electron concentration of 2.38×10^{20} cm^{-3}, close to the measured one for $Ba_8Ni_4Ge_{42}$ sample.

Zn, which is proportional with the contents of *d* states at band edges. More details will be given in our future work. The enhanced band edge DOS and decreased band gaps are the two main consequences from the TM substitutions, and the nearly zero band gaps of the Ni-, Pd-, Pt-, and Cu-substituted clathrates make them hard to be efficient TE materials at an elevated temperature.

The Seebeck coefficients and their temperature dependence reflect the contribution from the two conflicting factors—the increased band edge DOS and reduced band gaps (Figure 9.3c). By setting the theoretical electron concentrations to the experimental data of a $Ba_8Ni_4Ge_{42}$ sample, ~2.38×10^{20} cm^{-3}, estimated by Equation 9.18, the room-temperature Seebeck coefficient of $Ba_8Ni_4Ge_{42}$ is obviously higher than that of $Ba_8Ga_{16}Ge_{30}$; but the temperature corresponding to the peak of Seebeck coefficient is also shifted much lower due to the small gap. Seebeck coefficients going reversely with temperature above 450 K for $Ba_8Ni_4Ge_{42}$ comes from bipolar effect. Therefore, even with the Seebeck enhancement, reasonably large band gaps for TM-substituted type-I clathrates are still needed to realize improvement on their TE performance.

Solid solutions between clathrates, especially between the low-gap systems and the large-gap $Ba_8Ga_{16}Ge_{30}$, may help to optimize the properties in the TM-substituted systems. We systematically calculated the band structures of two solid solution systems: $Ba_8Ni_4Ge_{42} + Ba_8Ga_{16}Ge_{30}$ and $Ba_8Zn_8Ge_{38} + Ba_8Ga_{16}Ge_{30}$. Figure 9.4a plots the composition dependence of band gaps. These band gaps of the solid solutions follow roughly a linear way, and become larger with higher contents of $Ba_8Ga_{16}Ge_{30}$, implying that the band gaps can be adjusted by changing the ratio of parent compounds in type-I clathrates. Furthermore, the solid solutions also have tunable band edge DOS (Figure 9.3a) and enhanced Seebeck coefficients (Figure 9.3c). To confirm our theoretical results, we also conducted experiments for solid solutions.[28] Our experimental work confirmed that the compositions of solid solutions also follow the ZK rule. In other words, the multiple substituted type-I clathrates possess

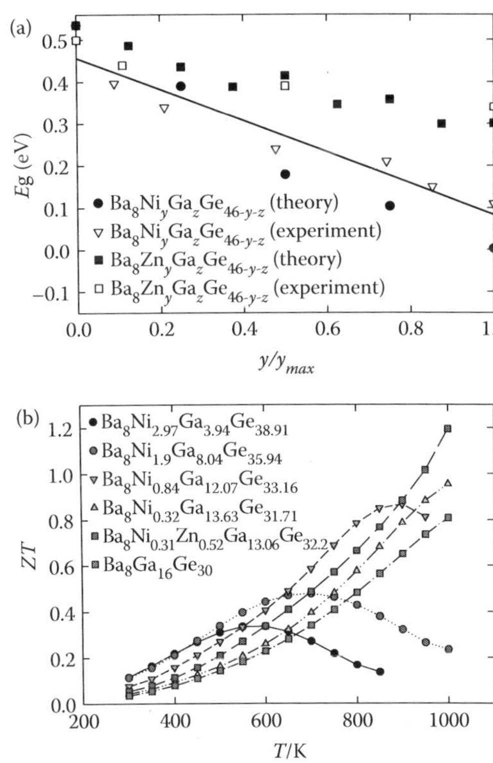

FIGURE 9.4 (a) Band gap variations of type-I clathrate solid solutions; (b) experimental *ZT* values of multiple-TM substituted type-I clathrates.

predictable compositions, favorable to material design. The experimental band gaps of the solid solutions, estimated by $E_g = 2eS_{max}T_{max}$ with S_{max} being the maximum Seebeck coefficient and T_{max} the corresponding temperature, are in fairly good agreement with our calculations (Figure 9.4a). One can get desired band gaps with careful composition control. Our experiments also showed that the solid solutions own reduced lattice thermal conductivities, together with the tunable band gaps and enhanced Seebeck coefficients, the clathrates with TM cross substitutions show improved *ZT* over the whole temperature range and the highest one reaches 1.2 at 1000 K for a triple substituted system (Figure 9.4b).

By studying the structural energetics and electronic structures of the type-I clathrates with TM substitutions for framework atoms, we found that all these compounds follow the Zintl–Klemm rule in their equilibrium states. This result helps the composition selection especially in the process of materials synthesis for multiple TM substitution. TMs also introduce localized *d* states into the band edge, and enhance DOS and thus Seebeck coefficients. On the other hand, single TM substitution also greatly reduces the band gaps, leading to the bipolar effect at relatively low temperature, and limits the performance enhancement. Theoretically, we showed that solid solutions of two or more clathrates, especially the one between single TM-substituted system and $Ba_8Ga_{16}Ge_{30}$, could be employed to tune band gaps to the expected values. Experimentally, tunable band gaps, stable compositions following ZK rule, and enhanced Seebeck coefficients at relatively low temperature are all realized, and *ZT* values are improved over the whole temperature range, especially in the intermediate temperatures. In principle, the cross substitution can be used not only in Ba-filled type-I clathrates, but also Sr- or Eu-filled clathrates, too. That might be a promising way to explore new high-performance type-I clathrates in the future. Furthermore, from our calculations, the TM substitutions have more influence on valence bands than on conduction bands since those *d* states are nearly fully filled. This implies that *p*-type type-I clathrates may also benefit from TM substitutions.

9.4 Half-Heusler Intermetallic Compounds

Ternary HH compounds of the MgAgAs structure type exhibit a wide variety of interesting physical properties. When the total valence electron count (VEC) in a primitive HH unit cell is 18, all the constituent atoms complete their electronic shells, and the HH becomes a semiconductor. Over the years, there have been a few semiconducting HHs investigated as potential TE materials, with the focus especially on MNiSn (M = Ti, Zr, and Hf) and related alloys.[31–35] Up to now, the best reliable maximum TE figure of merit ZT_{max} for MNiSn-related compounds is about 0.7,[32] These *ZT* values are close to those of state-of-the-art TE materials such as PbTe- and Bi_2Te_3-based materials.[36] Besides the extensively studied MNiSn-base alloys, only a few other HHs have been studied for TE purposes.[37–40] This is only a very small fraction of compounds that crystallize in HH structure. In fact there are more than 100 HHs that could be found in the *Pearson's Handbook*[41] and ICSD database.[29] Even using the criterion of VEC = 18 for semiconductors, there are still more than 30 HHs left. Using the DFT-based transport calculations discussed above, the electrical transport properties could be evaluated in order to find new HHs with promising properties.

Figure 9.5 shows the crystal structure of a typical HH compound. There are three atomic positions, A, B, and X in one primitive unit cell. For TE HHs, A is usually an early transition (IIIB or IVB) or rare-earth metal, X is a main group element (IVA or VA), and B is normally a TM between A and X in the periodic table. The structural information of HH compounds in this work is mainly taken from the ICSD database which has more than 100 HH compounds. We used the following rules to select HHs for further detailed investigation.

1. VEC equals to 18. Chemical bonding analysis showed that only HHs with 18 valence electrons may be semiconductors, which are suitable for TE purpose.
2. The HH compounds with lanthanides, except for La, are not included. Lanthanides contained HHs usually have no gaps, behave as metals, and therefore possess poor TE properties. Besides, unfilled *f* electrons usually need to be treated by *ab initio* method with strong-correlation effects included.

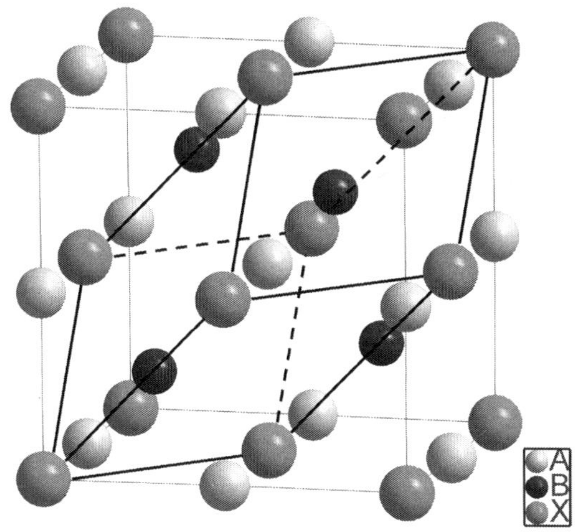

FIGURE 9.5 The crystallographic and primitive unit cell of a typical HH compound ABX.

Applying these two rules leads to those HH compounds listed in Table 9.1.

Electronic structures of the HHs are carefully calculated. All the HHs in Table 9.1 have TMs on their A, B sites, so *d* states are expected to become dominant at VBM and CBM. A-site elements vary from IIIB to VB group, while B-site elements are from Fe subgroup to IB group. TMs with different numbers of *d* electrons possess different atomic energy levels at CBM and VBM, and consequently, determine the band gaps. HHs in Table 9.1 are ordered by the groups of their A and B site elements. Band gaps for (IVB,

TABLE 9.1 TE Transport Properties of VEC = 18 HHs

Type	**IVB-(Ni, Pd)**					**IVB-Pt**				
Name	**TiNiSn**	**ZrNiSn**	**HfNiSn**	**ZrPdSn**	**HfPdSn**	**TiPtSn**	**ZrPtSn**	**HfPtSn**		
Gap(eV)	0.451	0.515	0.396	0.495	0.401	0.830	1.020	0.936		
Max PF(p)	25.13	21.93	22.03	19.62	19.83	24.91	21.11	21.46		
Max PF(n)	14.45	14.43	14.44	12.52	12.78	18.99	12.36	12.34		
p doping	−0.028	−0.019	−0.014	−0.017	−0.013	−0.033	−0.019	−0.014		
n doping	+0.013	+0.009	+0.009	+0.009	+0.008	+0.118	+0.014	+0.013		
Type	**IIIB-(Ni, Pd)**					**IIIB-Pt**		**IIIB-IB**		
Name	**ScNiSb**	**ScNiBi**	**YNiSb**	**YNiBi**	**LaPdBi**	**YPtSb**	**ScPtSb**	**YAuPb**	**ScAuSn**	
Gap(eV)	0.281	0.191	0.311	0.219	0.310	0.411	0.685	0.000	0.145	
Max PF(p)	15.82	14.54	13.18	11.98	8.54	12.56	15.90	10.52	18.53	
Max PF(n)	20.64	21.63	22.83	23.44	25.14	21.02	18.58	12.96	13.50	
p doping	−0.006	−0.006	−0.005	−0.004	−0.003	−0.004	−0.007	−0.016	−0.011	
n doping	+0.016	+0.016	+0.017	+0.015	+0.013	+0.034	+0.026	+0.016	+0.009	
Type	**IVB-(Co, Rh)**					**VB-(Co, Rh)**		**VB-(Fe, Ru)**		
Name	**TiCoSb**	**ZrCoBi**	**TiRhSb**	**ZrRhSb**	**HfRhSb**	**NbCoSn**	**NbRhSn**	**VFeSb**	**NbFeSb**	**TaRuSb**
Gap(eV)	1.05	0.992	0.773	1.215	1.145	0.987	0.665	0.324	0.529	0.655
Max PF(p)	65.95	52.88	21.92	18.51	18.42	67.10	56.12	38.03	35.49	28.79
Max PF(n)	22.57	23.58	10.86	22.51	22.26	24.12	7.85	9.35	7.59	7.08
p doping	−0.157	−0.152	−0.021	−0.012	−0.010	−0.084	−0.202	−0.052	−0.040	−0.025
n doping	+0.195	+0.038	+0.008	+0.121	+0.037	+0.045	+0.002	+0.007	+0.003	+0.003

Note: The maximum power factor is in unit of $\tau \times 10^{14}$ W cm^{-1} K^{-2} s^{-1} and the corresponding p- or n-type doping levels are in the unit of (electron/unit cell).

VB)-(Co, Rh)-containing HHs are about 0.8–1.2 eV; about 0.2–0.3 eV for IIIB-(Ni, Pd); about 0.4–0.6 eV for IVB-(Ni,Pd); about 0.3–0.6 eV for VB-(Fe, Ru); and about 0–0.2 eV for IIIB–IB. Usually, cross substitution between elements in the same column at the B sites does not affect the gap value substantially, except for Pt substituting for Ni or Pd (i.e., from ZrNiSn, ZrPdSn to ZrPtSn), which gives an enlarged band gap by several tenths of electron volt. Elements at the X sites could be approximately considered as accommodators for the donated valence electrons from atoms at A and B positions, leading to the formation of band gaps, and they themselves do not substantially affect the gap values.

As G. Slack discussed, that the Pauling electronegativity differences between the constitution elements had clear correlation with band gaps, especially for materials with similar structures in binary compounds. The fundamental correlation did improve the understanding of connections between elements and the macro properties of materials, such as weighted mobility,[13] and help the selection of new materials. Relationship between the effective electronegativity difference and the band gap is also studied in HH compounds (Figure 9.6a). Since A-site element is usually electropositive, while the electronegativities of B- and X-site element are similar, the formula ABX can be rewritten as $A^+[BX]^-$, and thus the effective electronegativity difference Δx of ABX is calculated by $|(x_B + x_X)/2 - x_A|$. Note that the band gap values could be roughly divided into three groups based on different A–B combinations as labeled in the inset of Figure 9.6a. HHs in the same group have similar band gaps with small variations, as mentioned above. Nevertheless, unlike binary compounds,[13] the band gaps of HHs show no clear trend with electronegativity difference. Even by taking main group elements Sn, Sb, or Bi as electron acceptor and thus defining $\Delta x = |(x_A + x_B)/2 - x_X|$, there is still no apparent relation between band gap and Δx. Figure 9.6b shows the electron density difference contour of typical HH compound ZrNiSn. Overall speaking, bonding in HHs possesses mixed characteristics, that is, partially covalent and partially metallic, and weak charge density accumulation is found between Zr and Ni since these two neighboring elements have relatively large electronegativity difference. The picture is quite different from the relatively simple chemical bonding in binary semiconductors. It would, therefore, be difficult to figure out simple rules to describe the dependence of band gap and transport property on the intrinsic properties of the constituent elements in HHs.

Semiconducting HHs need some levels of doping to reach their *p*-type or *n*-type power factor peaks. Estimating the optimal doping levels and related properties of a given material is very important for the experimental optimization of the systems. Due to the wide range of dopants and their small amounts in the whole systems, it is very difficult to calculate doped HHs and compare their electrical properties directly. Instead, we adopt the so-called rigid band approximation (RBA). It is a reasonably good

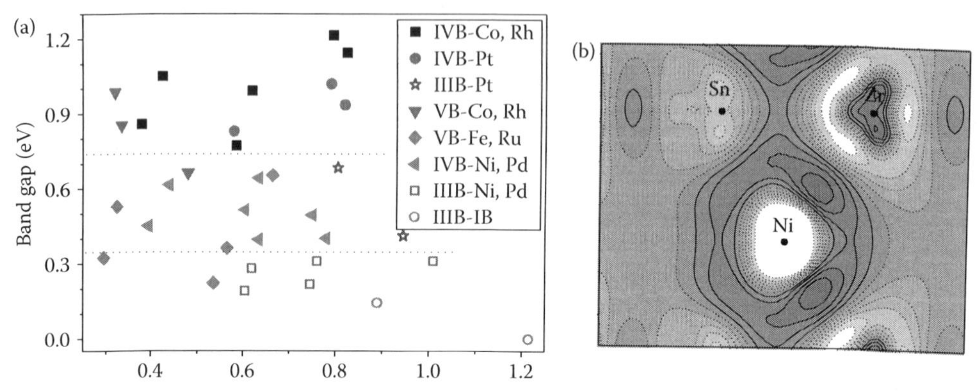

FIGURE 9.6 (a) Calculated band gap vs. effective electronegativity difference Δx in half-Heusler compound ABX. x_A, x_B, and x_X represents the electronegativity of elements A, B, and X, respectively. (b) Contour plot of the charge density difference of a typical half-Heusler compound ZrNiSn.

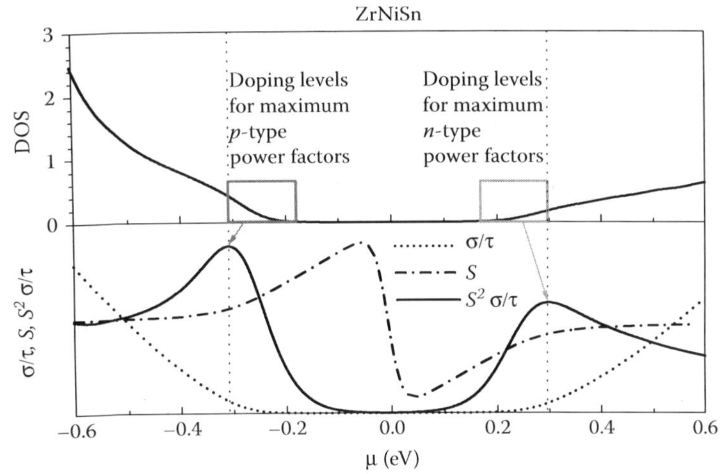

FIGURE 9.7 Procedure of evaluating *p*- or *n*-type optimal doping levels.

approximation if the doping level is not very high, and has been widely used for theoretical study of TE materials.[9,42] The optimal doping level is defined as the integral of DOS from band edge to the energy-level position corresponding to the maximum or the peak of power factor (Figure 9.7), as discussed in Section 9.2. For example, the optimal *n*-type doping level for ZrNiSn is estimated to be +0.009 e/u.c. This suggests that ZrNiSn reaches its maximum *n*-type power factor after 0.009 electrons are added per unit cell, which could be achieved by substituting 0.9% Sb for Sn. In Table 9.1, optimal doping levels and related power factors for every HH are listed. But before further analysis, we have to check the agreements between our calculated data and existing experimental ones, and the results are listed in Table 9.2. We give the dopants and the corresponding doping levels for HHs, and for comparison, the optimal doping levels and the corresponding Seebeck coefficients are calculated. Given the above approximations used, and the uncertainties in experiment data, the agreement is reasonably satisfactory and acceptable.

Table 9.1 contains the information about how to select good HHs from calculations. For *p*-type materials, the HHs with relatively high power factors usually contain Co, Fe, and Rh. Other IVB and IIIB containing compounds, usually having Ni subgroup and IB group as B-site atoms, are less promising *p*-type materials. Based on the estimated *n*-type power factors, IIIB-(Ni, Pd) HHs are expected to have better performance than the IVB-Ni-containing ones (e.g., ZrNiSn) which have been extensively studied

TABLE 9.2 Comparisons of Theoretical and Experimental Optimal Doping Levels, as Well as Corresponding Seebeck Coefficients

Systems	Experimental Doping Levels Seebeck (μV/K)	Theoretical Doping Levels Seebeck (μV/K)
ZrNiSn	1% Sb for Sn[32]	0.9%
n-type	−103	−138
HfPtSn	1% Co or Ir for Pt[35]	1.4%
p-type	120(Co)/130(Ir)	150
TiCoSb	15–30% Fe for Co[38]	15.7%
p-type	153.6	111
NbCoSn	6–10% Sb for Sn[40]	4.5%
n-type	−90	−113
ZrCoSb	10–15% Sn for Sb[39]	7%
p-type	150	129

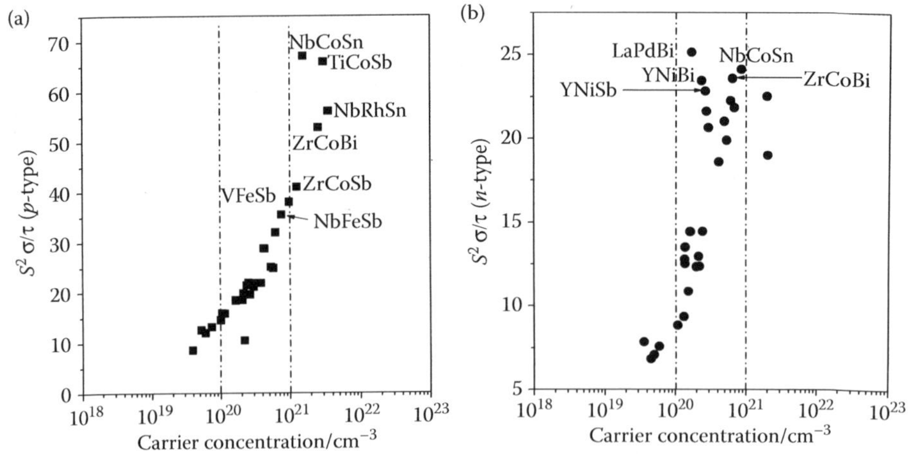

FIGURE 9.8 Maximum power factors versus the corresponding carrier concentrations for (a) *p*-type and (b) *n*-type cases. Compound names are indicated in the plots.

in recent years. LaPdBi has the highest *n*-type power factor. Furthermore, some Co-containing HHs also show reasonable *n*-type performance. Figure 9.8 plots the relationship between the maximum power factor and the corresponding carrier concentration (defined as Equation 9.18) for both *p*- and *n*-type HHs. Note that the best *n*-type HHs fall in the range of 10^{20}–10^{21} cm^{-3} and the best *p*-type over 10^{21} cm^{-3}, both correspond to heavily doped semiconductors. These results calculated from realistic and complicated band structures are interestingly in reasonable agreement with the optimal carrier concentration estimated based on a simple two-band semiconductor model proposed by Ioffe.[43] Figure 9.8 also gives the information that in *p*-type, high power factors require high carrier concentrations. But some IIIB-(Ni, Pd) compounds stand out for their *n*-type power factors, with relatively low carrier concentrations required. Band character analysis shows that *p* states from X-site elements are unneglectable at the CBM of these compounds, which enhances the group velocities of carriers. These contribute to high TDs and good power factors.

A systematical investigation of the electronic structures and electrical transport properties for over 30 well-selected HHs is quite interesting. Our work provides an overall picture of the electrical performance of the series compounds.[44] Theoretically, there are still many issues to be solved in predicting the exact values of electrical transport quantities. However, for the compounds with similar structures, the calculated results are expected to be comparable with each other, and therefore provide a systematic evaluation about their relative performance even with some uncertainties. Through the evaluations of optimal doping levels and corresponding power factors, several HH systems with potentially promising performance are predicted. Experimental works on these materials are encouraged.

9.5 *n*-Type Filled $CoSb_3$ Skutterudites

Filled skutterudites, close to be a type of TE materials satisfying the "phonon-glass-electron-crystal" paradigm, were recognized for their superior TE properties in the early 1990s.[3,45] Skutterudite $CoSb_3$ crystallizes with the body-centered-cubic structure in the space group *Im*3. The structure of $CoSb_3$ is composed of $CoSb_6$ octahedrons sharing corners with six neighboring octahedrons (Figure 9.9). Binary $CoSb_3$ exhibits high carrier mobility and relatively large electron effective mass, leading to reasonably high power factors ($S^2\sigma$). In one conventional crystallography unit cell, there are two intrinsic crystal voids that can be filled by electropositive metals. Filling the crystal voids of $CoSb_3$ with impurity atoms such as rare-earths (RE), alkaline-earths (AE), and alkaline metals (AM),[46–52] leads to the so-called

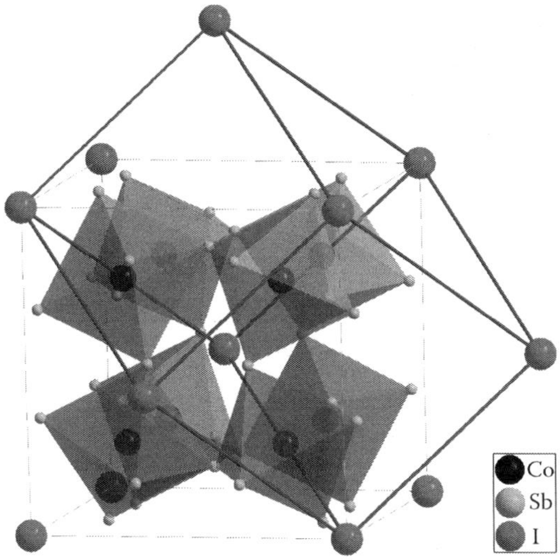

FIGURE 9.9 Crystal structure of filled skutterudite $I_yCo_4Sb_{12}$. I represents the void filler with a filling fraction y.

filled skutterudites that have been shown to effectively reduce the lattice thermal conductivity by introducing low-frequency phonon scattering centers, as well as to optimize electrical transport properties by tuning carrier concentrations. For most filled skutterudites, the experimentally reported power factors fall in the range of 40–60 μV/cm K^2, and the best ZT values are close to 1.7 at 800–850 K,[53] which makes the filled $CoSb_3$ skutterudites among the most promising bulk TE materials for power generation applications.

Earlier work on the maximum filling fraction, that is, the filling fraction limits for impurity partial filling in $CoSb_3$ led to the establishment of the so-called electronegativity rule for selecting possible fillers that could form stable filled $CoSb_3$. In this regard, we actually have proven that there exist only a few types of fillers to realize stable partial filling, including RE, AE, and AM atoms. Detailed discussions can be found in our publications[54–57] as well as in other chapters in this handbook, which is not the focus of the current paper. It was also clarified that the interaction between the fillers are mostly columbic, and the interaction depends critically on the effective valence charge states of the fillers. RE, AE, and AM show different valence charge states when filling into the lattice voids, but they similarly donate their outmost valence electrons to lattice and contribute to the change of carrier concentrations.

With the help of DFT calculations, we have developed an in-depth understanding of the electronic structures and the underlying relationship to electrical transport properties of the n-type filled skutterudites.[58] The main issue to be addressed is the optimization of electrical transport properties in these systems. We want to understand how filler atoms affect band structures especially in the area close to the CBM, and what the optimal filling levels for maximum power factors are for different filler atoms. To deal with the partially filled skutterudites, supercell calculations are directly used to study the transport properties, instead of using RBA. Our calculations are based on a $2 \times 2 \times 2$ supercell of Co_4Sb_{12} with eight crystallographic voids. We have calculated the properties of filled $CoSb_3$ with the filling fraction ranging from 1/8 to 7/8. Filler atoms are selected from AM, AE, and RE groups, and the results of Na-, K-, Ba-, Sr-, and Yb-filled $CoSb_3$ are presented.

Figure 9.10 shows the electronic structures of the Ba-filled skutterudite systems. Figure 9.10a is the band structure of $Ba_{0.25}Co_4Sb_{12}$. Comparing with the result of the unfilled Co_4Sb_{12} shown in Figure 9.1, the Fermi level in $Ba_{0.25}Co_4Sb_{12}$ moves upwards into the CB due to the electron donation from fillers. By integrating the DOS from CBM to the Fermi level, the valence charge states of various guest atoms are estimated to be around +1 for Na and K, +2 for Ba, Sr, and Yb, and +3 for La and Ce, consistent with experiments and our previous work. Another distinct feature derived from Figure 9.10a is that the band

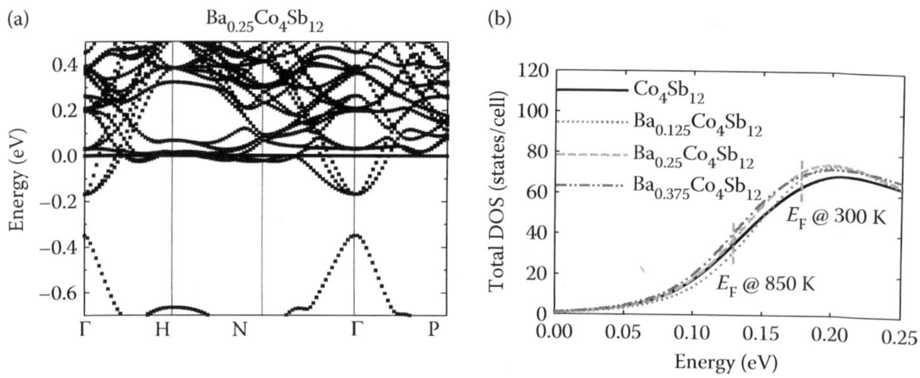

FIGURE 9.10 (a) Band structure of $Ba_{0.25}Co_4Sb_{12}$; (b) DOS close to the conduction band minima in several Ba-filled skutterudites and unfilled Co_4Sb_{12}.

structure in the area close to CBM for the Ba-filled system remains unchanged from that of $CoSb_3$. Figure 9.10b gives the total DOS plots close to CBM, and the DOS for the Ba-filled $CoSb_3$ in this energy range is shown to be very close to that of $CoSb_3$, despite the different filling fractions. All other filled systems we studied share the feature as the Ba-filled, that is, the DOS close to CBM is almost unaffected by the filling of the species of fillers and their filling fractions. Our earlier work for the AM-filled $CoSb_3$ shows the similar pattern, even with different calculation parameters.[57] The electronic states from the 3s,

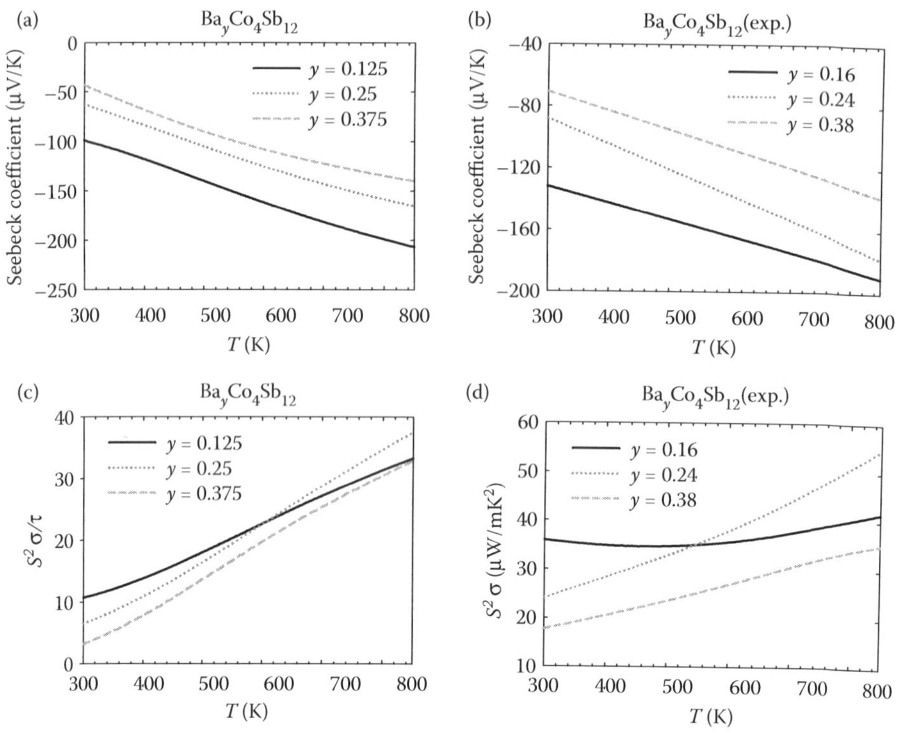

FIGURE 9.11 Calculated and measured temperature dependence of Seebeck coefficients and power factors for $Ba_yCo_4Sb_{12}$.

4s, and 6s electrons originating from fillers all go to the states well above the CBM. Figure 9.10b also indicates the Fermi level change with temperature for $Ba_{0.25}Co_4Sb_{12}$. Since all these are doped semiconductors, the Fermi levels move toward band edge at increasing temperature, leading to decreasing electrical conductivities and increasing Seebeck coefficients.

Seebeck coefficients and power factors as functions of temperature for the Ba-filled skutterudites are given in Figure 9.11, together with the experimental results.[46] Since filled skutterudites are typical heavily doped semiconductors, the electrical conductivities decrease with increasing temperature, which results in the increase of absolute Seebeck coefficients (Figure 9.11a and 9.11b). The temperature dependence and the magnitude of Seebeck coefficient (Figure 9.11a) agree reasonably well with the experimental data in comparable filling fraction range (Figure 9.11b). The experimentally observed trend of the temperature dependence of power factors is also reproduced reasonably well (Figure 9.11c and 9.11d) even with a constant approximation to the electron relaxation time in theory. The relative magnitude of the experimental power factors among various filling fractions can also be reproduced by the calculations. From Figure 9.11c we found that the filling fractions corresponding to maximum power factors vary with temperature, as mentioned in Section 9.2. At 300 K, $Ba_{0.125}Co_4Sb_{12}$ has the highest power factors; but at elevated temperature, $Ba_{0.25}Co_4Sb_{12}$ shows an even higher power factor than the samples with other filling fractions.

Generally speaking, Seebeck coefficients for all the *n*-type filled $CoSb_3$ investigated here have similar behaviors as $Ba_yCo_4Sb_{12}$. If the donated electrons from filler atoms are taken as *x*-axis, we get Seebeck coefficient variations in Figure 9.12a. Results at both 300 and 850 K are shown. All Seebeck coefficients decrease with the increasing filling levels. When given same donated electrons to the conduction bands of $CoSb_3$, filled skutterudites usually have very similar Seebeck coefficients. That is to say, *n*-type filled skutterudites follow a similar $S \sim n$ trend. This result is consistent with Bai et al.'s experimental collection

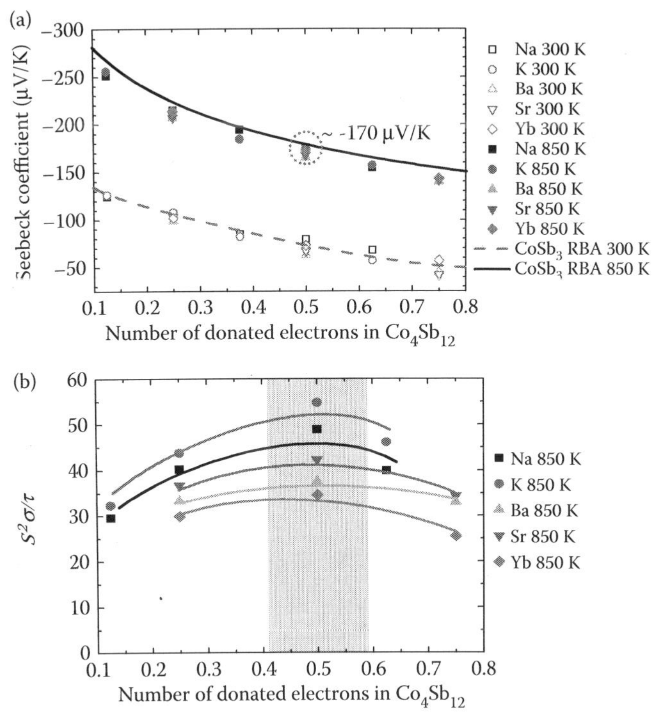

FIGURE 9.12 Seebeck coefficients (a) and power factors (b) for various *n*-type filled skutterudites investigated. In (a), Seebeck coefficients at both 300 and 850 K are shown. RBA results for unfilled Co_4Sb_{12} are also given for comparison. Only the power factor results at 850 K are given in (b).

of Seebeck coefficients, even for multiple filling.[59] In our calculations, the trend is found to be very close to the RBA results of Seebeck coefficients for the unfilled $CoSb_3$, both at room temperature and elevated temperature. All these imply the fact that the filling of the filler atoms has weak influence on conduction bands, and therefore Seebeck coefficients manifest similar tendency for pure and filled $CoSb_3$. At 300 K, the Seebeck coefficients are about −50 to −130 μV/K, which means that filled skutterudites are not very good TE materials at room temperature. But with the increase in temperature, Seebeck coefficients for all filled phases are gradually enhanced. Interestingly, when filling levels reach ~0.5 electrons in the unit of Co_4Sb_{12}, the Seebeck coefficients are around −170 μV/K. The value is very close to Ioffe's analysis ($2k_e/e$) for maximizing power factor.[43]

Figure 9.12b shows the power factors with respect to doping levels for different filled skutterudites at 850 K. A general feature is that power factor reaches the maximum when the filling fraction reaches ~0.5 electrons per unit cell. The results agree fairly well with many earlier reports and our own experimental measurements. Following-up study of the electronic structures and electrical transport properties confirmed that the picture is also valid for double- and multiple-filled $CoSb_3$ skutterudites,[51–59] since the underlying physics is the same. With that, we could easily combine different fillers with different charge states to realize the optimal doping levels for electrical transport. Together with the selection rule for the optimal combination of multiple fillers for the most effective thermal conductivity reduction,[60,61] the substantially enhanced ZT have been realized.[53] Details could be found in reference and another chapter in this handbook.

9.6 Concluding Remarks and Outlook

DFT-based *ab initio* methods have been widely used in understanding the fundamental properties of materials nowadays, covering structural energetics, electronic structures, and various functionalities. Good TE materials are traditionally considered to be narrow-gap semiconductors. From this point of view, DFT-based band structures can be greatly helpful to a rational search for novel materials. Such an approach has been followed by many people working in the TE community.

Combining DFT-based band structure calculations and Boltzmann transport theory, we try to expand the applicability and predictability of the fundamental theory. Our intention is to be of help to property optimization and materials exploratory, which has been a focus of recent efforts. In the present chapter, we reviewed our recent work by adopting the above approaches to understand the band structures and transport mechanisms of three different types of TE materials. For transition-metal-substituted type-I clathrates, we studied mostly the configurations and compositions by calculating the structural energetics, and TM-induced electronic structure change by looking into the bands. Our theoretical results reveal a new series of type-I clathrates after TM cross substitution for framework atoms. Those clathrates follow the Zintl–Klemm rule, and possess tunable band gaps and enhanced Seebeck coefficients as well, even in high temperature. By joining experiments, samples were synthesized and a relatively high ZT of 1.2 in triple substituted samples was obtained. For half-Heuslers, we applied 18 VEC as the rule to select from a large database, and used the combination of electronic structures and Boltzmann transport theory to select those compounds with potential high electrical transport performance. With the physically reasonable but simple approaches such as rigid band approximation and constant relaxation time, the theoretical results may not be exact; however, they are surely to be helpful to get rid of those systems with predicted bad performance that are hardly to be good in practice. In our work, several promising half-Heuslers are selected based on the relative magnitude of the calculated power factors. Our results also provide optimal carrier concentrations for the convenience of experimental optimization. For n-type filled skutterudites, the temperature dependence of electrical transport parameters is calculated to search for the optimal filling fractions in the operating temperature. The most interesting thing is that all the partially filled n-type $CoSb_3$ systems, including single, double, and triple filling, show generally similar band structures in the energy window close to the CBM. Electrical transport calculations confirmed that all systems reach their peak power factors at nearly a same carrier concentration. At 850 K, all of the filled systems reach maximum power factors after 0.5 electrons/unit cell are introduced into conduction bands, which has been confirmed by the following experimental data.[53,59] The conclusion is truly helpful for the performance optimization in the search for multiple filled systems with high *ZT*.

However, the whole package of methodology is far from being perfect. By skipping the underestimation of band gap in LDA for certain materials, the undetermined electron relaxation time is probably the most critical issue to be concerned for performance predictions and practical rational design of TEs. As mentioned in Section 9.2, the scattering rate is affected by many scattering mechanisms; most of them are hard to be solved precisely, especially from the atomic level. Among them, electron–phonon scattering is probably the one that is worth investigating in the first priority because it is the dominant scattering mechanism in many cases, especially at elevated temperature.[11] It also seems that electron–phonon scattering, especially the scattering due to the interaction between carriers and longitudinal acoustic phonons, is treatable under density functional perturbation theory (DFPT). After calculating the electron–phonon coupling matrix in DFPT, scattering of electronic states by phonon modes can be calculated to predict the final scattering rate, thus the energy-, and momentum-dependent carrier relaxation time. It is apparently a hard and very time-consuming task for today's computing technology. In recent years, electron scattering in simple semiconductors like Si and graphene has been investigated by quantum mechanical approaches,[62,63] but the progress is slow. Furthermore, it should also be pointed out that, even though the scattering from electron–phonon coupling could be handled in a reasonable way, scattering of carriers from various types of small and large defects are both difficult to be dealt with at atomic levels. It has been a challenge for decades. Even with so many challenges ahead, it is hopeful that in the near future, with more sophisticated transport theory, advanced supercomputers, and proper programs, *ab initio* transport calculations will become powerful tools in TE materials optimization and design.

References

1. Hohenberg, P. and Kohn, W. 1964. Inhomogeneous electron gas. *Phys. Rev.*, 136, B864–B871.
2. Kohn, W. and Sham, L. J. 1965. Self-consistent equations including exchange and correlation effects. *Phys. Rev.*, 140, 1133.
3. Singh, D. J. 2001. Theoretical and computational approaches for identifying and optimizing novel thermoelectric materials. In *Recent Trends in Thermoelectric Materials Research II*, ed. T. M. Tritt, pp. 125–178. San Diego: Academic Press.
4. Tse, J. S. and Klug, D. D. 2006. Recent trends for the design and optimization of thermoelectric materials—A theoretical perspective. In *Thermoelectrics Handbook: Macro to Nano*, ed. D. M. Rowe, Chapter 8. Boca Raton: CRC Press.
5. Martin, R. M. 2004. *Electronic Structure: Basic Theory and Practical Methods*. Cambridge: Cambridge University Press.
6. Ziman, J. M. 1960. *Electrons and Phonons: The Theory of Transport Phenomena in Solids*. London: Oxford University Press.
7. Singh, D. J. and Pickett, W. E. 1994. Skutterudite antimonides: Quasilinear bands and novel transport. *Phys. Rev. B*, 50, 11235.
8. Madsen, G. K. H. and Singh, D. J. 2006. BoltzTraP. A code for calculating band-structure dependent quantities. *Comput. Phys. Commun.*, 175, 67.
9. Scheidemantel, T. J., Ambrosch-Draxl, C., Thonhauser, T., Badding, J. V., and Sofo, J. O. 2003. Transport coefficients from first-principles calculations. *Phys. Rev. B*, 68, 125210.
10. Singh, J. 2003. *Electronic and Optoelectronic Properties of Semiconductor Structures*. Cambridge: Cambridge University Press.
11. Nag, B. R. 1972. *Theory of Electrical Transport in Semiconductors*. Oxford: Pergamon Press.

12. Allen, P. B., Pickett, W. E., and Krakauer, H. 1988. Anisotropic normal-state transport properties predicted and analyzed for high-T_c oxide superconductors. *Phys. Rev. B*, 37, 7482.
13. Slack, G. A. 1995. New materials and performance limits for thermoelectric cooling. In *CRC Handbook of Thermoelectrics*, ed. D. M. Rowe, Chapter 34. Boca Raton: CRC Press.
14. Mahan, G. D. 1998. Good thermoelectrics. In *Solid State Physics*, eds. H. Ehrenreich and F. Spaepen, pp. 81–157. New York: Academic Press.
15. Blöchl, P. E. 1994. Projector augmented-wave method. *Phys. Rev. B*, 50, 17953.
16. Kresse, G. and Furthmüller, J. 1996. Efficient iterative schemes for *ab initio* total-energy calculations using a plane-wave basis set. *Phys. Rev. B*, 54, 11169.
17. Rogl, P. 2006. Formation and crystal chemistry of clathrates. In *Thermoelectrics Handbook: Macro to Nano*, ed. D. M. Rowe, Chapter 8. Boca Raton: CRC Press.
18. Nolas, G. S. 2006. Structure, thermal conductivity, and thermoelectric properties of clathrate compounds. In *Thermoelectrics Handbook: Macro to Nano*, ed. D. M. Rowe, Chapter 8. Boca Raton: CRC Press.
19. May, A. F., Toberer, E. S., Saramat, A., and Snyder, G. J. 2009. Characterization and analysis of thermoelectric transport in *n*-type $Ba_8Ga_{16-x}Ge_{30+x}$. *Phys. Rev. B*, 80, 125205.
20. Wang, L., Chen, L. D., Chen, X. H., and Zhang, W. B. 2009. Synthesis and thermoelectric properties of n-type $Sr_8Ga_{16-x}Ge_{30-y}$ clathrates with different Ga/Ge ratios. *J. Phys. D: Appl. Phys.*, 42, 045113.
21. Schäfer, H. 1985. On the problem of polar intermetallic compounds: The stimulation of E. Zintl's work for the modern chemistry of intermetallics. *Annu. Rev. Mater. Sci.*, 15, 1.
22. Christensen, M. and Iversen, B. B. 2008. Host–guest coupling in semiconducting $Ba_8Zn_8Ge_{38}$. *J. Phys.: Condens. Matter*, 20, 104244.
23. Johnsen, S., Bentien, A., Madsen, G. K. H., Nygren, M., and Iversen, B. B. 2007. Crystal structure and transport properties of nickel containing germanium clathrates. *Phys. Rev. B*, 76, 245126.
24. Zhang, H., Zhao, J. T., Tang, M. B., Man, Z. Y., Chen, H. H., and Yang, X. X. 2009. Structure and low temperature physical properties of $Ba_8Cu_6Ge_{40}$. *J. Alloy Compd.*, 476, 1.
25. Blake, N. P., Bryan, D., Latturner, S., Møllnitz, L., Stucky, G. D., and Metiu, H. 2001. Structure and stability of the clathrates $Ba_8Ga_{16}Ge_{30}$, $Sr_8Ga_{16}Ge_{30}$, $Ba_8Ga_{16}Si_{30}$, and $Ba_8In_{16}Sn_{30}$. *J. Chem. Phys.*, 114, 10063.
26. Bentien, A., Iversen, B. B., Bryan, J. D., Stucky, G. D., and Palmqvist, A. E. C. 2002. Maximum entropy method analysis of thermal motion and disorder in thermoelectric clathrates $Ba_8Ga_{16}Si_{30}$. *J. Appl. Phys.*, 91, 5694.
27. Christensen, M., Lock, N., Overgaard, J., and Iversen, B. B. 2006. Crystal structures of thermoelectric *n*- and *p*-type $Ba_8Ga_{16}Ge_{30}$ studied by single crystal, multitemperature, neutron diffraction, conventional X-ray diffraction and resonant synchrotron X-ray diffraction. *J. Am. Chem. Soc.*, 128, 15657.
28. Shi, X., Yang, J., Bai, S., Yang, J., Wang, H., Chi, M., Salvador, J. R., Zhang, W., Chen, L., and Wong-Ng, W. 2010. On the design of high-efficiency thermoelectric clathrates through a systematic cross-substitution of framework elements. *Adv. Funct. Mater.*, 20, 755.
29. Inorganic Crystal Structure Database (ICSD). http://icsd.ill.eu/icsd/
30. Carrillo-Cabrera, W., Budnyk, S., Prots, Y., and Grin, Y. 2004. Ba_8Ge_{43} revisited: a $2a' \times 2a' \times 2a'$superstructure of the clathrate-I type with full vacancy ordering. *Z. Anorg. Allg. Chem.*, 630, 2267.
31. Uher, C., Yang, J., Hu, S., Morelli, D. T., and Meisner, G. P. 1999. Transport properties of pure and doped *M*NiSn ($M = $ Zr, Hf). *Phys. Rev. B*, 59, 8615.
32. Shen, Q., Chen, L., Goto, T., Hirai, T., Yang, J., Meisner, G. P., and Uher, C. 2001. Effects of partial substitution of Ni by Pd on the thermoelectric properties of ZrNiSn-based half-Heusler compounds. *Appl. Phys. Lett.*, 79, 4165.
33. Bhattacharya, B., Tritt, T. M., Xia, Y., Ponnambalam, V., Poon, S. J., and Thadhani, N. 2002. Grain structure effects on the lattice thermal conductivity of Ti-based half-Heusler alloys. *Appl. Phys. Lett.*, 81, 43.

34. Sakurada, S. and Shutoh, N. 2005. Effect of Ti substitution on the thermoelectric properties of (Zr, Hf)NiSn half-Heusler compounds. *Appl. Phys. Lett.*, 86, 082105.
35. Kimura, Y. and Zama, A. 2006. Thermoelectric properties of *p*-type half-Heusler compound HfPtSn and improvement for high performance by Ir and Co additions. *Appl. Phys. Lett.*, 89, 172110.
36. Nolas, G. S., Sharp, J., and Goldsmid, H. J. 2001. *Thermoelectrics: Basic Principles and New Materials Developments*. Berlin: Springer Press.
37. Mastronardi, K., Yong, D., Wang, C. C., Khalifah, P., Cava, R. J., and Ramirez, A. Z. 1999. Antimonides with the half-Heusler structure: New thermoelectric materials. *Appl. Phys. Lett.*, 74, 1415.
38. Wu, T., Jiang, W., Li, X. Y., Zhou, Y. F., and Chen, L. D. 2007. Thermoelectric properties of *p*-type Fe-doped TiCoSb half-Heusler compounds. *J. Appl. Phys.*, 102, 103705.
39. Sekimoto, T., Kurosaki, K., Muta, H., and Yamanaka, S. 2006. High-thermoelectric figure of merit realized in *p*-type half-Heusler compounds: $ZrCoSn_xSb_{1-x}$. *Jpn. J. Appl. Phys.*, 46, L673.
40. Ono, Y., Inayama, S., Adachi, H., and Kajitani, T. 2006. Thermoelectric properties of doped half-Heuslers $NbCoSn_{1-x}Sb_x$ and $Nb_{0.99}Ti_{0.01}CoSn_{1-x}Sb_x$. *Jpn. J. Appl. Phys.*, 45, 8740.
41. Villars, P. 1991. *Pearson's Handbook of Crystallographic Data for Intermetallic Phases*, 2nd ed., Ohio: ASM International Press.
42. Madsen, G. K. H. 2006. Automated search for new thermoelectric materials: The case of LiZnSb. *J. Am. Chem. Soc.*, 128, 12140.
43. Ioffe, A. F. 1957. *Semiconductor Thermoelements and Thermoelectric Cooling*, London: Infosearch Limited Press.
44. Yang, J., Li, H. M., Wu, T., Zhang, W., Chen, L. D., and Yang, J. 2008. Evaluation of half-Heusler compounds as thermoelectric materials based on the calculated electrical transport properties. *Adv. Funct. Mater.*, 18, 2880.
45. Uher, C. 2006. Skutterudite-based thermoelectrics. In *Thermoelectrics Handbook: Macro to Nano*, ed. D. M. Rowe, Chapter 34. Boca Raton: CRC Press.
46. Chen, L. D., Kawahara, T., Tang, X. F., Goto, T., Hirai, T., Dyck, J. S., Chen, W., and Uher, C. 2001. Anomalous barium filling fraction and *n*-type thermoelectric performance of $Ba_yCo_4Sb_{12}$. *J. Appl. Phys.*, 90, 1864.
47. Lamberton, Jr., G. A., Bhattacharya, S., Littleton IV, R. T., Kaeser, M. A., Tedstrom, R. H., Tritt, T. M., Yang, J., and Nolas, G. S. 2002. High figure of merit in Eu-filled $CoSb_3$-based skutterudites. *Appl. Phys. Lett.*, 80, 598.
48. Zhao, X. Y., Shi, X., Chen, L. D., Zhang, W., Bai, S. Q., Pei, Y. Z., Li, X. Y., and Goto, T. 2006. Synthesis of $Yb_yCo_4Sb_{12}/Yb_2O_3$ composites and their thermoelectric properties. *Appl. Phys. Lett.*, 89, 092121.
49. Puyet, M., Lenoir, B., Dauscher, A., Pécheur, P., Bellouard, C., Tobola, J., and Hejtmanek, J. 2006. Electronic, transport, and magnetic properties of $Ca_xCo_4Sb_{12}$ partially filled skutterudites. *Phys. Rev. B*, 73, 035126.
50. Pei, Y. Z., Chen, L. D., Zhang, W., Shi, X., Bai, S. Q., Zhao, X. Y., Mei, Z. G., and Li, X. Y. 2006. Synthesis and thermoelectric properties of $K_yCo_4Sb_{12}$. *Appl. Phys. Lett.*, 89, 221107.
51. Shi, X., Kong, H., Li, C.-P., Uher, C., Yang, J., Salvador, J. R., Wang, H., Chen, L., and Zhang, W. 2008. Low thermal conductivity and high thermoelectric figure of merit in *n*-type $Ba_xYb_yCo_4Sb_{12}$ double-filled skutterudites. *Appl. Phys. Lett.*, 92, 182101.
52. Pei, Y. Z., Yang, J., Chen, L. D., Zhang, W., Salvador, J. R., and Yang, J. 2009. Improving thermoelectric performance of caged compounds through light-element filling. *Appl. Phys. Lett.*, 95, 042101.
53. Shi, X., Yang, J., Salvador, J. R., Chi, M. F., Cho, J. Y., Wang, H., Bai, S. Q., Yang, J. H., Zhang, W., and Chen, L. 2011. Multiple-filled skutterudites: High thermoelectric figure of merit through separately optimizing electrical and thermal transports. *J. Am. Chem. Soc.*, 133, 7837.
54. Shi, X., Zhang, W., Chen, L. D., and Yang, J. 2005. Filling fraction limit for intrinsic voids in crystals: Doping in skutterudites. *Phys. Rev. Lett.*, 95, 185503.

55. Mei, Z. G., Zhang, W., Chen, L. D., and Yang, J. 2006. Filling fraction limits for rare-earth atoms in $CoSb_3$: An *ab initio* approach. *Phys. Rev. B*, 74, 153202.
56. Zhang, W., Shi, X., Mei, Z. G., Xu, Y., Chen, L. D., Yang, J., and Meisner, G. P. 2006. Predication of an ultrahigh filling fraction for K in $CoSb_3$. *Appl. Phys. Lett.*, 89, 112105.
57. Mei, Z. G., Yang, J., Pei, Y. Z., Zhang, W., Chen, L. D., and Yang, J. 2008. Alkali-metal-filled $CoSb_3$ skutterudites as thermoelectric materials: Theoretical study. *Phys. Rev. B*, 77, 045202.
58. Yang, J., Xi, L., Zhang, W., Chen, L. D., and Yang, J. 2009. Electrical transport properties of filled $CoSb_3$ skutterudites: A theoretical study. *J. Electron. Mater.*, 38, 1397.
59. Bai, S. Q., Pei, Y. Z., Chen, L. D., Zhang, W., Zhao, X., and Yang, J. 2009. Enhanced thermoelectric performance of dual-element-filled skutterudites $Ba_xCe_yCo_4Sb_{12}$. *Acta Mater.*, 57, 3135.
60. Xi, L., Yang J., Lu, C. F., Mei, Z. G., Zhang, W., and Chen, L. D. 2010. Systematic study of the multiple-element filling in caged skutterudite $CoSb_3$. *Chem. Mater.*, 22, 2384.
61. Yang, J., Zhang, W., Bai, S. Q., Mei, Z., and Chen, L. D. 2007. Dual-frequency resonant phonon scattering in $Ba_xR_yCo_4Sb_{12}$ (R = La, Ce, and Sr). *Appl. Phys. Lett.*, 90, 192111.
62. Restrepo, O. D., Varga, K., and Pantelides, S. T. 2009. First-principles calculations of electron mobilities in silicon: Phonon and Coulomb scattering. *Appl. Phys. Lett.*, 94, 212103.
63. Borysenko, K. M., Mullen, J. T., Barry, E. A., Paul, S., Semenov, Y. G., Zavada, J. M., Buongiorno, N. M., and Kim, K. W. 2010. First-principles analysis of electron–phonon interactions in graphene. *Phys. Rev. B*, 81, 121412(R).

10

Band Structure Guidelines for Higher Figure-of-Merit: Analytic Band Generation and Energy Filtering

Espen Flage-Larsen
University of Oslo

Ole Martin Løvvik
SINTEF Materials and Chemistry
University of Oslo

10.1 Introduction

The coefficients that determine the performance of thermoelectric materials are well established. Enhancements of the figure-of-merit are realized by improvements of the Seebeck coefficient α and the electrical conductivity σ. Both are limited by a prima facie inverse relation. Further improvements are possible if the thermal transport quantified by the thermal conductivity κ is reduced. However, the intricate correlation between σ, α, and κ confines the potential benefit of improving a single parameter.

One viable path to remedy the interdependence of these coefficients is to introduce effects that make it possible to simultaneously improve the coefficients and thus enhance the figure-of-merit further. One such scheme, which will be discussed here, involves a process which filters out the low-energy charge carriers from the electronic transport, breaking the strong interdependence, and thus opening up new possibilities to design materials with improved thermoelectric properties.

In order to properly assess the benefits of such an approach, we need to know the detailed fundamental parameters of the electronic transport, of which the electronic structure and scattering mechanisms are the most important. Consequently, we not only need to know how the transport coefficients depend on these parameters, but we also need detailed knowledge of the parameters.

The basic properties of the electronic and thermal transport have been studied by extensive experimental work on a wide range of materials. However, due to complicated measurement setups, processing conditions, and sample complexity it has been difficult to realize a general understanding of the correlation between the electronic structure and the transport coefficients of thermoelectric materials. Even though the experimental framework is continually improving and transport coefficients can be routinely measured with increased accuracy, it may nonetheless be difficult to reach conclusions without knowing the electronic structure, which is still a major task to achieve from experiments.

Accurate predictions of electronic structure from first principles would therefore be an attractive approach to determine the transport coefficients and obtain further knowledge of the system. This could not only assist the interpretation of experimental results but also lead to more in-depth knowledge about the electronic structure of well-performing thermoelectric materials. Such calculations could in addition facilitate screening of future materials or suggest alternative paths of optimizing current materials, thus improving the predictive power of the analysis toolbox. Electronic structure calculations are today possible for a wide selection of materials, but transport properties are unfortunately difficult to calculate for anything but the simplest systems with high symmetry and small unit cells. Additional complications arise due to the electron and lattice dynamics of the system which govern the scattering of the charge carriers. Elaborate knowledge about these properties is a prerequisite for rigorous first principle calculations of the transport properties and such calculations for real material system are thus to a large extent absent in the literature. However, by introducing simplified transport theories, for example, through the relaxation time approximation [1], the scattering properties can be parameterized through an effective relaxation time. Regardless of approach, determining the effective relaxation time or scattering properties in general is an exceedingly difficult task and the scattering properties are often simplified by energy-dependent models of the effective relaxation time [2]. By unifying first principle calculations and experimental measurements of the transport coefficients, it should be possible to improve the predictability and redundancy of the interpretations.

Transport measurements are commonly rationalized by imposing a variety of approximations in order to extract meaningful parameters such as the effective mass, the carrier concentration, and scattering properties. The single parabolic band approximation is often used to make tractable expressions for the electrical conductivity, Seebeck coefficient, Hall coefficient, and the electronic part of the thermal conductivity. These approximations have shortcomings related to the single-band approximation, the harmonicity of the bands, what kind of statistics is used to describe the charge carriers, and how the numerical implementation is tackled. Nevertheless, these approximations provide efficient means to calculate most of the fundamental properties in thermoelectric materials and thus also highlight the relationship between these properties.

A large portion of the pioneering work on the development of closed and analytic expressions for the transport coefficients and scattering effects is covered in the books of Fistul [2], Ioffe [3], Wilson [4], and Goldsmid [5]. Thermoelectric applications did not receive significant interest before the late 1950s and early 1960s. It became important to establish an upper limit to the figure-of-merit such that the future potential of thermoelectricity could be assessed. Rittner [6] was among the first to introduce such a limit based on a single-band model. This limit was generally thought to be too high for realistic material systems and subsequent work by Donahoe [7] and Price [8] tried to correct this by introducing an additional minority carrier band which consequently reduced the upper limit of the figure-of-merit. Simon [9] and Rittner and Neumark [10] later derived an upper bound of the figure-of-merit based on a general statistical two-band model with a result similar to the original work of Rittner. It was later pointed out [11,12] that there is no general analytical upper bound to the figure-of-merit, unless material-specific parameters are introduced, an insight that is still valid today. A material-dependent term of the figure-of-merit has been investigated in the literature [5,10,13,14]. It primarily contains the carrier effective mass, the carrier mobility, and the lattice thermal conductivity. The scattering mechanism of the carriers will modify the effective mass dependency of the mobility and from simplified models [5,14] an increase of the figure-of-merit is expected for ionized impurity scattering if there is an increase in the effective mass. The opposite is true for acoustic phonon scattering. It was also proposed [3,6] that ionized impurity scattering could be the most desirable scattering mechanism for high values of the figure-of-merit. Significant mathematical insight into the factors that improve the figure-of-merit was later gained by Mahan and Sofo [15]. Based on an analytic approach, they concluded that the transport distribution function should possess a delta function character in order to maximize the figure-of-merit. This was confirmed by Kim et al. [16] using Landauer [17] formalism. By assuming ballistic transport they pointed out that it is both the shape and magnitude of the distribution of the conducting density of states that determine a favorable transport distribution function. Enhancements of the figure-of-merit have also been predicted for systems with reduced dimensionality, for example, quantum wires and dots [18,19]. A recent parabolic band study by Pichanusakorn and Bandaru [20] suggests an optimal Seebeck coefficient for all types of materials complying with the parabolic approximation.

There are essentially three different, although somewhat related, strategies to hinder charge carriers with specific (low) energies to contribute to the conduction process. One possibility is to introduce a scattering mechanism which is much stronger for the low-energy carriers so that the electronic transport is biased toward higher energies. This can, for example, be induced by impurities and point defects. The second route is to introduce potential barriers at interfaces inside the material or between two dissimilar materials in the form of heterostructures. Similarly, the third route is to modify the band structure itself and introduce a potential barrier which the carriers need to cross in order to contribute to the conduction process. The second route can in principle be regarded as a subset of the third route. However, from an analytic point of view one could envisage modifications to the band structure which is not related to quantum confinement, heterostructures, or similar effects. In all strategies, the charge carrier filtering inhibits transmission of carriers with energies below the barrier or scattering threshold. Significant work has been performed for all strategies, both theoretically and experimentally; see, for example, recent reviews [21–24] and references therein. When we discuss energy filtering in this chapter, we will focus on an approach that combines strategy one and three. The concept of introducing a potential barrier was investigated early by Ravich et al. [25], who calculated the increase in Seebeck coefficient as a function of the barrier height, using a number of approximations [26]. Nishio and Hirano later calculated the optimal potential barrier height and distance between them, using analytic expressions based on similar assumptions [27]. Bian et al. [28] investigated the cross-plane Seebeck coefficient and Lorenz number of superlattices for different temperatures and thicknesses of quantum wells and potential barriers. They showed significant deviations from the Wiedemann–Franz law at low temperatures and an oscillatory behavior of the transport coefficients as a function of charge carrier concentration. Further investigations of the interface roughness and superlattice period length for InGaAs/InAlGaAs have been performed by Wang and Mingo [29]. The filtering concept has also been experimentally verified in numerous studies [28,30–32]. As an example, Kishimoto and Koyanagi demonstrated that reducing the grain size of a PbTe sample not only reduced the thermal conductivity but also increased the Seebeck coefficient [30]. Recent reviews show how this can be accomplished in a variety of materials [33,34]. Theoretical models have also recently been developed to demonstrate how filtering effects can be implemented using nano-inclusions as potential barriers for low-energy charge carriers [35,36]. However, it is important to point out a major obstacle for the experimental analysis of energy filtering; it is exceedingly difficult to isolate different scattering mechanisms and energy-filtering effects simultaneously. Such knowledge is necessary to obtain further insight into the potential of energy filtering.

There are several different schemes for theoretical calculations of the transport properties of materials, the two most used being the Boltzmann transport equations [1] and the Green–Kubo formalism [37,38]. The latter involves the calculation of response functions at the quantum level and is thus more complicated to implement. In this chapter, we will use the method developed from Boltzmann theory [1], which is based on a semiclassical derivation of the nonequilibrium distribution function and has been shown to accurately predict the semiclassical dynamics of a dilute system of particles close to equilibrium [39].

An advantage of Boltzmann theory is that it is under certain conditions fully dependent on the electronic band structure. Transport properties can thus be calculated based on a band structure, either numerical or analytical, as long as we assume a tractable model for the relaxation time. Consequently, more complex band structures can be analyzed and rigorous studies beyond the parabolic approximation are feasible. Based on an analysis of which band structures yield the most favorable conditions for a high figure-of-merit it should be possible to further relate these advantageous transport properties to the bonding and symmetry of the crystal due to the strong correlation between the band structure and geometric structure of the material. A wide range of structures could thus be efficiently screened and investigated without involving first principle frameworks. That is, with such insight, it may be possible from topology consideration to forecast promising thermoelectric materials that eventually need to be evaluated experimentally.

Boltzmann theory describing particle transport is semiclassically exact but requires accurate knowledge about the electronic structure and electron-scattering terms. Unfortunately, the scattering terms cannot be represented by any tractable expression. It is therefore common to introduce the relaxation time approximation, which simplifies these terms by introducing a scattering relaxation time [1]. This can in principle be calculated using an all-electron–phonon calculation involving the relaxation of electrons after electron–electron, electron–phonon, or other screened type of electron collisions. However, for real materials such a scheme is clearly out of reach with current techniques. Even though most of these calculations can be done with assistance of Monte Carlo procedures, their applications are still limited due to the lack of a proper description of the interacting electrons from first principles. Thus, further approximations are usually introduced to quantify the effect of scattering. We will come back to this in the following sections.

The Boltzmann transport equations provide considerable possibilities to predict a range of properties. To be able to explicitly calculate all transport coefficients and thus predict the figure-of-merit opens up for systematic studies of the correlation between fundamental band structures and transport properties. We will, in this chapter, present basic, but explicit connections between band structures and transport coefficients by solving the Boltzmann transport equations within the time relaxation approximation. Different band structures and scattering models are investigated to shed light on the complicated interdependence of the various transport coefficients which limit the figure-of-merit. This insight will be used to illustrate the potential of energy filtering by means of an energy-dependent scattering model, which will highlight the prospect of hindering the low-energy charge carrier transport.

10.2 Boltzmann Transport Equations

The derivation of the Boltzmann transport equation is already extensively covered in the literature; see, for example, the textbook of Ashcroft and Mermin [1]. However, for the sake of completeness we will in this section list the important transport tensor coefficients and how they relate to the thermoelectric parameters.

The electrical conductivity σ can be calculated from the band structure $\epsilon_n(\mathbf{k})$ from the following expression:

$$\sigma_{ij} = \frac{e^2}{\Omega} \sum_n \int \tau_n(\mathbf{k}) \left(-\frac{\partial f_n(\mathbf{k})}{\partial \epsilon} \right) \upsilon_i^n \upsilon_j^n \, d\mathbf{k}, \tag{10.1}$$

where f is the equilibrium distribution function for carriers in band n, e is the electronic charge, and $\mathbf{k}$ is the wave vector. For fermions, the Fermi distribution function should be used for f. The volume of integration is Ω, τ is the scattering relaxation time from the relaxation time approximation [1], and υ is the band velocity—the latter two will be defined later.

In order to define the Seebeck coefficient α, the χ tensor is introduced:

$$\chi_{ij} = \frac{e}{T\Omega} \sum_n \int \tau_n(\mathbf{k})(\epsilon_n(\mathbf{k}) - \mu) \left(-\frac{\partial f_n(\mathbf{k})}{\partial \epsilon} \right) \upsilon_i^n \upsilon_j^n \, d\mathbf{k}, \tag{10.2}$$

where μ is the chemical potential and T the temperature. The Seebeck coefficient α is then given by

$$\alpha_{kl} = (\sigma^{-1})_{ki}\chi_{il} \tag{10.3}$$

It is customary to transform the transport coefficients σ and χ to energy-dependent integrals to remove the wave vector integration in the following way:

$$\sigma_{ij} = e^2 \sum_n \int \Sigma_{ij}^{0,n}(\epsilon) \left(-\frac{\partial f_n(\mathbf{k})}{\partial \epsilon} \right) d\epsilon, \tag{10.4}$$

where

$$\Sigma_{ij}^{p,n}(\epsilon) = \frac{1}{n_k} \sum_{\mathbf{k}} \tau_n(\epsilon)(\epsilon_n(\mathbf{k}) - \mu)^p \upsilon_i^n \upsilon_j^n \frac{\delta(\epsilon - \epsilon_n(\mathbf{k}))}{d\epsilon}. \tag{10.5}$$

Here, n_k is the number of wave vectors in the energy slit. The χ tensor is similarly transformed as

$$\chi_{ij} = \frac{e}{T} \sum_n \int \Sigma_{ij}^{1,n}(\epsilon) \left(-\frac{\partial f_n(\mathbf{k})}{\partial \epsilon} \right) d\epsilon. \tag{10.6}$$

This approach is convenient when using the parabolic band approximation since the integrals are simplified due to the closed form of the energy and band velocity. These integrals are often referred to as Fermi integrals [2].

The electronic part of the thermal conductivity κ^e can also be calculated using

$$\kappa_{ij}^e = \kappa_{ij}^0 - T\chi_{ik}\alpha_{kj}, \tag{10.7}$$

where

$$\kappa_{ij}^0 = \frac{1}{e^2 T} \sum_n \int \Sigma_{ij}^{2,n}(\epsilon) \left(-\frac{\partial f_n(\mathbf{k})}{\partial \epsilon} \right) d\epsilon. \tag{10.8}$$

The seemingly simple implementation of the transport coefficients introduces numerous numerical challenges related to the sampling of the reciprocal space, the calculation of the band velocity, and the implementation of integration solvers, which are outside the scope of this chapter.

To be able to calculate the integrand of all transport coefficients, accurate knowledge of the band velocity υ_i^n is necessary. As an approximation, υ_i^n can be obtained by combining the velocity $\mathbf{v} = \nabla_{\mathbf{k}}\omega(\mathbf{k})$ of a Bloch wave packet with dispersion $\omega(\mathbf{k})$ from the de Broglie relation $\omega = E/\hbar$ as

$$\upsilon_i^n = \frac{1}{\hbar} \nabla_{k_i} \epsilon_n(\mathbf{k}). \tag{10.9}$$

This rather simple relation only requires knowledge of the band structure. It should be emphasized that this band velocity does not account for interband transitions or quantum effects that need a quantized calculation of the momentum transitions.

An important parameter for electronic transport is the effective mass. It is obtained by a Taylor expansion in $\mathbf{k}$ of the energy, which close to $\mathbf{k} = 0$ for quadratic bands can be shown to be

$$\frac{1}{m_{ij}^n} = \frac{1}{\hbar^2} \frac{\partial^2 \epsilon_n(\mathbf{k})}{\partial k_i \partial k_j}, \tag{10.10}$$

which illustrates the inverse relationship between the band velocity tensor and the effective mass tensor valid for parabolic or near-parabolic systems. A high effective mass is analogous to a small band velocity, implying a less curved band.

The thermoelectric dimensionless figure-of-merit is defined as

$$zT = \frac{\alpha^2\sigma}{\kappa}T. \tag{10.11}$$

The thermal conductivity κ is commonly separated into an electronic part κ^e and a lattice part κ^l, while the numerator of the figure-of-merit is referred to as the power factor, $W = \alpha^2\sigma$.

The electronic conductivity σ and the electronic part of the thermal conductivity κ^e are related through the Wiedemann–Franz ratio [40]. This defines the Lorenz function L as follows:

$$L = \frac{\kappa^e}{\sigma T}. \tag{10.12}$$

In general, both L and zT are tensors—here, we will only regard their isotropic values. The Lorenz function is often taken to be value, or the nondegenerate limit. It should be emphasized that these limits are in principle only valid for simple metals and insulators, and effort should be made to calculate L on a case-by-case basis as it varies from material to material.

By using the Lorenz function L, the figure-of-merit can be rewritten as follows:

$$zT = \frac{\alpha^2}{L(1+\kappa^l/\kappa^e)}. \tag{10.13}$$

Note that it is unreasonable to assume $\kappa^l = 0$. This would lead to a figure-of-merit of $zT = \alpha^2/L$, which diverges when the carrier concentration approaches zero (where the Seebeck coefficient diverges). It can, however, give an upper limit of zT when α is limited.

The figure-of-merit can thus within the relaxation time approximation be calculated purely from the band structure, provided we have access to an appropriate scattering model and the lattice thermal conductivity can be quantified. Unfortunately, neither of these quantities can be calculated directly from the electronic band structure. They also vary as a function of the chemical potential, carrier concentration and temperature. Nevertheless, we may gain important insight into the behavior of the figure-of-merit if κ^l is assumed to be constant or alternatively represented by an energy or wavevector-dependent model.

The relaxation, or scattering, time τ is generally difficult to calculate from first principles. To make the calculation of the transport coefficients possible, it is customary to assume that it is independent of the wave vector and either constant in energy or dependent on energy through analytic expressions from deducted scattering models [2]. For a constant relaxation time, the Hall and Seebeck coefficients are independent on the scattering due to the cancellation of the relaxation time. However, for the conductivities it does not cancel and needs to be specified. We will discuss the details regarding τ in the next section.

10.3 Technical Details

In this chapter, all integrals in the transport coefficients have been solved by Romberg integration in reciprocal space based on a controlled accuracy (cutoff at 10^{-3}) with on-the-fly three-dimensional modified Shepard interpolation [41].

10.3.1 Band Generation

We will throughout this chapter use the parabolic band approximation [1] to explicitly illustrate the band and scattering dependencies of the transport coefficients. These bands were generated according to the following equation:

$$\epsilon_n(\mathbf{k}) = \frac{\hbar^2 k^2}{2m_n^*} + \epsilon_s^n. \tag{10.14}$$

Here, ϵ_s^n is an energy shift determining the band onset and m_n^* is the effective mass for band n. To simplify we use the notation

$$m_n^* = a_n m_e, \tag{10.15}$$

where m_e is the electron mass and a_n is the ratio of the effective mass to the free electron mass. Note that a negative a_n applies to a valence band, while a positive a_n applies to a conduction band. Throughout this chapter we use negative values for a_n although all discussions and conclusions will be similar for positive a_n with a sign change where applicable.

From Equation 10.9, the parabolic band velocity is given by

$$\upsilon_i^n = \frac{\hbar k_i}{m_n^*}. \tag{10.16}$$

When multiple bands were studied, the different band energies $\epsilon_n(\mathbf{k})$ were generated according to the recipe above, with different input parameters a_n and ϵ_s^n, and combined in superposed band structures. Unless otherwise stated, all data in this chapter have been generated at a temperature of 300 K.

10.3.2 Chemical Potential and Charge Carrier Concentration

In this chapter, we present the transport coefficients as a function of both the chemical potential and the charge carrier concentration. To a certain degree this gives redundant information, but has the advantage that the behavior of the transport coefficients can be compared between the two. This is particularly convenient for coefficients which are shifted as a function of carrier concentration and yield what seems to be a change in magnitude at fixed carrier concentration.

To obtain the carrier concentration n_c for a single electron band, the following parabolic band relation [1] needs to be solved:

$$n_c(\mu,T) = \frac{1}{2\pi^3}\left(\frac{2am_e k_B T}{\hbar^2}\right)^{3/2} \int_0^\infty \epsilon^{1/2} f(\mu,T)\, d\epsilon \tag{10.17}$$

at a given temperature T. In order to calculate this, knowledge of the chemical potential μ is necessary. By summing the contribution from each band, the total carrier concentration is obtained. In this chapter, the chemical potential has been varied between −0.3 and 0.3 eV with corresponding values for the charge carrier concentrations. The first band onset is fixed at 0 eV.

10.3.3 The Conducting States

All transport coefficients within the Boltzmann formalism have some parts in common. One is the *pickup* integrands, which determine which states are included in the integration. The pickup P_n^i with band index n is defined as

$$P_n^i = (\epsilon_n - \mu)^i \frac{\partial f_n}{\partial \epsilon}, \tag{10.18}$$

where the constituents have been defined in preceding sections. The pickup integrands enter the electrical conductivity, the Seebeck coefficient, and the electrical part of the thermal conductivity with the exponent $i = 0$, 1, and 2, respectively.

It can be useful to visualize these pickup integrands at selected temperatures. In Figure 10.1, such a visualization is given as a function of $(\epsilon - \mu)$. In general, these functions determine which states are included in the transport coefficient integrals at the chemical potential μ, and thus which states contribute to the conduction process. A change in temperature mostly alters the width of the pickup function. The further away from the chemical potential the states are located, the less likely they will be included. Since the width increases with temperature, states lying higher in the conduction band or deeper in the valence band will contribute at elevated temperatures. As a consequence of the increased pickup width and changes to the chemical potential, minority carriers will often also be picked up at elevated temperatures. This yields a peak in the Seebeck coefficient as a function of temperature, causing further constraints on the optimal temperature range. For all reasonable temperatures, a width of 10 $k_B T$ from the chemical potential is sufficient to include all contributions. Bands outside this region will generally not contribute, except indirectly through, for example, a scattering process.

The symmetric behavior of P^0 around μ gives the electrical conductivity its simple band structure dependency. The pickup function of the Seebeck coefficient, however, is P^1, which is antisymmetric, divided by P^0 which is symmetric around μ. This has important implications for which band velocity configurations are favorable for the Seebeck coefficient; the transport distribution should be asymmetric around the chemical potential in order to maximize α [15].

It is worth noting that it is the total transport distribution $T_{ij}(\epsilon)$

$$T_{ij}(\epsilon) = N(\epsilon)\tau(\epsilon)\upsilon_i(\epsilon)\upsilon_j(\epsilon), \tag{10.19}$$

that should be asymmetric around the chemical potential, not necessarily all or just one of its constituents. $N(\epsilon)$ is here the density of states. The asymmetric behavior of the pickup function is precisely the reason why the Seebeck coefficient is very small in normal metals, while it is high for insulators (with the chemical potential in the band gap).

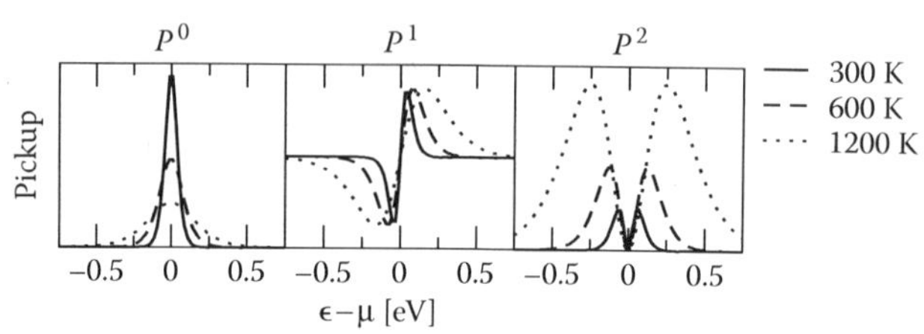

FIGURE 10.1 The pickup of conducting states (Equation 10.18) as a function of $\epsilon - \mu$ for three different temperatures: 300 K (solid curve), 600 K (dashed curve), and 1200 K (dotted curve).

It is interesting to compare the shapes of P^0 and P^2 in Figure 10.1, which constitute important parts of the electrical conductivity and the electrical part of the thermal conductivity, respectively. They will in general pick up different states for different temperatures and chemical potentials, which means that the dependency as a function of the chemical potential and carrier concentration can be different for σ and κ^e. It should thus be easy to convince the readers why the Wiedemann–Franz ratio generally yields a Lorenz *function* and not a Lorenz constant when the chemical potential is varied through different transport distributions. This has also been demonstrated in highly degenerate semiconductors [42] and superlattices [28], where the deviations of L from the Sommerfeld value are significant.

10.3.4 Scattering Models

To be able to predict the behavior of the transport coefficients for systems where the constant relaxation time is not applicable, energy-dependent scattering models have been included. Readers should consult the book by Fistul [2] and references therein for derivations and details of these models.

Three basic scattering models have been studied: acoustic phonon scattering from deformations, ionized impurity, and constant scattering. They can all be expressed mathematically in a similar manner, only distinguished by the scattering parameter r and different scattering prefactors τ_0:

$$\tau_r = \tau_0\left(\frac{\epsilon}{k_B T}\right)^{r-1/2}. \tag{10.20}$$

The band energy ϵ has its zero reference at the band onset, while the prefactor is in this chapter taken to be constant $\tau_0 = 1$ fs. The prefactor relies on the scattering model and can depend on the temperature, effective mass, and other band structure and deformation parameters [2]. These are highly system dependent, and outside the scope of this chapter. We would like to stress the fact that τ_0 depends on temperature, since this could weaken parts of the following analysis. However, defining a temperature dependence of τ_0 would be arbitrary without specifying the material. To simplify the discussions, we have redefined κ_l by dividing with the numeric value of τ_0 in units of fs. All discussions involving changes to κ_l in the following is thus a discussion of the effective κ_l, which can be modified by changes either to the lattice thermal conductivity or the scattering prefactor. The scattering parameter r distinguishes between the three models, and has been specified in Table 10.1.

Different scattering models are combined through the Matthiessen's rule. The total relaxation time can be built from separate contributions in the following manner:

$$\frac{1}{\tau} = \sum_i \frac{1}{\tau_i}, \tag{10.21}$$

where i specifies the scattering type.

TABLE 10.1 Scattering Parameter r for Three Different Scattering Models: Acoustic Phonon from Deformations, Ionized Impurity, and Constant Relaxation Time

	r
Acoustic phonon	0
Ionized impurity	2
Constant	$\frac{1}{2}$

10.4 Band Guidelines

With a well-defined formalism we will, in the following sections, investigate the basic trends of the band structure dependencies of the transport coefficients. Based on the limitations of the scattering models and the thermal conductivity it is important to note that direct comparisons with real systems should be avoided. We have therefore omitted the units for all transport coefficients and focussed on the trends.

10.4.1 Single Parabolic Band

The transport coefficients for single parabolic bands with three different effective masses are shown as a function of the chemical potential and the carrier concentration in Figure 10.2. Results for three different scattering models (constant relaxation time, acoustic phonon scattering, and ionized impurity scattering) have been included. The data for lower effective mass configurations are terminated due to the limited calculation window that was used for the chemical potential.

The electrical conductivity σ increases with increasing carrier concentration n_c, which is understood from the relation $\sigma = n_c e\mu$, where μ is the mobility. An increasing carrier concentration corresponds to a decreasing chemical potential for valence bands. It is furthermore easy to show that σ increases with increasing effective mass for parabolic bands at fixed temperature and chemical potential. The carrier concentration for a single parabolic band is proportional to $m^{*3/2}$, while the mobility is proportional to τm^{*-1}. The conductivity is the product of the mobility and carrier concentration, and is consequently proportional to $\tau m^{*1/2}$. This behavior is confirmed in Figure 10.2. The electrical conductivity increases with increasing effective mass not only for constant relaxation time but also for all the scattering processes. However, since the scattering prefactor also relies on the effective mass, we can expect modifications to these trends when studying real systems. The observed shifts as a function of carrier concentration are in compliance with Equation 10.17. While acoustic phonon scattering behaves quite similarly to the constant scattering model, ionized impurity scattering is less detrimental at high carrier energies for the electrical conductivity, assuming similar τ_0 for the processes. This is due to the $\tau \propto \epsilon^{3/2}$ dependence, which means that the ionized impurity scattering is only weakly present at high energies and primarily scatters charge carriers with low energies.

The Seebeck coefficient α has the opposite behavior; decreasing when the carrier concentration increases. A single parabolic band expression for the Seebeck coefficient can be obtained from Equation 10.3, such that

$$\alpha = \frac{1}{eT}\left(\frac{\int \tau\epsilon^{5/2}\frac{\partial f}{\partial \epsilon}\,d\epsilon}{\int \tau\epsilon^{3/2}\frac{\partial f}{\partial \epsilon}\,d\epsilon} - \mu\right). \tag{10.22}$$

This expression is independent of the band mass if given as a function of μ, which is also evident from Figure 10.2. However, as a function of the carrier concentration, the band mass is incorporated (through Equation 10.17) and thus the Seebeck coefficient is shifted when the effective mass varies, again in accordance with Figure 10.2. It should be emphasized that it is difficult to freely choose the chemical potential in a real system, but by, for instance, doping and substitutions it is possible to tune the Fermi level and thus the onset of the chemical potential at absolute zero. It is further important to realize that the Seebeck coefficient does not depend strongly on the scattering mechanism. This is largely due to the similar appearance of the denominator and numerator in the Seebeck coefficient, cancelling large parts of the effects of the scattering, leaving primarily the chemical potential dependency. However, for ionized impurity scattering, the ratio of the integrals is larger, yielding a slightly increased Seebeck coefficient, but with similar dependency on the chemical potential and effective mass. The effect is opposite for acoustic phonon scattering, although less, since the scattering factor r is larger for ionized impurity scattering and the constant scattering model.

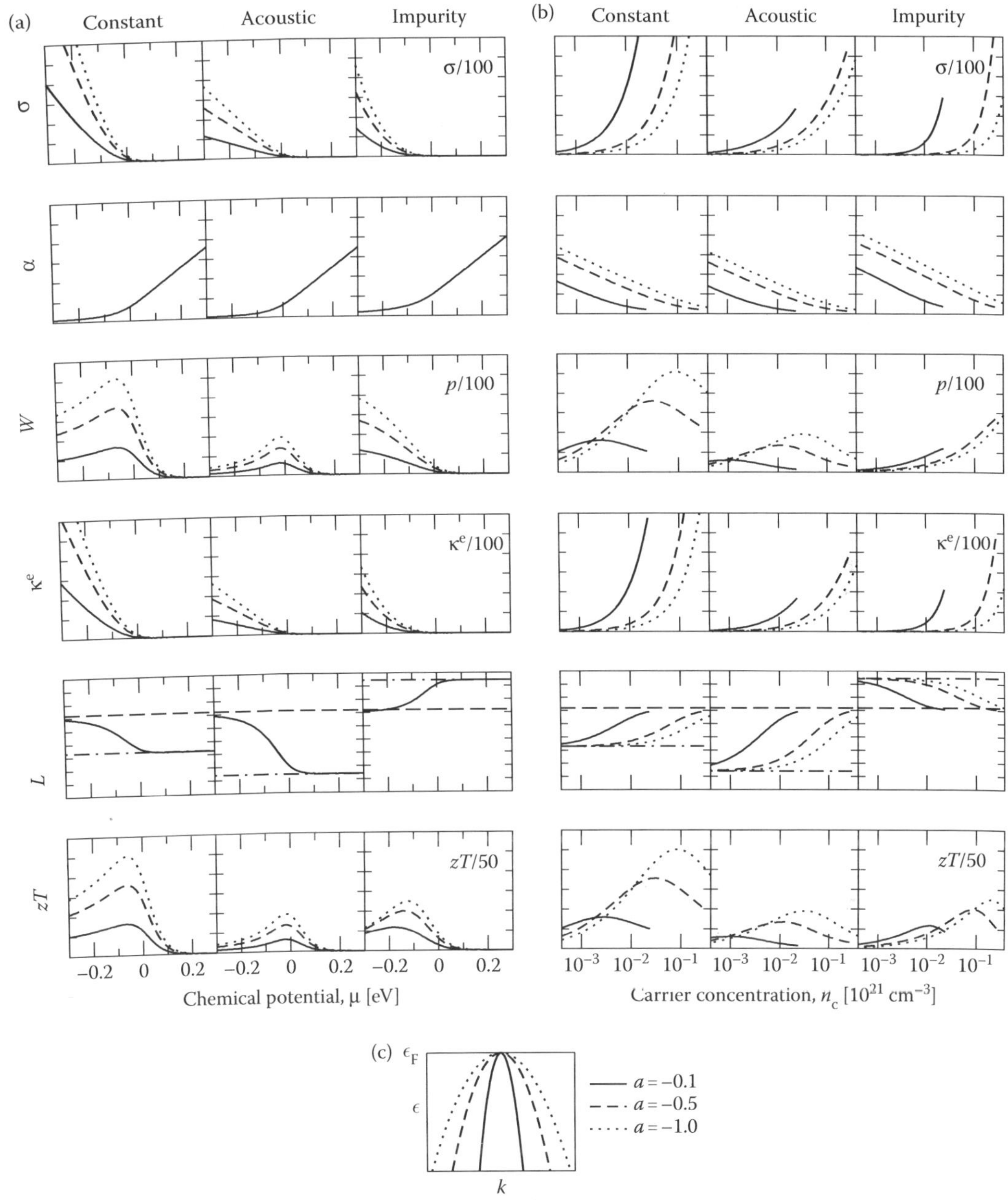

FIGURE 10.2 Transport coefficients, figure-of-merit, and Lorenz function for three different relative effective masses a as defined in Equation 10.15 ($a = -0.1$, solid line; $a = -0.5$, dashed line; and $a = -1.0$, dotted line) and three different scattering models given as a function of the chemical potential (a) and the carrier concentration (b). The plots show from top to bottom the electrical conductivity σ, the Seebeck coefficient α, the power factor $W = \alpha^2\sigma$, the electronic part of the thermal conductivity κ^e, the Lorenz function L, and the dimensionless figure-of-merit zT. From left to right, the scattering models are: constant relaxation time, acoustic phonon, and ionized impurity scattering (see Equation 10.20 for their definition). The figure-of-merit zT has been calculated using an effective $\kappa^l = 1\ \mathrm{W\ K^{-1}\ m^{-1}}$. The long-dashed and long-dashed-dotted lines represent the degenerate and nondegenerate limits of the Lorenz function. The legend and band structures are shown in (c).

The power factor W is calculated from $\alpha^2\sigma$. For a given chemical potential, it increases with increasing effective mass for all scattering processes. Since ionized impurity scattering is less detrimental for the electrical conductivity at high carrier concentration, and the Seebeck coefficient is slightly larger, the power factor is largest for the case with ionized impurity scattering. However, the magnitude of the power factor is dependent on the scattering prefactor and is thus heavily system dependent. Ionized impurity scattering mostly affects low-energy charge carriers and the peaks of the power factor for this type of scattering are consequently shifted to lower chemical potentials and higher carrier concentrations. Except for this, the power factor maximum aligns at similar chemical potentials for different effective masses. However, due to the mass dependence in Equation 10.17 this does not hold as a function of the carrier concentration. Consequently, the power factor peak shift is a function of carrier concentration.

The electronic part of the thermal conductivity κ^e is at first sight similar to the electrical conductivity. This is expected from the Wiedemann–Franz law when the Lorenz function and temperature are taken to be constant. However, as discussed above, there are important corrections to the Wiedemann–Franz law in the case of semiconductors or superlattices. These corrections can be seen in Figure 10.2 where the Lorenz function is visualized for all three scattering processes. The degenerate metallic limit, or Sommerfeld value of 2.44×10^{-8} V^2 K^{-2} is reached for all scattering processes at high carrier concentrations and low chemical potentials. However, there are significant corrections at lower carrier concentrations and higher chemical potentials. At the non-degenerate limit, the Lorenz function converges to the known values of 1.86, 1.48, and 2.97×10^{-8} V^2 K^{-2} for constant, acoustic phonon, and ionized impurity scattering, respectively. Assuming the Sommerfeld value for acoustic phonon scattering at low carrier concentrations, when the system is really dominated by ionized impurity scattering, can lead to a deviation of up to a factor of two. However, since most thermoelectric materials exhibit relatively high carrier concentrations it is unlikely that the deviations are this large for simple bulk systems.

Finally, the figure-of-merit is given in the lower panels of Figure 10.2. As expected, the behavior of the figure-of-merit is very similar to that of the power factor, with one exception. Relative to the power factor, ionized impurity scattering causes a shift of the zT peak to higher chemical potentials and lower carrier concentrations. This is caused by the compensating effects between κ^e and σ, that is, that both are reduced at low carrier concentrations. Ionized impurity scattering seems to be favorable to acoustic phonon scattering and the constant scattering model. However, it should be emphasized that an accurate quantification of τ_0 for the different scattering models should be done in order to reach definite conclusion on this manner. Such a study is system specific and out of the scope of this chapter. An important observation for carrier concentration optimization is the carrier concentration shift of the zT peak between the three different scattering models. Due to the energy dependence of the ionized impurity scattering, the significant part of the scattering affects low-energy charge carriers, where τ approaches zero, so that the electrical conductivity is reduced close to the band onset. Consequently, there is a shift of the figure-of-merit peak to higher carrier concentrations compared to the energy-independent constant scattering model. When acoustic phonon scattering dominates, the opposite occurs. Systems with predominantly acoustic phonon scattering will accordingly be optimized at a lower carrier concentration than what is expected for constant and ionized impurity scattering. However, modifications to this can arise for a system specific τ_0.

The manifestation of the strong interdependence between the transport coefficients is clear from the previous discussions. The parameter space is limited to the effective mass, the carrier concentration, and the lattice thermal conductivity for a single-band system. At this stage it seems difficult to simultaneously increase the Seebeck coefficient and the electrical conductivity due to their explicit inverse relation. However, as will be apparent later, there are ways of modifying this interdependence.

The majority of thermoelectric materials exhibit multiband behavior. Even though some of these effects can be approximated through a single parabolic band by adjusting the effective mass and scattering parameters, there are cases in which the multiband effects are nontrivial. We will devote the next section to a few representative multiband schemes to illustrate their ramification.

10.4.2 Multiple Parabolic Bands

The transport tensors in Equations 10.1 through 10.3 and 10.7 are well defined for multiple bands, and we will in this section study how this influences the transport properties. The multiband electrical conductivity is usually additive; there is thus, in principle, no need to study its explicit behavior for bands with similar band mass and scattering properties. However, for a combination of bands with different effective mass and different scattering properties there will be a modified dependence of the electrical conductivity on the chemical potential and carrier concentration. In addition, energy shifts between the bands will cause modifications related to the band pickup. The Seebeck coefficient needs particular attention, since it is necessary to sum the band contributions separately in the numerator and denominator before taking the ratio. As a function of the chemical potential for a single parabolic band, the Seebeck coefficient does not explicitly depend on the effective mass. However, perturbations of the Seebeck coefficient are expected for multiband scenarios with different effective mass, scattering properties, and energy onsets. Consequently, the power factor and figure-of-merit inherit changes based on this, and it is thus interesting to see the explicit multiband behavior of these coefficients.

In Figure 10.3, transport coefficients have been given for five different band structures, combining bands with different effective masses and energy shifts. Single parabolic bands with band masses $a = -0.1$ and -1.0 were compared to three combinations of bands: no energy shift $\epsilon_s = 0$ eV between the bands, the heavy band shifted by $\epsilon_s = -0.15$ eV, and the light band shifted similarly.

In a multiband scenario, the total electrical conductivity is composed of a sum of the electrical conductivities for each band and a transition region occurs for the band configurations, which includes a shifted band. This is also verified in Figure 10.3. The contributions from the energy-shifted bands have their onset at similarly shifted chemical potentials and carrier concentrations. Since the Seebeck coefficient is heavily dependent on the chemical potential and thus the carrier concentration, adding a shifted light band to a heavy band does not yield a large difference. However, when a shifted heavy band is added to a light band, a significant transition region can be observed. The Seebeck coefficient then follows an s-shaped behavior as a function of chemical potential or carrier concentration. Such a trend has indeed been observed in a number of experiments.

The power factor is accordingly modified; the location of the major peak follows the heaviest band, while local peaks or shoulders appear close to the onset of the lighter band. Depending on the energy shifts between these bands and their effective mass, the peaks or shoulders are more or less visible. From this it can be seen that the power factor can be improved over a wider range of the chemical potential if a favorable cascade of bands can be assembled—importantly, this is regardless of the scattering mechanism.

As for the single bands, the qualitative behavior of the electronic part of the thermal conductivity coincides with the electrical conductivity. The resulting figure-of-merit thus mimics the features of the power factor, except for the peak shift which occurs for ionized impurity scattering. Compared to the case of constant scattering, acoustic phonon and ionized impurity scattering change the zT magnitude and peak location between the nonshifted multiband and shifted multiband significantly. In particular, for ionized impurity scattering, the maximum magnitude of the figure-of-merit for the shifted heavy band is more similar to the other configurations than what is the case for acoustic phonon and constant scattering. This feature is expected from the energy dependency of the scattering models.

10.4.3 Thermal Dependencies

The thermal conductivity is rarely constant and depends on the dynamical lattice effects of the material. Even though the parameters determining the thermal conductivity are difficult to calculate and measure, it is worthwhile to illustrate how the figure-of-merit depends on the thermal conductivity and how the scattering mechanism affects this dependency. To investigate this, the effective $\kappa^l = 0.1$,

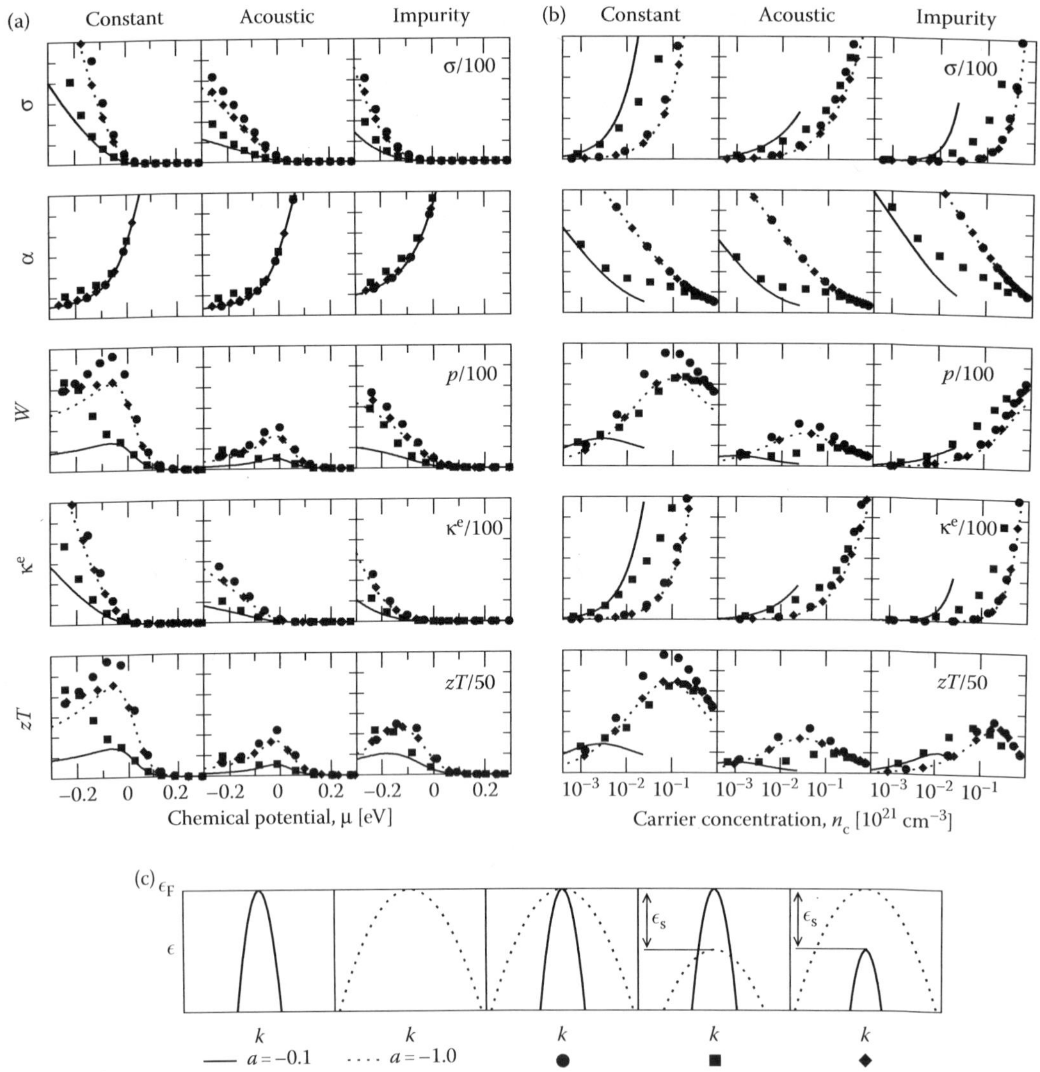

FIGURE 10.3 Transport coefficients for five different combinations of bands, effective masses, and energy shifts; solid line ($a = -0.1$), dashed line ($a = -1.0$), dots ($a_1 = -0.1$ and $a_2 = -1.0$), squares ($a_1 = -0.1$ and $a_2 = -1.0$, $\epsilon_s^2 = -0.15$ eV), and diamonds ($a_1 = -0.1$, $\epsilon_s^1 = -0.15$eV, and $a_2 = -1.0$), respectively, given as a function of the chemical potential (a) and carrier concentration (b). The legend with the band configurations is shown in (c).

1.0, and 10.0 W m^1 K^{-1} were chosen to give a representative selection of effective thermal conductivities. It is assumed that this thermal conductivity is independent of the chemical potential and temperature, which is a simplification that does not necessarily hold for real materials where for example the phonon dispersion changes with doping. The top of Figure 10.4 shows the figure-of-merit as a function of chemical potential and carrier concentration for the respective thermal conductivities. Note that the data have been scaled by the inverse of the effective κ^l (see Figure 10.4c). This has been done for the sake of comparison, based on the naive expectation that the figure-of-merit depends on the reciprocal thermal conductivity. Generally, as expected, a lower thermal conductivity indeed results in a higher figure-of-merit. However, important subtleties need particular attention; the peak is shifted to lower chemical potentials at high thermal conductivities for constant and ionized

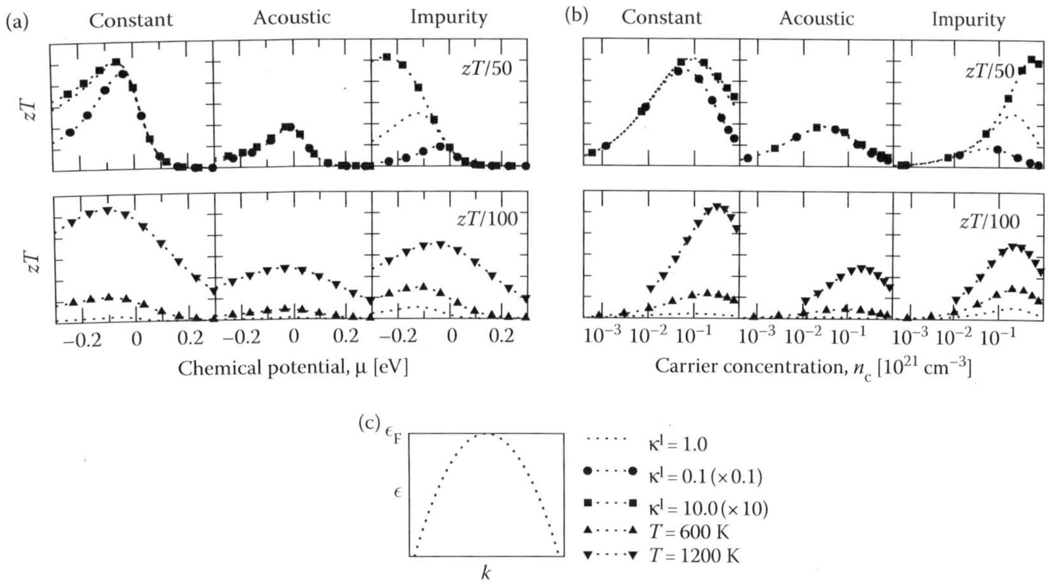

FIGURE 10.4 The figure-of-merit as a function of chemical potential (a) and carrier concentration (b). Three different values of κ^l at $T = 300$ K are shown in the top panels: dotted-dots ($\kappa^l = 0.1$ W K^{-1} m^{-1}), dotted ($\kappa^l = 1.0$ W K^{-1} m^{-1}), and dotted-squares ($\kappa^l = 10.0$ W K^{-1} m^{-1}). The bottom panels show the figure-of-merit for three different temperatures T with $\kappa^l = 1.0$ W K^{-1} m^{-1}: dotted ($T = 300$ K), dotted-up-triangles ($T = 600$ K), and dotted-down-triangles ($T = 1200$ K). An effective mass of $a = -1.0$ was used for all data sets. The legend of the configurations is shown in (c). Note the scaling for different κ^l.

impurity scattering, while there is no apparent shift for the acoustic phonon scattering. This can be explained by the weighting of κ^l/κ^e, determined by the size of κ^l. Any change in κ^e would be more visible for a larger κ^l. For ionized impurity scattering where κ^e is modified at low energies, there is a marked shift. On the opposite, for acoustic scattering this occurs at higher energies, and the figure-of-merit peaks are more aligned. Furthermore, the magnitude of the figure-of-merit scales well with the thermal conductivity for the acoustic phonon scattering. This is also partly true for the constant scattering, albeit only for the higher values of κ^l. For ionized impurity scattering, the relation between the peak magnitudes of the figure-of-merit for different thermal conductivities does not follow this behavior. We would like to stress the fact that in order to investigate these trends in a conclusive manner, a material specific estimate of τ_0 is needed. Detailed knowledge of the scattering properties in thermoelectric materials in general is thus of uttermost importance to reach the next level of understanding and performance.

In order to predict optimum operational temperature the temperature dependencies of the figure-of-merit are of particular importance. To shed light on this, temperatures of 300, 600, and 1200 K were chosen as a representative set. The figure-of-merit for each of those temperatures, at an effective $\kappa^l = 1.0$ W m^{-1} K^{-1}, is shown in the lower panels of Figure 10.4. It is immediately clear that the figure-of-merit increases with temperature, as expected from its explicit linear temperature dependency. However, there is an additional amplification due to the augmented pickup function which leads to increased electrical conductivity and Seebeck coefficient at elevated temperatures. Additional temperature dependencies enter through the scattering models and the relation between the chemical potential and carrier concentration. This makes for a complicated relationship, which is additionally burdened by the lack of more detailed knowledge of the scattering prefactors.

Even though multiband effects can be complicated, it is still possible to isolate and interpret the results if a proper modeling framework is used. To reach a new level of general understanding and ultimately higher performance, further knowledge on the detailed scattering properties is vital. Such knowledge can most likely be obtained by interwoven calculations and experiments. Furthermore, it is clear from the preceding sections that simultaneous improvements to the electrical conductivity and Seebeck coefficient seem to be a difficult task, given their explicit inverse relation. We will now use the preceding knowledge and investigate how this can be partly circumvented and consequently how the figure-of-merit can be improved by means of a concept known as energy filtering.

10.5 Energy Filtering

The interdependence between the transport coefficients emplace strict constraints on the optimization of thermoelectric materials, as demonstrated in the preceding sections. We thus need to pursue alternative paths to improve the electrical properties of thermoelectric materials. A possible approach involves reducing the number of low-energy charge carriers, a process that partly decouples the electrical conductivity and the Seebeck coefficient. This increases the asymmetry of the transport distribution function, ultimately improving the Seebeck coefficient without reducing the electrical conductivity deeper in the band. It is hence possible to improve the power factor and thus also the figure-of-merit. In this section, we will follow the route of "energy filtering" and investigate the concept by means of a simplified model.

The objective is to filter out low-energy charge carriers. A concept fulfilling this purpose is illustrated in Figure 10.5, where all charge carriers with energy less than the energy barrier ϵ_b have been removed from the conduction process. The result of this is illustrated in the lower part of Figure 10.5. As a consequence of the filtering, artificial asymmetry is introduced in the transport distribution tensor. This will

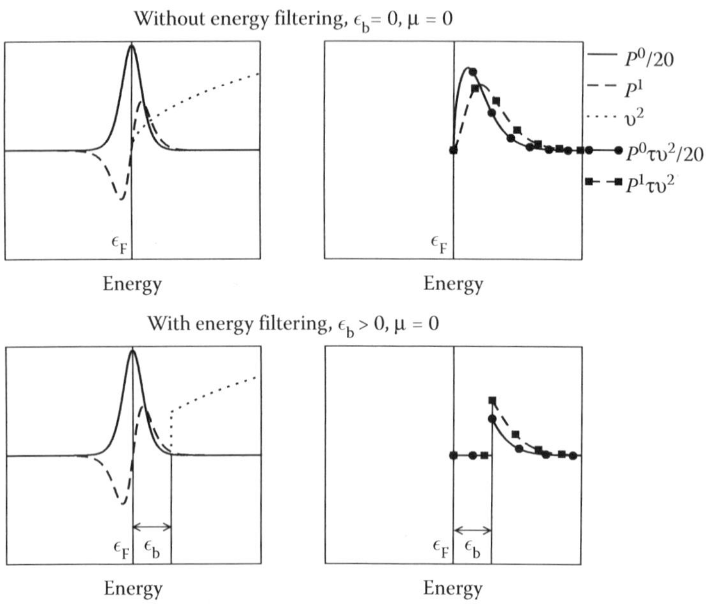

FIGURE 10.5 Schematics of the concept of energy filtering. The average trace of the band velocity tensor is drawn as a dotted line. The pickup functions for the conductivity (solid line), the Seebeck coefficient (dashed line), and the integrands in the denominator (solid-circle) and numerator (dashed-square) of the Seebeck coefficient are also shown. By removing the low-energy charge carriers from the conduction process through an energy barrier ϵ_b, a larger effective asymmetry in the transport distribution function can be obtained. This results in a change in the ratio between the integrals in the denominator and numerator of the Seebeck coefficient, ultimately leading to enhancements in the power factor and figure-of-merit.

increase the Seebeck coefficient, given the sensitivity to such asymmetry in its numerator integrand (see Equations 10.1 and 10.2 and Figure 10.5). More specifically, this means that the numerator integral of the Seebeck coefficient will be increased relative to the denominator integral, since their integrands have changed upon filtering. For the unfiltered case, the integral of $P^0\tau\upsilon^2$ is larger than $P^1\tau\upsilon^2$ by construction. Thus, by reducing the band velocity multiplied by the relaxation time at low energies, the $P^0\tau\upsilon^2$ integrand is more affected than $P^1\tau\upsilon^2$ and the ratio of the integrals in the numerator and denominator increases. This ultimately leads to an improvement of the Seebeck coefficient.

The carrier concentration in the filtering region is by definition zero at $T = 0$ K. This leads to a vanishing electrical conductivity, which at first sight seems to prevent improvements to the power factor and figure-of-merit. However, as will be apparent later, this approach will still yield elevated power factors and figure-of-merits due to the difference in the pickup functions of the Seebeck coefficient and electrical conductivity at higher temperatures.

A simple implementation of a filtering concept can be illustrated by introducing a completely inelastic charge carrier scattering model for the low-energy carriers. The relaxation time approximation is valid for elastic or isotropic inelastic scattering, but these effects can also surface from changes in the band velocities and the band structure. This approach is thus still valid for illustrative purposes. To this end, an energy-dependent relaxation time τ_f, based on the Heaviside step-function, can be introduced as follows:

$$\tau_f(\epsilon) = \tau_0 H\left(|\epsilon| - \epsilon_b\right), \tag{10.23}$$

where ϵ_b is the energy filtering barrier height and $\tau_0 = 1$ fs. The Heaviside step-function is defined as

$$H\left(|\epsilon| - \epsilon_b \begin{cases} 0 & |\epsilon| < \epsilon_b \\ \frac{1}{2} & |\epsilon| = \epsilon_b \\ 1 & |\epsilon| > \epsilon_b \end{cases}\right) \tag{10.24}$$

This filtering τ_f is combined with the constant, acoustic phonon, and ionized impurity scattering models through Mathiessen's rule (see Equation 10.21).

Transport coefficients for different ϵ_b and scattering mechanisms are shown in Figure 10.6. Note that the filtered and nonfiltered configurations are plotted along the same carrier concentration, which is an oversimplification. However, even though a shift is expected when correcting this, the rest of the discussion remains valid. The pickup function of the electrical conductivity is shifted to lower chemical potential and higher carrier concentrations due to the filtering. As expected, when the chemical potential is much lower than ϵ_b, at high carrier concentrations, the conductivity approaches the unfiltered limit. At the same time, the Seebeck coefficient is also shifted and experiences an increase in magnitude at a given chemical potential.

Since the power factor $W = \alpha^2\sigma$ and $\alpha \propto \chi/\sigma$, the power factor will be proportional to χ^2/σ. In general, the Seebeck coefficient is proportional to χ/σ, the extra χ factor in the power factor will make it even more sensitive to changes in the asymmetry due to, for example, filtering effects. We thus expect significant changes to the power factor upon energy filtering. Indeed, this can be seen in Figure 10.6, where there is a significant improvement in the power factor for all scattering processes. The magnitude and position of the peak depend on the filtering barrier height, the reduction of the electrical conductivity, the increase of the Seebeck coefficient, and the scattering mechanism. But the peak of the power factor is generally shifted to lower chemical potentials. The analytic relation between the chemical potential and carrier concentration is, in principle, not valid in the filtering region, and the resulting shift for the carrier concentration should be less than expected from Equation 10.17.

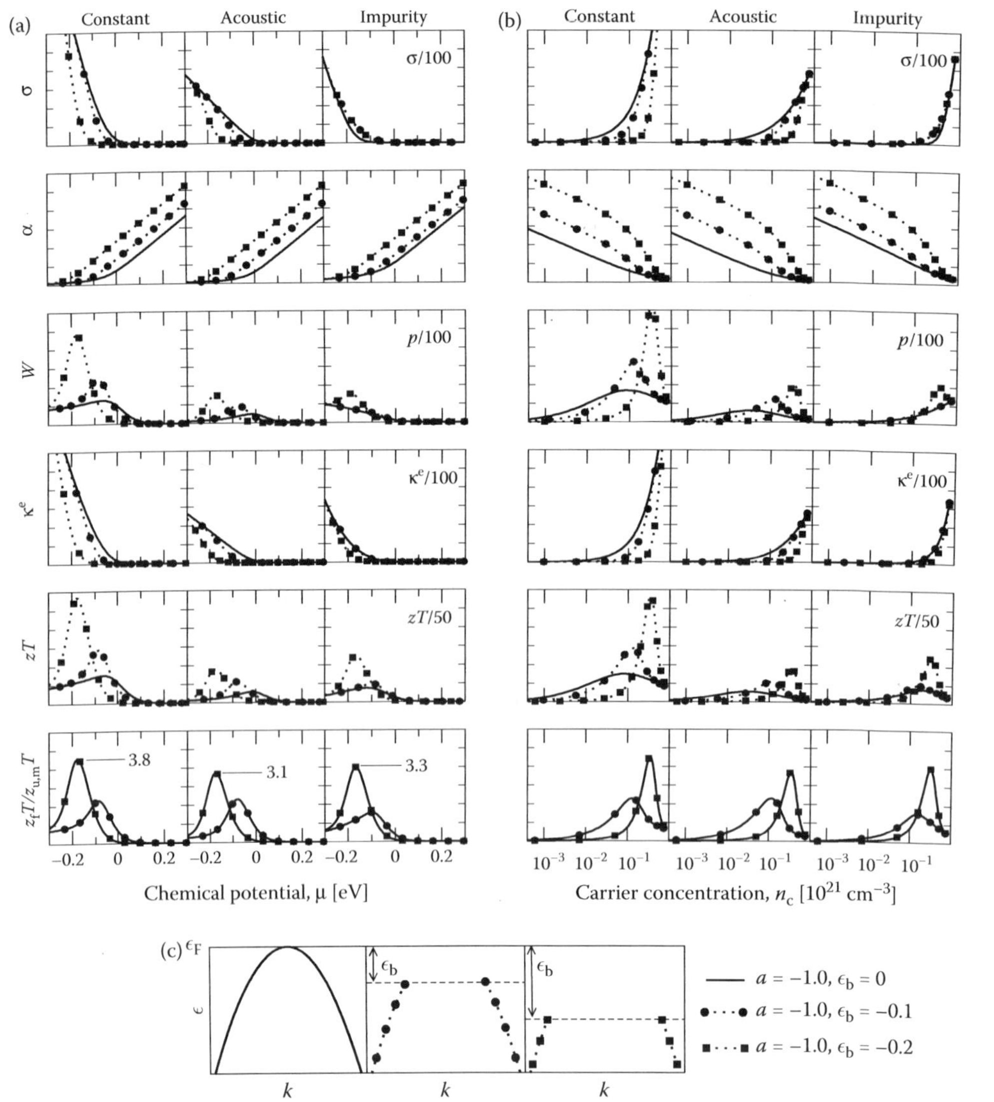

FIGURE 10.6 Transport coefficients as a function of chemical potential (a) and carrier concentration (b) for unfiltered (solid), filtered with $\epsilon_b = -0.1$ eV (dotted-dots) and filtered with $\epsilon_b = -0.2$ eV (dotted-squares). All configurations have a band mass of $a = -1.0$. For the band configuration legend, see (c). To demonstrate improvements to the figure-of-merit from energy filtering, the ratio $z_fT/z_{u,m}T$ of the maximum unfiltered $z_{u,m}T$ and the filtered z_fT has been plotted in the lower panels of (a) and (b).

Due to the similarities of κ^e and σ, the figure-of-merit sees improvements similar to those of the power factor. To better illustrate the impact of energy filtering on the figure-of-merit, the ratio $z_fT/z_{u,m}T$ of the filtered z_fT and the maximum unfiltered $z_{u,m}T$ is given explicitly in the lower panels of Figure 10.6. The maximum of the figure-of-merit may be increased significantly, typically a threefold increase for all scattering mechanisms. The peaks of the power factor and figure-of-merit are shifted to lower chemical potentials compared to the unfiltered case. Acoustic phonon scattering mainly affects carriers with high kinetic energy and modifications to the low-energy carriers become more important. Thus, for low barrier heights it is expected that the filtering concept is more effective for acoustic phonon scattering, while the trend changes at high barrier heights, where ionized impurity scattering is less important. This can be seen for the results for the smallest barrier height of 0.1 eV, where the improvement is largest for acoustic phonon scattering and smallest for ionized impurity scattering. At higher barrier heights, this is opposite. However, the detailed transition region will depend on the exact magnitude of τ_0.

In general, the maximum zT increases monotonously with the barrier height; also, such an increase is accompanied by shifting the peak to lower chemical potentials and higher carrier concentrations. Thus, there is an upper limit to the barrier height and thus the figure-of-merit determined by the achievable carrier concentrations in the material of interest. What is obvious is that energy filtering has significant potential to improve the figure-of-merit, in particular if acoustic phonon scattering dominates at low barrier heights or constant or ionized impurity scattering at higher barrier heights.

As an additional consequence of the shifted pickup, the peak of the power factor and figure-of-merit is shifted to a different chemical potential and carrier concentration. Energy filtering will thus not only improve the magnitudes of these peaks but might also serve as an additional tuning parameter in materials where it is otherwise difficult to tune the carrier concentration. By introducing energy filtering and controlling the barrier height, the figure-of-merit can potentially be optimized as a function of carrier concentration. This can serve as an alternative route to optimize carrier concentrations and thus further enhance thermoelectric materials.

As discussed in the introduction of this chapter, filtering effects have already been demonstrated in a few different materials [28,30–34]. The filtering was, for instance, achieved at interfaces in bulk samples, either between grains in a nanocrystalline sample [30] or between the host lattice and nanoinclusions [43].

Even though the energy filtering in this section is based on a very simple model, it nevertheless demonstrates the theoretical potential of filtering out low-energy charge carriers from the electronic conduction process. Whether such filtering can be implemented in materials remains to be seen, but the large potential of such an approach suggests that this should be investigated further.

10.6 Summary and Outlook

In this chapter, semiclassical Boltzmann theory in combination with the parabolic approximation were used to explicitly demonstrate how the transport coefficients depend on the band structure and three energy-dependent scattering mechanisms: constant, acoustic phonon, and ionized impurity scattering. An additional scattering model was incorporated to illustrate how these coefficients respond to the concept of hindering the low-energy charge carriers to contribute to the electronic conduction process, the concept usually referred to as "energy filtering."

In accordance with previous work, the figure-of-merit increases with the effective mass for all important scattering mechanisms. Acoustic phonon scattering tends to shift the peak of the figure-of-merit to lower carrier concentrations compared to the constant scattering model, while the opposite takes place for ionized impurity scattering. Assuming similar prefactors τ_0 for the scattering models, ionized impurity scattering potentially yields the highest figure-of-merit values. Important corrections to the Sommerfeld value of the Lorenz function were illustrated and a case-by-case calculations of the Lorenz function should be done to reach more accurate determination of the lattice thermal conductivity and interpretations of trends for specific materials. Furthermore, it was seen that the multiband behavior of the transport coefficients introduced important subtleties related to the peak position and peak width of the figure-of-merit as a function of chemical potential and carrier concentration. Thermal dependencies of the figure-of-merit were also investigated, but were limited by the lack of system specific scattering prefactors. We would like to emphasize the importance of estimating the prefactor. This is a particularly complicated task, but is important to reach conclusive predictions of future improvements.

The interdependence of the electrical conductivity and the Seebeck coefficient is indisputable. Modifications of this interdependence are however possible. Selective filtering of low-energy charge carriers is particularly promising, and optimistic accounts of the figure-of-merit indicates that it can be significantly increased, in our case by a factor of three. This effect stems from the electronic transport, and further improvements due to a reduced lattice thermal conductivity may be possible from nanostructuring.

As usual, the system needs to be optimized with respect to the charge carrier concentration in order to reach the maximum figure-of-merit. It is possible to predict the optimal carrier concentration if a barrier height is specified by obtaining the band structure from first principle calculations or through the parabolic approximation. If these carrier concentrations can be achieved with doping or substitutions, there is a possibility to simultaneously optimize the two quantities for optimum values of the figure-of-merit. Conversely, if doping of a material is difficult, energy filtering might provide a beneficial carrier concentration shift of the peak of the figure-of-merit, which is a novel route of optimizing this quantity. But the figure-of-merit peak is generally shifted deeper in the band, which will likely require more aggressive tuning of the carrier concentration to achieve peak alignment. In addition, it is expected that the largest possible energy barrier allowed by the feasible carrier concentration yields the highest value for the figure-of-merit.

Filtering out low-energy charge carriers creates a nonequilibrium distribution of carrier energies. If this ensemble is allowed to equilibrate entirely through additional scattering, the overall effect would be to limit the charge carriers, and little is gained. It is thus important to control the distance between scattering sites for the scattering controlling the filtering and possibly also for the equilibrating scattering mechanisms, in order to maximize the survival rates of the charge carriers. This has been investigated by Nishio et al. [27], but in general needs further studies.

Another issue which is a potential challenge is the transport of the filtered energy. Even though this energy is small it still needs to be dissipated, for example, as heat. This would be a particular problem for thermoelectric materials due to the resulting heat loss but few studies have so far investigated this issue.

The similarities between the impurity scattering and the filtering of low-energy charge carriers also need to be studied carefully. Even though they are analytically different, they could manifest themselves as equal in real samples making identification and control of the two processes difficult.

There is a large and unexploited terrain in this direction, and we can here only speculate where important breakthroughs may appear. It is obviously important to obtain control of the atomistic structure of the edifices constituting the energy barriers, whether they are grain boundaries, nanoinclusions, or something else. As an example, governing the level of various impurities may yield a variety of grain boundary compositions, particularly when the impurities are segregated toward the grain boundaries, which in turn can give different barrier heights and widths. However, isolating the filtering effects in these samples might be difficult. Another example is heterostructures where it might be easier to control the barrier height and separate the filtering effects from the other scattering mechanisms. Both examples need further studies.

The guidelines provided in this chapter depend on detailed and accurate knowledge of both band structure and input parameters. This illustrates the necessity of combining high-quality band structure calculations with precise and comprehensive experiments. Only when all this is under control will it be possible to perform directed development of novel thermoelectric systems.

Acknowledgments

We would like to express our deepest gratitude to Johan Taftø and Terje Finstad for fruitful discussions and proofreading. This work was supported by the Norwegian Research Council.

References

1. N. W. Ashcroft and N. D. Mermin, *Solid State Physics* (Thomson Learning, London, 1976).
2. V. I. Fistul, *Heavily Doped Semiconductors* (Translated from Russian by Albin Tybulewicz) (Plenum Press, New York, 1969).

3. A. F. Ioffe, *Semiconductor Thermoelements and Thermoelectric Cooling Information* (Infosearch, London, 1957).
4. A. H. Wilson, *The Theory of Metals* (Cambridge University Press, Cambridge, 1958).
5. H. J. Goldsmid, *Thermoelectric Refrigeration* (Plenum Press, New York, 1964).
6. E. S. Rittner, On the theory of the Peltier heat pump, *J. Appl. Phys.* **30**, 702, 1959.
7. F. J. Donahoe, *Elec. Eng.* **79**, 488, 1960.
8. P. J. Price, *Phys. Rev.* **104**, 1223, 1956.
9. R. Simon, Thermoelectric figure of merit of two-band semiconductors, *J. Appl. Phys.* **33**, 1830, 1962.
10. E. R. Rittner and G. F. Neumark, Theoretical bound on the thermoelectric figure of merit of two-band semiconductors, *J. Appl. Phys.* **34**, 2071, 1963.
11. H. Littman and B. Davidson, Theoretical bound on the thermoelectric figure of merit from irreversible thermodynamics, *J. Appl. Phys.* **32**, 217, 1961.
12. E. S. Rittner, Comment on "Theoretical bound on the thermoelectric figure of merit from irreversible thermodynamics," *J. Appl. Phys.* **33**, 2654, 1962.
13. R. P. Chasmar and R. J. Stratton, *J. Electron. Control* **7**, 52, 1959.
14. C. Wood, *Rep. Prog. Phys.* **51**, 459, 1988.
15. G. D. Mahan and J. O. Sofo, The best thermoelectric, *Proc. Natl. Acad. Sci.* **93**, 7436, 1996.
16. R. Kim, S. Datta, and M. Lundstrom, Influence of dimensionality on thermoelectric device performance, *J. Appl. Phys.* **105**, 034506, 2009.
17. R. Landauer, *IBM J. Res. Dev.* **1**, 223, 1957.
18. L. D. Hicks and M. S. Dresselhaus, Effect of quantum-well structures on the thermoelectric figure of merit, *Phys. Rev. B* **47**, 12727, 1993.
19. L. D. Hicks and M. S. Dresselhaus, Thermoelectric figure of merit of a one-dimensional conductor, *Phys. Rev. B* **47**, 16631, 1993.
20. P. Pichanusakorn and P. R. Bandaru, The best thermoelectric, *Proc. Natl. Acad. Sci.* **93**, 7436, 1996.
21. J. R. Sootsman, D. Y. Chung, and M. G. Kanatzidis, *Angew. Chem. Int. Ed.* **48**, 8616, 2009.
22. P. Pichanusakorn and P. Bandaru, Nanostructured thermoelectrics, *Mater. Sci. Eng. R* **67**(2–4), 19, 2010.
23. A. V. Dmitriev and I. P. Zvyagin, Current trends in the physics of thermoelectric materials, *Physics-Uspekhi* **53**(8), 789, 2010.
24. C. J. Vineis, A. Shakouri, A. Majumdar, and M. G. Kanatzidis, Nanostructured thermoelectrics: Big efficiency gains from small features, *Advanced Mater.* **22**(36), 3970, 2010.
25. Y. I. Ravich, B. A. Efimova, and V. I. Tamarche, Scattering of current carriers and transport phenomena in lead chalcogenides .1. Theory, *Phys. Stat. Sol. B* **43**(1), 11, 1971.
26. Y. I. Ravich, Selective carrier scattering in thermoelectric materials, in *CRC Handbook of Thermoelectrics*, edited by D. M. Rowe, Chapter 7, (CRC Press, 1995), pp. 67–73.
27. K. Nishio and T. Hirano, Improvement of the efficiency of thermoelectric energy conversion by utilizing potential barriers, *Japanese J. Appl. Phys. Part 1* **36**(1A), 170, 1997.
28. Z. Bian, M. Zebarjadi, R. Singh, Y. Ezzahri, A. Shakouri, G. Zeng, J.-H. Bahk, J. E. Bowers, J. M. O. Zide, and A. C. Gossard, *Phys. Rev. B* **76**, 205311, 2007.
29. S. Wang and N. Mingo, *Phys. Rev. B* **79**, 115316, 2009.
30. K. Kishimoto and T. Koyanagi, Preparation of sintered degenerate *n*-type PbTe with a small grain size and its thermoelectric properties, *J. Appl. Phys.* **92**(5), 2544, 2002.
31. J. M. O. Zide, D. Vashaee, Z. X. Bian, G. Zeng, J. E. Bowers, A. Shakouri, and A. C. Gossard, *Phys. Rev. B* **74**, 205335, 2006.
32. J. Martin, L. Wang, L. Chen, and G. S. Nolas, *Phys. Rev. B* **79**, 115311, 2009.
33. D. L. Medlin and G. J. Snyder, Interfaces in bulk thermoelectric materials: A review for current opinion in colloid and interface science, *Curr. Opin. Colloid Interf. Sci.* **14**(4), 226, 2009.
34. Y. Lan, A. J. Minnich, G. Chen, and Z. Ren, Enhancement of thermoelectric figure-of-merit by a bulk nanostructuring approach, *Advanced Func. Mater.* **20**(3), 357, 2010.

35. S. V. Faleev and F. Leonard, Theory of enhancement of thermoelectric properties of materials with nanoinclusions, *Phys. Rev. B* **77**(21), 214304, 2008.
36. A. Popescu, L. M. Woods, J. Martin, and G. S. Nolas, Model of transport properties of thermoelectric nanocomposite materials, *Phys. Rev. B* **79**(20), 205302, 2009.
37. H. S. Green, *J. Math. Phys.* **2**, 344, 1961.
38. R. Kubo, *J. Phys. Soc. Jpn.* **12**, 570, 1957.
39. P. T. Gressman and R. M. Strain, Global classical solutions of the Boltzmann equation with long-range interactions, *Proc. Nat. Acad. Sci.* **107**(13), 5744, 2010.
40. G. Wiedemann and R. Franz, *Ann. Phys.-Berlin* **165**, 497, 1853.
41. R. J. Renka, QSHEP3D: Quadratic Shepard method for trivariate interpolation of scattered data, *Acm. T. Math. Software* **14**, 151, 1988.
42. L. V. Prokof'eva, A. A. Shabaldin, V. A. Korchagin, S. A. Nemov, and Yu. I. Ravich, Lorentz number and Hall factor in degenerate semiconductors during resonance scattering of charge carriers, *Semiconductors* **42**(10), 1161, 2008.
43. Z. Xiong, X. Chen, X. Zhao, S. Bai, X. Huang, and L. Chen, Effects of nano-TiO_2 dispersion on the thermoelectric properties of filled-skutterudite $Ba_{0.22}Co_4Sb_{12}$, *Solid State Sci.* **11**(9), 1612, 2009.

11

Introduction to Modeling Thermoelectric Transport at High Temperatures

Andrew F. May
Oak Ridge National Laboratory

G. Jeffrey Snyder
California Institute of Technology

11.1 Introduction

The goal of this chapter is to provide experimentalists a basic outline for analyzing/modeling thermoelectric transport at high temperatures. The simplest model is that developed for itinerant conduction in a single parabolic band (SPB). The SPB model is a powerful analytical tool when investigating thermoelectric transport. For instance, it can be used to guide the enhancement and/or optimization of thermoelectric efficiency, or the analysis results may suggest the presence of more complex behavior affecting the thermoelectric efficiency. There are, of course, many cases when the SPB model fails due to either multiband effects [1,2] or nonparabolicity [3–7]. The interested reader has many texts available for studying the physical origin of thermoelectric transport and the models obtained from the Boltzmann transport equation [8–15].

A series of samples is not required to derive the parameters needed for an SPB model; indeed, it only takes data for a single sample. However, this modeling is best performed when a series of samples are available, as this allows the accuracy of the model to be tested. It is common for only a sample or two to be available, though, and in these cases the SPB model can suggest avenues to enhance the thermoelectric efficiency, such as increasing or decreasing the carrier density, and give a prediction for zT in an optimized sample. Alternatively, a model developed for a novel compound may suggest that a large figure of merit zT is not achievable in the system without the introduction (or possibly the discovery) of irregular features in the transport properties. Thus, the SPB model is something of a diagnostic tool for both overall thermoelectric efficiency and the existence of more complex transport behavior.

It is essential to develop the SPB model in a region where minority carrier transport is negligible. A simple check for this is that the magnitude of the Seebeck coefficient α and electrical resistivity ρ are increasing with increasing temperature. This is typically the case in heavily doped (degenerate) semiconductors at moderate temperatures, before the thermal excitation of electron–hole pairs influences transport at higher temperatures. Such excitations are clearly revealed via a maximum in the magnitude of the temperature-dependent Seebeck coefficient and electrical resistivity, and often through a minimum in the thermal conductivity due to the onset of bipolar thermal conduction. The effective onset temperature depends on the carrier density, where a low carrier density promotes the influence of this bipolar conduction at lower temperatures. Therefore, it may be worthwhile to weight the data collected on a high carrier concentration sample more than that measured on a low carrier concentration sample.

The analysis and modeling of thermoelectric transport at high temperatures can be performed rather easily, particularly if in addition to the transport coefficients that make up the figure of merit ($zT = \alpha^2 T/\rho\kappa$), the Hall coefficient R_H is known. The Hall coefficient R_H provides a means to characterize a sample's carrier density ($n_H = 1/R_H e$) and mobility ($\mu_H = R_H/\rho$). When Hall data are not available, a rough estimate of the carrier density may be obtained via chemical considerations—the counting of formal valences or basic doping considerations. It is generally useful to compare a chemically derived carrier density to one obtained via Hall measurements [16].

11.1.1 Summary Procedure for Analyzing Thermoelectric Data with a Single Parabolic Band Model

The following simplified steps are utilized in the analysis and modeling of transport data at high temperatures via the SPB model. The equations presented in these steps assume an SPB with conduction limited by acoustic phonon scattering. They employ the Fermi integrals F_j,

$$F_j(\eta) = \int_0^\infty f\epsilon^j d\epsilon = \int_0^\infty \frac{\epsilon^j d\epsilon}{1 + \mathrm{Exp}[\epsilon - \eta]}. \quad (11.1)$$

Relevant example calculations are tabulated in Table 11.1 for comparison.

- This SPB model assumes acoustic phonon scattering limits the carrier mobility. Verify this by plotting log[μ] versus log[T], which should give a slope between roughly −1 and −1.5. An acoustic phonon scattering assumption may be arguable for any slope <−0.5.
- Obtain the SPB effective mass for a sample and temperature of interest. At a given T, calculate the reduced electrochemical potential η from the magnitude of the Seebeck coefficient (consider Table 11.1).

TABLE 11.1 Summary of Chemical Potential Dependence of a Few Pertinent Properties for the Case of Acoustic Phonon Scattering

η (kT)	α (μV K^{-1})	L (10^{-8} W Ω K^{-2})	n (10^{19} cm^{-3})	r_H (−)	$\frac{F_{-1/2}}{2F_0}$ (−)	ψ (C K$^{-3/2}$ m^{-3})
−3	433	1.49	0.123	1.17	0.877	28.3
−2	350	1.51	0.324	1.17	0.863	73.9
−1	272	1.54	0.823	1.16	0.832	182
0	205	1.61	1.92	1.13	0.773	403
1	151	1.72	3.95	1.11	0.693	764
2	112	1.86	7.09	1.08	0.610	1240
3	86.1	1.99	11.3	1.05	0.539	1770
4	68.2	2.09	16.3	1.04	0.482	2340
6	46.9	2.24	28.7	1.02	0.403	3490
8	35.4	2.32	43.6	1.01	0.351	4660
10	28.3	2.36	60.4	1.01	0.315	5820

Note: A table of Fermi integrals is provided in Refs. [9,14]. These tables are useful as a starting point in analysis, or to verify a computation is being performed correctly. The values of n are provided for $m^*/m_e = 1$ and $T = 300$ K.

$$\alpha = \frac{k}{e}\left(\frac{2F_1}{F_0} - \eta\right) \tag{11.2}$$

Estimate m^* using η, the temperature and the carrier density (Table 11.1).

$$n = 4\pi\left(\frac{2m^*kT}{h^2}\right)^{3/2} F_{1/2} \tag{11.3}$$

The Hall carrier concentration ($n_H = 1/R_H e$) is related to the chemical carrier concentration n via $n_H = n/r_H$, where the Hall factor r_H for acoustic phonon scattering (see Table 11.1) is given by

$$r_H = \frac{3}{2}F_{1/2}\frac{F_{-1/2}}{2F_0^2}. \tag{11.4}$$

- Plot the magnitude of the Seebeck coefficient versus the Hall carrier density n_H at a given T (called by some a Pisarenko plot). Adjust the effective mass to obtain the best fit of the SPB model to the data for multiple samples, if available. This is the best stage to examine if the model is working well for the assumptions that m^* is independent of carrier concentration and multiple bands do not influence transport.
- Estimate the mobility parameter μ_0 at the temperature of interest. The Hall mobility $\mu_H = (\rho n_H e)^{-1}$ is typically used to estimate μ_0 from the SPB model. Nondegenerate (low carrier concentration and/or high T) semiconductors have $\mu_H = \left(\sqrt{\pi}/2\right)\mu_0$. The general SPB relationship utilizes μ_0 obtained from Equation 11.5 for μ_H (see Table 11.1), which assumes acoustic phonon scattering limits the carrier relaxation time. This equation can be used to explain the decrease in mobility with increasing carrier concentration at constant temperature using a plot similar to the Pisarenko plot for Seebeck coefficient.

$$\mu_H = \mu_0 \frac{F_{-1/2}}{2F_0} \tag{11.5}$$

If Hall data are not available at the temperature of interest, it may be reasonable to use the resistivity data with the assumption that the carrier concentration does not change with temperature.

- Calculate the Lorenz number L, using the following SPB equation or a plot of L versus α (Figure 11.5, Table 11.1).

$$L = \frac{k^2}{e^2}\frac{3F_0F_2 - 4F_1^2}{F_0^2} \tag{11.6}$$

Estimate the electronic contribution κ_e to the thermal conductivity. Subtract this from the total thermal conductivity and obtain an estimate for the lattice contribution κ_L,

$$\kappa_L = \kappa - \kappa_e = \kappa - L\sigma T. \tag{11.7}$$

If data for the phonon velocities are available (both the longitudinal υ_l and transverse υ_t velocities), compare the resultant κ_L to an estimation of κ_{min}. The following κ_{min} is developed by taking the high-temperature limit of the thermal conductivity calculated by Cahill et al. [17] for amorphous materials with an average volume per atom given by V.

$$\kappa_{\min} = \frac{1}{2}\left(\frac{\pi}{6}\right)^{1/3} kV^{-2/3}(2\upsilon_t + \upsilon_l) \tag{11.8}$$

- Repeat these steps for other temperatures to obtain m^*, μ_0, and κ_L versus temperature, making sure the data do not demonstrate the influence of minority carrier transport.
- Calculate a theoretical zT versus n to estimate the optimum carrier density at a particular temperature. Using the parameters m^*, μ_0, and κ_L at a temperature of interest, calculate the β parameter [10,18–20].

$$\beta = \frac{\mu_0 (m^*/m_e)^{3/2} T^{5/2}}{\kappa_L} \tag{11.9}$$

The function ψ also given in Table 11.1,

$$\psi = \frac{8\pi e}{3}\left(\frac{2m_e k}{h^2}\right)^{3/2} F_0, \tag{11.10}$$

is used to calculate the SBP prediction of zT from

$$zT = \frac{\alpha^2}{L + (\psi\beta)^{-1}}. \tag{11.11}$$

It is important to use the above method only for doping levels and temperatures where minority carrier effects would not be observed in the transport data. Thus, this method requires extrinsic samples and extrapolating within the extrinsic regime. In some circumstances, only an intrinsic sample is available; however, this can also be useful for assessing thermoelectric performance.

- The transport data of an intrinsic semiconductor may be utilized to estimate the thermal energy gap E_g and the lattice thermal conductivity κ_L.

  The thermal band gap E_g is generally estimated in one of two ways. A traditional method is to plot ln[1/ρ] versus $1/T$, which should be linear with a *slope* $= -E_g/2k$. The second method [21] is to utilize the maximum in the magnitude of the Seebeck coefficient α_{max} and the corresponding temperature T_{max}, as shown in Equation 11.12. This estimate for E_g at high T is very useful in thermoelectrics, as it quantifies the relative importance of minority carrier transport, which ultimately causes the decrease in zT at a high temperature. It should also be noted that E_g can vary strongly with T, such as in PbTe [3].

$$E_g = 2e\alpha_{max}T_{max} \tag{11.12}$$

The value of κ_L can often be assumed equal to κ for an intrinsic semiconductor, particularly at moderate T. When κ of an intrinsic sample is decreasing with decreasing T, roughly as T^{-1}, it is reasonable to set $\kappa = \kappa_L$. However, if κ rises or appears to flatten at high T, then the bipolar thermal conductivity must also be considered.

11.2 Developing an SPB Model

Most high-performance thermoelectric materials are heavily doped semiconductors with transport properties that appear more similar to metals than semiconductors. In general, electrical transport properties are determined by electrons (or holes) with energies near the electrochemical potential ζ, which is equal to the Fermi energy at 0 K. In thermoelectric materials, the carrier density n tends to be independent of temperature, though some thermal activation of electron–hole pairs occurs at higher temperatures. For the most part, the transport properties observed at moderate/high temperatures are easily understood in terms of classical semiconductor physics. The electrical resistivity ($\rho = 1/\sigma = 1/ne\mu$) increases with increasing T as the carrier mobility μ decreases due to electron–phonon interactions. The magnitude of the Seebeck coefficient increases with increasing T as the reduced electrochemical potential $\eta = \zeta/kT$ decreases with increasing kT as well as the associated broadening of the Fermi distribution. A decrease in the carrier density also results in a reduced η (fixed T), and thus the Seebeck coefficient increases with decreasing carrier density.

The SPB model is expected to break down at the extremes of carrier density (and temperature). Nonparabolic or multiband effects are more likely to be observed at higher carrier densities, when energies away from the band edge are probed. At low carrier densities, the deviation from the simple SPB model is more likely to be caused by a mobility edge, where a sharp decrease in μ may be observed due to carrier localization. Thermoelectric materials tend to optimize close by, but usually not in, the region where these features are significant [9].

11.2.1 Hall Mobility and Carrier Concentration

The carrier mobility provides insight into the nature of transport, and the temperature dependence of μ provides a strong indication of the dominant carrier scattering mechanism. To examine the theoretical temperature dependence of the Hall mobility μ_H, which is obtained via measurements of the Hall coefficient R_H and the electrical conductivity $\sigma = 1/\rho$, we consider the SPB model expression,

$$\mu_H = R_H\sigma = \frac{e}{m}\frac{\int_0^\infty \epsilon^{3/2}\tau^2(\partial f/\partial\epsilon)d\epsilon}{\int_0^\infty \epsilon^{3/2}\tau(\partial f/\partial\epsilon)d\epsilon}. \tag{11.13}$$

The carrier mobility is closely related to the carrier relaxation time, τ, the value of which is generally obtained by inverse summation of the τ values associated with different scattering mechanisms. In semiclassical transport models, such as the SPB model, the τ for various scattering mechanisms are commonly modeled by power laws in reduced carrier energy $\epsilon = E/kT$: $\tau = \tau_0\epsilon^{\lambda-1/2}$. The parameter λ is often called the scattering parameter and assumes the value of 0 for scattering by acoustic phonons; $\lambda = 2$ is often utilized for ionized impurity scattering. Note that in some representations λ may differ by the factor −1/2 [9]. Assuming this simple power law for τ, integration by parts reduces Equation 11.13 to

$$\mu_H = \mu_0\frac{(1/2+2\lambda)F_{2\lambda-1/2}}{(1+\lambda)F_\lambda}, \tag{11.14}$$

where $\mu_0 = e\tau_0/m^*$. Recall that F_j are dependent on η, which changes with T and thus the integrals F_j provide temperature dependence in the various transport properties. The term μ_0 is used in the model that allows the optimum carrier density to be predicted for a given temperature. To calculate μ_0 from Equation 11.5 using experimental transport data, one must compute η, and this is typically achieved using the Seebeck coefficient. In other formulations, such as that given by Goldsmid [9] expressions equivalent to Equation 11.5 are normalized so that μ_0 represents the nondegenerate mobility, or highest achievable mobility (when $\lambda = 0$) at the given T. The current definition for μ_0 utilizes the effective mass m^*, though it is more accurately defined as an inertial mass [13,10] which is separate from the commonly called "density of states effective mass" that is related to the carrier density (they are equal for the isotropic SPB). Note that the Hall mobility differs slights from the drift mobility, and the Hall factor r_H can be utilized to account for the minor difference (discussed below).

The temperature dependence of μ in an SPB arises due to two primary contributions (we consider only one relevant scattering mechanism). The first contribution is a power law in temperature introduced via the μ_0 term (more directly an associated τ_0). The second contribution is due to the η dependence of μ, which is hidden within the Fermi integrals; this introduces temperature dependence because η changes with T and thus the necessary integrals are functions of T. Power laws in T can often be utilized to model both the μ_0 term and a ratio of Fermi integrals [14]. Therefore, when a single scattering mechanism limits μ, the mobility often trends as roughly T^{-p} with an exponent $-p$ that may differ from the T-dependence observed (predicted) for μ_0. As such, a plot of log[μ] versus log[T] can be informative when assessing the dominant scattering mechanism.

When acoustic phonon scattering limits τ, the deformation potential theory (for SPB) provides $\tau = \tau_{0,ac}\epsilon^{-1/2}$. The $\tau_{0,ac}$ is given by

$$\tau_{0,ac} = \frac{\pi\hbar^4 \upsilon_l^2 d}{\sqrt{2}E_{def}^2 (m^* kT)^{3/2}}, \tag{11.15}$$

where E_{def} is the deformation potential that relates to changes in the electronic structure due to the presence of a phonon [14]. This theory predicts that the mobility decreases with increasing T as T^{-p} where $1 \le p \le 1.5$. The limit $p = 1.5$ is obtained in a nondegenerate electron gas (low doping levels and/or high T) and originates in the τ_0 term; the $p = 1$ case is observed for highly degenerate samples because $F_{-1/2}/F_0$ in Equation 11.5 goes as $\sim T^{0.5}$ for large η (the degenerate limit). A nice transition between the degenerate (metallic) $p = 1$ and semiconducting $p = 1.5$ is observed in $La_{3-x}Te_4$ [22].

It is important to note that significant deviations from this prediction are observed when material properties are temperature dependent. Considering Equation 11.15, it is clear that a temperature-dependent speed of sound (elastic properties), effective mass, or deformation potential can dramatically alter the temperature dependence of μ. For instance, a temperature-dependent m^* likely causes $\mu \propto T^{-2.5}$ in PbTe [3]. This is also observed in n-type $Ba_8Ga_{16-x}Ge_{30+x}$, as shown in Figure 11.1a, where the decay is closer to $T^{-1.9}$ at high temperatures. Generally, the observation of μ_H decreasing rapidly with increasing T suggests acoustic phonon scattering limits μ, and the model development can proceed.

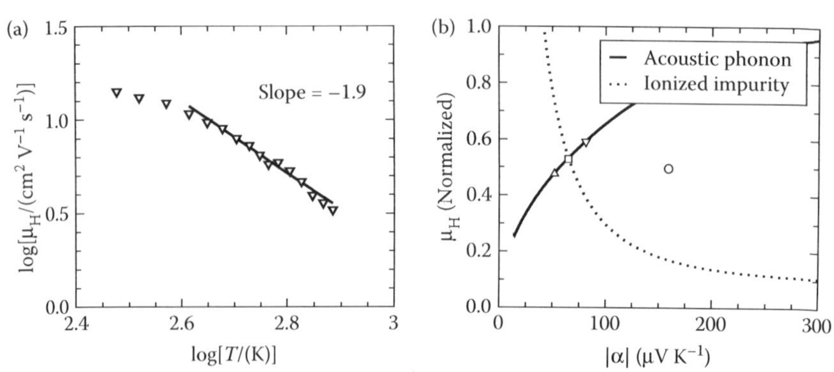

FIGURE 11.1 Hall mobility in n-type $Ba_8Ga_{16-x}Ge_{30+x}$, with experimental data shown by markers. (a) A generic power-law fit reveals that the mobility decays as roughly $T^{-1.9}$ at high temperatures. This behavior can be explained by acoustic phonon scattering when the effective mass increases with increasing T, as in PbTe. (Adapted from Y. I. Ravich, B. A. Efimova, I. A. Smirnov, *Semiconducting Lead Chalcogenides*: Plenum, New York, 1970.) (b) The Hall mobility versus Seebeck coefficient can be utilized to probe the dominant scattering mechanisms, as shown here at 300 K; the sample with the highest Seebeck coefficient (the lowest carrier density) shows a deviation from the simple theory for acoustic phonon scattering, suggesting other scattering mechanisms are active in this sample. Note this simple model for ionized impurity scattering utilizes $\lambda = 2$ and a constant μ_0.

The influence of additional scattering mechanisms may also be observed in the plot of log[μ] versus log[T]. For instance, a high residual resistance (from defects such as grain boundaries) reduces the overall p-value, as would scattering by ionized impurities ($\tau_0 \propto T^{3/2}$) and alloy scattering or optical phonon scattering ($\tau_0 \propto T^{-0.5}$). Indeed, in Figure 11.1a a lower p-value can be inferred from the data near room temperature, and it has been suggested that alloy scattering may be important in n-type $Ba_8Ga_{16-x}Ge_{30+x}$ [23]. However, due to their respective T dependences, these scattering mechanisms tend to be less influential at high T where they lead to long relaxation times.

The energy-dependent portion of Equation 11.13 allows plots of μ_H versus n_H, or μ_H versus α to be examined. Such plots can also be used to verify the assumption that acoustic phonon scattering limits τ, as shown in Refs. [24,25]. For instance, in Figure 11.1b, three of the four n-type $Ba_8Ga_{16-x}Ge_{30+x}$ samples follow the behavior predicted for acoustic phonon scattering, while one sample deviates from the theoretical curve with no clear explanation. Note, these plots require several samples to be present, as the plot only makes physical sense at a given temperature.

To plot a theoretical n dependence, the carrier's SPB effective mass m^* must be known. It is common to obtain this using the experimental Hall carrier density n_H and the η obtained from α at a given T. However, the Hall carrier density ($n_H = 1/R_H/e$) is not exactly equal to the chemical carrier density (n in Equation 11.3). For parabolic bands, the two are equal when the electron gas is highly degenerate—this happens at low T and high n (large η, regardless of λ). In the limit of a nondegenerate electron gas, the chemical n is roughly 18% larger than n_H for the scattering of carriers by acoustic phonons ($\lambda = 0$). A self-consistent model calculates a theoretical Hall coefficient R_H to obtain $n_H = 1/R_H/e$, or employs the Hall factor $r_H = n/n_H = \mu_H/\mu$ to account for the difference. The error in using the chemical n is not extremely large, though, especially considering the primary goals of this analysis (estimating the optimal carrier density and assessing the electronic structure).

$$R_H = \frac{3h^3}{8\pi e(2m^*kT)^{3/2}} \frac{\int_0^\infty \epsilon^{3/2}\tau^2(\partial f/\partial\epsilon)\,d\epsilon}{\left(\int_0^\infty \epsilon^{3/2}\tau(\partial f/\partial\epsilon)\,d\epsilon\right)^2} \tag{11.16}$$

For the simple power law description of the relaxation time $\tau = \tau_0\epsilon^{\lambda-1/2}$, this reduces to

$$R_H = \frac{3h^3}{8\pi e(2m^*kT)^{3/2}} \frac{(1/2+2\lambda)F_{2\lambda-1/2}}{(1+\lambda)^2F_\lambda^2}. \tag{11.17}$$

The Hall factor r_H is then

$$r_H = \frac{3}{2}F_{1/2}\frac{(1/2+2\lambda)F_{2\lambda-1/2}}{(1+\lambda)^2F_\lambda^2}. \tag{11.18}$$

11.2.2 The Seebeck Coefficient

When developing an SPB model, one must verify that the minority carriers are not influencing the transport properties. An easy check for this is to examine the temperature dependence of the Seebeck coefficient. The Seebeck coefficient increases with increasing temperature and/or decreasing carrier density in a simple, extrinsically doped semiconductor (n independent of T), because this T and/or n behavior corresponds to decreasing η. At high temperatures, the thermal activation of electron–hole pairs results in a compensation of the Seebeck coefficient, and a maximum in $|\alpha|$ is observed. Low extrinsic carrier concentrations promote the activation of electron–hole pairs due to lower η and a relative

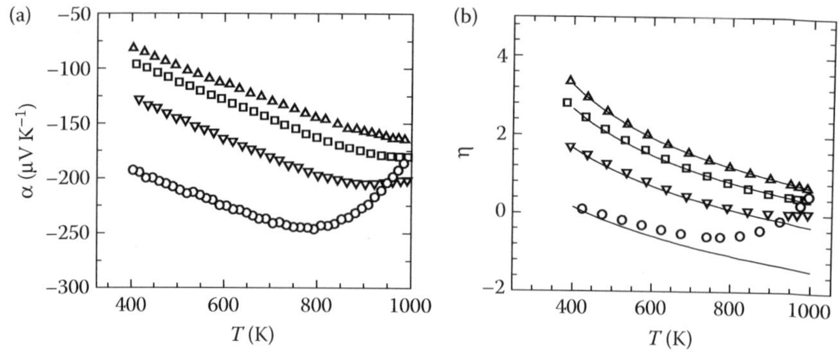

FIGURE 11.2 The temperature-dependent (a) Seebeck coefficient of n-type $Ba_8Ga_{16-x}Ge_{30+x}$ and the (b) reduced electrochemical potential η extracted from the data in (a) assuming an SPB model with electrons scattered by acoustic phonons. The result agrees well with the decay of η expected in an SPB with constant carrier density and band mass (solid curves). This is not true for the sample with the lowest carrier concentration, where the activation of minority carriers leads to a compensation of the Seebeck coefficient, and the rise of η at high T is an artifact of the SPB analysis. The carrier density of each sample can be inferred from Figure 11.3a where the symbols match.

change in the charge neutrality equation. Clearly, the SPB model is not valid at the temperatures and compositions where the influence of minority carrier transport is observed, and theoretical predictions for transport at low carrier concentrations should be considered carefully.

The Seebeck coefficient behavior typically observed in heavily doped semiconductors is shown in Figure 11.2a, where experimental Seebeck coefficients are plotted versus T for n-type $Ba_8Ga_{16-x}Ge_{30+x}$ [25]. The magnitude of α increases linearly with T at low/moderate T, then begins to lose linearity as the degeneracy of the electron gas decreases. In the sample with the largest room temperature $|\alpha|$ (the lowest carrier density), the expected maximum in $|\alpha|$ is observed at high temperatures where holes influence transport. A similar maximum is observed in the corresponding resistivity data, and the thermal conductivity data also suggest that bipolar conduction is important [25]. The E_g estimated via Equation 11.12 from the data in Figure 11.3a is $E_g = 0.4$ eV, which is rather close to that obtained from resistivity data on a near-insulating sample [25].

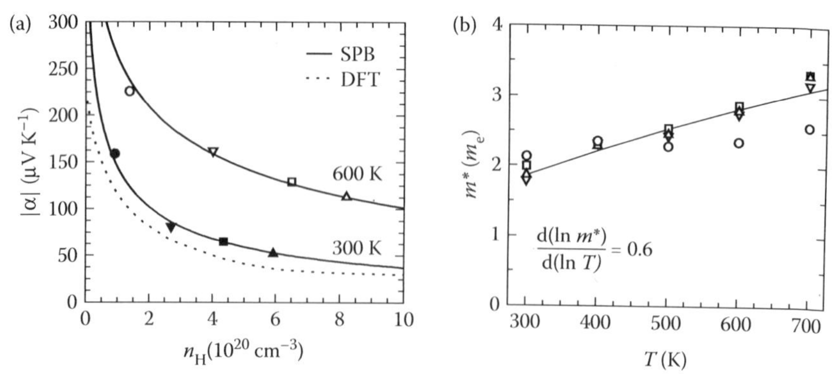

FIGURE 11.3 (a) The magnitude of the Seebeck coefficient in n-type $Ba_8Ga_{16-x}Ge_{30+x}$ is plotted versus the Hall carrier density and modeled by an SPB with m^*/m_e equal to 1.86 and 2.70 at 300 K and 600 K, respectively. DFT calculation results for 300 K are also shown, and the curvature agrees well with the SPB (DFT-based results extracted from literature and plotted in n). (b) The effective masses obtained from the experimental Hall and Seebeck coefficients increase with increasing temperature in a manner similar to n-type PbTe. Symbols in (b) correspond to carrier concentration with same symbol in (a). (Adapted from G. K. H. Madsen et al. *Physical Review B* **68**, 125212, 2003.)

To analyze the Seebeck coefficient data, we utilize the SPB description:

$$\alpha = \frac{k}{e}\left(\frac{\int_0^\infty \epsilon^{3/2}(\epsilon - \eta)\tau(\partial f/\partial \epsilon)d\epsilon}{\int_0^\infty \epsilon^{3/2}\tau(\partial f/\partial \epsilon)d\epsilon}\right), \tag{11.19}$$

which reduces to the following when $\tau = \tau_0 \epsilon^{\lambda-1/2}$:

$$\alpha = \frac{k}{e}\left(\frac{(2+\lambda)F_{\lambda+1}}{(1+\lambda)F_\lambda} - \eta\right). \tag{11.20}$$

For a SPB, the Seebeck coefficient only depends on η. That is, regardless of the band mass, all parabolic bands have the same Seebeck coefficient at a given η (Table 11.1). A larger band mass results in a lower η for a given n, T, and thus larger m^* provides larger α for a given n due to lower η.

The value of η (relative to the band edge) can be estimated from experimental α data using Equation 11.2 (assumes $\lambda = 0$). This allows the (SPB) effective mass to be calculated from the expressions provided for n or n_H. An estimate of m^* can be obtained using the linear region of the $\alpha(T)$ curve and the degenerate-limit equation for α (Equation 11.22), which contains n and m^* [26]. This can be achieved without invoking a computational program, and is therefore appealing in some circumstances. However, such limiting expressions should be avoided when possible; the expressions containing integrals should be used to estimate the transport parameters and the predicted optimal carrier density and corresponding zT.

In the limit of a nondegenerate electron gas, where Boltzmann statistics apply, the Seebeck coefficient in an SPB is related to the carrier density n and effective mass m^* via

$$\alpha = \frac{k}{e}(2+\lambda-\eta) = \frac{k}{e}\left(2+\lambda+\ln\left[\frac{2\left(2\pi m^* kT/h^2\right)^{3/2}}{n}\right]\right). \tag{11.21}$$

While in the limit of a degenerate electron gas (large η, which is approximately equal to the Fermi energy E_F), the SPB description reduces to

$$\alpha = (1+\lambda)\frac{\pi^2 k^2 T}{3eE_F} = (1+\lambda)\frac{8\pi^2 k^2 T m^*}{3eh^2}\left(\frac{\pi}{3n}\right)^{2/3}. \tag{11.22}$$

The reduced electrochemical potential $\eta(T)$ obtained by analyzing $\alpha(T)$ in n-type $Ba_8Ga_{16-x}Ge_{30+x}$ is plotted in Figure 11.2b. The solid curves in Figure 11.2b are the theoretical dependence of η on T in a parabolic band, and have been obtained via Equation 11.3. These temperature-dependent, theoretical curves are generated by aligning the η values at 300 K via minor adjustments in m^* at a given n, or vice versa, and then changing T with constant n, m^* to obtain $\eta(T)$. The agreement between the theoretical curves and the data points obtained via $\alpha(T)$ is very good, and strongly suggests that the conduction band is parabolic for the n, T examined.

A region where the SPB model for n-type $Ba_8Ga_{16-x}Ge_{30+x}$ clearly fails is indicated by the rise in $\eta(T)$ at high temperatures for the sample with a maximum in $|\alpha(T)|$. The calculated value of η rises because the minority carrier contribution has not been taken into account. Indeed, the actual η is most likely still decreasing with increasing temperature. In a sample with a lower carrier density, the artificial rise in η would occur at lower T, and thus the SPB model should not be utilized to consider low n data at high T.

It is worth noting that a relatively simple two band model (identical band properties for holes and electrons) can capture the maximum in $|\alpha|$ and produce the expected $\eta(T)$, though it does not fully capture the observed n_H behavior. Developing an accurate two-band model is very difficult, as the energy gap and its temperature dependence, as well as the transport properties of minority carriers are often difficult to accurately address. For this reason, the two band model mentioned here is not shown, but is noted simply to encourage the reader to attempt more challenging models.

The plot of α versus n_H is one of the best tools to assess a material's *effective* band structure. This is achieved by calculating α versus η, and n_H versus η. The calculation of n or n_H versus η only needs to be performed once, and then the values of T and m^* can be changed appropriately to describe the data; as such, data for $m^*/m_e = 1$ and $T = 300$ K are provided in Table 11.1. Once again, this reveals that n increases with increasing m^*, and that α is independent of band curvature for parabolic bands.

Similar to the comparisons shown in Figure 11.2b, agreement between the theoretical curves and the experimental data in Figure 11.3a strongly suggests conduction occurs in an SPB for the n-type $Ba_8Ga_{16-x}Ge_{30+x}$ system. The experimental data are well described by an SPB model at both 300 and 600 K (note that n_H increases slightly with T). Also shown on this plot is the result of a first principles calculation performed by Madsen and Singh [27], which possesses a similar curvature of $\alpha(n)$ at 300 K. Such a comparison to first principles calculations is often useful and may reveal interesting band features, as discussed below for $La_{3-x}Te_4$.

Interestingly, the SPB model suggests that the effective mass in $Ba_8Ga_{16-x}Ge_{30+x}$ increases with increasing temperature. The rise in m^* is similar to that observed in n-type PbTe, where $\Delta = (d\ln m^*/d\ln T) \sim 0.5$ is observed [3]. The increase in m^* is consistent with the strong temperature dependence of the Hall mobility. However, Figure 11.2b demonstrates that constant n, m^* is consistent with the Seebeck coefficient data. The apparent increase in m^* is due to a gradual increase in n_H with increasing T [25]. Such behavior is not expected in an extrinsic semiconductor, and thus the rise in m^* may be artificial. Therefore, despite the agreement between the data and the model, close inspection reveals that further investigation of the temperature dependence of the band features is warranted.

In some cases, such as in $La_{3-x}Te_4$, the SPB model clearly fails to describe the available data. At first glance, the electrical transport in $La_{3-x}Te_4$ is as expected for a heavily doped semiconductor, with the usual trends in α and σ being observed [24]. However, closer inspection of the Seebeck coefficient data reveals trends not consistent with conduction in a SPB [1]. The influence of different carrier scattering mechanisms is important to consider here, as the defect concentration is very high. However, the dependence of μ on n and T is as expected for acoustic phonon scattering, and even the sample with the highest lanthanum vacancy concentration can be described by the simple theory presented above for acoustic phonon scattering [24,28].

The clearest example of how $La_{3-x}Te_4$ deviates from an SPB model is shown in Figure 11.4a, where the magnitude of α is plotted versus the room temperature Hall carrier density. Two different parabolic band models are required to describe the Seebeck coefficient data, though four are shown to illustrate the trends with effective mass more clearly. A large effective mass is necessary to describe $\alpha(n)$ at high carrier densities, while a smaller m^* is required at low carrier densities. This effect is most pronounced at low/moderate temperatures, because significant thermal broadening at high temperatures leads to a situation that is described fairly well by an SPB.

The calculated, "SPB-equivalent" effective mass values are highly temperature dependent, and Figure 11.4b shows that they do not trend with T in a simple manner. For the most part, this SPB effective mass of the electrons at high carrier concentrations decreases with increasing T, while that of the low carrier concentration samples increases with T. This irregular behavior is a very strong indication that transport occurs in multiple bands.

Given the above irregularities, it is desirable to perform first principles calculations to provide insight into how the electronic structure may be influencing transport. Density functional theory (DFT) calculations on La_3Te_4 confirmed [1] the existence of multiple conduction bands within the energy range of interest for the given n, T values, and were thus an invaluable tool in understanding the transport in this system.

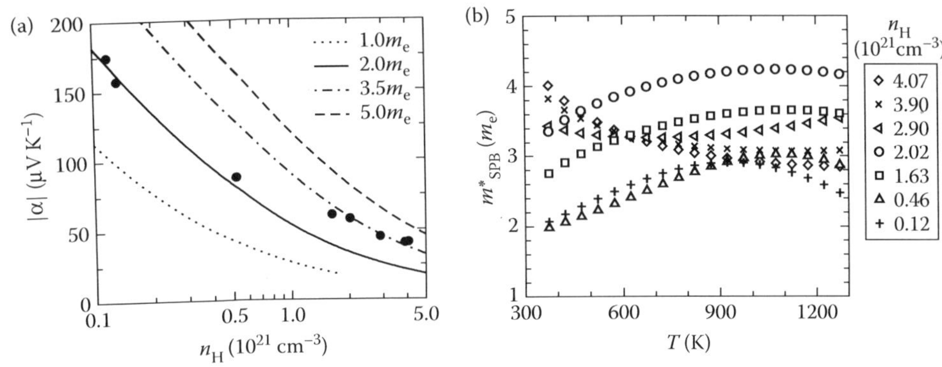

FIGURE 11.4 (a) In $La_{3-x}Te_4$, the carrier density dependence of $|\alpha|$ at 400 K suggests that multiple, parabolic bands contribute to the conduction process. The plot includes various band masses, which shift the calculated n_H to higher values with increasing m^*. (b) The single-band equivalent effective masses are heavily temperature dependent, and either increase or decrease with T, depending on the carrier concentration; this is another feature that suggests multiple bands contribute to conduction.

With the DFT results for band offsets and degeneracy as a starting point, a multiparabolic band model can be pursued. Such a procedure allows a semiempirical model guided by first principles to be developed, and thus a more reasonable model of the system results from interplay between classical modeling and fundamental calculations [1]. The initial analysis of the data using the SPB model was a necessary first step, though, as it suggested that a more complex electronic structure was influencing the transport properties.

11.2.3 Thermal Transport

The thermal conductivity κ is composed of two primary contributions, the electronic κ_e and the lattice κ_L. Reducing, or ideally eliminating, the lattice contribution is a primary goal of thermoelectric research, and thus determining κ_L is of great importance. It is customary to subtract an estimate for the electronic contribution from the measured κ and assume the remainder is the lattice contribution. Another method to determine κ_L is to plot κ versus σ (or σT) at a given temperature and obtain the value of κ_L from the intercept; this can only be done if transport data for a series of samples are available.

The electronic contribution is commonly calculated via the Wiedemann–Franz relationship, $\kappa_e = L\sigma T$, where L is the Lorenz number. While this method cannot account for electron–phonon interactions, it can provide a reasonable estimate for κ_L if an appropriate L is utilized. This method is only valid in the temperature range where minority carrier transport is negligible, or else the resulting κ_L may contain a fairly significant contribution from the bipolar thermal conductivity κ_b.

In the limit of a SPB with $\tau \propto \epsilon^{\lambda-1/2}$ the Lorenz number is

$$L = \frac{k^2}{e^2}\frac{(1+\lambda)(3+\lambda)F_\lambda F_{\lambda+2} - (2+\lambda)^2 F_{\lambda+1}^2}{(1+\lambda)^2 F_\lambda^2}. \tag{11.23}$$

The Lorenz number depends on η and the energy dependence of τ, as is the case for α. As such, it can be plotted as a function of α for a given scattering mechanism. In doing this for acoustic phonon scattering, as shown in Figure 11.5, we see that the degenerate limit of L is not reached in good thermoelectric materials. This is another way of viewing that the degenerate limit expressions are, in general, approximate but not accurate for thermoelectric materials. They do, however, provide insight into thermoelectric behavior and a decent starting point for understanding transport data.

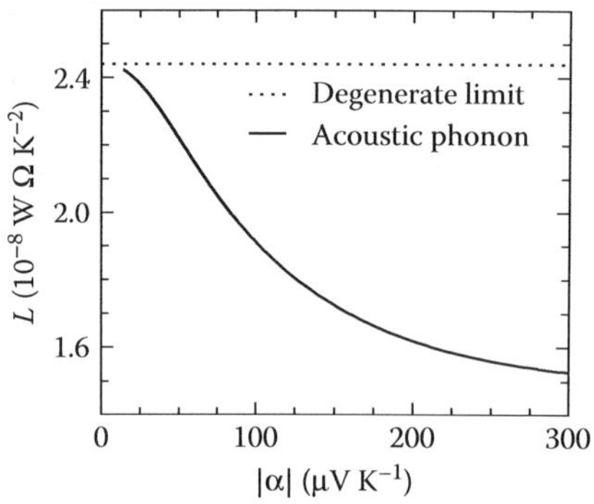

FIGURE 11.5 The Lorenz number expected in a SPB is plotted versus the Seebeck coefficient for acoustic phonon scattering (at all T). The dashed line represents the degenerate limit of L, which is rarely obtained in thermoelectric materials.

The degenerate limit of the Lorenz number is given by $L = \pi^2k^2/3/e^2 = 2.45 \times 10^{-8}$ W Ω K^{-2}, and is valid at high degeneracy regardless of the carrier scattering mechanism. This limit is often utilized in transport analysis. At high temperatures, the use of this value typically results in an overestimation of L (κ_e) and thus an underestimation of κ_L. In such cases, attention is often drawn to the "low" value of κ_L and incorrect conclusions can be drawn regarding the physics governing thermoelectric efficiency and the best methods for increasing efficiency. Indeed, using the degenerate limit of L can even result in an apparently negative κ_L! [24].

The results of SPB analysis of the thermal transport data for n-type $Ba_8Ga_{16-x}Ge_{30+x}$ are shown in Figure 11.6a,b for moderate temperatures. It is observed that the L values are significantly lower than the degenerate limit, as can be inferred from the magnitudes of α. At high temperatures, the SPB model breaks down where L is shown to artificially rise in the sample with the lowest carrier concentration. This is due to the compensation of the Seebeck coefficient and the corresponding rise (artificially) in η, which can be observed in Figure 11.2b. The calculated κ_L values are scattered about ~1 W/m/K at 300 K.

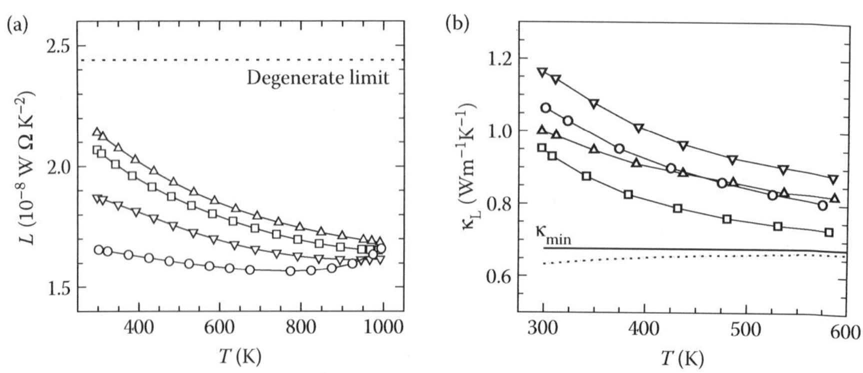

FIGURE 11.6 (a) The calculated Lorenz numbers for n-type $Ba_8Ga_{16-x}Ge_{30+x}$ are consistently lower than the degenerate limit (dashed line), and at high temperatures the SPB fails as the L values increase artificially (because η rises artificially). (b) The lattice thermal conductivity is plotted in the temperature range where the bipolar contribution is negligible. This shows that κ_L approaches the minimum thermal conductivity at high temperatures, and thus introducing additional scattering mechanisms are only likely to reduce κ_L at moderate T. The solid lower curve represents κ_{min} in the high-temperature limit (Equation 11.8) and the dashed curve represents the more detailed calculation of κ_{min} (Equation 11.24). (Adapted from A. F. May et al. *Physical Review B* **80**, 125205, 2009.)

The decrease in κ_L with increasing temperature is consistent with the *crystalline* behavior expected at high temperatures, where phonon–phonon interactions limit the phonon mean free path. As temperature rises and phonon populations increase linearly with T, phonon–phonon interactions increase and thus κ_L decreases as $\sim T^{-1}$. However, there is a minimum value for κ_L and thus the decay is not strictly at T^{-1} [25]. As with the carrier mobility, additional scattering mechanisms can also result in reduced temperature dependence in κ_L, as is often observed in alloys or amorphous materials.

It can be very useful to compare the calculated κ_L to an estimate for the minimum κ_L, termed κ_{min} here. In doing so, we find that $Ba_8Ga_{16-x}Ge_{30+x}$ approaches this lower limit at high T, and thus phonon–phonon scattering produces very low κ in this material. The relatively low κ_L is also related to the large number of atoms in the unit cell, which produces many optical phonons that generally have low group velocities [29].

Cahill and Pohl provided a formula for the estimation of κ_{min} by considering transport in amorphous materials (Equation 11.24) [17]. At high temperatures, this expression simplifies to Equation 11.8.

$$\kappa_{min} = \left(\frac{\pi}{6}\right)^{1/3} kV^{-2/3} \sum_i \upsilon_i \left(\frac{T}{\Theta_i}\right)^2 \int_0^{\Theta_i/T} \frac{x^3 e^x}{(e^x - 1)^2} dx \tag{11.24}$$

The summation is over the one longitudinal and two transverse modes, V represents the average volume per atom, $\Theta_i = \upsilon_i(\hbar/k_b)(6\pi^2/V)^{1/3}$, and υ_i is the appropriate sound velocity.

The bipolar thermal conductivity arises due to the creation and annihilation of electron–hole pairs [30]. The bipolar thermal conductivity is related to the partial conductivities σ_i and Seebeck coefficients α_i of electrons (subscript n) and holes (subscript p) [31], and tends to rise rapidly at high temperatures due to the thermal activation of minority carriers. An unaccounted for κ_b often results in an apparent rise in κ_L at high T or a premature flattening to an apparent amorphous limit that is higher than expected.

$$\kappa_b = \frac{\sigma_p \sigma_n}{\sigma_p + \sigma_n} (\alpha_p - \alpha_n)^2 T \tag{11.25}$$

11.2.4 The Optimization of *zT*

The net result of this process is a model for zT at moderate temperatures, where minority carriers do not significantly influence transport for the compositions of interest. As discussed above, the transport properties are strong functions of the reduced electrochemical potential η. The optimization process is thus an optimization in η, which is then correlated to an optimum carrier density n_{opt}. The optimum η occurs near the band edge (near $\eta = 0$), though its exact location depends on m^*, μ_0, and κ_L. Expressing zT as a function of energy-dependent terms α, L, and ψ allows the material-dependent terms to be grouped together in the factor β, shown in Equation 11.9.

It is important to note again that the SPB β hides the influence of band degeneracy in the effective mass m^* [10]. In this manner, this is truly an *effective* mass, as it averages out anisotropic effects, scattering influences, nonparabolicity, and the influence of multiple bands. When several bands are present, but have similar or identical energy minima, this method is particularly effective at providing a basic understanding of how zT optimizes as a function of n. A more accurate assessment of the band masses can be made by taking into account the presence of multiple bands, as discussed below regarding the number of equivalent valleys N_υ.

For the case of any relaxation time that can be modeled by a simple power law in energy ($\tau = \tau_0 \epsilon^{\lambda-1/2}$), the function ψ utilized in Equation 11.11 is given by

$$\psi = \frac{8\pi e}{3}\left(\frac{2m_e k}{h^2}\right)^{3/2} (1+\lambda) F_\lambda. \tag{11.26}$$

Enhancing zT can be seen as enhancing β, and it should be noted that a dependence on m^* is hidden within μ_0. In the limit of acoustic phonon scattering, the β factor is actually reduced when the effective mass increases, as μ_0 goes as $(m^*)^{-5/2}$ and thus the SPB zT decreases with increasing m^* for acoustic phonon scattering. Wood's review provides a nice discussion on the interplay between m^* and thermoelectric efficiency for various scattering mechanisms [18].

For the n-type $Ba_8Ga_{16-x}Ge_{30+x}$ example, performing the zT versus n_H calculation at 600 K suggests an optimum composition near $n_H = 2 \times 10^{20}$ cm^{-3}. This result is observed for models developed using data from two different carrier densities (solid and dashed curves), and a similar n_{opt} and maximum zT is also obtained from a model developed assuming $\lambda = 0.5$ [25]. While this model is likely to break down at much higher or lower n values, the region pertinent to thermoelectric application appears to be very well described, and thus the model achieves its primary task—to identify an optimum composition for application and identify means to enhance zT.

In many cases, only a single carrier density, or sample, is available. This is common in the exploration of novel compounds. It is useful to develop a basic model of transport to guide the optimization of performance. In some cases, this process indicates that large thermoelectric efficiency is unlikely to be achieved, and thus a simple model can save time in the exploration of novel compounds. The SPB model can also provide motivation to overcome synthetic difficulties. For instance, DFT calculations predicted n-type LiZnSb to have large thermoelectric efficiency [32]. While n-type samples are yet to be produced, the SPB model developed from two p-type samples suggested that the assumptions utilized in the first principles calculations were reasonable, and thus the model provides further incentive to explore synthetic routes to obtain n-type LiZnSb.

Finally, Figure 11.7b shows a theoretical calculation of $zT(n)$ for a parabolic band(s) semiconductor. It is observed that n_{opt} increases with increasing T, and thus it is difficult to optimize a thermoelectric material over a wide temperature range. For simplicity, consider Ioffe's result obtained via power factor optimization in a nondegenerate electron gas, where $n_{opt} \sim (m^*T)^{3/2}$ [8]. Even with this simplified approach to optimization, it is clear that *functionally graded materials* (with a spatially dependent concentration of

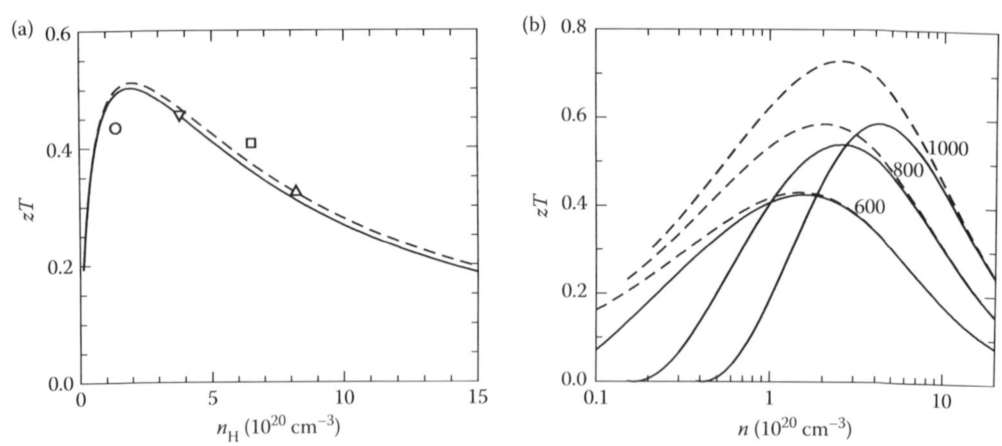

FIGURE 11.7 (a) The net result of the SPB is a theoretical zT versus n_H, shown here for 600 K in n-type $Ba_8Ga_{16-x}Ge_{30+x}$, with an optimum composition predicted near $x = 0.25$ regardless of the input parameters; dashed and solid curves are generated by analyzing data for two different samples assuming $\lambda = 0$. (b) Theoretical zT versus n curves for a simple semiconductor show how n_{opt} increases with T. Also shown is the effect minority carriers can have on the maximum zT and the corresponding n_{opt} via a multicarrier (2 band) model, which assumes similar μ_0 and m^* for electrons and holes. The dashed curves are the SPB models and the solid curves are for the two-band models, with temperatures indicated to the right of the curves. The parameters taken are similar to those in (a), but do not represent an accurate model for n-type $Ba_8Ga_{16-x}Ge_{30+x}$.

dopants) can provide significant enhancements in thermoelectric efficiency. See the chapter by E. Müller and G. J. Snyder in this text for further discussion on functionally graded materials.

Two sets of theoretical zT curves are plotted in Figure 11.7b. The dashed curves correspond to the SPB model shown in this chapter; the solid curves are for a simple semiconductor where electrons and holes have identical m^* and μ_0, and $E_g = 0.4$ eV. The effect of minority carriers is to reduce the overall zT due to compensation in α and the introduction of κ_b. To avoid the influence of these bipolar effects, the optimum carrier density shifts to higher values. For consistency with Figure 11.7a, the theoretical curves in Figure 11.7b are obtained using parameters similar to those for n-type $Ba_8Ga_{16-x}Ge_{30+x}$, but they do not represent an accurate model of transport in that system.

11.2.5 Expressions for Multiband Conduction

The following equations are utilized to calculate the transport coefficients when multiple bands contribute to transport [33]. The calculation of transport parameters for each band i must take into account the energy scale of electrons in that band; the Fermi integrals can be modified to take into account the various band edge energies. Also, the Seebeck and Hall coefficients must possess their appropriate sign (negative for electrons and positive for holes).

The expression for the electronic contribution to the thermal conductivity (κ_e) contains two terms. The term on the left is a summation over the contribution from each band where the SPB L_i are utilized with the correct adjustments for relative energy scales; for a single band conductor this is the "normal" electronic contribution. The term on the right exists due to transitions between different bands; in the limit that there is one electron band (n) and one hole band (p) this term reduces to the common expression for bipolar thermal conductivity (Equation 11.25).

$$n = \sum_i n_i \tag{11.27}$$

$$\sigma = \sum_i \sigma_i \tag{11.28}$$

$$R_H = \frac{\sum_i R_{H,i}\sigma_i^2}{\left(\sum_i \sigma_i\right)^2} \tag{11.29}$$

$$\alpha = \frac{\sum_i \alpha_i \sigma_i}{\sum_i \sigma_i} \tag{11.30}$$

$$\kappa_e = T(\sum_i L_i\sigma_i) + T\left(\sum_i \sigma_i\alpha_i^2 - \frac{\left(\sum_i \sigma_i\alpha_i\right)^2}{\sum_i \sigma_i}\right) \tag{11.31}$$

These expressions can be utilized to incorporate the influence of band degeneracy (multiple valleys with an equivalent energy minimum). In thermoelectrics, it is desirable to have a high-band degeneracy [10], as this promotes the simultaneous existence of large Seebeck and conductivity. The SPB model discussed here hides the influence of band degeneracy within the effective mass. If the band degeneracy is known, an estimate of m^* that is closer to the actual band mass can be obtained by incorporating this knowledge into the model. For N_υ equivalent valleys, the SPB effective mass m^* is related to the density of states band mass m_{DOS} via $(m^*)^{3/2} = N_\upsilon(m_{DOS})^{3/2}$ (consider Equations 11.3 and 11.27).

Acknowledgments

We would like to thank Yanzhong Pei and Eric Toberer for their comments on this chapter, as well as the many stimulating conversations over the last few years. We would also like to thank Michael Böttger for a careful review of this chapter. This chapter is the result of research sponsored by the Jet Propulsion Laboratory, California Institute of Technology under a contract with the National Aeronautics and Space Administration. We acknowledge funding from the Materials Sciences and Engineering Division, Office of Basic Energy Sciences, U.S. Department of Energy.

Symbols

d	density
E	particle energy
e	magnitude of charge of an electron or hole
E_{def}	deformation potential
F	Fermi distribution function
F_j	Fermi integral of order j
h	Planck constant
k	Boltzmann constant
L	Lorenz number
m_e	rest mass of electron
m^*	carrier effective band mass
n	carrier density
n_H	Hall carrier density ($=1/R_He$)
n_{opt}	optimum carrier density
P	generic power-law fit parameter for temperature dependence of mobility
R_H	Hall coefficient
r_H	Hall factor
T	absolute temperature
V	average volume per atom
υ_l	longitudinal speed of sound
υ_t	transverse speed of sound
zT	thermoelectric figure of merit (materials, dimensionless)
α	Seebeck coefficient
ϵ	reduced particle energy ($=E/kT$)
Z	electrochemical potential
η	reduced electrochemical potential ($=\zeta/kT$)
κ	thermal conductivity
κ_b	bipolar contribution to thermal conductivity
κ_e	electronic contribution to thermal conductivity
κ_L	lattice or phonon contribution to thermal conductivity
κ_{min}	minimum lattice thermal conductivity
Λ	scattering parameter related to energy dependence τ
μ	drift mobility
μ_H	Hall mobility
μ_0	SPB mobility parameter ($=e\tau_0/m^*$)
ρ	electrical resistivity
σ	electrical conductivity
τ	relaxation time for a charge carrier
τ_0	prefactor in power-law description of energy-dependent τ

References

1. A. F. May, D. J. Singh, G. J. Snyder, Influence of band structure on the large thermoelectric performance of lanthanum telluride, *Physical Review B* **79**, 153101, 2009.
2. D. J. Singh, I. I. Mazin, Calculated thermoelectric properties of La-filled skutterudites, *Physical Review B* **56**, R1650, 1997.
3. Y. I. Ravich, B. A. Efimova, I. A. Smirnov, *Semiconducting Lead Chalcogenides* (Plenum, New York, 1970).
4. D. J. Singh, W. E. Pickett, Skutterudite antimonides: Quasilinear bands and unusual transport, *Physical Review B* **50**, 11235–11238, 1994.
5. T. Caillat, A. Borshchevsky, J.-P. Fleurial, Properties of single crystalline semiconducting $CoSb_3$, *Journal of Applied Physics* **80**, 4442, 1996.
6. J. S. Dyck, W. Chen, C. Uher, L. Chen, X. Tang, T. Hirai, Thermoelectric properties of the *n*-type filled skutterudite $Ba_{0.3}Co_4Sb_{12}$ doped with Ni, *Journal of Applied Physics* **91**, 3698, 2002.
7. V. L. Kuznetsov, L. A. Kuznetsova, D. M. Rowe, Effect of partial void filling on the transport properties of $Nd_xDo_4Sb_{12}$ skutterudites, *Journal of Physics: Condensed Matter* **15**, 5035, 2003.
8. A. F. Ioffe, *Semiconductor Thermoelements and Thermoelectric Cooling* (Infosearch Ltd, London, 1957).
9. H. J. Goldsmid, *Applications of Thermoelectricity* (Butler & Tanner Ltd, London, 1960).
10. H. J. Goldsmid, *Introduction to Thermoelectricity* (Springer, New York, 2009).
11. A. H. Wilson, *The Theory of Metals* (The Syndics of the Cambridge University Press, London, 1953).
12. J. M. Ziman, *Electrons and Phonons. The Theory of Transport Phenomena in Solids* (Oxford University Press, London, 1963).
13 C. M. Bhandari, *CRC Handbook of Thermoelectrics* (CRC Press, Boca Raton, 1995).
14. V. I. Fistul, *Heavily Doped Semiconductors* (Plenum Press, New York, 1969).
15. N. W. Ashcroft, N. D. Mermin, *Solid State Physics* (Thomson Learning Inc, USA, 1976).
16. E. S. Toberer, A. F. May, G. J. Snyder, Zintl chemistry for designing high efficiency thermoelectric materials, *Chemistry of Materials* **22**, 624–634, 2010.
17. D. G. Cahill, S. K. Watson, R. O. Pohl, Low limit to the thermal conductivity of disordered crystals, *Physical Review B* **46**, 6131–6140, 1992.
18. C. Wood, Materials for thermoelectric energy conversion, *Reports on Progress in Physics* **51**, 459–539, 1988.
19. G. D. Mahan, *Solid State Physics*, Vol. 51 (Academic Press, New York, 1998).
20. F. DiSalvo, Thermoelectric cooling and power generation, *Science* **285**, 703, 1988.
21. H. J. Goldsmid, J. W. Sharp, Estimation of the thermal band gap of a semiconductor from Seebeck measurements, *Journal of Electronic Materials* **28**, 869–872, 1999.
22. L. Danielson, M. Alexander, C. Vining, R. Lockwood, C. Wood, Thermoelectric properties of $LaTe_y$, *Proceedings Seventh International Conference on Thermoelectric Energy Conversion*, p. 71 (University of Texas, Arlington, Texas, March 1988).
23. M. Hokazono, H. Anno, K. Matsubara, Effect of Cu substitution on thermoelectric properties of Ge clathrates, *Materials Transactions* **46**, 1485–1489, 2005.
24. A. F. May, J.-P. Fleurial, G. J. Snyder, Thermoelectric performance of lanthanum telluride produced via mechanical alloying, *Physical Review B* **78**, 125205, 2008.
25. A. F. May, E. S. Toberer, A. Saramat, G. J. Snyder, Characterization and analysis of thermoelectric transport in *n*-type $Ba_8Ga_{16-x}Ge_{30+x}$, *Physical Review B* **80**, 125205, 2009.
26. E. S. Toberer, G. J. Snyder, Complex thermoelectric materials, *Nature Materials* **7**, 105–114, 2008.
27. G. K. H. Madsen, K. Schwarz, P. Blaha, D. J. Singh, Electronic structure and transport in Type-I and Type-VIII clathrates containing strontium, barium, and europium, *Physical Review B* **68**, 125212, 2003.
28. A. F. May, E. Flage-Larsen, G. J. Snyder, Electron and phonon scattering in the high temperature thermoelectric $La_3Te_{4-z}M_z$, M = Sb,Bi, *Physical Review B* **81**, 125205, 2010.

29. M. Christensen, A. B. Abrahamsen, N. B. Christensen, F. Juranyi, N. H. Andersen, K. Lefmann, J. Andreasson, C. R. H. Bahl, B. B. Iversen, Avoided crossing of rattler modes in thermoelectric materials, *Nature Materials* **7**, 811–815, 2008.
30. H. J. Goldsmid, The thermal conductivity of bismuth telluride, *Proceedings of the Physical Society B* **69**, 203, 1956.
31. H. J. Goldsmid, *The Thermal Properties of Solids* (Dover Publications, New York, 1965).
32. G. K. H. Madsen, Automated search for new thermoelectric materials: The case for LiZnSb, *Journal of the American Chemical Society* **128**, 12140–12146, 2006.
33. E. H. Putley, Galvano- and thermo-magnetic coefficients for a multi-band conductor, *Journal of Physics C (Solid State Physics)* **8**, 1837–1840, 1975.

12

The Effect of Resonant Energy Levels on the Thermoelectric Power and Thermoelectric Power Factor

Joseph P. Heremans
The Ohio State University

12.1 Introduction

Most of the recent work aimed at increasing the thermoelectric (TE) figure of merit zT of modern materials has focused on reducing the lattice thermal conductivity using phonon scattering mechanisms, such as nanostructures or localized phonon modes. This approach has been very successful, leading to a doubling of zT, but further progress along these lines promises to come more slowly: Slack[1] teaches us that the phonon conductivity is limited in most bulk systems to what is known as the amorphous limit, that is, where the phonon mean free path is of the order of one interatomic distance. Comparatively less research effort has been aimed at increasing the thermoelectric power factor $P = S^2\sigma$, where S denotes the thermopower or Seebeck coefficient and σ the electrical conductivity. The quantity that needs optimization here is more specifically the product (S^2n), where n is the charge carrier concentration, because in doped semiconductors and metals, the thermopower $S(n)$ decreases with increasing n, a relation attributed to Pisarenko[2] and illustrated for p-type PbTe at room temperature as a full line in Figure 12.1. Combined with the electrical conductivity which is given to the first order by

$$\sigma = nq\mu \tag{12.1}$$

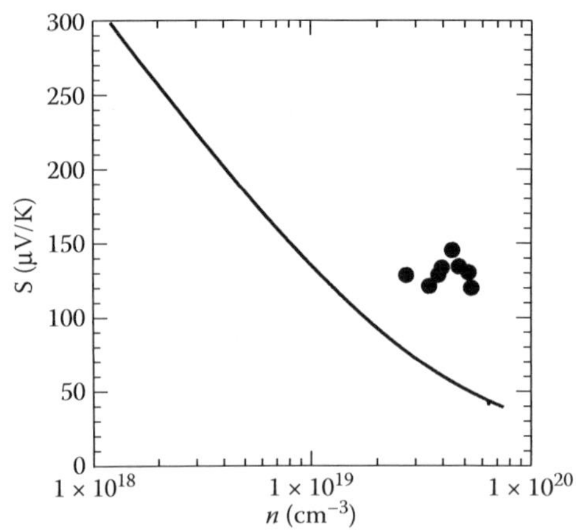

FIGURE 12.1 The Pisarenko relation between the thermopower S and the electron or hole concentration n in PbTe at 300 K (full line). The points are data for Tl-doped PbTe.

where q is the carrier charge and μ its mobility, the Pisarenko relation results in a power factor $P(n)$ that has a maximum for an optimum doping level that gives a value of n of the order of 10^{19}–10^{20} cm^{-3}. The higher the value of (S^2n), the higher the maximum power factor of optimally doped samples.

The product (S^2n) is a purely electronic property, dominated by the details of the electronic band structure and of the scattering mechanisms. We will show it to be a very strong function of the electronic density of states (DOS) $g(E)$, and somewhat less of the energy dependence of the scattering mechanism. Two main approaches to increasing S^2n have therefore been suggested, electron energy filtering[3,4] and distortions of the DOS near the Fermi energy. The two approaches are related: distortions in the DOS lead to resonant scattering, which has an energy filtering effect. Electron energy filtering has been illustrated using semiconductor superlattices grown by molecular beam epitaxy of metal–organic chemical vapor deposition techniques,[4] and several groups are currently trying to emulate these results using naturally occurring nanostructures in bulk alloys[5] or hybrid organic/inorganic structures and other self-assembling bulk materials. Here, we review the second approach: engineer the band structure to increase the DOS at the Fermi energy, $g(E_F)$. This was predicted to be beneficial to zT in the mid-1990s in three seminal papers, one by Mahan and Sofo,[6] and the others a set of two papers on the role of quantum confinement by Hicks and Dresselhaus[7] on quantum wells and quantum wires. In the next paragraph, we describe the band structure engineering tools used to create a favorable DOS.

The DOS, $g(E)$, in units of number of states per eV per cm^3 of material, is a measure of how many states are available to electrons in a unit volume of solid at energies between E and $E + \mathrm{d}E$. The dimensionality of the system is of primary importance in determining the energy dependence of the DOS. For a quasi-free three-dimensional electron with a quadratic dispersion relation $E(\mathbf{k})$, $g(E) \propto \sqrt{E}$. We will detail in the next section how S depends on $g(E = E_F)$, but here we start by paraphrasing Mahan and Sofo's[6] conclusion. The larger the g and the stronger its dependence on E, the higher the $S(n)$ will be for a given carrier concentration. The limiting case, where $g(E)$ is a delta function centered at $E = E_F$, is shown to give the optimal enhancement, but only if the background DOS under the delta function is zero. Such a situation does not occur in nature, and local variations in sample composition would broaden the delta function. The goal of research on the thermoelectric power factor is therefore to increase g as much as possible over and above a $g(E) \propto \sqrt{E}$ function near the Fermi level. The means to achieve such goal are summarized in Figure 12.2, and can be classified into four main directions:

1. One-dimensional quantum size effects in nanowires
2. Resonant impurity levels, which do not involve a magnetic exchange between the impurity and the host solid

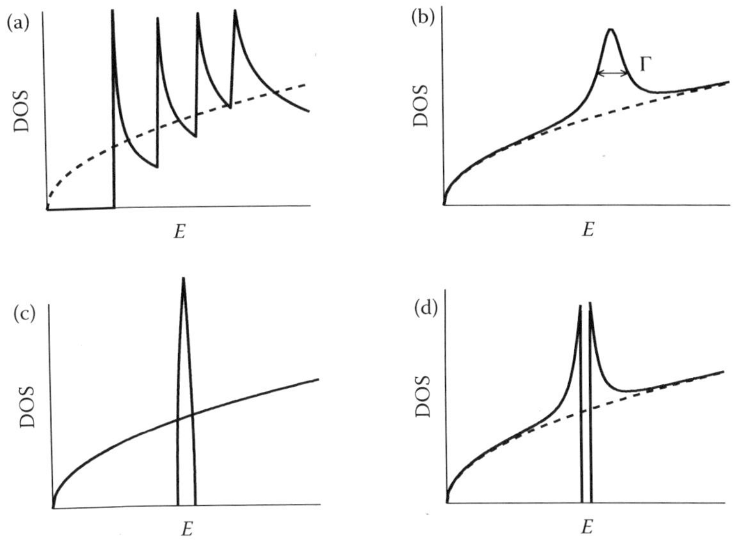

FIGURE 12.2 Density of states versus energy for different distorted bands. The mechanisms used to distort the density of states in solids and to increase their thermopower are: (a) Quantum size effects. (b) Resonant levels. (c) f-band of other narrow-band electrons. (d) Kondo systems.

3. The use of very narrow bands, ideally induced by *f*-levels in solids based on rare earth elements and displaying the physics of correlated electron systems; *d*-levels in carefully selected compounds may yield favorable results at a much lower cost in terms of the complexity of the physics involved
4. When the bands are narrow enough and hybridize with the normal extended states of the host solid, one reaches the case of Kondo systems, which are closely related to both (2) and (3) above

Hicks and Dresselhaus[7] predict that quantum wires could make excellent thermoelectric materials because $g(E)$ diverges at specific quantized energies in a true one-dimensional system by the very nature of the mathematical definition of $g(E)$ and the principles of statistical physics (Figure 12.2a). Quantum size effects exist in wires of a diameter smaller than the electron de Broglie wavelength of the material they are made of. In practice, the most promising systems in which this could be realized are elemental bismuth and BiSb alloys (the best TE materials below 150 K), in which the de Broglie wavelengths are of the order of 50 nm.[8] Experimental work on such effects in Bi showed that enhancements in S can exceed 100 mV/K.[9] This promises to increase zT below 150 K, and was reviewed previously.[10,11]

The use of electron energy bands that have a high DOS over a very narrow energy range is obviously advantageous. The Fermi energy of the well-known Cu–Ni thermocouple alloy constantan, for instance, lies at the root of a band related to the 3*d*-levels, giving rise to a thermopower that is indeed very large for a metal. The most promising of all would be the electronic states associated with *f*-electrons in rare earth elements: the atomic 4*f* orbitals are much more localized than even the highest occupied atomic *d*-levels in transition metals, and in an oversimplified picture (Figure 12.2c) would form extremely narrow bands. The physics of *f*-levels in solids is complicated by two factors. First, the independent electron approximation used in band structure calculations breaks down in systems that involve *f* electrons. Consequently, it is not accurate to draw a simple one-electron DOS picture for their $g(E)$ as done in Figure 12.2c: electron–electron interactions are strong, and the shape of $g(E)$ depends on their concentration. Second, the notion that a continuous transition exists between the extended states in the solid and the atomic levels on the atoms as the distance between the atoms is progressively increased also breaks down. As the nearest-neighbor separation between atoms is progressively increased, solids based on rare earth atoms can suddenly change from conductors to insulators at a critical interatomic separation value, a so-called Mott transition. This happens in many compound semiconductors based on rare earth elements, with the additional experimental difficulty that there are few solid solutions in their phase diagram, which complicates the tuning of their electronic properties. The band structure of even the elemental rare earth metals, none of which have particularly high thermopower, is not easy to calculate: the commonly accepted model is that the energy levels near the Fermi energy are due to fairly broad *s–p–d* bands, and that the *f*-levels split into two sharp peaks, one far below and the other far above the Fermi level.[12] Techniques to shift the Fermi level sufficiently to reach these levels are yet to be devised except in very few compounds. All of those that are known to date are impractical for use as thermoelectrics. The nomenclature covering these effects is "correlated electron physics." Figure 12.2c helps illustrate the relation between correlated systems and the other mechanisms, and is less inaccurate for transition metal atoms.

Kondo systems (Figure 12.2d) and resonant levels (RLs) (Figure 12.2b) are closely related to each other. Historically, RLs were identified in metal alloys such as Ag + 0.05 at% Mn in 1951 by Gerritsen and Korringa,[13] who calculated the hybridization between electrons on the dilute impurity atoms and those in the conduction of valence band (VB) of the host solid in metal alloys, pointing out that these resonances were analogous to those between α-radioactive nuclei and α-particles. That early paper accurately describes the properties of these levels that are important to transport. The concept was worked out more in detail in 1956 by Friedel.[14] In 1965, Jun Kondo[15] calculated the magnetic contributions to the coupling of dilute magnetic impurities in host metals (typically Cu, Ag, Au, but also Al), and successfully explained the anomalies in the temperature dependence of the resistivity and the strong enhancement in thermopower S observed at cryogenic temperatures in such systems.[16] This effect lies at the basis of cryogenic thermocouple alloys such as Au + 0.07 at% Fe, and is probably best labeled as *dilute Kondo systems.* Related physics is found in *Kondo insulators* such as $Ce_3Bi_4Pt_3$, which are compounds in which the transition or rare earth metal atoms are a major constituent. Similar are FeSi[17] or $CePd_3$,[18] and probably the interesting cryogenic system $FeSb_2$.[19] The energy bands in the solid that originate from the *f*- or *d*-electron shells of the transition metal atom are much narrower than those from *s*- and *p*-levels. If the two intersect at a certain energy, they hybridize (Figure 12.2d). This opens a gap, and forms an excess DOS on its edges; the system now is called a Kondo insulator.[17–20] Kondo insulators and dilute Kondo alloys are two limiting conditions, and intermediate situations occur at intermediate concentrations of transition metal or rare earth atoms.

12.2 Mechanisms for Thermopower Enhancements

DOS distortions such as shown in Figure 12.2, and RLs (Figure 12.2b) in particular, act on the thermoelectric transport properties in two ways:

1. The excess DOS $g(E)$ itself at the energy corresponding to the distortion creates an excess thermopower. The effect is purely due to details in the band structure, and has no more temperature dependence than the bands themselves, so that such effects can remain dominant at high temperatures.
2. Sharp excesses in DOS scatter conduction electrons in a way that is extremely sensitive to their energy, and the effect is known as "resonant scattering." Because phonon scattering dominates the high-temperature electronic transport properties of solids, the effects of resonant scattering are most often confined to cryogenic temperatures. Resonant scattering on virtual bound states (resonant impurity levels) was identified first in metals[21] and later applied to semiconductors by Ravich.[22]

Here, we first offer a simple phenomenological theory for both effects, and then suggest that magnetic fields can be used to separate their influences experimentally. We describe the experimental evidence in the subsequent section.

The phenomenological model is based on the Mott equation[23] that relates the thermopower S to the energy-dependent electrical conductivity $\sigma(E)$, in this case a scalar. In crystalline solids, both the energy-dependent $\ddot{\sigma}(E)$ and the total electrical conductivity $\ddot{\sigma}$ are tensors related by

$$\ddot{\sigma} = \int \ddot{\sigma}(E)\left(-\frac{\partial f}{\partial E}\right)\mathrm{d}E \tag{12.2}$$

where f is the equilibrium Fermi distribution function. One can demonstrate (Yibin Gao, private communication) from the Boltzmann transport equation using the relaxation time τ approximation that each component (μ, ν) of the conductivity tensor at energy E', $\ddot{\sigma}(E = E')$, is

$$\begin{aligned}\sigma_{\mu,\nu}(E') &= \frac{q^2}{4\pi^3}\frac{1}{\hbar^2}\oiint (\nabla_k E)\cdot\vec{\mathbf{e}}_\mu\,(\nabla_k E)\cdot\vec{\mathbf{e}}_\nu\,\tau(\vec{\mathbf{k}})\frac{dS}{|\nabla_k E|}\\ &= \frac{q^2}{4\pi^3}\frac{1}{\hbar^2}\iiint\left((\nabla_k\nabla_k E)_{\mu,\nu}d^3\mathbf{k}\right)\tau(E)\end{aligned} \tag{12.3}$$

Here, dS denotes an integral over the equienergetic surface at energy E' (the Fermi surface if $E' = E_F$), the volume integral is taken over its volume, and $\vec{\mathbf{e}}$ s are unit vectors. We assume that τ is a scalar. Mathematically, the DOS $g(E)$ is the function that enables the transformation of variables in the calculation of the transport integrals, from integrals of the type $(1/4\pi^3)\iiint_{E\le E'} F(\mathbf{k})d^3\mathbf{k}$ over momentum-space ($\mathbf{k}$-space) defined in a Brillouin zone of an n-dimensional solid, to scalar integrals of the type $\int_0^{E'} F(E)g(E)\mathrm{d}E$ over energy E (here $F(E)$ is an arbitrary function). The definition of the DOS is given by Ashcroft and Mermin,[12] and the total density of electrons at energies $E \le E'$ is

$$n(E') \equiv \frac{1}{4\pi^3}\iiint_{E\le E'} d^3\mathbf{k} = \int_0^{E'} g(E)\mathrm{d}E \tag{12.4}$$

Define the effective mass tensor $\ddot{\mathbf{m}}$ by its inverse $\ddot{\mathbf{m}}^{-1} \equiv \nabla_k\nabla_k E/\hbar^2$, the average inverse mass tensor for all electrons at energy $E \le E'$ is then

$$<\ddot{\mathbf{m}}^{-1}> \equiv \frac{1}{\hbar^2}\frac{\iiint_{E<E'}(\nabla_k\nabla_k E)d^3\mathbf{k}}{\iiint_{E<E'} d^3\mathbf{k}} \tag{12.5}$$

Equation 12.3 becomes

$$\ddot{\sigma}(E') = q^2 <\ddot{\mathbf{m}}^{-1}>\tau(E')n(E') = n(E')q\ddot{\mu}(E')(<\ddot{\mathbf{m}}^{-1}>:\ddot{\mathbf{m}}) \approx n(E')q\ddot{\mu}(E') \tag{12.6}$$

where $n(E')$ is the total density of charge carriers integrated from energy $E = 0$ up to energy $E = E'$. Assuming that the equienergetic surfaces maintain their shape, that is, if the elements $\partial^2E/\partial k_x^2$; $\partial^2E/\partial k_y^2$; $\partial^2E/\partial k_z^2$ at all values of $\vec{\mathbf{e}}$ have the same E'-dependence, the quantity $<\ddot{\mathbf{m}}^{-1}>:\ddot{\mathbf{m}}$ is a scalar. For parabolic bands, it is unity. The largest deviation of this factor from unity occurs in semiconductors like PbTe, which have Kane dispersions with an energy gap of the order of 0.15 eV, and are doped in the high-10^{19} cm^{-3} range in order to operate at high temperatures; even then, $<\ddot{\mathbf{m}}^{-1}>:\ddot{\mathbf{m}}$ remains within 25% of unity, and we also approximate it by 1 in Equation 12.6. The mobility tensor that applies to Equation 12.6 can then be defined at each value of energy E' by

$$\ddot{\mu}(E') \equiv q\tau(E')\ddot{\mathbf{m}}(E')^{-1} \tag{12.7}$$

Considering now the general form of the Mott relation[23] for the thermoelectric power, we can write for each component (μ, ν) of $\vec{\sigma}(E')$:

$$S = \frac{k_B}{q}\frac{1}{\sigma_{\mu,\nu}}\int_0^\infty \sigma_{\mu,\nu}(E)\left(\frac{E-E_F}{k_BT}\right)\left(-\frac{\partial f}{\partial E}\right)dE \tag{12.8}$$

Under the assumption that the equienergetic surfaces maintain their shape as the energy is changed, Equation 12.8 represents the thermopower as a scalar, a result consistent with the fact that the thermopower is closely related to the entropy of the electron divided by its charge. Such statement is strictly only applicable in the present case in which we considered only one pocket of the Fermi surface. The total thermopower of an anisotropic solid with several pockets in its Fermi surface is a tensor because it is the average of the partial thermopowers of each pocket weighted by the partial conductivities, which are tensors. A Bethe–Sommerfeld expansion of Equation 12.8 valid for degenerate statistics yields an approximate formula:

$$S = \frac{\pi^2}{3}\frac{k_B}{q}(k_BT)\left[\frac{1}{n(E)}\frac{dn(E)}{dE}+\frac{1}{\mu(E)}\frac{d\mu(E)}{dE}\right] = \frac{\pi^2}{3}\frac{k_B}{q}(k_BT)\left[\frac{g(E)}{n(E)}+\frac{1}{\mu(E)}\frac{d\mu(E)}{dE}\right] \tag{12.9}$$

using Equation 12.6, and defining μ as the tensor element $\mu_{\mu,\nu}$. The approximation comes from the fact that the thermopower is rigorously the logarithmic derivative of Equation 12.3, which contains the factor $(\nabla_k\nabla_k E)_{\mu,\nu}$ in the integrand. In spite of that, Equation 12.9 yields acceptable results in most practical semiconductors: for PbTe or Bi_2Te_3 doped to about 10^{19} cm^{-3} carriers and temperatures at or below 700 K, the error is less than 30%.

The presence of distortions in the DOS as shown in Figure 12.2 strongly affects Equation 12.9, which can be viewed as consisting of two terms, $g(E)/n(E)$ and $(1/\mu)$ $(d\mu/dE)$. The effects of distortions on the term $g(E)/n(E)$ correspond to the effect labeled (1) in the first paragraph of this section; that on the term $(1/\mu)$ $(d\mu/dE)$ corresponds to resonant scattering, the effect labeled (2) above. If the distortion is relatively slight, as it is in the case of metals, g/n in Equation 12.9 is small, and the impurities act mostly through the $1/\mu(d\mu/dE)$ resonant scattering term. In semiconductors, the g/n term is important because the background DOS and its integral over energy, $n(E)$, is much smaller than in metals.

12.3 Resonant Levels

12.3.1 Physical Origin

The RLs of interest in thermoelectric materials are those that exist in semiconductors. Consider first (Figure 12.3) the classical case of a donor impurity, such as P in Si. Phosphorous atoms share four electrons with the neighboring Si atoms, leaving one extra electron that remains weakly bound to the P. A small thermal activation energy R^* (approximately a few tens of meV) is sufficient to overcome the bond between the P atom and its fifth electron and releases that electron into the conduction band (CB) of the host material. This is similar to the case of the 1*s* electron on the hydrogen atom in vacuum, which can be released into vacuum by the addition of one Rydberg of energy. We label this the *hydrogen-like model*, and R^* is an *effective Rydberg*. We depict the energy level of this fifth electron as lying R^* below the CB edge and in the energy gap of the semiconductor.

In contrast, resonant donor impurities would have electronic energy levels for which there is no activation energy, but which lie at an energy E_D above the CB edge (or below the VB edge for *p*-type systems). The impurity level falls inside the CB and its energy coincides with energies of extended states (see Figure 12.3). Friedel labeled this as a "virtual bound state," now called a resonant state. As this state now would have the same energy as an extended state, which is forbidden by the Pauli principle, the two will resonate to build up two extended states of slightly different energies; these in turn would have the same energies as other

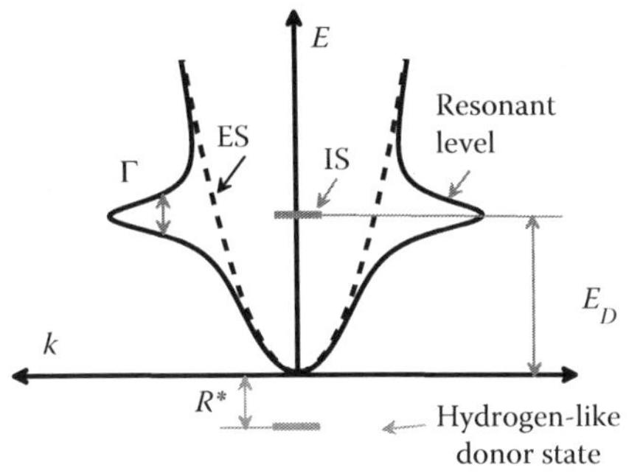

FIGURE 12.3 Schematic representation of a CB, a hydrogen-like impurity, and a resonant impurity.

extended states with whom they will resonate in turn, and so on. Consequently, the resonant state develops a certain width Γ. This warrants further discussion, for which we follow Daniel and Friedel.[24]

Since impurity atoms have multiple electron energy levels, some can be deeply bound, and others weakly bound or resonant. The Hamiltonian of an electron of the host solid unperturbed by the impurity has eigenvectors that are the propagation vectors **k** of the extended states (ES), schematically represented by the dashed parabola in Figure 12.3. In the spatial vicinity of the impurity, there is a disturbance of the periodic potential, on average by an amount *V*. In the simplest picture, the periodic potential of the host solid is averaged out, and we can treat the potential of the impurity on the basis of the simple one-dimensional quantum-mechanical picture for the particle-in-the-well of depth *V*. Electrons in this potential well have discrete eigenvalues and quantum numbers, just like electrons on isolated atoms, so we can label them atomic impurity states (IS). Those that fall inside the potential well correspond to bound states in the hydrogen-like model above (R^* in the gap in Figure 12.3). Those that fall above the edges of the well (E_D in the band in Figure 12.3) have an energy level already attributed to an extended state of the host bands. The extended states and those particular IS energy levels of the impurity can interact: the RL is a mixture of the two. Daniel and Friedel assume the impurity energy level to be a *d*-level, as they consider cases of transition metal impurities in metals with *sp* bands, such as Fe impurities in Au. They further solve the coupled Hamiltonian for the ES (assumed *s*-states) and the IS (assumed *d*-states) by *s*–*d* interactions. From this, they derive that the amount of impurity-induced state added to the ES (i.e., the difference between the dashed and the full line in Figure 12.3) has a Lorentzian energy dependence with a width given by

$$\Gamma = 2\pi \left| \overline{< O \mid V \mid \mathbf{k} >} \right|^2 g(E) \tag{12.10}$$

where $g(E)$ and **k** are the DOS and the wavevector of the ES, and *O* the orbital quantum number of the IS. A count of the electrons on the impurity and in the ES shows that the contribution to the total $g(E)$ is a Lorentzian, as shown in Figure 12.2b, with the same width Γ. The wave function of an electron on an RL is spatially concentrated in the atomic cell in which the impurity is located, but it is mixed with extended states. The simplest picture for this is to imagine that an electron in this mixed RL state spends a fraction of its time on the impurity, where it does not conduct, and a fraction of its time in the extended states, where it does. Therefore, conduction occurs through RLs even when the impurities are very dilute (ppm levels) and there is no overlap between the wavefunctions of the individual IS states. Electrons in RLs carry heat and charge, albeit with a decreased group velocity. If the impurities are at concentrations of the order of a few at.%, enough overlap exists between the wave functions of the different impurity atoms in neighboring cells, and the RLs can form an impurity band.

This discussion described the origin of the physical concepts necessary to exploit RLs in thermoelectric semiconductors: three criteria need to be met to design a semiconductor with the correct carrier density that optimizes the power factor.

1. The Fermi level must be located at the correct value E_F with respect to the band edge in order for the semiconductor to have the concentration of free carriers that maximizes $P(n)$. Since the Fermi level must also be in the RL in order for it to dominate transport, there is a second condition.
2. The RL energy E_D itself must fall appropriately with respect to the band edge.
3. Mahan and Sofo[6] calculate that the narrower the Γ, the higher the zT (as long as there is no underlying background DOS). In practice, experimentalists have only a limited degree of control over the sample and defect chemistry. This leads to criterion (12.3): experimentalists need a sufficient width, $\Gamma \sim$ 10–30 meV, to be able to locate E_F onto the RL.

The width Γ of the RL is quite different in metals from what it is in semiconductors, because of the factor $g(E)$ in the expression for Γ in Equation 12.10. In metals, where the bands are almost half-full, the background $g(E)$ filled up to E_F accommodates on the order of one electron per unit cell, which is very large. Consequently, in order to have a Γ sufficiently narrow to have a measurable effect on transport properties, the value of $<O|$ must be small. Indeed, RLs are observed in metals for *d*-electron impurities, while *s* or *p* electrons would result in RLs that are much too broad to be noticed. In contrast, in semiconductors, $g(E)$ integrated up to E_F accommodates only 10^{-4}–10^{-2} electrons per unit cell. While impurity atoms with *d*-levels certainly give RLs with very narrow Γ values in semiconductors, impurities with *p*-levels, such as those of the group III elements in PbTe or Sn in Bi_2Te_3, create RL's that are the most beneficial to zT.

12.3.2 Fermi Level Pinning

The excess DOS that results from the presence of a resonant impurity in a band often results in Fermi level pinning. Using the example of Ti doping of PbTe or Fe doping of HgSe,[25] the experimental observations that correspond to Fermi level pinning are as follows:

1. For an iron concentration $N_{Fe} < N_{Fe}^{CRIT}$, a critical value of 5×10^{18} cm^{-3} in HgSe:Fe, the concentration n of electrons is proportional to N_{Fe}.
2. For $N_{Fe} \geq N_{Fe}^{CRIT}$, $n \approx N_{Fe}^{CRIT}$.
3. At a low temperature, where ionized impurity scattering dominates, the mobility of the electrons varies nonmonotonically with N_{Fe}. For $N_{Fe} < N_{Fe}^{CRIT}$, $\mu \propto N_{Fe}^{-1}$, for $N_{Fe} \geq N_{Fe}^{CRIT}$, μ is much less sensitive to N_{Fe} and can even increase slightly with it. Finally, at a higher concentration yet ($N_{Fe} \sim 2 \times 10^{19}$ cm^{-3}), μ decreases again with increasing Fe concentration. The temperature dependence of the mobility is also different in the three regimes.

These effects can all be understood in the light of Figure 12.2b, where the excess DOS in the example above originates from the Fe atoms on the Hg sites. These can have two valences, Fe^{2+} which is neutral, and Fe^{3+} which is a donor. As long as the Fermi level lies below the RL, all Fe atoms are ionized into the 3+ state; the more Fe atoms are added, the more electrons are released into the extended states. When the Fermi level reaches the level E_D from the band edge, it gets pinned: every Fe atom added above the concentration $\approx N_{Fe}^{CRIT}$ goes into the lattice as Fe^{2+}. The value of N_{Fe}^{CRIT} corresponds to the electron concentration at which the Fermi level equals E_D: $N_{Fe}^{CRIT} = n(E_F = E_D)$. The concentration and temperature dependences of the mobility are consistent with this picture, as the Fe^{3+} ions result in ionized impurity scattering of the free carriers, while the Fe^{2+} ions do not.

The experimental observation of Fermi level pinning is not a condition for an impurity to give a thermopower above the Pisarenko relation. Indeed, Fermi level pinning is based solely on the total DOS, and does not imply that the electrons pinned contribute to transport, and thus have a high S.

12.3.3 Resonant Scattering

Here, we estimate the effect of RLs on the second of the two terms in Equation 12.9. With $g(E)$ shown in Figure 12.2b, and $n(E)$ its integral up to the Fermi energy, the first term is clear. We can treat the second term, resonant scattering, using the simple one-dimensional particle-in-the-well picture above and following the didactic derivation developed by Popescu et al.[26] Given a well width a and depth V, the one-dimensional Schrödinger equation for an electron of mass m and energy E above the well edges gives a transmission coefficient $T(E)$, as reported in textbooks on modern physics. Considering now a number N of these wells a distance L apart, and taking the limit for $N \to \infty$ gives the mean free path for scattering over these wells as a function of $T(E)$:

$$\text{mfp}(E) = \sum_{j=1}^{N\to\infty} T(E)^j (1 - T(E)^j L = \frac{T(E)}{1-T(E)} L = \frac{E(E+V)}{V^2} \frac{L}{\sin^2\left[\sqrt{2m(E+V)}\dfrac{a}{\hbar}\right]} \tag{12.11}$$

which diverges not only when $V = 0$, but also when $E + V = i^2\pi^2\hbar^2/2ma^2$ $(i = 0,1,\ldots)$.

Other parallel mechanisms also limit the mean free path, and, if we assume resonant scattering to act independently from other background scattering mechanisms characterized by a relaxation time τ_a, the total relaxation time including the effect of resonant scattering (τ_{RES}) can be obtained from adding the scattering frequencies and is thus described by

$$\tau = \left[\frac{1}{\tau_a} + \frac{1}{\tau_{RES}}\right]^{-1} = \left[\frac{1}{\tau_a} + \frac{v_g}{\text{mfp}(E)}\right]^{-1} \tag{12.12}$$

where v_g is the carrier group velocity. This quantity does not diverge, but can be strongly nonmonotonic. At energies where

$$E = i^2 \frac{\pi^2\hbar^2}{2ma^2} - V \tag{12.13}$$

the mean free path is limited to a value of order of L and a, whereas at energies different from Equation 12.13, the mean free path is that due to phonon scattering. Such strongly energy-dependent behavior provides a very effective energy-filtering mechanism that affects the thermopower. When τ is strongly energy dependent, so is μ, and the term $(1/\mu(E))(d\mu(E)/dE)$ in Equation 12.9 is large. At high temperatures, where $\tau_a \ll \text{mfp}(E)/v_g$, the effect of resonant scattering disappears.

We illustrate schematically the various influences of the effect of a spike in DOS and of resonant scattering in Figure 12.4. Here, we use a parabolic dispersion for the band, with an effective mass $m^* = 0.2m_e$ (m_e = the free electron value), and represent the distortion using a "piecemeal parabolic" model with $m^* = 1m_e$, in order to avoid singularities in evaluating $\nabla_k\nabla_k E$ which would occur if we used a Lorentzian shape for the DOS distortion. The background τ_a is assumed to be limited by acoustic phonon scattering, and the resonant scattering is calculated from Equations 12.11 and 12.12 assuming $V = 2.8$ eV, $a = 0.8$ nm, and $L = 4$ nm; these numbers were chosen so as to show both effects at 300 K. The thermopower S is then further obtained from Equation 12.9 under the main assumption that the effect of RLs on conductivity σ_r and thermopower S_r are independent of those of the bands of the host solid (σ_b and S_b), so that we write $S = (S_r\sigma_r + S_b\sigma_b)/(\sigma_r + \sigma_b)$ (this is a very coarse approximation that does not hold for a strongly interacting system, and is solely used for didactic purposes). We show S for three cases: (1) a simple parabolic band with only acoustic phonon scattering (dashed line), (2) acoustic phonon scattering dominating, but adding the effect of the excess DOS (full line), and (3) resonant scattering dominating on the simple parabolic band. It is seen that the effects of the DOS and of resonant scattering in this case counteract each other. Resonant scattering increases S at lower doping levels, but decreases S when the Fermi energy is located inside the excess DOS. This result is certainly not universal; for different

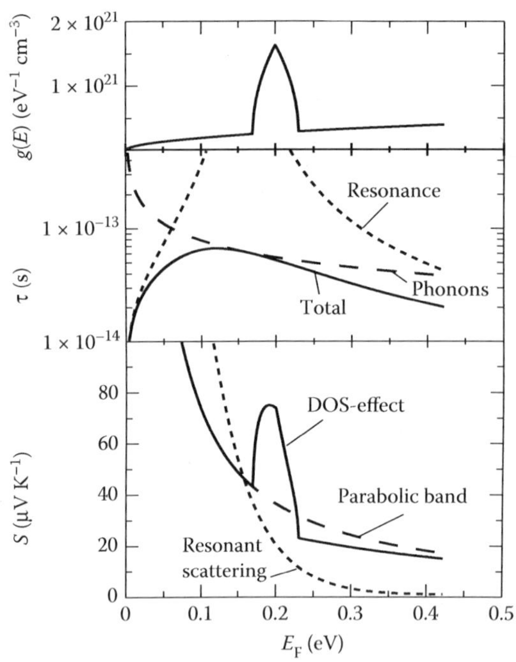

FIGURE 12.4 Schematic representation, as a function of Fermi energy, of the effect of a resonant impurity on the DOS, and of the effect of resonant scattering (middle frame) on the relaxation time τ. The background τ_a (dashed line) was assumed to be limited by acoustic phonon scattering to an $E^{-1/2}$ power law, the calculated τ_{RES} (short dashes) limited by resonant scattering diverges at 0.15 eV. The thermopower S is shown in the bottom frame. S calculated for the background parabolic band assuming only acoustic phonon scattering is shown as a dashed line, S including the effect of the DOS but still with acoustic phonon scattering as a full line, and S calculated neglecting the effect of the DOS but assuming resonant scattering as a dotted line.

values of V, the two effects can add. Since neither V nor a is accurately know for practical systems, experimental observation of these effects is in order.

Thermomagnetic phenomena are useful to separate experimentally the contributions from band structure effects from those of scattering effects in the thermopower. Two different experimental techniques can be used: low-field transverse Nernst–Ettingshausen measurements, or high-field measurements of the thermopower. The scale of the magnetic field B is set by the mobility μ of the sample: a magnetic field is defined as low when the dimensionless product $\mu B < 1$ and high when $\mu B > 1$.

The high-field limit of the thermoelectric power, $S_\infty = \lim_{B\to\infty}[S(B)]$, is independent of the scattering mechanism.[27] Very schematically, the product of the cyclotron frequency ω_c and the relaxation time τ is $\omega_c\tau = \mu B$. Therefore, when $\mu B > 1$, the carriers move on a cyclotron orbit before they undergo a collision, and their thermopower is unaffected by scattering. Note that this image cannot be extended to the magnetoresistance, which obviously must remain finite. Therefore, while both scattering and DOS contribute to S at zero field (Equation 12.14), the $\mu B > 1$ high-field thermopower has only the contribution of the first term g/n in Equation 12.9, and the effects of DOS and of resonant scattering can be separated experimentally.

In the low-field $\mu B < 1$ limit, and assuming that there is only one type of carrier present in the solid, a technique can be applied that we labeled the "method of the four coefficients."[28,29] In this method, four electronic transport properties are measured on each sample at each temperature. These are (1) the electrical resistivity ρ, (2) the thermopower S, (3) the low-field Hall coefficient R_H, and (4) the low-field isothermal

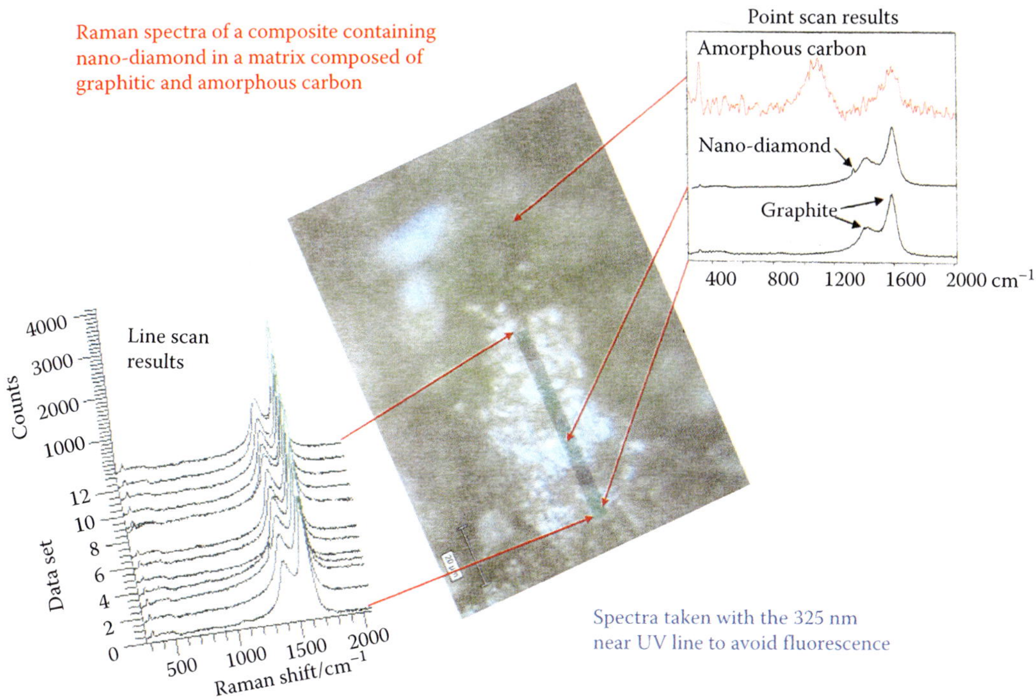

FIGURE 8.1 Raman spectra of nanocarbon ensembles containing UNCD cores surrounded by graphitic and amorphous carbon shells.

HR-TEM of hollow polyhedral nanographite particles

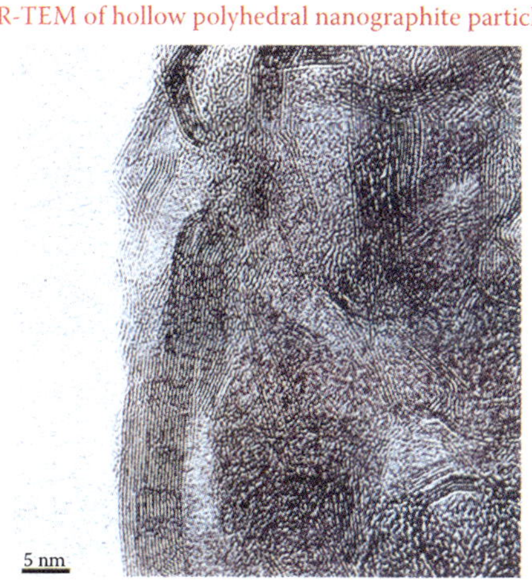

TEM micrographs for the 5% B_4C doped samples of nanographite annealed at 2500 K

FIGURE 8.2 High-resolution transmission electron micrograph of highly electrically conducting hollow polyhedral NG structures formed at annealing temperatures of 2500 K.

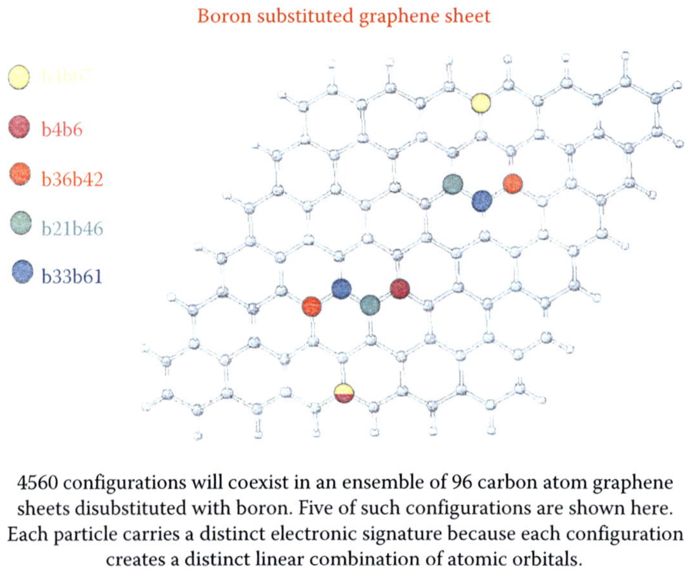

FIGURE 8.5 Five distinct configurations of doubly substituted borons in a graphene sheet of about 100 carbon atoms.

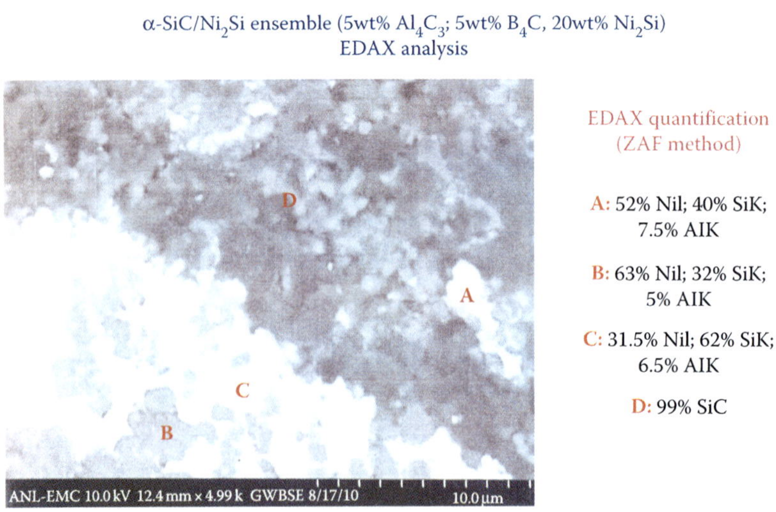

FIGURE 8.12 SEM and EDAX analysis of SiC/Ni_2Si SPS-processed ensemble.

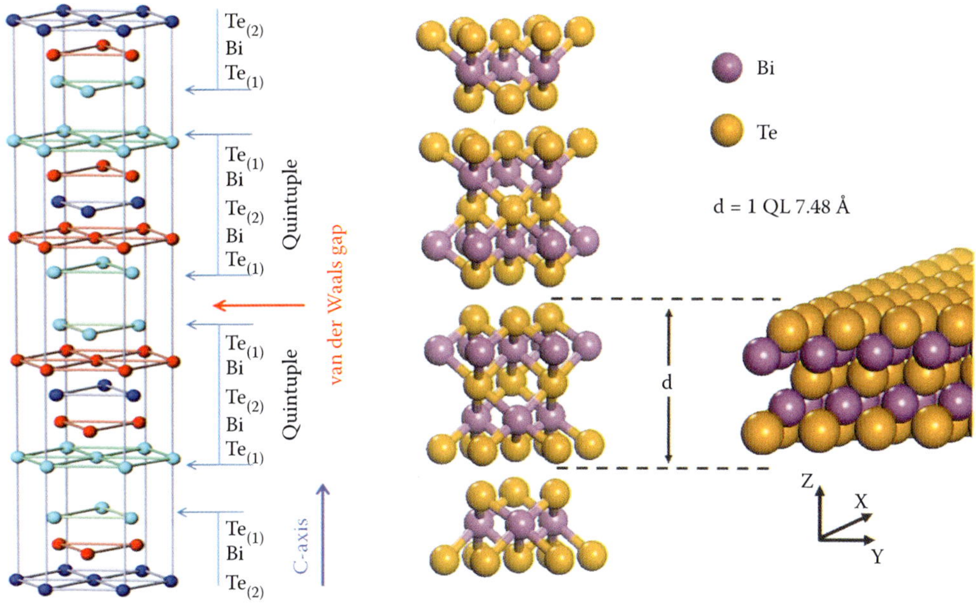

FIGURE 13.1 Schematic of Bi_2Te_3 crystal structure of D_{3d}^5 $-R(-3)m$ space group showing quintuple layers and location of van der Waals gaps (left panel). The $Te^{(1)}$–$Te^{(1)}$ bond is the weakest while Bi–$Te^{(1)}$ bond is the strongest. The mechanical exfoliation results in breaking the $Te^{(1)}$–$Te^{(1)}$ van der Waals bond and formation of few-quintuple films or individual quintuple layers. The atomic layer model used in the *ab initio* simulations of thermoelectric properties of individual quintuple. (Adapted from Zahid, F., Lake, R., *Appl. Phys. Lett.*, 97, 212102, 2010; Teweldebrhan, D., Goyal, V., Balandin, A.A., *Nano Lett.*, 10, 1209, 2010.)

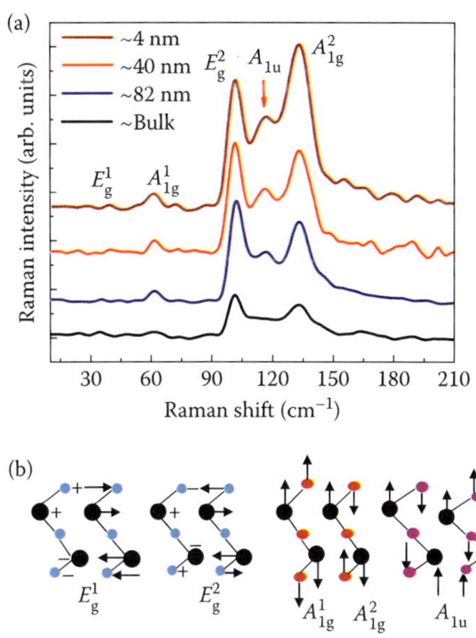

FIGURE 13.4 (a) Raman spectra of reference bulk Bi_2Te_3 crystal and few-quintuple films. Note that A_{1u} mode at ~117 cm^{-1} becomes Raman active only in the thin films. (b) Schematic of the main lattice vibrations in {–$Te^{(1)}$–Bi–$Te^{(2)}$–Bi–$Te^{(1)}$–} quintuples. The large circles are Bi atoms. The small circle in the middle is $Te^{(2)}$ atom, which is the center of the inversion symmetry. The "+" and "–" signs in the schematics indicate atomic motions toward and from the observer. In all panels, the letters "E" and "A" indicate the in-plane and out-of-plane (c axis) lattice vibrations, respectively. The subscript "g" denotes Raman active while "u" stands for IR-active modes. (Adapted from Shahil, K.M.F. et al., *Appl. Phys. Lett.*, 96, 153103, 2010.)

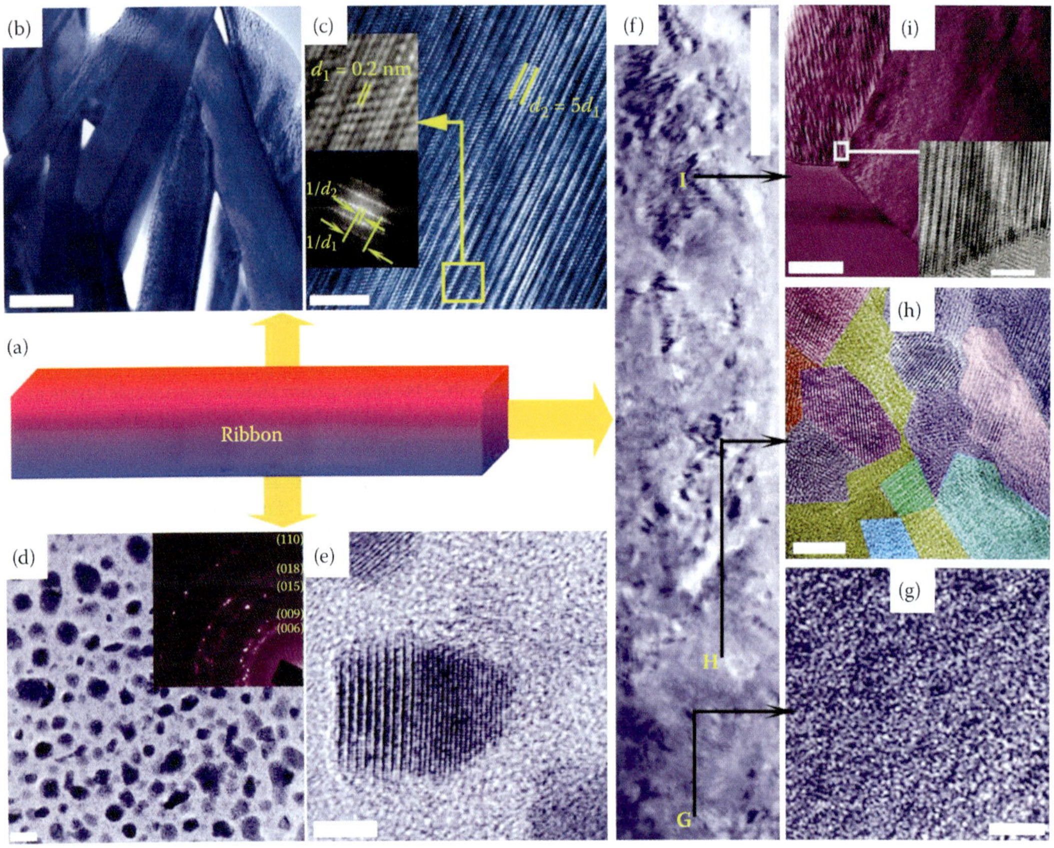

FIGURE 16.3 Microscopy images of the melt spun ribbon. (a) Schematic figure of ribbon sample; (b) TEM image of the free surface that is typified by microsize dendritic crystals, and (c) HRTEM image of nanoscale modulations in the dendritic crystal shown in (b); (d) TEM image of the near contact surface; (e) HRTEM image of nanocrystalline domains; (f) TEM image of the cross-section of the ribbon, the bottom and the top of figure are images near the contact surface and the free surface, respectively; (g) HRTEM image of the amorphous phase near the contact surface; (h) HRTEM image of nanocrystalline domains with mostly coherent boundaries; (i) TEM image of micron-sized crystals, with the inset displaying the HRTEM image of the nanoscale modulation, the scale bar for the inset is 2 nm; The scale bars for (b), (c), (d), (e), (f), (g), (h), and (i) are 200 nm, 5 nm, 20 nm, 5 nm, 0.5 μm, 10 nm, 5 nm, and 50 nm, respectively. (Reprinted with permission from Xie, W. J. et al., Identifying the specific nanostructures responsible for the high thermoelectric performance of $(Bi,Sb)_2Te_3$ nanocomposites, *Nano Lett.*, 10, 3283, Copyright 2010, American Chemical Society.)

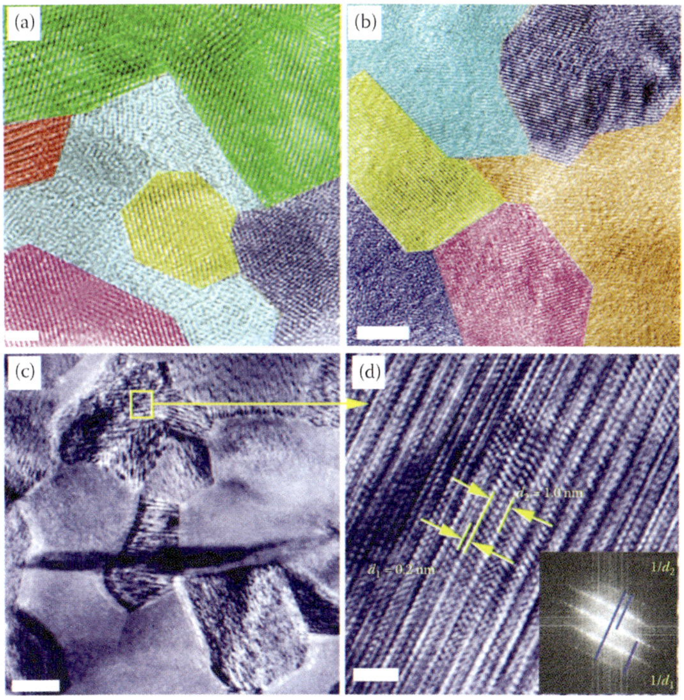

FIGURE 16.4 (a)–(d) are representative TEM images of SE–MS–SPS (single element–melt spun–SPS) sample; (a) and (b) show HRTEM images of typical nanostructures consisting of 10–20 nm nanocrystals embedded in the bulk matrix. (c) and (d) show the HRTEM images of nanoscale modulations. The scale bars for (a), (b), (c), and (d), are 2, 5, 100, and 2 nm, respectively. (Reprinted with permission from Xie, W. J. et al., Identifying the specific nanostructures responsible for the high thermoelectric performance of $(Bi,Sb)_2Te_3$ nanocomposites, *Nano Lett.*, 10, 3283, Copyright 2010, American Chemical Society.)

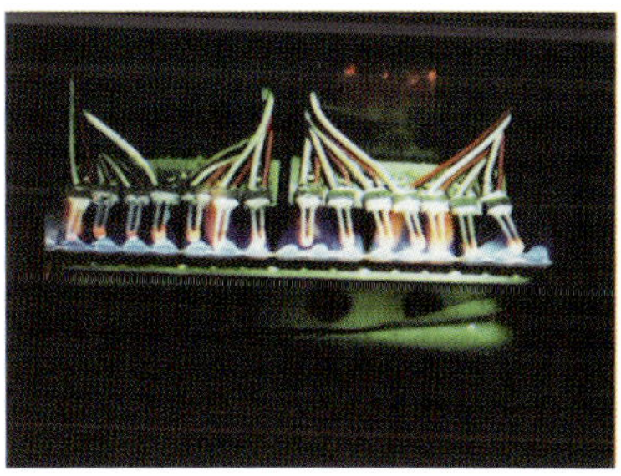

FIGURE 18.22 The durability testing of β-$FeSi_2$ thermocouples in the town gas flame.

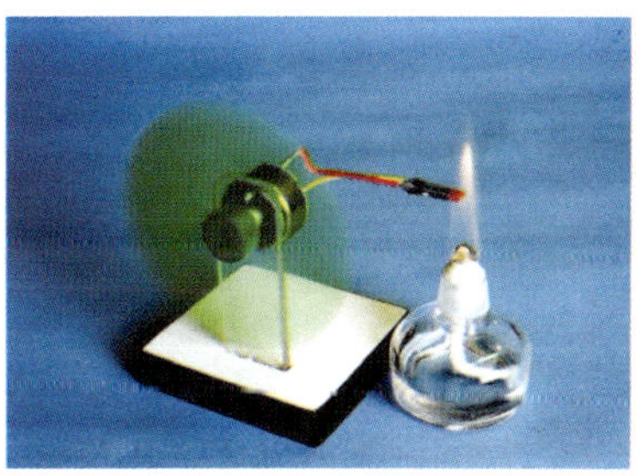

FIGURE 18.26 The simplest thermoelectric generator system made by using a pair of β-$FeSi_2$ thermocouples. (From I. A. Nishida; *Thermoelectrics—Principles and Applications*, edited by M. Sakata, Realize Inc., Tokyo, 2000, 199–209. With permission.)

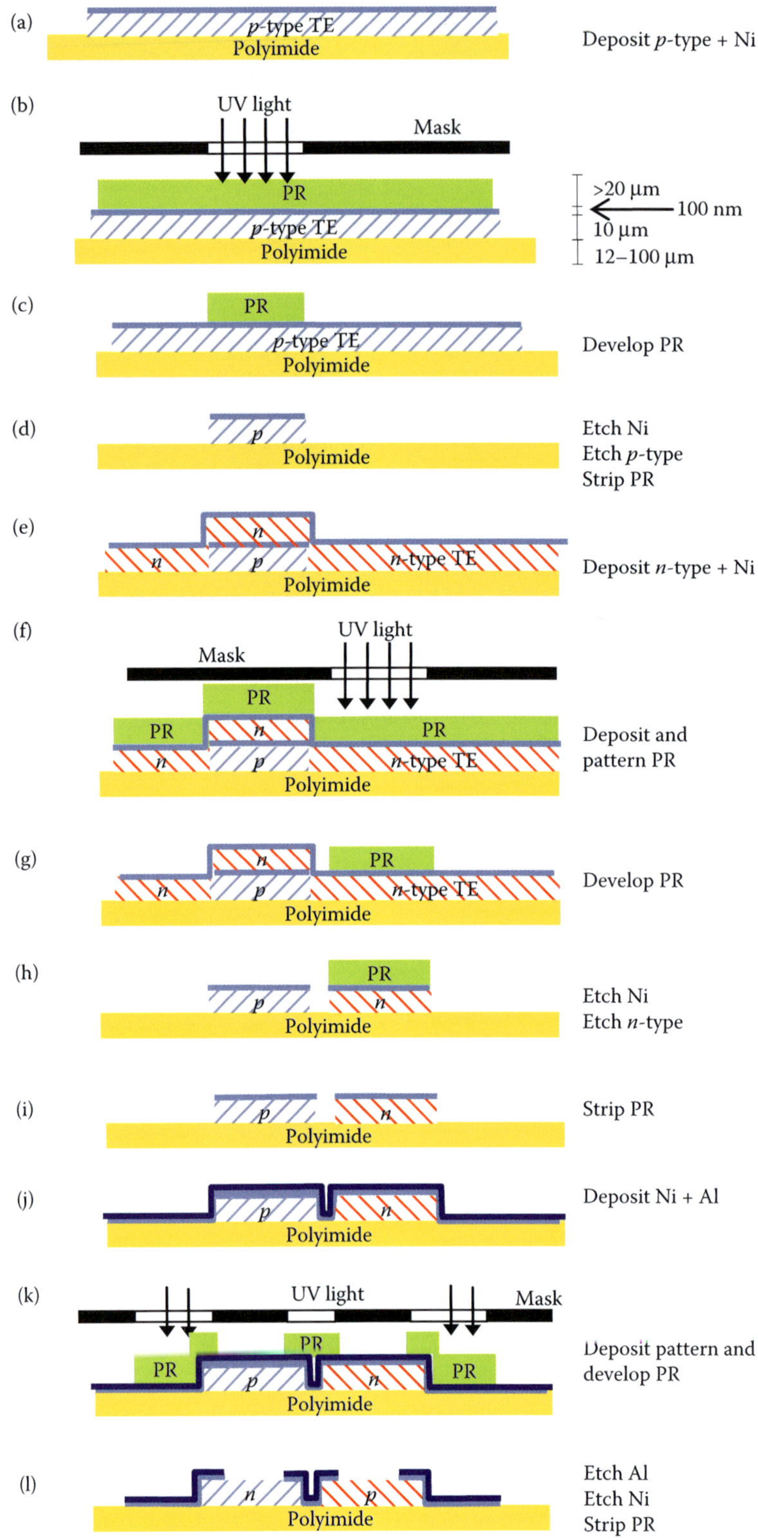

FIGURE 19.14 Fabrication steps of thermoelectric converters. (Adapted from L.M. Goncalves et al., *Journal of Micromechanics and Microengineering*, 18, 2008, 064008.)

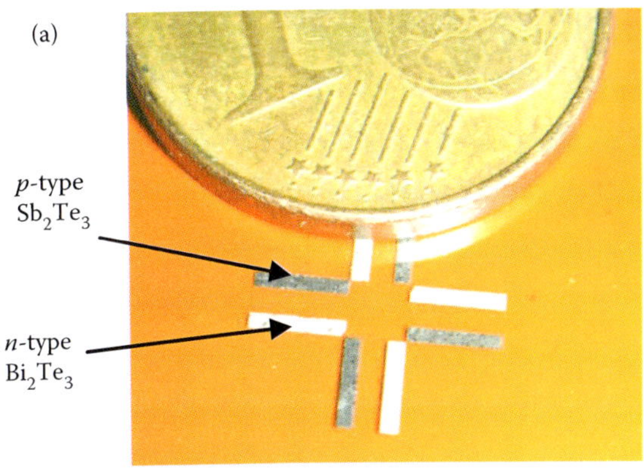

FIGURE 19.15 (a) Photo of *n*-type and *p*-type elements, before deposition of top contacts. (b) Photo of microcooler with eight pairs of thermoelectric elements, fabricated with bottom contacts. (Adapted from L.M. Goncalves et al., *Journal of Micromechanics and Microengineering*, 18, 2008, 064008.)

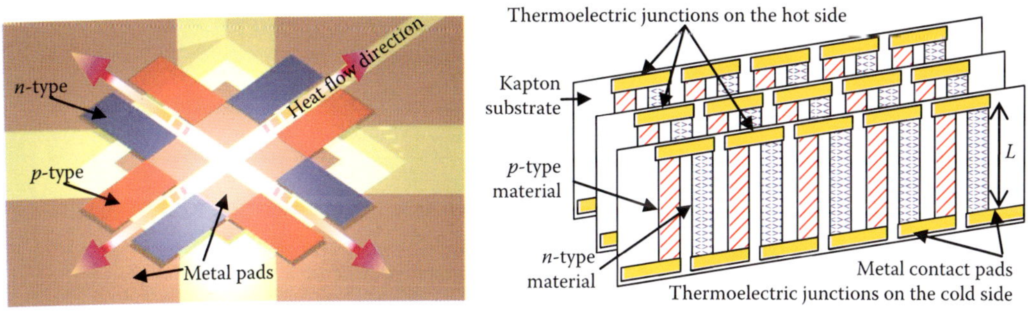

FIGURE 19.16 Planar thermoelectric device (left) and array of thermoelectric devices, fabricated in Kapton foils.

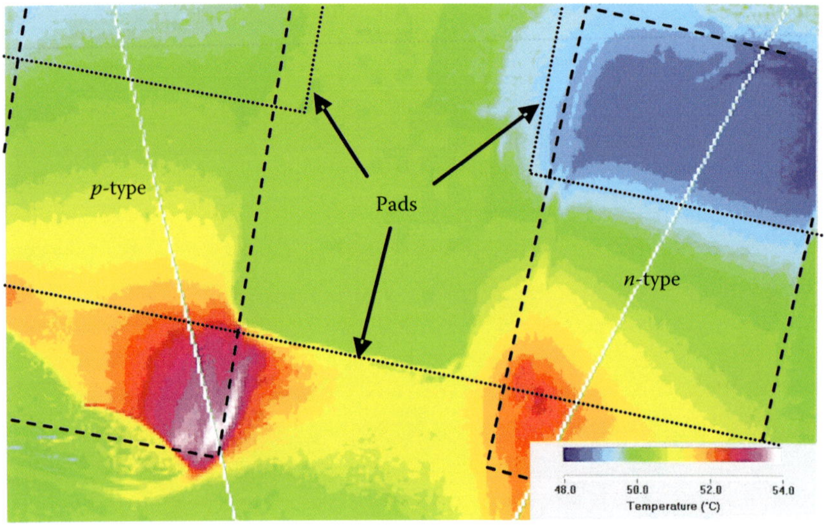

FIGURE 19.18 Thermal image of *n*-type and *p*-type thermoelectric elements, powered with 4 mA current, under vacuum. (Adapted from L.M. Goncalves et al., *Journal of Micromechanics and Microengineering*, 17, 2007, S168–S173.)

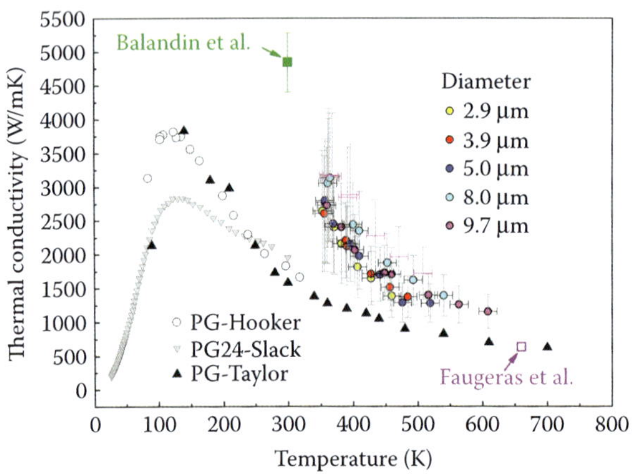

FIGURE 21.12 Thermal conductivity of the suspended CVD graphene (Chen et al. 2011) as a function of the measured temperature of the graphene monolayer suspended in vacuum over holes of various diameters. Shown in comparison are the literature thermal conductivity data of pyrolytic graphite samples as a function of temperature (Hooker et al. 1965; Taylor 1966; Slack 1962), and the thermal conductivity data of mechanically exfoliated graphene monolayer reported in Balandin et al. (2008) and Faugeras et al. (2010). The figure was prepared by Shanshan Chen.

transverse Nernst–Ettingshausen coefficient N. The latter parameter is very sensitive to scattering; indeed, in the high-field limit where we showed that transport is scattering independent, $\lim_{\mu B \to \infty}(N) = 0$. If the material is isotropic, for instance, in crystals with a cubic structure, the measurements can be carried out on either polycrystals or single crystals; if the material is anisotropic, for instance, rhombohedral Bi_2Te_3, the measurements are easier to interpret when made along principal axes, such as with the current and voltages in the trigonal plane and the magnetic field along the trigonal axis of Bi_2Te_3. In the method of the four coefficients, it is further assumed that the relaxation time has an energy dependence that is a power law:

$$\tau(E) = \tau_0 E^{\lambda} \tag{12.14}$$

where λ, the scattering exponent, is indicative of the dominant mechanism that scatters electrons, and is almost directly measured by N. For scattering of electrons on acoustic or optical phonons with electron–phonon coupling mediated by the deformation potentials, $\lambda = -1/2$. For electron scattering on grain boundaries or macroscopic uncharged defects, $\lambda = 0$. Electron scattering on polar optical phonons gives $\lambda = +1/2$, and finally for electrons scattering on ionized impurities, such as ionized donor or acceptor impurity atoms, we have $\lambda = +3/2$.[30] If there is only one type of carrier in the solid, either in one single pocket of the Fermi surface or in several exactly degenerate pockets, then the properties of such electrons are completely determined by four parameters: (1) the electron or hole concentration n, (2) the electron mobility μ, (3) the scattering exponent λ, and (4) either the Fermi energy E_F or the DOS $g(E_F)$ or, if the band is assumed to have a known dispersion relation like a parabola or a Dirac dispersion, the density-of-states effective mass m_{DOS}. In such calculation, g and m_{DOS} are uniquely related, and n, g, and E_F are as well, so they amount to only one unknown. Therefore, at each temperature, one can exactly solve the four measured quantities (ρ, S, R_H, N) for the four electronic parameters (n, μ, λ, m_{DOS}). If an increase in S related in presence of an RL is due to the DOS effect (g/n in 9), this is revealed by an increase in m_{DOS} over its value in the host solid; if it is due to resonant scattering ($d\mu/dE$ in 9), a strong increase in λ is expected, and can be seen in N. This method will be applied to the measurements described next.

12.4 Examples of Resonant Impurity Levels in Semiconductors

In principle, RLs exist in all semiconductors, including III–V and II–VI compound semiconductors, but here we focus on IV–VI and V_2–VI_3 compounds. For completion, we mention two special cases first, size-quantized structures and semimetals. First, impurities behave differently in quantum wells or superlattices, and in bulk semiconductors. In a size-quantized structure, the E_D of an impurity does depend on the width of a quantum well. Impurities like Si in GaAs that are hydrogen-like in bulk can become resonant in size-quantized structures;[31] this will not be discussed further. Second, L. A. Fal'kovskii[32] points out, on a theoretical basis, that RLs are quite likely to exist in many of the very narrow-gap semiconductors, particularly those with a strongly nonparabolic dispersion relation, and also in semimetals like elemental bismuth. At first sight, all electrically active dopants could become resonant impurities in zero-gap semiconductors and in semimetals, since these have a band overlap or a negative band gap, yet no RLs have been identified to date that significantly increase $S(n)$ in the group V semimetals. This proves that arguments based exclusively on the energy of the impurity level with respect to the bands are insufficient because they ignore the coupling (Equation 12.10) between the impurity and the extended states that is necessary for the impurity level to contribute to electrical and thermoelectric transport.

Impurity states that are merely located inside a band but are not coupled to it can absorb or emit electrons depending on the location of the Fermi level with respect to the impurity level, but if those electrons spend all their time on the impurity rather than a fraction of their time in the extended states of the solid, they do not contribute to transport. Many more examples of the latter exist than of RLs that increase $S(n)$; much recent work seems to indicate that Cd,[33] Sb,[34] Ti, and Cr belong to this class in PbTe. A plot of the DOS as a function of energy for a given impurity in a solid, as is often found in the literature, can therefore give only a necessary but not a sufficient condition for an impurity level to be resonant.

12.4.1 Rocksalt IV–VI Compounds

Alloys of the group IV–VI compound semiconductors that crystallize in the rocksalt structure are the mainstay of thermoelectric technology aimed at operating near 500°C, where combustion processes typically provide the heat. The binary semiconductors are PbTe, PbSe, PbS, SnTe, and GeTe in its high-temperature phase. Excellent high-zT thermoelectric materials based on them are PbTe, PbS_xTe_{1-x}, $Pb_{1-x}Sn_xTe$, PbTe + $AgSbTe_2$ (LAST), and GeTe + $AgSbTe_2$ (TAGS). Divalent metal M^{2+} substitutions on the Pb site in $Pb_{1-x}M_xTe$ and $Pb_{1-x}M_xSe$ alloys can change the energy gap but do not transfer electrons to the system. True dopants, that is, atoms that transfer charge to the semiconductor lattice, are (1) off-stoichiometric compositions; (2) classical donors (halogens, group V atoms on the metal site) and acceptors (alkali, Ag, Sb on the chalcogen site[34]); (3) transition metals Fe[35], Cr[36,37]; (4) rare earths[38–40] and uranium[41]; and (5) group III atoms, reviewed hereunder.

Rosenberg and Wald[42] first studied the electrical properties of $Pb_{1-x}In_xTe$ in 1965, found that In added electrons to the system, and gave the first indications of Fermi level pinning. This was more clearly identified in 1971,[43] based on the anomalous temperature dependence of the Hall coefficient. These studies were extended to In-doped $Pb_{1-x}Sn_xTe$.[44] Persistent photoconductivity effects were observed,[45] and the suspicion arose that a cubic-rhombohedral phase transition similar to that in GeTe at 700 K might occur in In-doped PbSnTe at a low temperature. Soon thereafter the work was extended to the other group III elements, Tl and Ga. Several of these group III elements were recognized as being resonant impurities, by the observation and calculation of resonant scattering at a low temperature.[46] Excellent reviews are given by the Moscow State[47] and St Petersburg[48] groups.

Band structure calculations reveal the physical nature of the RLs induced by the group III elements.[49] A systematic *ab initio* study of several types of defects in PbTe, Pb, and Te vacancies, monovalent (Na, K, Rb, Cs, Cu, and Ag), divalent (Zn, Cd, Sn, and Ge), and trivalent (Ga, In, Tl, As, Sb, and Bi) impurities on the Pb site, and Se, S, and I on the Te site,[50] was extended to other IV–VI compounds.[51] Of special interest here is the detailed study of the deep levels induced by the group III elements Tl, In, Ga,[52] and more recently Al (K. Hoang and S. D. Mahanti, private communication). The results are shown in the partial charge density maps projected in the <100> plane (Figure 12.5). The group III elements (the example used is In) substituting for a Pb atom in PbTe, and shown as a blue dot in the middle of the frame, form two energy levels. The first, shown in the left panel, is a hyperdeep defect state (HDS), filled with two electrons per **k**-point. It is predominantly of In-5*s* character; the neighboring states, Te-5*p* states are localized about 0.5 nm from them. The HDS state is full and electrons in it are localized. The second level is shown in the right frame: it is called a deep defect state (DDS), which has more contributions of the neighboring Te-5*p* states, and is only half-filled. The Fermi level is pinned in the middle of this band: As long as the impurity–impurity interactions are weak and no impurity band forms, the position of the Fermi level does not change with impurity concentration. The DDS forms a band, but its dispersion is similar to that of a free electron (m^*Nm_e); the HDS has a much smaller bandwidth yet. The DDS is split off from neither the top of the VB nor the bottom of the CB but is the result of the strong hybridization between the impurity-*s*-states and the entire VB, which is predominantly Te-*p* in character. When the In concentration becomes of the order of 5%, an impurity band forms, and the electrical conductivity becomes dominated by hopping effects.[53]

The calculations identify the nature of the RL and of the interactions that give rise to them, but they are not precise enough to predict the energy values E_D necessary to optimize thermoelectric performance, or the temperature dependence of the entire picture, which is crucial for thermoelectric applications.

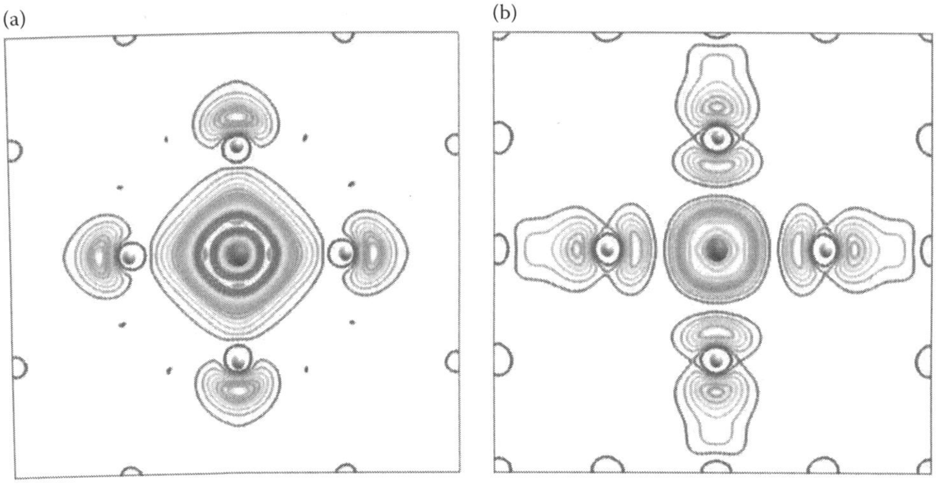

FIGURE 12.5 Partial charge density plot of an In atom on a Pb site in a PbTe lattice, showing the bonding (a, the HDS) and antibonding (b, the DDS) states of the In-5*s* state and its neighboring Te-5*p* states. The HDS state has a very strong In-5*s* character, is very narrow, and is filled. The DDS state has a mixed In-5*s* and Te-5*p* character, and is less narrow: that state is the resonant level. It is partially filled, which accounts for the Fermi level pinning effects, and is strongly hybridized with the entire VB. (After S. K. Hoang, D. Mahanti, and M. G. Kanatzidis, *Phys. Rev B.* **81**, 115106, 2010.)

Summarizing the published experimental data in the binary rocksalt V–VI compounds, we have the following values for E_D:

1. In PbTe, Tl forms a DDS starting from the band edge and peaking 60 meV into the VB. There might be a dependence of E_D on Tl concentration. The excess DOS leads to superconductivity[54] and Kondo-like transport properties.[55] Indium gives $E_D \sim 90$ meV into the CB at $T = 0$ K. Gallium appears to display two different levels depending on the Ga concentration and the preparation method.[46] Aluminum is calculated to create an RL near the CB edge, but none is observed experimentally (C. M. Jaworski and J. P. Heremans, private communication).
2. In PbSe, Tl displays Fermi level pinning at a hole concentration of 1×10^{20} cm^{-3}.[56] The In level in PbSe lies considerably deeper into the CB of PbSe than in PbTe, with $E_D \sim 300$ meV[57] at 77 K.
3. In PbS and ternary alloys of sulfides and the heavier chalcogens, it becomes difficult to follow the Tl level. As the direct L-point gap opens with sulfur concentration, the distance between the light VB maximum at L-points and the heavy VB maximum at Σ-points decreases, so that at a certain S concentration, the Tl level becomes degenerate with the heavy VB. Because the L-point gap is temperature dependent, the cross-over point also is.
4. In SnTe, indium forms an RL in the VB,[58] but the dominance of Sn vacancies in this system, and the fact that they act as acceptors, makes the Fermi level pinning behave almost inversely to what it does in the other RL systems. For each concentration of In in SnTe, there is a range of concentrations of excess Te where the acceptor effect of the Sn vacancy is not observed. As a result, the Fermi level is pinned at the low values of vacancy concentrations, and released above a threshold value, which depends on the In concentration. This is modeled by assuming that E_D is a relatively strong function of In concentration. Thermopower increases are observed above Pisarenko at low temperature,[58] and are attributed to resonant scattering.
5. Both PbTe and SnTe are direct-gap semiconductors that form complete solid solutions, but their CB and VB are inverted[59] (Figure 12.6a). Consequently, alloys at $x \sim 34\%$ are exactly zero-gap semiconductors at $T = 0$ K, with a pure Dirac dispersion. The bands have very strong temperature

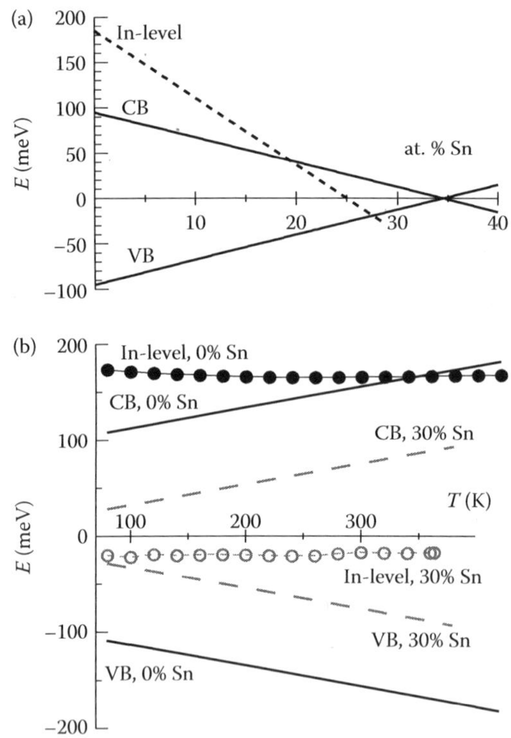

FIGURE 12.6 (a) Composition dependence of the CB and VB edges, and of the In level in $Pb_{1-x}Sn_xTe$ alloys at low temperature ($T \leq 80$ K). (b) Temperature dependence of the In level for $x = 0$ and $x = 30$ at.%. The In level moves into the gap around room temperature, where ceases to be an RL and becomes a trap. (Redrawn from data in J. O. Dimmock, I. Melngailis, and A. J. Strauss, *Phys. Rev. Lett.* **16**, 1193, 1966.)

dependences, so that the concentration at which the gap is zero varies with T. The indium level in $Pb_{1-x}Sn_xTe$ alloys[44] moves from the CB to the VB with x, as shown in Figure 12.6a. There is also a very strong temperature dependence to E_D in these alloys, shown in Figure 12.6b.[60] In pure PbTe, the In level moves into the energy gap near 250 K, where it acts as a trap, and it loses its usefulness in thermoelectric performance above that temperature.

While it has long been known that resonant scattering increases the thermoelectric power of PbTe:Tl at a low temperature,[22,47] the effect of the Tl RL at room temperature and above is important for applications, because it nearly doubles the zT of the p-type PbTe materials[61] compared to literature values. Figure 12.1 shows as data points the thermopower at 300 K of PbTe samples doped with various amounts of Tl, and they are all pinned at about 125–150 μV/K, considerably above the Pisarenko plot. Figure 12.7 shows the thermoelectric performance of Tl-doped PbTe, and compares the thermoelectric figure of merit zT of Tl-doped samples compared to similarly doped Na-doped PbTe. Measurements of the transverse Nernst–Ettingshausen coefficient and of the Hall coefficient make it possible to use the method of the four coefficients described in Section 12.3.3, using those two measurements along that of resistivity and thermopower to determine the hole concentration (3×10^{19} to 6×10^{19} cm^{-1}), mobility (90 to 70 cm^2/Vs), scattering exponent (Equation 12.14 becomes $\tau(E) = \tau_0 E^{-1/2}$ indicative of acoustic phonon scattering), and hole density-of-states $g(E)$. Below ~800 K, the Fermi surface of degenerately doped p-type PbTe consists of four ellipsoidal pockets at the L-points of the Brillouin zone along the $|\Gamma L|$ direction.[62] The DOS $g(E)$ includes the degeneracy factor of 4, and is further parameterized in terms of a DOS effective mass m_{DOS} as shown in Figure 12.8, defined as the inverse of the curvature of a parabola drawn through

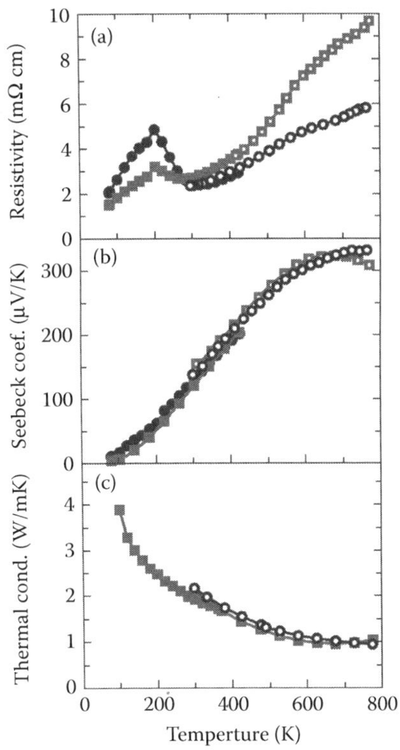

FIGURE 12.7 (a) Electrical resistivity ρ, (b) thermopower S, and (c) calculated figure of merit zT (the thermal conductivity κ is shown as insert) versus temperature of two samples of PbTe doped with (1) 1 at% Tl, and (2) 2 at% Tl. The open and closed dots correspond to measurements made in two different instruments in different temperature ranges. (After V. Jovovic et al. *J. Appl. Phys.* **103**, 053710, 2008.)

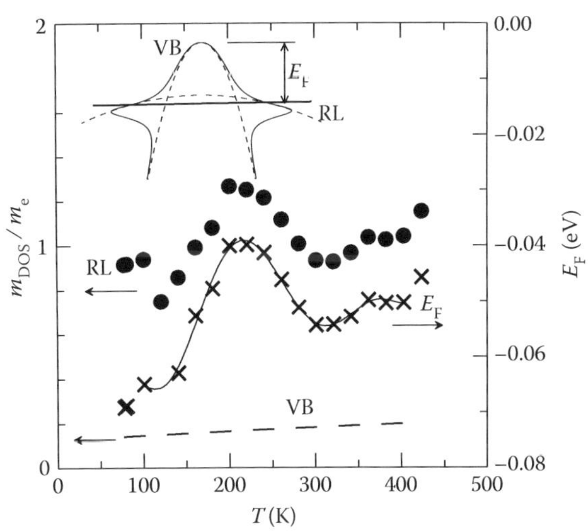

FIGURE 12.8 Temperature dependence of the Fermi energy E_F and of the density-of-states effective mass (left) of the VB (dashed line, VB) and of the Tl-doped samples. The mass m_{DOS} is an approximation based on drawing a parabolic band as a dashed line through the RL.

the RL at the Fermi level. For the Tl-doped samples, Figure 12.8 shows how the Fermi level falls near the maximum DOS of the RL, which is known to be 60 meV below the VB edge, and how the effective mass $m_{DOS} \sim 0.8$ to $1m_e$, is roughly 4 to 5 times higher than the known mass[62] of the VB of conventionally doped PbTe at the same hole concentration $m_{DOS} \sim 0.15$ to $0.2m_e$ (the latter includes the DOS degeneracy factor of 4, and the effects of nonparabolicity). The value of 0.8 to $1m_e$ is consistent with the DOS obtained from measurements of the electronic specific heat.[63]

The conclusion of this work is that the enhancement in S, and ultimately of zT, is due to an increase in the local DOS, and not to an increase in the energy dependence of the scattering by resonant scattering. Had the latter been at the origin, one could have expected the effect of the RL to decrease with increasing temperature where acoustic and optical phonon scattering become ever more dominating, precluding the use of RLs in high-temperature applications such as electrical power generators.

12.4.2 Tetradymite V_2–VI_3 Compounds

Bi_2Te_3 is the binary semiconductor that is the basis for commercial thermoelectric alloys used for Peltier cooling near room temperature. The upper valence bands (UVB) of degenerately doped materials consist of six ellipsoidal Fermi surfaces in **k**-space,[64] centered in the mirror plane of the Brillouin zone lying 40 ± 10% of the zone edge along |ΓX| direction. They have an integral DOS effective mass $m_{DOS} = 0.35m_e$.[62] A lower valence band (LVB), consisting of six ellipsoids, exists 20.5 meV below the UVB in **k**-space at 35 ± 5% along |ΓA|.[64] Bi_2Te_3 can be doped p-type with group IV atoms Ge and Pb. Tin forms an RL with $E_D \sim 15$ meV[65,66] below the top of the UVB at 2 K and "stabilizes" the thermopower of single crystals of Bi_2Te_3[67,68] at cryogenic temperatures.

Jaworski et al.[69] use the Sn RL to strongly increase the thermopower above the Pisarenko relation at 300 K, as shown in Figure 12.9. The work was carried out on single crystals in the direction perpendicular to the trigonal axis, but for moderately heavily doped material, the sample's thermopower is

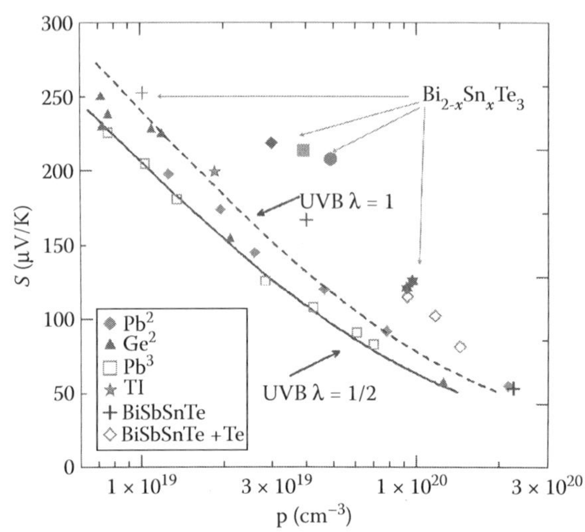

FIGURE 12.9 Pisarenko relation (thermopower versus carrier concentration) at 300 K for p-type Bi_2Te_3. The lines are calculated using a classical parabolic band with the effective mass for the UVB. The data points are obtained for the different acceptor impurities as marked, those with superscript 3 by C. M. Jaworski et al. and those with superscript 2 by others (see text). The points where Sn is used as a dopant differ significantly from the others and from the calculated lines. (Adapted from C. M. Jaworski, V. A. Kulbachinskii, and J. P. Heremans, *Phys. Rev. B.* **80**, 233201, 2009.)

isotropic and equals the scalar partial hole coefficient $S(p)$,* which is calculated for p-type Bi_2Te_3 as lines in Figure 12.9. The calculation assumed a relaxation time given by Equation 12.14 for polar optical phonon scattering ($\lambda = +1/2$) and also for $\lambda = 1$ to illustrate the effect of an additional contribution of ionized impurity scattering for which $\lambda = +3/2$. It further assumes $m_{DOS} = 0.35m_e$. Figure 12.9 shows literature data for Pb- and Ge-doped samples[69–71] at 300 K. The data measured on $Bi_{2-x}Te_3Sn_x$ sample depart from the conventional Pisarenko relation, giving strong experimental indication for the presence of an RL. One sample with $x = 0.015$ has double the S of similarly doped samples without Sn. The power factors of such samples is also enhanced. Measurements of the period of the Shubnikov–de Haas oscillations[69] on the same samples reinforce the thermopower measurements. Finally, a measurement of the Nernst and Hall coefficients, and the application of the method of the four coefficients shows that the effect is partially due to resonant scattering at 100 K, but that it is due to the enhancement of the DOS itself at 300 K. One salient difference between the Sn level in Bi_2Te_3 and the Tl level in PbTe can be seen by comparing Figures 12.1 and 12.9, where in PbTe:Tl $S(p)$ remains steady at 130–150 μV/K for increasing Tl atom and hole concentrations, in Bi_2Te_3:Sn $S(p)$ falls back to the Pisarenko line for $p > 10^{20}$ cm^{-3}. It is possible to overdope in Bi_2Te_3 with Sn, whereas overdoping does not occur in PbTe presumably within the solubility limit of Tl.

12.5 Conclusions

Distortions in the dispersion relations of thermoelectric semiconductors can strongly improve the thermoelectric power factor and figure of merit zT. These distortions can be classified schematically into four types: they can be induced by quantum size effects, by doping and alloying with select impurities that form RLs, Kondo systems, or hybridize with the bands. The last three mechanisms are related. The band structure distortions can work either by the excess DOS they induce or at low temperature by inducing resonant scattering. A brief review on RLs in compound semiconductors is also given, from the point of view of their influence on thermoelectric properties. A more extensive review of RL's in other compound semiconductor is given in Ref. 72.

Acknowledgments

This work is supported by grants from NSF-CBET-0754023, DOE EFRC "Revolutionary materials for solid-state energy conversion" 61-32128 and AFOSR "MURI: Cryogenic Peltier cooling" FA9550-10-1-0533.

References

1. G. A. Slack, New materials and performance limits for thermoelectric cooling, p. 407, in *CRC Handbook of Thermoelectrics*, D. M. Rowe, Editor, CRC Press LLC, Boca Raton, FL, 1995.
2. A. F. Ioffe, *Physics of Semiconductors*, Academic Press, New York, 1960 (translated from Russian, "Fizika Poluprovodnikov", Russian Academy of Sciences, Moscow, 1957).
3. D. Vashaee and A. Shakouri, Improved thermoelectric power factor in metal-based superlattices, *Phys. Rev. Lett.* 106103, **92**, 2004.
4. Z. Bian, H. Wang, and A. Shakouri, Maximum cooling temperature and uniform efficiency criterion for inhomogeneous thermoelectric materials, *Phys. Rev. B.* **75**, 245208, 2007.

* The thermopower of Bi_2Te_3 is anisotropic in general, while the partial thermopowers of each electron or hole pocket of the Fermi surface are scalars. The anisotropy of the total thermopower arises from the fact that the total thermopower is an average of the partial coefficients of each pocket weighted by the partial conductivities of those pockets, which are anisotropic. In degenerately doped material with all pockets identical, the total thermopower equals the partial thermopowers of each pocket and is thus scalar.

5. J. P. Heremans, C. M. Thrush, and D. T. Morelli, Thermopower enhancement in PbTe with Pb precipitates, *J. Appl. Phys.* **98**, 063703, 2005.
6. G. D. Mahan and J. O. Sofo, The best thermoelectric, *Proc. Natl. Acad. Sci. USA* **93**, 7436, 1996.
7. L. D. Hicks and M. S. Dresselhaus, Effect of quantum-well structures on the thermoelectric figure of merit, *Phys. Rev. B.* **47**, 12727-31, 1993, and Thermoelectric figure of merit of a one-dimensional conductor, *Phys. Rev. B.* **47**, 16631-4, 1993.
8. Z. B. Zhang, X. Sun, M. S. Dresselhaus, J. Y. Ying, and J. Heremans, Electronic transport properties of single-crystal bismuth nanowire arrays, *Phys. Rev. B.* **61**, 4850, 2000; J. Heremans, C. M. Thrush, Y-M. Lin, S. Cronin, Z. Zhang, M. S. Dresselhaus, and J. F. Mansfield, Bismuth nanowire arrays: Synthesis and galvanomagnetic properties, *Phys. Rev. B.* **61**, 2921, 2000.
9. J. P. Heremans, C. M. Thrush, D. T. Morelli, and M-C. Wu, Thermoelectric power of bismuth nanocomposites, *Phys. Rev. Lett.* **88**, 216801, 2002.
10. J. P. Heremans, Low-dimensional thermoelectricity, *Acta Phys. Pol. A.* **108**, 609, 2005.
11. M. S. Dresselhaus and J. P. Heremans, Recent developments in low-dimensional thermoelectric materials, Chapter 39, pp. 1–24, in *CRC Thermoelectrics Handbook*, D. M. Rowe, Editor, CRC Press, Boca Raton, FL, 2006.
12. N. W. Ashcroft and N. D. Mermin, *Solid State Physics*, Holt, Rhinehart and Winston, New York, 1976.
13. A. N. Gerritsen and J. Korringa, Anomalous resistance of noble metals containing paramagnetic ions, *Phys. Rev.* **84**, 604, 1951; J. Korringa and A. N. Gerritsen, The cooperative electron phenomenon in dilute alloys, *Physica.* **19**, 457, 1953.
14. J. Friedel, On some electrical and magnetic properties of metallic solid solutions, *Can. J. Phys.* **34**, 1190, 1956.
15. J. Kondo, Giant thermo-electric power of dilute magnetic alloys, *Progr. Theoret. Phys.* **34**, 372, 1965.
16. F. J. Blatt, P. A. Schroeder, C. L. Foiles, and D. Greig, *Thermoelectric Power of Metals*, Plenum Press, New York, 1976.
17. B. C. Sales, E. C. Jones, B. C. Chakoumakos, J. A. Fernandez-Baca, H. E. Harmon, J. W. Sharp, and H. E. Volckmann, Magnetic, transport, and structural properties of $Fe_{1-x}Ir_xSi$, *Phys. Rev. B.* **50**, 8207, 1994.
18. Z. Fisk, J. L. Sarrao, and J. D. Thompson, *Current Opinion in Solid State Material Science*, **142**, 1996.
19. A. Bentien, S. Johnson, G. K. H. Madsen, B. B. Iversen, and F. Steglich, Colossal Seebeck coefficient in strongly correlated semiconductor $FeSb_2$, *Europhys. Lett.* **80**, 17008, 2007.
20. Ph. Nozières, Some comments on Kondo lattices and the Mott transition, *Eur. Phys. J. B.* **6**, 447, 1998.
21. P. de Faget de Casteljau and J. Friedel, Etude de la résistivité électrique et du pouvoir thermoélectrique des impuretés dissoutes dans les métaux noble, *J. Phys. Radium* **17**, 27, 1956; A. Blandin and J. Friedel, Propiétés magnétiques des alliages dilués: interactions magnétiques et antiferromagnétisme dans les alliages du type metal noble–metal de transition, *J. Phys. Radium* **20**, 160, 1959 (in French).
22. Y. I. Ravich, Selective carrier scattering in thermoelectric materials, in *CRC Handbook of Thermoelectrics*, D. M. Rowe, Editor, CRC Press, Boca Raton, FL, p. 67, 1995.
23. N. F. Mott and E. A. Davis, *Electronic Conduction in Non-Crystalline Materials*, Clarendon Press, Oxford, 1979.
24. E. Daniel and J. Friedel, Localized states in dilute alloys, in *Low-Temperature Physics* LT9, J. G. Daunt et al., Editors, Plenum Press, New York, pp. 933–954, 1965.
25. I. M. Tsidilkovskii, Zero-gap semiconductors with magnetic impurities forming resonance donor states, *Sov. Phys. Usp.* **35**, 85, 1992; J. D. König et al. *Phys. Rev. B.* 2011.
26. A. Popescu, L. M. Woods, J. Martin, and G. S. Nolas, Model of transport properties of thermoelectric nanocomposite materials, *Phys. Rev. B.* **79**, 205302, 2009.

27. W. Zawadzki and R. Lassnig, Magnetization, specific heat, magnetothermal effect and thermoelectric power of two-dimensional electron gas in a quantizing magnetic field, *Surf. Sci.* **142**, 225, 1984.
28. J. P. Heremans, C. M. Thrush, and D. T. Morelli, Thermopower enhancement in lead telluride nanostructures, *Phys. Rev. B.* **70**, 115334, 2004.
29. J. P. Heremans, C. M. Thrush, and D. T. Morelli, Thermopower enhancement in PbTe with Pb precipitates, *J. Appl. Phys.* **98**, 063703, 2005.
30. S. Ahmad and S. D. Mahanti, Energy and temperature dependence of relaxation time and Wiedemann-Franz law on PbTe, *Phys. Rev. B.* **81**, 165203, 2010.
31. T. A. Perry, R. Merlin, B. V. Shanabrook, and J. Comas, Observation of resonant impurity states in semiconductor quantum-well structures, *Phys. Rev. Lett.* **54**, 2623, 1985.
32. L. A. Falkov'skii, Impurity states in substances with narrow energy gaps, *Sov. Phys. JETP.* **41**, 767, 1976.
33. K. Ahn, M. K. Han, J. He, J. Androulakis, S. Ballikaya, C. Uher, V. P. Dravid, and M. G. Kanatzidis, Exploring resonance levels and nanostructuring in the PbTe–CdTe system and enhancement of the thermoelectric figure of merit, *J. Am. Chem. Soc.* **132**, 5227-35, 2010.
34. C. M. Jaworski, J. Tobola, E. M. Levin, K. Schmidt-Rohr, and J. P. Heremans, Antimony as an amphoteric dopant in lead telluride, *Phys. Rev. B.* **80**, 125208, 2009.
35. D. T. Morelli, J. P. Heremans, and C. M. Thrush, Magnetic and thermal properties of iron-doped lead telluride, *Phys. Rev. B.* **67**, 035206, 2003.
36. E. P. Skipetrov, N. A. Pichugin, E. I. Slyn'ko, and V. E. Slyn'ko, Electronic structure of chromium-doped lead telluride-based diluted magnetic semiconductors, *Low Temp. Phys.* **37**, 210, 2011.
37. M. D. Nielsen, E. M. Levin, C. M. Jaworski, K. Schmidt-Rohr, and J. P. Heremans, Chromium as resonant donor impurity in PbTe, *Phys. Rev. B.* **85**, 045210, 2012.
38. A. Bruno, J. P. Lascaray, M. Averous, G. Filion, and J. F. Dumas, Determination of the Gd^{3+}–Gd^{3+} exchange constants from susceptibility measurements on $Pb_{1-x}Gd_xTe$, *Phys. Rev. B.* **37**, 1186, 1988.
39. D. L. Partin, Lead telluride doped with rare-earth elements, *J. Appl. Phys.* **57**, 1997, 1985.
40. V. Jovovic, S. J. Thiagarajan, J. West, J. P. Heremans, T. Story, Z. Golacki, W. Paszkowicz, and V. Osinniy, Transport and magnetic properties of dilute rare-earth-PbSe alloys, *J. Appl. Phys.* **102**, 043707 1-6, 2007.
41. S. Isber, C. Fau, S. Charar, M. Averous, and Z. Golacki, Uranium diluted in IV–VI compounds, *Mater. Sci. Forum* **182–184**, 657, 1995.
42. A. J. Rosenberg and F. Wald, Solid solutions of CdSe and InSe in PbSe, *J. Phys. Chem. Solids* **26**, 1079, 1965.
43. A. A. Averkin, V. I. Kaidanov, and R. B. Melnik, *Sov. Phys. Semicond.* **5**, 75, 1971.
44. B. A. Akimov, R. S. Vadnkhva, V. P. Zlomanov, L. I. Ryabova, and S. M. Chudinov, *Sov. Phys. Semicond.* **11**, 637, 1977.
45. B. A. Akimov, N. B. Brandt, S. A. Bogoslovskii, L. I. Ryabova, and S. M. Chudinov, *JETP Lett.* **29**, 9, 1979; B. A. Akimov, N. B. Brandt, S. O. Klimonskiy, L. I. Ryabova, and D. R. Khokhlov, Dynamics of the semiconductor-metal transition induced by infrared illumination in $Pb_{1-x}Sn_xTe$ (In) alloys, *Phys. Lett.* **88A**, 483, 1982.
46. V. I. Kaidanov, S. A. Nemov, and Yu. I. Ravich, Resonant scattering of carriers in IV–VI semiconductors, *Soc. Phys. Semicond.* **26**, 113, 1992.
47. B. A. Volkov, L. I. Ryabova, and D. R. Khokhlov, Mixed-valence impurities in lead telluride-based solid solutions, *Physics-Uspekhi.* **45**, 819, 2002.
48. V. I. Kaidanov and Yu. I. Ravich, Deep and resonance states in AIV BVI semiconductors, *Sov. Phys. Usp.* **28**, 31, 1985.
49. S. Ahmad, K. Hoang, and S. D. Mahanti, *Ab initio* study of deep defect states in narrow band-gap semiconductors: Group III impurities in PbTe, *Phys. Rev. Lett.* **96**, 056403, 2006.
50. S. Ahmad, S. D. Mahanti, K. Hoang, and M. G. Kanatzidis, *Ab initio* studies of the electronic structure of defects in PbTe, *Phys. Rev. B.* **74**, 155206, 2006.

51. S. K. Hoang, D. Mahanti, and M. G. Kanatzidis, Impurity clustering and impurity-induced bands in PbTe-, SnTe-, and GeTe-based bulk thermoelectrics, *Phys. Rev B.* **81**, 115106, 2010.
52. K. Hoang and S. D. Mahanti, Electronic structure of Ga-, In-, and Tl-doped PbTe: A supercell study of the impurity bands, *Phys. Rev B.* **78**, 085111, 2008.
53. Yu. I. Ravich and S. A. Nemov, Hopping conduction via strongly localized impurity states of indium in PbTe and its solid solutions, *Semiconductors* **36**, 3, 2002.
54. Y. Matsushita, H. Blum, T. H. Geballe, and I. R. Fisher, Evidence for charge kondo effect in superconducting Tl-doped PbTe, *Phys. Rev. Lett.* **94**, 157002, 2005.
55. A. S. Ericson, N. P. Breznay, E. A. Nowadnick, T. H. Geballe, and I. R. Fisher, Correlation of anomalous normal state properties with superconductivity in $Pb_{1-x-y}Tl_xIn_yTe$, *Phys. Rev. B*, **13**, 2010-06-04, 2010.
56. A. N. Veis, et al. *Sov. Phys. Semicond.* **11**, 588, 1977.
57. V. I. Kaidanov, R. B. Melnik, and N. V. Germanas, *Sov. Phys. Semicond.* **6**, 627, 1972.
58. G. S. Bushmarina, B. F. Gruzinov, I. A. Drabkin, E. Ya. Lev, and V. M. Yuneev, Characteristics of the effects of In as a dopant in SnTe, *Sov. Phys. Semicond.* **18**, 1374, 1984.
59. J. O. Dimmock, I. Melngailis, and A. J. Strauss, Band structure and laser action in $Pb_xSn_{1-x}Te$, *Phys. Rev. Lett.* **16**, 1193, 1966.
60. V. Jovovic, S. J. Thiagarajan, J. P. Heremans, T. Kommisarova, D. Khokhlov, and A. Nicorici, Low temperature thermal, thermoelectric, and thermomagnetic transport in indium rich $Pb_{1-x}Sn_xTe$ alloys, *J. Appl. Phys.* **103**, 053710, 2008.
61. J. P. Heremans, V. Jovovic, E. S. Toberer, A. Saramat, K. Kurosaki, A. Charoenphakdee, S. Yamanaka, and G. J. Snyder, Enhancement of thermoelectric efficiency in PbTe by distortion of the electronic density of states, *Science* **321**, 554–558, 2008.
62. O. Madelung (Ed.), The integrated density-of-states mass is the density of states mass of each pocket of the Fermi surface, multiplied by the number of pockets to the power 2/3, in *Landolt–Bornstein Numerical Data and Functional Relationships in Science and Technology*, Vol. **17**, Subvol. f, Springer-Verlag, Berlin, 1983.
63. Y. Matsushita, P. A. Wianecki, A. T. Sommer, T. H. Geballe, and I. R. Fisher, Type II superconducting parameters of Tl-doped PbTe determined from heat capacity and electronic transport measurements, *Phys. Rev. B.* **74**, 134512, 2006.
64. H. Kohler, Parabolicity of the highest valence band of Bi_2Te_3 from Shubnikov-de Haas effect, *Phys. Status Solidi (b)* **74**, 591, 1976.
65. V. A. Kulbachinskii, N.B. Brandt, P.A. Cheremnykh, S.A. Azou, J. Horak, and P. Lostak, Magnetoresistance and Hall effect in $Bi_2Te_3\langle Sn\rangle$ in ultrahigh magnetic fields and under pressure, *Phys. Status Solidi (b)* **150**, 237, 1988.
66. V. A. Kulbachinskii, et al. Valence-band energy spectrum of solid solutions of narrow-gap-semiconductor $Bi_{2-x}Sn_xTe_3$ single crystals, *Phys. Rev. B.* **50**, 16921, 1994.
67. V. A. Kulbachinskii, et al. Thermoelectric power and scattering of carriers in $Bi_{2-x}Sn_xTe_3$ with layered structure, *Phys. Status Solidi (b)* **199**, 505, 1997.
68. M. K. Zhitinskaya, S. A. Nemov, T. E. Svechnikova, Peculiarities of electrophysical properties of Bi_2Te_3 doped with Sn, *Proc. 16th International Conférence on Thermoelectricity*, p. 97, IEEE, 1997.
69. C. M. Jaworski, V. A. Kulbachinskii, and J. P. Heremans, Tin forms a resonant level in Bi_2Te_3 that enhances the room temperature thermoelectric power, *Phys. Rev. B.* **80**, 233201, 2009.
70. G. Bergmann, Untersuchnungen über die dotiereigenschaften der elemente germanium und blei in wismuttellurid Bi_2Te_3, *Z. Natuforsch.* **18a**, 1169, 1963.
71. T. Plecháček, J. Navrátil, J. Horák, and P. Lošťák, Defect structure of Pb-doped Bi_2Te_3 single crystals, *Philos. Mag.* **84**, 2217, 2004.
72. J. P. Heremans, B. Wiendlocha, and A. M. Chamoire, Resonant levels in bulk thermoelectric semiconductors, *Energy Environ. Sci.* 2011, DOI: 10.1039/c1ee01895g.

13

Graphene-Like Exfoliated Quasi-2D Thermoelectric Crystals

Alexander A. Balandin
University of California, Riverside

13.1 Introduction

Since the discovery of its extraordinary thermoelectric properties, bismuth–telluride (Bi_2Te_3) has become a vital component for thermoelectric industry [1–3]. Bulk Bi_2Te_3 materials are known to have some of the highest thermoelectric figure of merit, $ZT \sim 1.14$ at room temperature (RT). The thermoelectric figure of merit is defined as $ZT = S^2\sigma T/K$, where $S = -\Delta V/\Delta T$ is the Seebeck coefficient (ΔV is the voltage difference corresponding to a given temperature difference ΔT), σ is the electrical conductivity, and K is the thermal conductivity, which has contributions from electrons and phonons. It is clear from ZT definition that in order to improve thermoelectric figure of merit one should increase $S^2\sigma$ and decrease the thermal conductivity. Different approaches have been tried in order to enhance the thermoelectric properties of bulk Bi_2Te_3 or its alloys. These approaches included the composition change from its stoichiometry, the use of polycrystalline materials with different grain sizes, intentional introduction of structural defects and incorporation of different dopants, for example, Sb or Se, into Bi_2Te_3 lattice. The optimization of bulk Bi_2Te_3 led to incremental improvements but no breakthrough enhancement in ZT was achieved. More promising results ($ZT \sim 2.4$ for p-type material at RT) were achieved with Bi_2Te_3/Sb_2Te_3 superlattices produced by low-temperature deposition [4]. A recent study indicated that the low-dimensional structuring of BiSbTe alloys [5] also allows for ZT enhancement to ~1.5 at RT. But still higher ZT values are needed for major practical impact. It has been shown that ZT above 3 or 4 at RT are needed in order to make thermoelectric cooling and power generation competitive with conventional methods [6].

It follows from many theoretical predictions that a drastic improvement in ZT can be achieved in low-dimensional structures where electrons and holes are strongly confined in one or two dimensions [7]. Hicks and Dresselhaus [8] predicted that ZT can be increased in Bi_2Te_3 quantum well by a factor of ~13 over the bulk value. This would require a complete carrier confinement in a quantum well with a width H of the order of ~1 nm. According to Dresselhaus et al. [7,8], quantum confinement of charge carriers in quantum wells leads to a drastic ZT improvement due to the increase in the carrier density of states near the Fermi level and corresponding increase in the thermopower. The crucial condition for such mechanism is quantum confinement of carriers in quantum wells, which is only possible if materials are *crystalline* and essentially free of defects. The thickness of the thin film required to achieve the quantum confinement conditions has to be of the order of few atomic layers. Note that in the superlattices commonly used in thermoelectric studies the charge carriers are only partially confined, if confined at all due to the small potential barrier height and relatively low material quality. The barrier height has a pronounced effect on ZT. Broido and Reinecke [9] have shown theoretically that $ZT = 3$ can be achieved in Bi_2Te_3 superlattices with infinite potentials when the quantum well width (i.e., thickness of the thin film) is $H \sim 3$ nm. In the structures with incomplete quantum confinement the maximum ZT decreases to ~2.5 and the required width becomes as small as ~2 nm.

Balandin and Wang [10,11] proposed a different strategy for increasing ZT in low-dimensional structures by reducing its thermal conductivity via spatial confinement of acoustic phonons, which carry bulk of heat in thermoelectric materials. The improvement of thermoelectric properties via phonon engineering [10,11] also can be achieved in thin films or nanowires with the thickness of just few atomic layers and high quality of interfaces. In nanostructures with rough interfaces, the thermal conductivity can be reduced due to phonon scattering on boundaries and defects [12–14]. At the same time, defects and disorder can also lead to electron mobility degradation limiting the improvement.

Thus, in order to explore the full strength of the low-dimensional confinement effects for improving thermoelectric figure of merit either via electron band-structure or phonon engineering one needs to produce quasi two-dimensional (2D) crystalline structures with a few-atomic layer thickness and high-quality interfaces. Conventional chemical vapor deposition (CVD), electrochemical or other means are not capable for producing such quality structures. Even the best superlattice structures grown by the molecular beam epitaxy (MBE) have limitations owing to the small conduction and valence band discontinuity resulting in the low height of the confining potential for electrons and holes. The epitaxially grown films have to be lattice matched to the substrate, which leads to a small acoustic impedance mismatch and, as a result, the absence of strong confinement effects for the acoustic phonons [10,11]. These considerations create strong motivations for the search of alternative approaches to fabrication of quasi-2D crystals made of thermoelectric materials.

Recent developments renewed interest to Bi_2Te_3, and opened up a completely different strategy for ZT enhancement in Bi_2Te_3 thin films. It was discovered that Bi_2Te_3 family of materials are topological insulators (TIs) [15–19]. TIs are materials with a bulk insulating gap and conducting surface states that are topologically protected against scattering by the time-reversal symmetry [15–19]. The TIs and graphene constitute a new class of materials—termed as "Dirac materials"—which reveal Dirac cones, for example, regions in the reciprocal space with the linear dispersion relation for electrons. The Dirac cones in TI materials are located on their surfaces. It was shown theoretically that ZT can be strongly enhanced in Bi_2Te_3 thin-film TIs provided that the Fermi level is tuned to ensure that the surface transport regime and the films are thin enough to open a gap in the Dirac cone dispersion on the surface [20]. This is a principally new mechanism for ZT enhancement than any considered previously.

In order to explain the new approach for improving the thermoelectric properties we need to review the basics of crystal structure of Bi_2Te_3. Bi_2Te_3 has the rhombohedral crystal structure of the space group D_{3d}^5–$R(-3)m$ with five atoms in one unit cell. The lattice parameters of the hexagonal cells of Bi_2Te_3 are $a_H = 0.4384$ nm and $c_H = 3.045$ nm. Its atomic arrangement can be visualized in terms of the layered sandwich structure (see Figure 13.1). Each sandwich is built up by five mono-atomic sheets, referred to as *quintuple* layers, along the c_H axis with the sequence–[$Te^{(1)}$-Bi-$Te^{(2)}$-Bi-$Te^{(1)}$] – [$Te^{(1)}$-Bi-$Te^{(2)}$-Bi-$Te^{(1)}$]–. Here superscripts (1) and (2) denote two different chemical states for the anions. The outmost atoms $Te^{(1)}$ are strongly bound to three planar $Te^{(1)}$ and three Bi metal atoms of the same quintuple layers and weakly bound to three $Te^{(1)}$ atoms of the next sandwich. The binding between adjacent $Te^{(1)}$ layers originates mostly from the weak van der Waals forces although other long-range Coulomb

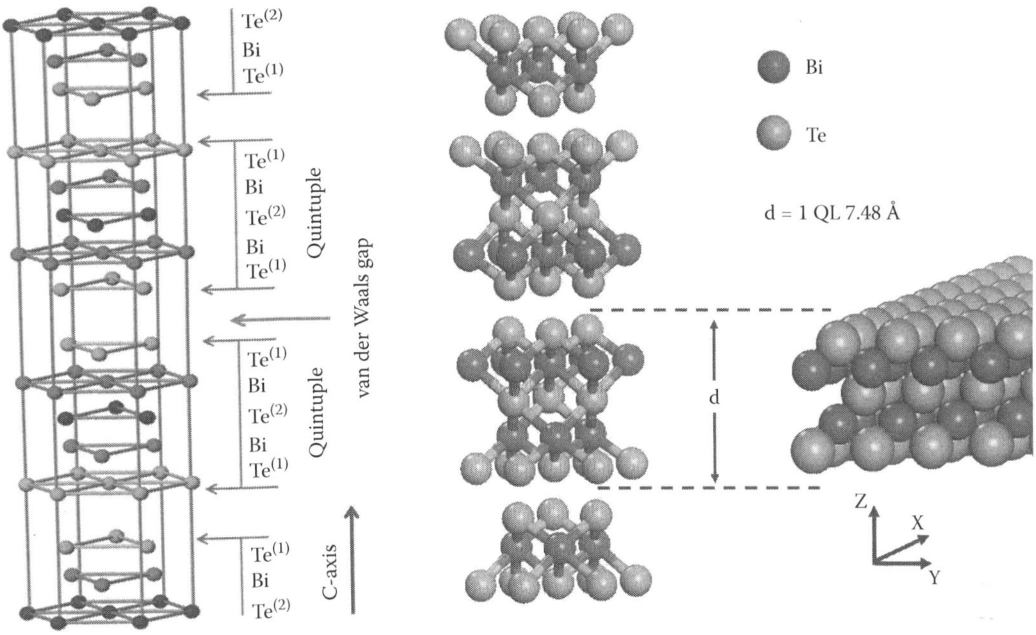

FIGURE 13.1 **(See color insert.)** Schematic of Bi_2Te_3 crystal structure of $D_{3d}^5 - R(-3)m$ space group showing quintuple layers and location of van der Waals gaps (left panel). The $Te^{(1)}$–$Te^{(1)}$ bond is the weakest while Bi–$Te^{(1)}$ bond is the strongest. The mechanical exfoliation results in breaking the $Te^{(1)}$–$Te^{(1)}$ van der Waals bond and formation of few-quintuple films or individual quintuple layers. The atomic layer model used in the *ab initio* simulations of thermoelectric properties of individual quintuple. (Adapted from Zahid, F., Lake, R., *Appl. Phys. Lett.*, 97, 212102, 2010; Teweldebrhan, D., Goyal, V., Balandin, A.A., *Nano Lett.*, 10, 1209, 2010.)

forces play role in the bonding. The stronger bonds inside the quintuple layers are of the covalent or partially ionic nature. The presence of the van der Waals gap between the quintuples allows for cleavage between the adjacent $Te^{(1)}$ layers.

Zahid and Lake [20] calculated the thermoelectric figure of merit ZT of a single quintuple layer, and found that ZT increases by a factor of 10 to the RT value of 7.15 (see Figure 13.2). The extremely large enhancement in ZT results from the change in the distribution of the valence band density of modes brought about by the quantum confinement in the thin film. The theoretical model used the *ab initio* electronic structure calculations with the full quantum–mechanical structure relaxation combined with a Landauer formalism for the linear-response transport coefficients [20]. Independently, Ghaemi et al. [21] studied electronic transport in quintuples of TIs, which are known to be good thermoelectric materials in bulk form for example, Bi_2Te_3 and Bi_2Se_3. These materials were classified as TIs because the recent photoemission experiments have shown that their surfaces have an unusual electronic structure with a single Dirac cone [15–19]. In the single quintuple or few-quintuple films, the surface states from top and bottom surfaces hybridize, and conventional diffusive transport predicts that the tunable hybridization-induced band gap leads to increased thermoelectric performance at low temperatures [21]. One should note here that confinement and hybridization of the Dirac cones in the single or few quintuple Bi_2Te_3 films changes the character of the electron transport on the surface. In this sense such films are no longer "true" TIs where the scattering is suppressed by the time-reversal symmetry. At the same time, the ZT enhancement mechanisms predicted for the single or few-quintuple Bi_2Te_3 films in the TI context are related to their surface states and, as such, are completely different from those discussed previously [7–9]. For this reason it makes sense to talk about thermoelectric properties of TI materials.

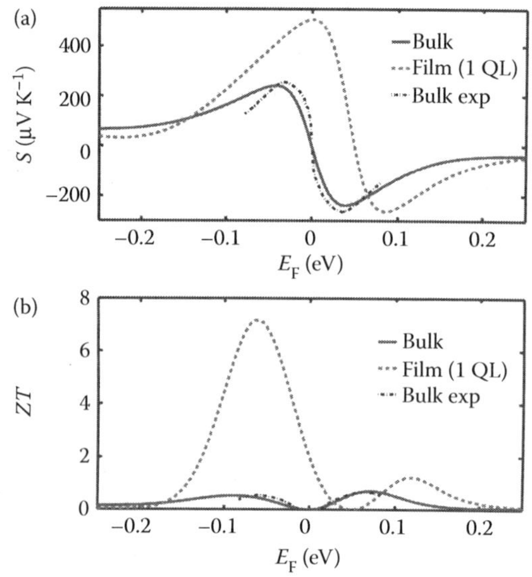

FIGURE 13.2 (a) Seebeck coefficient *S* and (b) *ZT* at RT as a function of the Fermi energy E_F for bulk and thin film of the one-quintuple thickness. The dotted black line shows the experimental data for bulk. The electron and hole mean free paths 14 and 8 nm, respectively, which are obtained by fitting our results to the bulk experimental data. The zero in the energy scale represents the midgap energy. (The figure is courtesy of Roger Lake with the results after Zahid, F., Lake, R., *Appl. Phys. Lett.*, 97, 212102, 2010.)

13.2 Mechanical Exfoliation of Quasi-2D Crystals of Bi_2Te_3

We have recently demonstrated that separate Bi–Te atomic layers can be mechanically exfoliated from bulk Bi_2Te_3 in a *graphene-like* fashion [22–24]. The presence of the van der Waals gap allowed us to disassemble Bi_2Te_3 crystal into its quintuple building blocks—five mono-atomic sheets of $Te^{(1)}$–Bi-$Te^{(2)}$–Bi-$Te^{(1)}$, which have the thickness of ~1 nm. The resulting quasi-2D crystals retain their good electrical conduction and poor thermal conduction properties important for thermoelectric applications. The exfoliated atomically thin films have nearly *infinite* confining potential barriers for both electrons as holes as well as for acoustic phonons, which carry the heat. Thus, the graphene-like exfoliated films allow one to achieve the strong quantum confinement regime for charge carriers and phonons.

In order to isolate Bi_2Te_3 quintuples and break them into atomic planes we employed a method used for exfoliation of single-layer graphene [25,26]. Through a mechanical cleavage process we separated thin films from crystalline bulk Bi_2Te_3. The process was repeated several times to obtain the layers with just few-atomic planes. Owing to the specific structure of Bi_2Te_3 crystal along c_H direction ($c_H = 3.045$ nm is very large lattice constant as compared with other materials) we were able to verify the number of layers using the optical inspection combined with the atomic force microscopy (AFM) and the scanning electron microscopy (SEM). The thickness of the atomic quintuple is $H \sim 1$ nm. The step-like changes in the cleaved layers highest of ~1 nm can be distinguished well with AFM.

The produced atomic layers were placed on Si substrates containing a 300 nm-thick SiO_2 capping layer. The SiO_2 thickness was selected to visualize the atomic films through light interference in the same way as it is done for graphene on Si/SiO_2. We then isolated and separated individual crystal planes which exhibited high crystal quality with little to no structural defects. For detailed material characterization, we suspended some of the Bi–Te atomic films over trenches in SiO_2/Si substrates. The trenches were fabricated by the reactive ion etching. They had a depth of ~300 nm and widths ranging from 1 to 5 μm. By suspending the ultra-thin films over these trenches we reduced the coupling to the substrate.

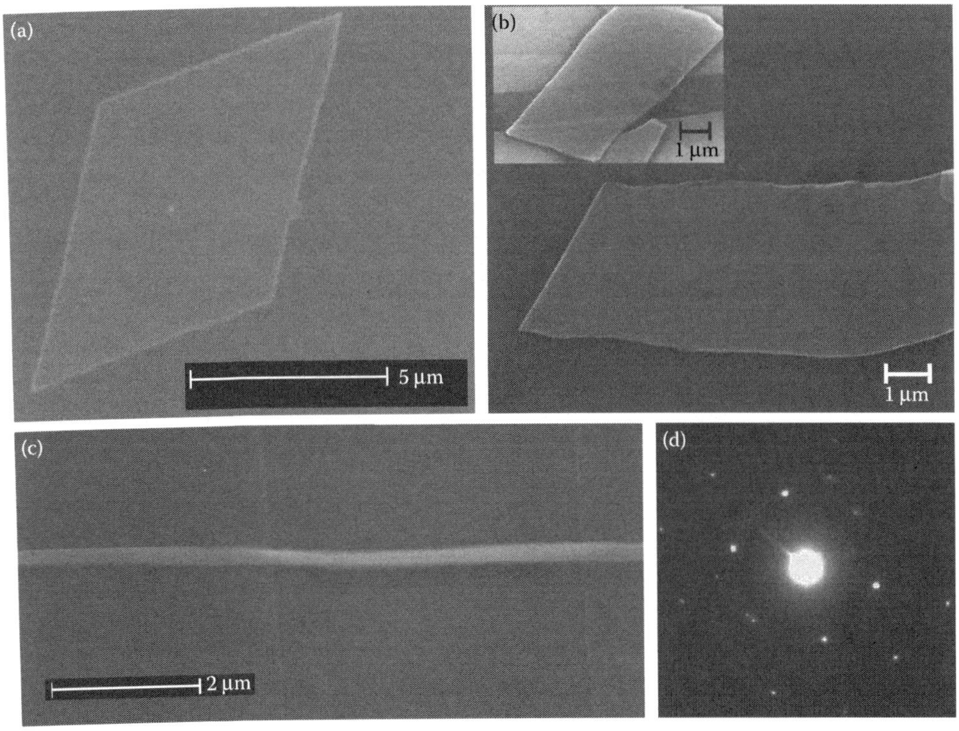

FIGURE 13.3 (a) SEM image of a Bi_2Te_3 quintuple of a regular geometrical shape suggesting a single-crystal structure. (b) SEM image of the few-quintuple layer with the inset showing the layer suspension over a trench in the Si/SiO_2 wafer. (c) SEM image of the few-quintuple Bi_2Te_3 ribbon suspended across a trench in the wafer. The ribbon geometry allows for strong spatial confinement in two directions. (d) The electron diffraction pattern indicating that quasi-2D Bi_2Te_3 films are crystalline. (Adapted from Teweldebrhan, D. et al., *Appl. Phys. Lett.*, 96, 053107, 2010; Teweldebrhan, D., Goyal, V., Balandin, A.A., *Nano Lett.*, 10, 1209, 2010.)

The latter allowed us to achieve better understanding of the intrinsic properties of the atomically thin layers (see Figure 13.3).

The graphene-like exfoliation of the single or few quintuple films open up a possibility of changing the "effective" atomic composition in the crystalline ultra-thin films. It is known that the thermoelectric properties of Bi_2Te_3, that a small variation of ~0.1% in the Bi to Te atomic ratio can change the properties of the material from *p*-type to *n*-type [27]. Intentional deviation from stoichiometry during synthesis of Bi_2Te_3 compounds and alloys has been conventionally used for "doping" this type of materials [28]. Our atomically thin films are crystalline with the stoichiometric composition. But due to their small thickness one may achieve "stoichiometric doping" through variation in the film thickness. This approach is different from that implemented in the polycrystalline or disordered alloy Bi_2Te_3 materials. In this sense, the quasi-2D crystals may open up a new way for doping and tuning the properties of Bi_2Te_3 materials.

13.3 Raman Spectroscopic Nanometrology of Quasi-2D Thermoelectric Crystals

Thermoelectric applications of the exfoliated films require a tool for the control of the number of atomic planes and crystalline quality. We found that several interesting features in Raman spectra of few-nm-thick Bi_2Te_3 films create a possibility for using micro-Raman spectroscopy as a nanometrology tool for thermoelectric and TI applications. All Raman spectra in our study were excited at RT with laser light ($\lambda = 488$ nm) and recorded in backscattering configuration through a 50× objective. An 1800-lines/mm grating provided a spectral resolution of ~1 cm^{-1}, which was software-enhanced to ~0.1 cm^{-1}. Since Bi_2Te_3 has very low thermal conductivity and a low melting point, special care was taken to avoid local heating and melting during the measurements.

Since bulk Bi_2Te_3 is a semiconductor with a five-atom primitive cell, which belongs to the space group $R\bar{3}m(D^5_{3d})$ [29], it reveals 15 lattice vibration modes (phonon polarization branches). Three of these branches are acoustic and 12 are optical phonons. According to the group theory classification, 12 optical branches have $2A_{1g}$, $2E_g$, $2A_{1u}$, and $2E_u$ symmetry [30]. Due to the inversion symmetry of the crystal these phonon modes are exclusively either Raman or infrared (IR) active [30]. Figure 13.4 shows typical Raman spectra of bulk Bi_2Te_3 crystal and three few-quintuple layers as their thickness H decreases from ~82 to ~4 nm. The frequencies of the observed peaks and their assignment are listed in Table 13.1

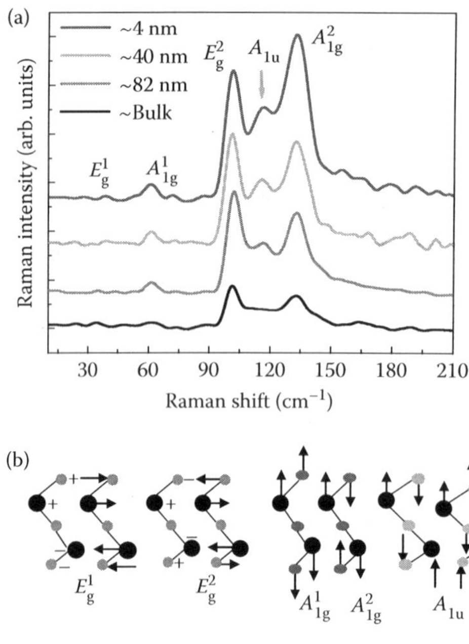

FIGURE 13.4 **(See color insert.)** (a) Raman spectra of reference bulk Bi_2Te_3 crystal and few-quintuple films. Note that A_{1u} mode at ~117 cm^{-1} becomes Raman active only in the thin films. (b) Schematic of the main lattice vibrations in {–Te$^{(1)}$–Bi–Te$^{(2)}$–Bi–Te$^{(1)}$–} quintuples. The large circles are Bi atoms. The small circle in the middle is Te$^{(2)}$ atom, which is the center of the inversion symmetry. The "+" and "–" signs in the schematics indicate atomic motions toward and from the observer. In all panels, the letters "E" and "A" indicate the in-plane and out-of-plane (c axis) lattice vibrations, respectively. The subscript "g" denotes Raman active while "u" stands for IR-active modes. (Adapted from Shahil, K.M.F. et al., *Appl. Phys. Lett.*, 96, 153103, 2010.)

TABLE 13.1 Raman Peaks in Bi_2Te_3 Crystals and Few-Quintuple Films

	E^1_g	A^1_{1g}	E^2_g	A_{1u}	A^2_{1g}	$I(A^2_{1g})/I(E^2_g)$	$I(A_{1u})/I(E^2_g)$	Comments
Bulk	34.4	62.1	101.7	—	134.0	0.75	—	Ref. [35]
82 nm	35.8	61.5	101.9	116.9	132.7	0.83	0.62	
40 nm	38.9	61.3	101.3	116.2	133.0	0.92	0.76	
4 nm	38.9	60.9	101.4	116.7	132.9	1.30	0.86	
Bulk	36.5	62.0	102.3	—	134.0	—	—	Ref. [29]
Bulk	—	62.3	103.7	—	134.2	—	—	Ref. [32]

together with previously reported data for bulk Bi_2Te_3. One can see that our results for four Brillouin zone (BZ) center Raman-active modes, E_g^1 (TO), A_{1g}^1 (LO), E_g^2 (TO), and A_{1g}^2 (LO), are consistent with literature [29–32]. In this nomenclature, TO and LO are transverse and longitudinal optical phonons, respectively. Both E_g and A_{1g} modes are twofold degenerate: in E_g, the atoms vibrate in the basal plane, while in A_{1g}, the atoms vibrate along c_H. The A_{1g}^1 and Eg^1 vibrations occur at lower frequencies than A_{1g}^2 and E_g^2. The latter modes, where the outer Bi and $Te^{(1)}$ atoms move in opposite phase, are mainly affected by the forces between Bi and $Te^{(1)}$ atoms. In E_g^1 and A_{1g}^1 modes the outer Bi–$Te^{(1)}$ pairs move in phase. Thus, the Bi–$Te^{(2)}$ bonding forces are primarily involved in these vibrations. In crystals with inversion symmetry, the IR-active modes (A_{1u}) must be odd parity while the Raman-active modes (E_g, A_{1g}) must be even parity under inversion. The odd-parity phonons (IR active) do not show up in Raman spectra for bulk samples as long as crystal retains its symmetry.

It is interesting to notice in Figure 13.4 that an additional peak appears at ~117 cm^{-1} in few-quintuple layers (FQLs). Its intensity, normalized to the intensity of E_g^2 (the most pronounced feature in the spectrum), grows with decreasing FQL thickness (see Table 13.1). This peak was identified as A_{1u} mode composed of TO phonons at the BZ boundary (Z point). The A_{1u} peak is IR active but not Raman active in bulk Bi_2Te_3 [30,33,34]. We attribute its appearance in FQL to breaking the crystal symmetry due to the presence of two interfaces. A single quintuple is inversely symmetric, which suggests that the identified crystal symmetry breaking is likely related to the loss of infinite crystal periodicity due to interfaces and corresponding relaxation of the phonon wave vector $q = 0$ selection rule. Since E_g^2 (TO) is a regular BZ-center peak originating in the "bulk" of the film, the $I(A_{1u})/I(E_g^2)$ ratio increases with decreasing H because of the decreasing interaction volume $V = S \times H$ (S is the cross-sectional area of the laser spot), which defines $I(E_g^2)$ for H smaller than the light penetration depth in a given material. We estimated the light penetration depth in our samples to be 60–110 nm for 488 nm laser depending on the carrier concentration. This values correlates well with H when the 117 cm^{-1} peak appears in FQL's spectrum.

Another intriguing observation from this figure is an evolution of the $I(A_{1g}^2)/I(E_g^2)$ ratio as one goes from bulk crystal to FQLs. In bulk, the measured intensity of in-plane vibrations $I(E_g^2)$ is higher than that of the out-of-plane vibrations $I(A_{1g}^2)$, which is in agreement with literature for bulk Bi_2Te_3 [29–32]. But in the atomically thin films the intensity of the out-of-plane vibrations increases with decreasing thickness and at some H exceeds that of $I(E_g^2)$. In FQL with $H = 4$ nm, the ratio $I(A_{1g}^2)/I(E_g^2) > 1$. It is reasonable to assume that the out-of-plane vibrations will be less restrained in a four-quintuple film than in bulk, which may lead to larger amplitudes of vibrations. We used the developed Raman nanometrology technique to select the "graphene-like" exfoliate Bi_2Te_3 flakes of the proper thickness and crystal quality [24].

13.4 Thermoelectric Properties of the "Dirac Materials"

The mechanically exfoliated few-quintuple films have a number of benefits compared to the CVD- or MBE-grown thin films. They are perfectly crystalline and have an essentially infinite potential barrier for electrons and holes. The latter is a drastic difference from Bi_2Te_3-based superlattices grown by MBE or other techniques. The ultimately high-potential barriers together with a few-quintuple thickness and possibility of electrical back-gating can ensure the strong quantum confinement for electrons and E_F fine-tuning. We investigated a possibility of ZT enhancement in the mechanically exfoliated Bi_2Te_3 films via reduction of the thermal conductivity [35]. The thermoelectric applications require a sufficient quantity of material, that is, "bulk," that is, the single quintuples would hardly be practical. For this reason, we studied the stacks of the exfoliated films, which were put on top of each other and subjected to thermal treatment.

The "graphene-like" exfoliated films were transferred to a substrate and mechanically put on top of each other. The stacks were annealed at ~250°C to reduce the air gaps between the layers and improve structural stability. Figure 13.5 shows the cross-sectional SEM image of a stacked sample. One can see

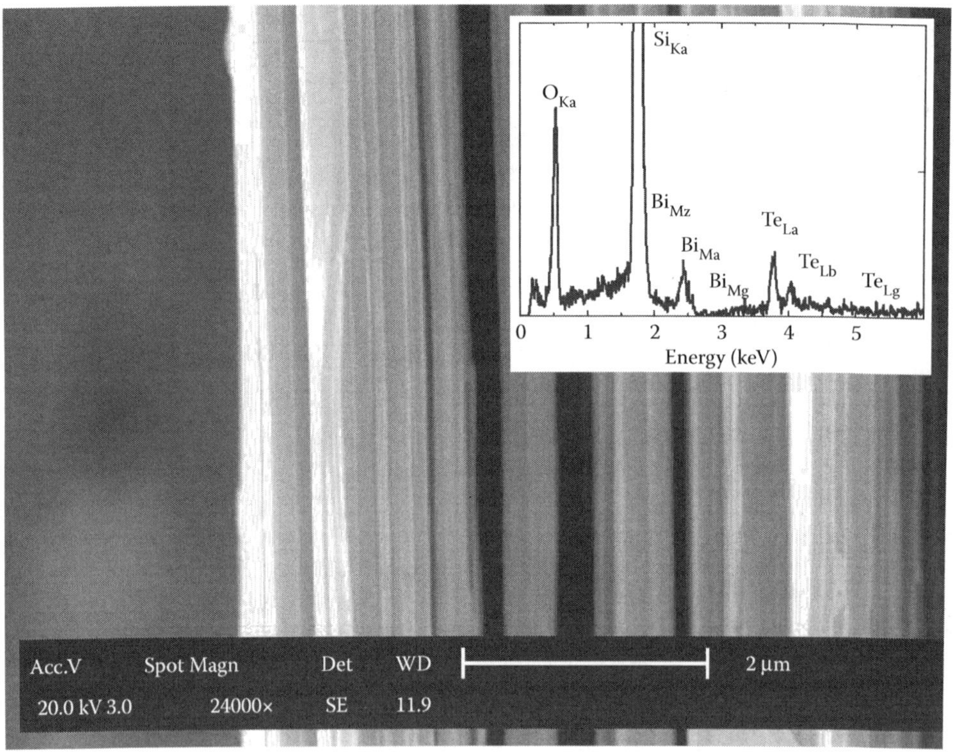

FIGURE 13.5 SEM image of the stacked "pseudo-superlattice" of the mechanically exfoliated Bi_2Te_3 films. The inset shows EDS spectrum of the exfoliated films indicating their atomic composition and transparency for the electron beam. (Adapted from Goyal, V., Teweldebrhan, D., Balandin, A.A., *Appl. Phys. Lett.*, 97, 133117, 2010.)

that the films of different thicknesses are assembled into a nonperiodic "pseudo-superlattice" (also referred to as compositionally modulated structures or "random superlattices"). The nonperiodic "superlattices" may have certain benefits for thermoelectric applications owing to flexibility for tuning the phonon transport. The films elemental composition and stoichiometry were verified with the energy dispersive spectroscopy (EDS). The presence of prominent Si and O peaks indicate that the individual films, which make up the stacks, are transparent to the electron beams.

The measurements of the thermal conductivity, K, were performed by two different experimental techniques to obtain its in-plane and cross-plane components. The first technique was the transient plane source (TPS) "hot disk" technique, which can measure the average in-plane thermal conductivity. The second technique was the optical "laser flash" technique (LFT), which measures the average cross-plane K. In the TPS technique, a short electric pulse is passed through an electrically insulated sensor sandwiched between two pieces of the sample under investigation. The sensor acts both as a heat source and a thermometer to determine the temperature rise, ΔT, in the sample in response to heating. The thermal diffusivity, α, and specific heat, required for determining K value, were obtained by measuring ΔT as a function of time. Our LFT instrument (Netzsch NanoFlash LFA 447) was equipped with a xenon flash lamp which heated the sample from one end by light shots. The temperature rise was determined at the back side of the samples with the nitrogen-cooled InSb IR detector. Using the thermal-wave travel time, we measured α and determined K from the equation $K = \alpha\rho C_p$, where C_p is the heat capacity and ρ is the mass density of the material.

Figure 13.6 gives the results of our thermal measurements for three representative "pseudo-superlattices" (samples A with $W \sim 0.4$ mm, B with $W \sim 0.3$ mm, and C with $W \sim 0.1$ mm), reference bulk samples,

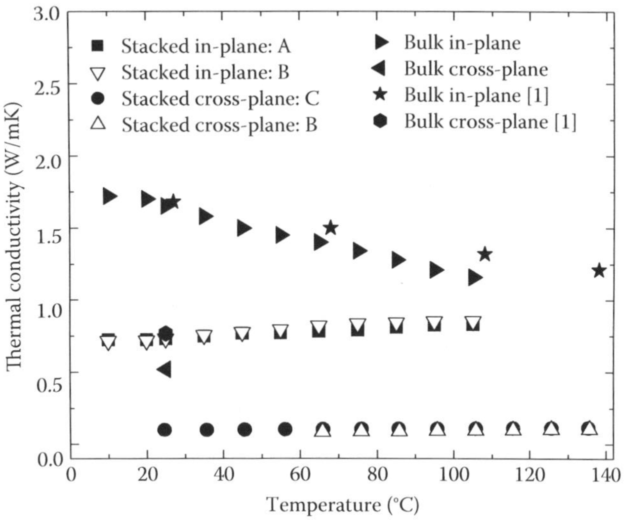

FIGURE 13.6 Thermal conductivity as a function of temperature for the stacked "pseudo-superlattices" and reference bulk Bi_2Te_3 crystals. The literature values for bulk Bi_2Te_3 are also shown for comparison.

and literature data for bulk Bi_2Te_3. Our data for the in-plane K in bulk (denoted with the black–white triangles) are in good agreement with literature values [2] marked by the stars. The monotonic K decrease with T is characteristic for semiconductor materials where thermal transport is limited by the crystal inharmonicity. The data for the bulk cross-plane K are also consistent. The good agreement for bulk attests for accuracy of our measurements.

One can see in Figure 13.6 that there is a strong decrease in the in-plane thermal conductivity of the "pseudo-superlattices" as compared to the bulk. The in-plane K for the stacked Bi_2Te_3 superlattices is ~0.7 W/mK, which is a reduction by a factor of ~2.4 from the bulk value of ~1.7 W/mK. The RT cross-plane K of the stacked films is ~0.14 W/mK, which is a significant drop, by a factor of 3.5, from the bulk cross-plane value of ~0.5 W/mK. The "pseudo-superlattice" K is only weakly dependent on temperature. This is expected for the thermal transport limited by the phonon-boundary scattering. The cross-plane K reduction is much stronger possibly owing to the large thermal boundary resistance between the exfoliated films. The thermal conduction in our samples was mostly by the acoustic phonons as estimated from the Wiedemann–Franz law. Another observation is that K values obtained for the stacks with different thicknesses are nearly the same. It means that the thermal conductivity was limited by the phonon scattering at the interfaces between the individual Bi_2Te_3 layers rather than by the scattering on the outside boundaries of the samples. The overall decrease of K in our stacks is exceptional. It is on the higher end or exceeds that reported for Bi_2Te_3 nanoparticles [36], alloy films [37], and highly textured materials [38]. The cross-plane K in our "pseudo-superlattices" approaches the theoretical *minimum* value predicted for the disordered crystals [39].

We analyzed the effect of annealing on K. Annealing at 350°C for 30 s led to ~5% increase in the thermal conductivity at RT. The latter was attributed to stronger bonds between the exfoliated Bi_2Te_3 layers and further decrease in the air gaps. The annealing at 450°C for the same time led to the approximately the same decrease in K. We explained it by the onset of interdiffusion between the layers and increased disorder. The latter was confirmed by the cross-plane SEM studies. The melting point for Bi_2Te_3 is rather low (T_m ~ 570°C), which supports our conclusion. The Seebeck measurements (MMR SB100) gave S values in the range from 231 to 247 μV/K. The higher values were obtained for thinner samples. We also investigated the current–voltage characteristics (Signatone and HP4142) of the "pseudo-superlattices" with different thicknesses. The electrical measurements revealed electrical resistivity of the order of ~10^{-4} Ω m. The strong decrease in the thermal conductivity with preserved electrical properties translates to ~140–250% increase in *ZT* at RT.

It is important to emphasize that the estimated increase is achieved entirely via reduction in the thermal conductivity. The Fermi level in the "pseudo-superlattices" was not optimized. The carrier transport in the films was likely of the mixed volume and surface nature. The further increase in crystal quality, thinning of the films, and gating for achieving the pure surface transport is expected to result in additional strong *ZT* increase predicted theoretically [20–21].

13.5 Conclusions

We discussed thermoelectric properties of individual quintuples and stacks of the graphene-like exfoliated single-crystal quasi-2D Bi_2Te_3 films. We showed that *ZT* in such "pseudo-superlattices" of the stacked films can be substantially increased via reduction of the in-plane and cross-plane thermal conductivity with simultaneously preserved good electrical properties in the films with smooth interfaces. It was recently predicted theoretically and demonstrated experimentally that Bi_2Te_3 and related materials are TIs, which reveal unusual surface states with the Dirac type of electron dispersion. Theoretical considerations suggest that this class of materials can have strongly enhanced thermoelectric properties over a wide temperature range. Although it can be difficult to scale up the mechanical exfoliation technique to the industrial scale, it allows one to investigate possible new routes for *ZT* increase, which can be eventually implemented with a different production technology.

Acknowledgments

This work was supported, in part, by SRC–DARPA through the FCRP Center on Functional Engineered Nano Architectonics (FENA), and DARPA Defense Microelectronics Activity DMEA under agreement number H94003-10-2-1003. The author is indebted to Roger Lake for providing figures with the calculated ZT in single quintuple Bi_2Te_3.

References

1. Ioffe, A.F., *Semiconductor Thermoelements* (Nauka, Moscow, 1956) (in Russian); or Ioffe, A.F., *Semiconductor Thermoelectric and Thermoelectric Cooling* (Infosearch, London, 1957).
2. Goldsmid, H.J., Douglas, R.W., The use of semiconductors in thermoelectric refrigeration. *Br. J. Appl. Phys.*, 5, 458, 1954.
3. Wright, D.A., Thermoelectric properties of bismuth telluride and its alloys. *Nature*, 181, 834, 1958.
4. Venkatasubramanian, R., Siivola, E., Colpitts, T., O'Quinn, B., Thin-film thermoelectric devices with high room-temperature figures of merit. *Nature*, 413, 597–602, 2001.
5. Xie, W., Tang, X., Yan, Y., Zhang, Q., Tritt, T.M., High thermoelectric performance BiSbTe alloy with unique low-dimensional structure. *J. Appl. Phys.*, 105, 113713, 2009.
6. DiSalvo, F.J., Thermoelectric cooling and power generation. *Science*, 285, 5428, 703, 1999.
7. Dresselhaus, M.S., Dresselhaus, G., Sun, X., Zhang, Z., Cronin, S.B., Koga, T., Low-dimensional thermoelectric materials. *Phys. Solid State*, 41(5), 679–682, 1999.
8. Hicks, L.D., Dresselhaus, M.S., Thermoelectric figure of merit of a one-dimensional conductor. *Phys. Rev. B.*, 47, 12727, 1993.
9. Broido, D.A., Reinecke, T.L., Comment on use of quantum well superlattices to obtain high figure of merit from nonconventional thermoelectric materials'. *Appl. Phys. Lett.*, 67, 100, 1995.
10. Balandin, A.A., Wang, K.L., Significant decrease of the lattice thermal conductivity due to phonon confinement in a free-standing semiconductor quantum well. *Phys. Rev. B.*, 58, 1544, 1998.
11. Balandin, A., Wang, K.L., Effect of phonon confinement on the thermoelectric figure of merit of quantum wells. *J. Appl. Phys.*, 84, 6149, 1998.

12. Chen, G., Borca-Tasciuc, T., Yang, B., Song, D., Liu, W. L., Zeng, T., Achimov, D.A., Heat conduction mechanisms and phonon engineering in superlattice structures. *Therm. Sci. Eng.*, 7, 43–51, 1999.
13. Borca-Tasciuc, T., Achimov, D., Liu, W.L., Chen, G., Ren, H.-W., Lin, C.-H., Pei, S.S., Thermal conductivity of InAs/AlSb superlattices. *Microscale Thermophys. Eng.*, 5, 225–231, 2001.
14. Feutelais, Y., Legendre, B., Rodier, N., Agafonov, V.A., Study of the phases in the bismuth–tellurium system. *Mater. Res. Bull.*, 28, 591–596, 1993.
15. Bernevig, B.A., Hughes, T.L., Zhang, S.C., Quantum spin Hall effect and topological phase transition in HgTe quantum wells. *Science,* 314, 1757, 2006.
16. Konig, M., Wiedmann, S., Brune, C., Roth, A., Buhmann, H., Molenkamp, L., Qi, X.L., Zhang, S.C., Quantum spin Hall insulator state in HgTe quantum wells. *Science,* 318, 766, 2007.
17. Hsieh, D., Qian, D., Wray, L., Xia, Y., Hor, Y.S., Cava, R.J., Hasan, M.Z., A topological Dirac insulator in a quantum spin Hall phase. *Nature,* 452, 970, 2008.
18. Zhang, H., Liu, C., Qi, X., Dai, X., Fang, Z., Zhang, S., Topological insulators in Bi_2Se_3, Bi_2Te_3 and Sb_2Te_3 with a single Dirac cone on the surface. *Nat. Phys.*, 5, 438, 2009.
19. Xia, Y., Qian, D., Hsieh, D., Wray, L., Pal, A., Lin, H., Bansil, A., Grauer, D., Hor, Y.S., Cava, R.J., Observation of a large-gap topological-insulator class with a single Dirac cone on the surface. *Nat. Phys.,* 5, 398, 2009.
20. Zahid, F., Lake, R., Thermoelectric properties of Bi_2Te_3 atomic quintuple thin films. *Appl. Phys. Lett.*, 97, 212102, 2010.
21. Ghaemi, P., Mong, R.S.K., Moore, J.E., In-plane transport and enhanced TE in thin films of the TIs Bi_2Te_3 and Bi_2Se_3. *Phys. Rev. Lett.*, 105, 166603, 2010.
22. Teweldebrhan, D., Goyal, V., Rahman, M., Balandin, A.A., Atomically-thin crystalline films and ribbons of bismuth telluride. *Appl. Phys. Lett.*, 96, 053107, 2010.
23. Teweldebrhan, D., Goyal, V., Balandin, A.A., Exfoliation and characterization of bismuth telluride atomic *quintuples* and quasi-two-dimensional crystals. *Nano Lett.*, 10, 1209, 2010.
24. Shahil, K.M.F., Hossain, M.Z., Teweldebrhan, D., Balandin, A.A., Crystal symmetry breaking in few-quintuple Bi_2Te_3 films: Applications in nanometrology of topological insulators. *Appl. Phys. Lett.*, 96, 153103, 2010.
25. Novoselov, K.S., Geim, A.K., Morozov, S.V., Jiang, D., Zhang, Y., Dubonos, S.V., Grigorieva, I.V., Firsov, A.A., Electric field effect in atomically thin carbon films. *Science*, 306, 666, 2004.
26. Novoselov, K.S., Geim, A.K., Morozov, S.V., Jiang, D., Katsnelson, M.I., Grigorieva, I.V., Dubonos. S.V., Firsov, A.A., Two-dimensional gas of massless Dirac fermions in graphene. *Nature*, 438, 197, 2005.
27. Jones, P., Hubera, T.E., Melngailisb, J., Barry, J., Ervin, M.H., Zhelevac, T.S., Nikolaevad, A., Konopko, L., Graf, M., Electrical contact resistance of bismuth telluride nanowires, *ICT '06. 25th International Conference on Thermoelectrics, 2006.* ICT '06, 693–696, 2006, doi: 10.1109/ICT.2006.331236.
28. Golia, S., Arora, M., Sharma, R.K., Rastogi, A.C., Electrochemically deposited bismuth telluride thin films. *Curr. Appl. Phys.*, 3, 195, 2003.
29. Kullmann, W., Geurts, J., Richter, W., Lehner, N., Rauh, H., Steigenberger, U., Eichhorn, G., Geick, R., Lattice dynamics and phonon dispersion in the narrow gap semiconductor Bi_2Te_3 with sandwich structure. *Phys. Status Sol. (b)*, 125, 131, 1984.
30. Richter, W., Kohler, H., Becker, C.R., A Raman and far-infrared investigation of phonons in the rhombohedral V_2–VI_3 compounds Bi_2Te_3, Bi_2Se_3, Sb_2Te_3 and $Bi_2(Te_{1-x}Se_x)3$ $(0 < x < 1)$, $(Bi_{1-y}Sb_y)_2Te_3$ $(0 < y < 1)$. *Phys. Status Sol. (b),* 84, 619, 1977.
31. Russo, V., Bailini, A., Zamboni, M., Passoni, M., Conti, C., Casari, C.S., Li Bassi, A., Bottani, C.E., *Raman* spectroscopy of Bi-Te thin films, *J. Raman Spec.*, 39, 205, 2008.
32. Goncalves, L.M., Couto, C., Alpuim, P., Rolo, A.G., Völklein, F., Correia, J.H., Fabrication of thermoelectric devices by applying microsystems technology. *Thin Solid Films,* 518, 2816, 2010.
33. Yu, P.Y., Cardona, M., *Fundamentals of Semiconductors* (Springer, New York, 1999).
34. Wagner, V., Dolling, G., Powell, B.M., Landwehr, G., Lattice vibrations of Bi_2Te_3. *Phys. Stat. Sol. (b),* 85, 311, 1978.

35. Goyal, V., Teweldebrhan, D., Balandin, A.A., Mechanically-exfoliated stacks of thin films of Bi_2Te_3 topological insulators with enhanced thermoelectric performance. *Appl. Phys. Lett.*, 97, 133117, 2010.
36. Dirmyer, M.R., Martin, J., Nolas, G.S., Sen, A., Badding, J.V., Thermal and electrical conductivity of size-tuned bismuth telluride nanoparticles. *Small*, 5, 933, 2009.
37. Chiritescu, C., Mortensen, C., Cahill, D.G., Johnson, D., Zschack, P., Lower limit to the lattice thermal conductivity of nanostructured Bi_2Te_3-based materials. *J. Appl. Phys.*, 106, 073503, 2009.
38. Ben-Yehuda, O., Shuker, R., Gelbstein, Y., Dashebsky, Z., Dariel, M.P., Highly textured Bi_2Te_3-based materials for thermoelectric energy conversion. *J. Appl. Phys.*, 101, 113707, 2007.
39. Cahill, D., Watson, S., Pohl, R., Lower limit to the thermal conductivity of disordered crystals. *Phys. Rev. B*, 46, 6131, 1992.

14

The Bottom-Up Approach to Bulk Thermoelectric Materials with Nanoscale Domains

Anuja Datta
University of South Florida

Adrian Popescu
University of South Florida

Lilia Woods
University of South Florida

George S. Nolas
University of South Florida

14.1 Introduction

Many of the properties that directly relate the potential of a material for thermoelectric (TE) applications are given by the materials' dimensionless figure of merit, ZT,

$$ZT = \frac{S^2\sigma}{\kappa}T \tag{14.1}$$

where S is the Seebeck coefficient, σ is the electrical conductivity ($= 1/\rho$, where ρ is the electrical resistivity), and κ is the total thermal conductivity (where $\kappa = \kappa_L + \kappa_e$, the lattice and electronic contributions, respectively).[1] The power factor, $S^2\sigma$, is usually optimized as a function of carrier concentration (typically ~10^{19} carriers/cm^3) through doping in order to achieve the largest ZT.[1] An enhancement in ZT by a factor of two or more over existing TE materials currently in use is necessary for TEs to fully compete with other technologies for refrigeration or power conversion applications.[2]

Often, the failure in achieving excellent TE properties in seemingly superior TE materials can be directly attributed to the lack of "tunability" of the material systems. Low-dimensional materials offer avenues for the manipulation of electron and phonon transport. Hicks and Dresselhaus[3] first demonstrated that nanometer-scale TE materials exhibit improved TE properties as compared to that of the bulk, thus offering a new avenue for TE properties "tunability." More recently, nanoscale TE materials research has been at the core of many of the recent advances in TEs.[4–7] The two mechanisms that led to this success were a strong reduction in κ_L and an increase in $S^2\sigma$. The reduction in κ_L can be understood as a function of phonon mean free path.[1,5–9] Much of the work in improving the TE properties by introducing nanoscale domains in bulk materials has been focused on this approach.[5–7,10,11] Quantum confinement results in a strongly increased S due to the increase in the density of states (DOS) of the carriers.[3] This observation was at the origin of the interest in nanoscale thermoelectricity and experimentally observed on bismuth nanowires.[12] However, TE technology requires materials in large quantities, typically in bulk form. This leaves open the question: Is enhancement of $S^2\sigma$ possible in bulk materials?

Carrier scattering at the boundaries that inhibits the conduction of low-energy charge carriers thus causing an overall increase in S was first proposed theoretically,[13,14] and was experimentally investigated in III–V heterostructures.[15] Bulk materials that incorporate nanoscale features, or domains, effectively increase the number of interfaces per unit volume. Thus, nanocomposites that involve the incorporation of nanograins, or nanoscale inclusions, within a bulk matrix are gaining interest. There are, however, several different approaches to synthesizing bulk materials with nanoscale features or domains. These range from solution-phase to mechanical grinding techniques. There are also different avenues for densification, from simply cold pressing (CP) and annealing to employing spark plasma sintering (SPS). In this chapter, we provide an overview of these avenues of research while focusing on our work on a "bottom-up" approach for the synthesis of bulk materials with nanograin domains that exhibit enhanced TE properties.

14.2 Physical Preparative Methods for Nanoscale TE Materials

TE properties depend largely on crystal structure, composition, microstructure, defects, and impurities, and methods for materials preparation require appropriate tuning of the synthesis parameters for optimization. This is critically important for TE materials with nanoscale grains or inclusions. TE materials currently in use are prepared by a variety of physical processes that typically result in bulk materials in powder form or as ingots.[1,16,17] The synthesis of TE materials that contain nanoscale domains are, to a large extent, prepared by physical processes that can basically be described as "top-down" approaches. However, preparing nanoscale TE materials with finite grain size distributions has often been a challenge for these methods due to the thermodynamics and kinetics of the reactions. Despite these shortcomings, several bulk processes have been successfully applied in recent years to create stable nanocomposites with improved properties over those of their bulk counterparts. In this section, we review a few examples of these established physical preparation methods.

14.2.1 Melt Spinning

TE materials are typically compounds or solid solutions of two or more constituent elements. Hence, in order to prepare TE materials from a melt, proper understanding of solid–liquid phase diagrams are necessary. The preparation of TE materials from the melt begins with melting of the elements and subsequent quenching the product, thus producing homogeneous and uniform phases. Due to the rapid temperature decrease from quenching, there is limited growth while producing a relatively uniform grain size distribution. Nanostructured TE materials can thus be prepared by rapidly cooling a liquid melt to room temperature or below, a process known as melt spinning (MS). In MS, a rotating wheel is cooled internally, usually by water or liquid nitrogen, and rotated. A thin stream of melt then falls onto the wheel and is cooled, in an Ar atmosphere, causing rapid solidification (Figure 14.1a).[18,19]

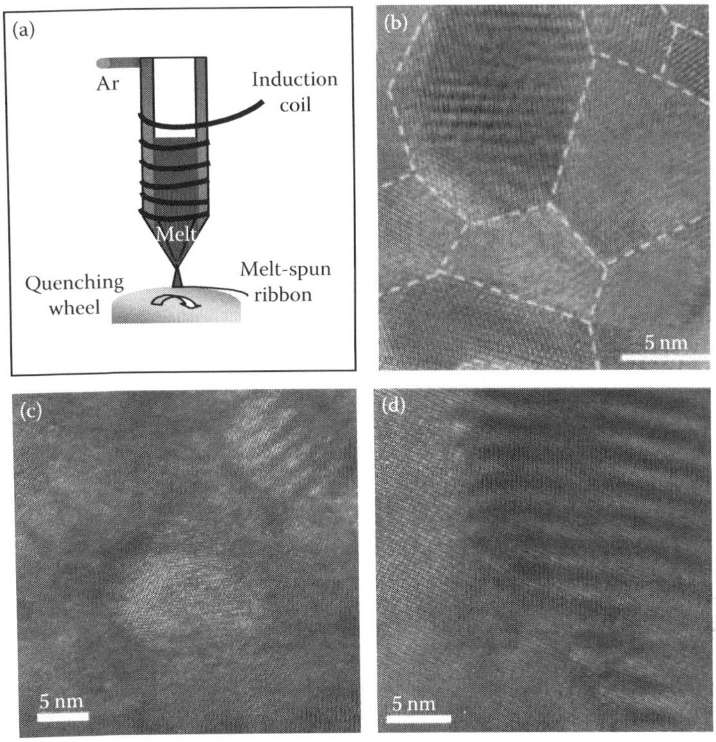

FIGURE 14.1 (a) Schematic diagram showing the MS process. (b) High resolution transmission electron microscopy (HRTEM) image of MS–SPS-prepared *p*-type $Bi_{0.52}Sb_{1.48}Te_3$ showing coherent interfaces between the nanograins. (Reprinted with permission from Xie, W. J. et al., Unique nanostructures and enhanced thermoelectric performance of melt-spun BiSbTe alloys, *Appl. Phys. Lett.*, 94, 102111, Copyright 2009, American Institute of Physics.) (c) HRTEM image indicating the presence of nanograins inside the $Bi_{0.52}Sb_{1.48}Te_3$ nanocomposite prepared by ball milling and hot pressing. (Reprinted with permission from Ma, Y. et al., Enhanced thermoelectric figure-of-merit in *p*-type nanostructured bismuth antimony tellurium alloys made from elemental chunks, *Nano Lett.*, 8, 2580, Copyright 2008, American Chemical Society.) (d) HRTEM image of $(Pb_{0.95}Sn_{0.05}Te)_{0.92}(PbS)_{0.08}$ grown by spinoidal decomposition. The parallel lines correspond to periodic compositional fluctuations of PbS (bright) and PbTe (dark) phases. (Reprinted with permission from Androulakis, J. et al., Spinodal decomposition and nucleation and growth as a means to bulk nanostructured thermoelectrics: Enhanced performance in $Pb_{1-x}Sn_xTe$-PbS, *J. Am. Chem. Soc.*, 129, 9780, Copyright 2007, American Chemical Society.)

The cooling rate achievable by MS can reach up to 10^6 K/s.[20,21] SPS is often used to densify the product after MS (MS–SPS). MS has successfully produced bulk Bi_2Te_3 and $Bi_{0.52}Sb_{1.48}Te_3$ TE materials with nanoscale features.[22–24] Microstructures generated in these MS materials are sensitive to their position in relation to their location in the MS chamber.[22–24] With the MS technique, there always exists a contact surface and a free surface, each having significantly different cooling rates. Different cooling rates at different positions relative to the rotating wheel result in variations in microstructure in a single specimen. This variation in microstructure is also related to the variation in the degree of crystallinity of the specimen. Electron microscopic studies on MS–SPS-generated nanocomposites indicate the presence of coherent interfaces between the grains in $Bi_{0.52}Sb_{1.48}Te_3$ (Figure 14.1b) and Bi_2Te_3. These interfaces were reported to influence the electrical and thermal transport properties of these materials. Room temperature *ZT* values of ~1.6 and ~1.4 were reported for *p*-type $Bi_{0.52}Sb_{1.48}Te_3$ and Bi_2Te_3, respectively.[22,23]

14.2.2 Ball Milling

In a typical ball milling process, ingots or elemental chunks together with steel balls are added to a jar made of cemented carbide materials. The mixture is then rotated so that the steel balls mechanically impact on the enclosed material. Different types of milling processes, such as vibration ball milling, rotation ball milling, planetary ball milling, and attrition milling, are used for the preparation of fine powders. The resulting mixture is then hot pressed (HP) at an appropriate temperature and pressure. With proper adjustment of the HP pressure, temperature, and time, a relatively dense bulk material with nanosize grains results. This material was referred to as "random nanostructured bulk"[25,26] due to the presence of random nanograins in the bulk material (Figure 14.1c). Recently, this technique has been used to synthesize Si–Ge and $Bi_xSb_{2-x}Te_3$ alloys.[25–28] So far, the highest ZT reported for bulk p-type $Bi_{0.5}Sb_{1.5}Te_3$ ($ZT \sim 1.4$ at 373 K), was prepared by ball milling followed by direct-current hot pressing.[25,28] The TE enhancement is largely from a reduction in κ_L, together with an increase in $S^2\sigma$ as compared to bulk p-type $Bi_xSb_{2-x}Te_3$.

Nanocomposite Si–Ge alloys were also prepared by ball milling and HPing.[10,26,27] These nanostructured alloys were reported to exhibit an increase in $S^2\sigma$ due to an enhanced S compared to bulk Si–Ge alloys over the temperature range of 300–1000 K. A $ZT \sim 0.72$ at 1000 K was reported, as compared to $ZT \sim 0.6$ for bulk Si–Ge alloys. The increased $S^2\sigma$ for the nanostructured bulk Si–Ge alloy is consistent with what was observed for the $Bi_{0.5}Sb_{1.5}Te_3$ alloy.

14.2.3 Thermal Processing Techniques: Nanoscale Precipitates, Matrix Encapsulation, and Spinoidal Decomposition

Thermal processing techniques include a particular or a combination of metallurgical methods in which different compounds or elements are sealed in ampoules and thermally treated under Ar atmosphere. Materials are chosen for this process based on the solubility between the constituent phases. The constituent phases are generally in the liquid state. The major phase commonly has a higher melting point than the minor phase (precipitate).[29] Rapid quenching of the melt, and subsequent tempering or annealing, leads to the formation of nanoscale precipitates within a bulk matrix. These are observed as inclusions encapsulated by the major phase. Polycrystalline PbTe containing 30–40 nm size precipitates of excess Pb was prepared using this process by Heremans et al.[30] They achieved an enhancement in S over that of bulk PbTe. Nanostructured PbTe–Pb–Sb composites were also prepared in a similar manner by first melting elemental Pb and Te, then adding excess Pb and Sb to the melt.[29,31] Subsequent melting of PbTe, Pb, and Sb, and rapid quenching, generated Pb and Sb nanodots within the bulk. This process, referred to as "synergistic nanostructuring,"[31] led to an improvement in the TE properties in this system. Enhanced $S^2\sigma$ for certain carrier concentrations above 600 K was reported.[29,31] In addition, a decrease in κ_L resulted in a $ZT \sim 1.4$ for the composition PbTe–Pb (0.5%)–Sb (2%) at 700 K.

Other than direct thermal precipitation of elements or compounds, spinodal decomposition is used to prepare metastable solid solutions of two different phases. By this process, metastable solid solutions are thermodynamically stabilized by phase segregation, thus nanostructuring occurs due to nucleation and growth. Typically, two isomorphous phases are chosen for spinoidal decomposition, such as PbTe–PbS (Figure 14.1d).[32] This type of coherent nanoscale metastable solid solution heterogeneities, such as PbTe–PbS, results in a decrease in κ_L while maintaining a high electron mobility (>100 cm^2/V s at 700 K). A $ZT \sim 1.5$ was achieved at 642 K for $(Pb_{0.95}Sn_{0.05}Te)_{0.92}(PbS)_{0.08}$ prepared by spinoidal decomposition.

A very similar process to that described above was used for the preparation of several different solid solutions in the PbTe system, namely $AgPb_mSbTe_{2+m}$,[33–38] $AgPb_mSn_nSbTe_{2+m+n}$,[39] $NaPb_mSbTe_{2+m}$,[40,41] and KPb_mSbTe_{2+m}.[5,42] To synthesize these materials, nominal quantities of the respective elements, calculated based on the stoichiometry of the compounds, were sealed under vacuum in carbon-coated quartz tubes and allowed to melt completely. The homogeneity of the liquid phase was maintained by keeping

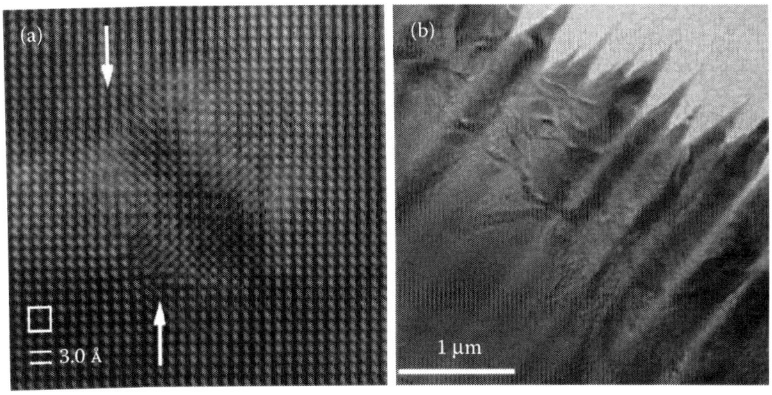

FIGURE 14.2 (a) HRTEM image of $AgPb_mSbTe_{2+m}$ showing a nanoparticulate precipitate inside the bulk. (Reprinted with permission from Quarez, E. et al., Nanostructuring, compositional fluctuations, and atomic ordering in the thermoelectric materials $AgPb_mSbTe_{2+m}$. The myth of solid solutions, *J. Am. Chem. Soc.*, 127, 9177, Copyright 2005, American Chemical Society.) (b) Lamellar nanostructuring inside $NaPb_{13}Sn_5BiTe_{20}$. (Reprinted with permission from Gueguen, A. et al., Thermoelectric properties and nanostructuring in the *p*-type materials $NaPb_{18-x}Sn_xMTe_{20}$ (M = Sb, Bi), *Chem. Mater.*, 21, 1683, Copyright 2009, American Chemical Society.)

the melt at temperature for 4–12 h followed by 2–6 h of nucleation and crystal growth. Cooling for 12–84 h resulted in the formation of nanostructures within the phase-pure materials.[5,33,35,36,40–42] It was proposed that complex chemical compositions in these material systems may result in the sporadic enrichment of heavy elements within the bulk. This may also result in the formation of nanostructures and concomitant interface-induced matrix perturbations. Nanostructures varying in morphology, from dots (precipitates) (Figure 14.2a) to lamellae (Figure 14.2b), observed by transmission electron microscopy (TEM) analyses, were reported to contribute to the reduction in κ_L in all these materials. The nanostructure and interface-induced matrix elastic perturbations were reported to play an important role in scattering the phonons resulting in reduced κ in these systems. A $ZT \sim 1.7$ at 700 K was obtained from $AgPb_mSbTe_{2+m}$ with $m \sim 18$–22.[35] $AgPb_mSbTe_{2+m}$ specimens prepared by MA–SPS[37,38] were also reported to have similar nanostructuring and enhanced TE properties. In *p*-type Na alkali metal containing $NaPb_mSbTe_{2+m}$, a low κ resulted in $ZT \sim 1$ near 475 K and $ZT \sim 1.7$ at 650 K, the widest temperature ranges reported in which a single material exhibits $ZT > 1$.[40]

The outline presented in this section serves to demonstrate the successful application of the top-down approach toward nanostructured TE materials processing. Issues relating to scalability and cost, as well as an interest in investigating the fundamental effects on the TE properties by nanoscale domains incorporated within bulk TE materials, lead us to investigate a bottom-up approach. In the next section, we will review the bottom-up approach in terms of understanding the role of chemical synthesis and densification on the TE properties of known TE materials with nanoscale domains.

14.3 Fabricating Bulk Dimensional Nanocomposite TE Materials by Chemical Processing: The Bottom-Up Approach

High-temperature solid-state or liquid-phase reactions were traditionally employed in the synthesis of TE materials. When nanoscale domains are of interest, this approach does not typically allow for control of the size or composition of the nanoscale grains or inclusions. Moreover, it may be difficult to establish a set of predictable rules by which a particular material or composite can be synthesized. A possible solution to this drawback lies in adopting a bottom-up chemical synthesis approach that will not only offer a solution to the problem of reproducibility but also has the added advantages of (1) scalability, (2) cost-effectiveness, (3) controllability of grain size, shape, and size distribution, (4) allowing for the study of a broad range of chemical compositions, and (5) allowing for a way to control the carrier concentration through chemical doping. A bottom-up synthesis approach allows for the manipulation of matter at the molecular level and for good chemical homogeneity at the nanoscale. The benefits of bottom-up synthetic processing in synthesizing TE compounds with nanoscale inclusions have now been widely accepted.[43–52] After preparation of the desired materials in the form of nanocrystals, another vital step is to densify the nanocrystals into a dense polycrystalline solid. A popular method to prepare dense polycrystalline materials with nanoscale inclusions is to densify the prepared nanocrystals and bulk powder together, mixed in appropriate ratios.[53–55] An alternative method is to densify only the nanocrystals by a suitable compaction technique.[50–52,56] This approach can result in dense polycrystalline compounds with nanoscale grains. These types of "dimensional nanocomposites," where all grains are of similar composition but different size, can result in a reduced κ (due to phonon scattering at grain boundaries), and an increased $S^2\sigma$ (due to charge carrier scattering at the grain boundaries) in certain cases.[50–52,56] Major concerns with this approach include compaction with limited grain growth, preventing agglomeration, and providing highly dense materials. In this section, we describe an approach that addresses several of these issues.

14.3.1 Solution Phase Synthesis

Although chemical processes are favorable over solid-state reaction techniques, a major concern with several chemical synthesis routes is obtaining phase pure materials after complete removal of any surface binding molecules used in the synthesis process. These impurities, if not removed, may have a detrimental effect on the desired TE properties after densification. Therefore, the selection of rational synthetic tools is of major importance. In addition, entrapment of impurities in the final product must be avoided, or at least minimized, before densification. In this section, we discuss a few chemical processes that are feasible in terms of the quality of the nanocrystals that are obtained, ease of preparation, cost-effectiveness, yield, and allow for a wide range of potential compounds.

14.3.1.1 Direct Chemical Precipitation

In a typical direct chemical precipitation reaction, soluble ions in separate solutions are mixed together to form an insoluble compound that precipitates from the reaction medium as a solid. This precipitate is then collected after repeated washing. For a direct precipitation reaction the chemicals should be soluble in the solvent. The rate and efficiency of a precipitation reaction also depend on the "solubility rules" of the chemicals. Precipitation reactions are typically water based and therefore can be carried out at low temperatures. This also makes them attractive economically because of their scalability for large yields. A wide range of crystal sizes, from 5 nm to over 1 µm, can be synthesized by direct precipitation. An essential aspect of this approach is its simplicity. For binary systems, this involves reactions of only one set of cation and anions; however, the system becomes more complicated for ternary, quaternary, and when doping. In a solution containing two dissolved solutes that contain the same ions (common-ion effect), the reaction rate may be suppressed. The type of reactant and pH also determines the reaction product. For example, simple ionic precursors or small organic molecular salts are generally chosen for direct precipitation due to the ease of ion release in the solvent. A variety of materials such as oxides, sulfides, tellurides, and selenides can be precipitated by this process.[50,57–62] This has been demonstrated with lead telluride (PbTe) and bismuth telluride (Bi_2Te_3) derivatives.[50,51,60–62] As a tellurium (Te) precursor, metallic Te easily dissolves in highly alkaline aqueous solutions. A cation precursor solution is then added to prepare the desired telluride compound. It is possible to vary the rate of the reaction by varying the pH of the solution. This directly controls the growth rate of the nanocrystals, hence the nanocrystal size, as shown in the TEM and scanning electron microscope (SEM) images in Figure 14.3. The PbTe nanocrystals shown in Figure 14.3a through c increase in diameter from 50 to 200 nm when the pH is varied from 9 to 12. As no capping

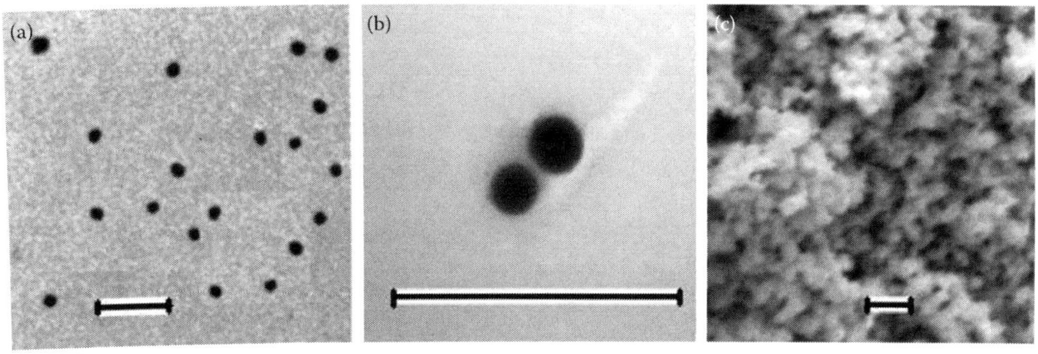

FIGURE 14.3 TEM images of PbTe nanocrystals synthesized at a pH of (a) 9 and (b) 10, and SEM image of PbTe nanocrystals synthesized at pH (c) 12. The scale bar corresponds to 500 nm.

or surface active molecules are used in this process, other than the reactant precursors, the products do not need any further purification or annealing treatments before densification.

14.3.1.2 Emulsion Process

In an emulsion, typically using two immiscible liquids, one liquid (the dispersed phase) is dispersed into the other (the continuous phase). These two phases give rise to a colloidal solution that can be used as a "nanoreactor."[57,63] When the colloidal solution is kept undisturbed it typically tends to separate into two constituent phases. In most cases, a surfactant is used to stabilize the reactor droplet forming "microemulsions" that consist of isotropic liquid mixtures of oil, that is, liquid immiscible phase, water, and surfactant. Pileni and coworkers[63–68] were the first to develop this technique. Depending on the dispersed and dispersant media two basic types of microemulsions can be prepared, direct ("oil in water") and reverse ("water in oil"). The interfaces of these two liquid phases are stabilized by surfactant molecules. Surfactants are usually organic compounds that contain both hydrophobic groups ("tails") and hydrophilic groups ("heads"). They are therefore soluble in both organic solvents and water. In microemulsion systems, the surfactant molecules at the interface between the oil and water, with the "tails" dissolved in the oil phase and the "head" groups in the aqueous phase, protect the droplets of the dispersed phase and thus control the size of the nanocrystals. This is illustrated schematically in Figure 14.4. As in binary systems (water/surfactant or oil/surfactant), self assembled structures of different types can

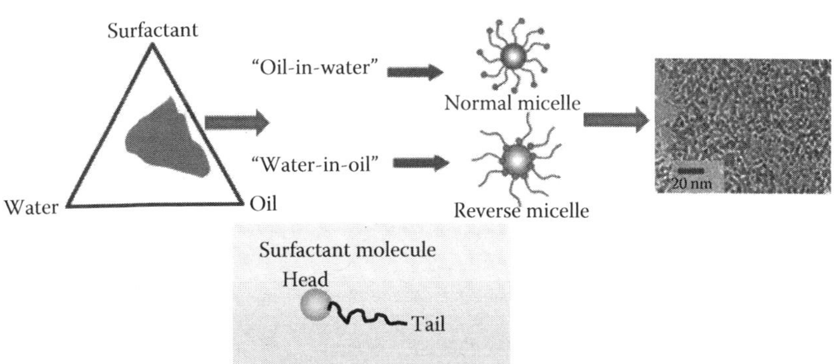

FIGURE 14.4 Schematic diagram illustrating the formation of nanocrystals by microemulsion synthesis. Also shown are the constituent parts of a surfactant molecule. The "head" of the surfactant is hydrophobic and the "tail" is hydrophilic.

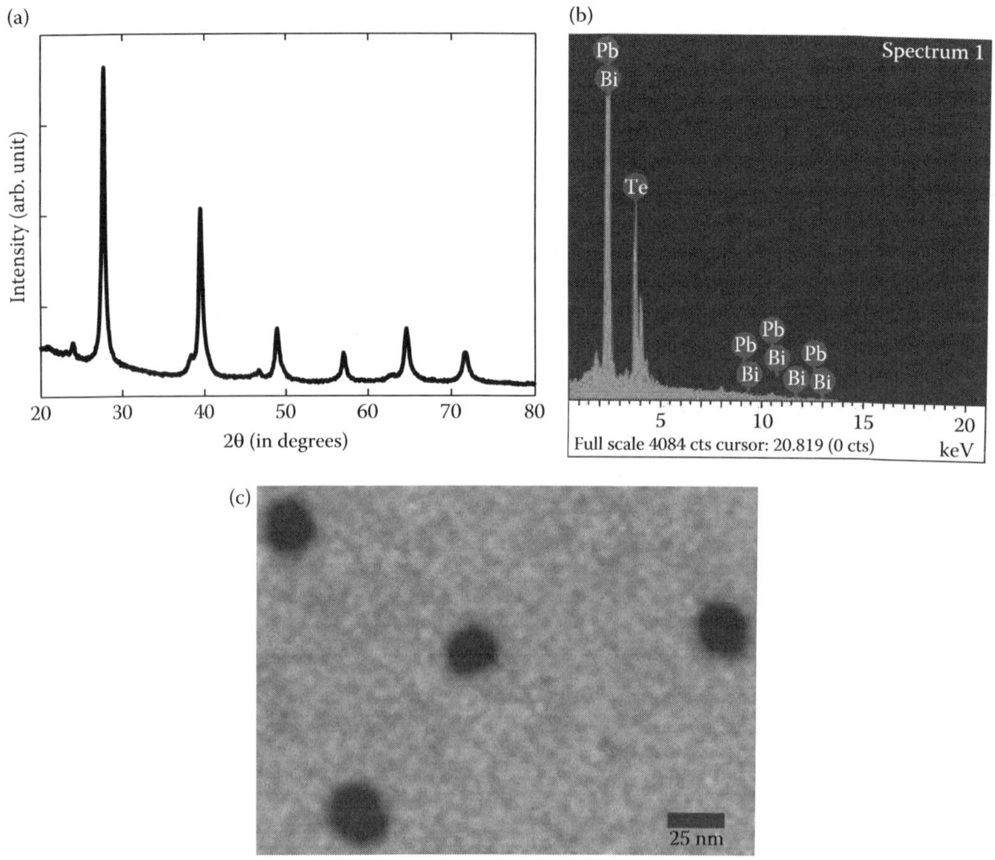

FIGURE 14.5 Bi-doped PbTe nanocrystals of ~20–30 nm in diameter prepared by a microemulsion synthesis process using a w/o ratio of 1/20 and w/AOT ratio of 25 as shown by (a) XRD, (b) EDS, and (c) TEM.

form with morphologies that vary, for example, from spherical and cylindrical micelles to lamellar phases. Nanocrystals of II–VI semiconductors are typically synthesized by microemulsion.[57,68–73] The size dependence was carefully studied as a function of water/surfactant ratio by several groups.[57,68–73] The process is also attractive for synthesizing nanocrystals of TE materials such as PbTe and Bi_2Te_3.[46,74–76] Recently, our group has prepared *p*- and *n*-type PbTe nanocrystals by a microemulsion process. We were able to vary the carrier concentration in densified nanocomposites by doping the nanocrystals before densification. The nanocrystals were synthesized by employing a water/surfactant ratio of 20, where sodium 2-bis(2-ethylhexyl)sulfosuccinate (AOT) was used as a surfactant. As much as 40% Bi substitution, or doping, for Pb was achieved by this process with no secondary phase observed in the x-ray diffraction (XRD) spectra (Figure 14.5a). Bi doping was also confirmed by energy dispersive spectroscopic (EDS) analysis, suggesting this to be a very convenient synthesis route for these types of TE materials (Figure 14.5b). Nanocrystals in the range of 20–30 nm in diameter were prepared (Figure 14.5c).

14.3.1.3 Solvothermal Synthesis

Solvothermal processing typically involves the use of a sealed reaction vessel under a temperature above the boiling point of the solvent used. Under these conditions, pressure is developed due to vaporization of the solvent at temperature, and thus pressure increases rapidly with temperature.[77] The percentage fill is defined as the volume of the solvent inside the container with respect to the volume of the container. Figure 14.6 demonstrates the rapid rise in pressure when a high percentage fill in reaction vessel is used.

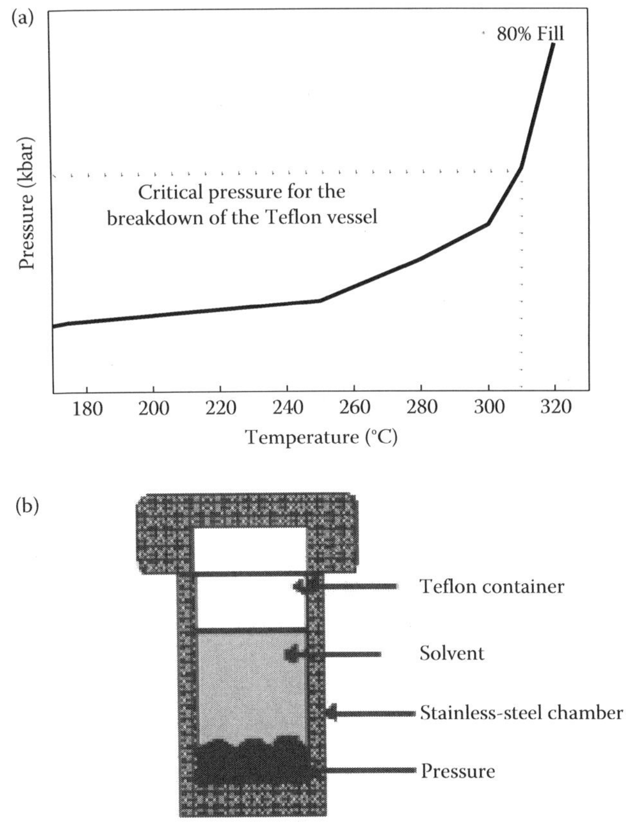

FIGURE 14.6 (a) An illustrative graph indicating the rapid rise in the pressure in a sealed Teflon-lined solvothermal vessel as a function of temperature for a particular percentage fill (80%) of a solvent. (b) Diagram showing the solvent and reactants in the sealed vessel. The solvent fill percentage is typically maintained at not more than 80% because of the potential for vessel rapture or leakage at higher concentrations.

Below the critical point of the solvent, and even below 200°C, a high percentage fill (≥80%) allows access to pressures of several millibar. A solvothermal process can be specified depending on the solvent used, for example, hydrothermal, ethanothermal, or benzothermal. Nearly, all inorganic substances at elevated temperatures and pressures are soluble in a majority of solvents. Crystallization of the desired product occurs at elevated pressures and temperatures. In addition to the percentage fill, temperature, and solvent nature, the relation between precursor and product is also dependent on the precursor types as well as the time of reaction associated with the solvothermal reaction process. The particular precursor material used depends on the material to be synthesized. It is necessary for the precursors to be soluble in the solvent at an elevated temperature so that the reaction can continue until completion. Solvothermal synthesis is commonly performed using Teflon-lined, stainless-steel reactors, as shown in Figure 14.6. The pressures developed in solvothermal reactions necessitate the use of a sealed reactor. An inert teflon liner is typically used to protect the stainless-steel outer shell from corrosive reagents. This process is advantageous because it allows for the proper mixing of the reagents, incorporates stable kinetic phases, allows for control of particle size, and is a single-step process. The solvothermal approach is widely used for nanocrystals of metals, oxides, sulfides, tellurides, selenides, and nitrides.[52,59,78] Moreover, doping of the nanocrystals is readily achieved. Table 14.1 shows the precursor, solvent, reaction conditions, product, and grain size for typical syntheses of some TE materials. PbTe and Bi_2Te_3 nanocrystals of 5–100 nm can be prepared by this method in large yield.

TABLE 14.1 Solvothermal Synthesis Parameters of Some Chalcogenide TE Materials

Composition	Precursors	Solvent	T (°C)	Morphology	Dimension (nm)
PbTe	Lead nitrate, sodium tellurite, hydrazine hydrate	Water	200	Nanoparticles	$d = \sim 30$
Bi-doped PbTe	Lead nitrate, sodium tellurite, bismuth nitrate, hydrazine hydrate	Water	200	Nanoparticles	$d = \sim 30$
Bi_2Te_3	Bismuth nitrate, sodium tellurite, hydrazine hydrate	Ethylene glycol	230	Nanoparticles	$d = \sim 15$
Se-doped Bi_2Te_3	Bismuth nitrate, sodium tellurite, selenium di-oxide, hydrazine hydrate	Ethylene glycol	230	Nanoparticles	$d = \sim 15$
Sb_2Te_3	Antimony tri-chloride, sodium tellurite, hydrazine hydrate	Ethylene glycol	230	Nanoflakes	$t = \sim 5$, $l = 1–2$ μm
Se-doped Sb_2Te_3	Antimony tri-chloride, selenium di-oxide, sodium tellurite, hydrazine hydrate	Ethylene glycol	230	Nanoflakes	$t = \sim 5$, $l = 1–2$ μm
$Bi_{0.5}Sb_{1.5}Te_3$	Bismuth nitrate, antimony tri-chloride, sodium tellurite, hydrazine hydrate	Ethylene glycol	230	Nanoparticles	$d = \sim 18$
Se-doped $Bi_{0.5}Sb_{1.5}Te_3$	Bismuth nitrate, antimony tri-chloride, sodium tellurite, selenium di-oxide, hydrazine hydrate	Ethylene glycol	230	Nanoparticles	$d = \sim 18$

Note: In all the syntheses the reaction time was 12 h.

We have synthesized nanocrystals of TE alloys $Bi_{2-x}Sb_xTe_{3-y}Se$ in the 5–20 nm range by an ethylene glycol-mediated solvothermal process. Figure 14.7a through f are SEM images of the undoped and Se-doped Bi_2Te_3, Sb_2Te_3, and $Bi_{0.5}Sb_{1.5}Te_3$ nanocrystals. The inset images show the compositional analyses by EDS. Solvothermal processing of TE materials have proven to be an attractive tool in the synthesis of nanocrystalline TE materials.[76,79–86]

14.3.1.4 Polyol Process

The polyol method is a low-temperature process that is environmentally benign as compared with other solution-based synthesis approaches. Reactions carried out in polyalcohols, or polyols, having high boiling points, such as ethylene glycol, tend to yield more monodispersed products than the processes described above. The advantage of this process is the ability of polyols to effectively act as bidentate chelating agents for the solvated metal cations. The polyols also serve as reducing and stabilizing agents once the produced nanoparticles are precipitated. Fievet et al.[87] developed a polyol process for the synthesis of submicron-size metal particles. The process was modified by Cable et al.[88] by adding a strong reducing agent to produce size- and shape-controlled nanocrystals of a variety of metals, alloys, and intermetallics. Many colloidal metals and compounds are prepared in gram scale by the polyol process or modified polyol processes.[89–93] Polyol reduction processes are often modified by adding a stabilizing agent, such as poly(*N*-vinyl-2-pyrrolidone), which stabilizes the nanocrystals. So far, direct or modified polyol processes have been used for the synthesis of nanocrystalline TE compounds such as Bi,[94] Bi_2Te_3,[95] PbTe,[96,97] Sb_2S_3,[98] and $CoSb_3$,[88,99] making this process potentially useful for synthesizing a wide range of TE materials in large quantities.

14.3.1.5 Organometallic Synthesis

As described above, controlling size, phase purity, and obtaining nanocrystals in high yield are major requirements in synthesizing nanocrystals for TE applications. Among the various chemical synthesis routes for size controllability, perhaps the most effective is the organometallic synthesis route. As

FIGURE 14.7 SEM images of Bi_2Te_3 (a and b), Sb_2Te_3 nanoflakes (c and d), and $Bi_{0.5}Sb_{1.5}Te_3$ (e and f). Arrows in (c) indicate that two superimposed flakes are visible due to the very small thickness (~5 nm) of the nanoflakes. Corresponding EDS spectra of the doped specimen are also shown as insets. (Reprinted with permission from Datta, A. et al., Facile chemical synthesis of nanocrystalline thermoelectric alloys based on Bi-Sb-Te-Se, *Cryst. Growth Design*, 10, 3983, Copyright 2010, American Chemical Society.)

shown by LaMer and Dinegar,[100] the production of monodisperse colloids requires a sequentially discrete nucleation event followed by slower controlled growth on the existing nuclei. Rapid addition of reagents to the reaction vessel raises the precursor concentration above the nucleation threshold so that the nanocrystals begin to grow. The monodispersity of the produced nanocrystals is then dependent on the rate of the growth of the nanocrystals and the rate of consumption of the reagent. Since the growth rates of all the nanocrystals are the same, the initial size distribution is largely determined by the time in which the nuclei are formed and begin to grow. If the percentage of nanocrystal growth during the nucleation period is small compared with subsequent growth, the nanocrystals can become more uniform over time. In the case of compound semiconductor nanocrystals, the requisite supersaturation and subsequent nucleation can be triggered by rapid injection of metal–organic precursors into a vigorously stirred flask containing a hot (~150–350°C) coordinating solvent. Typically, the solvents are mixtures of long-chain alkylphosphines R_3P, alkylphosphine oxides R_3PO (R being butyl or octyl), or alkylamines. In the case of chalcogens, metal alkyls and organophosphine chalcogenides or bistrimethylsilylchalcogenides are often chosen as precursors. This process has been largely adopted in the preparation of precise size- and shape-selective chalcogenide nanocrystals.[101–103] As a result of the reduced probability of carrier trapping and nonradiative recombinations in these nanocrystals, high-luminescence quantum efficiencies were achieved.[101–103] However, for TE materials, this process has a few drawbacks. First, the organic ligands typically remain adhered to the nanocrystals and are difficult to remove. Upon densification, even a thin organic layer may have a detrimental effect on the electrical conductivity by forming an insulating layer. Second, yield is typically small for this process. Several batches of reactions are therefore necessary in order to prepare nanocrystals in quantities sufficient for bulk specimen formation. In recent years, modifications to organometallic colloidal synthesis have been employed in synthesizing nanocrystals of Bi_2Te_3–Sb_2Te_3-based alloys.[47,48,62,104–107] In these cases, however, a mixed-metal ion polymer-like composite or a mixed organic ligand composite is typically formed. Additional steps are thus needed in the process in order to remove the carbonaceous materials.[107] Subsequent annealing can cause growth of the nanocrystals even before densification. Provided approaches for the removal of organic ligands can be improved and scalability issues can be addressed, this approach will develop into an effective way of producing nanocrystals of TE materials.

14.3.2 Densification to Bulk Materials

Densification of TE materials is an important step in obtaining dense polycrystalline specimens. However, a major challenge with densifying nanocrystals or nanoscale powders is obtaining highly dense specimens while maintaining nanoscale domains or grains. Traditionally, densification involves enhancing the grain-to-grain contact by using static pressure and subsequent sintering of the pressed powders at elevated temperatures in order to obtain dense, bulk materials. Relatively long sintering times are often necessary to densify small grain powders that have a high surface-to-volume ratio. This typically promotes grain growth. Grain growth will also occur at elevated temperatures. Selecting a densification technique is therefore a key goal toward preparing dense bulk TE material with nanoscale domains. In general, there are three widely used densification techniques, CP and annealing, hot pressing (HP), and SPS.

14.3.2.1 Cold Pressing and Annealing

CP is used mainly for preforming of powders prior to annealing.[108] Compaction of the powder is carried out inside a die and punch assembly by applying uniaxial pressure. The press is usually mechanical or hydraulic. Annealing essentially increases the density of the compact by increasing the grain-to-grain contact. A major concern with this technique is cracking and distortion of the pellets. CPing nanocrystals alone usually does not allow for the formation of dense pellets. Thus, it is typically accompanied by annealing. This results in grain growth, as shown in Figure 14.8a. After CP at 10 MPa and annealing at 540°C for 24 h, grain growth is observed within the densified polycrystalline PbTe specimen. Although, high-density pellets may be obtained through CP and annealing for certain TE materials, an ideal compaction technique will require preservation of the nanoscale features while reducing porosity for all types of materials.

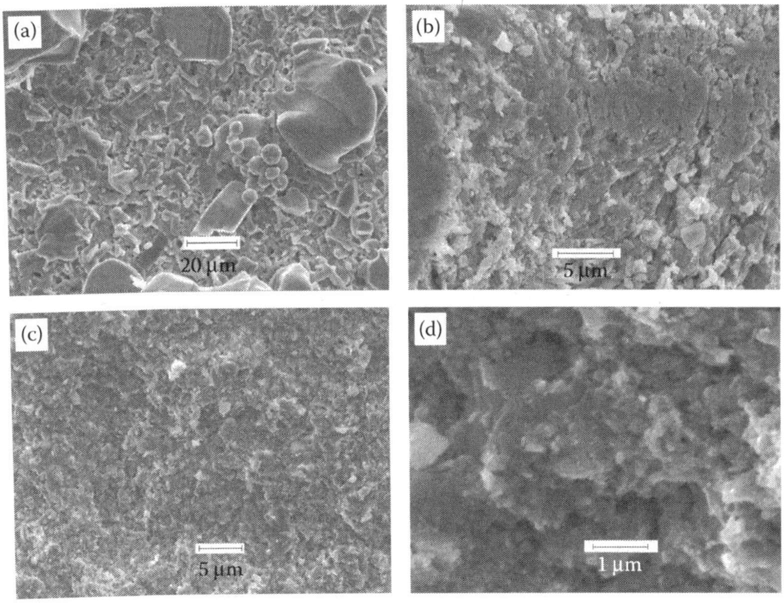

FIGURE 14.8 SEM images of PbTe nanocrystals densified by different techniques. (a) CP at 10 MPa and annealing at 540°C for 24 h. The control of grain growth by this process is typically minimal. (b) Nanocrystals densified by uniaxial hot pressing at 2000 kg and 540°C for 30 min. The plastic deformation of the grains is useful in obtaining dense pellets although some of the structural features are lost. (c) and (d) Fractured surfaces after densification by SPS at 60 MPa, 300°C for 10 min. The density obtained using SPS is more than 95%. Nanograins are preserved within the dense pellets.

14.3.2.2 Hot Pressing

HP is the application of heat and pressure simultaneously resulting in the densification of powders into a polycrystalline pellet. As in CP, a die and punch assembly (typically graphite or a refractory material) is used and heated resistively or inductively. This method can be used to form a simple shape (pellet or disc) or more elaborate shapes. HP is typically done at a temperature close to but below the melting point of the material to be densified. For many applications, this allows for the formation of dense polycrystalline pellets. For TE materials with nanoscale domains the HP technique is widely used.[5,11,43,49,75,101,109] However, there is always a chance of plastic deformation for relatively long sintering times under pressure. Care must be taken to reduce grain growth as shown in Figure 14.8b.

14.3.2.3 Spark Plasma Sintering

Although HP is widely used for compacting nanocrystalline powders prepared by ball milling, or mixtures of nanocrystals with bulk powders,[25,28,53,110] for many materials systems it may not be the most effective densification technique. SPS is a variant of field-assisted processing. In little more than a decade it has become an established consolidation method for fabrication of dense polycrystalline specimens.[111–113] The SPS process is similar to conventional HP; however, instead of heating the die and powder inductively or resistively, a direct electrical current, or a pulse of electric current, is applied to the graphite punch-and-die set which passes through the specimen so that it is heated both from the inside and from the outside. This process generates internal localized heating thus allowing for the rapid, low-temperature, densification of fine-grained powders. SPS minimizes the temperature gradients inside the specimen by guarantying electrical contacts between all parts. Since, only the surface temperature of the nanopowders rises rapidly by self-heating, grain growth is controlled. A sintered compact is densified in a much shorter time as compared to HP. The general speed of the process, therefore, avoids grain growth. Our approach in preparing dense TE materials with nanoscale domains by SPS[50–52] was initiated due to the inherent limitations associated with other densification techniques. Densifying bulk pellets from nanocrystals results in dimensional nanocomposites as shown in the SEM images in Figure 14.8c and d and modifies the transport properties of these TE materials as compared with that of the bulk.[6,10,11,30,50–52,114]

14.3.3 State-of-the-Art TE Materials as Nanocomposites

Several new potential TE materials have been reported over the past few years.[2,7,11] It is not sufficient to investigate a TE material solely on the basis of a high S, a low ρ, and a low κ. Mechanical properties, thermal cycling, bonding, and expense incurred in manufacturing should also be considered. However, the fundamental properties of nanostructured TE materials is a current interest in evaluating the potential of this technique for the optimization of materials' TE properties before research into mechanical properties optimization is explored. While a variety of TE material systems with nanoscale domains and their processing methods are now being investigated, we focused on binary state-of-the-art TE materials prepared by solution-based synthesis processes in order to explore the fundamental phenomena associated with the addition of these domains, as well as for proof-of-principle studies.[50–52,115] In this section, we outline recent developments in this field as well as our latest results.

14.3.3.1 Skutterudites

Compounds within the skutterudite crystal structure are among the most promising materials for TE applications, and are good examples of the phonon-glass electron-crystal concept proposed by Slack.[116] Atoms present in the voids of the skutterudite crystal structure create phonon-scattering centers, and therefore reduce κ without substantially affecting the electrical properties.[1,116–118] Nanograin polycrystalline materials also provide a way for achieving a lower κ in skutterudites, resulting in improved TE properties.[53,54,119,120] Chemical synthesis of skutterudites, adopted for the preparation of nanoscale $CoSb_3$ powders, and their filled variants, have also been employed because of the simplicity of processing and control of grain size.[53,88,120–123]

Polycrystalline $CoSb_3$ with nanoscale grains has been reported to reduce κ.[123,124] Stiewe et al.[123] reported improved TE properties of Ni and Te cosubstituted nanograined $CoSb_3$ with spherical nanoparticles of 10–20 nm in diameter. They used a chemical alloying route to precipitate $Co_{1-x}Ni_xSb_{3-y}Te_y$ nanoparticles from a stoichiometric solution of the components at a pH in the range of 1–3. After densification, the grain size was 150–200 nm as a result of grain growth during HP. They reported a ZT value of 0.65 at 680 K, a value that is much higher than that of bulk $CoSb_3$. In another approach, Mi et al.[120] synthesized nanocrystals of $CoSb_3$ by a solvothermal method using $CoCl_2$ and $SbCl_3$ as the precursors and $NaBH_4$ as the reductant in an ethanol solution. These $CoSb_3$ nanocrystals were densified by SPS together with micron-sized grains of partially filled $Yb_{0.15}Co_4Sb_{12}$, prepared by the direct reaction of the constituent elements. The composite made up of nanograins of $CoSb_3$ within bulk $Yb_{0.15}Co_4Sb_{12}$ showed a reduction in κ as a result of enhanced phonon scattering due to the nanograins. Additionally, S was reported to be enhanced due to potential charge carrier scattering at the grain boundaries. A ZT value of 0.95 was reported at 700 K. In a unique study, in order to further reduce κ in filled skutterudites, nanoscale skutterudite compounds were preferentially introduced at the grain boundaries of filled skutterudites.[53] In investigating the $La_{0.9}CoFe_3Sb_{12}$–$CoSb_3$ nanocomposite system, a hydrothermal nanoparticle-plating technique was employed in order to grow a layer of $CoSb_3$ nanoparticles on the surface of $La_{0.9}CoFe_3Sb_{12}$ micron-sized grains. In the reaction process, the precursors used to prepare $CoSb_3$ were mixed with an appropriate amount of $La_{0.9}CoFe_3Sb_{12}$, ethanol, and reductant, and were then heat treated solvothermally at 240°C for 80 h. The nanoparticles of $CoSb_3$, with size ranging from 30 to 40 nm, coated the filled skutterudite. This component was then densified by HP and the TE properties were measured. The $La_{0.9}CoFe_3Sb_{12}$–$CoSb_3$ nanocomposite was reported to display a systematic decrease in κ_L with increasing wt% of the nanoparticles. A 15% higher ZT than that of the bulk specimen was reported.

These examples illustrate the breadth of the ongoing research on skutterudites with nanoscale inclusions or domains.

14.3.3.2 Lead Chalcogenides

Lead chalcogenide compounds (PbE with E = S, Se, Te) comprise an entire class of TE materials. Among these, PbTe-based materials are among the most popular for potential TE applications. The TE properties of PbTe are determined by many factors, including doping, defects, and deviations from stoichiometry. These factors are dependent on the synthesis approach and conditions. From a cursory investigation of the PbTe phase diagram, PbTe-based compounds have a relatively wide range of homogeneous compositions extending from the Te-rich to the Pb-rich region.[125,126] PbTe-based bulk compounds have been in focus as potential TE materials for quite some time,[16,127–129] while PbTe-based TE materials with nanoscale inclusions or domains are one of the most well-studied systems in recent years.[6,10,50,51,115] The reported TE enhancement in two-dimensional *n*-type PbTe quantum well superlattice structures[130] resulted in a renewed and increased interest in this material system. Several important investigations on polycrystalline nanostructured PbTe have been made over the past few years. *S* enhancement in PbTe with grain sizes of the order of 30–50 nm was reported by Heremans et al.[131] In addition, polycrystalline PbTe containing nanometer-sized precipitates of Pb metal were reported to have higher *S* relative to bulk PbTe[30] resulting in an enhancement in the power factor for *p*-type PbTe.

We began a study of the TE properties of PbTe dimensional nanocomposites prepared by a bottom-up approach.[50,51] The PbTe nanocomposites were obtained by densifying solely PbTe nanocrystals by SPS. One to two grams of 100–150 nm PbTe nanocrystals were synthesized by an aqueous solution-phase direct precipitation process, as shown in the TEM image of Figure 14.9a. After SPS, the grain size distribution was in between 100 nm and 1 μm, with a large amount of the 100 nm grains within larger grains (Figure 14.9b). The densified PbTe specimens also had minor oxide impurity phases. In nanocrystalline materials, this chemisorption of oxygen may result in increased trapping of carriers at grain boundaries, forming energy barriers that impede the conduction of carriers between grains.[51,132] A large room temperature *S* value of 325 μV/K was observed for undoped *p*-type PbTe nanocomposites (Figure 14.9c). This was ~23% higher than that of the bulk with a similar carrier concentration. The ρ values, however, were higher in the nanocomposite due to grain boundary scattering of the charge carriers originating from the smaller average grain size (Figure 14.9c). In order to modify the carrier concentration of *p*-type PbTe nanocomposites, doping of the nanocrystals was done with Ag and Na prior to densification. Adjusting the carrier concentrations in the *p*-type nanocomposite specimens allowed for a decrease in the resistivity as compared to undoped PbTe nanocomposites. Furthermore, in *p*-type nanocomposites, it was observed that ρ increased below room temperature with a peak occurring below 100 K (Figure 14.9d). Two competing scattering mechanisms, long-wavelength acoustic phonon scattering and grain boundary scattering, are the reasons behind the temperature dependence observed in ρ.[7] The *S* versus *n* plot, the so-called Pisarenko relation[133] (Figure 14.9e), indicated an enhancement in *S* values for the nanocomposites compared to that of bulk PbTe with similar carrier concentrations. These larger *S* values result in an improved room temperature power factor of up to a factor of 2 (inset of Figure 14.9e). Temperature-dependent κ measurements (Figure 14.9f) showed similar values to that of bulk polycrystalline PbTe.[129,134,135] This $S^2\sigma$ enhancement, in addition to similar κ values, resulted in a room temperature *ZT* of up to a factor of 2 higher than that of bulk PbTe.

In continuing our previous work,[50,51] it was of interest to explore the TE properties of *n*-type PbTe. *n*-Type PbTe dimensional nanocomposites were prepared by doping the PbTe nanocrystals with Bi.[129,136] The nanocrystals were prepared in different sizes and with different concentrations of Bi by different synthesis processes. Microemulsion produced ~20–30 nm diameter nanoparticles (Figure 14.5) whereas direct precipitation resulted in crystals of different sizes by varying the pH of the reaction media (Figure 14.3). All the processes yielded PbTe nanocrystals in large yields. Different doping concentrations of Bi using $Bi(NO_3)_3.5H_2O$ as the precursor were introduced into PbTe. Bismuth-doped *n*-type PbTe nanocomposites were prepared by densifying the nanocrystals via SPS. The grain size after densification was

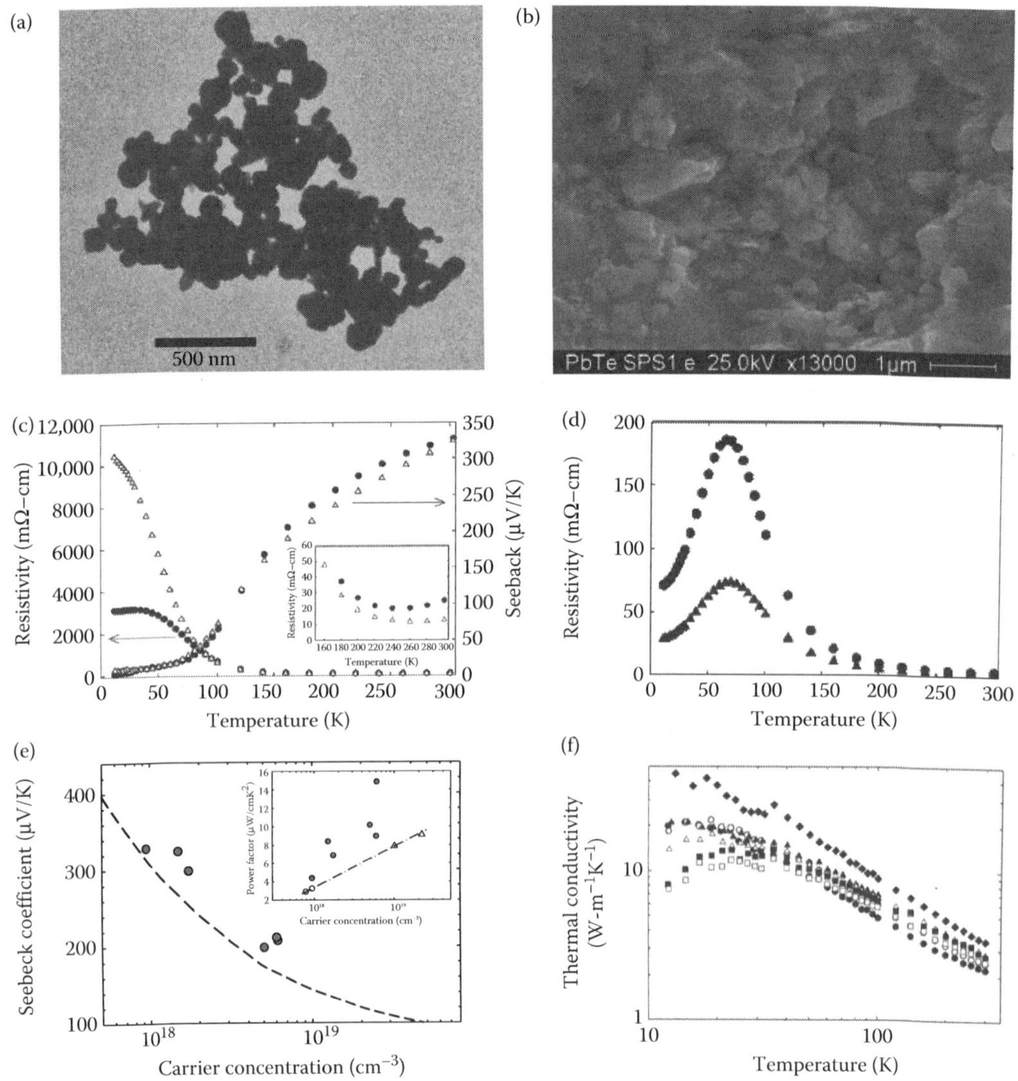

FIGURE 14.9 (a) TEM image of PbTe nanocrystals prepared by a direct precipitation process. (b) SEM image of a fractured surface of a representative PbTe nanocomposite specimen. (Reprinted with permission from Martin, J. et al., PbTe nanocomposites synthesized from PbTe nanocrystals, *Appl. Phys. Lett.*, 90, 222112, Copyright 2007, American Institute of Physics.) (c) Temperature-dependent ρ and *S* of undoped PbTe nanocomposites with different Pb-to-Te ratios. (Reprinted with permission from Martin, J. et al., PbTe nanocomposites synthesized from PbTe nanocrystals, *Appl. Phys. Lett.*, 90, 222112, Copyright 2007, American Institute of Physics.) (d) Temperature-dependent ρ of Ag-doped PbTe nanocomposites with carrier concentrations of 5.1×10^{18} cm^{-3} and 6.2×10^{18} cm^{-3}. (Reprinted with permission from Martin, J. et al., Enhanced Seebeck coefficient through energy-barrier scattering in PbTe nanocomposites, *Phys. Rev. B*, 79, 115311, Copyright 2009, American Physical Society.) (e) Pisarenko relation for PbTe nanocomposites indicate *S* values for the nanocomposites are enhanced as compared to bulk. The dotted line is taken from Crocker et al.[133] and represents the bulk line. The inset shows $S^2\sigma$ versus *n* for the nanocomposites. Δ and O are bulk PbTe data, and the dashed-dotted line is a guideline for the eye. (f) Temperature-dependent κ for the PbTe nanocomposites shown in (e). Room temperature κ for the nanocomposites are similar to that of the bulk.

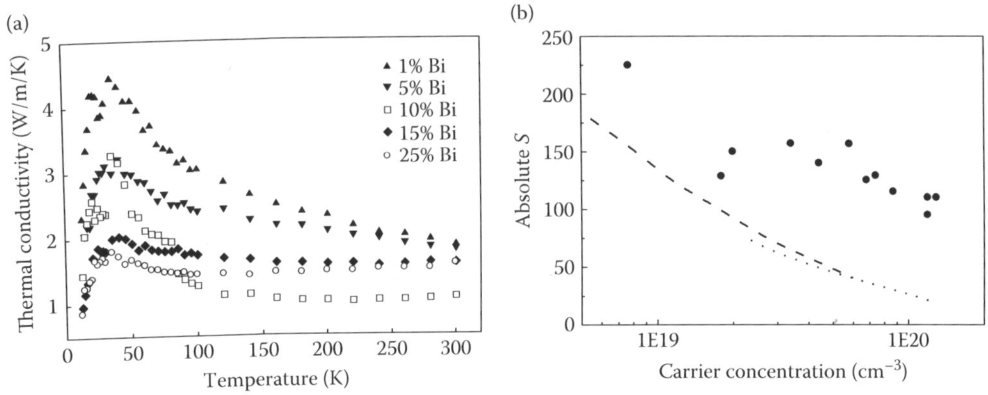

FIGURE 14.10 (a) Temperature-dependent κ for *n*-type Bi-doped PbTe nanocomposites with grain sizes in the 50–100 nm range. (b) The Pisarenko relation for *n*-type PbTe together with *n*-type PbTe nanocomposites are shown. The dotted and dashed lines are the Pisarenko relation from Refs. [144] and [145], respectively.

estimated to be ~30–50 nm, ~50–100 nm, and ~100–300 nm, depending on the starting crystal sizes. A higher Bi concentration resulted in an expected decrease in absolute *S*, $|S|$, and ρ. In addition, the peak in κ at ~40 K decreased with increasing Bi concentration (Figure 14.10a). Room-temperature Hall measurements confirmed that Bi doping increases the electron concentration by up to two orders of magnitude. The Pisarenko relation[137,138] together with the room temperature $|S|$ values for *n*-type PbTe specimens is shown in Figure 14.10b. *ZT* ~ 0.3 at room temperature was obtained for PbTe with the highest electron concentration, $n = 1.2 \times 10^{20}$ cm^{-3}. Higher carrier concentrations (higher Bi doping) and smaller grain sizes for κ reduction, as well as an investigation into the mechanical properties as a function of grain size, are potential future directions for this work.

14.3.3.3 Bi_2Te_3-Based Alloys

At present, Bi_2Te_3-based compounds are the most widely used materials in state-of-the-art TE devices. In spite of decades of research on these materials in devices, *ZT* ~1 at room temperature.[1] Bi_2Se_3 and Sb_2Te_3 have similar crystal structure and chemical properties with Bi_2Te_3. These materials therefore form isomorphous solid solutions.[139] Significant progress in enhancing *ZT* in Bi_2Te_3-based materials have been made by incorporating nanoscale grains in polycrystalline bulk materials.[10,11,114] Notable examples include the work on cross-plane transport in Bi_2Te_3/Sb_2Te_3 superlattices where phonon scattering perpendicular to the superlattice layers reduced κ more than σ.[140,141] A room temperature *ZT* of 2.4 was reported for these superlattice structures grown by chemical-vapor deposition.[114] The extraordinarily high *ZT* value in these systems originates from the low κ_L values (0.24 Wm^{-1}K^{-1} at 300 K) in the direction perpendicular to the superlattice layers. Improvements in the TE properties have been reported for nanograined bulk Bi_2Te_3-based alloys prepared by top-down approaches.[10,22,23,25,28,110,142,143] Bottom-up syntheses approaches have also been employed for the preparation of nanocrystals of Bi_2Te_3- and Sb_2Te_3-based TE alloys.[47,48,81,83,106] In order to synthesize these nanocrystals, organic capping molecules are typically used.[47,48,81,83,106] However, the use of organic molecules during the synthesis may result in the formation of resistive impurities in the densified bulk TE material, as discussed in Section 14.3.1.5. Nevertheless, a facile and scalable chemical process using nontoxic and less expensive chemicals would be desirable. So far, the synthesis of size-selective Bi_2Te_3-based alloys employing scalable processes has remained a challenge. Furthermore, high-density nanostructured Bi_2Te_3 based alloys are difficult to achieve due to the inherent structural anisotropy in these materials.[48,106] Dimensional nanocomposites prepared from chemically synthesized nanocrystals may provide a unique solution to these issues. A high yield and readily scalable glycol-based solvothermal synthesis approach was adopted to prepare

Bi_2Te_3 and Sb_2Te_3 nanocrystals in the 5–18 nm range.[52] The simplified process was also extended to modify the carrier concentration of Bi_2Te_3 and Sb_2Te_3 nanocrystals with Se doping. SEM and TEM images of the Bi_2Te_3 and Sb_2Te_3 nanocrystals are shown in Figure 14.7a through d and Figure 14.11a through d, respectively. The undoped and Se-doped Bi_2Te_3 nanocrystals where spherical and 20 ± 5 nm in diameter (Figure 14.11a and b), whereas undoped and Se-doped Sb_2Te_3 nanocrystals were typically flake-like having a thickness as low as 5 nm with lengths in the range of 100–1.5 μm (Figures 14.7c, 14.7d, and 14.11c). Densification by SPS resulted in bulk nanocomposites with densities of at least 95% of the theoretical values for both Bi_2Te_3 and Sb_2Te_3. The Bi_2Te_3 nanocomposites were *n*-type. The room temperature *n* value for the undoped Bi_2Te_3 was relatively high ($n = 4.8 \times 10^{19}$ cm^{-3}) and increases with increasing Se doping ($n = 6.6 \times 10^{19}$ cm^{-3} for 10% Se doping and $n = 8.8 \times 10^{19}$ cm^{-3} for 20% Se doping). The ρ and *S* values decrease with increasing *n*, as expected. The room temperature variation of $|S|$ with *n* showed an enhancement in $|S|$ for the doped polycrystalline bulk nanocomposites over theoretical predictions[127] and bulk polycrystalline specimens.[144,145] In the case of *p*-type Sb_2Te_3 nanocomposites, the ρ values were low and decreased with increased Se doping. The flake-like nanostructures of the Sb_2Te_3 nanocomposites may have some influence in lowering ρ in these specimens, but the exact role is not yet clearly understood. The κ values of Sb_2Te_3 nanocomposites were also found to be a factor of 2 lower than that reported for bulk polycrystalline specimens at 300 K.[146] A room-temperature $ZT \sim 0.17$ was

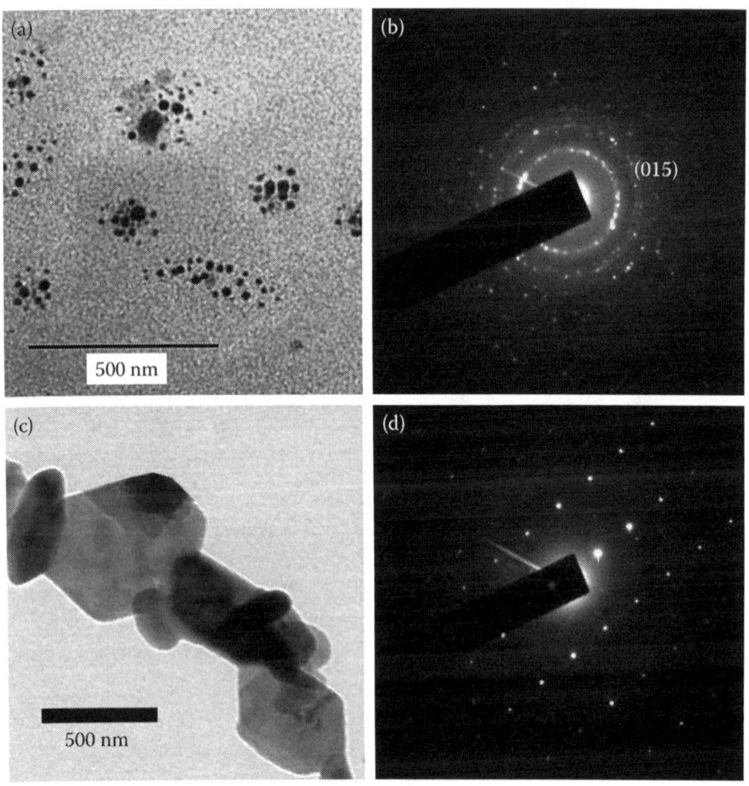

FIGURE 14.11 (a) Bi_2Te_3 nanocrystals in aggregate. (b) Polycrystalline selected area electron diffraction (SAED) pattern of the Bi_2Te_3 nanocrystals. The diffraction ring is indexed to the (015) plane of rhombohedral Bi_2Te_3. (Reprinted with permission from Datta, A. et al., Facile chemical synthesis of nanocrystalline thermoelectric alloys based on Bi-Sb-Te-Se, *Cryst. Growth Design*, 10, 3983, Copyright 2010, American Chemical Society.) (c) Bright field TEM image of Sb_2Te_3 nanoflakes. Superimposed flakes are very thin and devoid of structural defects. (d) Single crystalline SAED pattern of a Sb_2Te_3 nanoflake. The diffraction spots indexed to the corresponding planes indicates the rhombohedral structure.

obtained for 20% Se-doped Sb_2Te_3 nanocomposites. Further investigations are needed in order to optimize these TE materials; however, these data illustrate that the bottom-up synthesis approach is a promising method in terms of scalability, cost, and reproducibility of TE materials.

In addition to experimental efforts, theoretical investigations are also needed in order to explore different ways of improving the TE properties of various TE materials. First-principles calculations and analytical models are indispensable in the design of materials characteristics at their optimum regimes. A review including some of our recent results is given in the next section.

14.4 Theoretical Modeling of Transport in TE Composites

14.4.1 Boltzmann Equation Formulation of the Transport Properties

The discussion in this section is based on carrier transport obtained from the classical Boltzmann equation (BE). The BE relates the evolution of charge carriers in terms of their nonequilibrium distribution function under perturbations originating from the applied electric fields and temperature gradients. It is given by the expression[147]

$$\left(\frac{\partial f_l}{\partial t}\right)_{sc} = \frac{\partial f}{\partial t} + \boldsymbol{v}\cdot\nabla f - \frac{e\boldsymbol{\Omega}}{\hbar}\cdot\nabla_k f_l \tag{14.2}$$

where $\boldsymbol{\Omega}$ is a small electric field, $f_l(\mathbf{k},\mathbf{r},t)$ is the nonequilibrium distribution function for the carriers in energy band l with wave vector $\mathbf{k}$, and $\boldsymbol{v}$ is the associated group velocity. The standard approach is to solve the BE by requiring that $\partial f/\partial t = 0$ within the carrier relaxation time $\tau(E_k)$ ansatz with E_k, the wave vector-dependent carrier energy.[147] Carriers are considered to be in quasi-equilibrium so that $(\partial f/\partial t)_{sc} = -(f-f_0)/\tau(E_k)$. The solution to Equation 14.2 is then expressed in terms of the equilibrium Fermi–Dirac distribution function f_0.

The carrier transport is further characterized by the charge and thermal currents $\mathbf{J}$ and $\mathbf{J}_Q$, respectively. Within the linear response theory, they are given by the expressions

$$\begin{aligned}\mathbf{J} &= \sigma\boldsymbol{\Omega} + L\nabla T\\ \mathbf{J}_Q &= M\boldsymbol{\Omega} + N\nabla T\end{aligned} \tag{14.3}$$

where ∇T is the temperature gradient. σ, L, M, and N are coefficients found by inserting the relaxation time ansatz into Equation 14.2. These are usually expressed through their relations to σ, S, and κ_e of the material, and are given by

$$\sigma = e^2\int dE\left(-\frac{\partial f_0}{dE}\right)\Sigma(E) \tag{14.4}$$

$$S = -\sigma^{-1}L = \frac{e}{\sigma T}\int dE\left(-\frac{\partial f_0}{dE}\right)\Sigma(E)(E-\mu) \tag{14.5}$$

$$\kappa_e = M\sigma^{-1}L - N = \frac{1}{T}\int dE\left(-\frac{\partial f_0}{dE}\right)\Sigma(E)(E-\mu)^2 - T\sigma S^2 \tag{14.6}$$

where

$$\Sigma(E) = \int\frac{2d^3k}{(2\pi)^3}\boldsymbol{v}^2(k)\tau(k)\delta(E-E_k)$$

is the transport distribution function.

The BE approach is widely used for modeling TE properties because it provides satisfactory agreement with experimental results even when nanostructures are involved. The success of this method has been related to the fact that many of the effects of nanostructuring, such as confinement and tunneling, can be taken into consideration through the density of states (DOS), while those originating from scattering can be accounted for in the relaxation time parameters.

14.4.2 Bulk Electronic Structure Characteristics and Scattering Mechanisms

The important properties that enter into the TE transport coefficients, σ, S, κ_e, are related to the materials' electronic structure: the DOS and carriers group velocity. Although the electronic structure of any material is complex, the part located in the vicinity of the Fermi level is typically responsible for the transport. This gives the flexibility to use simplified models that nevertheless capture the important phenomena limiting the transport, while providing an ease of use for further studies.

As examples of this approach, we focus on the nanocomposite PbTe, $CoSb_3$, and $Si_{1-x}Ge_x$. These materials have attracted attention due to their enhanced TE performance as compared to that of the bulk.[26,50,115,131,148–150] This is partly due to their nonparabolic energy band structure and energy-dependent properties, such as band gaps and effective masses. The band structure for these materials can be described by the standard two-band Kane model[115]

$$E(1+\alpha E) = \frac{\hbar^2 k_l^2}{2m_l^*} + 2\times\frac{\hbar^2 k_t^2}{2m_t^*} \tag{14.7}$$

where k_l and k_t are the carrier momentum, m_l^* and m_t^* are the effective mass along the longitudinal and transverse directions at the band edge, respectively, and α is a nonparabolicity factor. For small band gap semiconductors, $\alpha = 1/E_g$, where E_g is the energy gap. For PbTe and $CoSb_3$ $m^* = \beta^{2/3}(m_l^* m_t^{*2})^{1/3}$, where β is the degeneracy of the Fermi surfaces containing more than one pocket.[150] For Si–Ge, this is not appropriate, thus m_l^* and m_t^* must be kept separately in Equation 14.7.[1,150]

In addition to the electronic structure properties, the relevant scattering mechanisms need to be identified for each bulk material. Within the relaxation time approximation, each mechanism is independent of the other and can be described by its own relaxation time. Using the Mathiessen's rule the total relaxation time is $\tau^{-1} = \sum_j \tau_j^{-1}$, where j sums over all contributing mechanisms. For bulk PbTe and $CoSb_3$, the most relevant relaxation mechanisms are due to electron-acoustic phonon, electron-optical phonon, and electron-charged impurity scattering mechanisms.[128,151] They can be described as $\tau_j(E) = a_j(T)E^{r_j}$, where $r_j = -1/2$, 1/2, and 3/2 for scattering by acoustic phonons, nonpolar optical phonons, and ionized impurities, respectively. The parameters a_j are temperature dependent and their expressions with relevant parameters are given in Ref. [152]. Since $Si_{1-x}Ge_x$ is not a polar material, the polar scattering mechanism is not available. Expressions for the relaxation times for $Si_{1-x}Ge_x$ due to acoustic phonon and charged impurity scatterings can be obtained from various sources.[150,152]

In order to have a full set of tools to model the TE transport properties in bulk materials, a description of κ_L is also needed. Reliable results can be obtained using the Holland–Callaway model. This approach gives an approximate expression for κ_L found from the phonon BE[153–155] within the relaxation time approximation and variational techniques:

$$\kappa_L = \frac{k_B}{2\pi^2 v_{ph}}\left(\frac{k_B T}{\hbar}\right)^3 \int_0^{\theta_D/T} \tau_{ph}(x)\frac{x^4 e^x}{(e^x-1)^2}dx \tag{14.8}$$

where v_{ph} is the average phonon group velocity, θ_D is the Debye temperature, and $x = \hbar\omega/k_B T$ is dimensionless with $\hbar$ being Planck's constant, k_B the Boltzmann constant, and ω the frequency of excitations.

Also, τ_{ph} is the total phonon relaxation time given by $\tau_{ph}^{-1} = \sum_i \tau_{ph,i}^{-1}$ within the Mathiessen's rule. Each $\tau_{ph,i}$ corresponds to a phonon-scattering mechanism. Relevant mechanisms determining $\tau_{ph,i}$ originate from umklapp processes, normal processes, spinodal decomposition, phonon–phonon processes, strains, precipitates, point defects, boundaries, and dislocations.[156] Here, we concentrate on the contribution from umklapp scattering, $\tau_{ph,U}^{-1} = BT\omega^2$, since this is the dominant mechanism at elevated temperatures.[128] B is a constant which can be determined from available experimental data for the specific material. The expressions for other scattering times for PbTe, $CoSb_3$, and $Si_{1-x}Ge_x$ are also available.[157]

14.4.3 Increasing the Power Factor and Figure of Merit beyond Bulk Values

14.4.3.1 Energy Filtering Scattering

There are different routes to increase the TE performance of materials. According to Equations 14.4 through 14.8, these involve modification of the total DOS, the carrier scattering time, and/or the phonon scattering time. Nanocomposites offer the possibility of changing the carrier and phonon properties. When nanoinclusions are added, the grain boundaries in granular composites and the boundaries in heterostructures introduce additional carrier and phonon-scattering mechanisms. This interface scattering can lead to a reduction in κ_L or enhancement of S. Proper design of the nanocomposite characteristics is then needed in order to ensure optimum enhancement of ZT.

To describe the carrier/interface scattering one can model each interface as a potential barrier. These can be described as a collection of planar barriers of average height, V_g, distance between them, L, and width, w, (Figure 14.12a) for granular composites, or as randomly distributed spherical formations of radius R and barrier height V_i for composites with nanoinclusions (Figure 14.12b). Using standard quantum mechanical considerations, the relevant carrier scattering times may be calculated. For example, granular interface scattering is described as $\tau_g = \lambda/v$, where the mean free path λ is found by assuming that the incident carriers are plane-wave-like.[115] The nanoinclusion scattering can be described as $\tau_i^{-1} = n_i v \sigma_t$, where n_i is the concentration of inclusions and σ_t is the total scattering cross section found using the Born approximation or the phase shift method.[158,159] τ_g and τ_i correspond to filtering of the low-energy carriers, while only the high-energy carriers contribute to the transport. Using perturbation techniques for the differences in the mass and bond stiffness between the host medium and the nanoinclusions, analytical expressions for the phonon/interface scattering time are also derived.[160]

From Equations 14.4 through 14.8, the transport characteristics are calculated in order to determine the optimum regime of operation in terms of the composite physical characteristics for a specific

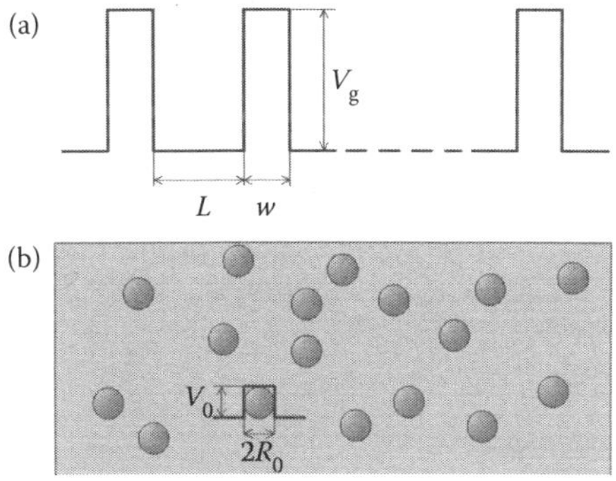

FIGURE 14.12 Schematic illustration of (a) grain regions limited by rectangular potential barriers where the height of the barrier is V_g, the width is w, and the grain size is L, and (b) spherical nanoinclusions embedded in a host bulk matrix. V_0 and R_0 represent the nanoinclusions potential height and radii, respectively.

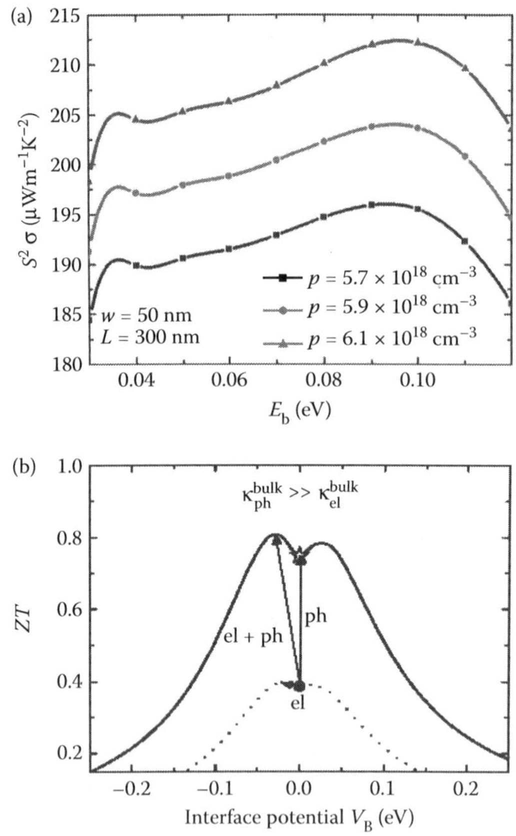

FIGURE 14.13 (a) Power factor as a function of the potential barrier height at room temperature for different carrier concentrations. (Reprinted with permission from Popescu, A. et al., Model of transport properties of thermoelectric nanocomposite materials, *Phys. Rev. B*, 79, 205302, Copyright 2009, American Physical Society.) (b) Calculated ZT for PbTe illustrating the relative effects of electronic and phonon scattering from nanoinclusions. (Reprinted with permission from Faleev, S.V. and Leonard, F., Theory of enhancement of thermoelectric properties of materials with nanoinclusions, *Phys. Rev. B*, 77, 214304, Copyright 2008, American Physical Society.)

material. It has been demonstrated that $S^2\sigma$ and ZT are maximum for certain values of V_g and L.[115,150,158] Using the infinite number barrier model,[115] $S^2\sigma$ for granular p-doped PbTe is seen to exhibit maximum values as a function of V_g indicating that the decrease in σ is not as significant as the increase in S (Figure 14.13a). Enhancement in ZT for similar values of V_g is also found for n-type PbTe with nanoinclusions[158] (Figure 14.13b). Similar results have been observed for $Si_{0.7}Ge_{0.3}$ nanocomposites[150] as indicated in Figure 14.14. The TE enhancement is attributed to the energy dependence of τ. Further analysis shows that a regime can be found in terms of the characteristics of the nanograins or nanoinclusions (size, distribution, and barrier heights) where S and κ are increased, but the σ is not significantly degraded.[115,150,158] This regime is also dependent on the properties of the material itself, such energy bands, effective masses, band gaps, and doping.

14.4.3.2 Total DOS Modifications

Another potential avenue for TE properties enhancement is associated with localized modifications in the total DOS of bulk materials. This idea was suggested by Mahan and Sofo,[161] where it was shown that if there is a δ function at the Fermi level, S increases leading to $ZT > 2$ as shown in Figure 14.15. The model used is a generic representation of the DOS for materials with parabolic energy bands:

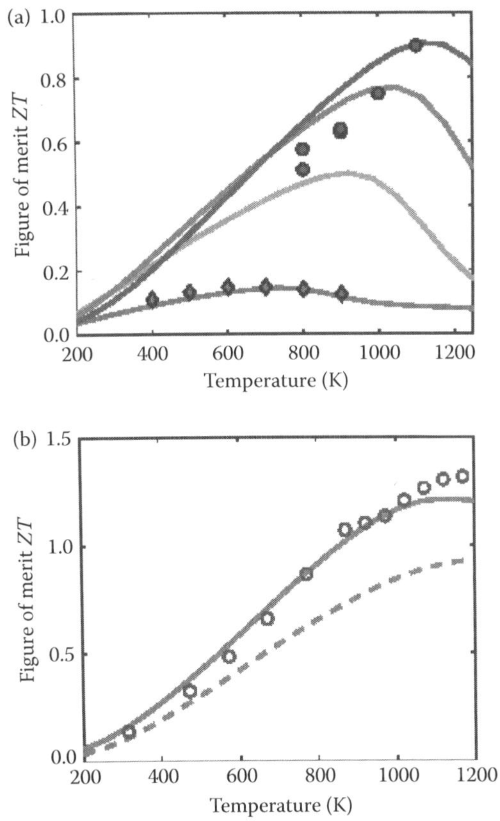

FIGURE 14.14 Experimental (symbols) and computed (solid lines) TE properties of bulk *n*-type $Si_{0.7}Ge_{0.3}$ (a) at different doping concentrations ($\times 10^{19}$ cm^{-3}): ◊ = 0.22; ○ = 17 and (b) for the nanocomposite with ○ = 0.22 (the dashed line is computed without considering grain boundary scattering). (Reprinted with permission from Minnich, A.J. et al., Modeling study of thermoelectric SiGe nanocomposites, *Phys. Rev. B*, 80, 155327, Copyright 2009, American Physical Society.)

$$g(E) = n_i\left[\delta(E - bk_BT) + \frac{1}{W}\right] \tag{14.9}$$

where n_i is the concentration of the energy levels, b indicates the position of the peak with respect to the Fermi level, and W is the bandwidth. Changing the position b and the background k_BT/W, one finds that it is most beneficial to have the δ peak at the Fermi level with less than 1% contribution from the background. Recently, this idea has experienced renewed interest due to studies demonstrating that resonant impurity levels in PbTe and ZnSe achieved by appropriate doping can be used for such an increase in S.[162–165] Experimental evidence has also been reported indicating that such locally distorted DOS is responsible for TE properties enhancement in PbTe at elevated temperatures.[166]

The success of these studies is based on the prediction by Hjalmarson et al.[167] that the incorporation of isoelectronic impurities in semiconductors leads to highly localized impurity levels. When the impurities are highly electronegative, their energy levels are located close to the conduction band edge. This has been shown for different classes of materials, such as GaAs with substitutional nitrogen,[168] various dopants in PbTe and PbSe,[163] and oxygen impurities in ZnSe.[164] Ahmad et al.[162] have presented a microscopic understanding of the fact that it is important to achieve a resonant impurity state which will contribute to the transport in order to obtain changes in the TE properties (Figure 14.16).

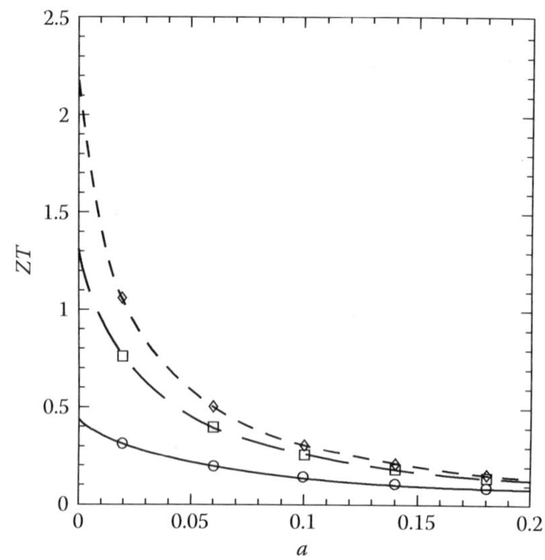

FIGURE 14.15 *ZT* calculated for a generic material with total DOS composed of a δ-function centered at point b and a background versus the intensity of the background a–$g(E) \sim \delta(E - b) + a$. (Reprinted from Mahan, G.D. and Sofo, J., The best thermoelectric, *Proc. Natl. Acad. Sci. USA*, 93, 7436, Copyright 1996, National Academy of Sciences.)

The outline presented above indicates that a locally modulated DOS is a promising way of achieving enhancement of the TE properties in bulk materials. This approach may be possible in a variety of TE materials. However, further investigations are needed in order to understand the role of the resonant states on the transport properties by taking into account a realistic electronic structure with temperature-dependent parameters. Efforts are also needed to search for appropriate dopants in different host materials, and experimentally investigate their TE properties.

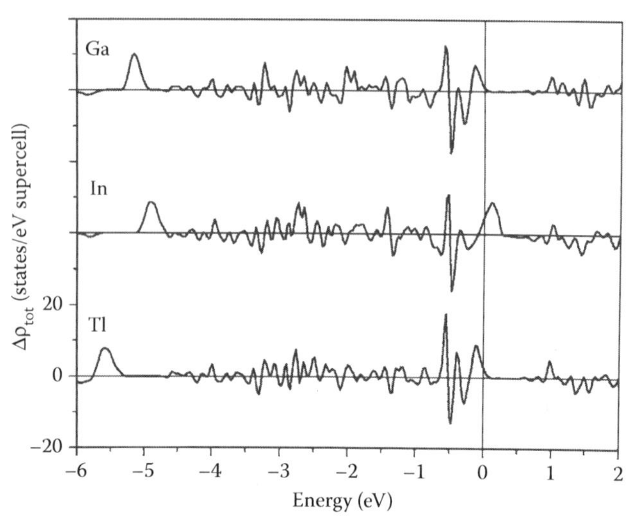

FIGURE 14.16 The difference in the total DOS between PbTe with and without trivalent group III impurities Ga, In, and Tl. (Reprinted with permission from Ahmad, S. et al., *Ab initio* studies of the electronic structure of defects in PbTe, *Phys. Rev. B*, 74, 155205, Copyright 2006, American Physical Society.)

14.5 Summary

In this chapter, we have provided an overview of different methods for processing bulk materials with nanoscale domains that result in enhanced TE properties. We described different approaches, many from the authors' recent work undertaken for the bottom-up approach, and outlined several examples of top-down approaches that have currently been reported. We also outlined our recent theoretical analyses in investigating the fundamental issues related to transport of bulk materials with nanoscale domains or inclusions. A combined approach that includes theory and experiment allows for a more complete investigation into the phenomena that directly affects the TE properties. It is clear that further development in materials research will lead to the understanding of new fundamental phenomena that can be used to develop into new materials with superior properties for TE applications. We believe that further work in the field will not only provide the potential for significant *ZT* enhancement but will also lead to an understanding of the fundamental physics that may have implications, and perhaps direct impact, on other technologically significant fields.

Acknowledgments

This work was supported by the U.S. Army Medical Research and Materiel Command under Grant No. W81XWH-07-1-0708 and the National Science Foundation under Grant Nos. CBET-0932526 and CMMI-0927637. We thank Professor Lidong Chen of Shanghai Institute of Ceramics for SPS processing and Dr. Joshua Martin of NIST for Hall measurements.

References

1. Nolas, G.S., Sharp, J.W., and Goldsmid, H.J., *Thermoelectrics: Basic Principles and New Materials Developments*, Springer-Verlag, Heidelberg, 2001.
2. Nolas, G.S., Poon J., and Kanatzidis, M.G., Recent developments in bulk thermoelectric materials, *Mater. Res. Soc. Bull.*, 31, 199, 2006.
3. Hicks, L.D. and Dresselhaus, M.S., Effect of quantum-well structures on the thermoelectric figure of merit, *Phys. Rev. B*, 47, 12727, 1993.
4. Yang, J., Nolas, G.S., Koumoto, K., and Grin, Y. (eds), *Materials and Technologies for Direct Thermal-to-Electrical Energy Conversion*, Mater. Res. Soc. Proc. Vol. 1166, Cambridge University Press, NY, 2009.
5. Kanatzidis, M.G., Nanostructured thermoelectrics: The new paradigm? *Chem. Mater.*, 22, 648, 2010.
6. Minnich, A.J., Dresselhaus, M.S., Ren, Z.F., and Chen, G., Bulk nanostructured thermoelectric materials: Current research and future prospects, *Energy Environ. Sci.*, 2, 466, 2009.
7. Snyder, G.J. and Toberer, E.S., Complex thermoelectric materials, *Nat. Mater.*, 7, 105, 2008.
8. Sharp, J.W., Poon, S.J., and Goldsmid, H.J., Boundary scattering of phonons and thermoelectric figure of merit, *Phys. Status Solidi A*, 187, 507, 2001.
9. Nolas, G.S. and Goldsmid, H.J., The figure of merit in amorphous thermoelectrics, *Physica B*, 194, 271, 2002.
10. Dresselhaus, M.S., Chen, G., Tang, M.Y., Yang, R., Lee, H., Wang, D., Ren, Z.F., Fleurial, J.-P., and Gogna, P., New directions for low-dimensional thermoelectric materials, *Adv. Mater.*, 19, 1043, 2007.
11. Sootsman, J.R., Chung, D.Y., and Kanatzidis, M.G., New and old concepts in thermoelectric materials, *Angew. Chem., Int. Ed.*, 48, 8616, 2009.
12. Heremans, J.P., Thrush, C.M., Morelli, D.T., and Wu, M., Thermoelectric power of bismuth nanocomposites, *Phys. Rev. Lett.*, 88, 216801, 2002.
13. Whitlow, L.W. and Hirano, T., Superlattice applications to thermoelectricity, *J. Appl. Phys.*, 78, 5460, 1995.

14. Moyzhes, B. and Nemchinsky, V., Thermoelectric figure of merit of metal-semiconductor barrier structure based on energy relaxation length, *Appl. Phys. Lett.*, 73, 1895, 1998.
15. Vashaee, D. and Shakouri, A., Improved thermoelectric power factor in metal-based superlattices, *Phys. Rev. Lett.*, 92, 106103, 2004.
16. Rowe, D.W. and Bhandari, C.M., *Modern Thermoelectrics*, Holt Saunders, London, 1983.
17. Disalvo, F.J., Thermoelectric cooling and power generation, *Science*, 285, 703, 1999.
18. Greer, A.L., Metallic glasses, *Science*, 267, 1947, 1995.
19. Chen, H., He, Y., Shiflet, G.J., and Poon, S.J., Deformation-induced nanocrystal formation in shear bands of amorphous alloys, *Nature (London)*, 367, 541, 1994.
20. Tkatch, V.I., Limanovskii, A.I., Denisenko, S.N., and Rassolov, S.G., The effect of the melt-spinning processing parameters on the rate of cooling, *Mater. Sci. Eng. A.*, 323, 91, 2002.
21. Tkatch, V.I., Denisenko, S.N., and Beloshov, O.N., Direct measurements of the cooling rates in the single roller rapid solidification technique, *Acta Mater.*, 45, 2821, 1997.
22. Tang, X.F., Xie, W.J., Li, H., Zhao, W.Y., Zhang, Q.J., and Niino, M., Preparation and thermoelectric transport properties of high-performance *p*-type Bi_2Te_3 with layered nanostructure, *Appl. Phys. Lett.* 90, 012102, 2007.
23. Xie, W.J., Tang, X.F., Yan, Y.G., Zhang, Q.J., and Tritt, T.M., Unique nanostructures and enhanced thermoelectric performance of melt-spun BiSbTe alloys, *Appl. Phys. Lett.*, 94, 102111, 2009.
24. Xie, W.J., Tang, X.F., Yan, Y.G., Zhang, Q.J., and Tritt, T.M., High thermoelectric performance BiSbTe alloy with unique low-dimensional structure, *J. Appl. Phys.*, 105, 113713, 2009.
25. Poudel, B., Hao, Q., Ma, Y., Lan, Y., Minnich, A., Yu, B., Yan, X. et al., High-thermoelectric performance of nanostructured bismuth antimony telluride bulk alloys, *Science*, 320, 634, 2008.
26. Joshi, G., Lee, H., Lan, Y., Wang, X., Zhu, G., Wang, D., Gould, R. W. et al., Enhanced thermoelectric figure-of-merit in nanostructured *p*-type silicon germanium bulk alloys, *Nano Lett.*, 8, 4670, 2008.
27. Wang, X.W., Lee, H., Lan, Y.C., Zhu, G.H., Joshi, G., Wang, D.Z., Yang, J. et al., Enhanced thermoelectric figure of merit in nanostructured *n*-type silicon germanium bulk alloy, *Appl. Phys. Lett.*, 93, 193121, 2008.
28. Ma, Y., Hao, Q., Poudel, B., Lan, Y., Yu, B., Wang, D., Chen, G., and Ren, Z.F., Enhanced thermoelectric figure-of-merit in *p*-type nanostructured bismuth antimony tellurium alloys made from elemental chunks, *Nano Lett.*, 8, 2580, 2008.
29. Sootsman, J.R., Pcionek, R.J., Kong, H., Uher, C., and Kanatzidis, M.G., Strong reduction of thermal conductivity in nanostructured PbTe prepared by matrix encapsulation, *Chem. Mater.*, 18, 4993, 2006.
30. Heremans, J.P., Thrush, C.M., and Morelli, D.T., Thermopower enhancement in PbTe with Pb precipitates, *J. Appl. Phys.*, 98, 063703, 2005.
31. Sootsman, J.R., Kong, H., Uher, C., D'Angelo, J.J., Wu, C.-I., Hogan, T.P., Caillat, T., and Kanatzidis, M.G., Large enhancements in the thermoelectric power factor of bulk PbTe at high temperature by synergistic nanostructuring, *Angew. Chem., Int. Ed.*, 47, 8618, 2008.
32. Androulakis, J., Lin, C.-H., Kong, H.-J., Uher, C., Wu, C.-I., Hogan, T., Cook, B.A., Caillat, T., Paraskevopoulos, K.M., and Kanatzidis, M.G., Spinodal decomposition and nucleation and growth as a means to bulk nanostructured thermoelectrics: Enhanced performance in $Pb_{1-x}Sn_xTe$-PbS, *J. Am. Chem. Soc.*, 129, 9780, 2007.
33. Cook, B.A., Kramer, M.J., Harringa, J.L., Han, M.K., Chung, D.Y., and Kanatzidis, M.G., Analysis of nanostructuring in high figure-of-merit $Ag_{1-x}Pb_mSbTe_{2+m}$ thermoelectric materials, *Adv. Funct. Mater.*, 19, 1254, 2009.
34. Bilc, D.I., Mahanti, S.D., and Kanatzidis, M.G., Electronic transport properties of PbTe and $AgPb_mSbTe_{2+m}$ systems, *Phys. Rev. B*, 74, 125202, 2006.
35. Hsu, K.F., Loo, S., Guo, F., Chen, W., Dyck, J.S., Uher, C., Hogan, T., Polychroniadis, E.K., and Kanatzidis, M.G., Cubic $AgPb_mSbTe_{2+m}$: Bulk thermoelectric materials with high figure of merit, *Science*, 303, 818, 2004.

36. Quarez, E., Hsu, K.F., Pcionek, R., Frangis, N., Polychroniadis, E., and Kanatzidis, M.G., Nanostructuring, compositional fluctuations, and atomic ordering in the thermoelectric materials $AgPb_mSbTe_{2+m}$. The myth of solid solutions, *J. Am. Chem. Soc.*, 127, 9177, 2005.
37. Wang, H., Li, J.-F., Nan, C.-W., Zhou, M., Liu, W., Zhang, B.-P., and Kita, T., High-performance $Ag_{0.8}Pb_{18+x}SbTe_{20}$ thermoelectric bulk materials fabricated by mechanical alloying and spark plasma sintering, *Appl. Phys. Lett.*, 88, 092104, 2006.
38. Zhou, M., Li, J.-F., and Kita, T., Nanostructured $AgPb_mSbTe_{m+2}$ system bulk materials with enhanced thermoelectric performance, *J. Am. Chem. Soc.*, 130, 4527, 2008.
39. Androulakis, J., Hsu, K.F., Pcionek, R., Kong, H., Uher, C., DAngelo, J.J., Downey, A., Hogan, T., and Kanatzidis, M.G., Nanostructuring and high thermoelectric efficiency in *p*-type $Ag(Pb_{1-y}Sn_y)_mSbTe_{2+m}$, *Adv. Mater.*, 18, 1170, 2006.
40. Poudeu, P.F.P., D'Angelo, J., Downey, A.D., Short, J.L., Hogan, T.P., and Kanatzidis, M.G., High thermoelectric figure of merit and nanostructuring in bulk *p*-type $Na_{1-x}Pb_mSb_yTe_{m+2}$, *Angew. Chem., Int. Ed.*, 45, 3835, 2006.
41. Gueguen, A., Poudeu, P.F.P., Li, C.P., Moses, S., Uher, C., He, J.Q., Dravid, V., Paraskevopoulos, K.A., and Kanatzidis, M.G., Thermoelectric properties and nanostructuring in the *p*-type materials $NaPb_{18-x}Sn_xMTe_{20}$ (M = Sb, Bi), *Chem. Mater.*, 21, 1683, 2009.
42. Poudeu, P.F.P., Guéguen, A., Wu, C.-I., Hogan, T., and Kanatzidis, M.G., High figure of merit in nanostructured *n*-type KPb_mSbTe_{m+2} thermoelectric materials, *Chem. Mater.*, 22, 1046, 2010.
43. Zhang, S. and He, J., Thermoelectric nanocomposites—A new paradigm, *J. South Carolina Acad. Sci.*, 6, 14, 2008.
44. Ji, X., Tritt, T.M., Zhao, X., and Kolis, J.W., Solution chemical synthesis of nanostructured thermoelectric materials, *J. South Carolina Acad. Sci.*, 6, 1, 2008.
45. Zhou, W., Zhu, J., Li, D., Hng, H.H., Boey, F.Y.C., Ma, J., Zhang, H., and Yan, Q., Binary-phased nanoparticles for enhanced thermoelectric properties, *Adv. Mater.*, 21, 3196, 2009.
46. Purkayastha, A., Kim, S., Gandhi, D.D., Ganesan, P.G., Borca-Tasciuc, T., and Ramanath, G., Molecularly protected bismuth telluride nanoparticles: microemulsion synthesis and thermoelectric transport properties, *Adv. Mater.*, 18, 2958, 2006.
47. Scheele, M., Oeschler, N., Meier, K., Kornowski, A., Klinke, C., and Weller, H., Synthesis and thermoelectric characterization of Bi_2Te_3 nanoparticles, *Adv. Funct. Mater.*, 19, 3476, 2009.
48. Zhao, Y., Dyck, J.S., Hernandez, B.M., and Burda, C., Enhancing thermoelectric performance of ternary nanocrystals through adjusting carrier concentration, *J. Am. Chem. Soc.* 132, 4982, 2010.
49. Zhang, G., Yu, Q., and Li, X., Wet chemical synthesis and thermoelectric properties of V–VI one- and two-dimensional nanostructures, *Dalton Trans.*, 39, 993, 2010.
50. Martin, J., Nolas, G.S., Zhang, W., and Chen, L., PbTe nanocomposites synthesized from PbTe nanocrystals, *Appl. Phys. Lett.*, 90, 222112, 2007.
51. Martin, J., Wang, L., Chen, L., and Nolas, G.S., Enhanced Seebeck coefficient through energy-barrier scattering in PbTe nanocomposites, *Phys. Rev. B*, 79, 115311, 2009.
52. Datta, A., Paul, J., Kar, A., Patra, A., Sun, Z., Chen, L., Martin, J., and Nolas, G.S., Facile chemical synthesis of nanocrystalline thermoelectric alloys based on Bi-Sb-Te-Se, *Cryst. Growth Design*, 10, 3983, 2010.
53. Gothard, N., Ji, X., He, J., and Tritt, T.M., Thermoelectric and transport properties of *n*-type Bi_2Te_3 nanocomposites, *J. Appl. Phys.*, 103, 054314, 2008.
54. Alboni, P.N., Ji, X., He, J., Gothard, N., and Tritt, T.M., Thermoelectric properties of $La_{0.9}CoFe_3Sb_{12}$–$CoSb_3$ skutterudite nanocomposites, *J. Appl. Phys.*, 103, 113707, 2008.
55. Mi, J.L., Zhao, X.B., Zhu, T.J., and Tu, J.P., Improved thermoelectric figure of merit in *n*-type $CoSb_3$ based nanocomposites, *Appl. Phys. Lett.*, 91, 172116, 2007.
56. Nolas, G.S., Bulk dimensional nanocomposites for thermoelectrics applications, Patent pending, Application no. 12417326, August 16, 2010.
57. Cushing, B.L., Kolesnichenko, V.L., and O'Connor, C.J., Recent advances in the liquid-phase syntheses of inorganic nanoparticles, *Chem. Rev.*, 104, 3893, 2004.

58. Ring, T.A., *Fundamentals of Ceramic Powder Processing and Synthesis*, Academic Press: San Diego, CA, 1996.
59. Nielsen, A.E., *Kinetics of Precipitation*, Pergamon Press: Oxford, 1964.
60. Burda, C., Chen, X., Narayanan, R., and El-Sayed, M.A., Chemistry and properties of nanocrystals of different shapes, *Chem. Rev.*, 105, 1025, 2005.
61. Zhang, W., Zhang, L., Cheng, Y., Hui, Z., Zhang, X., Xie, Y., and Qian, Y., Synthesis of nanocrystalline lead chalcogenides PbE (E = S, Se, or Te) from alkaline aqueous solutions, *Mater. Res. Bull.*, 35, 2009, 2000.
62. Lu, W., Ding, Y., Chen, Y., Wang, Z., Fang, J., Bismuth telluride hexagonal nanoplatelets and their two-step epitaxial growth, *J. Am. Chem. Soc.*, 127, 10112, 2005.
63. Pileni, M.P., Nanosized particles made in colloidal assemblies, *Langmuir*, 13, 3266, 1997.
64. Petit, C. and Pileni, M.P., Synthesis of cadmium sulfide *in situ* in reverse micelles and in hydrocarbon gels, *J. Phys. Chem.*, 92, 2282, 1988.
65. Pileni, M.P., Reverse micelles as microreactors, *J. Phys. Chem.*, 97, 6961, 1993.
66. Pileni, M.P., Nanocrystal self-assemblies: Fabrication and collective properties, *J. Phys. Chem.*, 100, 1867, 1996.
67. Pileni, M.P., The role of soft colloidal templates in controlling the size and shape of inorganic nanocrystals, *Nat. Mater.*, 2, 145, 2003.
68. Pileni, M.P., II–VI semiconductors made by soft chemistry: Syntheses and properties, *Catal. Today*, 58, 151, 2000.
69. Viau, G., Brayner, R., Poul, L., Chakroune, N., Lacaze, E., Fievet-Vincent, F., Fievet, F., Ruthenium nanoparticles: Size, shape, and self-assemblies, *Chem. Mater.*, 15, 486, 2003.
70. Wu, S.-H. and Chen, D.-H., Synthesis and characterization of nickel nanoparticles by hydrazine reduction in ethylene glycol, *J. Colloid Interface Sci.*, 259, 282, 2003.
71. Hou, Y. and Gao, S., Monodisperse nickel nanoparticles prepared from a monosurfactant system and their magnetic properties, *J. Mater. Chem.*, 13, 1510, 2003.
72. Cotton, F.A. and Wilkinson, G., *Advanced Inorganic Chemistry*, 5th edition; Wiley: New York, 1988.
73. Schwuger, M.-J. and Stickdornt, K., Microemulsions in technical processes, *Chem. Rev.*, 95, 849, 1995.
74. Vafaei, S., Purkayastha, A., Jain, A., Ramanath, G., and Borca-Tasciuc, T., The effect of nanoparticles on the liquid–gas surface tension of Bi_2Te_3 nanofluids, *Nanotechnology*, 20, 185702, 2009.
75. Foos, E.E., Stroud, R.M., and Berry, A.D., Synthesis and characterization of nanocrystalline bismuth telluride, *Nano Lett.*, 1, 693, 2001.
76. Zou, G., Liu, Z., Wang, D., Jiang, C., and Qian, Y., Selected-control solvothermal synthesis of nanoscale hollow spheres and single-crystal tubes of PbTe, *Eur. J. Inorg. Chem.*, 22, 4521, 2004.
77. Rabeneau, A., The role of hydrothermal synthesis in preparative chemistry, *Angew. Chem., Int. Ed.*, 24, 1026, 1985.
78. Walton, R.I., Subcritical solvothermal synthesis of condensed inorganic materials, *Chem. Soc. Rev.*, 31, 230, 2002.
79. Shi, S., Cao, M., and Hu, C., Controlled solvothermal synthesis and structural characterization of antimony telluride nanoforks, *Cryst. Growth Design*, 9, 2057, 2009.
80. Shi, W., Yu, J., Wang, H., and Zhang, H., Hydrothermal synthesis of single-crystalline antimony telluride nanobelts, *J. Am. Chem. Soc.*, 128, 16490, 2006.
81. Shi, W., Zhou, L., Song, S., Yang, J., and Zhang, H., Hydrothermal synthesis and thermoelectric transport properties of impurity-free antimony telluride hexagonal nanoplates, *Adv. Mater.*, 20, 1892, 2008.
82. Wang, W., Poudel, B., Yang, J., Wang, D.Z., and Ren, Z.F., High-yield synthesis of single-crystalline antimony telluride hexagonal nanoplates using a solvothermal approach, *J. Am. Chem. Soc.*, 127, 13792, 2005.

83. Wang, W., Yan, X., Poudel, B., Ma, Y., Hao, Q., Yang, J., Chen, G., and Ren, Z., Chemical synthesis of anisotropic nanocrystalline Sb_2Te_3 and low thermal conductivity of the compacted dense bulk, *J. Nanosci. Nanotechnol.*, 8, 452, 2008.
84. Zhang, G., Wang, W., Lu, X., and Li, X., Solvothermal synthesis of V–VI binary and ternary hexagonal platelets: The oriented attachment mechanism, *Cryst. Growth Design*, 9, 145, 2009.
85. Poudel, B., Wang, W.Z., Wang, D.Z., Huang, J.Y., and Ren, Z.F., Shape evolution of lead telluride and selenide nanostructures under different hydrothermal synthesis conditions, *J. Nanosci. Nanotechnol.*, 6, 1050, 2006.
86. Wang, W.Z., Poudel, B., Wang, D., and Ren, Z.F., Synthesis of PbTe nanoboxes using a solvothermal technique, *Adv. Mater.*, 17, 2110, 2005.
87. Fievet, F., Lagier, J.P., Blin, B., Beaudoin, B., and Figlarz, M., Homogeneous and heterogeneous nucleations in the polyol process for the preparation of micron and submicron size metal particles, *Solid State Ion.*, 32, 198, 1989.
88. Cable, R.E. and Schaak, R.E., Low-temperature solution synthesis of nanocrystalline binary intermetallic compounds using the polyol process, *Chem. Mater.*, 17, 6835, 2005.
89. Zhao, Y., Zhu, J.J., Hong, J.M., Bian, N., and Chen, H.Y., Microwave-induced polyol-process synthesis of copper and copper oxide nanocrystals with controllable morphology, *Eur. J. Inorg. Chem.*, 20, 4072, 2004.
90. Komarneni, S., Katsuki, H., Li, D., and Bhalla, A.S., Microwave–polyol process for metal nanophases, *J. Phys., Condens. Matter*, 16, S1305, 2004.
91. Silvert, P.Y., Herrera-Urbinab, R., and Tekaia-Elhsissen, K., Preparation of colloidal silver dispersions by the polyolprocess, *J. Mater. Chem.*, 7, 293, 1997.
92. Wan, J., Cai, W., Meng, X., and Liu, E., Monodisperse water-soluble magnetite nanoparticles prepared by polyol process for high-performance magnetic resonance imaging, *Chem. Commun.*, 5004, 2007.
93. Peng, Z., Jiang, Y., Song, Y., Wang, C., and Zhang, H., Morphology control of nanoscale PbS particles in a polyol process, *Chem. Mater.*, 20, 3153, 2008.
94. Wang, Y. and Xia, Y., Bottom-up and top-down approaches to the synthesis of monodispersed spherical colloids of low melting-point metals, *Nano Lett.*, 4, 2047, 2004.
95. Zhou, B., Zhao, Y., Pu, L., and Zhu, J.J., Microwave-assisted synthesis of nanocrystalline Bi_2Te_3, *Mater. Chem. Phys.*, 96, 192, 2006.
96. Yan, Q., Chen, H., Zhou, W., Hng, H.H., Boey, F.Y.C., and Ma, J., A simple chemical approach for PbTe nanowires with enhanced thermoelectric properties, *Chem. Mater.*, 20, 6298, 2008.
97. Kerner, R., Palchik, O., and Gedanken, A., Sonochemical and microwave-assisted preparations of PbTe and PbSe. A comparative study, *Chem. Mater.*, 13, 1413, 2001.
98. Zhang, R., Chen, X., Mo, M., Wang, Z., Zhang, M., Liu, X., and Qian, Y., Morphology-controlled growth of crystalline antimony sulfide via a refluxing polyol process, *J. Cryst. Growth*, 262, 449, 2004.
99. Yang, L., Hng, H.H., Cheng, H., Sun, T., and Ma, J., Synthesis of $CoSb_3$ by a modified polyol process, *Mater. Lett.*, 62, 2483, 2008.
100. LaMer, V.K. and Dinegar, R.H., Theory, production and mechanism of formation of monodispersed hydrosols, *J. Am. Chem. Soc.*, 72, 4847, 1950.
101. Talapin, D.V., Lee, J., Kovalenko, M.V., and Shevchenko, E.V., Prospects of colloidal nanocrystals for electronic and optoelectronic applications, *Chem. Rev.*, 110, 389, 2010.
102. Cozzoli, P.D., Pellegrino, T., and Manna, L., Synthesis, properties and perspectives of hybrid nanocrystal structures, *Chem. Soc. Rev.*, 35, 1195, 2006.
103. Buhro, W.E. and Colvin, V.L., Semiconductor nanocrystals: Shape matters, *Nat. Mater.*, 2, 138, 2008.
104. Ritter, J.J. and Maruthamuthu, P., Synthesis of polycrystalline bismuth telluride by a metal-organo complex method, *Inorg. Chem.*, 34, 4278, 1995.105.
105. Garje, S.S., Eisler, D.J., Ritch, J.S., Afzaal, M., O'Brien, P., and Chivers, T., A new route to antimony telluride nanoplates from a single-source precursor, *J. Am. Chem. Soc.*, 128, 3120, 2006.

106. Dirmyer, M.R., Martin, J., Nolas, G.S., Sen, A., and Badding, J.V., Thermal and electrical conductivity of size-tuned bismuth telluride nanoparticles, *Small*, 5, 933, 2009.
107. Zhao, Y., Dyck, J.S., Hernandez, B.M., and Burda, C., Improving thermoelectric properties of chemically synthesized Bi_2Te_3-based nanocrystals by annealing, *J. Phys. Chem. C*, 114, 11607, 2010.
108. Scoville, A.N., Powder metallurgy techniques, in *CRC Thermoelectrics Handbook*, Rowe, D.M. (ed.), CRC Press, Boca Raton, Florida, 101, 1995.
109. Zhao, Y. and Burda, C., Chemical synthesis of $Bi_{0.5}Sb_{1.5}Te_3$ nanocrystals and their surface oxidation properties, *ACS Appl. Mater. Interfaces*, 1, 1259, 2009.
110. Lan, Y., Poudel, B., Ma, Y., Wang, D., Dresselhaus, M.S., Chen, G., and Ren, Z., Structure study of bulk nanograined thermoelectric bismuth antimony telluride, *Nano Lett.*, 9, 1419, 2009.
111. Tokita, M., Trends in advanced SPS spark plasma sintering systems and technology, *J. Soc. Powder Tech. Jpn.*, 30, 790, 1993.
112. Omori, M., Sintering, consolidation, reaction and crystal growth by the spark plasma system (SPS), *Mater. Sci. Eng. A*, 287, 183, 2000.
113. Schmidt, J., Boehling, M., Burkhardt, U., and Grin, Y., Preparation of titanium diboride TiB_2 by spark plasma sintering at slow heating rate, *Sci. Technol. Adv. Mater.*, 8, 376, 2007.
114. Venkatasubramanian, R., Siivola, E., Colpitts, T., O'Quinn, B., Thin-film thermoelectric devices with high room-temperature figures of merit, *Nature*, 413, 597, 2001.
115. Popescu, A., Woods, L.M., Martin, J., and Nolas, G.S., Model of transport properties of thermoelectric nanocomposite materials, *Phys. Rev. B*, 79, 205302, 2009.
116. Slack, G.A., New materials and performance limits for thermoelectric cooling, in *CRC Thermoelectrics Handbook*, Rowe, D.M. (ed.), CRC Press, Boca Raton, Florida, p. 407, 1995.
117. Uher, C., Skutterudites: Prospective novel thermoelectrics, in *Semiconductors and Semimetals*, Vol. 69, Tritt, T.M. (ed.), Academic Press, New York, pp. 139–253, 2001.
118. Nolas, G.S., Morelli, D.T., and Tritt, T.M., Skutterudites: A phonon-glass-electron crystal approach to advanced thermoelectric energy conversion applications, *Annu. Rev. Mater. Sci.*, 29, 89, 1999.
119. Toprak, M.S., Stiewe, C., Platzek, D., Williams, S., Bertini, L., Muller, E., Gatti, C., Zhang, Y., Rowe, M., and Muhammed, M., The impact of nanostructuring on the thermal conductivity of thermoelectric $CoSb_3$, *Adv. Funct. Mater.*, 14, 1189, 2004.
120. Mi, J.L., Zhao, X.B., Zhu, T.J., and Tu, J.P., Thermoelectric properties of $Yb_{0.15}Co_4Sb_{12}$ based nanocomposites with $CoSb_3$ nano-inclusion, *J. Phys. D: Appl. Phys.*, 41, 205403, 2008.
121. Mi, J.L., Zhao, X.B., Zhu, T.J., Tu, J.P., and Cao, G.S., Solvothermal synthesis and electrical transport properties of skutterudite $CoSb_3$, *J. Alloys Compd.*, 417, 269, 2006.
122. Mi, J.L., Zhu, T.J., Zhao, X.B., and Ma, J., Nanostructuring and thermoelectric properties of bulk skutterudite compound $CoSb_3$, *J. Appl. Phys.*, 101, 054314, 2007.
123. Stiewe, C., Bertini, L., Toprak, M., Christensen, M., Platzek, D., Williams, S., Gatti, C., Müller, E., Iversen, B. B., Muhammed, M., and Rowe, M., Nanostructured $Co_{1-x}Ni_x(Sb_{1-y}Te_y)_3$ skutterudites: Theoretical modeling, synthesis and thermoelectric properties, *J. Appl. Phys.*, 97, 044317, 2005.
124. Yang, J., Zhang, W., Bai, S.Q., Mei, Z., and Chen, L.D., Dual-frequency resonant phonon scattering in $Ba_xR_yCo_4Sb_{12}$ (R = La, Ce, and Sr), *Appl. Phys. Lett.*, 90, 192111, 2007.
125. Brebrick, R.F., Non-stoichiometry in binary semiconductor compounds, in *Progress in Solid State Chemistry*, Reiss, H. (ed.), Vol. 3, Pergamon Press, New York, pp. 213–264, 1966.
126. Gomez, M.P., Stevenson, D.A., and Huggins, R.A., Self-diffusion of Pb and Te in lead telluride, *J. Phys. Chem.*, 32, 335, 1971.
127. Ioffe, A.F., *Physics of Semiconductors*, Academic Press, New York, 1960.
128. Ravich, Y.I., Efimova, E.A., and Smirnov, I.A., *Semiconducting Lead Chalcogenides*, Plenum Press, New York, 1970.
129. Fano, V., Lead telluride and its alloys, in *CRC Thermoelectrics Handbook*, Rowe, D.M., CRC Press, Boca Raton, Florida, p. 257, 1995.

130. Hicks, L.D., Harman, T.C., Sun, X., and Dresselhaus, M.S., Experimental study of the effect of quantum-well structures on the thermoelectric figure of merit, *Phys. Rev. B*, 53, R10493, 1996.
131. Heremans, J.P., Thrush, C.M., and Morelli, D.T., Thermopower enhancement in lead telluride nanostructures, *Phys. Rev. B*, 70, 115334, 2004.
132. Zyubina, T.S., Neudachina, V.S., Yashina, L.V., and Shtanov, V.I., XPS and *ab initio* study of the interaction of PbTe with molecular oxygen, *Surf. Sci.*, 574, 52, 2005.
133. Crocker, A.J. and Rogers, L.M., Interpretation of the Hall coefficient, electrical resistivity and Seebeck coefficient of *p*-type lead telluride, *Br. J. Appl. Phys.*, 18, 563, 1967.
134. Scanlon, W.W., Polar semiconductors, in *Solid State Physics*, Seitz, F. and Turnbull, D. (eds), Academic Press, New York, 9, pp. 83–137, 1959.
135. Dughaish, Z.H., Lead telluride as a thermoelectric material for thermoelectric power generation, *Physica B*, 322, 205, 2002.
136. Borisova, L.D., Thermoelectric properties of impurity doped PbTe, *Phys. Status Solidi A*, 53, K19, 1979.
137. Ahn, K., Li, C., Uher, C., and Kanatzidis, M.G., Improvement in the thermoelectric figure of merit by La/Ag cosubstitution in PbTe, *Chem. Mater.*, 21, 1361, 2009.
138. Kishimoto, K. and Koyanagi, T., Preparation of sintered degenerate *n*-type PbTe with a small grain size and its thermoelectric properties, *J. Appl. Phys.*, 92, 2544, 2002.
139. Scherrer, H. and Scherrer, S., Thermoelectric properties of bismuth antimony telluride solid solutions, in *CRC Thermoelectrics Handbook*, Rowe, D.M. (ed.), CRC Press, Boca Raton, Florida, pp. 27-1–27-16, 2006.
140. Venkatasubramanian, R., Phonon blocking electron transmitting superlattice structures as advanced thin film thermoelectric materials, in *Semiconductors and Semimetals: Recent Trends in Thermoelectric Materials Research III*, Tritt, T.M. (ed.), Vol. 71, Academic Press, San Diego, CA, Chapter 4, pp. 175–201, 2001.
141. Yang, B. and Chen, G., Phonon heat conduction in superlattices, in *Chemistry, Physics, and Materials Science for Thermoelectric Materials: Beyond Bismuth Telluride*, Eds. Kanatzidis, M.G., Hogan, T.P., and Mahanti, S.D. (eds), Kluwer Academic/Plenum Publishers, New York, pp. 147–167, 2003.
142. Zhao, X.B., Ji, X.H., Zhang, Y.H., Zhu, T.J., Tu, J.P., and Zhang, X.B., Bismuth telluride nanotubes and the effects on the thermoelectric properties of nanotube-containing nanocomposites, *Appl. Phys. Lett.*, 86, 062111, 2005.
143. Xiao, X.B., Zhu, T.J., and Ji, X.H., in *Thermoelectrics Handbook: Macro to Nano*, Rowe, D.M. (ed.), CRC, Boca Raton, Florida, 25, 2006.
144. Kim, D.H. and Mitani, T., Thermoelectric properties of fine-grained Bi_2Te_3 alloys, *J. Alloys Compd.*, 399, 14, 2005.
145. Zhao, L.D., Zhang, B.-P., Liu, W.S., Zhang, H.L., and Li, J.-F., Effects of annealing on electrical properties of *n*-type Bi_2Te_3 fabricated by mechanical alloying and spark plasma sintering, *J. Alloys Comp.*, 467, 91, 2009.
146. Touloukian, Y.S., *Thermal Conductivity: Metallic Elements and Alloys*, IFI/Plenum, New York, 1970.
147. Butcher, P., March, N.H., and Tosi, M.P. (eds), Theory of electron transport in low-dimension semiconductor structures, in *Physics of Low-Dimensional Semiconductor Structures*, Plenum, New York, Chapter 3, p. 95, 1993.
148. Lan, Y., Minnich, A.J., Chen, G., and Ren, Z., Enhancement of thermoelectric figure-of-merit by a bulk nanostructuring approach, *Adv. Funct. Mater.*, 20, 357, 2010.
149. Dresselhaus, M.S., Chen, G., Tang, M.Y., Lee, H., Wang, D.Z., Ren, Z.F., Fleural, J.P., and Gogna, P., New directions for nanoscale thermoelectric materials research, *Mater. Res. Soc. Symp. Proc.*, 886, 3, 2006.
150. Minnich, A.J., Lee, H., Wang, X.W., Joshi, G., Dresselhaus, M.S., Ren, Z.F., Chen, G., and Vashaee, D., Modeling study of thermoelectric SiGe nanocomposites, *Phys. Rev. B*, 80, 155327, 2009.

151. Vineis, C.J., Harman, T.C., Calawa, S.D., Walsh, M.P., Reeder, R.E., Singh, R., and Shakouri, A., Carrier concentration and temperature dependence of the electronic transport properties of epitaxial PbTe and PbTe/PbSe nanodot superlattices, *Phys. Rev. B*, 77, 235202, 2008.
152. Ravich, Y.I., Efimova, B.A., and Tamarchenko, V.I., Scattering of current carriers and transport phenomena in lead chalcogenides II experiment, *Phys. Status Solidi B*, 43, 453, 1971.
153. Callaway, J., Model for lattice thermal conductivity at low temperatures, *Phys. Rev.*, 113, 1046, 1959.
154. Holland, M.G., Analysis of lattice thermal conductivity, *Phys. Rev.*, 132, 2461, 1963.
155. Srivastava, G.P., *The Physics of Phonons*, Taylor & Francis Group, New York, 1990.
156. Callaway, J. and von Bayer, H.C., Effect of point imperfections on lattice thermal conductivity, *Phys. Rev.*, 120, 1149, 1960.
157. Steigmeier, E.F. and Abeles, B., Scattering of phonons by electrons in germanium-silicon alloys, *Phys. Rev.*, 136, A1149, 1964.
158. Faleev, S.V. and Leonard, F., Theory of enhancement of thermoelectric properties of materials with nanoinclusions, *Phys. Rev. B*, 77, 214304, 2008.
159. Landau, L.D. and Lifshitz, E.M., *Quantum Mechanics (Non-Relativistic Theory)*, 3rd edition, Vol. 3, Pergamon Press, New York, 1977.
160. Kim, W. and Majumdar, A., Phonon scattering cross section of polydispersed spherical nanoparticles, *J. Appl. Phys.*, 99, 084306, 2006.
161. Mahan, G.D. and Sofo, J., The best thermoelectric, *Proc. Natl. Acad. Sci. USA*, 93, 7436, 1996.
162. Ahmad, S., Hoang, K., and Mahanti, S.D., Erratum: *Ab initio study* of deep defect states in narrow band-gap semiconductors: Group III impurities in PbTe, *Phys. Rev. Lett.*, 96, 169907(E), 2006.
163. Ahmad, S., Mahanti, S.D., Hoang, K., and Kanatzidis, M.G., Ab initio studies of the electronic structure of defects in PbTe, *Phys. Rev. B*, 74, 155205, 2006.
164. Lee, J.H., Wu, J., and Grossman, J.C., Enhancing the thermoelectric power factor with highly mismatched isoelectronic doping, *Phys. Rev. Lett.*, 104, 016602, 2010.
165. Popescu, A. and Woods, L., Enhanced thermoelectricity in composites by electronic structure modifications and nanostructuring, *Appl. Phys. Lett.*, 97, 052102, 2010.
166. Heremans, J.P., Jovovic, V., Toberer, E.S., Saramat, A., Kurosaki, K., Charoenphakdee, A., Yamanaka, S., and Snyder, G.J., Enhancement of thermoelectric efficiency in PbTe by distortion of the electronic density of states, *Science*, 321, 554, 2008.
167. Hjalmarson, H.P., Vogl, P., Wolford, D.J., and Dow, J.D., Theory of substitutional deep traps in covalent semiconductors, *Phys. Rev. Lett.*, 44, 810, 1980.
168. Harris, J.S., GaInNAs long-wavelength lasers: Progress and challenges, *Semicond. Sci. Technol.*, 17, 860, 2002.

15

Surface and Interface Effects on Thermoelectric Behavior in Crystalline Nanowires

Frédéric Sansoz
The University of Vermont

15.1 Introduction

Intense research efforts have been devoted to understanding how surfaces and interfaces influence properties at the nanoscale level in crystalline nanowires (NWs) from semiconductors and metals. For instance, material properties such as mechanical strength, chemical sensing, photovoltaic conversion, and electrical transport have been dramatically improved by controlling the size, shape, and microstructure of NWs (Bauer et al. 2004; Baxter and Aydil 2005; Deng and Sansoz 2009; Wu et al. 2004, 2005). In this chapter, the effects of sample size, grain boundaries, and surface morphology on the thermoelectric behavior of semiconductor NWs are addressed with particular focus on Si NWs, which are pertinent for power generation in microscale- and nanoscale-integrated devices. Future directions in this research field are also discussed.

The efficiency of a thermoelectric material is determined by the dimensionless figure-of-merit defined as $ZT = S^2\sigma T/\kappa$, where S is the thermoelectric power (Seebeck coefficient), σ is the electrical conductivity, κ is the thermal conductivity, which is partitioned into an electronic contribution (κ_e) and a phonon contribution (κ_{ph}), and T is the (average) temperature. Maximizing ZT is challenging because optimizing one physical parameter often adversely affects another in bulk materials. The best commercially available thermoelectric devices made of Bi_2Te_3 alloys have a ZT of 1 (Hicks and Dresselhaus 1993a), although a ZT value of ~3 is required to replace current compressor-based refrigeration and generator technologies (Majumdar 2004; Mingo 2004). The ideal thermoelectric material is assumed to be *an electronic crystal* and *a phonon glass*, that is, a material with high cooling power factor $S^2\sigma$, but low thermal conductivity κ such as in heavily doped semiconductors (Snyder et al. 2004).

Semiconductors have a lower density of carriers leading to larger S, and κ is dominated by phonons because the electrical and thermal conductivities are somewhat decoupled in this type of materials (Boukai et al. 2008; Hochbaum et al. 2008). On the one hand, bulk thermoelectric semiconductors can be engineered to achieve ZT values above 2 (Hsu et al. 2004); however, the materials (Bi, Te, Pb, Sb, Ag) and processes used are not often easy to scale to practically useful dimensions (Hochbaum et al. 2008). On the other hand, Si, which is a ubiquitous and low-cost semiconductor, could make possible the integration of thermoelectric materials into microelectronic devices (Chowdhury et al. 2009; Snyder et al. 2003); however, bulk Si is known to exhibit poor thermoelectric behavior due to its high thermal conductivity. Semiconductor nanostructures, including low-dimensional materials like Si NWs, hold great promise for thermoelectric applications. In low-dimensional systems, the density of states has sharp peaks, leading to high S, although harnessing this electronic effect to produce high-ZT materials has only had limited success. Another strategy is to optimize phonon dynamics and heat transport in Si NWs, which has yielded remarkable results as shown later. In semiconductor NWs, charge carriers are naturally pushed away from the surface. Consequently, surface scattering is expected to reduce κ in greater proportion than σ and S (Martin et al. 2010). Therefore, current research focuses on finding new ways to fabricate Si NWs with specific defects and sizes in order to reduce thermal conductivity in this semiconductor.

To illustrate the size dependence of κ in Si, Figure 15.1 summarizes the thermal conductivity of pure Si NWs at 300 K, as a function of size (or diameter), from reports in the literature using either experiments or computer simulations/theory. Two important conclusions can be drawn from this figure. First, the thermal conductivity of Si is found to decrease by more than two orders of magnitude via size reduction. Applying the semiempirical Matthiessen's rule the expected NW thermal conductivity can be estimated as a function of NW diameter D (Doerk et al. 2010) such as

$$\kappa = \frac{\kappa_{bulk} \times D}{l_{bulk} + D}, \tag{15.1}$$

where κ_{bulk} is the bulk Si thermal conductivity [~150 W m^{-1} K^{-1} at room temperature (Touloukian 1970)] and l_{bulk} is the effective mean free path in bulk Si, which is approximately 300 nm as obtained

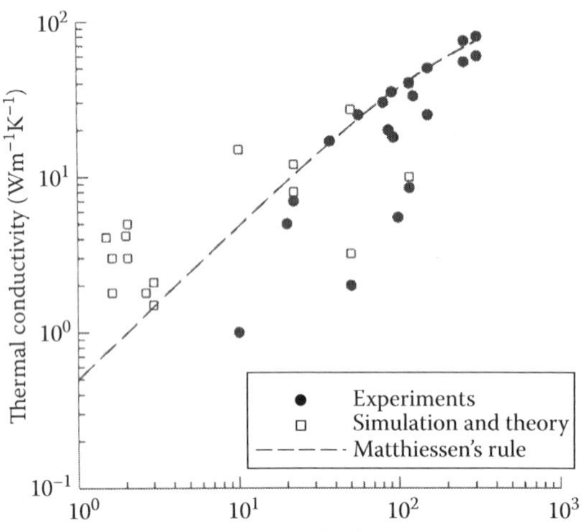

FIGURE 15.1 Size effects on thermal conductivity of single-crystalline Si nanowires at 300 K from past experiments (Boukai et al. 2008; Doerk et al. 2010; Hippalgaonkar et al. 2010; Hochbaum et al. 2008; Kim et al. 2010; Li et al. 2003) and computer simulation/theoretical studies (Chan et al. 2010; Donadio and Galli 2009; Donadio and Galli 2010; Liu and Chen 2010; Lu and Chu 2006; Martin et al. 2010; Moore et al. 2008; Papanikolaou 2008; Volz and Chen 1999). The predicted curve assuming only boundary scattering via Matthiessen's rule (dashed line) is included for comparison.

from measurements on thin films (Doerk et al. 2010; Hochbaum et al. 2008). This trend is closely followed by both experimental and theoretical data for $D > 50$ nm. However, the semiempirical Matthiessen's rule is found to underestimate κ for smaller NWs in comparison with more accurate theoretical results, which is consistent with an earlier report for in-plane thermal conductivity in thin Si films (Turney et al. 2010). Nevertheless, the strong decrease of κ with sample size reduction points to the importance of free surfaces on heat transport at the nanoscale level as the surface-to-volume ratio increases with decreasing *D*.

Second, Figure 15.1 shows excellent agreement between experiments and computer simulations and theory, including first-principles calculations and molecular dynamics (MD) simulations. However, thermal conductivities measured experimentally are usually smaller in the diameter range between 10 and 50 nm as compared with theoretical predictions. This discrepancy is due to the fact that different surface effects other than sample size can also profoundly affect thermal conductivity at the nanoscale level. Three major factors influencing lattice thermal conductivity in Si NWs are described next: sample size, grain boundary superlattices, and surface morphology. We also show that direct insights into the basic mechanisms of thermal conductivity in Si NWs can be achieved using atomic-scale computer simulations. This chapter is divided as follows. Conventional atomic-scale simulation techniques for computing thermal conductivities in NWs are described in Section 15.2. Section 15.3 provides an overview of size effects on thermal conductivity of Si NWs with or without a native oxide layer. The roles played by grain boundary superlattices and surface morphology on nanoscale heat transport are addressed in Sections 15.4 and 15.5, respectively. Finally, Section 15.6 presents a discussion on the effects of sawtooth surface faceting and alloy scattering on thermal conductivity in Si-based NWs, which opens up new opportunities for future research in nanoscale thermoelectrics.

15.2 Computational Methods

15.2.1 *Ab Initio* Density Functional Theory Calculations

Ab initio density functional theory (DFT) calculations have been used routinely to compute electronic and thermal properties in Si NWs. In particular, it is possible to measure σ, *S*, and κ_e from the electronic band structure predicted by *ab initio* DFT-based technique in conjunction with the solution of one-dimensional Boltzmann equation given by Shi et al. (2009):

$$\sigma = \Lambda^{(0)}, \tag{15.2}$$

$$\kappa_e = \frac{1}{e^2T}\left[\Lambda^{(2)} - \Lambda^{(1)}\left(\Lambda^{(0)}\right)^{-1}\Lambda^{(1)}\right], \tag{15.3}$$

$$S = \frac{1}{eT}\left(\Lambda^{(0)}\right)^{-1}\Lambda^{(1)}, \tag{15.4}$$

with

$$\Lambda(n) = e^2\tau\frac{2}{m^*}\sum_{E_k}\Delta E\left[\frac{\beta\exp\left[\beta(E_k-\mu)\right]}{\left\{1+\exp\left[\beta(E_k-\mu)\right]\right\}^2}\right]D(E_k)E_k(E_k-\mu)^n, \tag{15.5}$$

where *e* is the charge of carriers, *T* is the temperature, E_k is the electron energy, τ is the relaxation time, m^* is the effective mass of the charge carrier, μ is the electron chemical potential, and $D(E_k)$ is the density of states.

Both conventional DFT and DFT-derived tight-binding (DFTB) methods have been utilized to predict thermoelectric properties in <110> and <111> oriented Si NWs with diameters up to 7 nm. Various effects have been investigated, including carrier concentration (Shi et al. 2009; Vo et al. 2008), cross-section geometry (Yao et al. 2008), surface decoration (Chen et al. 2009; Markussen et al. 2009), surface roughness (Persson et al. 2008), and NW diameter (Shi et al. 2009). The DFTB method has higher computational efficiency for the simulation of large atomistic systems (>1000 atoms) than conventional DFT, and achieves similar accuracy. For instance, it has been shown that the accuracy of DFTB to predict the band gap in bulk Si is very close to the experimental value of 1.17 eV (Yao et al. 2008), and is therefore superior to that calculated with commonly used DFT methods, which underestimate band gaps.

15.2.2 Molecular Dynamics Simulation

The determination of κ_{ph} in pristine Si NWs at finite temperatures has been conducted by classical MD simulation technique for more than one decade (Volz and Chen 1999). Two common MD approaches are the Green–Kubo method and the direct method (Schelling et al. 2002; Zhou et al. 2009). The Green–Kubo formalism relates the ensemble average of the autocorrelation of the heat flux J to κ. This technique has largely been used for thermal conductivity calculations in bulk systems with three-dimensional periodic boundary conditions. However, the direct method is more robust for computing thermal conductivity in systems with free surfaces such as NWs (Schelling et al. 2002). The direct method proposed by Muller-Plathe (Muller-Plathe 1997) reverses the cause and effect since J is directly imposed on the system while the time-averaged temperature gradient $\partial T/\partial x$, with x the coordinate along the wire axis ($x < L$), is computed from the simulation. κ is then obtained from Fourier's law such as $\kappa = -J/(\partial T/\partial x)$. The Muller-Plathe method is implemented in MD simulations as follows. In order to impose a heat flux and to calculate a temperature profile, the simulation box is divided into N slabs perpendicular to the NW axis. J is added and removed along the hot and cold regions, which are carefully set in the simulation box. By way of illustration, in Figure 15.2a, slab 0 is defined as the "cold" slab and slab $N/2$ as the "hot" slab. J is generated

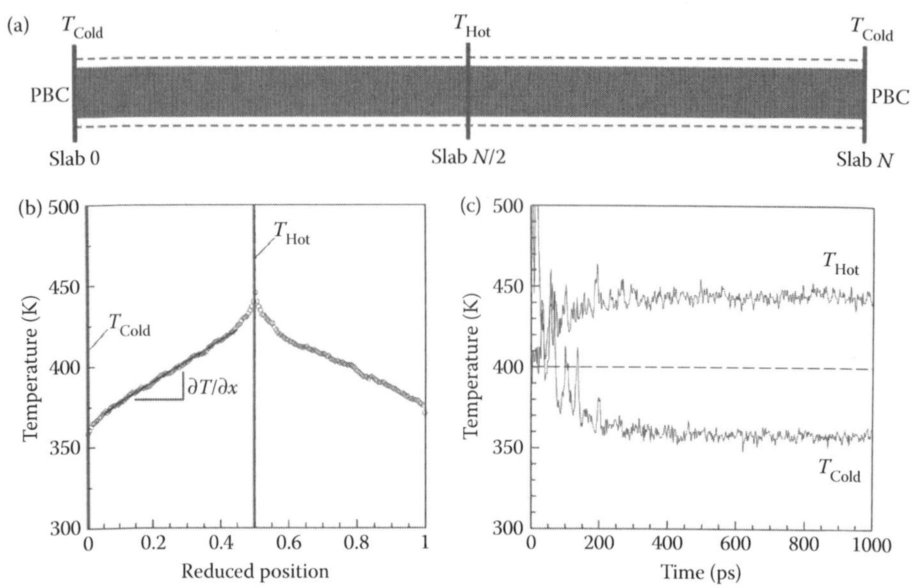

FIGURE 15.2 Atomic-level simulation for computing thermal conductivity in crystalline nanowires using the Muller-Plathe method. (a) Atomistic model setup for a <111> oriented Si nanowire. The system includes periodic boundary conditions (PBC). (b) Typical temperature profile. (c) Time-averaged temperature for the cold and hot sides corresponding to thin slabs in the model at temperature T_{Hot} and T_{Cold}, respectively.

by exchanging the velocity vectors between the hottest atom in the cold slab and the coldest one in the hot slab in a way that the temperature increases in the hot slab and decreases in the cold slab. This mechanism produces an energy transfer from the cold slab to the hot slab leading to a linear temperature gradient in the intervening regions as shown in Figure 15.2b. However, a challenge is that large system sizes must be simulated to increase the system's length (L) and therefore avoid imposing very high temperature gradients. These issues are typically solved by using massively parallel computations.

Furthermore, the interatomic forces and energies between Si atoms can be best described by using either a three-body empirical potential from Stillinger and Weber (1985) or that by Tersoff (1988). More accurate modified embedded-atom-method (MEAM) potentials have also been developed for Si (Lee and Baskes 2000). The thermal conductivity predicted by MD simulations in pristine Si NWs with the Tersoff potential was found to agree well with more elaborate DFT calculations (Markussen et al. 2008). Also, MD predictions of thermal conductivity in bulk Si with the Stillinger–Weber potential have been found consistent with experimental data (Schelling et al. 2002). Such atomistic computations can be carried out using freely available MD parallel codes such as LAMMPS (large-scale atomic molecular massively parallel simulator, Sandia National Laboratories) (Plimpton 1995), where these potentials have been implemented.

Here, the Muller-Plathe method was applied to compute the lattice thermal conductivity of a cylindrical <111> Si NW with a circular cross section of 9.8 nm in diameter using the Stillinger–Weber potential. For diamond cubic Si, NWs have preferential growth along the <111> direction (Arbiol et al. 2008). Also, it was found that <111> oriented Si NWs exhibit lower thermal conductance than <110> Si NWs (Markussen et al. 2008); thereby, thermoelectric devices can make better use of Si NWs with <111> crystal orientation. Figure 15.2c shows that a steady state was reached in this system in less than 1 ns, which was attained by performing one million MD steps of 1 fs each. Furthermore, finite-size effects are known to arise in simulation when the periodic length of the simulation cell L_z is not significantly longer than the phonon mean free path (Poetzsch and Bottger 1994; Oligschleger and Schon 1999; Schelling et al. 2002), although a dimension scaling theory has recently been proposed to bridge the length scales between simulated and actual nanostructures (Zhou et al. 2010). This effect results from scattering that occurs at the interfaces with the heat source and sink. For a sample with length smaller than the mean free path in an infinite system, κ is limited by the system size. Using an analysis rooted in kinetic theory, Schelling et al. (2002) have established that κ depends on the inverse mean free paths for scattering such as

$$\frac{1}{\kappa} = \frac{a^3}{4k_B v}\left(\frac{1}{l_{\text{inf}}} + \frac{2}{L_z}\right), \tag{15.6}$$

where a is the lattice constant, k_B is the Boltzmann constant, v is the group velocity for an acoustic branch, and l_{inf} is the mean free path in an infinite system. More importantly, Equation 15.6 suggests that a plot of $1/\kappa$ versus $1/L_z$ should be linear, and that the thermal conductivity of an infinite system can be obtained by extrapolating to $1/L_z \sim 0$. Therefore, it is critically important to quantitatively assess the dependence of κ predicted by MD as a function of L_z. For that purpose, Figure 15.3 displays the system size dependence of $1/\kappa$ as predicted from MD simulations at an average temperature of 100 K. For <100> bulk Si as shown in Figure 15.3b, Figure 15.3c predicts that κ for a system of infinite size is equal to 488 W m^{-1} K^{-1} at 100 K when $1/L_z \sim 0$. This MD prediction is of the same order of magnitude than the value of ~900 W m^{-1} K^{-1} reported experimentally for natural bulk Si at the same temperature (Glassbrenner and Slack 1964), which therefore supports the validity of this computational scheme.

15.3 Size Effects

All evidence proved in the past that NWs with small diameters are preferred to bulk materials for thermoelectric applications (Hicks and Dresselhaus 1993b). With cross-sectional area decreasing, σ decreases slowly, while S increases remarkably, thus leading to a quick increase in the cooling power

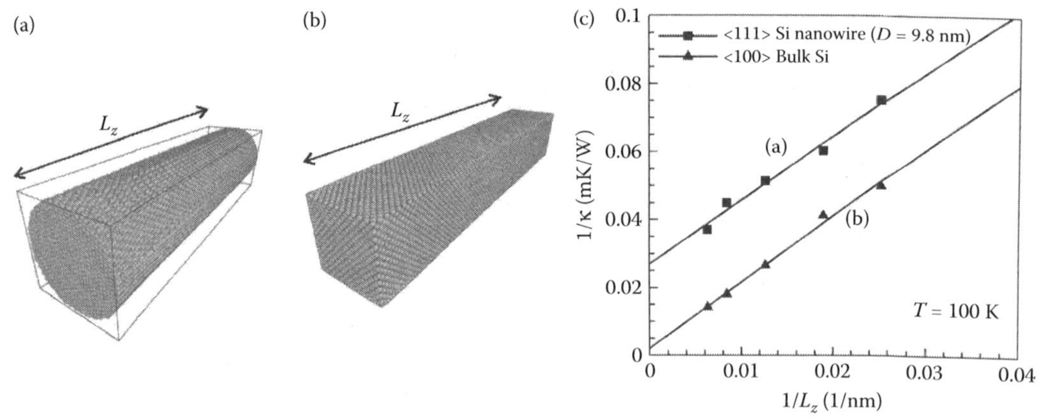

FIGURE 15.3 Atomistic model for (a) <111> cylindrical Si nanowire with circular cross-section and (b) <100> bulk Si. (c) System size dependence of $1/\kappa$ on $1/L_z$ at 100 K for both models.

factor (Shi et al. 2009). In small Si NWs, however, the primary effect of sample size on thermoelectric performance is a pronounced reduction of κ_{ph}, because the mean free path for phonons is significantly larger than that for electrons (Hochbaum et al. 2008). In bulk form, the thermal conductivity of Si is ~150 W m^{-1} K^{-1} at 300 K (Glassbrenner and Slack 1964). For comparison, both experimental and theoretical studies presented in Figure 15.1 clearly show that thermal conductivity in Si NWs can be more than 100 times smaller than that of bulk Si at this temperature. Calculations of κ from MD simulations with the same system size (L_z = 156.4 nm) for temperatures between 100 and 500 K are presented in Figure 15.4. This figure reveals that introducing free surfaces in smooth Si NWs causes a threefold decrease in lattice thermal conductivity compared to that of bulk Si, regardless of extrinsic factors such as surface roughness or the presence of a native oxide layer. Figure 15.3 also shows that the thermal conductivity for a system of infinite size is found to decrease by more than one order of magnitude from

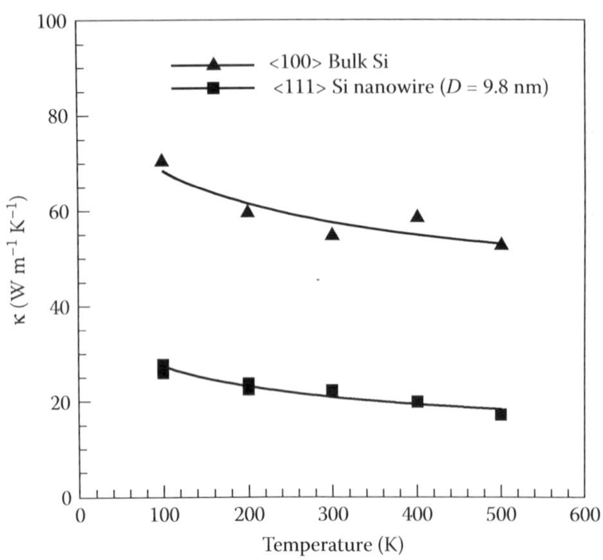

FIGURE 15.4 Temperature effects on thermal conductivity of <100> bulk Si and <111> Si nanowires with a diameter of 9.8 nm. The system size corresponds to L_z = 156.4 nm in all atomistic models.

488 to 37 W m^{-1} K^{-1} from <100> bulk to <111> Si NWs at 100 K. Free surface effects on κ can be attributed to significant decrease in mean free path for phonon scattering due to boundaries (Hochbaum et al. 2008). In Figure 15.3, it is observed that l_{inf} decreases from 475 and 35 nm between bulk Si and smooth circular Si NWs, respectively, which supports this hypothesis.

As discussed earlier, spatial confinement is not the sole process influencing thermal transport in Si NWs. In particular, Donadio and Galli (2009) noted that an amorphous SiO_2 layer causes smaller thermal conductivity and increases the significance of size effects in smooth Si NWs. The system can be viewed as a core–shell structure consisting of a crystalline core and an amorphous shell. It is worth noting here that the amorphous SiO_2 layer shows very similar thermal properties than amorphous Si (Donadio and Galli 2009). In addition, Donadio and Galli (2010) have found that the presence of a native oxide layer in thin Si NWs eliminates the temperature dependence of κ, which is in contrast to the thermal behavior of oxide-free Si NWs as shown in Figure 15.4. These authors have hypothesized that the crystalline core–amorphous shell structure in oxidized Si NWs promotes the presence of nonpropagating vibrational modes similar to those contributing to heat transport in amorphous Si. The diffusivity of these nonpropagating modes at the crystalline–amorphous interface does not depend on the temperature. Experimental observations by transmission electron microscopy have shown that Si NWs grown by vapor–liquid–solid (VLS) technique do promote an oxide layer between 1 and 3 nm in thickness (Hochbaum et al. 2008). In the case of the VLS-grown Si NW of 22 nm diameter, it was determined that κ is reduced to 7 W m^{-1} K^{-1} at room temperature (Li et al. 2003).

15.4 Grain Boundary Superlattice Effects

Earlier studies have successfully demonstrated that thermoelectric superlattice-based films could significantly enhance thermoelectric power in microelectronic devices; for example, to cool very local “hot spots” on computer chips (Chowdhury et al. 2009). The idea of using two-dimensional multilayered superlattices has long been proposed to create quantum-well structures for thermoelectric applications (Hicks and Dresselhaus 1993b). This technique was used to approach the *ZT* value of 2.5 with bulk alloys made of different nanolayers such as Bi_2Te_3/Sb_2Te_3 superlattice (Venkatasubramanian et al. 2001) and PbSeTe-based superlattice (Harman et al. 2002). Compared to superlattices in bulk semiconductors, however, theoretical calculations in Si/Ge superlattice NWs have shown that the additional boundary scattering in nanoscale superlattice heterostructures could further reduce κ by a factor of 2 or more (Dames and Chen 2004). It was also argued that the formation of cracks and dislocations at interfaces between nanolayers in bulk thermoelectric superlattices is detrimental, while this issue may be resolved in NWs because free surfaces can more easily relax stresses than plastic deformation at the interfaces (Gillet et al. 2009).

For low-dimensional systems like Si NWs, Becker et al. (2006) have used MD simulations to gain fundamental insights into the scattering processes of a wave packet at an interface in an Si NW. They predicted that the presence of gaps in the phonon spectrum of thin NWs could result in a complete reflection of phonons at the interfaces. This theory relies on interfaces between semiconductors of different masses such as Si/Ge NWs. In parallel, other investigations have focused on interfacial effects in single-crystalline NWs with Si/SiGe superlattices because these structures can be grown easily by the VLS method (Wu et al. 2002). In this case, the interface scattering effect predominates in the reduction of thermal conductivity while the intrinsic size effects on average phonon velocity and phonon mean free path also appear to be critical (Yang et al. 2008). Moreover, enhanced electrical conductivity has been measured in undoped Si/SiGe heterostructure NWs whose electrical properties were found to be more strongly influenced by free surfaces than pure Si NWs (Das Kanungo et al. 2009).

Recently, control over ordering phenomena at the atomic scale have enabled the fabrication of well-defined coherent twinning superlattices in a vast range of semiconductor NWs (Arbiol et al. 2008; Algra et al. 2008; Banerjee et al. 2006; Caroff et al. 2009; Davidson et al. 2007; Wang et al. 2008). Although VLS-grown Si NWs are generally single crystalline due to the large twin formation energy, which is about half of the stacking fault energy (Lindefelt et al. 2003), advances in synthesis have made it possible to grow <111>

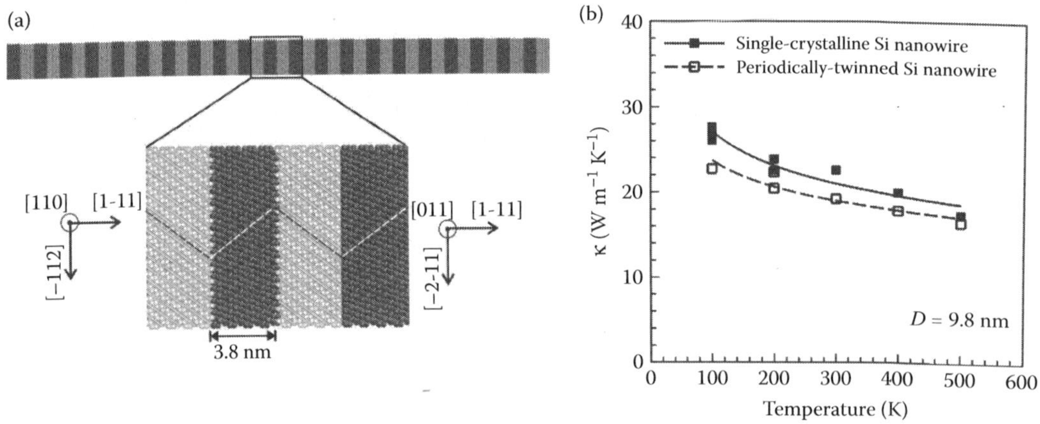

FIGURE 15.5 Effects of a periodically twinned superlattice on thermal conductivity of <111> Si nanowires. (a) Atomistic model of a circular NW with a coherent twinning superlattice of 3.8 nm in grain size. The {111} crystallographic planes are indicated by dashed line in the close-up. (b) Thermal conductivities predicted by direct atomistic simulation technique as a function of average temperature.

oriented twinned Si NWs from Cu catalysts (Arbiol et al. 2008). Twin interfaces are unique, low-energy grain boundaries with mirror-symmetry and perfect crystal structure. In Si NWs, additional thermoelectric performance could be gained from introducing twin boundaries to promote both surface and interface phonon scattering, while the material could be expected to retain the same electrical properties than those in twin-free single crystals (Lu et al. 2004). Therefore, coherent twinning superlattices in NW semiconductors are of considerable interest for thermoelectric applications.

In the present study, MD simulations were conducted to characterize the intrinsic effect of a coherent twinning superlattice in pure Si NWs. Figure 15.5 presents simulated thermal conductivities in the 100–500 K range for <111> Si NWs with a periodically twinned superlattice of 3.8 nm in grain size. This figure shows relatively smaller thermal conductivities in periodically twinned NWs than in single-crystalline ones with the same orientation. Optimization of twin boundary spacing may enhance this effect. However, because twinning is usually found associated with the formation of special surface facets in semiconductor NWs (Arbiol et al. 2008; Algra et al. 2008; Caroff et al. 2009; Li et al. 2009; Sansoz et al. 2008; Shim et al. 2008; Wang et al. 2008), twinning-induced surface faceting may have stronger impact on nanoscale heat transport than interface scattering due to twin boundaries as discussed later.

15.5 Surface Morphology Effects

In theory, phonon–phonon, impurity, boundary, and surface roughness scattering processes should be simultaneously considered for accurate predictions of thermal conductivity in Si NWs (Martin et al. 2009, 2010). Experimental evidence has shown however that efficient reduction of κ can be best accomplished through surface scattering of the phonon spectrum by the introduction of nanostructures at different length scales (diameter, roughness, and defects) (Hochbaum et al. 2008). Figure 15.6 schematically illustrates how lattice thermal conductivity and thermoelectric figure-of-merit ZT can be controlled in pure Si by modifications of surface morphology at difference scales. Diffuse surface scattering is partly responsible for the thermal conductivity reduction from bulk Si to pristine NWs, especially when the NW diameter is below 50 nm. As discussed in Section 15.3, the addition of a native oxide layer on the surface of Si NWs causes further decrease in thermal conductivity due to additional diffusivity taking place at the Si and SiO_2 interface. Here, the structure of a crystalline core with an oxide shell was found to cause a decrease of propagating mode lifetimes at the crystalline–amorphous interface,

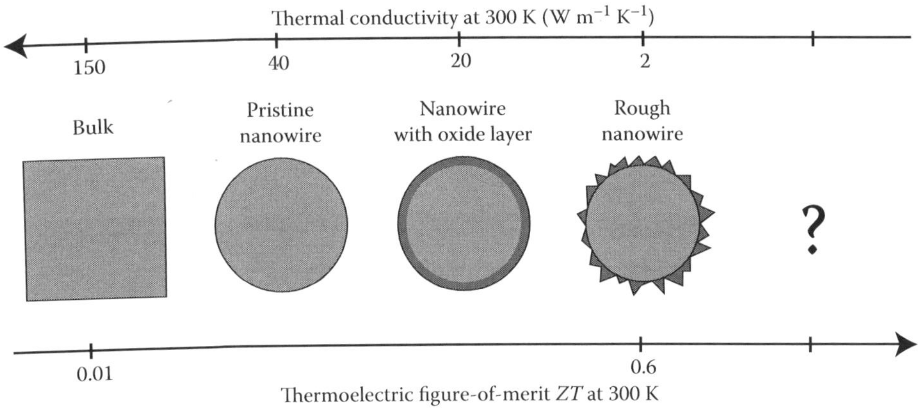

FIGURE 15.6 Schematic illustrating the effects of surface morphology on lattice thermal conductivity and thermoelectric figure-of-merit *ZT* in Si nanowires at 300 K.

combined with the presence of nonpropagating, diffusive modes similar to those found in amorphous Si (Donadio and Galli 2009). The exact *ZT* in Si NWs with a native oxide and its dependence on oxide thickness, however, remains elusive.

Furthermore, κ can be reduced by introducing scattering at additional length scales beyond the NW diameter. Significant progress was made in this area by Hochbaum et al. (2008) who showed that surface roughness has a profound impact on κ in Si NWs. These authors have experimentally measured that roughened Si NWs promote low κ values down to ~1.6 W m^{-1} K^{-1}, with the phonon contribution close to the amorphous limit and without significant alteration in the cooling power factor. As such, 52-nm-diameter-roughened Si NWs have been found to exhibit *ZT* as small as 0.6 at room temperature. MD simulations in oxide-free Si NWs have proved that roughness phonon scattering is an intrinsic mechanism, which can be controlled by patterning periodic ripples on the NW surface (Liu and Chen 2010). In the case of rough Si NWs, it was assumed that the roughness at the NW surface behaves like secondary scattering phases, and acts to augment the rates of diffuse reflection or backscattering of phonons at the boundary (Hochbaum et al. 2008; Moore et al. 2008). Moreover, using a new approach based on perturbation theory, Martin et al. (2009) have predicted a quadratic dependence of κ on both diameter and roughness in $(D/\Delta)^2$, where Δ is the mean square-root of roughness heights, which was in excellent agreement with κ measured experimentally in rough Si NWs. They also confirmed the secondary scattering hypothesis, and established a new type of phonon scattering, where phonons "see" a rough NW as a series of constrictions along their propagation direction. Recently, this theory has been successfully extended to other semiconductor NW systems such as Ge NWs (Martin et al. 2010).

15.6 Outlook for Future Research

This chapter has shown that introducing defects to scatter phonons at different length scales can significantly reduce κ and thus dramatically improve *ZT* in Si NWs via boundary diffuse scattering and roughness scattering processes. To date, fabricating roughened single-crystalline Si NWs of 52 nm in diameter has made it possible to achieve the amorphous Si limit of κ ~ 1 W m^{-1} K^{-1} and to enable *ZT* of 0.6 in Si NWs. However, practically useful applications would require producing materials with *ZT* above 3 and κ nearly 10 times smaller than that of roughened Si NWs. Therefore, an outlook for future research in nanoscale thermoelectricity with crystalline Si NWs can be summarized by asking two fundamental questions:

- Can surface roughness scattering be controlled and maximized by introducing sawtooth surface faceting? Coherent twinning superlattices have the potential to suppress thermal conductivity in

NWs in two ways, that is, grain boundary superlattice effects and twinning-induced surface faceting as found in various Si-based NW systems. For example, SiC NWs synthesized with perfect coherent twinning superlattices exhibit zigzag structures with {111} surface facets (Wang et al. 2008). Shim et al. (2008) explained this effect based on the energy minimization of interfaces and edges near twin boundaries to account for the {111} faceted-twinning structure. Similarly, sawtooth faceting with predominantly {111}/{100} facets and {111}/({113} facets has been observed in VLS-grown Si NWs with or without twin boundaries (Arbiol et al. 2008; Li et al. 2009; Oehler et al. 2010; Ross et al. 2005; Xu et al. 2010). Li et al. (2009) have demonstrated that the size and type of surface facets can be accurately controlled by axially varying *p*- and *n*-type doping during growth. It is possible that such controlled surface faceting, similar to that encountered in periodically twinned SiC NWs, is likely to increase phonon-surface dispersion via phonon backscattering process. In theory, it has been shown that the backscattering process would operate in the same way whether the surface of Si NWs presents random roughness or periodic sawtooth faceting (Moore et al. 2008).

- Is alloying the necessary route for optimizing thermoelectric performance in Si-based NWs? Much attention has recently been paid to the thermoelectric behavior of SiGe alloy NWs for two reasons. First, it has long been predicted that Ge NWs should have a smaller thermal conductivity than their Si counterparts (Mingo et al. 2003). Compared to rough Si NWs, Martin et al. (2010) have predicted a remarkably low κ of 0.1 W m^{-1} K^{-1} in Ge NWs with similar roughness and diameter. Second, theoretical studies have already shown the enhanced thermoelectric properties of Si/Ge NWs with either superlattice structure (Dames and Chen 2004) or core–shell structure (Chen et al. 2010). However, recent experimental and computational studies tend to indicate that alloy scattering in SiGe NWs is likely to promote larger thermoelectric improvements than these two types of nanoscale heterostructures. Using a genetic algorithm optimization technique with coarse-grained cluster expansion, Chan et al. (2010) have predicted atomistic configurations that significantly decrease κ in $Si_{1-x}Ge_x$ NWs. In particular, configurations comprised of atomically thin planes of Ge atoms in crystalline Si NWs showed drastically low κ below 1 W m^{-1} K^{-1}. Experimentally, the thermal conductivities of $Si_{1-x}Ge_x$ NWs synthesized with different Ge concentrations have been observed to decrease by almost one order of magnitude as the Ge concentration increases to 8%, compared to pure Si NWs with corresponding diameters (Kim et al. 2010). Remarkably, the thermal conductivity of SiGe alloy NWs was found less dependent on the NW diameter, which confirms that the scattering process is dominated by alloy scattering in this system, instead of boundary phonon scattering (Kim et al. 2010; Wang and Mingo 2010).

To conclude, the critical role of phonon transport in thermoelectric semiconductor devices is well known. This overview has shown that this role is likely to become increasingly important when such devices incorporate low-dimensional semiconductor nanostructures like Si NWs. Current fundamental research shows that phonon transport can be efficiently tailored at the nanoscale level by controlling the size, microstructure, and surface morphology of Si NWs. Furthermore, theoretical studies harnessing the power of atomic-scale computer simulations sheds light on new scattering processes in NW systems, which will also motivate further experimental progress to understand nanoscale thermoelectricity.

The author gratefully acknowledges the computational resources provided by the Vermont Advanced Computing Center (NASA grant NNX06AC88G).

References

Algra, R. E., M. A. Verheijen, M. T. Borgstrom et al. 2008. Twinning superlattices in indium phosphide nanowires. *Nature* 456(7220):369–372.

Arbiol, J., A. F. I. Morral, S. Estrade et al. 2008. Influence of the (111) twinning on the formation of diamond cubic/diamond hexagonal heterostructures in Cu-catalyzed Si nanowires. *Journal of Applied Physics* 104(6):064312.

Banerjee, R., A. Bhattacharya, A. Genc, and B. M. Arora. 2006. Structure of twins in GaAs nanowires grown by the vapour-liquid-solid process. *Philosophical Magazine Letters* 86(12):807–816.

Bauer, L. A., N. S. Birenbaum, and G. J. Meyer. 2004. Biological applications of high aspect ratio nanoparticles. *Journal of Materials Chemistry* 14(4):517–526.

Baxter, J. B. and E. S. Aydil. 2005. Nanowire-based dye-sensitized solar cells. *Applied Physics Letters* 86(5):053114.

Becker, B., P. K. Schelling, and S. R. Phillpot. 2006. Interfacial phonon scattering in semiconductor nanowires by molecular-dynamics simulation. *Journal of Applied Physics* 99(12):123715.

Boukai, A. I., Y. Bunimovich, J. Tahir-Kheli, J. K. Yu, W. A. Goddard, and J. R. Heath. 2008. Silicon nanowires as efficient thermoelectric materials. *Nature* 451(7175):168–171.

Caroff, P., K. A. Dick, J. Johansson, M. E. Messing, K. Deppert, and L. Samuelson. 2009. Controlled polytypic and twin-plane superlattices in III–V nanowires. *Nature Nanotechnology* 4:50–55.

Chan, M. K. Y., J. Reed, D. Donadio et al. 2010. Cluster expansion and optimization of thermal conductivity in SiGe nanowires. *Physical Review B* 81(17):174303.

Chen, X., Y. C. Wang, and Y. M. Ma. 2010. High thermoelectric performance of Ge/Si core-shell nanowires: First-principles prediction. *Journal of Physical Chemistry C* 114(19):9096–9100.

Chen, X., Y. C. Wang, Y. M. Ma, T. Cui, and G. T. Zou. 2009. Origin of the high thermoelectric performance in Si nanowires: A first-principle study. *Journal of Physical Chemistry C* 113(31):14001–14005.

Chowdhury, I., R. Prasher, K. Lofgreen et al. 2009. On-chip cooling by superlattice-based thin-film thermoelectrics. *Nature Nanotechnology* 4(4):235–238.

Dames, C. and G. Chen. 2004. Theoretical phonon thermal conductivity of Si/Ge superlattice nanowires. *Journal of Applied Physics* 95(2):682–693.

Das Kanungo, P., A. Wolfsteller, N. D. Zakharov, P. Werner, and U. Gosele. 2009. Enhanced electrical properties of nominally undoped Si/SiGe heterostructure nanowires grown by molecular beam epitaxy. *Microelectronics Journal* 40(3):452–455.

Davidson, F. M., D. C. Lee, D. D. Fanfair, and B. A. Korgel. 2007. Lamellar twinning in semiconductor nanowires. *Journal of Physical Chemistry C* 111(7):2929–2935.

Deng, C. and F. Sansoz. 2009. Near-ideal strength in gold nanowires achieved through microstructural design. *ACS Nano* 3(10):3001–3008.

Doerk, G. S., C. Carraro, and R. Maboudian. 2010. Single nanowire thermal conductivity measurements by Raman thermography. *ACS Nano* 4(8):4908–4914.

Donadio, D. and G. Galli. 2009. Atomistic simulations of heat transport in silicon nanowires. *Physical Review Letters* 102(19):195901.

Donadio, D. and G. Galli. 2010. Temperature dependence of the thermal conductivity of thin silicon nanowires. *Nano Letters* 10(3):847–851.

Gillet, J. N., Y. Chalopin, and S. Volz. 2009. Atomic-scale three-dimensional phononic crystals with a very low thermal conductivity to design crystalline thermoelectric devices. *Journal of Heat Transfer-Transactions of the ASME* 131(4):043206.

Glassbrenner, C. J. and G. A. Slack. 1964. Thermal conductivity of silicon and germanium from 3°K to the melting point. *Physical Review B* 134:A1058–A1069.

Harman, T. C., P. J. Taylor, M. P. Walsh, and B. E. LaForge. 2002. Quantum dot superlattice thermoelectric materials and devices. *Science* 297(5590):2229–2232.

Hicks, L. D. and M. S. Dresselhaus. 1993a. Effect of quantum-well structures on the thermoelectric figure of merit. *Physical Review B* 47(19):12727–12731.

Hicks, L. D. and M. S. Dresselhaus. 1993b. Thermoelectric figure of merit of a one-dimensional conductor. *Physical Review B* 47(24):16631–16634.

Hippalgaonkar, K., B. Huang, R. Chen, K. Sawyer, P. Ercius, and A. Majumdar. 2010. Fabrication of microdevices with integrated nanowires for investigating low-dimensional phonon transport. *Nano Letters* 10:4341–4348.

Hochbaum, A. I., R. K. Chen, R. D. Delgado et al. 2008. Enhanced thermoelectric performance of rough silicon nanowires. *Nature* 451(7175):163-U5.

Hsu, K. F., S. Loo, F. Guo et al. 2004. Cubic AgPbmSbTe2 + m: Bulk thermoelectric materials with high figure of merit. *Science* 303(5659):818–821.

Kim, H., I. Kim, H. J. Choi, and W. Kim. 2010. Thermal conductivities of $Si_{1-x}Ge_x$ nanowires with different germanium concentrations and diameters. *Applied Physics Letters* 96(23):233106.

Lee, B. J. and M. I. Baskes. 2000. Second nearest-neighbor modified embedded-atom-method potential. *Physical Review B* 62:8564–8567.

Li, D. Y., Y. Y. Wu, P. Kim, L. Shi, P. D. Yang, and A. Majumdar. 2003. Thermal conductivity of individual silicon nanowires. *Applied Physics Letters* 83(14):2934–2936.

Li, F., P. D. Nellist, and D. J. H. Cockayne. 2009. Doping-dependent nanofaceting on silicon nanowire surfaces. *Applied Physics Letters* 94(26):263111.

Lindefelt, U., H. Iwata, S. Oberg, and P. R. Briddon. 2003. Stacking faults in 3C-, 4H-, and 6H-SiC polytypes investigated by an *ab initio* supercell method. *Physical Review B* 67(15):155204.

Liu, L. and X. Chen. 2010. Effect of surface roughness on thermal conductivity of silicon nanowires. *Journal of Applied Physics* 107(3):033501.

Lu, L., Y. F. Shen, X. H. Chen, L. H. Qian, and K. Lu. 2004. Ultrahigh strength and high electrical conductivity in copper. *Science* 304(5669):422–426.

Lu, X. and J. Chu. 2006. Lattice thermal conductivity in a silicon nanowire with square cross section. *Journal of Applied Physics* 100(1):014305.

Majumdar, A. 2004. Thermoelectricity in semiconductor nanostructures. *Science* 303(5659):777–778.

Markussen, T., A. P. Jauho, and M. Brandbyge. 2008. Heat conductance is strongly anisotropic for pristine silicon nanowires. *Nano Letters* 8(11):3771–3775.

Markussen, T., A. P. Jauho, and M. Brandbyge. 2009. Surface-decorated silicon nanowires: A route to high-ZT thermoelectrics. *Physical Review Letters* 103(5):055502.

Martin, P., Z. Aksamija, E. Pop, and U. Ravaioli. 2009. Impact of phonon-surface roughness scattering on thermal conductivity of thin Si nanowires. *Physical Review Letters* 102(12):125503.

Martin, P. N., Z. Aksamija, E. Pop, and U. Ravaioli. 2010. Reduced thermal conductivity in nanoengineered rough Ge and GaAs nanowires. *Nano Letters* 10(4):1120–1124.

Mingo, N. 2004. Thermoelectric figure of merit and maximum power factor in III–V semiconductor nanowires. *Applied Physics Letters* 84(14):2652–2654.

Mingo, N., L. Yang, D. Li, and A. Majumdar. 2003. Predicting the thermal conductivity of Si and Ge nanowires. *Nano Letters* 3(12):1713–1716.

Moore, A. L., S. K. Saha, R. S. Prasher, and L. Shi. 2008. Phonon backscattering and thermal conductivity suppression in sawtooth nanowires. *Applied Physics Letters* 93(8):083112.

Muller-Plathe, F. 1997. A simple nonequilibrium molecular dynamics method for calculating the thermal conductivity. *Journal of Chemical Physics* 106(14):6082–6085.

Oehler, F., P. Gentile, T. Baron, P. Ferret, M. Den Hertog, and J. Rouviere. 2010. The importance of the radial growth in the faceting of silicon nanowires. *Nano Letters* 10(7):2335–2341.

Oligschleger, C. and J. C. Schon. 1999. Simulation of thermal conductivity and heat transport in solids. *Physical Review B* 59(6):4125–4133.

Papanikolaou, N. 2008. Lattice thermal conductivity of SiC nanowires. *Journal of Physics-Condensed Matter* 20(13):135201.

Persson, M. P., A. Lherbier, Y. M. Niquet, F. Triozon, and S. Roche. 2008. Orientational dependence of charge transport in disordered silicon nanowires. *Nano Letters* 8(12):4146–4150.

Plimpton, S. 1995. Fast parallel algorithms for short-range molecular-dynamics. *Journal of Computational Physics* 117(1):1–19.

Poetzsch, R. H. H. and H. Bottger. 1994. Interplay of disorder and anharmonicity in heat-conduction–molecular-dynamics study. *Physical Review B* 50(21):15757–15763.

Ross, F. M., J. Tersoff, and M. C. Reuter. 2005. Sawtooth faceting in silicon nanowires. *Physical Review Letters* 95(14):146104.

Sansoz, F., H. C. Huang, and D. H. Warner. 2008. An atomistic perspective on twinning phenomena in nano-enhanced fcc metals. *JOM* 60(9):79–84.

Schelling, P. K., S. R. Phillpot, and P. Keblinski. 2002. Comparison of atomic-level simulation methods for computing thermal conductivity. *Physical Review B* 65(14):144306.

Shi, L. H., D. L. Yao, G. Zhang, and B. W. Li. 2009. Size dependent thermoelectric properties of silicon nanowires. *Applied Physics Letters* 95(6):063102.

Shim, H. W., Y. F. Zhang, and H. C. Huang. 2008. Twin formation during SiC nanowire synthesis. *Journal of Applied Physics* 104(6):063511.

Snyder, G. J., M. Christensen, E. Nishibori, T. Caillat, and B. B. Iversen. 2004. Disordered zinc in Zn_4Sb_3 with phonon-glass and electron-crystal thermoelectric properties. *Nature Materials* 3(7):458–463.

Snyder, G. J., J. R. Lim, C. K. Huang, and J. P. Fleurial. 2003. Thermoelectric microdevice fabricated by a MEMS-like electrochemical process. *Nature Materials* 2(8):528–531.

Stillinger, F. H. and T. A. Weber. 1985. Computer-simulation of local order in condensed phases of silicon. *Physical Review B* 31(8):5262–5271.

Tersoff, J. 1988. New empirical-approach for the structure and energy of covalent systems. *Physical Review B* 37(12):6991–7000.

Touloukian, Y. S. 1970. Thermal conductivity: Metallic elements and alloys. In *Thermophysical Properties of Matter*, edited by Y. S. Touloukian, R. W. Powell, C. Y. Ho, and P. G. Klemens. New York: IFI/Plenum.

Turney, J. E., A. J. H. McGaughey, and C. H. Amon. 2010. In-plane phonon transport in thin films. *Journal of Applied Physics* 107(2):024317.

Venkatasubramanian, R., E. Siivola, T. Colpitts, and B. O'Quinn. 2001. Thin-film thermoelectric devices with high room-temperature figures of merit. *Nature* 413(6856):597–602.

Vo, T. T. M., A. J. Williamson, V. Lordi, and G. Galli. 2008. Atomistic design of thermoelectric properties of silicon nanowires. *Nano Letters* 8(4):1111–1114.

Volz, S. G. and G. Chen. 1999. Molecular dynamics simulation of thermal conductivity of silicon nanowires. *Applied Physics Letters* 75(14):2056–2058.

Wang, D. H., D. Q. Wang, Y. J. Hao, G. Q. Jin, X. Y. Guo, and K. N. Tu. 2008. Periodically twinned SiC nanowires. *Nanotechnology* 19(21):215602.

Wang, Z. and N. Mingo. 2010. Diameter dependence of SiGe nanowire thermal conductivity. *Applied Physics Letters* 97(10):101903.

Wu, B., A. Heidelberg, and J. J. Boland. 2005. Mechanical properties of ultrahigh-strength gold nanowires. *Nature Materials* 4(7):525–529.

Wu, Y., J. Xiang, C. Yang, W. Lu, and C. M. Lieber. 2004. Single-crystal metallic nanowires and metal/semiconductor nanowire heterostructures. *Nature* 430(6995):61–65.

Wu, Y. Y., R. Fan, and P. D. Yang. 2002. Block-by-block growth of single-crystalline Si/SiGe superlattice nanowires. *Nano Letters* 2(2):83–86.

Xu, T., J. P. Nys, A. Addad et al. 2010. Faceted sidewalls of silicon nanowires: Au-induced structural reconstructions and electronic properties. *Physical Review B* 81(11):115403.

Yang, C. C., J. Armellin, and S. Li. 2008. Determinants of thermal conductivity and diffusivity in nanostructural semiconductors. *The Journal of Physical Chemistry B* 112(5):1482–1486.

Yao, D. L., G. Zhang, and B. W. Li. 2008. A universal expression of band gap for silicon nanowires of different cross-section geometries. *Nano Letters* 8(12):4557–4561.

Zhou, X. W., S. Aubry, R. E. Jones, A. Greenstein, and P. K. Schelling. 2009. Towards more accurate molecular dynamics calculation of thermal conductivity: Case study of GaN bulk crystals. *Physical Review B* 79(11):115201.

Zhou, X. W., R. E. Jones, and S. Aubry. 2010. Molecular dynamics prediction of thermal conductivity of GaN films and wires at realistic length scales. *Physical Review B* 81(15):155321.

II

Materials Preparation and Measurement

16

High-Performance Nanostructured Thermoelectric Materials Prepared by Melt Spinning and Spark Plasma Sintering

Xinfeng Tang
Wuhan University of Technology

Wenjie Xie
Wuhan University of Technology

Han Li
Wuhan University of Technology

Baoli Du
Wuhan University of Technology

Qingjie Zhang
Wuhan University of Technology

Terry M. Tritt
Clemson University

Ctirad Uher
University of Michigan

16.1 Introduction and Background

The demand for alternative energy technology to reduce our global reliance on fossil fuels and the greenhouse gas emissions leads to new and important regimes of research, including that of direct thermal-to-electrical energy conversion via thermoelectricity [1–4]. Thermoelectric (TE) materials and devices possess several merits such as an all solid-state assembly, lightweight structure and compactness, rapid responsiveness, operation with no moving mechanical parts or hazardous working fluids, and the added attribute for potential miniaturization. However, despite the obvious merits of thermoelectricity, the efficiency of TE conversion processes is less than that of equivalent mechanical systems by a factor of 2–4. For this reason, thermoelectricity has been long limited to *niche* applications such as providing the power sources for deep space missions (using radioisotope TE generators) and unattended terrestrial systems, where the availability and reliability of the power supply are more of a concern than the efficiency. Hence, improving the conversion efficiency is the key for the development of large-scale applications of thermoelectricity.

The efficiency of TE energy conversion depends primarily on the performance of the TE material, which is gauged by the dimensionless figure of merit, ZT. This parameter is defined as

$$ZT = \frac{\alpha^2 \sigma T}{(\kappa_e + \kappa_{ph})} \tag{16.1}$$

where σ is the electrical conductivity, α is the thermopower (i.e., Seebeck coefficient), κ is the thermal conductivity (often separated into the lattice thermal conductivity, κ_{ph}, and the carrier thermal conductivity, κ_e), and T is the temperature in Kelvin.

As is apparent from Equation 16.1, the material's transport properties are the key to achieving a high TE figure of merit. Thus, searching for materials that possess a high ZT value along with a relatively temperature-independent compatibility factor is the goal of much of the research in this field [3]. Referring to Equation 16.1, the high ZT material should be what has come to be known as a "phonon-glass electron-crystal" (PGEC) [5] that is, a material that simultaneously possesses a high Seebeck coefficient (often called the thermopower), high electrical conductivity (weak electron scattering), and a low lattice thermal conductivity (strong phonon scattering). It is difficult to satisfy these criteria in a simple crystalline bulk material since all three quantities (σ, α, and κ) that govern ZT are interrelated, and a modification of any of these quantities often adversely affects the others, so classes of materials that are good TE materials are relatively rare.

The traditional way to improve TE performance is to investigate new TE materials that possess open crystal structures, which have good electronic transport properties or/and they also possess a low lattice thermal conductivity. Skutterudites are an excellent example in this respect [5,6]. Skutterudites (e.g., $CoSb_3$) are cubic materials with the unit cell can be thought of as comprising eight cubes with Co situated at the cube corners. Six of the cubes possess near-square rings of Sb atoms and two of the cubes are empty. Thus, the formula unit can be written as $\Upsilon_2Co_8Sb_{24}$ where Υ represents an empty void site. An important aspect of the skutterudite structure are these two icosahedron-shaped voids in the unit cell that can be filled with foreign species such as rare earths [6–13], alkali metals [14], and alkaline earths [15,16]. Slack postulated that such *filled skutterudites* may, indeed, be the "perfect candidate materials" for his "PGEC" paradigm, that is, they might possess a very low glass-like thermal conductivity while exhibiting good crystalline character in their electronic transport properties [5]. Subsequently, intensive research efforts focused on filled skutterudite compounds with many different filling elements in order to lower the lattice thermal conductivity. This, indeed, succeeded and the lattice thermal conductivity was lowered by over an *order of magnitude* relative to the parent unfilled material [6–16]. Other novel candidate TE materials are SiGe alloys [17–22], half-Heusler compounds [23–30], lead antimony silver tellurium compounds [31–36], and Bi_2Te_3 alloys [37–50]. Clathrate compounds with their cage structure similar to that of skutterudites proved to be yet another very good example of a PGEC material with extremely low lattice thermal conductivity when the voids are filled [51–59]. Type I clathrates in particular are a prospective class of TE materials for high-temperature power generation applications due to their very low thermal conductivity values, which originate from the static and dynamic disorder of the guest atom inside the cage structure. There are also other interesting TE materials such as $NaCo_2O_4$-based oxides [60–65], Zn_4Sb_3 [66–70], $Yb_{14}MnSb_{11}$ [71–75], and $La_{3-x}Te_4$ [76–79], all of them attracting interest as possible power generation TEs.

Over the past decade, the efforts of pursuing high ZT materials have culminated in another new paradigm: a TE bulk nanocomposite material [80–82]. In a TE nanocomposite material, the presence of multiple length-scales down to a few nanometers provides new avenues for phonon scattering for example, from mass fluctuation alloying, grain boundary, and strain fields, which can occur in parallel and thus each adds to the process according to the Matthiessen's rule with the shortest scattering time (process) dominating. Therefore, all these various scattering processes cover a wide range of phonon wavelengths in order to effectively reduce the lattice thermal conductivity. Meanwhile, the multiscale complexity can be tuned such that the nanostructures scatter phonons more effectively than they scatter electrons. There is accumulating evidence supporting the view that "nanostructuring" is indeed an effective method to achieve a high *ZT* in TE materials and is considered a new paradigm in TE materials research, much like the earlier PGEC concept. One well-known example is the $AgPb_mSbTe_{2+m}$ [the so-called lead antimony silver tellurium (LAST-m)] system. It has been found that Ag_2Te nanodomains that form *in situ* in the matrix of PbTe and that have been identified by electron microscopy studies, are at the heart of high *ZT* values of ~1.7 in the temperature range 600–700 K in the LAST-m material [31].

Several novel techniques have been reported recently for preparing zero-dimensional, one-dimensional, and two-dimensional TE materials, such as molecular-beam epitaxy for superlattice films [40], electrochemical deposition for nanowires [83–86] and superlattice nanowires [87–89], hydrothermal methods for nanoparticles [90], and other methods for nanoplatelets [91] and nanorods [92]. However, these techniques are not suitable for fabrication of *bulk* nanocomposite TE materials. Hence, it is important to explore and develop new preparation techniques for obtaining high performance bulk materials that exhibit the attributes of the lower-dimensional structures. Recently, we reported on our results employing a melt-spinning (MS) technique combined with a subsequent spark plasma sintering (SPS) process (MS–SPS) in order to obtain high-performance bulk materials, such as Bi_2Te_3-based alloys [43,45–47,93,94], filled $CoSb_3$-based skutterudites [95–98], clatherates [99], Zn_3Sb_4 [100,101], InSb [102], $MnSi_{1.75}$ [103], and $AgSbTe_2$ [104]. Compared with other preparation techniques, the MS–SPS process not only optimizes TE transport properties by introducing unique low-dimensional nanostructures, but also improves mechanical strength. Moreover, an important attribute of this MS–SPS processing is a significantly shorter synthesis time and accompanying energy saving.

In this chapter, we present the details of an innovative route of a MS technique combined with a subsequent SPS process that we developed and applied to the synthesis of high-performance bulk TE materials with unique nanostructures, such as bulk *p*-type $(Bi,Sb)_2Te_3$, *n*-type filled $CoSb_3$-based skutterudites, and *p*-type $AgSbTe_2$ bulk materials. This route allows the nanostructure of ribbons and bulk materials to be controlled by adjusting the synthesis parameters during the MS and SPS processing, and we designate such samples as MS–SPS. Subsequently, we found that the electrical and thermal transport properties are significantly affected by the unique low-dimensional nanostructure such samples contain, and the resulting values of the TE figure of merit *ZT* reach a maximum value of *ZT* ~1.5 at 390 K, *ZT* ~1.3 at 800 K, and *ZT* ~1.65 at 570 K for the *p*-type $(Bi,Sb)_2Te_3$, $Yb_{0.3}CoSb_3$, and $AgSbTe_2$ MS–SPS bulk materials, respectively. The chapter is divided into three parts as follows: First, we introduce the MS and SPS techniques; second, we summarize our recent results on high performance bulk TE materials prepared by MS in combination with SPS; and, finally, we offer our perspective and outlook for the MS technique followed by SPS.

16.2 Introduction to MS and SPS Synthesis

16.2.1 MS to Achieve Unique Microstructures

The MS technique has been widely used to prepare amorphous alloys [105,106]. The main characteristic of the MS technique is that during the MS process, the molten alloy is ejected onto a high-speed rotating copper drum under a protective Ar gas atmosphere. The rapidly cooled melt forms an amorphous or nanocrystalline ribbon-like material due to extremely large cooling rates which can reach ~10^6 K/s [107,108]. The schematic of the MS technique is shown in Figure 16.1. Recently, we reported on some of our preliminary studies using the MS technique aiming to synthesize TE nanocomposite materials [43,45–47,94–104]. Prior to these reports, the MS technique had not been considered as a viable means of preparing TE materials with either an amorphous structure or a fine nanostructure. We have since found that the MS technique gives us considerable control over the resulting nanostructure.

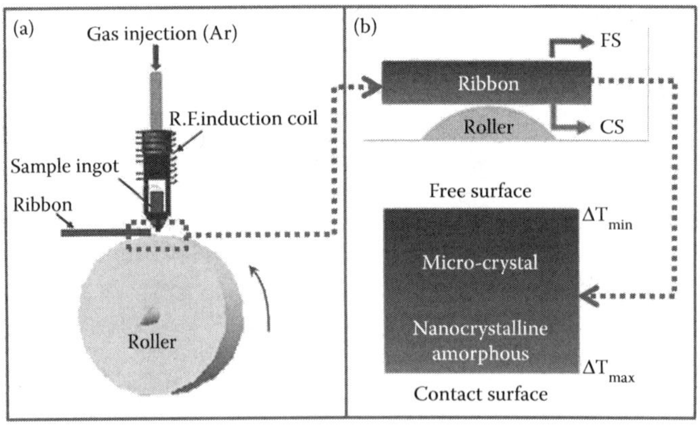

FIGURE 16.1 (a) Diagram of the MS process, (b) designation of the free surface (FS) and the contact surface (CS), and schematic distribution of the crystalline state across the thickness of the ribbon. (Reprinted with permission from Xie, W. J. et al., Identifying the specific nanostructures responsible for the high thermoelectric performance of $(Bi,Sb)_2Te_3$ nanocomposites, *Nano Lett.*, 10, 3283, Copyright 2010, American Chemical Society.)

16.2.2 SPS to Achieve Fully Densified Bulk Materials

Currently, there are many solidification/sintering techniques used to obtain bulk TE materials, among them hot pressing [109], high-pressure sintering [110], hot isostatic pressing [111], shock consolidation [112], SPS [113], and so on. However, for nanostructured materials, SPS has its unique advantages: rapid densification at relatively low temperature, preservation of metastable nanostructures as well as internal interfaces, and even its use in the crystal growth [114–118]. SPS is a high-speed powder consolidation/sintering technology capable of processing conductive as well as nonconductive materials. Theories concerning the SPS process vary, but the most commonly accepted viewpoint is the microspark/plasma concept, which is based on the electrical spark discharge phenomenon wherein a high-energy, low-voltage pulse current momentarily generates spark plasma at high temperatures (many thousands of °C) in fine local areas between particles [119]. The schematic of SPS is shown in Figure 16.2.

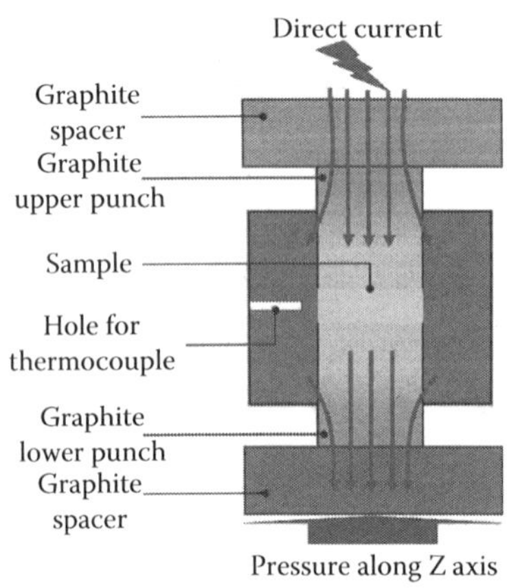

FIGURE 16.2 Diagram of the SPS process. A sample is placed in a graphite die and small initial force is applied via two graphite rods. Pulsed current is injected into the sample which heats through self-Joule heating. At a set temperature the force is increased and the relative displacement of the rods is monitored. Upon sintering, the changing displacement of the rods indicates the progress of sintering and when the rods cease to move further the sintering process is essentially complete. (Adapted from Wenjie, X., et al., *J. Mater. Res.*, 26, 1791–1799, 2011, doi:10.1557/jmr.2011.170. With permission.)

Therefore, in our case, we adopt the MS technique to prepare TE materials with multiscale nanostructures, and then use SPS to preserve thermoelectrically favorable nanostructures in the bulk materials. Bulk nanocomposite samples can be produced with densities of 99% or greater of the theoretical density. Another advantage of the SPS processing is its rapid nature (on the order of minutes) that should be easily adaptable to large-scale manufacturing.

16.3 Microstructure and TE Properties of MS–SPSed Nanocomposites

Combining the advantages of MS and SPS, we have been able to prepare TE nanocomposites. We can adjust the composition of starting materials and also the cooling rate in the MS process to control the growth of nanostructures desirable for different materials. We then employ SPS to prepare fully densified bulk materials in which the nanostructure is preserved. The entire process takes only a few minutes. In order to adjust the MS–SPS process and control the size of the nanometer-scale inclusions, we have systemically investigated the correlation between the nanostructure and the resulting TE transport properties. Table 16.1 summarizes some reported TE nanocomposites prepared by the MS technique. In the following sections, we will illustrate three successful variations of the MS–SPS processing as applied to the formation of Bi_2Te_3, $CoSb_3$, and $AgSbTe_2$ nanocomposite materials.

16.3.1 Bi_2Te_3-Based Materials Prepared by MS–SPS

Bi_2Te_3-based bulk materials are the most widely used TEs that operate in the temperature range of 200–400 K. The maximum figure of merit of commercial Bi_2Te_3 ingots is about ZT ~1.0 (p-type) and ZT ~0.8 (n-type) and has remained essentially unchanged since their discovery in the mid-1950s. It has been reported recently that high ZT p-type $(Bi,Sb)_2Te_3$ nanocrystalline bulk materials have been prepared from a commercial ingot and elemental chunks by a combined ball milling and hot-pressing procedure in an inert atmosphere [48,49]. We also have reported on a novel two-stage, time- and cost-efficient MS–SPSed synthesis of high-performance p-type $(Bi,Sb)_2Te_3$ nanocomposite TE materials, which achieved the highest ZT of 1.5 at ~390 K [47]. Besides, the ZT value of MS–SPSed n-type $(Bi,Sb)_2(Te,Se)_3$ also achieved $ZT \approx 1.0$ around 460 K, and it remains above 0.9 at temperatures from 360 to 500 K [93].

TABLE 16.1 TE Nanocomposites Prepared by MS

Nanocomposites	Reference	Carrier Type	ZT_{max}	Method[a]
$Bi_{0.52}Sb_{1.48}Te_3$	[45,46]	p	1.56 (300 K)	ZM + MS + SPS
$Bi_{0.48}Sb_{1.52}Te_3$	[47]	p	1.5 (390 K)	MS + SPS
$(Bi_{0.85}Sb_{0.15})_2(Te_{0.83}Se_{0.17})_3$	[93]	n	0.96 (360 K)	MC + MS + SPS
$(Bi_{0.95}Sb_{0.05})_2(Te_{0.85}Se_{0.15})_3$	[94]	n	1.0 (460 K)	MC + MS + SPS
$Yb_{0.2}Co_4Sb_{12.3}$	[96]	n	1.26 (800 K)	MQA + MS + SPS
$Yb_{0.3}Co_4Sb_{12.3}$	[98]	n	1.3 (800 K)	MQA + MS + SPS
$Yb_{0.3}Co_4Sb_{12}$	[97]	n	1.22 (800 K)	MQA + MS + SPS
$Ce_{0.3}Fe_{1.5}Co_{2.5}Sb_{12}$	[95]	p	0.55 (750 K)	MQA + MS + SPS
$Mn_{1.75}Si$	[103]	p	0.62 (800 K)	IM + SPS
InSb	[102]	n	0.49 (700 K)	MC + MS + SPS
$AgSbTe_2$	[104]	p	1.65 (570 K)	MQA + MS + SPS
Zn_4Sb_3	[100]	p	1.16 (700 K)	MC + MS + SPS
$Ba_8Ga_{12}Zn_2Ge_{12}$	[99]	n	0.9 (900 K)	MC + MS + SPS

[a] ZM, zone melting; MS, melt spinning; SPS, spark plasma sintering; MC, melt cooling; MQA, melt quench annealing; IM, induction melting.

16.3.1.1 *p*-Type $(Bi,Sb)_2Te_3$ Nanocomposites Prepared by MS–SPS

Recently, we succeeded in the preparation of *p*-type BiSbTe materials directly from single elements and without any prior reaction such as by zone-melting the ingot. We simply weigh powders of the starting elements (Bi, Sb, and Te) in the appropriate proportions and place them into a quartz tube with a tapered end. The material is then inductively melted and ejected under pressure on a cold copper drum as described previously. We call this synthesis procedure the single-element, melt-spinning and spark plasma sintering (SE–MS–SPS) process [47]. Surprisingly, thus prepared materials have very similar properties as the MS–SPS *p*-type material [45,46]. In fact, the nanostructures of these SE–MS–SPS materials and their TE properties were all quite similar to the MS–SPS materials. This has important implications as far as the cost and efficiency of the manufacturing process are concerned. The *ZT* of MS–SPSed $(Bi,Sb)_2Te_3$ nanocomposites has reached values of 1.5 at 390 K, which represents an increase of ~50% in the temperature range of 270–450 K compared to commercial ingots [47].

The MS process creates ribbons with two distinct surfaces: the surface that contacts the copper drum is called the contact surface and the other surface we call the free surface, Figure 16.1a. Significantly different cooling rates on the contact and free surfaces result in multiscale nanostructures. Going from the contact surface of the ribbon towards the free surface we observe an evolution from an amorphous structure, to 5–10 nm fine nanocrystalline regions, to dendritic crystals with nanoscale modulation. Due to the rapid solidification and low thermal conductivity of melts in the MS process, the contact surface and the free surface are expected to have significantly different temperatures, critical radii for nucleation, and growth speeds for nucleation sites. In order to verify these speculations, high-resolution transmission electron microscopy (HRTEM) was used to investigate the nanostructures of the two surfaces and the cross-section of SE–MS ribbons. The results show that the contact surface is characterized by 5–10 nm nanocrystalline domains embedded in the amorphous phase (Figure 16.3b and 16.3c), whereas the free surface features micron-sized dendritic crystals (Figure 16.3d). More interestingly, we have identified nanoscale modulations within the dendritic crystals (Figure 16.3e), and the Fourier analysis yielded periodicity close to 1.0 nm—about 5 times that of the $d_1 = d_{0015}$ spacing. The nanostructures observed on the cross-section of ribbons are shown in Figure 16.3f through 16.3i. Note, a gradient of multiscale nanostructures forming across the thickness of ribbons during the MS process (Figure 16.3f through 16.3i): The selected area electron diffraction confirmed that the nanocrystalline domains can be indexed as Bi_2Te_3 and the orientation of domains appeared to be random.

A question that often arises relates to the stability of multiscale nanostructures induced by MS during the sintering process. The SPS process (heat treatment and sintering process) enables the nuclei to grow into nanocrystalline regions after nucleation and growth. However, owing to the rapid sintering process, the amorphous structure and 5–10 nm nanocrystallites cannot grow too much, and therefore are preserved and embedded in the resulting ingot. We have examined the impact of the SPS process on these multiscale nanostructures. As shown in Figure 16.4a and 16.4b, the typical nanostructure following the SPS process contains 10–20 nm nanocrystals embedded in the MS–SPSed bulk matrix. These 10–20 nm nanocrystals are probably derived from the amorphous phase and the 5–10 nm nanocrystals that are shown in Figure 16.3a and 16.3b. We note that there is no clear grain boundary between nanocrystalline regions shown in Figure 16.4a and 16.4b, which means that the interfaces are coherent. Moreover, the nanoscale modulation structure is found in large crystals with sizes of several hundred nm, shown in Figure 16.4c and 16.4d. According to results shown in Figure 16.4, it appears that the multiscale nanostructures in melt-spun ribbon samples can be preserved during the rapid SPS process.

To supplement the microscopy study of the nanostructures, we have subsequently performed small angle neutron scattering (SANS) measurements. SANS is a well-established technique for characterizing nanostructures on length-scales from a few nm to a few hundreds nm, which greatly complements the electron microscopy study. In Figure 16.5 we present the SANS intensity (*I*) as a function of momentum transfer (*Q*) for the SE–MS–SPS and zone melted (ZM) samples. The plot is in a *double-logarithmic* scale. Due to the complexity of nanostructures, we shall restrict the data analysis to a comparison of the

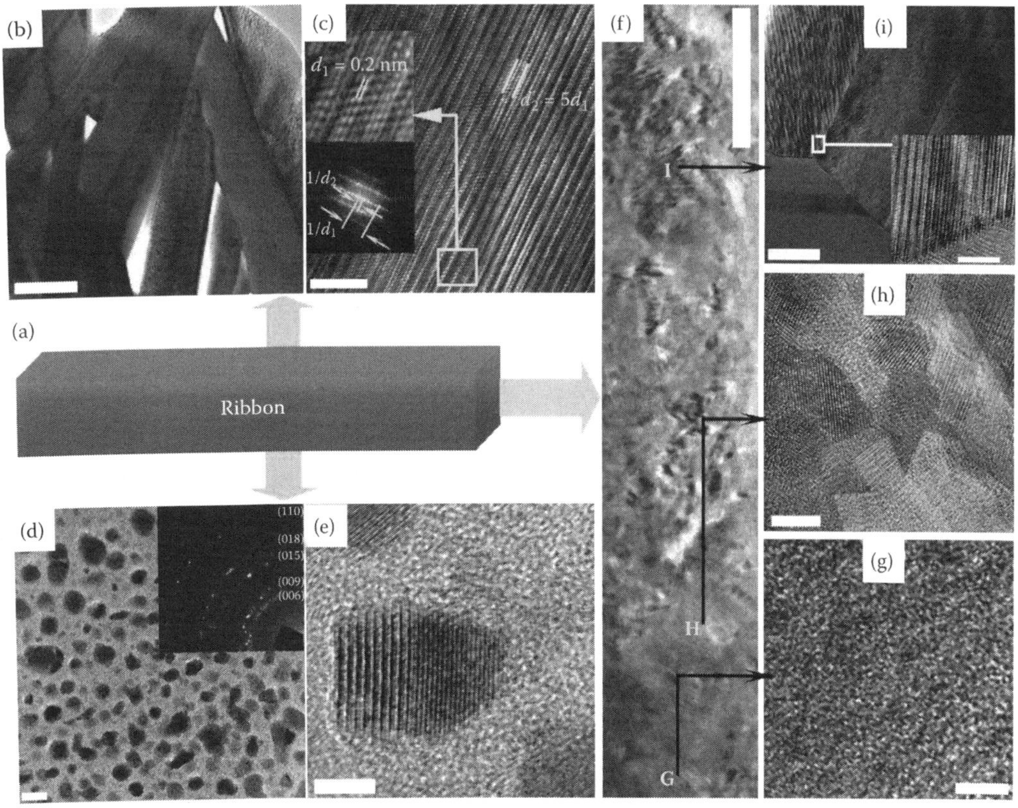

FIGURE 16.3 (**See color insert.**) Microscopy images of the melt spun ribbon. (a) Schematic figure of ribbon sample; (b) TEM image of the free surface that is typified by microsize dendritic crystals, and (c) HRTEM image of nanoscale modulations in the dendritic crystal shown in (b); (d) TEM image of the near contact surface; (e) HRTEM image of nanocrystalline domains; (f) TEM image of the cross-section of the ribbon, the bottom and the top of figure are images near the contact surface and the free surface, respectively; (g) HRTEM image of the amorphous phase near the contact surface; (h) HRTEM image of nanocrystalline domains with mostly coherent boundaries; (i) TEM image of micron-sized crystals, with the inset displaying the HRTEM image of the nanoscale modulation, the scale bar for the inset is 2 nm; The scale bars for (b), (c), (d), (e), (f), (g), (h), and (i) are 200 nm, 5 nm, 20 nm, 5 nm, 0.5 μm, 10 nm, 5 nm, and 50 nm, respectively. (Reprinted with permission from Xie, W. J. et al., Identifying the specific nanostructures responsible for the high thermoelectric performance of $(Bi,Sb)_2Te_3$ nanocomposites, *Nano Lett.*, 10, 3283, Copyright 2010, American Chemical Society.)

Q-dependence of scattered intensity in the two respective samples. First of all, the scattered intensity for the SE–MS–SPS sample exhibits a salient power-law behavior with an exponent, $a = 3.7$ between $Q_{min} = 0.001$ Å^{-1} and $Q_{max} = 0.1$ Å^{-1} that is, $I \sim Q^{-a}$ with $a = 3.70 \pm 0.01$. This observation is consistent with the existence of a broad distribution of nanostructures over a decade or more in length-scale, which is consistent with the results concerning multiscale nanostructures in the electron microscopy study (Figures 16.3 and 16.4). In addition, the observed power-law exponent is close to $a \approx 4$, which indicates that the interfaces among the nanostructures and the bulk matrix in the SE–MS sample are mostly homogeneous, or *coherent*, in accord with the HRTEM results.

Temperature dependences of the electrical resistivity (ρ) and Seebeck coefficient (α) are presented in Figure 16.6a and 16.6b, respectively. For comparison, a sample cut directly from the commercial ZM ingot along the ZM direction is designated as ZM. Both the ZM and SE–MS–SPS samples exhibit a typical semimetallic behavior in their electrical resistivity and Seebeck coefficient. Despite their similar resistivity behavior, they display different behaviors in their Seebeck coefficient. The SE–MS–SPS sample has

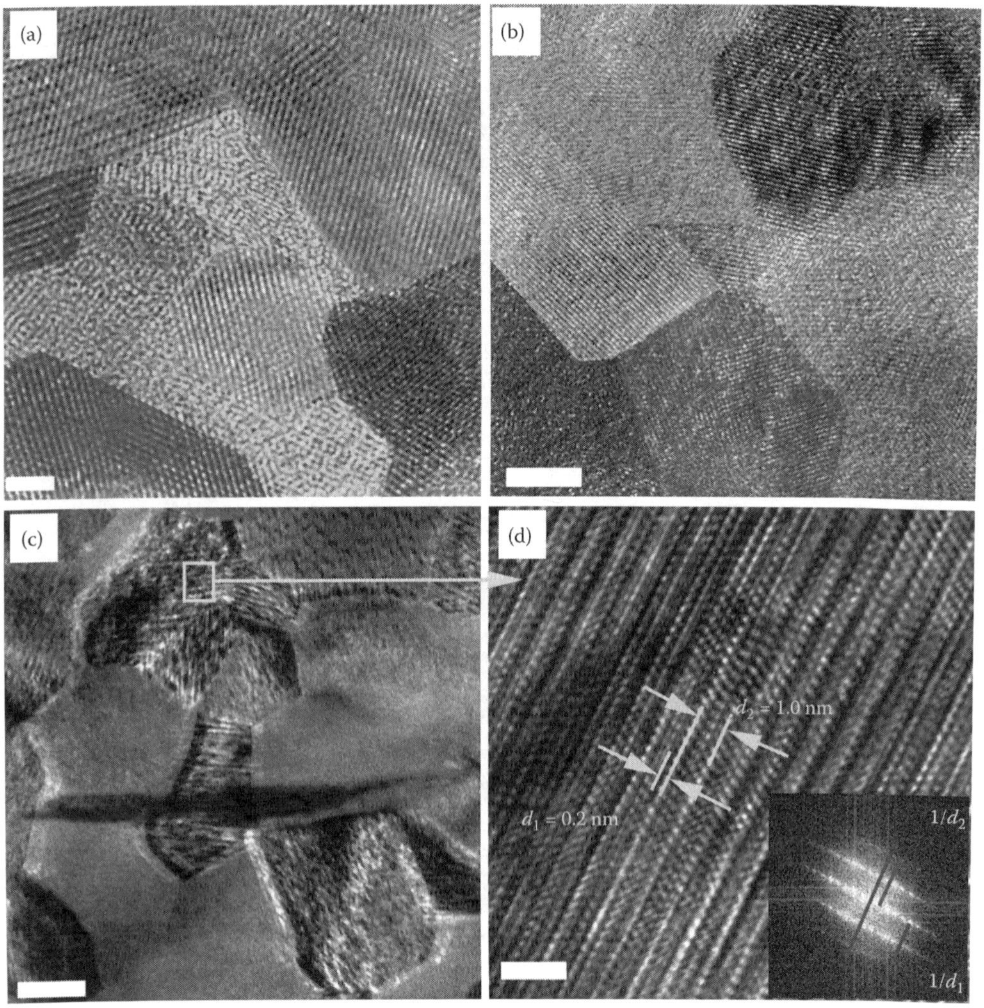

FIGURE 16.4 **(See color insert.)** (a)–(d) are representative TEM images of SE–MS–SPS (single element–melt spun–SPS) sample; (a) and (b) show HRTEM images of typical nanostructures consisting of 10–20 nm nanocrystals embedded in the bulk matrix. (c) and (d) show the HRTEM images of nanoscale modulations. The scale bars for (a), (b), (c), and (d), are 2, 5, 100, and 2 nm, respectively. (Reprinted with permission from Xie, W. J. et al., Identifying the specific nanostructures responsible for the high thermoelectric performance of $(Bi,Sb)_2Te_3$ nanocomposites, *Nano Lett.*, 10, 3283, Copyright 2010, American Chemical Society.)

as much as 10% higher Seebeck coefficient than the ZM sample between $T \approx 270$ K and 540 K. Furthermore, the Seebeck coefficient of the SE–MS–SPS sample peaks at $T \approx 450$ K, the temperature that is significantly higher than that of the ZM sample. The results of Hall coefficient measurements indicate that both samples have a carrier concentration on the order of 10^{19} cm^{-3} and within ±10% of each other between 5 and 300 K. Moreover, the $T^{-3/2}$ dependence of the Hall mobility is consistent with a predominant electron–phonon scattering near room temperature. The concomitant enhancement and the upshift of the Seebeck coefficient peak with temperature strongly suggest that the nanostructuring process induced a favorable change in the Fermi surface topology. This argument is further supported by the temperature dependence of the lattice thermal conductivity. As shown in the main panel and inset of Figure 16.6c, the minima of the lattice thermal conductivity (κ_L) and the total thermal conductivity (κ) are close to the temperature where the Seebeck coefficient peaks, a signature of bipolar thermal

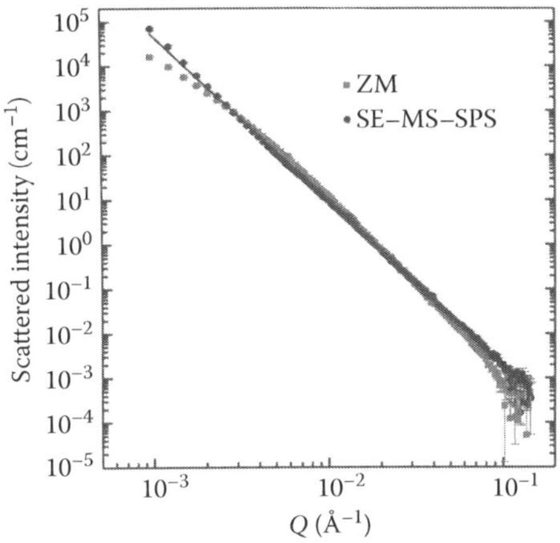

FIGURE 16.5 SANS intensities as a function of momentum transfer Q in a *double-logarithmic* plot. The SE–MS–SPS sample is fitted with the Q^{-a} dependence for a full range of Q values with $a = 3.7$ (black line). (Reprinted with permission from Xie, W. J. et al., Identifying the specific nanostructures responsible for the high thermoelectric performance of $(Bi,Sb)_2Te_3$ nanocomposites, *Nano Lett.*, 10, 3283, Copyright 2010, American Chemical Society.)

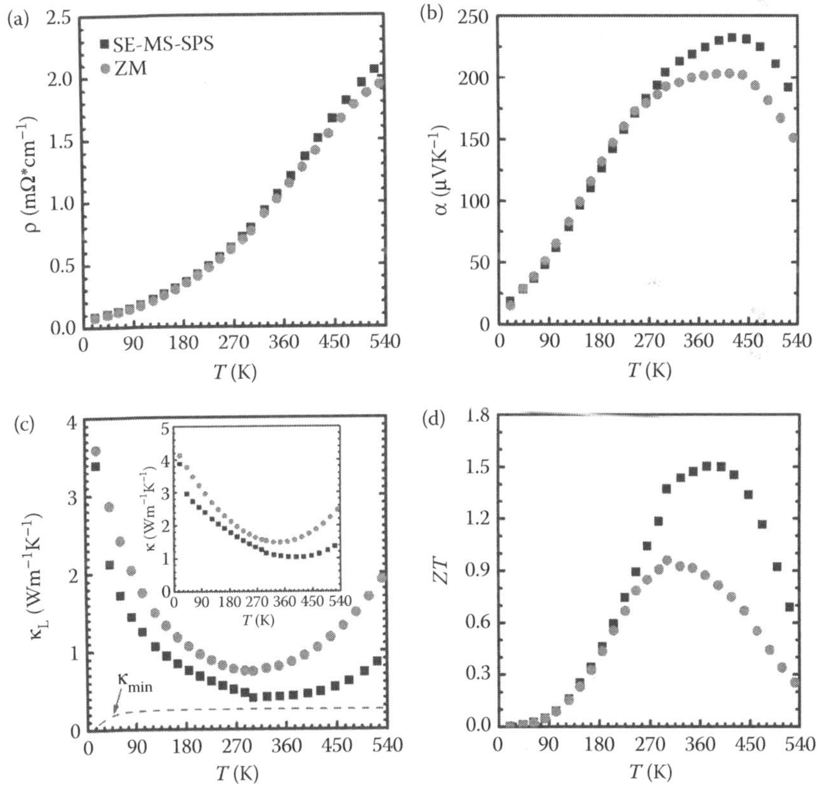

FIGURE 16.6 TE properties of ZM and SE–MS–SPS bulk materials: (a) electrical resistivity; (b) Seebeck coefficient; (c) lattice thermal conductivity with an insert displaying the total thermal conductivity; (d) figure of merit ZT values. (Reprinted with permission from Xie, W. J. et al., Identifying the specific nanostructures responsible for the high thermoelectric performance of $(Bi,Sb)_2Te_3$ nanocomposites, *Nano Lett.*, 10, 3283, Copyright 2010, American Chemical Society.)

conduction. The MS–SPS nanostructuring procedure thus alters the band structure and shifts the impact of the detrimental ambipolar effects to higher temperatures. (*Ambipolar effects become apparent when temperatures are high enough in a given material so that minority carriers start to contribute to the conduction process.*) As a result, the adopted two-stage nanostructuring procedure leads to a slightly higher resistivity, a moderately enhanced Seebeck coefficient, and a significantly reduced lattice thermal conductivity in the SE–MS–SPS sample. In particular, the SE–MS–SPS sample has a lattice thermal conductivity, $\kappa \approx 0.4–0.5$ W/m-K between 300 K and 400 K, as compared to a $\kappa \approx 0.8–0.9$ W/m-K of the ZM sample. An increase of 50% or more in the ZT values between 280 and 540 K over the ZM sample has been attained, and the maximum ZT value is ≈1.5 at 390 K (Figure 16.6d).

Detailed analysis of nanostructures by the TEM technique combined with the neutron scattering analysis [47] yielded valuable information pertaining to the origin of the high ZT observed in the p-type $(Bi,Sb)_2Te_3$ nanocomposite. In general, nanostructures of the as-prepared MS–SPSed sample are characterized by closely packed micron-sized crystals with nanoscale modulations, a small amount of an amorphous phase, and a large population of 10–20 nm nanocrystalline domains with coherent boundaries. The characteristic length-scales of these nanostructures extend down to a few nanometers, comparable to the estimated wavelength of heat-carrying phonons, while the mostly coherent grain boundaries cause a minimal degradation of the carrier mobility. This is the reason why these nanostructures give rise to only a slight increase in resistivity but significantly reduce the lattice thermal conductivity. In other words, the boundaries between the nanocrystalline domains and the bulk matrix scatter phonons more effectively than they scatter electrons.

16.3.1.2 n-Type $(Bi,Sb)_2(Te,Se)_3$ Nanocomposites Prepared by MS–SPS

We have found that n-type $(Bi,Sb)_2(Te,Se)_3$ nanocomposites can also be prepared by MS in combination with SPS [93]. The morphology of n-type $(Bi,Sb)_2(Te,Se)_3$ nanocomposites is very similar to that of p-type $(Bi,Sb)_2Te_3$ nanocomposites [45–47,93]. By combining compositional optimization with nanostructuring, the MS–SPSed n-type $(Bi_{0.95}Sb_{0.05})_2(Te_{0.85}Se_{0.15})_3$ nanocomposite with a layered nanostructure achieves low lattice thermal conductivity, resulting in the highest ZT value of 1.0 at 460 K shown in Figure 16.7d. Furthermore, the ZT value of MS–SPSed n-type $(Bi_{0.95}Sb_{0.05})_2(Te_{0.85}Se_{0.15})_3$ nanocomposites remains above 0.9 in the temperature range from 360–500 K. These n-type Bi_2Te_3 nanocomposites not only exhibit relatively good TE performance, but they also possess very good mechanical properties as compared with those of commercial zone-melted ingots.

The above results indicate that, compared with the p-type Bi_2Te_3 material, it is very difficult to significantly improve ZT for n-type Bi_2Te_3 bulk materials by introducing nanostructures via the MS process. We also note that essentially the same results were obtained for n-type materials by other researchers who used ball milling and sintering techniques [48–50]. Currently, our focus has been more on methods with which to decrease the lattice thermal conductivity. In future, the focus should shift towards improving the power factor for n-type Bi_2Te_3 and especially achieving a higher Seebeck coefficient.

16.3.2 n-Type Filled Skutterudites Prepared by MS–SPS

Skutterudites are the highest pnicogen (P, As, Sb) concentration species that form stable compounds with Co, Rh, and Ir. The skutterudite structure—the ε-phase $CoSb_3$ in the case of cobalt antimonides—is obtained in a peritectic reaction at 873°C from the δ-phase ($CoSb_2$) and the liquid phase [120,121]. Because of very slow kinetics, the peritectic reaction is rarely completed and long annealing times are required to achieve the $CoSb_3$ phase. To assure full densification, the ingots are crushed and hot pressed with a subsequent further annealing below the peritectic temperature. The process is time and energy consuming and typically takes a week to 10 days to prepare a fully densified single-phase skutterudite structure [6,15,122–125].

The skutterudite lattice contains structural voids that can be filled by foreign species such as rare earth, alkaline earth, and other atoms, forming what is known as filled skutterudites [126]. Viewed originally as one of the examples of the "PGEC" concept [5], filled skutterudites have stimulated much

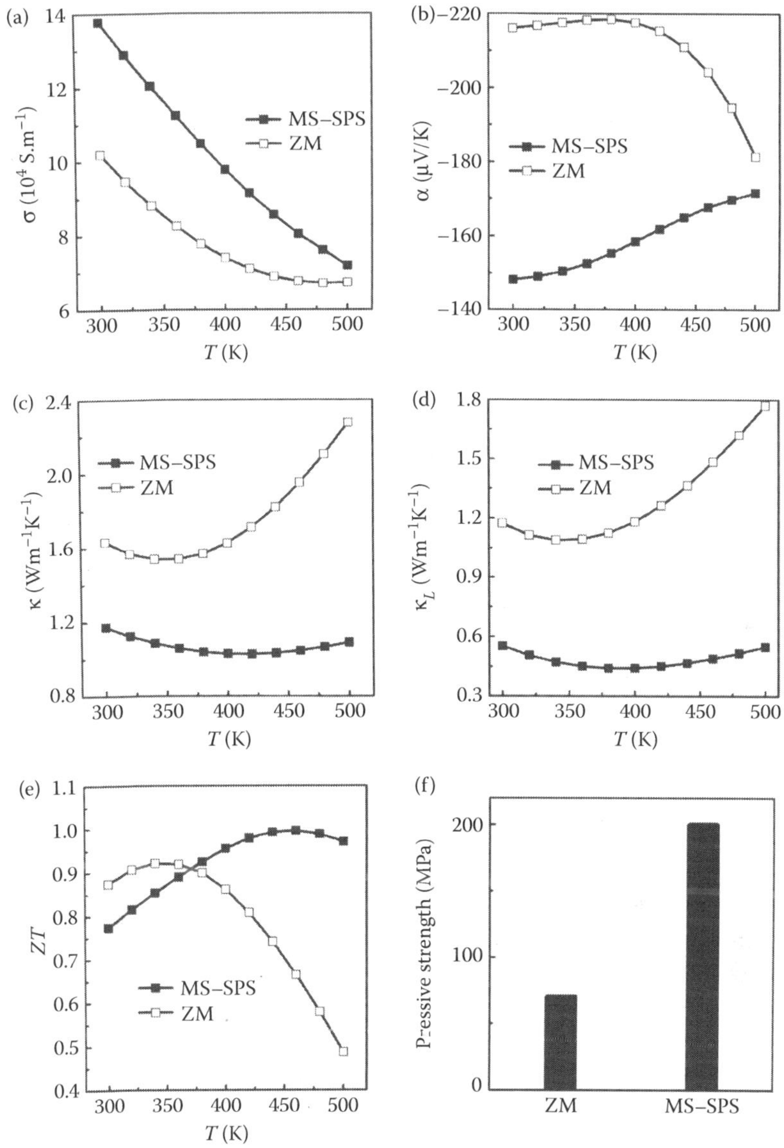

FIGURE 16.7 TE transport properties and mechanical properties of zone-melted $Bi_2(Se_{0.07}Te_{0.93})_3$ + 0.08wt%TeI_4 and MS–SPSed *n*-type $(Bi_{0.95}Sb_{0.05})_2(Te_{0.85}Se_{0.15})_3$ bulk materials: (a) electrical resistivity; (b) Seebeck coefficient; (c) total thermal conductivity; (d) lattice thermal conductivity; (e) figure of merit *ZT* values; and (f) compressive strength. (Adapted from Wang, S. Y. et al., *J. Phys. D, Appl. Phys.*, 43, 335404, 2010. With permission.)

scientific interest worldwide as one of the premier novel TE materials. Since the filler species enter the structure as electropositive ions, filling the voids of $CoSb_3$ is in fact a strong *n*-type doping mechanism. Although, such *n*-type filled skutterudites have a significantly reduced thermal conductivity in comparison with $CoSb_3$ on account of the fillers "rattling" in the oversized icosahedron-shaped voids, the thermal conductivity is nevertheless still too high and limits the magnitude of *ZT*. Research over the last decade has revealed that when ions of the rare earths, alkali metals, and alkaline earths [6–16] with large atomic mass and small ionic radius are used as fillers (especially species such as Yb, La, Ce, Eu, Ba), the thermal conductivity is particularly strongly affected.

Since theoretical arguments [127] suggest that lower-dimensional structures might possess considerably enhanced TE properties, it is desirable to introduce nanometer-scale features into the bulk skutterudite matrix. Nanostructures that are properly chosen should preferentially scatter heat-carrying phonons without significantly hindering charge carrier transport. In fact, there might even be favorable circumstances where the presence of nanostructures enhances the Seebeck coefficient [128]. Regarding skutterudites, Anno et al. [129] and Toprak et al. [130] reported on the improved TE performance when the grain size in $CoSb_3$ was reduced.

In this section, we describe the synthesis route that is combining MS with SPS and that results in a rapid formation of nanostructured bulk skutterudites with the nearly full theoretical density. Using single-filled $Yb_xCo_4Sb_{12}$ ($x = 0.1$, 0.2, 0.3) as an example, we have demonstrated that the MS–SPS processing technique achieves excellent n-type skutterudites. Subsequently, $Yb_xCo_4Sb_{12+y}$ nanocomposites with Sb nanoinclusions have also been prepared by the MS–SPS process, and transport property measurements show that the multiscale nanostructure significantly improves the TE performance.

16.3.2.1 Microstructure and TE Properties of MS–SPS Prepared $Yb_xCo_4Sb_{12}$

In this research, we found that the exact Yb filling fraction has a minor influence on the nanostructure and phase composition of the ribbon samples following MS and on bulk materials after SPS processing. Thus, for convenience, we take the $Yb_{0.2}Co_4Sb_{12}$ compound as a representative example of the entire series of $Yb_xCo_4Sb_{12}$ ($x = 0.1$, 0.2, 0.3) compounds when we discuss nanostructures and phase composition.

Figure 16.8 shows the phase composition of ribbons prepared under different MS parameters. All melt-spun ribbons exhibit a complex phase composition and the diffraction peaks are broad due to a high cooling rate of the melt during MS processing and the formation of nanostructures in the ribbons (see Figure 16.9). When the copper drum speed $V = 10$ m/s, the main phase is that of a skutterudite as shown in Figure 16.8a. When V increases to 30 m/s (see Figure 16.8b), the fraction of the skutterudite phase decreases notably and the main phase in the ribbons is $CoSb_2$. We note that the diffraction peaks broaden further and the diffraction intensity decreases. With respect to the MS technique, the cooling rate increases with the increasing linear speed V of the spinning copper drum and the crystalline phases are therefore composed mainly of high-temperature phases such as $CoSb_2$ and CoSb, and the size of the crystallites becomes much finer (see Figure 16.9b and 16.9b-1).

The field emission scanning electron microscopy (FESEM) photographs of a ribbon's free face after MS are shown in Figure 16.9. As depicted in Figure 16.9a and 16.9a-1, when $V = 10$ m/s, the grain size changes from 100 to 200 nm and the crystals agglomerate into dendritic patterns. Energy-dispersive x-ray (EDX) analysis reveals that the free face exhibits a shortage of Sb and Yb. When $V = 30$ m/s, more

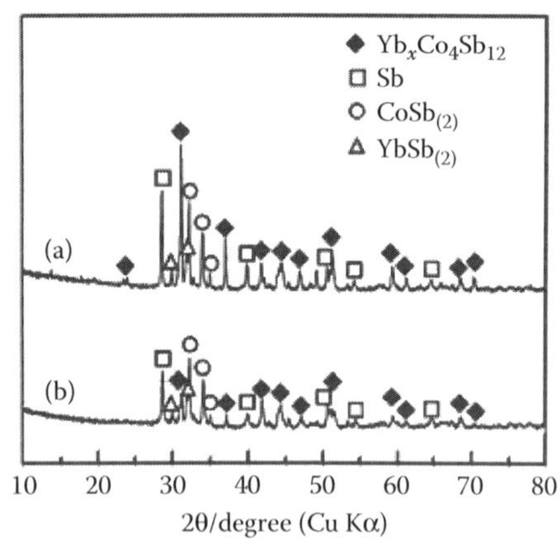

FIGURE 16.8 XRD patterns of ribbons after MS processing. (a) $V = 10$ m/s; (b) $V = 30$ m/s. (Adapted from Li, H. et al., *J. Phys. D, Appl. Phys.*, 42, 145409, 2009. With permission.)

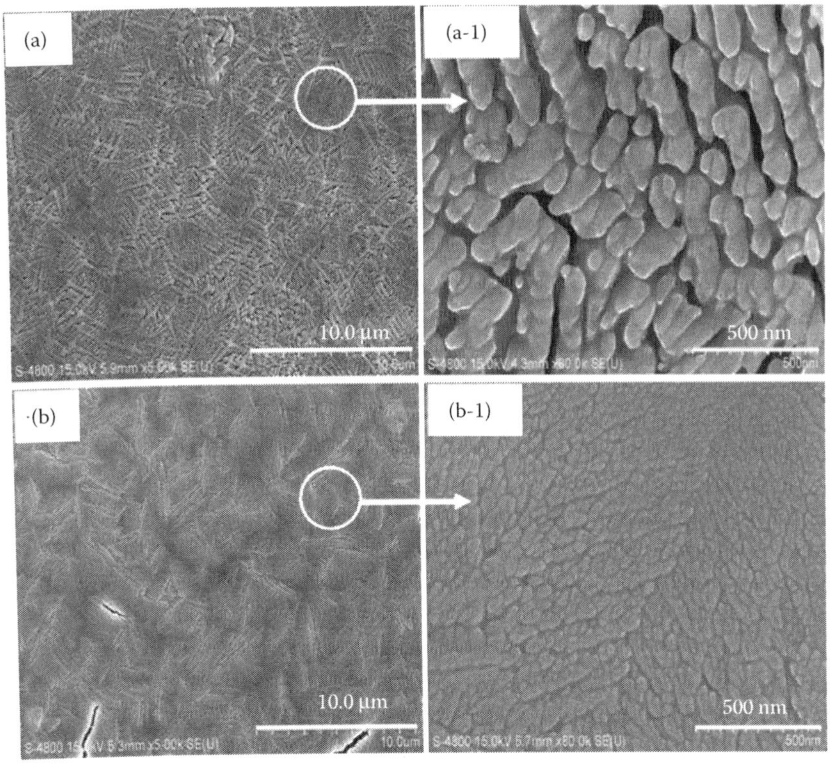

FIGURE 16.9 FESEM photographs of the ribbon's free face after MS processing: (a) $V = 10$ m/s, (a-1) partially enlarged detail of (a); (b) $V = 30$ m/s, (b-1) partially enlarged detail of (b). (Adapted from Li, H. et al., *J. Phys. D, Appl. Phys.*, 42, 145409, 2009. With permission.)

evenly distributed nanocrystals of about 20–50 nm size can be observed on the free face of the ribbons, see Figure 16.9b and 16.9b-1. EDX analysis shows homogeneously distributed elements, the stoichiometry of which is approximately in accord with the nominal elemental ratio.

In comparison with the free face of the ribbons, the contact face is formed by the rapidly freezing melt that contacts directly with the chilled copper drum, so the cooling rate is much higher. Figure 16.10 shows the FESEM photographs of the ribbon's contact face after MS. EDX analysis indicates a minute

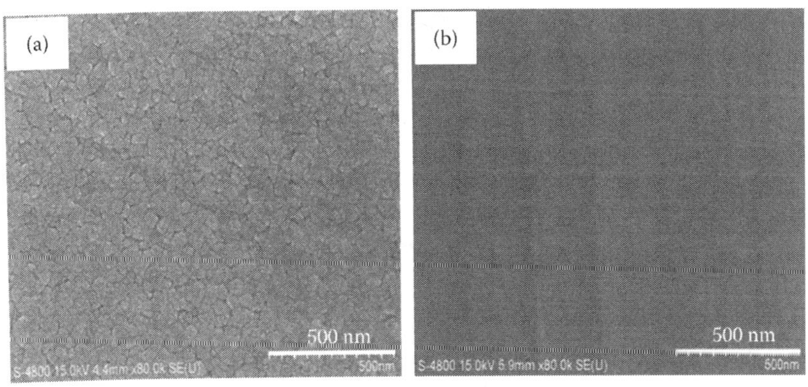

FIGURE 16.10 FESEM photographs of the ribbon's contact face after MS processing: (a) $V = 10$ m/s; (b) $V = 30$ m/s. (Adapted from Li, H. et al., *J. Phys. D, Appl. Phys.*, 42, 145409, 2009. With permission.)

amount of excess Sb and Yb when $V = 10$ m/s. At the same linear speed V, the grain size on the contact face is much smaller than that on the free face. This is especially so at the higher speed $V = 30$ m/s where the contact face appears as a uniform amorphous structure with no discernable details. To identify the finer nanostructure of the ribbons when $V = 30$ m/s, the ribbon is thinned down using a single-side ion-milling method for observation of the contact face and the free face by HRTEM. HRTEM images of the ribbon after MS with $V = 30$ m/s are shown in Figure 16.11. The high-resolution photographs (Figure 16.11a and 16.11a-1) and electron diffraction pattern (the insert of Figure 16.11a-1) of the contact face verify our suspicion that the amorphous structure has formed on the contact face. Figure 16.11b and 16.11b-1 show a compact arrangement of the nanoscale crystals with the grain size of 30 nm. Overall, the results of the HRTEM analysis are in good agreement with the results of FESEM.

Bulk materials are obtained by sintering the ribbons using the SPS technique. The XRD patterns of bulk materials are shown in Figure 16.12. The most *remarkable* feature of these data is the formation of a *single-phase skutterudite phase* after the ribbons are sintered for just 5 min in the SPS process. In the traditional synthesis route, it takes long annealing times (>100 h) to obtain the single-phase skutterudite phase [6,15,122–125]. The rapid formation of the skutterudite phase is likely due to a homogeneous distribution of the constituents and a very fine nanostructure of the ribbons prepared by MS. It should be mentioned that, when $V = 10$ m/s, a small amount of $CoSb_2$ impurity is detected (see Figure 16.12a). However, this impurity phase is beyond the detection limit when the linear speed is increased to 30 m/s and one is left with an essentially pure skutterudite phase as shown in Figure 16.12b. It may be concluded that the more homogeneous distribution of the elements the ribbon possesses, the more complete and rapid the formation of the skutterudite phase. Consequently, carrying out MS under higher speeds should benefit the phase purity of the resulting bulk material.

Nanostructures of the starting material and of bulk samples prepared by the combined MS + SPS processing are shown in Figure 16.13. It is clear from this figure that the MS + SPS method is very

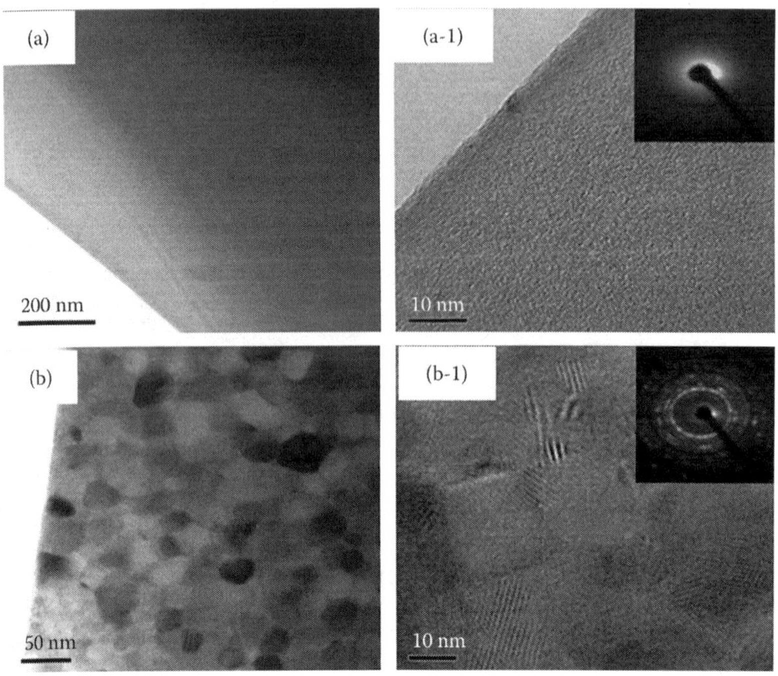

FIGURE 16.11 HRTEM image of a ribbon after MS processing with $V = 30$ m/s: (a) contact face, (a-1) high-resolution image of (a); (b) free face, (b-1) high-resolution image of (b). (Adapted from Li, H. et al., *J. Phys. D, Appl. Phys.*, 42, 145409, 2009. With permission.)

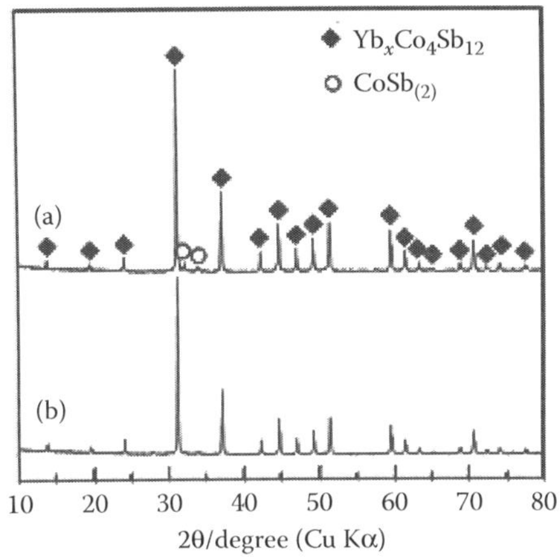

FIGURE 16.12 XRD patterns of bulk $Yb_{0.2}Co_4Sb_{12}$: (a) $V = 10$ m/s; (b) $V = 30$ m/s. (Adapted from Li, H. et al., *J. Phys. D, Appl. Phys.*, 42, 145409, 2009. With permission.)

effective in decreasing the grain size in bulk skutterudites. When $V = 10$ m/s, the grain size is approximately on the scale of 300 nm; when V is increased to 30 m/s, the grain size decreases to about 100 nm range and the crystallites are more evenly distributed. Looking at Figures 16.9 and 16.13, it is apparent that the size of nanocrystals in the ribbons has a great influence on the ultimate grain size of bulk skutterudite samples. It is worthwhile to mention that all our bulk skutterudite compounds are very dense materials with the density approaching 100%.

FIGURE 16.13 FESEM images of bulk $Yb_{0.2}Co_4Sb_{12}$ after SPS processing: (a) starting material; (b) $V = 10$ m/s; (c) $V = 30$ m/s. (Adapted from Li, H. et al., *J. Phys. D, Appl. Phys.*, 42, 145409, 2009. With permission.)

When we inspect and compare growth features in Figures 16.9 and 16.10, it is obvious that there are differences between the growth patterns developed on the contact face and on the free face of the ribbons with the former resulting in a much finer nanostructure than the latter. The question arises whether these two different morphologies corresponding to the contact and free faces of the ribbons are reflected in the nanocrystalline pattern of the final product—the SPS-processed bulk skutterudite samples? Because of its limited resolution, an FESEM photograph of a bulk skutterudite in Figure 16.13c cannot provide an unequivocal answer and we need to have images with higher resolution. This we obtained using HRTEM on our bulk nanostructured $Yb_{0.2}Co_4Sb_{12}$ sample. In Figure 16.14a, we note nanocrystalline regions with two distinct length-scales, those of typically 100 nm size and areas of much finer nanocrystals. A region of the finer nanostructure is magnified in Figure 16.14b where one clearly sees nanocrystals of typically 10–20 nm size. The likely origin of this landscape is the contact face region of the ribbons that was originally entirely amorphous (see Figure 16.10b) and has evolved into a fine nanostructure as a result of SPS processing. Thus, the bulk sample contains a composite nanostructure with two distinct length-scales, regions with crystallites of typically 100 nm size and regions of a much finer nanostructure. It is this very fine, 10–20 nm nanostructure that we expect to have a significant impact on the scattering of phonons as it is of comparable size to the phonon wavelength. Moreover, Figure 16.14c shows the high-resolution image of a grain boundary region and we note that the interfaces are substantially coherent with no signs of the presence of any grain-boundary phase. Such a structure, in addition to providing for strong phonon scattering, should at the same time enable smooth, unimpeded electron transport and therefore support high power factors.

Transport parameters and their temperature dependence are displayed in Figure 16.15. Electrical conductivity (σ) and Seebeck coefficient (α) for the MS–SPS-prepared nanostructured bulk $Yb_xCo_4Sb_{12}$ ($x = 0.1$,

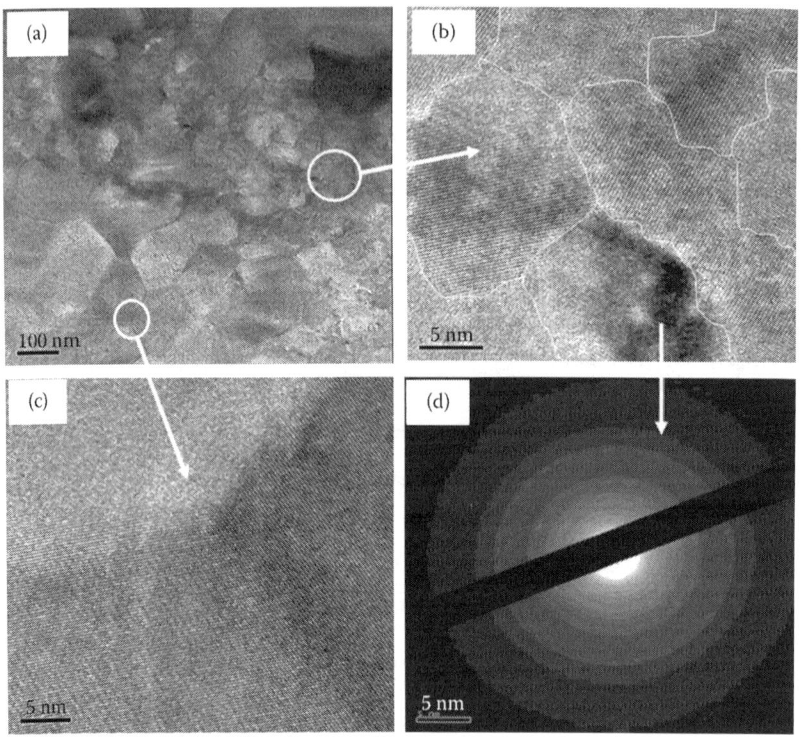

FIGURE 16.14 HRTEM images of (a) bulk $Yb_{0.2}Co_4Sb_{12}$ after SPS processing with $V = 30$ m/s; (b) finer nanocrystals area; (c) high resolution image of a grain boundary; (d) electron diffraction pattern of (b). (Adapted from Li, H. et al., *J. Phys. D, Appl. Phys.*, 42, 145409, 2009. With permission.)

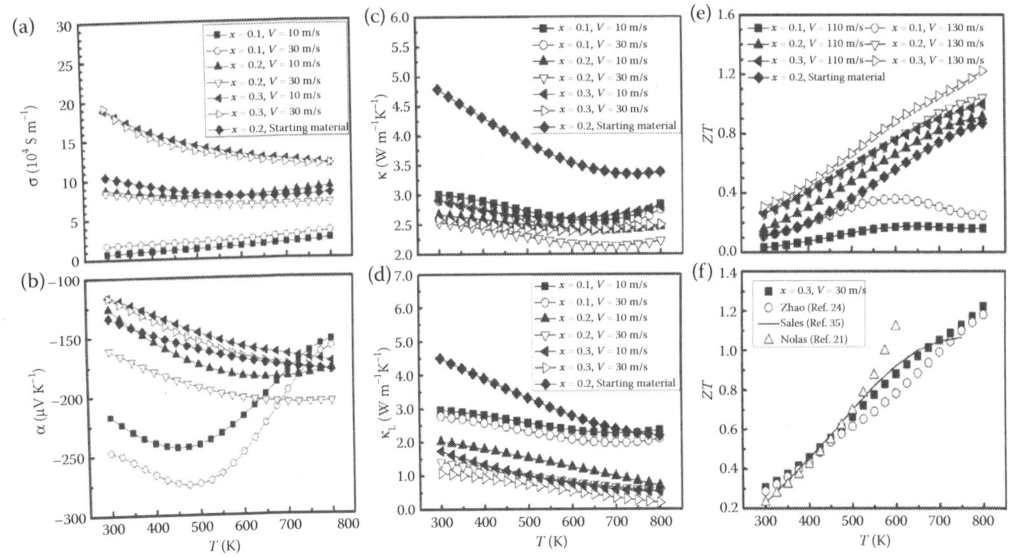

FIGURE 16.15 Temperature dependence of (a) electrical conductivity, (b) Seebeck coefficient, (c) total thermal conductivity, (d) lattice thermal conductivity, and (e) ZT for $Yb_xCo_4Sb_{12}$ ($x = 0.1, 0.2, 0.3$). Dimensionless figure of merit ZT of Yb single-filled skutterudites from the literature and our data for $Yb_{0.3}Co_4Sb_{12}$ are shown in part (f). (Adapted from Li, H. et al., *J. Phys. D, Appl. Phys.*, 42, 145409, 2009. With permission.)

0.2, 0.3) compounds between 300 and 800 K are plotted in Figure 16.15. The electrical conductivity increases and changes its character from that of a nondegenerate to a strongly degenerate semiconductor as the filling fraction of Yb increases [123,124,131]. The linear speed of the copper drum during the MS-formation of ribbons seems to have very little effect on the magnitude of the electrical conductivity and shows no distinct trend. For $x = 0.2$, we can make a comparison with a sample of similar Yb content ($Yb_{0.19}Co_4Sb_{12}$) prepared by the traditional method of synthesis [132]. Electrical conductivity is about 9.5×10^4 S/m for this sample at 600 K and thus is somewhat higher than the conductivity of both our $x = 0.2$ filled skutterudite samples which falls in the range of $7–8 \times 10^4$ S/m at 600 K. As shown in Figure 16.15b, and as expected, the Seebeck coefficient is strongly dependent on the filling fraction x and decreases with the increasing Yb content. However, we also note that with the increasing filling fraction of Yb the extrinsic regime of conduction is greatly extended. The dependence of the Seebeck coefficient on the linear speed of the copper drum, that is, on the size and extent of the nanostructure, is very surprising and important. We observe a large enhancement in the Seebeck coefficient with the reduced size of the nanocrystals as the linear speed of the copper wheel is increased. This is clearly seen in Figure 16.15b for $x = 0.1$ and 0.2, and although the effect is less pronounced for $x = 0.3$, it still accounts for a 5% enhancement at 800 K.

The most dramatic influence of the nanostructure on the transport properties comes via its impact on the thermal conductivity. In Figure 16.15c, we plot the behavior of the measured total thermal conductivity. For comparison, we include a plot of the total thermal conductivity of a sample with $x = 0.2$ prepared by the traditional synthesis—our so called starting material. The data reveal not only the obvious advantage of the MS processing in reducing the thermal conductivity but also a clear trend that the smaller the nanocrystalline grains in the MS–SPS-prepared skutterudites, the more effective the scattering of phonons on the interfaces. Using the Wiedemann–Franz relationship and taking the Lorenz constant at its Sommerfeld value of 2.45×10^{-8} V^2/K^2, we subtract the electronic thermal conductivity from the total thermal conductivity and the resulting lattice thermal conductivity is displayed in Figure 16.15d. Except for the fact that the lattice thermal conductivity is quite low, one should not pay too much attention to its exact value since it depends critically on the value of the Lorenz constant chosen to subtract the electronic thermal conductivity. Since, nobody knows what the actual Lorenz value of highly degenerate filled skutterudites is, it makes sense, for comparative purposes, to opt for the Sommerfeld value, that is, treating the system as fully degenerate. The redeeming feature of this approach is the fact that a vast majority of the published data on lattice thermal conductivity of skutterudites assumes precisely this to be the case and thus, one has a chance to compare the respective lattice thermal conductivities without any confusion concerning what is or is not the Lorenz value assumed. Again, the influence of the nanostructure is clearly evident in very low values of the lattice thermal conductivity displayed in Figure 16.15d. Comparing the MS–SPS prepared $Yb_{0.2}Co_4Sb_{12}$ with our starting material of the same composition, we note a reduction of the lattice thermal conductivity by 60% at 300 K and by 67% at 800 K. The currently record-high *n*-type double-filled skutterudite of Ref. [124] that was prepared by the traditional method of synthesis has the lattice thermal conductivity comparable to our single-filled MS–SPS prepared $Yb_{0.3}Co_4Sb_{12}$ and should clearly benefit from applying the MS–SPS processing to its synthesis. Figure 16.8d further attests to the fact that a fine nanostructure such as achieved with higher linear speeds of the copper drum will reduce the lattice thermal conductivity more effectively. Although we do not have a direct evidence, it is possible that the filling of Yb at $x = 0.3$ exceeds the filling fraction limit believed to be about 0.22 for this filler. In that case, a minute amount of Yb might have precipitated in our $Yb_{0.3}Co_4Sb_{12}$ sample and formed Yb_2O_3 just as was the case in samples of Refs. [123 and 124]. On the other hand, it is also possible that the rapid synthesis of the skutterudite phase that takes place in our MS–SPS processing method enhances the filling fraction limit of Yb and no ytterbium is segregated. This issue will have to be pursued in future research.

The dimensionless TE figure of merit ZT is calculated based on the transport data presented and is displayed in Figure 16.15e. Except for samples with $x = 0.1$ that have clearly a too low filling fraction to make them useful TE materials, skutterudites with the Yb content of $x = 0.2$ and 0.3 approach or exceed the ZT value of unity. The overall benefit of a fine nanostructure (samples with $V = 30$ m/s) is again clearly manifested in the form of high ZT values with the highest $ZT = 1.2$ achieved in $Yb_{0.3}Co_4Sb_{12}$ at 800 K. For comparison, alongside with our data, Figure 16.8f displays some other prominent values of ZT for Yb single-filled skutterudites. Our $Yb_{0.3}Co_4Sb_{12}$ compound fares very favorably and the results indicate that the MS–SPS processing method holds a great promise for achieving even higher values of the figure of merit especially when the starting composition is chosen to take advantage of multifilling and the purity of Co is better than 99.9%, [124] the material used in this work.

The application of the technique to skutterudites results in a fully dense material that possesses extensive nanostructure on the length-scale as short as 10–20 nm. The homogeneous nature of the MS-prepared ribbons leads to a remarkably rapid and efficient phase transformation of the strata into a near fully dense, single-phase, nanocrystalline skutterudite structure under a brief SPS processing. Moreover, HRTEM images of the grain boundary regions of the resulting bulk skutterudites indicate phase coherence and no secondary phases precipitating at the grain boundaries. Both of these attributes (nanostructural features and the phase coherence of the grain boundaries) are highly beneficial to the TE properties as the high density of fine nanocrystals strongly impedes phonon transport while the coherent grain boundaries support good electronic properties. Moreover, we have actually observed an enhancement in the Seebeck coefficient at a level of at least 10% when the nanograins become smaller as a result of a higher cooling rate of the ribbons when the linear speed of the spinning cooper wheel is increased.

16.3.2.2 Rapid Preparation and Improved TE Properties of Sb: Overstoichiometric $Yb_xCo_4Sb_{12+y}$ Compounds Prepared by MS–SPS

As shown in the preceding section, the manipulation of the nanostructure in skutterudites has a significant influence on their TE properties, especially, nanostructures in bulk materials can markedly increase the concentration of the grain boundaries, and the lattice thermal conductivity may be reduced due to enhanced scattering effects on phonons [123,129,130,133–135]. However, this scattering mechanism might have a similar effect on charge carriers too, so the electrical conductivity may be depressed to a certain degree. For the *n*-type skutterudites, which exhibit metallic behavior [122,123,132], electrical transport properties can be effectively improved by introducing a small excess of Sb [136,137].

Based on the results of the preceding section, we know that when single-phase skutterudites are used as starting ingots for the MS process, the resulting ribbon samples show obvious signatures of phase separation. Hence, whether the starting ingot is or is not a single phase skutterudite may not be so important. We have therefore modified the MS–SPS synthetic route. The technique relies on a combination of melt-quench with melt spinning and spark plasma sintering (MMS) and cuts the preparation time to <2 days. A moderate excess of Sb is considered an effective way to adjust the electrical properties for *n*-type skutterudites. Moreover, the highly nonequilibrium nature of the synthesis process allows for a higher filling fraction of Yb than is otherwise possible via the traditional synthesis routes. The outcome is an efficient synthesis process that yields high-quality samples with outstanding TE properties.

Figure 16.16 shows the XRD patterns of all bulk $Yb_{0.3}Co_4Sb_{12+y}$ samples after SPS. The XRD pattern of a sample with no Sb excess that was prepared by the traditional method (sample $Yb_{0.3}Co_4Sb_{12}$) is shown in Figure 16.16a. Although, this sample went through a 168 h annealing process, a small quantities of YbSb and $CoSb_2$ impurity were detected. This is due to the Yb filling fraction of 0.3 that exceeds the theoretically and experimentally estimated critical filling fraction of 0.2 in this near-equilibrium prepared sample [138]. Figure 16.16b and 16.16c show the XRD patterns of bulk $Yb_{0.3}Co_4Sb_{12.3}$ and $Yb_{0.3}Co_4Sb_{12.6}$ samples synthesized by the MMS method in this research. There are no traces of YbSb and $CoSb_2$ impurities detected by XRD in these samples. However, a small amount of pure Sb is observed in the XRD pattern of the $Yb_{0.3}Co_4Sb_{12.6}$ sample. Electron probe micro-analyzer (EPMA) analysis shows that the actual composition of the $Yb_{0.3}Co_4Sb_{12.3}$ and $Yb_{0.3}Co_4Sb_{12.6}$ samples is $Yb_{0.285}Co_4Sb_{12.268}$ and $Yb_{0.289}Co_4Sb_{12.583}$, respectively, that is, the Yb content indeed exceeds the theoretical and experimental filling fraction limit 0.2. The results also suggest that the skutterudite structure cannot accommodate Sb excess greater than about $y = 0.3$. The vital difference between the MMS and the traditional method of preparation is that the material prior to SPS processing are supercooled ribbons made by MS (a highly nonequilibrium synthesis with the cooling rate on the order of 10^5 K/s) rather than skutterudite powders prepared by long time annealing. The ribbons contain a large fraction of nanocrystalline and amorphous phases and the rapid formation of the skutterudite structure is likely due to this fine and homogeneously distributed nanostructure. The reason why there is a 50% enhancement in the critical filling fraction of Yb awaits further detailed structural studies.

FESEM photographs of the bulk $Yb_{0.3}Co_4Sb_{12+y}$ samples after SPS are shown in Figure 16.17. Samples prepared by MMS show very high density approaching the theoretical 100%. As seen in Figure 16.17a and 16.17b, the grain size of the samples varies between 250 and 300 nm. The sample with $y = 0.3$

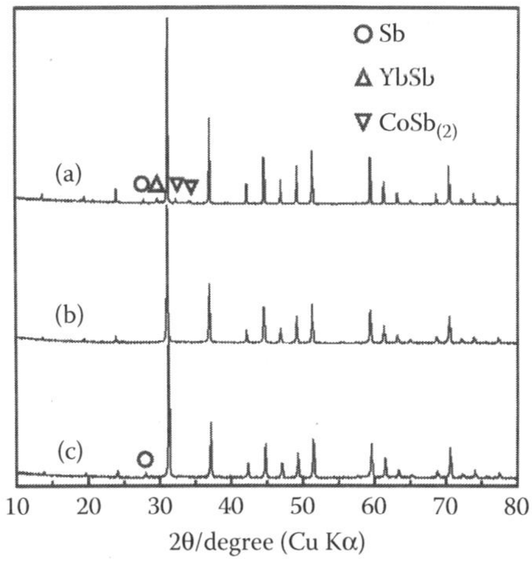

FIGURE 16.16 XRD patterns of bulk $Yb_{0.3}Co_4Sb_{12+y}$: (a) $y = 0$, the sample prepared by the traditional method of synthesis; (b) $y = 0.3$; (c) $y = 0.6$. (Reprinted with permission from Li, H. et al., Preparation and thermoelectric properties of high-performance Sb additional $Yb_{0.2}Co_4Sb_{12+y}$ bulk materials with nanostructure, *Appl. Phys. Lett.*, 92, 202114, Copyright 2008, American Institute of Physics.)

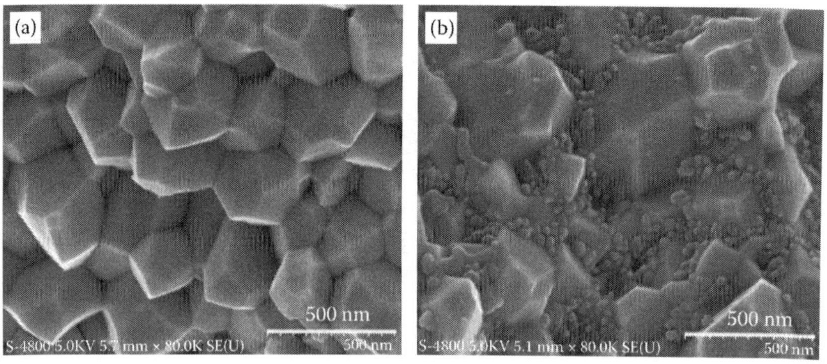

FIGURE 16.17 FESEM photographs of bulk nanostructured $Yb_{0.3}Co_4Sb_{12+y}$ after SPS: (a) $y = 0.3$; (b) $y = 0.6$. (Reprinted with permission from Li, H. et al., Preparation and thermoelectric properties of high-performance Sb additional $Yb_{0.2}Co_4Sb_{12+y}$ bulk materials with nanostructure, *Appl. Phys. Lett.*, 92, 202114, Copyright 2008, American Institute of Physics.)

possesses evenly distributed grain size and clear and sharp grain boundaries. However, when $y = 0.6$, the excess of Sb is too large and antimony segregates in the form of nanoparticles (~30 nm) at the grain boundaries, as confirmed by EDX analysis.

The temperature dependence of electrical conductivity, Seebeck coefficient and power factor for $Yb_{0.3}Co_4Sb_{12+y}$ compounds are shown in Figure 16.18a and 16.18b, respectively. Both $Yb_{0.3}Co_4Sb_{12+y}$ compounds show a metallic conduction behavior and σ increases with the increasing excess of Sb. The absolute value of the Seebeck coefficient α increases with the increasing temperature and decreases with the increasing excess of Sb. Compared with other literature values of the Seebeck coefficient for high Yb filling fraction samples [123,132], because of the presence of nanostructure, α values in this research are higher [128]. Another reason is that the carrier concentrations of our $Yb_{0.3}Co_4Sb_{12.3}$ and $Yb_{0.3}Co_4Sb_{12.6}$ are 7.2×10^{20} and 7.9×10^{20} cm^{-3}, respectively, which are lower than the carrier concentration of 8.4×10^{20} cm^{-3} for the $Yb_{0.25}Co_4Sb_{12.69}/Yb_2O_3$ composite material [123]. Overall, electrical transport properties are markedly improved by higher Yb filling fraction and nanostructural features in the bulk matrix.

Figure 16.18c displays temperature dependence of the thermal conductivity and the lattice thermal conductivity of $Yb_{0.3}Co_4Sb_{12+y}$ compounds. Compared with $Yb_{0.3}Co_4Sb_{12}$ prepared by the traditional method [6,125], κ of the nanostructured $Yb_{0.3}Co_4Sb_{12+y}$ samples is decreased remarkably. Taking as an example $Yb_{0.3}Co_4Sb_{12.3}$, the thermal conductivity is reduced by 30% at room temperature and by 22% at 800 K. These values of thermal conductivity obtained on our bulk nanostructured $Yb_{0.3}Co_4Sb_{12+y}$ samples are comparable with those measured on double-filled $Ba_{0.08}Yb_{0.09}Co_4Sb_{12}$ [124], which suggests that even lower thermal conductivity might be achieved by preparing nanostructured double-filled skutterudite compounds. Moreover, the over-stoichiometric amount of Sb does not seem to increase thermal conductivity significantly. Using the Wiedemann–Franz relationship and taking the Lorenz number at its Sommerfeld value of 2.45×10^{-8} V^2/K^2, we subtract the electronic thermal conductivity from the total thermal conductivity and the resulting lattice thermal conductivity is plotted in the inset of Figure 16.18c. It should be noted that the sample with $y = 0.6$ possesses lower lattice thermal conductivity compared with the $y = 0.3$ sample. We attribute this to enhanced scattering of phonons on the nanometer-scale precipitates of Sb at the grain boundaries of the sample with $y = 0.6$, see Figure 16.10b. The value of κ_L is comparable to that of the $Yb_{0.25}Co_4Sb_{12}/Yb_2O_3$ composite material [123].

The dimensionless TE figure of merit ZT of our n-type $Yb_{0.3}Co_4Sb_{12+y}$ is calculated based on the above transport data and is presented in Figure 16.18d together with other representative values of ZT for Yb single-filled skutterudites. As is evident from Figure 16.14b, the bulk nanostructured $Yb_{0.3}Co_4Sb_{12+y}$ ($y = 0.3$ and 0.6) prepared by the MMS method show higher ZT values, and $(ZT)_{MAX}$ reaches values of at least 1.3 at 800 K for the $Yb_{0.3}Co_4Sb_{12.3}$ compound. It is worth mentioning that the ZT of $Yb_{0.3}Co_4Sb_{12.3}$

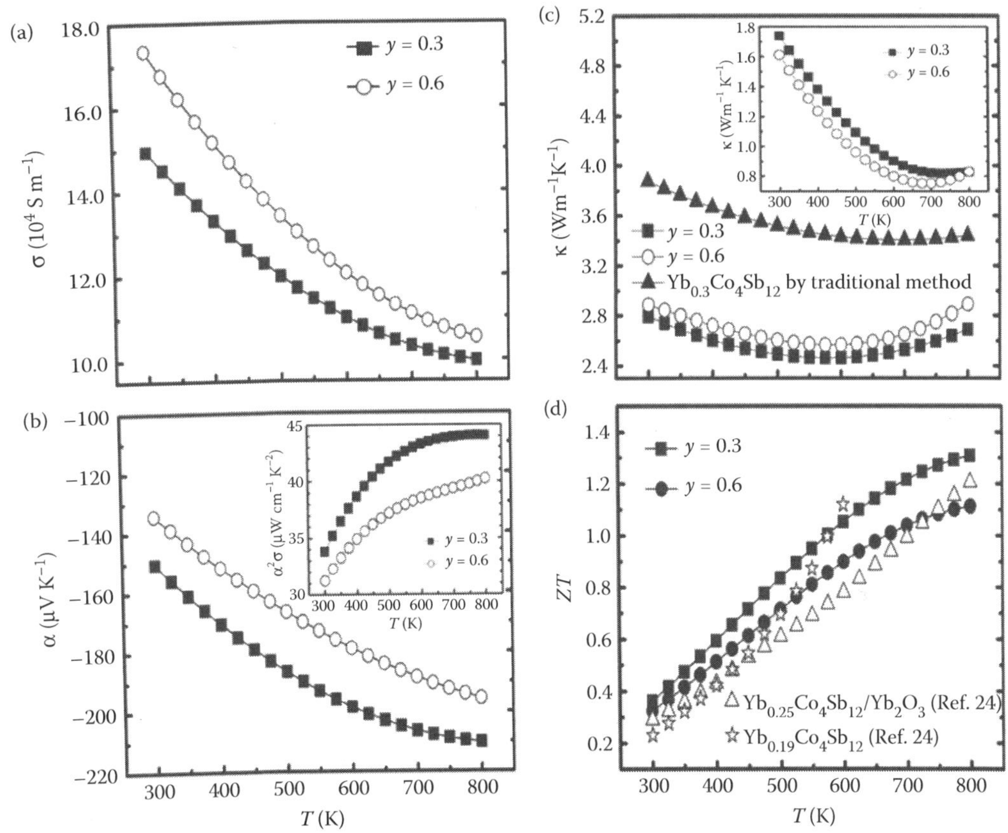

FIGURE 16.18 Temperature dependence of (a) electrical conductivity, (b) Seebeck coefficient, (c) thermal conductivity, (d) *ZT* for $Yb_{0.3}Co_4Sb_{12+y}$, and some representative Yb single-filled skutterudites from the literature. (Reprinted with permission from Li, H. et al., Preparation and thermoelectric properties of high-performance Sb additional $Yb_{0.2}Co_4Sb_{12+y}$ bulk materials with nanostructure, *Appl. Phys. Lett.*, 92, 202114, Copyright 2008, American Institute of Physics.)

exceeds the value of unity already at 570 K and keeps rising with the increasing temperature. For comparison, so far the best *n*-type skutterudite—the double-filled $Ba_{0.08}Yb_{0.09}Co_4Sb_{12}$ compound [125]—exceeds the value of unity at 600 K. Thus, the bulk nanostructured $Yb_{0.3}Co_4Sb_{12.3}$ material seems to show high performance over a marginally wider temperature range in spite of being merely single filled.

A rapid processing method that combines melt-quenching with MS and SPS cuts the synthesis time of high-performance filled skutterudites to <2 days yet yields superior quality bulk nanostructured materials. The highly nonequilibrium nature of the synthesis process provides an additional benefit by significantly enhancing the critical filling fraction that is beneficial to the TE properties. Much reduced lattice thermal conductivity due to a significantly smaller grain size structure contributes to the high figure of merit of the compounds that reach values of $ZT = 1.3$ at 800 K. This processing route holds great promise for commercial scale production of skutterudite-based TE materials. Applied to multiple-filled skutterudites and using even faster cooling rates to reduce the grain size to a few tens of nanometers, the technique should yield higher-performance filled skutterudites with considerable savings in time and energy.

16.3.3 *p*-Type $AgSbTe_2$ Compounds Prepared by MS–SPS

$AgSbTe_2$ has been studied extensively as a *p*-type medium-temperature TE material [139–142] and as an end component of the two classes of high-performance TE material—$(AgSbTe_2)_{1-x}(GeTe)_x$ [143,144] and

$(AgSbTe_2)_{1-x}(PbTe)_x$ [31,35]. The thermal conductivity of the $AgSbTe_2$ compound is limited to a very low value (only 0.6–0.7 $Wm^{-1}K^{-1}$) by its complex crystal structure, Ag/Sb disorder on cation sites and intrinsically strong scattering of acoustic phonon modes by optic modes [145,146]. However, according to the Ag_2Te–Sb_2Te_3 pseudo-binary phase diagram [147,148], $AgSbTe_2$ is a thermodynamic high-temperature phase. So there are small amounts of unexpected impurities (such as Ag_2Te or Sb_2Te_3) in samples prepared by the melt and slow-cooling method (MC) [141] and other synthesis methods [128,142,149,150]. In 2008, Wang et al. [149] prepared $AgSbTe_2$ compound with a ZT of 1.59 by mechanical alloying (MA). The high ZT value is attributed to the extremely low thermal conductivity resulting from the low relative density (~80%) and the large pore percentage (~13.9%). Here, we present and discuss nanostructures observed in the polycrystalline bulk p-type $AgSbTe_2$ compound prepared by MS combined with SPS, and we document their effect on TE transport properties.

We have found that the nanostructure of melt-spun $AgSbTe_2$ ribbons is totally different from those of melt-spun $(Bi,Sb)_2Te_3$ and filled skutterudites. Figure 16.19a through 16.19c shows FESEM photographs of (a) the free surface, (b) the contact surface, and (c) the cross-section of the ribbons. Nanograins sized 400–600 nm with fine surface texture (20–40 nm) are observed on the free surface of the ribbons. On the contact surface, the size of nanograins spans from 20 to 100 nm. As shown in Figure 16.19c, the grain size increases monotonously from the contact surface to the free surface, due to changes in the cooling rate. Owing to the ultrahigh nucleation rate without a timely replenishment of the melt during the MS process, large amounts of pores and gully-like structures are formed on both the free side and the drum side of the ribbons. It is also interesting to think about how those nanograins and fine surface textures induced by MS will be altered after the SPS. After the obtained melt-spun $AgSbTe_2$ ribbon were hand-ground into powders, the powders of melt-spun ribbons were sintered by SPS [104]. Nanostructures of MS–SPSed $AgSbTe_2$ were also observed by FESEM, and the results are presented in Figure 16.20. As

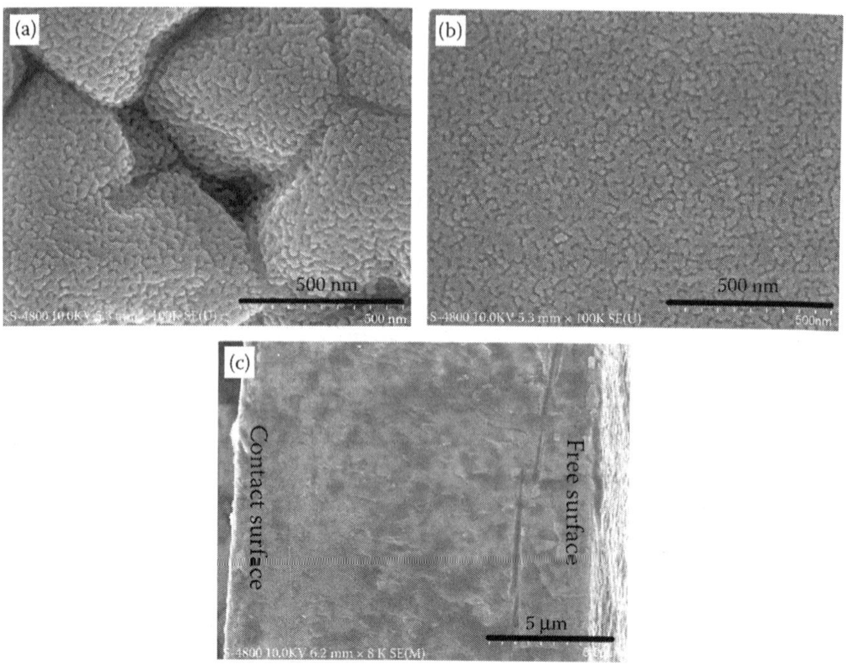

FIGURE 16.19 FESEM photographs of a melt-spun $AgSbTe_2$ ribbon: (a) the free surface; (b) the contact surface; (c) the cross-section. (Reprinted from *J. Solid State Chem.*, Du, B. L. et al., Enhanced thermoelectric performance and novel nanopores in $AgSbTe_2$ prepared by melt spinning, doi:10.1016/j.jssc.2010.10.036, Copyright 2010, with permission from Elsevier.)

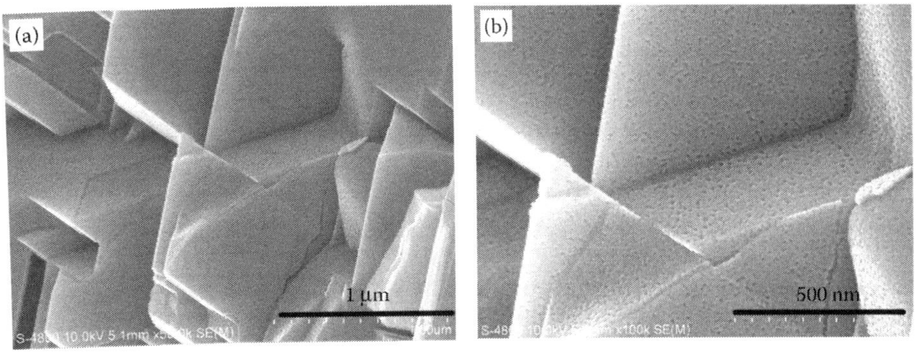

FIGURE 16.20 FESEM photographs of a MS–SPSed $AgSbTe_2$ bulk material: (a) low magnification; (b) high magnification. (Reprinted from *J. Solid State Chem.*, Du, B. L. et al., Enhanced thermoelectric performance and novel nanopores in $AgSbTe_2$ prepared by melt spinning, doi:10.1016/j.jssc.2010.10.036, Copyright 2010, with permission from Elsevier.)

shown in Figure 16.20b, uniformly distributed 5–10 nm pores were obtained on the surface of the $AgSbTe_2$ matrix in the MS–SPS sample, corresponding to the nanostructure characteristics of the ribbons. The detail formation mechanism of nanopores was discussed in the Ref. [104].

By having a thin layer of nanopores coated on $AgSbTe_2$ matrix grains, we expect additional electron and phonon scattering mechanisms to play a role in the MS–SPSed $AgSbTe_2$ system. Beyond reducing the thermal conductivity by strong scattering of heat-carrying phonons, selective scattering affecting charge carriers may also introduce energy filtering effects that could increase the Seebeck coefficient. Figure 16.21 shows the TE transport properties of the MS–SPSed $AgSbTe_2$ bulk material. To aid a comparison, the TE properties of a melt-quench-SPS $AgSbTe_2$ sample with the same SPS treatment and some reference data from the literature are also plotted in Figure 16.21. As shown in Figure 16.21a, the electrical conductivity of the MS–SPS sample increases with the increasing temperature and reaches a maximum of 2.0×10^4 Sm^{-1} at ~665 K, a value much higher than that of the MA sample [149] with low relative density and higher than that of the SPS sample above 380 K. However, it is lower than that of the samples prepared by the traditional MC method, which may be ascribed to the impurities with high electrical conductivity (such as Ag_2Te, Sb_2Te_3) in the MC sample [141]. The MS–SPS sample exhibits positive Seebeck coefficient (shown in Figure 16.21b) as does the SPS sample with nano-Ag_2Te evenly distributed on the matrix grain surface [142]. The value of α is >255 μV K^{-1} in the whole measured temperature range and reaches its maximum of 285 μV K^{-1} at ~390 K. At room temperature, the Seebeck coefficient of the MS–SPS sample is 2.5 times higher than that of the MC (75 μV K^{-1}) sample [141], and is similar to that of the MA sample (260 μV K^{-1}) [149].

In general, Seebeck coefficient α of degenerate semiconductors as a function of the energy dependence of the electrical conductivity $\sigma(E)$ can be given by the Mott relation [150]:

$$\alpha = \frac{\pi^2}{3}\frac{k_B^2}{e}\left\{\frac{d[\ln](\sigma(E))}{dE}\right\}_{E=E_F} \tag{16.2}$$

where the electrical conductivity is $\sigma = \sigma(E_F)$. E, e, and k_B are carrier energy, electron charge, and Boltzmann constant, respectively. In addition, for solids with parabolic bands, the energy dependence of the relaxation time τ is customarily introduced as $\tau = \tau_0 E^{\lambda-1/2}$, where λ is scattering factor [128]. Combining with the equations $\sigma = ne\mu = ne^2\tau/m^*$ (n, μ, and m^*, are carrier density, Hall mobility, and the density-of-states effective mass, respectively), the Seebeck coefficient for a given carrier density increases as the scattering factor increases. According to the Hall measurement, the MS–SPS sample has the similar carrier density (6×10^{19} cm^{-3}) as the MC sample (5×10^{19} cm^{-3}). Thus, the increased Seebeck coefficient should be attributed

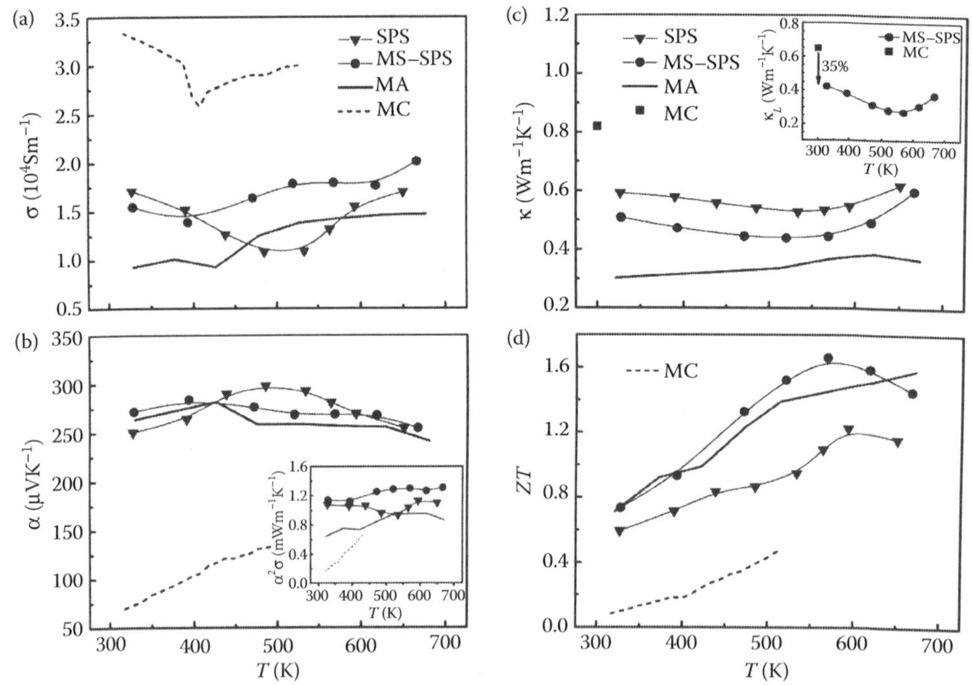

FIGURE 16.21 Temperature dependence of (a) electrical conductivity, (b) Seebeck coefficient, power factor (inset in Figure 4b, (c) thermal conductivity, lattice thermal conductivity (inset in Figure 5a, and (d) figure of merit *ZT* for MS–SPS and reference samples. Lines are plotted as a guide to the eye. (Reprinted from *J. Solid State Chem.*, Du, B. L. et al., Enhanced thermoelectric performance and novel nanopores in $AgSbTe_2$ prepared by melt spinning, doi:10.1016/j.jssc.2010.10.036, Copyright 2010, with permission from Elsevier.)

to an increased scattering factor, presumably from the potential barrier scattering effect [151] associated with the 5–10 nm pores, similar to the increase in the Seebeck coefficient in the SPS sample caused by evenly distributed 20–50 nm Ag_2Te nanograins [142,152]. Therefore, the nanopores embedded in the matrix can introduce an additional scattering mechanism which preferentially scatters low-energy electrons, minimizing their contribution to transport properties and thus increasing the Seebeck coefficient [153]. The inset in Figure 16.21b presents the temperature dependence of the power factor $\alpha^2\sigma$ for the MS–SPS sample and some reference samples. The value of $\alpha^2\sigma$ for the MS–SPS sample reaches 1.3 mW K^{-2} m^{-1} at ~665 K, which is much higher than the values obtained with SPS, MA, and MC samples. The above results indicate that the MS–SPS sample can exhibit high electrical conductivity without sacrificing the Seebeck coefficient.

Temperature dependence of the thermal conductivity κ and the lattice thermal conductivity κ_L for the MS–SPS sample and reference samples are plotted in Figure 16.21c and 16.21d. The thermal conductivity of the MS–SPS sample is 20–60% lower than those of SPS and MC samples, respectively. Moreover, though the MS–SPS sample possesses nearly the theoretical density, the value of κ does not increase too much compared with that of the MA sample with a relatively low density. To further investigate the influence of nanopores on thermal transport properties, the lattice thermal conductivity κ_L is calculated using the Wiedemann–Franz law

$$\kappa_L = \kappa - L\sigma T \tag{16.3}$$

where $L = 0.7L_0$ [140,154], $L_0 = \pi^2/3(k_B/e)^2 = 2.45\times10^{-8}V^2/K^2$ (shown in inset of Figure 16.21c). At room temperature, the lattice thermal conductivity of the MS–SPS sample (0.42 $Wm^{-1}K^{-1}$) is 35% lower than that of the MC sample (0.65 $Wm^{-1}K^{-1}$), and a minimum of 0.30 $Wm^{-1}K^{-1}$ is reached at ~570 K, approaching the minimum theoretical thermal conductivity (~0.3 $Wm^{-1}K^{-1}$) calculated from formulas reported by Cahill et al. [155].

In the traditional diffusion-approximation-based Maxwell–Garnett's model [156,157], the bulk thermal conductivity of porous materials is only a function of the pore volume fraction ϕ:

$$\kappa_{eff} \approx \kappa_m[2(1-\phi)/(2+\phi)] \tag{16.4}$$

where κ_m is the bulk thermal conductivity of the host medium. Assuming $\phi = 0$ for the MC sample [141], the lattice thermal conductivity of the MS–SPS sample is as low as that of the sample with porosity $\phi = 27\%$ prepared by the traditional MC method. So, the diffusion-approximation-based effective medium models cannot be used to predict the thermal conductivity of the MS–SPS sample, since the pore size can be comparable to the mean free path of phonons [158]. Strictly speaking, the Boltzmann transport equation must be solved to capture the ballistic nature of thermal transport and phonon size effects due to the small pore diameter. Therefore, the 5–10 nm pores can act as scattering centers of heat-carrying phonons resulting in the reduction of the thermal conductivity.

The MS–SPSed $AgSbTe_2$ sample exhibits similar Seebeck coefficient and higher electrical conductivity compared with samples prepared by the melt-quench-SPS and MA method. The thermal conductivity is greatly reduced due to the strong phonon scattering effects resulting from 5–10 nm pores. As a result, the TE performance of the MS-SPS sample improves remarkably and a *ZT* of 1.65 is achieved at ~570 K shown in Figure 16.21d, which increases ~35% and ~200% compared with samples prepared by melt-quench-SPS and traditional melting and slow-cooling methods. Our study suggests that nanopores obtained by MS introduce a new scattering mechanism and have great impact on electron and phonon transport properties in the $AgSbTe_2$ system. Therefore, MS combined with SPS is a promising method for designing and optimizing $AgSbTe_2$-based TE materials.

16.4 Conclusions and Outlook

High-performance nanostructured TE materials have been successfully produced using a unique nanostructuring approach: MS followed by SPS. Because nanostructures of melt-spun ribbons are determined not only by the MS process but also by physical properties of the starting materials, the nanostructures for different material systems (including Bi_2Te_3, $CoSb_3$, and $AgSbTe_2$) are totally different compared with each other. The unique nanostructures of nanocomposite materials greatly depress the lattice thermal conductivity by increasing phonon scattering for all materials mentioned in this work. At the same time, for Bi_2Te_3 and $CoSb_3$ systems, coherent boundaries in the nanocrystallites scatter phonons more effectively than they scatter electrons, which results in relatively good power factors. Therefore, the *ZT* values of MS–SPSed nanocomposites are enhanced due to the presence of nanostructures formed during the MS process. The level of improvement varies in different systems. For example, the peak *ZT* is increased to 1.5 in *p*-type $(Bi,Sb)_2Te_3$ nanocomposites compared to *ZT* ~ 1.0 in commercial zone-melted ingots. In a single-filled *n*-type skutterudite $Yb_{0.3}Co_4Sb_{12.3}$ with Sb nanoinclusions the *ZT* of MS samples reaches values of 1.3 compared to a value of 1.05 achieved with samples of similar composition but prepared by traditional synthesis routes. In the case of *p*-type $AgSbTe_2$, the MS–SPS process results in a nanostructure consisting of fine nanopores that coat the grains of the matrix and the figure of merit of these structures reaches values of 1.65 that are to be compared to values of 1.2 achieved with melt-quenched SPS-fabricated bulk materials. The MS–SPS processing route clearly holds great promise for commercial scale production of *p*-type Bi_2Te_3, *n*-type skutterudite, and *p*-type $AgSbTe_2$ TE materials with high TE performance.

Due to the fact that TE applications need both *p*- and *n*-types legs to prepare TE modules, further investigations seeking improvements in the performance of *n*-type Bi_2Te_3 and *p*-type $CoSb_3$ are urgently needed if these materials are indeed going to be used for commercial applications. Besides significantly reducing the lattice thermal conductivity by introducing nanostructures, increasing the power factor is also a very important route to improve *ZT*, for example, by introducing resonant states [159]. In future work, we hope to more fully understand how we can independently optimize the three constituent transport parameters of *ZT*, so as not only to degrade heat transport but also enhance the Seebeck coefficient and perhaps even the electrical conductivity. Nanostructures introduced either by ball milling or MS (and possibly other new synthesis processes) along with the utilization of SPS will certainly play a prominent role in the development and production of a new generation of TE materials with the enhanced conversion efficiency.

Acknowledgments

The authors are thankful to Dr. Yonggao Yan, Shanyu Wang, Xianli Su, Dekui Qi, Song Zhu, Dale Hitchcock, Tim Holgate, Professor Jian He, and Professor Hye Jung Kang for their contributions to this work. We acknowledge the support of the National Basic Research Program of China (Grant No. 2007CB607501) and the Natural Science Foundation of China (Grant No. 50731006 and 50672118) along with 111 Project (Grant No. B07040). The work at Clemson University is supported by DOE/EPSCoR Implementation Grant (#DE-FG02-04ER-46139), and the SC EPSCoR cost sharing program. The work at the University of Michigan is supported by a grant from the University Research Corridor and by the Center for Solar and Thermal Conversion funded by DOE, Office of Basic Energy Sciences (Grant DESC0000957). W.J. Xie would also like to thank the China Scholarship Council (CSC) for support in the form of a partial fellowship (No. 2008695022).

References

1. Tritt, T. M. and Subramanian, M. A. (eds). Harvesting energy through thermoelectrics: Power generation and cooling, *MRS Bull.*, 31, 2006, special issue.
2. Tritt, T. M., Bottner, H., Chen, L. D., Thermoelectrics: Direct solar thermal energy conversion, *MRS Bull.*, 33, 366, 2008.
3. Snyder, G. J., Toberer, E. S., Complex thermoelectric materials, *Nature Mater.*, 7, 105, 2008.
4. Bell, L. E., Cooling, heating, generating power, and recovering waste heat with thermoelectric systems, *Science*, 321, 1457, 2008.
5. Slack, G. A., New materials and performance limits for the thermoelectric cooling, In *CRC Handbook of Thermoelectrics*, D. M. Rowe (ed.), pp. 407–440. Boca Raton, FL: CRC Press, 1995.
6. Sales, B. C., Mandrus, D., Williams, R. K., Filled skutterudite antimonides: A new class of thermoelectric materials, *Science*, 272, 1325, 1996.
7. Tang, X. F., Zhang, L. M., Yuan, R. Z., Chen, L. D., Goto, T., Hirai, T., Dyck, J. S., Chen, W., and Uher, C., *J. Mater. Res.*, High-temperature thermoelectric properties of *n*-type $Ba_yNi_xCo_{4-x}Sb_{12}$, 16, 3343, 2001.
8. Kuznetsov, V. L., Kuznetsova, L. A., Rowe, D. M., Effect of partial void filling on the transport properties of $Nd_xCo_4Sb_{12}$ skutterudites, *J. Phys., Condens. Matter.*, 15, 5035, 2003.
9. Nolas, G. S., Cohn, J. L., Slack, G. A., Effect of partial void filling on the lattice thermal conductivity of skutterudites, *Phys. Rev. B*, 58, 164, 1998.
10. Lamberton, G. A., Bhattacharya, Jr. S., Littleton, IV. R. T., Kaeser, M. A., Tedstrom, R. H., Tritt, T. M., Yang, J., Nolas, G. S., High figure of merit in Eu-filled $CoSb_3$-based skutterudites, *Appl. Phys. Lett.*, 80, 598, 2002.
11. Morelli, D. T., Meisner, G. P., Chen, B. X., Hu, S. Q., Uher, C., Cerium filling and doping of cobalt triantimonide, *Phys. Rev. B*, 56, 7376, 1997.
12. Yang, J., Morelli, D. T., Meisner, G. P., Chen, W., Dyck, J. S., Uher, C., Effect of Sn substituting for Sb on the low-temperature transport properties of ytterbium-filled skutterudites, *Phys. Rev. B*, 67, 165207, 2003.
13. Sales, B. C., Mandrus, D., Chakoumakos, B. C., Keppens, V., Thompson, J. R., Filled skutterudite antimonides: Electron crystals and phonon glasses, *Phys. Rev. B*, 56, 15081, 1997.

14. Pei, Y. Z., Chen, L. D., Zhang, W., Shi, X., Bai, S. Q., Zhao, X. Y., Mei, Z. G., Li, X. Y., Synthesis and thermoelectric properties of $K_yCo_4Sb_{12}$, *Appl. Phys. Lett.*, 89, 221107, 2006.
15. Chen, L. D., Kawahara, T., Tang, X. F., Goto, T., Hirai, T., Dyck, J. S., Chen, W., Uher, C., Anomalous barium filling fraction and *n*-type thermoelectric performance of $Ba_yCo_4Sb_{12}$, *J. Appl. Phys.*, 90, 1864, 2001.
16. Puyet, M., Lenoir, B., Dauscher, A., Pècheur, P., Bellouard, C., Tobola, J., Hejtmanek, J., Electronic, transport, and magnetic properties of $Ca_xCo_4Sb_{12}$ partially filled skutterudites, *Phys. Rev. B*, 73, 035126, 2006.
17. Bhandari, C. M., Rowe, D. M., Silicon–germanium alloys as high-temperature thermoelectric materials, *Contemp. Phys.*, 21, 219, 1980.
18. Rowe, D. M., Recent advances in silicon-germanium alloy technology and an assessment of the problems of building the modules for a radioisotope thermo-electric generator, *J. Power Sources*, 19, 247, 1987.
19. Slack, G. A., Hussain, M. A., The maximum possible conversion efficiency of silicon–germanium thermoelectric generators, *J. Appl. Phys.*, 70, 2694, 1991.
20. Vining, C. B., Silicon germanium, In *CRC Handbook of Thermoelectrics*, D. M. Rowe (ed.), pp. 329–337. Boca Raton, FL: CRC Press, 1995.
21. Wang, X. W., Lee, H., Lan, Y. C., Zhu, G. H., Joshi, G., Wang, D. Z., Yang, J. et al., Enhanced thermoelectric figure of merit in nanostructured *n*-type silicon germanium bulk alloy, *Appl. Phys. Lett.*, 93, 193121, 2008.
22. Joshi, G., Lee, H., Lan, Y., Wang, X., Zhu, G., Wang, D., Gould, R. W. et al., Enhanced thermoelectric figure-of-merit in nanostructured *p*-type silicon germanium bulk alloys, *Nano Lett.*, 8, 4670, 2008.
23. Shen, Q., Chen, L., Goto, T., Hirai, T., Yang, J., Meisner, G. P., Uher, C., Effects of partial substitution of Ni by Pd on the thermoelectric properties of ZrNiSn-based half-Heusler compounds, *Appl. Phys. Lett.*, 79, 4165, 2001.
24. Nolas, G. S., Poon, J., Kanatzidis, M., Recent developments in bulk thermoelectric materials, *MRS Bull.*, 31, 199, 2006.
25. Uher, C., Yang, J., Hu, S., Morelli, D. T., Meisner, G. P., Transport properties of pure and doped MNiSn (M = Zr, Hf), *Phys. Rev. B*, 59, 8615, 1999.
26. Culp, S. R., Poon, S. J., Hickman, N., Tritt, T. M., Blumm, J., Effect of substitutions on the thermoelectric figure of merit of half-Heusler phases at 800°C, *Appl. Phys. Lett.*, 88, 042106, 2006.
27. Yu, C., Zhu, T. J., Shi, R. Z., Zhang, Y., Zhao, X. B., He, J., High-performance half-Heusler thermoelectric materials $Hf_{1-x}Zr_xNiSn_{1-y}Sb_y$ prepared by levitation melting and spark plasma sintering, *Acta Mater.*, 57, 2757, 2009.
28. Xie, W. J., Jin, Q., Tang, X. F., The preparation and thermoelectric properties of $Ti_{0.5}Zr_{0.25}Hf_{0.25}Co_{1-x}Ni_xSb$ half-Heusler compounds, *J. Appl. Phys.*, 103, 043711, 2008.
29. Xie, W. J., He, J., Zhu, S., Su, X. L., Wang, S. Y., Holgate, T., Graff, J. W., Ponnambalam, V., Poon, S. J., Tang, X. F., Zhang, Q. J., Tritt, T. M., Simultaneously optimizing the independent thermoelectric properties in (Ti,Zr,Hf)(Co,Ni)Sb alloy by *in situ* forming InSb nanoinclusions, *Acta Mater.*, 58, 4705, 2010.
30. Qiu, P. F., Yang, J., Huang, X. Y., Chen, X. H., Chen, L. D., Effect of antisite defects on band structure and thermoelectric performance of ZrNiSn half-Heusler alloys, *Appl. Phys. Lett.*, 96, 152105, 2010.
31. Hsu, K. F., Loo, S., Guo, F., Chen, W., Dyck, J. S., Uher, C., Hogan, T., Polychroniadis, E. K., Kanatzidis, M. G., Cubic $AgPb_mSbTe_{2+m}$: Bulk thermoelectric materials with high figure of merit, *Science*, 303, 818, 2004.
32. Quarez, E., Hsu, K. F., Pcionek, R., Frangis, N., Polychroniadis, E. K., Kanatzidis, M. G., Nanostructuring, compositional fluctuations, and atomic ordering in the thermoelectric materials $AgPb_mSbTe_{2+m}$. The myth of solid solutions, *J. Am. Chem. Soc.*, 127, 9177, 2005.
33. Poudeu, P. F. R., D'Angelo, J., Downey, A. D., Short, J. L., Hogan, T. P., Kanatzidis, M. G., High thermoelectric figure of merit and nanostructuring in bulk *p*-type $Na_{1-x}Pb_mSb_yTe_{m+2}$, *Angew. Chem. Int. Ed.*, 45, 3835, 2006.

34. Wang, H., Li, J. F., Nan, C. W., Zhou, M., Liu, W. S., Zhang, B. P., Kita, T., High-performance $Ag_{0.8}Pb_{18+x}SbTe_{20}$ thermoelectric bulk materials fabricated by mechanical alloying and spark plasma sintering, *Appl. Phys. Lett.*, 88, 092104, 2006.
35. Ke, X. Z., Chen, C. F., Yang, J. H., Wu, L. J., Zhou, J., Li, Q., Zhu, Y. M., Kent, P. R. C., Microstructure and a nucleation mechanism for nanoprecipitates in PbTe-$AgSbTe_2$, *Phys. Rev. Lett.*, 103, 145502, 2009.
36. Zhou, M., Li, J. F., Kita, T., Nanostructured $AgPb_mSbTe_{m+2}$ system bulk materials with enhanced thermoelectric performance, *J. Am. Chem. Soc.*, 130, 4527, 2008.
37. Anisworth, R. S., Scanlon, W. W., Mobility of electrons and holes in PbS, PbSe, and PbTe between room temperature and 4.2°K, *Phys. Rev.*, 111, 1029, 1958.
38. Goldsmid, H. J., *Thermoelectric Refrigeration*, Pion Ltd., London, 1986.
39. Scherrer, H., Scherrer, S., Bismuth telluride, antimony telluride, and their solid solutions, In *CRC Handbook of Thermoelectrics*, D. M. Rowe (ed.), pp. 211–237. Boca Raton, FL: CRC Press, 1995.
40. Venkatasubramanian, R., Siivola, E., Colpitts, T., O'Quinn, B., Thin-film thermoelectric devices with high room-temperature figures of merit, *Nature*, 413, 597, 2001.
41. Hicks, L. D., Dresselhaus, M. S., Effect of quantum-well structures on the thermoelectric figure of merit, *Phys. Rev. B*, 47, 12727, 1993.
42. Zhao, X. B., Ji, X. H., Zhang, Y. H., Zhu, T. J., Tu, J. P., Zhang, X. B., Bismuth telluride nanotubes and the effects on the thermoelectric properties of nanotube-containing nanocomposites, *Appl. Phys. Lett.*, 86, 062111, 2005.
43. Tang, X. F., Xie, W. J., Li, H., Zhao, W. Y., Zhang, Q. J., Niino, M., Preparation and thermoelectric transport properties of high-performance *p*-type Bi_2Te_3 with layered nanostructure, *Appl. Phys. Lett.*, 90, 012102, 2007.
44. Cao, Y. Q., Zhao, X. B., Zhu, T. J., Zhang, X. B., Tu, J. P., Syntheses and thermoelectric properties of Bi_2Te_3/Sb_2Te_3 bulk nanocomposites with laminated nanostructure, *Appl. Phys. Lett.*, 92, 143106, 2008.
45. Xie, W. J., Tang, X. F., Yan, Y. G., Zhang, Q. J., Tritt, T. M., Unique nanostructures and enhanced thermoelectric performance of melt-spun BiSbTe alloys, *Appl. Phys. Lett.*, 94, 102111, 2009.
46. Xie, W. J., Tang, X. F., Yan, Y. G., Zhang, Q. J., Tritt, T. M., High thermoelectric performance BiSbTe alloy with unique low-dimensional structure, *J. Appl. Phys.*, 105, 113713, 2009.
47. Xie, W. J., He, J., Kang, H. J., Tang, X. F., Zhu, S., Laver, M., Wang, S. Y., Copley, J. R. D., Brown, C. M., Zhang, Q. J., Tritt, T. M., Identifying the specific nanostructures responsible for the high thermoelectric performance of $(Bi,Sb)_2Te_3$ nanocomposites, *Nano Lett.*, 10, 3283, 2010.
48. Poudel, B., Hao, Q., Ma, Y., Lan, Y., Minnich, A., Yu, B., Yan, X., Wang, D., Muto, A., Vashaee, D., Chen, X., Liu, J., Dresselhaus, M. S., Chen, G., Ren, Z., High-thermoelectric performance of nanostructured bismuth antimony telluride bulk alloys, *Science*, 320, 634, 2008.
49. Ma, Y., Hao, Q., Poudel, B., Lan, Y., Yu, B., Wang, D., Chen, G,. Ren, Z., Enhanced thermoelectric figure-of-merit in *p*-type nanostructured bismuth antimony tellurium alloys made from elemental chunks, *Nano Lett.*, 8, 2580, 2008.
50. Yan, X., Poudel, B., Ma, Y., Liu, W. S., Joshi, G., Wang, H., Lan, Y., Wang, D., Chen, G., Ren, Z. F., Experimental studies on anisotropic thermoelectric properties and structures of *n*-type $Bi_2Te_{2.7}Se_{0.3}$, *Nano Lett.*, 10, 3373, 2010.
51. Nolas, G. S., Cohn, J. L., Slack, G. A., Schujman, S. B., Semiconducting Ge clathrates: Promising candidates for thermoelectric applications, *Appl. Phys. Lett.*, 73, 178, 1998.
52. Saramat, A., Svensson, G., Palmqvist, A. E. C., Stiewe, C., Mueller, E., Platzek, D., Williams, S. G. K., Rowe, D. M., Bryan, J. D., Stucky, G. D., Large thermoelectric figure of merit at high temperature in Czochralski-grown clathrate $Ba_8Ga_{16}Ge_{30}$, *J. Appl. Phys.*, 99, 023708, 2006.
53. Toberer, E. S., Christensen, M., Iversen, B. B., Snyder, G. J., High temperature thermoelectric efficiency in $Ba_8Ga_{16}Ge_{30}$, *Phys. Rev. B*, 77, 075203, 2008.

54. Deng, S. K., Tang, X. F., Zhang, Q. J., Synthesis and thermoelectric properties of *p*-type $Ba_8Ga_{16}Zn_xGe_{30-x}$ type-I clathrates, *J. Appl. Phys.*, 102, 043702, 2007.
55. Tang, X. F., Li, P., Deng, S. K., Zhang, Q. J., High temperature thermoelectric transport properties of double-atom-filled clathrate compounds $Yb_xBa_{8-x}Ga_{16}Ge_{30}$, *J. Appl. Phys.*, 104, 013706, 2008.
56. Deng, S. K., Tang, X. F., Zhang, Q. J., High temperature thermoelectric transport properties of *p*-type $Ba_8Ga_{16}Al_xGe_{30-x}$ type-I clathrates with high performance, *J. Appl. Phys.*, 103, 073503, 2008.
57. May, A. F., Toberer, E. S., Saramat, A., Snyder G. J., Characterization and analysis of thermoelectric transport in *n*-type $Ba_8Ga_{16-x}Ge_{30+x}$, *Phys. Rev. B*, 80, 125205, 2009.
58. Deng, S. K., Saiga, Y. Suekuni, K. Takabatake T., Enhancement of thermoelectric efficiency in type-VIII clathrate $Ba_8Ga_{16}Sn_{30}$ by Al substitution for Ga, *J. Appl. Phys.*, 108, 073705, 2010.
59. Shi, X., Yang, J., Bai, S. Q.,Yang, J. H., Wang, H., Chi, M. F., Salvador, J. R., Zhang, W. Q., Chen, L.D., Wong-Ng, W., On the design of high efficiency thermoelectric clathrates through a systematic cross-substitution of main-frame elements, *Adv. Funct. Mater.*, 20, 755, 2010.
60. Terasaki, I., Sasago, Y., Uchinokura, K., Large thermoelectric power in $NaCo_2O_4$ single crystals, *Phys. Rev. B*, 56, 12685, 1997.
61. Kawata, T., Iguchi, Y., Itoh, T., Takahata, K., Terasaki, I., Na-site substitution effects on the thermoelectric properties of $NaCo_2O_4$, *Phys. Rev. B*, 60, 10584, 1999.
62. Takahata, K., Iguchi, Y., Tanaka, D., Itoh, T., Terasaki, I., Low thermal conductivity of the layered oxide $(Na,Ca)Co_2O_4$: Another example of a phonon glass and an electron crystal, *Phys. Rev. B*, 61, 12551, 2000.
63. Motohashi, T., Naujalis, E., Ueda, R., Isawa, K., Karppinen, M., Yamauchi, H., Simultaneously enhanced thermoelectric power and reduced resistivity of $Na_xCo_2O_4$ by controlling Na nonstoichiometry, *Appl. Phys. Lett.*, 79, 1480, 2001.
64. Terasaki, I., Tsukada, I., Iguchi, Y., Impurity-induced transition and impurity-enhanced thermopower in the thermoelectric oxide $NaCo_{2-x}Cu_xO_4$, *Phys. Rev. B*, 65, 195106, 2002.
65. Wang, Y. Y., Rogado, N. S., Cava, R. J., Ong, N. P., Spin entropy as the likely source of enhanced thermopower in $Na_xCo_2O_4$, *Nature*, 423, 425, 2003.
66. Caillat, T., Fleurial, J. P., Borshchevsky, A., Preparation and thermoelectric properties of semiconducting Zn_4Sb_3, *J. Phys. Chem. Solids*, 58, 1119, 1997.
67. Snyder, G. J., Christensen, M., Nishibori, E., Caillat, T., Iversen, B. B., Disordered zinc in Zn_4Sb_3 with phonon-glass and electron-crystal thermoelectric properties, *Nature Mater.*, 3, 458, 2004.
68. Nylen, J., Andersson, M., Lidin, S., Haussermann, U., The structure of α-Zn_4Sb_3: Ordering of the phonon-glass thermoelectric material β-Zn_4Sb_3, *J. Am. Chem. Soc.*, 126, 16306, 2004.
69. Mozharivskyj, Y., Pecharsky, A. O., Bud'ko, S., Miller, G. J., A Promising thermoelectric material: Zn_4Sb_3 or $Zn_{6-d}Sb_5$. Its composition, structure, stability, and polymorphs. structure and stability of $Zn_{1-d}Sb$, *Chem. Mater.*, 16, 1580, 2004.
70. Tsutsui, M., Zhang, L. T., Ito, K., Yamaguchi, A., Effects of in-doping on the thermoelectric properties of β-Zn_4Sb_3, *Intermetallics*, 12, 809, 2004.
71. Brown, S. R., Kauzlarich, S. M., Gascoin, F., Snyder, G. J., $Yb_{14}MnSb_{11}$: New high efficiency thermoelectric material for power generation, *Chem. Mater.*, 18, 1873, 2006.
72. Kauzlarich, S. M., Brown, S. R., Snyder, G. J., Zintl phases for thermoelectric devices, *Dalton Trans.*, 21, 2099, 2007.
73. Brown, S. R., Toberer, E. S., Ikeda, T., Cox, C. A., Gascoin, F., Kauzlarich, S. M., Snyder, G. J., Improved thermoelectric performance in $Yb_{14}Mn_{1-x}Zn_xSb_{11}$ by the reduction of spin-disorder scattering, *Chem. Mater.*, 20, 3412, 2008.
74. Toberer, E. S., Cox, C. A., Brown, S. R., Ikeda, T., May, A. F., Kauzlarich, S. M., Snyder, G. J., Traversing the metal-insulator transition in a zintl phase: Rational enhancement of thermoelectric efficiency in YbMnAlSb, *Adv. Funct. Mater.*, 18, 2795, 2008.
75. Toberer, E. S., May, A. F., Snyder, G. J., Zintl chemistry for designing high efficiency thermoelectric materials, *Chem. Mater.*, 22, 624, 2010.

76. May, A. F., Fleurial, J., Snyder, G. J., Thermoelectric performance of lanthanum telluride produced via mechanical alloying, *Phys. Rev. B*, 78, 125205, 2008.
77. May, A. F., Singh, D. J., Snyder, G. J., Influence of band structure on the large thermoelectric performance of lanthanum telluride, *Phys. Rev. B*, 79, 153101, 2009.
78. May, A. F., Flage-Larsen, E., Snyder, G. J., Electron and phonon scattering in the high-temperature thermoelectric $La_3Te_{4-z}M_z$ (M = Sb,Bi), *Phys. Rev. B*, 81, 125205, 2010.
79. May, A. F., Fleurial, J., Snyder, G. J., Optimizing thermoelectric efficiency in $La_{3-x}Te_4$ via Yb substitution, *Chem. Mater.*, 22, 2995, 2010.
80. Dresselhaus, M. S., Chen, G., Tang, M. Y., Yang, R. G., Lee, H., Wang, D. Z., Ren, Z. F., Fleurial, J. P., Gogna, P., New directions for low-dimensional thermoelectric materials, *Adv. Mater.*, 19, 1043, 2007.
81. Kanatzidis, M. G., Nanostructured thermoelectrics: The new paradigm, *Chem. Mater.*, 22(3), 648, 2010.
82. Sootsman, J. R., Chung, D. Y., Kanatzidis, M. G., New and old concepts in thermoelectric materials, *Angew. Chem., Int. Ed.*, 48, 8616, 2009.
83. Menke, E. J., Li, Q., Penner, R. M., Bismuth telluride (Bi_2Te_3) nanowires synthesized by cyclic electrodeposition/stripping coupled with step edge decoration, *Nano Lett.*, 2004, 4, 2009.
84. Sander, M. S., Prieto, A. L., Gronsky, R., Sands, T., Stacy, A. M., Fabrication of high-density, high aspect ratio, large-area bismuth telluride nanowire arrays by electrodeposition into porous anodic alumina templates, *Adv. Mater.*, 2002, 14, 665.
85. Martín-González, M., Prieto, A. L., Gronsky, R., Sands, T., Stacy, A. M., High-density 40 nm diameter Sb-Rich $Bi_{2-x}Sb_xTe_3$ nanowire arrays, *Adv. Mater.*, 2003, 15, 1003.
86. Trahey, L., Becker, C. R., Stacy, A. M., Electrodeposited bismuth telluride nanowire arrays with uniform growth fronts, *Nano Lett.*, 2007, 7, 2535.
87. Xue, F. H., Fei, G. T., Wu, B., Cui, P., Zhang, L. D., Direct electrodeposition of highly dense Bi/Sb superlattice nanowire arrays, *J. Am. Chem. Soc.*, 2005, 127, 15348.
88. Yoo, B., Xiao, F., Bozhilov, K. N., Herman, J., Ryan, M. A., Myung, N. V., Electrodeposition of thermoelectric superlattice nanowires, *Adv. Mater.*, 2007, 19, 296.
89. Wang, W., Lu, X. L., Zhang, T., Zhang, G. Q., Jiang, W. J., Li, X. G., Bi_2Te_3/Te multiple heterostructure nanowire arrays formed by confined precipitation, *J. Am. Chem. Soc.*, 2007, 129, 6702.
90. Purkayastha, A., Kim, S., Gandhi, D. D., Ganesan, P. G., Borca-Tasciuc, T., Ramanath, G., Molecularly protected bismuth telluride nanoparticles: microemulsion synthesis and thermoelectric transport properties, *Adv. Mater.*, 2006, 18, 2958.
91. Lu, W. G., Ding, Y., Chen, Y. X., Wang, Z. L., Fang, J. Y., Bismuth telluride hexagonal nanoplatelets and their two-step epitaxial growth, *J. Am. Chem. Soc.*, 2005, 127, 10112.
92. Purkayastha, A., Lupo, F., Kim, S., Borca-Tasciuc, T., Ramanath, G., Low-temperature templateless synthesis of single-crystal bismuth telluride nanorods, *Adv. Mater.*, 2006, 18, 496.
93. Wang, S. Y., Xie, W. J., Li, H., Tang, X. F., High performance *n*-type $(Bi,Sb)_2(Te,Se)_3$ for low temperature thermoelectric generator, *J. Phys. D, Appl. Phys.*, 43, 335404, 2010.
94. Wang, S. Y., Xie, W. J., Li, H., Tang, X. F., Microstructures and thermoelectric properties of *n*-type melting spun $(Bi_{0.85}Sb_{0.15})_2(Te_{1-x}Se_x)_3$ compounds, *Acta Phys. Sin.*, 59, 605, 2010.
95. Guo, Q. S., Li, H., Su, X. L., Tang, X. F., Microstructure and themoelectric properties of *p*-type filled skutterudite $Ce_{0.3}Fe_{1.5}Co_{2.5}Sb_{12}$ prepared by melt-spinning method, *Acta Phys. Sin.*, 59, 6666, 2010.
96. Li, H., Tang, X. F., Su, X. L., Zhang, Q. J., Preparation and thermoelectric properties of high-performance Sb additional $Yb_{0.2}Co_4Sb_{12+y}$ bulk materials with nanostructure, *Appl. Phys. Lett.*, 92, 202114, 2008.
97. Li, H., Tang, X. F., Su, X. L., Zhang, Q. J., Uher, C., Nanostructured bulk $Yb_xCo_4Sb_{12}$ with high thermoelectric performance prepared by the rapid solidification method, *J. Phys. D, Appl. Phys.*, 42, 145409, 2009.

98. Li, H., Tang, X. F., Zhang, Q. J., Uher, C., Rapid preparation method of bulk nanostructured $Yb_{0.3}Co_4Sb_{12+y}$ compounds and their improved thermoelectric performance, *Appl. Phys. Lett.*, 93, 252109, 2008.
99. Cao, W. Q., Deng, S. K., Tang, X. F., Li, P., The effects of melt spinning process on microstructure and thermoelectric properties of Zn-doped type-I clathrates, *Acta Phys. Sin.*, 58, 612, 2009.
100. Qi, D. K., Tang, X. F., Li, H., Yan, Y. G., Zhang, Q. J., Improved thermoelectric performance and mechanical properties of nanostructured melt-spun beta-Zn_4Sb_3, *J. Electron. Mater.*, 39, 1159, 2010.
101. Qi, D. K., Yan, Y. G., Li, H., Tang, X. F., Effects of rapid solidification method on thermoelectric and mechanical properties of beta-$Zn_{4+x}Sb_3$ materials, *J. Inorg. Mater.*, 25, 603, 2010.
102. Su, X. L., Tang, X. F., Li, H., Effects of melt spinning process on microstructure and thermoelectric properties of *n*-type InSb compounds, *Acta Phys. Sin.*, 59, 2860, 2010.
103. Luo, W. H., Li, H., Yan, Y. G., Lin, Z. B., Tang, X. F., Zhang, Q. J., Uher, C., Rapid synthesis of high thermoelectric performance higher manganese silicide with *in-situ* formed nano-phase of MnSi, *Intermetallics*, 19, 404, 2011.
104. Du, B. L., Li, H., Xu, J. J., Tang, X. F., Uher, C., Enhanced thermoelectric performance and novel nanopores in $AgSbTe_2$ prepared by melt spinning, *J. Solid State Chem.*, 184, 109, 2011.
105. Greer, A.L., Metallic glasses, *Science*, 267, 1947, 1995.
106. Chen, H., He, Y., Shiflet, G. J., Poon, S. J., Deformation-induced nanocrystal formation in shear bands of amorphous alloys, *Nature*, 367, 541, 1994.
107. Tkatch, V. I., Limanovskii, A. I., Denisenko, S. N., Rassolov, S. G., The effect of the melt-spinning of processing parameters on the rate cooling, *Mater. Sci. Eng. A*, 323, 91, 2002.
108. Tkatch, V. I., Denisenko, S. N., Beloshov, O. N., Direct measurements of the cooling rates in the single roller rapid solidification technique, *Acta. Mater.*, 45, 2821, 1996.
109. Oh, T. S., Hyun, D., Kolomoets, N. V., Thermoelectric properties of the hot-pressed $(Bi,Sb)_2(Te,Se)_3$ alloys, *Scr. Mater.*, 42, 849, 2000.
110. Hahn, H., Microstructure and properties of nanostructured oxides, *Nanostruct. Mater.*, 2, 251, 1993.
111. Larker, H. T., Recent advances in hot isostatic pressing processes for high performance ceramics, *Mater. Sci. Eng.*, 71, 329, 1985.
112. Jin, Z. Q., Rockett, C., Liu, J. P., Hokamoto, K., Thadhani, N. N., Shock compaction of bulk nanocomposite magnetic materials, *Mater. Sci. Forum*, 465, 93, 2004.
113. Tokita, M., Mechanism of spark plasma sintering, *J. Soc. Powder Technol. Jpn.*, 30, 790, 1993.
114. Zhan,G. D., Kuntz, J., Wan, J., Garay, J., Mukherjee, A. K., Spark-plasma-sintered $BaTiO_3/Al_2O_3$ nanocomposites, *Mater. Sci. Eng. A*, 356, 443, 2003.
115. Nygren, M., Shen, Z., Spark plasma sintering: Possibilities and limitations, *Key Eng. Mater.*, 264–268, 719, 2004.
116. Omori, M., Sintering, consolidation, reaction and crystal growth by the spark plasma system (SPS), *Mater. Sci. Eng. A*, 287, 183, 2000.
117. Anselmi-Tamburini, U., Garay, J. E., Munir, Z. A., Fast low-temperature consolidation of nanometric ceramic materials, *Scr. Mater.*, 54, 823, 2006.
118. Liu, W., Naka M., *In situ* joining of dissimilar nanocrystalline materials by spark plasma sintering, *Scr. Mater.*, 48, 1225, 2003.
119. Aalund, R., Spark plasma sintering, *Ceramic Industry*, May 1, 2008.
120. Feschotte, P., Lorin, D., Les systemes binaires Fe-Sb, Co-Sb et Ni-Sb, *J. Less-Common Met.*, 155, 255, 1989.
121. Geng, H. Y., Ochi, S., Guo, J. Q., Solidification contraction-free synthesis for the $Yb_{0.15}Co_4Sb_{12}$ bulk material, *Appl. Phys. Lett.*, 91, 022106, 2007.
122. Sales, B. C., Chakoumakos, B. C., Mandrus, D., When does a crystal conduct heat like a glass? *MRS Symposia No. 626*, Materials Research Society, Pittsburgh, 2001, p. z7.11.

123. Zhao, X. Y., Shi, X., Chen, L. D., Zhang, W. Q., Bai, S. Q., Pei, Y. Z., Li, X. Y., Goto, T., Synthesis of $Yb_yCo_4Sb_{12}/Yb_2O_3$ composites and their thermoelectric properties, *Appl. Phys. Lett.*, 89, 092121, 2006.
124. Shi, X., Kong, H., Li, C. P., Uher, C., Yang, J., Salvador, J. R., Wang, H., Chen, L., Zhang, W., Low thermal conductivity and high thermoelectric figure of merit in *n*-type $Ba_xYb_yCo_4Sb_{12}$ double-filled skutterudites, *Appl. Phys. Lett.*, 92, 182101, 2008.
125. Tang, X. F., Li, H., Zhang, Q. J., Synthesis and thermoelectric properties of double-atom-filled skutterudite compounds $Ca_mCe_nFe_xCo_{4-x}Sb_{12}$, *J. Appl. Phys.*, 100, 123702, 2006.
126. Jeitschko, W., Braun, D., $LaFe_4P_{12}$ with filled $CoAs_3$-type structure and isotypic lanthanoid-transition metal polyphosphides, *Acta Crystallogr. B, Struct. Crystallogr. Cryst. Chem.*, 33, 3401, 1977.
127. Hicks, L. D., Harman, T. C., Dresselhaus, M. S., Use of quantum-well superlattices to obtain a high figure of merit from nonconventional thermoelectric materials, *Appl. Phys. Lett.*, 63, 3230, 1993.
128. Heremans, J. P., Thrush, C. M., Morelli, D. T., Thermopower enhancement in PbTe with Pb precipitates, *J. Appl. Phys.*, 98, 063703, 2005.
129. Anno, H., Hatada, K., Shimizu, H., Structural and electronic transport properties of polycrystalline *p*-type $CoSb_3$, *J. Appl. Phys.*, 83, 5270, 1998.
130. Toprak, M. S., Stiewe, C., Platzek, D. Williams, S., Bertini, L., Müller, E., Gatti, C., Zhang, Y., Rowe, M., Muhammed, M., The impact of nanostructuring on the thermal conductivity of thermoelectric $CoSb_3$, *Adv. Funct. Mater.*, 14, 1189, 2004.
131. Li, H., Tang, X. F., Su, X. L., Zhang, Q. J., Preparation and thermoelectric properties of high-performance Sb additional $Yb_{0.2}Co_4Sb_{12+y}$ bulk materials with nanostructure, *Appl. Phys. Lett.*, 92, 202114, 2008.
132. Nolas, G. S., Kaeser, M., Littleton IV, R. T., Tritt, T. M., High figure of merit in partially filled ytterbium skutterudite materials, *Appl. Phys. Lett.*, 77, 1855, 2000.
133. Yu, B. L., Tang, X. F., Qi, Q., Zhang, Q. J., Effect of grain size on thermoelectric properties of $CoSb_3$ compound, *Acta Phys. Sin.*, 53, 3130, 2004.
134. Williams, J. R., Smalley, A. L. E., Sellinschegg, H., Daniels-Hafer, C., Harris, J., Johnson, M. B., Johnson, D. C., Synthesis of crystalline skutterudite superlattices using the modulated elemental reactant method, *J. Am. Chem. Soc.*, 125, 10335, 2003.
135. Savchuk, V., Boulouz, A., Chakraborty, S., Schumann, J., Vinzelberg, H., Transport and structural properties of binary skutterudite $CoSb_3$ thin films grown by dc magnetron sputtering technique, *J. Appl. Phys.*, 92, 5319, 2002.
136. He, Z., Stiewe, C., Platzek, D., Karpinski, G., Müller, E., Li, S. H., Toprak, M., Muhammed, M., Effect of ceramic dispersion on thermoelectric properties of nano-$ZrO_2/CoSb_3$ composites, *J. Appl. Phys.*, 101, 043707, 2007.
137. Katsuyama, S., Watanabe, M., Kuroki, M., Maehata, T., Ito, M., Effect of NiSb on the thermoelectric properties of skutterudite CoSb, *J. Appl. Phys.*, 93, 2758, 2003.
138. Shi, X., Zhang, W., Chen, L. D., Yang, J., Filling fraction limit for intrinsic voids in crystals: Doping in skutterudites, *Phys. Rev. Lett.*, 95, 185503, 2005.
139. Ye, L. H., Hoang, K., Freeman, A. J., Mahanti, S. D., He, J., Tritt, T. M., Kanatzidis, M. G., First-principles study of the electronic, optical, and lattice vibrational properties of $AgSbTe_2$, *Phys. Rev. B*, 77, 245203, 2008.
140. Jovovic, V., Heremans, J. P., Measurements of the energy band gap and valence band structure of $AgSbTe_2$, *Phys. Rev. B*, 77, 245204, 2008.
141. Wojciechowski, K. T., Schmidt, M., Structural and thermoelectric properties of $AgSbTe_2$-$AgSbSe_2$ pseudobinary system, *Phys. Rev. B*, 79, 184202, 2009.
142. Xu, J., Li, H., Du, B., Tang, X., Zhang, Q., Uher, C., High thermoelectric figure of merit and nanostructuring in bulk $AgSbTe_2$, *J. Mater. Chem.*, 20, 6138, 2010.
143. Rowe, D. M., *Handbook of Thermoelectrics Macro to Nano*, CRC press, Boca Raton, FL, 2006.
144. Yang, S. H., Zhu, T. J., Sun, T., He, J., Zhang, S. N., Zhao, X. B., Nanostructures in high-performance $(GeTe)_x(AgSbTe_2)_{100-x}$ thermoelectric materials, *Nanotechnology*, 19, 245707, 2008.

145. Barabash, S. V., Ozolins, V., Wolverton, C., First-principles theory of competing order types, phase separation, and phonon spectra in thermoelectric $AgPb_mSbTe_{m+2}$ alloys, *Phys. Rev. Lett.*, 101, 155704, 2008.
146. Morelli, D. T., Jovovic, V., Heremans, J. P., Intrinsically minimal thermal conductivity in cubic I-V-VI_2 semiconductors, *Phys. Rev. Lett.*, 101, 035901, 2008.
147. Petzow, G., Effenberg, G., *Ternary Alloys*, John Wiley & Sons, New York, 2, 554, 1988.
148. Marin, R.-M., Brun, G., Tedenac, J.-C., The ternary system silver-antimony-tellurium. Study of the subternary Sb_2Te_3-Ag_2Te-Te, *J. Mater. Sci.*, 20, 730, 1985.
149. Wang, H., Li, J. F., Zou, M. M., Sui, T., Synthesis and transport property of $AgSbTe_2$ as a promising thermoelectric compound, *Appl. Phys. Lett.*, 93, 202106, 2008.
150. Mott, N. F., Davis, E. A., *Electronic Processes in Non-Crystalline Materials*, Clarendon, Oxford University Press, USA, 1979.
151. Bytenskii, L. I., Gudkin, T. S., Iordanishvili, E. K., Kaz'min, S. A., Kaidanov, V. I., Nemov, S. A., Ravich, Y. I., Influence of potential barriers on thermoelectric properties of lead chalcogenides films, *Semiconductors*, 11, 894, 1977.
152. Du, B., Xu, J., Zhang, W., Tang, X., Impact of *in situ* generated Ag_2Te nanoparticles on the microstructure and thermoelectric properties of $AgSbTe_2$ compounds, *J. Electron Mater.*, 40, 1249, 2011.
153. Minnich, A. J., Dresselhaus, M. S., Ren, Z. F., Chen, G., Bulk nanostructured thermoelectric materials: Current research and future prospects, *Energy Environ. Sci.*, 2, 466, 2009.
154. Zhang, S. N., Zhu, T. J., Yang, S. H., Yu, C., Zhao, X. B., Phase compositions, nanoscale microstructures and thermoelectric properties in $Ag_{2-y}Sb_yTe_{1+y}$ alloys with precipitated Sb_2Te_3 plates, *Acta Mater.*, 58, 4160, 2010.
155. David, G. C., Watson, S. K., Pohl, R. O. Lower limit to the thermal conductivity of disordered crystals, *Phys. Rev. B*, 46, 6131, 1992.
156. Szelagowski, H., Arvanitidis, I., Seetharaman, S., Effective thermal conductivity of porous strontium oxide and strontium carbonate samples, *J. Appl. Phys.*, 85, 193, 1999.
157. Maxwell-Garnett, J., Philos, C., A comprehensive treatise on inorganic and theoretical chemistry, *Tran. Roy. Soc.*, 203, 385, 1904.
158. Prasher, R., Transverse thermal conductivity of porous materials made from aligned nano- and microcylindrical pores, *J. Appl. Phys.*, 100, 064302, 2006.
159. Heremans, J. P., Jovovic, V., Toberer, E. S., Saramat, A., Kurosaki, K., Charoenphakdee, A., Yamanaka, S., Snyder, G. J., Enhancement of thermoelectric efficiency in PbTe by distortion of the electronic density of states, *Science*, 321, 554, 2008.
160. Wenjie, X., Jian, H., Song, Z., Tim, H., Shanyu, W., Xinfeng, T., Qingjie, Z., Terry, M. T., Investigation of the sintering pressure and thermal conductivity anisotropy of melt-spun spark-plasma-sintered $(Bi,Sb)_2Te_3$ thermoelectric materials, *J. Mater. Res.*, 26, 1791–1799, 2011, doi:10.1557/jmr.2011.170.

17

Fabrication Routes for Nanostructured TE Material Architectures

Muhammet S. Toprak
Royal Institute of Technology

Shanghua Li
Royal Institute of Technology

Mamoun Muhammed
Royal Institute of Technology

17.1 Introduction

Nanomaterials have been an emerging field of research due to the novel properties exhibited when the size of building blocks is reduced below 100 nm. Several size-dependent phenomena make nanomaterials attractive in terms of potential applicability compared to their larger-sized counterparts, justifying the importance and attention of this research.[1–3] For thermoelectric research, nanomaterials are of great interest due to the possibility of decoupling electrical and thermal transport properties which may help attain higher ZT values for the currently available materials.[4,5]

The commonly accepted nomenclature of nanomaterials is based on the number of degrees of freedom of charge carriers for the description of their dimensionalities. Thus, a thin film or superlattice is confined in one dimension, but is a 2D nanomaterial. Similarly, a nanowire is confined in two dimensions, hence is a 1D nanomaterial, and finally a nanoparticle is confined in three dimensions and is therefore a 0D nanomaterial. In the case of larger nanocrystals (not quantum confined), the size and shape of the nanocrystals also affect their properties due to differences in surface-to-volume ratio. Bulk nanostructured (NS) TE materials are fabricated using a bulk process rather than a nanofabrication process, which has the important advantage of being produced in large quantities and in a form that is compatible with commercially available devices.[6] They are different from the advanced low-dimensional TE materials as they are consolidated under high T and P, resulting in larger-sized (40–200 nm) grains. The advanced low-dimensional TE materials are usually in quantum size and have well-defined shapes such as rod-like or tube-like structures. The nanoeffects of bulk NS TE materials will not be as strong as in the case of advanced low-dimensional TE materials, but since the material is closer to bulk material, it can be easily handled the same way as bulk TE materials using conventional TE module/device technology for further steps toward direct applications. So far, of all the NS materials, only bulk NS materials have been produced in enough quantity to be used in this manner.

The nanostructuring approach offers the possibility of engineering power factor and thermal conductivity separately, thus allowing improvement in the *ZT* of the TE materials. There are several reviews on the enhancement of thermoelectric figure of merit by nanostructuring approach.[4,6–8] These papers have mostly focused on the review of performance of the TE materials, not much detailing the processes of their fabrication. In this chapter, we aim at providing an understanding of the physical and chemical fabrication processes by giving a brief description about each technique followed by examples of TE material architectures synthesized using them.

Traditionally, TE materials have been made by solid-state reactions or melt crystallization routes. However, these techniques generally produce materials with very large grain size due to prolonged exposure to higher temperatures. In the last few decades, a variety of gas phase and solution-based methodologies have been introduced and implemented for the fabrication of TE nanomaterials. The strategies can be mainly divided into two groups as physical and chemical routes; physical routes include physical vapor deposition (PVD) techniques, melt-alloying, solid-state synthesis, and mechanochemical alloying, while chemical routes include chemical vapor deposition (CVD), solvothermal, coprecipitation, sol–gel, electrochemical synthesis methods, as well as postprocessing of bulk powders.

Another classification of synthesis strategies is based on the routes of forming nanostructures. Top-down strategies are those based on forming a bulk and then crashing it down by various routes while bottom-up strategies can tailor the size and morphology of the structures formed controlling the critical parameters during the synthesis. Chart 17.1 summarizes material architectures and the strategies that can be utilized to fabricate them.

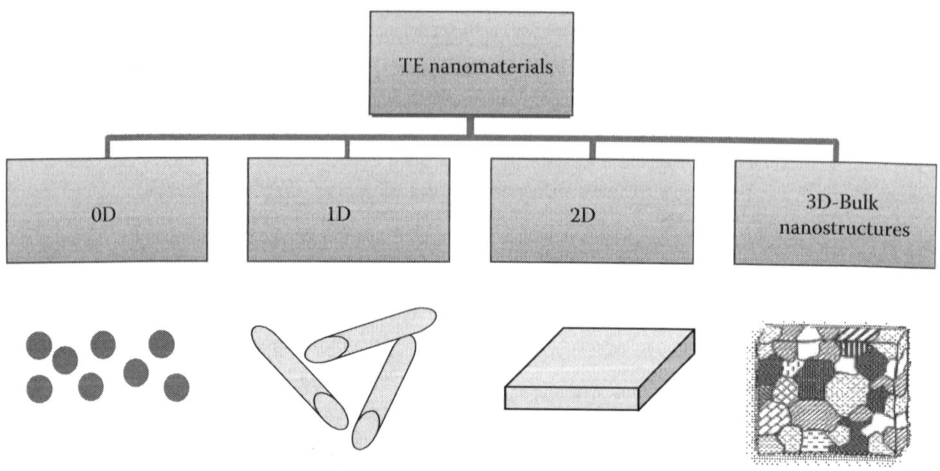

Bottom-up strategies			
Molecular beam epitaxy (MBE)	Chemical/physical vapor deposition (CVD/PVD)	Chemical/physical vapor deposition (CVD/PVD)	Co-precipitation
Thermolysis	Electrochemical deposition	Atomic layer deposition	Electrochemical deposition
Solution reduction	Solution reduction	Solution reduction	Mechanical alloying
Sol-gel synthesis	Hydrothermal synthesis	Molecular beam epitaxy	Sol-gel synthesis
Hydrothermal reduction		Sputtering	Melt alloying
Sonochemical synthesis			
Top-down strategies			
Ball-milling	Electrochemical etching	Electrochemical etching	Ball-milling
Post alkali processing	Melt spinning	Plasma etching	Post alkali processing
			Melt-alloying + Ball milling
			Solid-state synthesis + Ball-milling

CHART 17.1 Nanomaterials: Classifications and strategies used for their fabrication.

17.2 Physical Routes of Fabrication

Solid-state synthesis reactions are commonly applied for the fabrication of TE materials. In this process, a stoichiometric, or near stoichiometric, mixture of high-purity elemental components, in loose or compacted form, are heated for extended periods in a sealed container under vacuum or inert gas at a temperature below the melting point of the components to form the desired TE phase.

Various solid-state synthesis techniques, including sintering of cold compacts followed by hot pressing, direct synthesis by hot pressing, and sinter-forging, have been used in order to produce high-efficiency Zn_4Sb_3 bulk specimens.[9] Elemental mixtures used in this work were slightly rich in Zn, yet resulted in stoichiometric composition of Zn_4Sb_3. Solid-state synthesis utilizing the hot consolidation process has been suggested to offer a potential processing route to produce bulk Zn_4Sb_3.

Ni-substituted skutterudite compound has been synthesized by a direct reaction between the elements in a carbon-coated silica tube, under vacuum. All the handling took place in a glove box under a continuously purified argon atmosphere. The ampule was rapidly heated to 1000°C, kept at this temperature for 48 h, water quenched, then annealed at 700°C for 4 days.[10]

Some ZrNiSn-based half-Heusler compounds have been successfully fabricated by the method of solid-state reaction starting from high-purity elemental components as starting materials.[11]

Due to high temperatures and long reaction times used, materials synthesized via solid-state methods exhibit tens or hundreds of micrometers particle/grain sizes. Solid-state synthesis is generally combined with ball-milling in order to achieve nanostructuring. NS bismuth antimony telluride bulk alloys have been reported by Poudel et al. using a combination of solid-state synthesis of the bulk material, ball-milling, and hot-pressing of nanoparticles into bulk ingots.[12]

Solid-state syntheses are mostly performed using fine-powdered raw materials to enhance diffusion and to achieve better homogeneity in the sample. Skutterudite compound with the formula $CoSb_3$ has been prepared from high-purity elemental powders by ball-milling the powder to fine grade followed by sealing and reacting in an evacuated quartz tube at 800°C for 2 days.[13]

NS thermoelectric oxides with low thermal conductivity have been synthesized by both solid-state reactions and a soft chemistry process from water-soluble precursors. Powders of substituted cobaltates (p-type) and manganates (n-type) were synthesized.[14,15]

Melt-alloying is very similar to solid-state reactions where the components are heated at temperatures above the melting points and then are slowly cooled down to grow crystals of the desired phase.

$CoSb_3$ powders have been prepared by vacuum melting and annealing using high-purity metallic elemental rectants.[16] Spark plasma sintering (SPS) is a sintering method where direct current is passed through the sample during compaction, reaching very high densification values with minimum exposure to elevated temperatures and pressures. High-performance NS bulk skutterudites $Yb_xCo_4Sb_{12}$ have been prepared by combining the melt-spinning technique with SPS. Prepared polycrystalline ingots with desired stoichiometry using the traditional method (melt + quench + annealing + SPS) were inductively melted in a quartz tube and the melt is ejected on a chilled Cu wheel where it rapidly solidified. Formed ribbons were then collected and loaded into a graphite die and sintered by the SPS method. As a result, NS materials with grain size of 100–200 nm and finer structure with nanocrystals of 10–20 nm were obtained as can be seen from the micrographs presented in Figure 17.1.[17]

Melt-alloying route was also reported for the fabrication of NS bulk Tl-doped PbTe by direct reaction of appropriate amounts of Pb, Te, and Tl_2Te in a fused-silica tube sealed under a vacuum. Each sample was briefly melted at 1273 K for 24 h and lightly shaken to ensure homogeneity of the liquid, then furnace-cooled to 800 K and annealed for 1 week. The obtained ingot was crushed into fine powder and hot-pressed at 803 K for 2 h under a flowing 4% H_2–Ar atmosphere. Results showed an average ZT value of 1.5 ± 0.3 and authors anticipate that deliberately engineered impurity-induced band-structure distortions will be a generally applicable route to enhanced S and ZT in all TE materials.[18]

Mechanical alloying (MA), which is an alloy formation in the solid phase by means of repeated mechanical impact during high-energy ball-milling of high-purity-powdered elemental components,

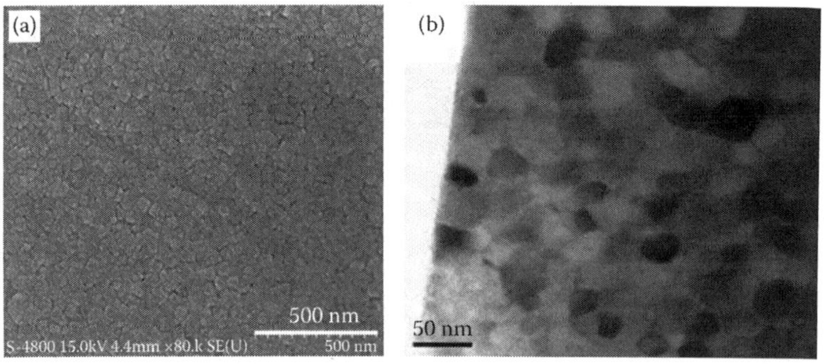

FIGURE 17.1 (a) FESEM photograph and (b) HRTEM image of $Yb_xCo_4Sb_{12}$ ribbon. (From H. Li et al., Nanostructured bulk $Yb_xCo_4Sb_{12}$ with high thermoelectric performance prepared by the rapid solidification method, *J. Phys. D: Appl. Phys.* **42**, 145409, 2009. With permission.)

has proven to be an advantageous method for the synthesis of semiconducting thermoelectric compounds or solid solutions. The method even allows the formation of those alloys which are—due to an unfavorable phase diagram—difficult to obtain via solidification processes.[19]

NS bismuth antimony tellurium alloys have been prepared by ball-milling elemental components. Resultant powder showed an average particle size of 10 nm. Nanostructure has been maintained even after hot-pressing, resulting in a high *ZT* value for the NS material. The process of synthesizing nanocrystalline alloys from elemental chunks is reported to be also more environment friendly compared to alloyed crystalline ingots as it requires less energy than forming ingots first and then ball-milling.[20]

Synthesis of $CoSb_3$ skutterudite by MA has been reported by Yang et al.[21] Elemental powder mixtures with an atomic ratio of Co:Sb = 1:3 were mechanically alloyed by high-energy ball-milling in a planetary ball mill. It is found that the $CoSb_3$ compound was synthesized after milling for 10 h, and its concentration increased with prolonged milling time. The single-phase $CoSb_3$ skutterudite could not be obtained even on further milling for 50 h. It is shown that single-phase $CoSb_3$ could be synthesized easily by annealing the as-mechanical alloyed powders in vacuum at 700°C, and the annealing time could be as short as 1 h for the powders milled for over 10 h.

17.3 Chemical Routes of Fabrication

Chemical fabrication routes are bottom-up strategies that allow a better control on the size and morphology of TE nanostructures as various parameters can be adjusted/tailored during the fabrication process.

Solution coprecipitation is a method that has been recently implemented for the fabrication of TE materials. In this technique, various components are simultaneously precipitated out of a solution by pH modifications or introduction of a precipitating agent into the mixture. These powders can then be exposed to thermochemical processing toward the formation of the desired TE phase.

Various schemes are reported for the precipitation of the components and conversion of these coprecipitated powders to Bi_2Te_3 powder through hydrogen reduction. Ritter et al. has reported a room-temperature coprecipitation process of bismuth telluride precursor (Bi_2O_3, $3TeO \cdot xH_2O$) in aqueous media, which is then converted to polycrystalline Bi_2Te_3 via hydrogen reduction at 275°C for 12 h.[22] Toprak et al. adapted a similar approach, replacing bismuth oxide with bismuth oxalate that has higher reactivity and thus could reduce the time required for hydrogen reduction significantly down to 2 h at 350°C producing 40 nm nanocrystals of Bi_2Te_3.[23] This is a powerful method and is proven to be successful in the synthesis of bulk quantities of TE powder with the desired composition. The use of metal-organo complexes as a route to mixed cation oxide powders has also been utilized for the synthesis of various TE materials. For instance, in the synthesis of Bi_2Te_3, Ritter et al. reported an organic precursor prepared by a reaction

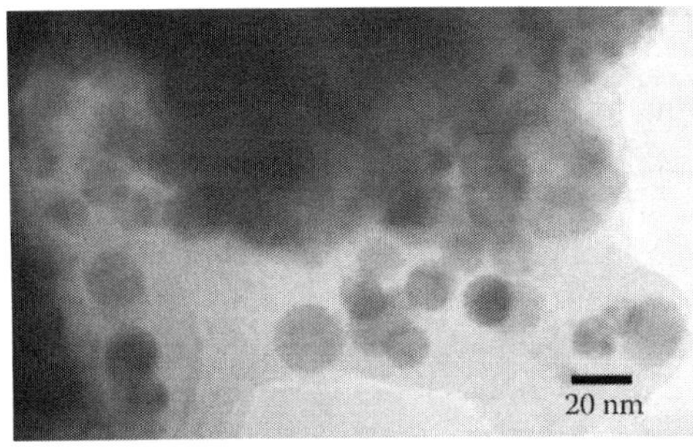

FIGURE 17.2 TEM micrograph of skutterudite powder prepared using solution coprecipitation route. (Reprinted with permission from C. Stiewe et al., Nanostructured $Co_{1-x}Ni_x(Sb_{1-y}Te_y)_3$ skutterudites: Theoretical modeling, synthesis and thermoelectric properties, *J. Appl. Phys* **97**(4), 044317-1–044317-7, Copyright 2005, American Institute of Physics.)

mixture of Bi_2O_3, ethylene glycol, metallic Te, and DL-tartaric acid and nitric acid. This precursor was then calcined and reduced under hydrogen to form nanocrystalline Bi_2Te_3 powder.[24]

Skutterudites have also been prepared using solution coprecipitation, also called *chemical alloying* techniques.[25] In a series of papers by Toprak et al., the use of thermodynamic modeling for the selection of reaction conditions and the fabrication of doped $CoSb_3$-based skutterudites are reported. Undoped ($CoSb_3$),[26] Ni-doped ($Ni_xCo_{1-x}Sb_3$),[27] Fe-doped ($Fe_yCo_4Sb_{12}$),[28] Ni- and Te-doped ($Co_{1-x}Ni_x(Sb_{1-y}Te_y)_3$)[29–31] compositions have successfully been synthesized via oxalate and/or oxide precursor coprecipitation. A TEM micrograph of skutterudite powder prepared using solution coprecipitation route is shown in Figure 17.2.

Sol–gel technique is commonly used for the fabrication of NS oxide thermoelectric. This route involves precursor materials dissolved in water containing some gelling agents and complex-forming ions. By pH adjustment, the metal ions are bound forming complexes, and temperature increase allows a slow evaporation of water resulting in a gel structure in which all the components are trapped in the pores of the gel. This can then be calcined for the decomposition and removal of all organic reagents and the formation of the desired TE phase.

Sol–gel technique provides the advantage of mixing the constituting precursors at the molecular level and gives more homogeneous cation stoichiometry. Furthermore, it avoids the necessity of the intermediate grindings which are required to overcome the diffusion problem encountered in the solid-state reaction. Liu et al. have prepared $NaCo_2O_{4-\delta}$ at temperature as low as 700°C via aqueous sol–gel methods, using starting materials $NaNO_3$, $Co(NO_3)_3$ in deionized water, and urea.[32]

$ZnO–TiB_2$ composite samples have been prepared by a sol–gel method, which were then hot pressed to obtain the composite where ZnO and TiB_2 are chosen as a TE phase and a highly conducting phase, respectively, to construct two-phase ball-shell-type composites.[33]

Solvothermal reactions are also used for the fabrication of NS TEs. In this route, precursor materials are mixed in a small volume of a solvent, and a reducing agent, such as sodium borohydride, is added into this mixture. The reaction is performed at slightly elevated temperatures in a sealed pressure container, during which the metal ions coming from the dissolution of the precursor materials are precipitated by the reducing agent, thus forming the desired TE phase.

Nano-sized $CoSb_3$ powders with particle size of ~100 nm were synthesized by a solvothermal method using $CoCl_2$ and $SbCl_3$ as the precursors and $NaBH_4$ as the reducing agent in an ethanol solution.[34] Yang et al. reported a rapid, one-step synthesis of nanocrystalline $CoSb_3$ with high purity via a modified polyol process. In this modified process, a stronger reducing agent $NaBH_4$ was introduced. The synthesis of pure $CoSb_3$ was realized from precursors consisting of a well-defined mixture of $CoCl_2 \cdot 6H_2O$ and $SbCl_3$. Polyvinylpyrrolidone (PVP) and tetraethylene glycol (TEG) were used as stabilizer and solvent, at temperatures as low as 180°C.[35] Nano-sized thermoelectric skutterudite $Co_{4-x}Fe_xSb_{12}$ with a single phase has been synthesized via a solvothermal method, starting from analytical grade $SbCl_3$, $CoCl_2 \cdot 6H_2O$, $FeCl_3 \cdot 6H_2O$, $NaBH_4$, in triethlyene glycol solution at 290°C for 12 h.[36]

Bi_2Te_3 nanowires prepared by aqueous chemical reduction has been reported by Sun et al. from the starting materials of tellurium powder or Na_2TeO_3, $BiCl_3$, and $NaBH_4$ in a highly basic aqueous solution at 65°C. Structure was directed by the use of surfactant SDS.[37] Nano-sized powders of $La_xBi_{(2-x)}Te_3$ has been synthesized solvothermally at 120°C, 150°C, and 180°C, starting from the precursors $LaCl_3$, $BiCl_3$, and Te powder, ethanol as the solvent, $NaBH_4$ as the reducing agent, and NaOH for controlling the pH.[38]

A biomolecule-assisted hydrothermal approach has been developed for the fabrication of Bi_2Te_3 thermoelectric nanomaterials with a nanostring-cluster hierarchical structure which is composed of ordered and aligned platelet-like crystals. Initially, Te nanorods, shown in Figure 17.3, are formed using alginic acid as both reductant and template. Subsequently, Bi_2Te_3 grows in a certain direction on the surface of the Te rods, resulting in the nanostring structure. The nanostrings further recombine side-by-side with each other to achieve the ordered nanostring clusters. The particle size and morphology can be controlled by adjusting the concentration of NaOH, which plays a crucial role in the formation mechanism of Bi_2Te_3.[39]

Ao et al. reported on the synthesis of nano-sized $AgPb_{18}SbTe_{20}$ by hydrothermal method at 180°C for 48 h for the first time, using starting materials $AgNO_3$, $Pb(CH_3COO)_2 \cdot 3H_2O$, $SbCl_3$, Te powder, $NaBH_4$, and KOH. Morphology of particles obtained was spherical and rod-like, which is suggested to improve the ZT due to the quantum effect at the crystal boundaries.[40]

Thermolysis, decomposition of metallo-organic precursors in organic liquids, is capable of producing metallic/intermetallic nanoparticles under inert atmosphere. Weigang et al. reported on the synthesis of discrete, single-crystal, defect-free, and hexagonal Bi_2Te_3 nanoplatelets using a high-temperature organic solution approach. They also demonstrated a two-step epitaxial growth of the cylindrical strings of Bi_2Te_3 nanoplatelets on the surface of the Te rod by packing them along *c*-axis in a top–bottom–top–bottom sequence.[41] In the synthetic process, they used bismuth-2-ethylhexanoate and trioctylphosphine-telluride (TOP-Te) in phenylether solution at 150°C.

PbSe nanocrystals have been synthesized by reacting lead oleate with tri-*n*-octylphosphine selenide in squalane in the presence of oleic acid capping ligands, resulting in the size of the nanocrystals from ~4.3 to ~8.6 nm with the size distribution well below 10%. Quantum dot superlattice (QDSL) films of monodisperse PbSe nanocrystals were deposited by dropcasting solutions on substrates with preformed parallel Cr/Au electrodes (Figure 17.4), resulting in film thickness of 50–200 nm that exhibit a significant thermopower enhancement of several hundred microvolts per K relative to bulk PbSe.[42]

Synthesis of Bi_2Te_3 and $Bi_{0.5}Sb_{1.5}Te_3$ nanocrystals via thermolysis has been reported by Zhao et al.[43,44] Starting materials were bismuth acetate-dodecanethiol, antimony chloride-dodecanethiol dissolved in phenyl ether, and TOP-Te. The reaction was performed by heating the solution to 150°C for 60 min under argon gas.

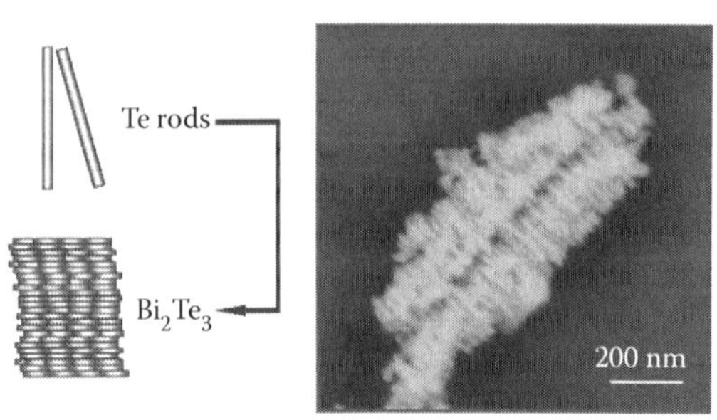

FIGURE 17.3 Scheme of assembly and TEM image of a multistring cluster hierarchical Te rod/Bi_2Te_3 platelet structure synthesized by biomolecule-assisted hydrothermal approach. (Reprinted with permission from J.L. Mi et al., Biomolecule-assisted hydrothermal synthesis and self-assembly of Bi_2Te_3 nanostring-cluster hierarchical structure, *ACS Nano* **4**(5), 2523–2530, Copyright 2010, American Chemical Society.)

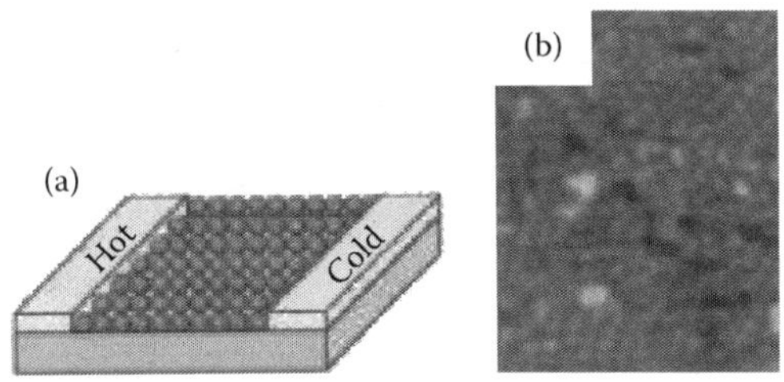

FIGURE 17.4 (a) The measurement device consisting of a glass wafer with two Cr/Au electrodes. (b) Typical high-resolution scanning electron microscopy image of a conductive PbSe nanocrystal film. (Reprinted with permission from R.Y. Wang et al., Enhanced thermopower in PbSe nanocrystal quantum dot superlattices, *Nano Lett.* **8**(8), 2283–2288, Copyright 2008, American Chemical Society.)

Organic solvents and precursors sometimes are extremely reactive and therefore should be handled with great care; furthermore, the quantities of materials in these processes are also limited making them incompatible for large-scale production. An aqueous precursor method could be a safe-to-handle alternative to the organic precursor method. Dhak et al. reported on the synthesis of nanocrystalline Bi_2Te_3 using water-soluble EDTA complexes of Bi and Te. Addition of a strong reducing agent KBH_4 to this highly basic mixture resulted in the formation of Bi_2Te_3 nanocrystals.[45]

Microemulsion synthesis technique makes use of a dispersed phase in a continuous phase as reactors for size-controlled synthesis of nanoparticles. Microemulsions are thermodynamically stable, isotropic liquid mixtures of oil (*o*), water (*w*), and surfactant, in which one of the liquids is dispersed in the other; based on the their respective distribution there are two categories as normal (oil in water *o*/*w*) and inverse/reverse (water in oil *w*/*o*) microemulsions.[46] Two or more separate microemulsions can be mixed together for the particle formation via intermicellar exchange of the reactants.

Nanocrystalline Bi_2Te_3 with particle sizes <10 nm was synthesized by microemulsion technique in which the synthesis involves the reaction of $BiOClO_4 \cdot xH_2O$ with $(Me_3Si)_2Te$ in a reverse micelle formed by sodium dioctylsulfosuccinate and water in hexane.[47]

Kanatzidis et al. reported the synthesis of nanoparticles with compositions $AgSbTe_2$, $AgPbSbTe_3$, and $AgPb_2SbTe_4$ using reverse micellar synthesis coupled with a sodium borohydride reduction. In this route, all metal ions, Ag^+, Pb^{2+}, Sb^{3+}, including Te^{4+} (in Na_2TeO_3), are probably simultaneously reduced to the elemental nanoparticles which combine to form the quaternary phases. The process has been described as convenient, general, low-cost, and environmentally friendly, by employing common water-soluble metal salts.[48]

Ultrasonic synthesis technique has drawn a lot of attention due to its rapid reaction rate, controllable reaction conditions, and the ability to form nanoparticles with uniform shapes, narrow size distributions, and high purity. Sonochemistry arises from acoustic cavitation, the formation, growth, and implosive collapse of bubbles in a liquid. The collapse of bubbles generates localized hot spots with transient temperatures of 5000 K and cooling rates in excess of 10^{10} K/s.[49,50] A typical sonochemical synthesis apparatus is shown in Figure 17.5.

Sonochemistry has been used for the synthesis of Bi_2Te_3 nanocrystals and Bi–Se–Ti alloys as well as other selenide and telluride nanocrystals, such as Ag_2Se, Ag_2Te, Cu_5Se_4, and PbSe.[51] For instance, in Bi_2Te_3 synthesis, $Bi(NO_3)_3$, Na_2TeO_3, $N_2H_4 \cdot H_2O$, and deionized water have been mixed and exposed to high-intensity ultrasonic irradiation in air at an ambient temperature for 1–4 h. Nanocrystalline Bi_2Te_3 thermoelectric compounds have been synthesized by sonochemical methods at 70°C using Te and $BiCl_3$ as the reactants, $NaBH_4$ as the reductant, and NaOH as the pH-value controller.[52] Bi_2Se_3 nanobelts with a typical width of 20–80 nm, thickness of 8–10 nm, and length of several micrometers have been successfully synthesized using a simple coreduction method under ultrasonic irradiation for 15 h at room temperature.[53] The starting chemicals were bismuth chloride, selenious acid, and

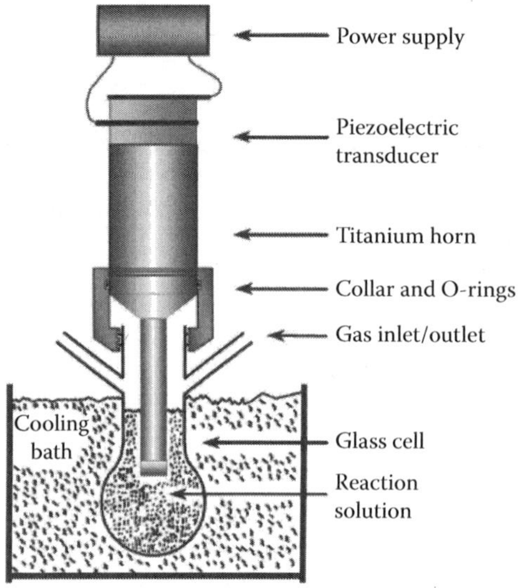

FIGURE 17.5 A typical sonochemical apparatus with direct immersion ultrasonic horn. Ultrasound can be easily introduced into a chemical reaction with good control of temperature and ambient atmosphere. (With kind permission from Springer Science+Business Media: *Sonochemistry and Sonoluminescence*, Applications of sonochemistry to materials synthesis, eds. Crum, L.A. et al., Kluwer Publishers: Dordrecht, Netherlands, 1999, pp. 291–320, K.S., Suslick et al.)

hydrazine where morphology and size of the nanostructures were found to vary with synthesis temperature.

Electrochemical synthesis routes have been used for the fabrication of TE nanowires/nanorods and films. A solution of metal ions are first screened by cyclic voltammetry in order to find out the regions where oxidation and reduction processes take place as a function of voltage or current sweeping. Then, the optimum conditions are chosen for the fabrication of the desired material phase. This route is laborious as the initial solution concentration is not reflected in the final deposit. Therefore, there are a lot of adjustment steps before optimum conditions are obtained. An advantage of this technique is access to very narrow channels, thus making it possible to synthesize nanorods of TEs in the pores of alumina membranes. It is also important to mention that electrochemical methods do not require expensive equipment and are also scalable.

Bismuth telluride nanowires have been prepared by several groups using electrochemical deposition techniques. Usually, porous alumina membranes have been used as they provide well-defined cylindrical channels for the deposition process.[54–57] Bi_2Te_3 nanowire array in porous alumina membrane and a single nanowire prepared by A. L. Prieto et al. is presented in Figure 17.6a and 17.6b, respectively. Sb_xTe_y thin films and nanowires have also been potentiostatically electrodeposited at room temperature and a strong dependence of the morphology, film composition, and thickness of the deposits on the applied potential has been observed.[58]

Free-standing nanowires are difficult to handle when it comes to integration onto electronic components for active cooling. An interesting route to TE nanowires is electrochemical etching process. Seo et al.[59] reported on the fabrication of precisely controlled vertical arrays of silicon nanowires (Figure 17.6c) and cones using a combined treatment strategy of electrochemical and chemical etching. In the electrochemical process, current density plays an important parameter in determining the wire diameter and wire spacing.

Lead telluride (PbTe) nanofilms have been fabricated by electrochemical atomic layer deposition (ALD) method.[60] Deposits were formed by using an ALD cycle on Au substrates depositing one atomic layer at a time making use of surface-limited reactions.

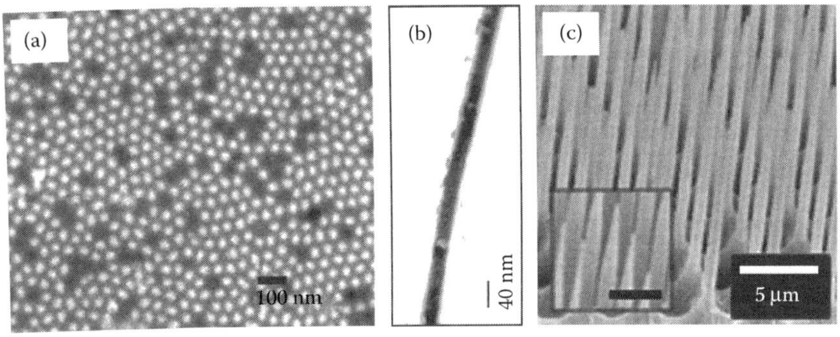

FIGURE 17.6 (a) SEM image of Bi_2Te_3 nanowire array composite. The bright regions are the filled pores. (b) Bright-field TEM image of a single Bi_2Te_3 nanowire. (Reprinted with permission from A.L. Prieto et al., Electrodeposition of ordered Bi_2Te_3 nanowire arrays, *J. Am. Chem. Soc.* **123**(29), 7160–7161, Copyright 2001, American Chemical Society.) (c) Si nanostructures fabricated by simple electrochemical etching at a current density of 65 mA/cm^2. (Reprinted from *Mater. Lett.* **63**, H.-S. Seo et al., Fabrication of precisely controlled silicon wire and cone arrays by electrochemical etching, 2567–2569, Copyright 2009, with permission from Elsevier.)

Bismuth telluride nanoparticles have also been synthesized by electrodeposition on a Cu substrate from a nitrate solution of the constituents. Clusters of ~400 nm made of particles with ~80 nm diameter were obtained.[61]

Galvanic displacement can be considered as a simple and fast process to deposit TE thin films on a substrate and to tailor the morphology, crystal structure, and composition by modifying reaction parameters. Chang et al. reported on the formation of Bi_xTe_y nanotubes,[62] as schematically described in Figure 17.7, followed by a systematic study on the Bi and Te films from sacrificial metals.[63] In order to synthesize Bi and Te thin films by galvanic displacement, three different sacrificial thin films made of Ni, Fe, and Co were electrodeposited and substrates were dipped into the solutions of Bi- or Te-containing electrolytes. Figure 17.8 shows thickness and microstructure of Te thin film made by galvanostatic reaction with a sacrificial Fe layer.

Controllable synthesis of Bi_2Te_3 intermetallic compounds with hierarchical nanostructures via the electrochemical deposition route was recently reported by Li et al.[64] The concentration ratio of $Bi(NO_3)_3$ to Na_2TeO_3 was a key factor that affected the morphologies and structures of Bi_2Te_3 intermetallic compounds reported in this work.

Film devices are required to allow for localized cooling at points of interest in many TE applications, such as thermochemistry on a chip, biothermoelectric chips, and active cooling for microelectronic

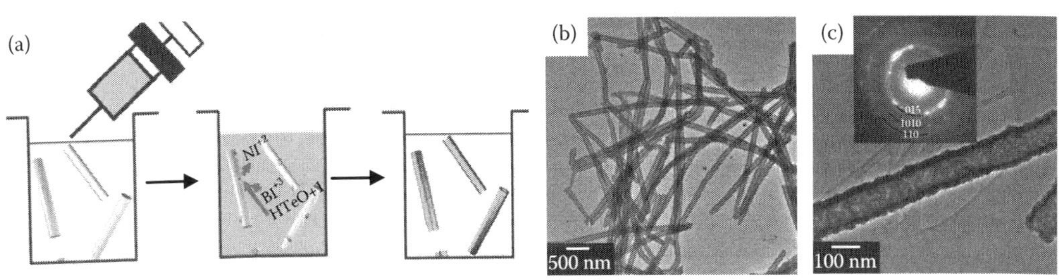

FIGURE 17.7 (a) Schematic illustrations of Bi_2Te_3 nanotube synthesis: TEM images and SAED pattern of high aspect ratio Bi_2Te_3 nanotubes (b,c) synthesized from nickel nanowire. (Reprinted with permission from F. Xiao et al., Synthesis of Bi_2Te_3 nanotubes by galvanic displacement, *J. Am. Chem. Soc.* **129**, 10068, Copyright 2007, American Chemical Society.)

FIGURE 17.8 (a) Te thin film deposited by galvanic displacement with the sacrificial Fe thin film (3 μm thick) for 1 h, (b) Te thin film deposited by galvanic displacement with the sacrificial Fe thin film (20 μm thick) for 1 h, (c) top view of Te thin film in (b). (Reprinted from *Electrochim. Acta* **55**(3), C.H. Chang et al., Bi and Te thin films synthesized by galvanic displacement from acidic nitric baths, 743–752, Copyright 2010, with permission from Elsevier.)

processors. Thick film-based devices have technological advantages over conventional TE module technology, for which there are practical problems in fabricating pellets with lengths of a few hundred micrometers. Li et al. reported on the synthesis of NS bismuth telluride thick films using electrochemical deposition technique.[65–67]

Dresselhaus et al.[68,69] reported on the preparation of Bi nanowires using porous alumina host materials filled with liquid Bi by a high-pressure technique. This process resulted in solid single-crystal Bi nanowires. In the high-pressure liquid technique, the pressure that is needed to force liquid Bi into the pores of the membrane is inversely proportional to the wire diameter. Therefore, the smallest diameter of single-crystal Bi nanowire prepared from the melt is 65 nm.

Heremans et al. reported a new vapor phase method for the synthesis of Bi nanowires with diameters between 7 and 200 nm. The method is based on vapor deposition in a diffusion-pumped liquid nitrogen-trapped vacuum evaporator to fill the pores of nanoporous alumina membrane and the setup is shown in Figure 17.9. This method produced single-crystal wires of Bi. As it relies on the diffusion of vapor, it

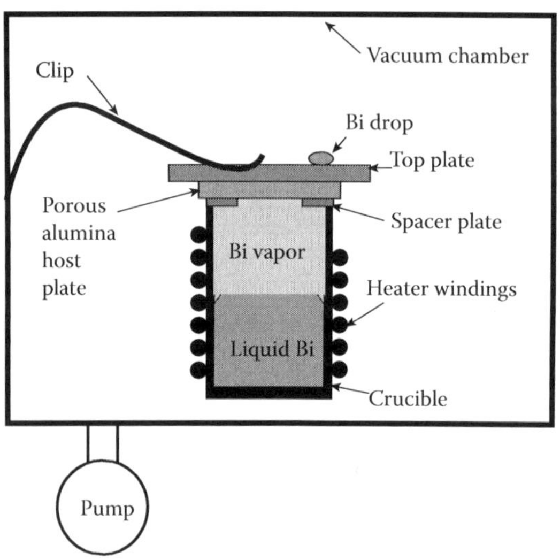

FIGURE 17.9 Schematic diagram of the vacuum chamber containing the vapor-phase sample-preparation technique. (Reprinted with permission from J. Heremans et al., *Phys. Rev. B* **61**, 2921–2930, Copyright 2000 by the American Physical Society.)

circumvents the limitation of high-pressure liquid technique used previously, and allows the preparation of wires with smaller diameter.[70]

17.4 Vapor Deposition Techniques

Physical vapor deposition (PVD) techniques are commonly used for the fabrication of thin-film structures of various compositions. The target materials could be elemental targets or with the desired TE composition that are evaporated using various sources such a laser, magnetron, and so on, and deposited on a heated substrate to form the desired TE phase.

Magnetron sputtering has been used for the fabrication of Zn_4Sb_3 thin films by codeposition of Zn and Sb. The as-deposited films have been annealed to crystallize in the form of β-Zn_4Sb_3. Microstructure of the deposited films have been controlled by changing sputtering power and deposition time.[71]

Molecular beam epitaxy (MBE) is a technique for epitaxial growth via the interaction of one or several molecular or atomic beams that occur on a surface of a heated crystalline substrate. A typical setup of MBE systems is presented in Figure 17.10. The solid-source materials are placed in evaporation cells to provide an angular distribution of atoms or molecules in a beam. The substrate is heated to the necessary temperature and, when needed, continuously rotated to improve the growth homogeneity.

Self-assembled quantum dot nanomaterials represent one of a number of new approaches being investigated in order to improve the ZT. MBE has been used for the fabrication of TE QDSLs. Ge QDSLs were grown by solid-source MBE on *p*-type Si (100) wafers. The growth temperature was kept at 540°C and the growth rates were 1 and 0.2 Å/s for Si and Ge, respectively. Each Ge quantum dot layer was formed by self-assembly via the Stranski–Krastanov growth mode with deposition of 15 Å Ge. A sample consisting of 10 bilayers, in which Ge quantum dot layers were separated by 20-nm-thick Si spacer layers is shown in Figure 17.11, where dot base diameter and height were determined to be 75 and 7 nm, respectively.[72]

PbSeTe/PbTe QDSL have been fabricated on BaF_2 substrates with a growth rate of 1–2 μm/h growth rate.[73] A TE device has been demonstrated using this QDSL that showed improved performance to the conventional TE cooling material $(Bi,Sb)_2(Se,Te)_3$.[74]

Chemical vapor deposition (CVD) is a vapor transport mechanism in which the gaseous reactants decompose and recombine to form some desired thin film or other nanoscale features. Lead telluride (PbTe) nanocrystals have been synthesized using a CVD technique via a size-selective precipitation mechanism which enables relatively good control of the particle size distribution by variation of the heating temperature, the Ar gas flow rate, and an admixture with Au particles. However, the yield is in the order of several hundred milligrams.[75]

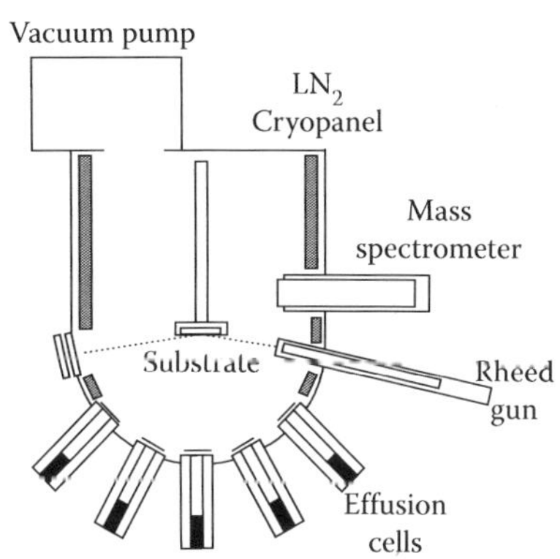

FIGURE 17.10 A typical setup of MBE apparatus.

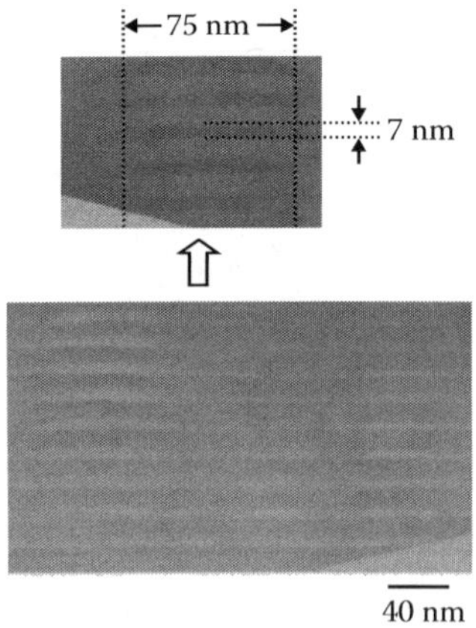

FIGURE 17.11 Cross-sectional HREM image of the 10-period quantum dot superlattice sample. The typical dot base diameter and height were determined to be 75 and 7 nm, respectively. (Reprinted from *J. Cryst. Growth*, **227–228**, J.L. Liu et al., Growth of Ge quantum dot superlattices for thermoelectric applications, 1111–1115, Copyright 2001, with permission from Elsevier.)

17.5 Nanocomposite Processing Routes

Bulk thermoelectric material which exhibits or contains nanoscale substructure within the material has been an intensive topic of interest as they might allow decoupling of electronic and thermal transport components.[5,6] These materials bring together low-dimensional and bulk materials for thermoelectric applications.

In general, bulk crystals with embedded nanocrystals represent a fascinating set of NS materials whose scope extends beyond the field of TEs especially when the properties of guest/matrix are chosen for specific functions. By using the melt-alloying approach nanoscale precipitates have been formed in several materials systems.[76–79] By choosing the appropriate compounds and subjecting the material to a thermal processing procedure, a metastable solid solution of different materials can be obtained. Some materials are soluble in others in liquid state but not in the solid state. When a mixture of molten liquids is rapidly cooled, the insoluble minority phase will precipitate forming nanoparticles embedded in the host phase.[78] Some examples fabricated via this process are *p*-type $Na_{1-x}Pb_mSb_yTe_{m+2}$,[76] *n*-type $Pb_{1-x}Sn_xTe/PbS$,[77] and *n*-type $AgPbTe_2/PbTe$[79] which all showed much lower thermal conductivity values than the host material.

Postprocessing of bulk TE materials is a recently reported route which is very effective in introducing nanoscale features. The process is based on alkali hydrothermal treatment of bulk TE powder in the presence of Na/K hydroxide, Na/K fluoride, and Na/K borohydride in a sealed reactor at ~150°C. The complex series of reactions taking place might be simplified as the dissolution of the bulk powder at elevated temperatures and highly basic conditions via attacks by Na/K ions and subsequent reduction to their original state by the strongly reducing borohydride ions. Tritt et al. reported examples of materials in which nanoscale fractal grain boundaries are generated via such innovative treatments. Nanoscale grain boundaries have been successfully generated in polycrystalline $Pb_{0.75}Sn_{0.25}Te$ system via this treatment, a representative micrograph is presented in Figure 17.12. The sample showed significant reduction in thermal conductivity due to the control of micromorphology of the intergrain boundary.[80] In this report, $Pb_{0.75}Sn_{0.25}Te$ has been grown by solid-state reaction of high-purity elemental components. The flexibility of the technique is the ability to produce these nanoscale fractals independent of the route the material is originally made.

Using a hydrothermal growth route, 2D-coated nanostructures into bulk thermoelectric nanocomposites has been demonstrated by forming a layer of $Pb_{0.75}Sn_{0.25}Se$ with typical thickness of 30 nm on the

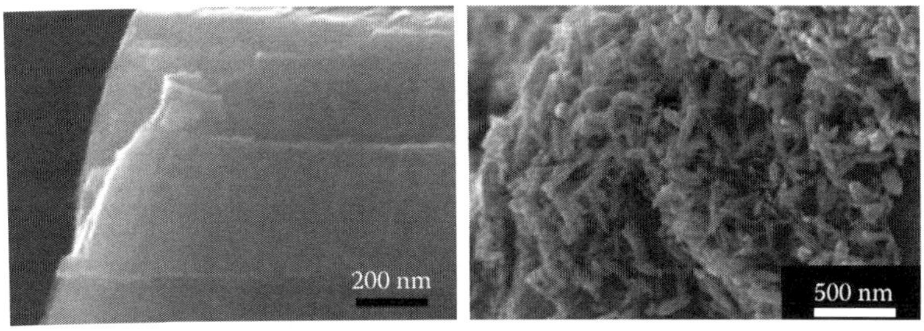

FIGURE 17.12 Micrographs of bulk reference sample with no treatment and after Na-hydrothermal treatment. (X. Ji et al., Nanoscale granular boundaries in polycrystalline $Pb_{0.75}Sn_{0.25}Te$: An innovative approach to enhance the thermoelectric figure of merit, *Phys. Status Solidi A* 2009, **206**, 221–228. Copyright Wiley-VC H Verlag GmbH & Co. KGaA. Reproduced with permission.)

surfaces of $Pb_{0.75}Sn_{0.25}Te$ micron-sized seed particles.[81] The surface of the seed particles are roughened after hydrothermal treatment while the size and shape of the seed particles are preserved as shown in Figure 17.13. The seed particles were doped PbTe nanoparticles that have been synthesized via a high yield and size-selective CVD technique.

$CoSb_3$ nanocomposites with different contents of $CoSb_3$ nanoparticles have been prepared by mixing nano-sized $CoSb_3$ particles obtained via polyol method and micron-sized particles obtained via a solid-state reaction. Resultant nanocomposites showed improved TE performance over the bulk material.[82] Highly dense *n*-type Bi_2Te_3 dispersed with various vol% nano-SiC particles thermoelectric materials have been fabricated by MA and SPS method. The maximum dimensionless figure of merit (ZT_{max}) was improved from 0.99 for Bi_2Te_3 sample to 1.04 for 0.1 vol% SiC-dispersed nanocomposite.[83]

Ceramic-dispersed thermoelectrics are investigated due to the low thermal conductivity of the ceramic phase, and it is expected to further lower the thermal conductivity of the TE material by amplifying phonon scattering. $ZrO_2/CoSb_3$ nanocomposites, with different ZrO_2 content, have been prepared by ball-milling followed by hot-pressing nanopowders of $CoSb_3$ and commercially available 25 nm ZrO_2 nanoparticles. ZrO_2 has been chosen due to its very low thermal conductivity and it has distributed itself uniformly decorating the surfaces of $CoSb_3$ particles as observed from the SEM micrographs taken on hot-pressed samples.[84–86]

Developing advanced TE materials can impact on conversion technologies of waste heat to electrical power. The introduction of low-dimensional material architectures concept into the field of thermoelectricity has stimulated a lot of activities on innovative synthetic approaches for their fabrication. It is evident that there are significant improvements in the synthetic strategies, yet the TE materials have not reached high *ZT* values to make a big impact on the energy conversion/recovery field.

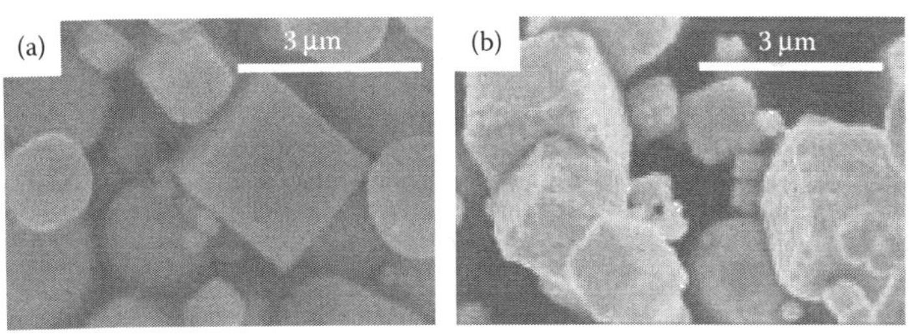

FIGURE 17.13 SEM images of $Pb_{0.75}Sn_{0.25}Te$ are shown before and after the coating process, where, in the right figure, a layer of unknown composition and thickness has formed on the surfaces of matrix particles. (Reprinted with permission from B. Zhang et al., Controlled two-dimensional coated nanostructures for bulk thermoelectric composites. *J. Appl. Phys. Lett.* **89**, 163114, Copyright 2006, American Institute of Physics.)

References

1. M. Muhammed, Engineering of nanostructured materials, in *Nanostructures: Synthesis, Functional Properties and Applications*, eds. T. Tsakalakos et al., Kluwer Academic Publishers, Netherlands, pp. 37–79, 2003.
2. M. Muhammed and T. Tsakalakos, Nanostructured materials and nanotechnology: Overview, *J. Korean Chem. Soc.* **40**(11), 1027–1046, 2003.
3. H. Gleiter, Nanostructured materials: Basic concepts and microstructure, *Acta Mater.* **48**(1), 1–29, 2000.
4. M.S. Dresselhaus, G. Chen, M.Y. Tang, R. Yang, H. Lee, D. Wang, Z. Ren, J.-P. Fleurial, and P. Gogna, New directions for low-dimensional thermoelectric materials, *Adv. Mater.* **19**, 1043–1053, 2007.
5. M.S. Dresselhaus and J.P. Heremans (review), Recent developments in low-dimensional thermoelectric materials, in *CRC Thermoelectrics Handbook*, ed. D.M. Rowe, CRC Press, Boca Raton, FL, pp. 1–24, 2006.
6. A.J. Minnich, M.S. Dresselhaus, Z.F. Ren, and G. Chen, Bulk nanostructured thermoelectric materials: Current research and future prospects, *Energy Environ. Sci.* **2**, 466–479, 2009.
7. Y. Lan, A.J. Minnich, G. Chen, and Z. Ren, Enhancement of thermoelectric figure-of-merit by a bulk nanostructuring approach, *Adv. Func. Mater.* **20**, 357–376, 2010.
8. P. Pichanusakorn and P. Bandaru, Nanostructured thermoelectrics, *Mater. Sci. Eng. R* **67**, 19–63, 2010.
9. S.C. Ur, P. Nash, and I.H. Kim, Solid-state syntheses and properties of Zn Sb thermoelectric materials, *J. Alloys Compd.* **361**, 84–91, 2003.
10. L. Chapon, D. Ravot, and J.C. Tedenac, Nickel-substituted skutterudites: Synthesis, structural and electrical properties, *J. Alloys Compd.* **282**, 58–63, 1999.
11. Q. Shen, L. Zhang, L. Chen, T. Goto, and T. Hirai, Thermoelectric properties of ZrNiSn-based half-Heusler compounds, by solid state reaction method, *J. Mater. Sci. Lett.* **20**, 2197–2199, 2001.
12. B. Poudel, Q. Hao, Y. Ma, Y. Lan, A. Minnich, B. Yu, X. Yan, et al., High-thermoelectric performance of nanostructured bismuth antimony telluride bulk alloys, *Science* **320**, 634, 2008.
13. P.N. Alboni, X. Ji, J. He, N. Gothard, J. Hbbard, and T.M. Tritt, Synthesis and thermoelectric properties of "nano-engineered" $CoSb_3$ skutterudite materials, *J. Electron. Mater.* **36**(7), 711–715, 2007.
14. A. Weidenkaff, R. Robert, M.H. Aguirre, L. Bocher, and L. Schlapbach, Nanostructured thermoelectric oxides with low thermal conductivity, *Phy. Stat. Sol. (RRL)* **6**, 247–249, 2007.
15. L. Bocher, M.H. Aguirre, R. Robert, M. Trottman, D. Logvinovich, P. Hug, and A. Weidenkaff, Chimie douce synthesis and thermochemical characterization of mesoporous perovskite-type titanate phases, *Thermochim. Acta* **457**, 11, 2007.
16. J.L. Mi, X.B. Zhao, and J.P. Tu, Improved thermoelectric figure of merit in *n*-type $CoSb_3$ based nanocomposites, *Appl. Phys. Lett.* **91**, 172116, 2007.
17. H. Li, X. Tang, X. Su, Q. Zhang, and C. Uher, Nanostructured bulk $Yb_xCo_4Sb_{12}$ with high thermoelectric performance prepared by the rapid solidification method, *J. Phys. D: Appl. Phys.* **42**, 145409, 2009.
18. J.P. Heremans, V. Jovovic, E.S. Toberer, A. Saramat, K. Kurosaki, A. Charoenphakdee, S. Yamanaka, and G.J. Snyder, Enhancement of thermoelectric efficiency in PbTe by distortion of the electronic density of states, *Science* **321**, 554–557, 2008.
19. J. Schilz, M. Riffel, K. Pixius, and H.-J. Meyer, Synthesis of thermoelectric materials by mechanical alloying in planetary ball mills, *Powder Technol.* **105**(1–3), 149–154, 1999.
20. Y. Ma, Q. Hao, B. Poudel, Y. Lan, B. Yu, D. Wang, G. Chen, and Z. Ren, Enhanced thermoelectric figure-of-merit in *p*-type nanostructured bismuth antimony tellurium alloys made from elemental chunks, *Nano Lett.* **8**(8), 2580–2584, 2008.
21. J. Yang, Y. Chen, J. Peng, X. Song, W. Zhu, J. Su, and R. Chen, Synthesis of $CoSb_3$ skutterudite by mechanical alloying, *J. Alloys Compd.* **375**, 229–232, 2004.
22. J.J. Ritter, A novel synthesis of polycrystalline bismuth telluride, *Inorg. Chem.* **33**, 6419, 1994.

23. M. Toprak, Y. Zhang, and M. Muhammed, Chemical alloying and characterization of nanocrystalline bismuth telluride, *Mater. Lett.* **57**, 3976–3982, 2003.
24. J.J. Ritter and P. Maruthamuthu, Synthesis of polycrystalline bismuth telluride by a metal-organo complex method, *Inorg. Chem.* **34**, 4278–4280, 1995.
25. M.S. Toprak, Y. Zhang, Y. Jo, D. K. Kim, and M. Muhammed, Nanocrystalline thermoelectric alloys: Processing, chemistry and characterization, *Diffusion and Defect Data Pt. B: Solid State Phenomena*, **101–102**, 197–204, 2005.
26. M.S. Toprak, C. Stiewe, D. Platzek, S. Williams, L. Bertini, E. Müller, C. Gatti, Y. Zhang, M. Rowe, and M. Muhammed, The impact of nanostructuring on the thermal conductivity of thermoelectric $CoSb_3$, *Adv. Func. Mater.* **14**(12), 1189–1196, 2004.
27. L. Bertini, C. Stiewe, M. Toprak, S. Williams, D. Platzek, A. Mrotzek, Y. Zhang, C. Gatti, E. Müller, M. Muhammed, and M. Rowe, Nanostructured $Co_{1-x}Ni_xSb_3$ skutterudites: Synthesis, thermoelectric properties, and theoretical modeling, *J. Appl. Phys.* **93**, 438–447, 2003.
28. M. Christensen, B.B. Iversen, L. Bertini, C. Gatti, M. Toprak, M. Muhammed, and E. Nishibori, Structural study of Fe doped and Ni substituted thermoelectric skutterudites by combined synchrotron and neutron powder diffraction and *ab initio* theory, *J. Appl. Phys.* **96**(6), 3148–3157, 2004.
29. C. Stiewe, L. Bertini, M. Toprak, M. Christensen, D. Platzek, S. Williams, C. Gatti, E. Müller, B. B. Iversen, M. Muhammed, and M. Rowe, Nanostructured $Co_{1-x}Ni_x(Sb_{1-y}Te_y)_3$ skutterudites: Theoretical modeling, synthesis and thermoelectric properties, *J. Appl. Phys.* **97**(4), 044317-1–044317-7, 2005.
30. V.Y. Kodash, J.R. Groza, G. Aldica, M.S. Toprak, S. Li, and M. Mamoun, Field-activated sintering of skutterudites, *Scr. Mater.* **57**(6), 509–511, 2007.
31. M. Muhammed and M. Toprak, Nanostructured skutterudites, in *Thermoelectrics Handbook: Macro to Nano*, Chapter 41, ed. D.M. Rowe, CRC Press, Boca Raton, Florida, ISBN: 0849322642, 2005.
32. C.-J. Liu, J.-Y. Liao, T.-W. Wu, and B.-Y. Jen, Preparation and transport properties of aqueous sol-gel synthesized $NaCo_2O_{4-\delta}$, *J. Mater. Sci.* **39**, 4569–4573, 2004.
33. K. F. Cai, E. Muller, C. DraSar, and A. Mrotzek, Preparation and thermoelectric properties of ZnO-TiB_2 composites, in *Proc. 21st Int. Conf. Thermoelectronics*, 142–146, 2002.
34. J.L. Mi, X.B. Zhao, T.J. Zhu, J.P. Tu, and G.S. Cao, Solvothermal synthesis and electrical transport properties of skutterudite $CoSb_3$, *J. Alloys Compd.* **417**, 269–272, 2006.
35. L. Yang, H.H. Hng, H. Chenga, T. Sun, and J. Ma, Synthesis of $CoSb_3$ by a modified polyol process, *Mater. Lett.* **62**, 2483–2485, 2008.
36. J.Q. Li, X.W. Feng, W.A. Sun, W.Q. Ao, F.S. Liu, and Y. Du, Solvothermal synthesis of nano-sized skutterudite $Co_{4-x}Fe_xSb_{12}$ powders, *Mater. Chem. Phys.* **112**, 57–62, 2008.
37. T. Sun, X.B. Zhao, T.J. Zhu, and J.P. Tu, Aqueous chemical reduction synthesis of Bi_2Te_3 nanowires with surfactant assistance, *Mater. Lett.* **60**, 2534–2537, 2006.
38. X.B. Zhao, Y.H. Zhang, and X.H. Ji, Solvothermal synthesis of nano-sized $La_xBi_{(2-x)}Te_3$ thermoelectric powders, *Inorg. Chem. Commun.* **7**, 386–388, 2004.
39. J.L. Mi, N. Lock, T. Sun, M. Christensen, M. Søndergaard, P. Hald, H.H. Hng, J. Ma, and B.B. Iversen, Biomolecule-assisted hydrothermal synthesis and self-assembly of Bi_2Te_3 nanostring-cluster hierarchical structure, *ACS Nano* **4**(5), 2523–2530, 2010.
40. W.Q. Ao, W.A. Sun, J.Q. Li, F.S. Liu, and Y. Du, Hydrothermal synthesis of nano-sized $AgPb_{18}SbTe_{20}$ thermoelectric powders, *J. Alloy Compd.* **475**, L22–L24, 2009.
41. W. Lu, Y. Ding, Y. Chen, Z.L. Wang, and J. Fang, Bismuth telluride hexagonal nanoplatelets and their two-step epitaxial growth, *J. Am. Chem. Soc.* **127**(28), 10112–10116, 2005.
42. R.Y. Wang, J.P. Feser, J.-S. Lee, D.V. Talapin, R. Segalman, and A. Majumdar, Enhanced thermopower in PbSe nanocrystal quantum dot superlattices, *Nano. Lett.* **8**(8), 2283–2288, 2008.
43. Y. Zhao, J.S. Dyck, B.M. Hernandez, and C. Burda, Improving thermoelectric properties of chemically synthesized Bi_2Te_3-based nanocrystals by annealing, *J. Phys. Chem. C* **114**, 11607–11613, 2010.

44. S. Li, S. Zhang, Z. He, M. Toprak, C. Stiewe, M. Muhammed, and E. Müller, Novel solution route synthesis of low thermal conductivity nanocrystalline bismuth telluride, *J. Nanosci. Nanotechnol.* **10**, 1–5, 2010.
45. D. Dhak and P. Pramanik, Characterization of nanocrystalline bismuth telluride (Bi_2Te_3) synthesized by a novel approach through aqueous precursor method, *J. Am. Ceram. Soc.* **89**(2), 534–537, 2006.
46. M.A. López-Quintela, Synthesis of nanomaterials in microemulsions: Formation mechanisms and growth control, *Curr. Opin. Colloid.* **8**, 137–144, 2003.
47. E.E. Foos, R.M. Stroud, and A.D. Berry, Synthesis and characterization of nanocrystalline bismuth telluride, *Nano Lett.* **1**(12), 693–695, 2001.
48. A.J. Karkamkar and M.G. Kanatzidis, Chemical routes to nanocrystalline thermoelectrically relevant $AgPb_mSbTe_{m+2}$ materials, *J. Am. Chem. Soc.* **128**, 6002–6003, 2006.
49. K.S. Suslick, M. Fang, and T. Hyeon, Sonochemical synthesis of iron colloids, *J. Am. Chem. Soc.* **118**, 11960–11961, 1996.
50. K.S. Suslick, M.M. Fang, T. Hyeon, and M.M. Mdleleni, Applications of sonochemistry to materials synthesis, in *Sonochemistry and Sonoluminescence*, eds. L.A. Crum, T.J. Mason, J. Reisse, K.S. Suslick, Kluwer Publishers: Dordrecht, Netherlands, pp. 291–320, 1999.
51. J.P. Ge and Y.D. Li, Ultrasonic synthesis of nanocrystals of metal selenides and tellurides, *J. Mater. Chem.* **13**, 911–915, 2003.
52. Y.Y. Zheng, T.J. Zhu, X.B. Zhao, J.P. Tu, and G.S. Cao, Sonochemical synthesis of nanocrystalline Bi_2Te_3 thermoelectric compounds, *Mater. Lett.* **59**, 2886–2888, 2005.
53. H. Cui, H. Liu, J. Wang, X. Lia, F. Hana, and R.I. Boughton, Sonochemical synthesis of bismuth selenide nanobelts at room temperature, *J. Cryst. Growth* **271**, 456–461, 2004.
54. R. Mannam, M. Agarwal, A. Roy, V. Singh, K. Varahramyan, and D. Davis, Electrodeposition and thermoelectric characterization of bismuth telluride nanowires, *J. Electrochem. Soc.* **156**(8), B871–B875, 2009.
55. A.L. Prieto, M.S. Sander, M.S. Martín-González, R. Gronsky, T. Sands, and A.M. Stacy, Electrodeposition of ordered Bi_2Te_3 nanowire arrays, *J. Am. Chem. Soc.* **123**(29), 7160–7161, 2001.
56. S. Li, Y. Liang, J. Qin, M. Toprak, and M. Muhammed, Template electrodeposition of ordered bismuth telluride nanowire arrays, *J. Nanosci. Nanotechnol.* **8**, 1–5, 2008.
57. E.J. Menke, M.A. Brown, Q. Li, J.C. Hemminger, and R.M. Penner, Bismuth telluride (Bi_2Te_3) nanowires: Synthesis by cyclic electrodeposition/stripping, thinning by electrooxidation, and electrical power generation, *Langmuir* **22**, 10564–10574, 2006.
58. K. Park, F. Xiao, Y. Rheem, and N.V. Myung, Electrochemical deposition of thermoelectric Sb_xTe_y thin films and nanowires, *J. Alloy Compd.* **485**, 362–266, 2009.
59. H.-S. Seo, X. Li, H.D. Um, B. Yoo, J.H. Kim, K.P. Kim, Y.W. Cho, and J.H. Lee, Fabrication of precisely controlled silicon wire and cone arrays by electrochemical etching, *Mater. Lett.* **63**, 2567–2569, 2009.
60. D.O. Banga, R. Vaidyanathan, L. Xuehai, J. L. Stickney, S. Cox, and U. Happeck, Formation of PbTe nanofilms by electrochemical atomic layer deposition (ALD), *Electrochim. Acta* **53**(23), 6988–6994, 2008.
61. Y.X. Gan, J. Sweetman, and J.G. Lawrence, Electrodeposition and morphology analysis of Bi–Te thermoelectric alloy nanoparticles on copper substrate, *Mater. Lett.* **64**(3), 449–452, 2010.
62. F. Xiao, B.Y. Yoo, K.-H. Lee, and N.V. Myung, Synthesis of Bi_2Te_3 nanotubes by galvanic displacement, *J. Am. Chem. Soc.* **129**, 10068–10069, 2007.
63. C.H. Chang, Y. Rheem, Y.H. Choa, D.H. Shin, D.Y. Park, and N.V. Myung, Bi and Te thin films synthesized by galvanic displacement from acidic nitric baths, *Electrochim. Acta* **55**(3), 743–752, 2010.
64. G.-R. Li, F.-L. Zheng, and Y.X. Tong, Controllable synthesis of Bi_2Te_3 intermetallic compounds with hierarchical nanostructures via electrochemical deposition route, *Cryst. Growth Des.* **8**(4), 1226–1232, 2008.

65. S. Li, M.S. Toprak, H.M.A. Soliman, J. Zhou, M. Muhammed, D. Platzek, and E. Müller, Fabrication of nanostructured thermoelectric bismuth telluride thick films by electrochemical deposition, *Chem. Mater.* **18**(16), 3627–3633, 2006.
66. S. Li, H.M.A. Soliman, J. Zhou, M.S. Toprak, M. Muhammed, D. Platzek, P. Ziolkowski, and E. Müller, Effects of annealing and doping on nanostructured bismuth telluride thick films, *Chem. Mater.* **20**(13), 4403–4410, 2008.
67. J. Zhou, S. Li, H.M.A. Soliman, M.S. Toprak, M. Muhammed, D. Platzek, and E. Müller, Synthesis and Seebeck coefficient of nanostructured phosphorus-alloyed bismuth telluride thick films, *Phys. Stat. Sol.* (c) **5**(11), 3453–3457, 2008.
68. Z. Zhang, X. Sun, M.S. Dresselhaus, J.Y. Ying, and J.P. Heremans, Magnetotransport investigations of ultrafine single-crystalline bismuth nanowire arrays, *Appl. Phys. Lett.* **73**, 1589–1591, 1998.
69. Z. Zhang, J.Y. Ying, and M.S. Dresselhaus, Bismuth quantum-wire arrays fabricated by a vacuum melting and pressure injection process, *J. Mater. Res.* **13**, 1745–1748, 1998.
70. J. Heremans, C.M. Thrush, Yu-Ming Lin, S. Cronin, Z. Zhang, M.S. Dresselhaus, and J.F. Mansfield, Bismuth nanowire arrays: Synthesis and galvanomagnetic properties, *Phys. Rev. B* **61**, 2921–2930, 2000.
71. L.T. Zhang, M. Tsutsui, K. Ito, and M. Yamaguchi, Thermoelectric properties of Zn_4Sb_3 thin films prepared by magnetron sputtering, *Thin Solid Films* **443**, 84–90, 2003.
72. J.L. Liu, A. Khitun, K.L. Wang, T. Borca-Tasciuc, W.L. Liu, G. Chen, and D.P. Yu, Growth of Ge quantum dot superlattices for thermoelectric applications, *J. Cryst. Growth* **227–228**, 1111–1115, 2001.
73. T.C. Harman, M.P. Walsh, B.E. Laforge, and G.W. Turner, Nanostructured thermoelectric materials, *J. Electron. Mater.* **34**, 5 L19–L22, 2005.
74. T.C. Harman, P.J. Taylor, M.P. Walsh, and B.E. Laforge, Quantum dot superlattice thermoelectric materials and devices, *Science* **297**, 2229–2232, 2002.
75. B. Zhang, J. He, and T.M. Tritt, Size-selective high-yield growth of lead telluride (PbTe) nanocrystals using a chemical vapor deposition technique, *Appl. Phys. Lett.* **88**, 043119–043119-3, 2006.
76. P.F.P. Podeu, J.D. Angelo, A.D. Downey, J.L. Short, T.P. Hogan, and M. Kanatzidis, High thermoelectric figure of merit and nanostructuring in bulk *p*-type $Na_{1-x}Pb_mSb_yTe_{m+2}$, *Angew. Chem., Int. Ed.* **45**, 3835–3839, 2006.
77. J. Androulakis, C.H. Lin, H.J. Kong, C. Uher, C.I. Wu, T. Hogan, B.A. Cook, T. Caillat, T.M. Paraskevopoulos, and M. Kanatzidis, Spinodal decomposition and nucleation and growth as a means to bulk nanostructured thermoelectrics: Enhanced performance in $Pb_{1-x}Sn_xTe$ – PbS, *J. Am. Chem. Soc.* **129**, 9780 9788, 2007.
78. R. Sootsman, R.J. Pcionek, H. Kong, C. Uher, and M.G. Kanatzidis, Strong reduction of thermal conductivity in nanostructured PbTe prepared by matrix encapsulation, *Chem. Mater.* **18**, 4993–4995, 2006.
79. K.F. Hsu, S. Loo, F. Guo, W. Chen, J.S. Dyck, C. Uher, T. Hogan, E.K. Polychroniadis, and M.G. Kanatzidis, Cubic $AgPb_mSbTe_{2+m}$: Bulk thermoelectric materials with high figure of merit, *Science* **303**, 818–821, 2004.
80. X.Ji, B. Zheng, Z. Su, T. Holgate, J. He, and T.M. Tritt, Nanoscale granular boundaries in polycrystalline $Pb_{0.75}Sn_{0.25}Te$: An innovative approach to enhance the thermoelectric figure of merit, *Phys. Status Solidi A* **206**, 221–228, 2009.
81. B. Zhang, J. He, X. Ji, T.M. Tritt, and A. Kumbhar, Controlled two-dimensional coated nanostructures for bulk thermoelectric composites, *J. Appl. Phys. Lett.* **89**, 163114-163114-3, 2006.
82. L. Yang, H.H. Hng, D. Li, Q.Y. Yan, J. Ma, T.J. Zhu, X.B. Zhao, and H. Huang, Thermoelectric properties of *p*-type $CoSb_3$ nanocomposites with dispersed $CoSb_3$ nanoparticles, *J. Appl. Phys.* **106**(1), 013705-013705-6, 2009.
83. L.D. Zhao, B.P. Zhang, J.F. Li, M. Zhou, W.S. Liu, and J. Liu, Thermoelectric and mechanical properties of nano-SiC-dispersed Bi_2Te_3 fabricated by mechanical alloying and spark plasma sintering, *J. Alloys Compd.* **455**, 259–264, 2008.

84. C. Stiewe, Z. He, D. Platzek, G. Karpinski, E. Müller, S. Li, M. Toprak, and M. Muhammed, Control of thermoelectric properties in $ZrO_2/CoSb_3$ nano-dispersed composites, *Mat. Wiss. U. Werkstofftech.* **38**(9), 773–776, 2007.
85. Z. He, C. Stiewe, D. Platzek, G. Karpinski, E. Müller, S. Li, M. Toprak, and M. Muhammed, Nano $ZrO_2/CoSb_3$ composites with improved thermoelectric figure of merit, *Nanotechnol.* **18**(23), 235602 (5p.), 2007.
86. Z. He, C. Stiewe, D. Platzek, G. Karpinski, E. Müller, S. Li, M. Toprak, and M. Muhammed, Effect of ceramic dispersion on thermoelectric properties of nano-$ZrO_2/CoSb_3$ composites, *J. Appl. Phys.* **101**(4), 043707 (7p.), 2007.

18 Preparation and Thermoelectric Properties of Iron Disilicide

Yukihiro Isoda
National Institute for Materials Science

Haruhiko Udono
Ibaraki University

18.1 Introduction

The stoichiometric compound iron disilicide ($FeSi_2$) is composed of environment-friendly, nontoxic, and abundantly available elements. According to the equilibrium phase diagram established by Piton and Fay,[1] $FeSi_2$ causes the metal–semiconductor transition at the peritectoid temperature of 1259 K, as shown in Figure 18.1.[2–6] The low-temperature phase β-$FeSi_2$, which belongs to the *Cmca* space group of the orthorhombic crystalline structure,[7] possesses stable semiconducting properties up to 1259 K. Above this peritectoid temperature, the metallic phase is a eutectic alloy composed of the tetragonal α-phase[8,9] and the cubic ε-phase.[10,11] The crystallographical notation of the α-phase is $FeSi_2$ or $Fe_{1-Υ}Si_2$ (Υ: vacancies of the iron sublattice); however, it exists within a rather wide range of compositions through the accommodation of vacancies in the iron sublattice. The compound with 20% vacancies is represented by the chemical formula Fe_2Si_5. The ε-phase is iron monosilicide FeSi with space group $P2_13$. The eutectic and peritectoid temperatures of the eutectic alloy (Fe-66.7 at% Si) are 1485 and 1259 K, respectively. The β-$FeSi_2$ phase transforms into the α-Fe_2Si_5 phase and ε-FeSi phase above 1259 K as follows:

$$\beta\text{-FeSi}_2 \rightarrow \alpha\text{-Fe}_2\text{Si}_5 + \varepsilon\text{-FeSi}.$$

Below 1223 K, it transforms into β-$FeSi_2$ from the eutectic alloy of the high-temperature phase according to the eutectoid and peritectoid reactions as follows:

$$\alpha\text{-Fe}_2\text{Si}_5 \rightarrow \beta\text{-FeSi}_2 + \text{Si}\ \ (\text{eutectoid decomposition}),$$

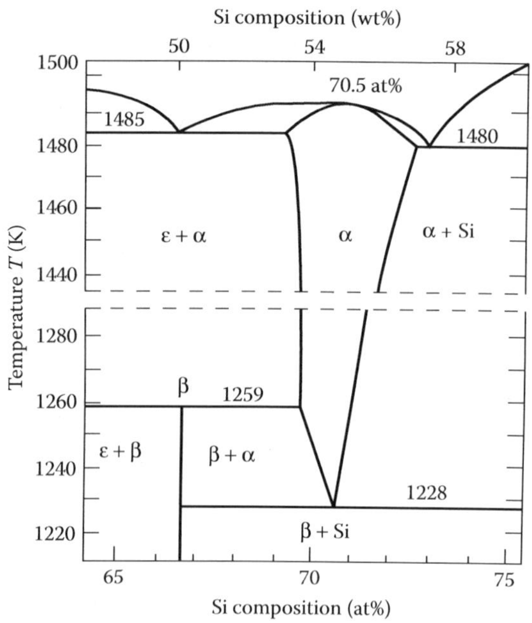

FIGURE 18.1 The equilibrium phase diagram for the system Fe–Si by Piton and Fay. α-phase:$FeSi_2$, β-phase:$FeSi_2$, ε-phase:FeSi.

$$\varepsilon\text{-FeSi} + \text{Si} \rightarrow \beta\text{-FeSi}_2 \quad \text{(peritectoid reaction)}.$$

The growth of β-$FeSi_2$ phase with a lamellar structure in the cast specimen is not completed even after a prolonged annealing time of 500 h. Hence, the semiconducting single phase of β-$FeSi_2$ cannot be grown from the melt. Consequently, the general preparation of bulk β-$FeSi_2$ is based on powder metallurgical methods. These methods include the slip casting method,[12] cold-pressing and sintering method,[13–18] hot-pressing method,[19–21] and spark plasma sintering (SPS) method, which is the same as the plasma-activated sintering (PAS) method.[22–24] A single crystal of β-$FeSi_2$ has been grown by using the chemical vapor transport (CVT),[25–27] horizontal freezing gradient (HFG),[28,29] and temperature gradient solution growth (TGSG)[30–34] techniques.

β-$FeSi_2$ is called the "environment-friendly thermoelectric semiconductor" because of its environment-friendly elements, high heat and oxidation resistance, and relatively high thermoelectromotive force. Nondoped β-$FeSi_2$ is an intrinsic semiconductor. However, nondoped β-$FeSi_2$ prepared with 4 N purity of source materials exhibits *p*-type conduction, and nondoped single crystals prepared with 5 N purity exhibit *n*-type.[35,36] For these reasons, the conduction type of β-$FeSi_2$ can be controlled by doping various elements. *p*-Type conductivity is exhibited by manganese,[6,13,16–19] chromium,[37,38] vanadium, titanium,[39] and aluminum,[40–42] whereas *n*-type conductivity is exhibited by cobalt,[13,14,16,40] nickel,[37,43] platinum,[37] and boron.[44,45]

In this chapter, the preparation of a single crystal of β-$FeSi_2$ by the and TGSG techniques and the preparation of bulk polycrystalline β-$FeSi_2$ by the cold-pressing and sintering method will be discussed. Moreover, the thermoelectric properties of the obtained β-$FeSi_2$ will be discussed. In the last section of this chapter, it will be shown that β-$FeSi_2$ is a good candidate for many civil applications.

18.2 Single Crystal Growth of ß-$FeSi_2$

Several studies on the single crystal growth of semiconducting iron disilicide (ß-$FeSi_2$) have been carried out by using the CVT[25–27] and solution growth (flux growth)[30–34] method, which can grow a crystal below the melting point because the solidus line of this phase does not coexist with the melt[1] (Figure 18.1). The high-temperature phase of α-$FeSi_2$, which is thermodynamically metastable at higher temperatures, transforms into the low-temperature ß-$FeSi_2$ phase according to the eutectoid reaction.[8–11] The ß-$FeSi_2$ phase is stable below the transition temperature. Therefore, ordinary growth methods from the melt such as the Bridgmann, Czochralski, floating zone, and horizontal gradient freeze methods are difficult to use for the bulk single crystal growth of ß-$FeSi_2$.

Wandji et al. reported the single crystal growth of ß-$FeSi_2$ by the CVT method using I_2 as a transport agent.[25] They obtained needle-like crystals of the ß-$FeSi_2$ phase with a length of 2–3 mm during the growth period of 196 h. So far, I_2 is the most preferred transport agent for the CVT growth of ß-$FeSi_2$, even though other transport agents such as Br_2 and Cl_2 can be used owing to their higher transport ability for the Fe species. Figure 18.2 shows a schematic illustration of a typical CVT system and the grown crystals of ß-$FeSi_2$.[46] The temperature gradient of the growth ampoule was stabilized between 750°C and 1050°C during the growth period (7–30 days). Depending on the growth period, crystal growth is encouraged; however, twin growth frequently occurs in the large-sized crystals (Figure 18.2c). Therefore, the size of a single crystal is usually limited to ~$10 \times 2 \times 0.5$ mm^3.[26,27,46]

Single crystal growth of ß-$FeSi_2$ is also reported by the TGSG method using Ga,[31,33] Zn,[47,48] Sn,[49] and Sb[32] as solvents. The source of the $FeSi_2$ ingot, which is presynthesized by melting highly pure Fe and Si, is dissolved into the solvent metal under the gradient temperature condition. The dissolved solute is transported by diffusion and convection from a higher-temperature position to a lower-temperature position, and crystal growth then occurs by supersaturation at the lower-temperature position (Figure 18.3).

The crystal habits of the solution-grown ß-$FeSi_2$ change depending on the solvent (Figure 18.3). Most of the grown crystals have a polyhedral shape when using the Ga, Zn, and Sb solvents, whereas most of the crystals have a needle-like shape when using the Sn solvent. Depending on the growth period, crystal growth is also encouraged by the TGSG method. Twin growth, however, is frequently the same as in the case of CVT growth. Typical dimensions of a single crystal without twins for the growth period of 1 week are $(2–5) \times (2–7) \times (1–3)$ mm^3 for the Ga solvent, $(1–3) \times (1–5) \times (0.5–2)$ mm^3 for the Zn solvent, and $(0.5–1) \times (0.5–1) \times (3–4)$ mm^3 for the Sn solvent. The differences in the crystal sizes are mainly due to the $FeSi_2$ source solubility against the solvents. In addition, plate-like bulk crystals having a diameter of 10 mm and thickness of 0.3–0.5 mm were successfully grown on a graphite template using the Ga solvent.[50] This method can be potentially adopted to obtain large bulk ß-$FeSi_2$ single crystals.

There are several preferable growth directions for ß-$FeSi_2$. In the CVT-grown crystal, Irmuscher et al. found that their needle-like crystals have a twin grain boundary formed by the contact of (001) and (010)

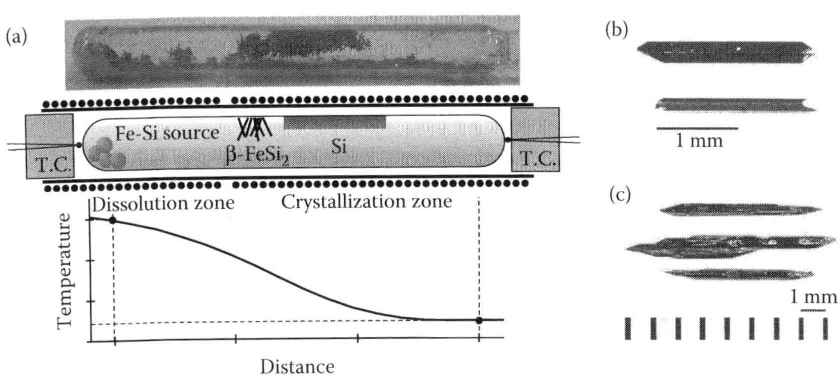

FIGURE 18.2 (a) Schematic illustration of growth ampoule and temperature profile of CVT method. A number of needle-like crystals are observed in the as-grown ampoule. Microphotographs of (b) CVT-grown ß-$FeSi_2$ single crystals without twin, and (c) large-sized CVT-grown ß-$FeSi_2$ crystals. (Courtesy of Professor Y. Hara, Ibaraki College of Technology.)

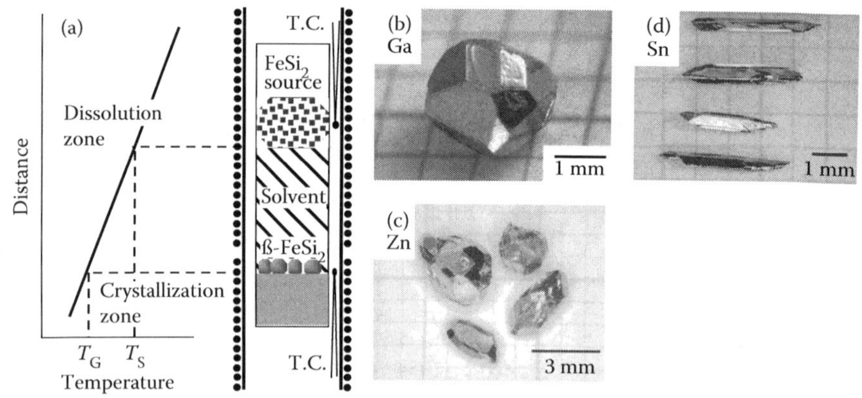

FIGURE 18.3 (a) Schematic illustration of growth ampoule and temperature profile of TGSG method. Microphotographs of typical ß-$FeSi_2$ single crystals grown from (b) Ga solvent, (c) Zn solvent, and (d) Sn solvent.

planes, as well as (100) planes rotated 90° to each other without requiring large reconstruction, and both grains have [010] and [001] parallel to the preferred growth directions.[51]

On the other hand, Udono et al. found two different types of needle-like crystals in their solution-grown crystals using the Sn solvent.[49] One has a <010> preferred growth direction and the other has a <011> preferred growth direction. The former crystal has good crystalline quality without twins, while the latter one frequently has twins. Both types of crystals have clear growth facets (Table 18.1). The crystals grown toward the <011> direction have 10 growth facets belonging to the symmetrically equivalent faces of {100}, {111}, and {311}, whereas those grown toward the <010> direction have eight growth facets belonging to those of {100}, {001}, and {101}. These Miller indexes of the growth facets are also observed in both solution-grown and CVT-grown ß-$FeSi_2$ single crystals.[33,47,51]

Since the silicon dioxide is formed on the surface in ambient atmosphere, single crystalline ß-$FeSi_2$ has especially high chemical stability and corrosion resistance compared with those of a sintered ß-$FeSi_2$ polycrystal.[52,53] Tables 18.2 and 18.3 show corrosion resistance against several acids and alkalis and etch rates dependent on the surface orientation of the single crystalline ß-$FeSi_2$, respectively. The single crystalline ß-$FeSi_2$ does not dissolve in HCl, HNO_3, H_2SO_4, and aqua regia, while sintered ß-$FeSi_2$ polycrystal usually dissolves in them because it usually remains in the vicinity of α-$FeSi_2$ and ε-FeSi phases, and both of them dissolve in those acids.

TABLE 18.1 List of Growth Facets Observed in the Needle-Like Crystals Grown toward the <011> and <010> Directions

Needle	Observed Facet Angle (°)															
Axis	*a*	*b*	*c*	*d*	*e*	*f*	*g*	*h*	*I*	*j*	*k*	*l*	*m*	*n*	*o*	*p*
<011>	0	31	—	63	—	120	—	147	179	211	—	241	—	298	—	330
<010>	0	—	47	—	94	—	137	—	176	—	226	—	270	—	319	—
M-Index	(100)	$31\bar{1}$	(101)	$11\bar{1}$	(001)	$\bar{1}1\bar{1}$	$\bar{1}01$	$\bar{3}\bar{1}1$	$\bar{1}00$	$\bar{3}\bar{1}1$	$\bar{1}0\bar{1}$	$\bar{1}\bar{1}1$	$00\bar{1}$	$1\bar{1}1$	$10\bar{1}$	$3\bar{1}\bar{1}$

Note: The facet planes are parallel to the needle-like axis. The angle accuracy is within 5°.

TABLE 18.2 Corrosion Resistance of Single Crystalline ß-$FeSi_2$ for Acid and Alkali Solutions

Conditions	HCl (36%)	HNO_3 (60%)	H_2SO_4 (96%)	Aqua Regia (1:3)	Dilute HF (5%)	NaOH (60%)
22°C, 24 h (wt%)	0.0	0.0	0.0	0.0	30	0.0
50°C, 5 h (wt%)	0.0	0.0	0.0	0.0	—	0.7

TABLE 18.3 Etch Rate of *p*-Type ß-$FeSi_2$ {100}, {001}, and {111} Faces at 22°C

	Etch Rate (μm/min) at 22°C			
	HF(49%):HNO_3(60%):H_2O			
Surface	1:1:2	1:1:4	1:1:8	5% HF
{111}	1.57 ± 0.2	0.53 ± 0.05	0.08 ± 0.02	0.07 ± 0.02
{100}	1.43 ± 0.2	0.43 ± 0.05	0.05 ± 0.01	0.05 ± 0.01
{001}	1.09 ± 0.1	0.33 ± 0.04	0.03 ± 0.01	0.02 ± 0.01

A diluted HF solution is a good chemical etchant for removing the damaged surface layer that was introduced by mechanical polishing during sample preparation. The etch rate can be increased by adding oxidizing acid (HNO_3). There are also some reports about etch pits of ß-$FeSi_2$. Etch pits with a specific shape depend on crystal orientation and appear on the surface when etching with the diluted HF and HF:HNO_3:H_2O solutions. These pits are convenient for identifying the surface orientation of the growth facets.

18.3 Thermoelectric Properties of Single Crystalline ß-$FeSi_2$

Studies on single crystalline ß-$FeSi_2$ significantly extend our understanding of the fundamental thermoelectric properties of this material. In the CVT-grown single crystal, the thermoelectric properties of Cr-doped *p*-type ß-$FeSi_2$ and Co-doped *n*-type ß-$FeSi_2$ are mainly studied.[54,55] Figure 18.4 shows the reported thermoelectric power and resistivity of CVT-grown single crystals. The thermoelectric power of both *p*- and *n*-type crystals is more than 400 μV/K between 200 and 600 K. These values are relatively large compared with conventional thermoelectric materials. However, the electrical resistivity of these crystals is quite high owing to the difficulty in doping control when using the CVT method. Thus, the thermoelectric power factor is limited in a CVT-grown single crystal.

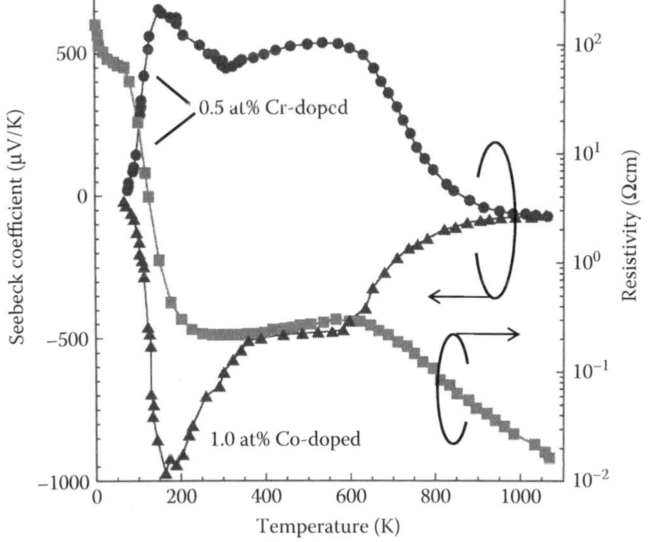

FIGURE 18.4 Temperature dependence of Seebeck coefficient and resistivity of CVT-grown ß-$FeSi_2$ single crystals. Positive Seebeck coefficient and resistivity are measured from 0.5 at% Cr-doped ß-$FeSi_2$ and negative Seebeck coefficient from 1.0 at% Co-doped one.

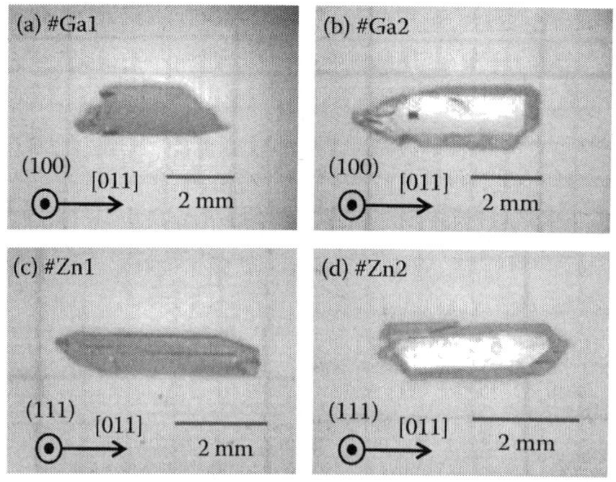

FIGURE 18.5 Single crystalline ß-$FeSi_2$ samples grown by the temperature gradient solution growth (TGSG) method. Samples are grown from Ga solvent: (a), (b) and Zn solvent: (c), (d). The surface are (100) face for #Ga1 and #Ga2 and (111) for #Zn1 and #Zn2. Longitudinal axis of all samples are [011] direction.

In contrast to the CVT method, impurity doping is much easier in the solution growth method. Suzuki et al. and Udono et al. studied the temperature dependence of carrier concentration, Hall mobility, electrical conductivity, and thermoelectric power of solution-grown single crystals at a low temperature.[56–58] Figure 18.5 shows the typical ß-$FeSi_2$ samples, which were prepared from TGSG crystals with anisotropic crystal habits, used for thermoelectric measurements. Their growth conditions and purity of source metals are listed in Table 18.4. The samples grown from Ga solvent show *p*-type conductivity due to Ga impurity incorporated from the solvent, whereas samples grown from Zn solvent show *n*-type conductivity due to the unintentionally doped boron and nickel impurity that incorporated in the $FeSi_2$ source. (Crystals grown from Zn solvent without boron and nickel doping show *p*-type conductivity due to Zn impurity incorporated from the solvent.) The samples are single crystals without twins and their sizes are $(4–6) \times (1–2) \times 0.3$ mm^3. The orientation of the longitudinal axis is [011] for all samples. Figure 18.6 demonstrates the temperature dependence of the carrier concentration and Hall mobility of the samples. The carrier concentrations of *p*-type (Ga solvent) and *n*-type (Zn solvent) samples are $\sim 1 \times 10^{19}$ and 5×10^{18} cm^{-3} at 300 K, respectively. The temperature dependence of the carrier concentration and the Hall mobility of the Ga solvent samples are almost independent of the $FeSi_2$ source purity, whereas those of the Zn solvent samples are affected by the source purity because the impurity concentration of Ga is much higher than that of Zn. Therefore, the difference of Hall mobility μ between #Zn-1 (9 cm^2/Vs) and #Zn-2 (2 cm^2/Vs) would be mainly due to the difference of dominant impurity and their concentration level.

The temperature dependence of Seebeck coefficient *S* measured along the [011] direction is plotted in Figure 18.7. The samples #Ga-1 and #Ga-2 have positive values, whereas the samples #Zn-1 and #Zn-2 have negative values. The values of *S* for the *p*- and *n*-type samples are ~350 and 700 μV/K at 300 K, respectively. These values are comparable to the *S* of Mn-doped sintered ß-$FeSi_2$, and Co-doped CVT-ß-$FeSi_2$,

TABLE 18.4 Growth Conditions of ß-$FeSi_2$ Samples for Thermoelectric Measurements

Sample #	$FeSi_2$ Source Starting Material and Dopant		T_G/T_S (°C)	Growth Period (h)
#Ga1	3N-Fe, 4Nup-Si	—	880/920	336
#Ga2	4Nup-Fe, 5Nup-Si	Boron		
#Zn1	3N-Fe, 4Nup-Si	—		
#Zn2	4Nup-Fe, 5Nup-Si	Boron		

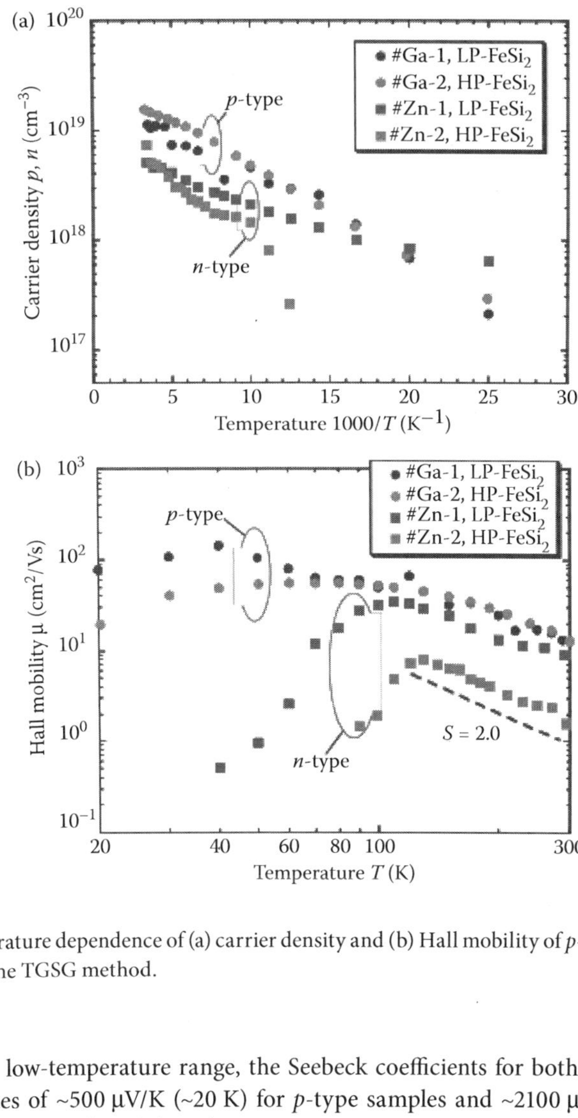

FIGURE 18.6 Temperature dependence of (a) carrier density and (b) Hall mobility of *p*- and *n*-type single crystalline ß-FeSi$_2$ grown by the TGSG method.

respectively.[54,62] At a low-temperature range, the Seebeck coefficients for both *p*- and *n*-type samples have maximum values of ~500 μV/K (~20 K) for *p*-type samples and ~2100 μV/K (~70 K) for *n*-type samples. As shown by the Hall mobility, the temperatures at maximum *S* are slightly lower than those of the maximum Hall mobility. Therefore, the large *S* values at a low-temperature range might be due to the phonon drag effect.

Figure 18.8 shows the temperature dependence of resistivity ρ. The resistivity of both *p*-type samples is almost same and the values are 0.03–0.04 Ωcm at 300 K. This value is ~1/2–1/5 times lower than that of CVT and sintered ß-FeSi$_2$ crystals. In the lower temperature range between 300 and 50 K, the resistivity of *p*-type samples is almost constant, and thus the resistivity are more than one order of magnitude lower than that of CVT and sintered ß-FeSi$_2$ crystals.

In the *n*-type samples, the resistivity at 300 K is ~0.16 and 0.62 Ωcm for #Zn-1 and #Zn-2, respectively. The values are more than one order of magnitude higher than those of the *p*-type samples due to the lower electron concentration and Hall mobility, but they are comparable to those of Co-doped CVT and sintered samples. The resistivity is almost constant between 300 and 130 K similar to the *p*-type samples. Below this temperature range, resistivity increases rapidly as the Hall mobility decreases.

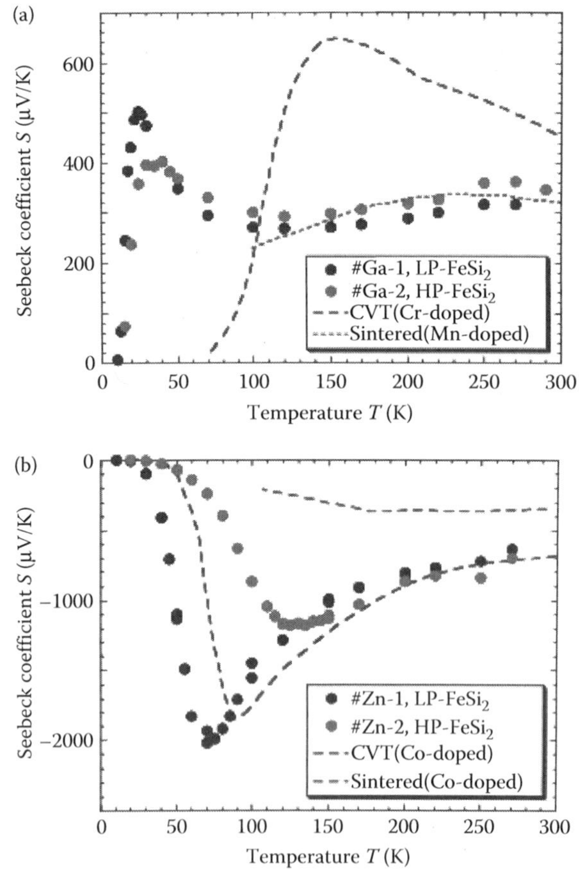

FIGURE 18.7 Temperature dependence of Seebeck coefficient *S* of TGSG-grown single crystalline ß-$FeSi_2$. (a) *p*-Type and (b) *n*-type samples.

The temperature dependence of the power factor, $P = S^2/\rho$ is plotted in Figure 18.9. The values at 300 K are 4.4×10^{-6} (#Ga-2) and 2.0×10^{-6} $Wcm^{-1}K^{-2}$ (#Zn-1). They increase as the temperature decreases and reaches the maximum values of 6.0×10^{-6} $Wcm^{-1}K^{-2}$ (#Ga-2) at 170 K and 23×10^{-6} $Wcm^{-1}K^{-2}$ (#Zn-1) at 100 K. These values are relatively high as compared with those of a sintered polycrystal and a CVT-grown single crystal.[54,62] Thus, the power factors of both *p*- and *n*-type ß-$FeSi_2$ are greatly improved in the low-temperature range by using the solution-grown single crystals.

Doping control of the solution-grown crystal was also studied by using a combination of the Ga–Al solvent and a nondoped $FeSi_2$ source for the *p*-type crystal and the Ga solvent and a Co-doped $FeSi_2$ source for the *n*-type crystal. By using the Ga–Al solvent and nondoped $FeSi_2$ source, the carrier concentration of the *p*-type crystal increases up to 2.7×10^{19} cm^{-3} and its resistivity decreases to 0.02 Ωcm at 300 K. However, the power factor *P* was not improved effectively due to rapid decrease of the Seebeck coefficient.

In the *n*-type crystal, Co-doping effectively improves the thermoelectric property near 300 K. The power factor increases to ~8×10^{-6} $Wcm^{-1}K^{-2}$ near room temperature by the doping of 1 wt% Co in the $FeSi_2$ source. Furthermore, thermal conductivity is also improved by the Co-doping. Figure 18.10 shows the thermal conductivity of single crystalline ß-$FeSi_2$ measured by the laser flush method. The thermal conductivity of a Co-doped *n*-type single crystal decreases to ~0.06 W/cmK. This value is half that of a typical single crystal grown using the Ga solvent and a nondoped $FeSi_2$ source.

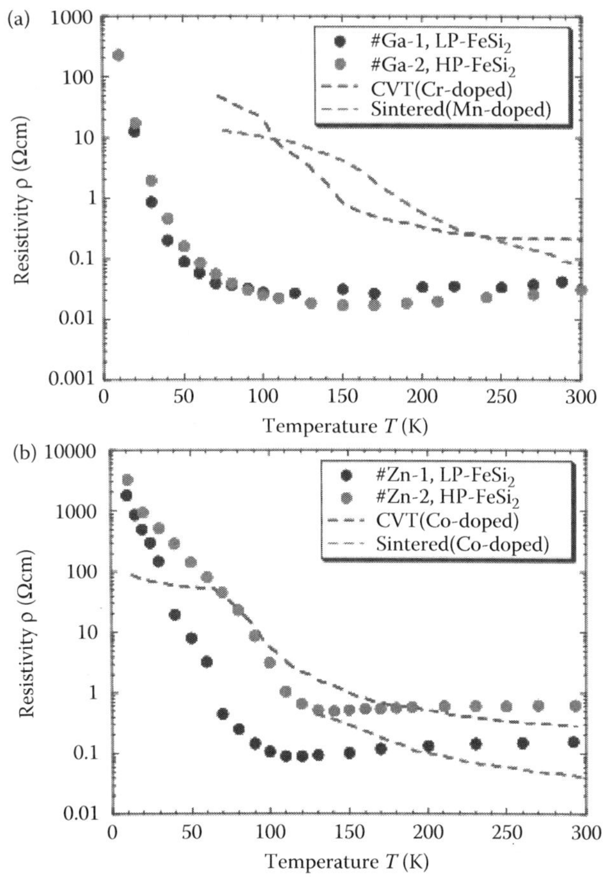

FIGURE 18.8 Temperature dependence of resistivity of TGSG-grown single crystalline ß-$FeSi_2$. (a) *p*-Type and (b) *n*-type samples.

18.4 Preparation of β-$FeSi_2$ Thermocouple by the Cold-Pressing and Sintering Method

The preparation by the powder metallurgy method has the following advantages:

- Improvement of mechanical strength
- Improvement of machinability
- Improvement of yield
- Improvement of bondability
- Easier control of composition
- Reduction of thermal conductivity

Moreover, the cold-pressing and sintering method simply prepares a thermocouple by directly joining the *p–n* junction. This method results in higher mass productivity and lower cost. Figure 18.11 shows the preparation of a β-$FeSi_2$ thermocouple by the cold-pressing and sintering method.[13–17] This method is carried out by the following process: weighing capacity, smelting, powderization, granulation, sieving, cold-pressing, sintering, solid solution treatment, and annealing. The weighed raw materials are smelted under argon or an Ar + H_2 atmosphere in quarts or ceramic crucibles using resistance or induction heating. In the experiment that produced only a little β-$FeSi_2$, arc melting was used. The melted ingots made by the

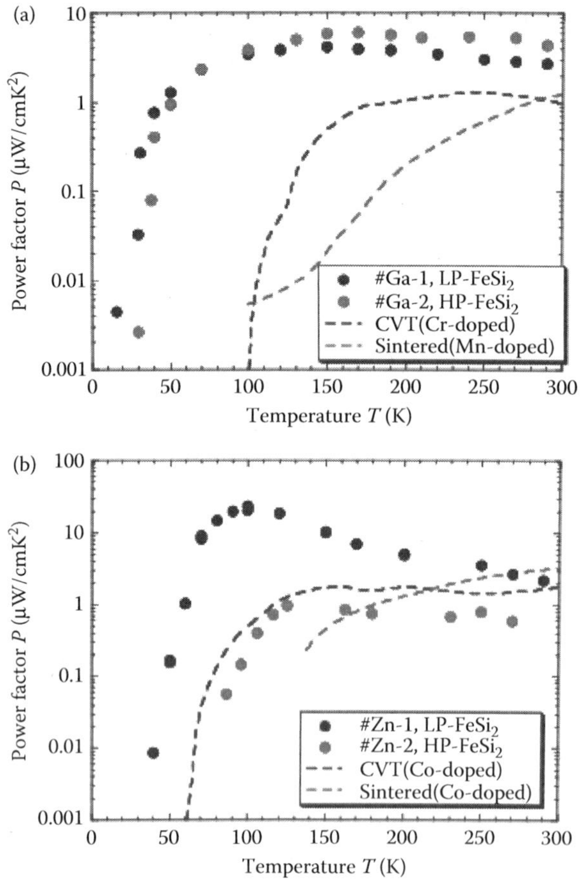

FIGURE 18.9 Temperature dependence of power factor *P* of TGSG-grown single crystalline ß-$FeSi_2$. (a) *p*-Type and (b) *n*-type samples.

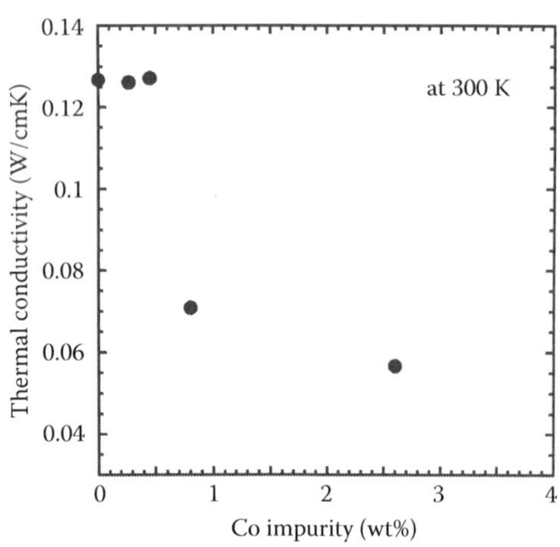

FIGURE 18.10 Thermal conductivity of TGSG-grown single crystalline ß-$FeSi_2$ measured by laser flush method at 300 K. Thermal conductivity decreases with doping the Co impurity.

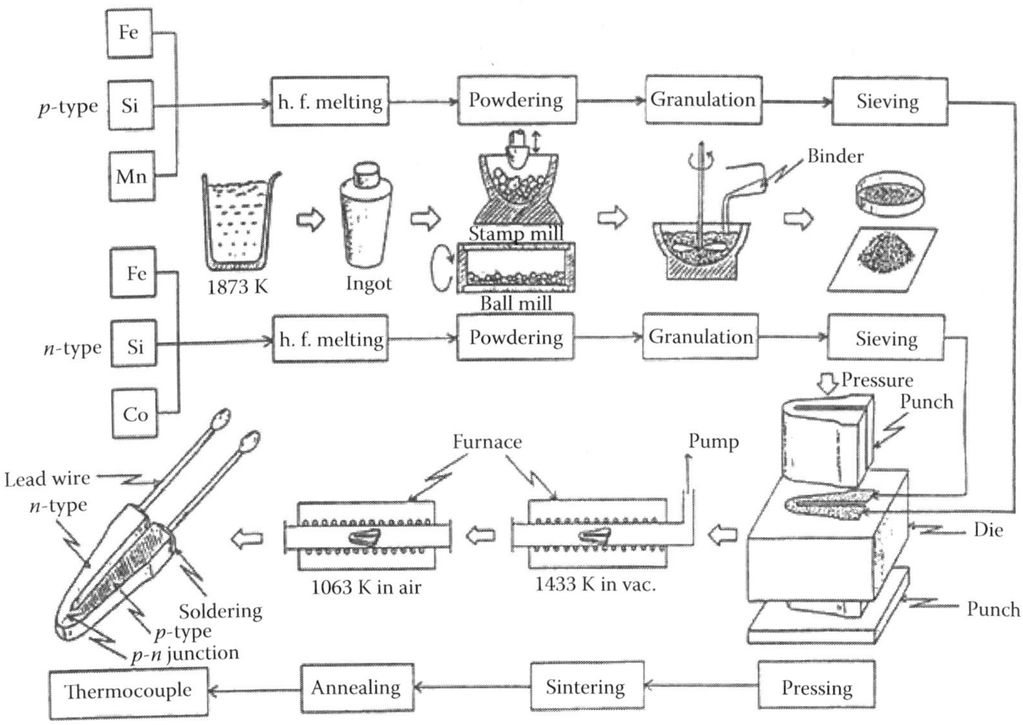

FIGURE 18.11 The preparation process of β-$FeSi_2$ thermocouple by the cold-pressing and sintering method. (From I. A. Nishida; *Thermoelectrics—Principles and Applications*, edited by M. Sakata, Realize Inc., Tokyo, 2000, 199–209. With permission.)

above-mentioned process were crushed to powder with mean particle sizes of 100 μm or less by a steel stamp mill, and they are powdered to 1.1–2.2 μm in diameter by an iron ball mill. The ground powders need granulation for easier handling. Polyvinyl alcohol and colloidal paraffin were used during granulation as a binder and lubricant, respectively. After granulation, the powders were sieved by particle size, and *p*- and *n* type $FeSi_2$ powders were simultaneously poured into steel U-shaped dies. These powders were cold-pressed under a pressure of 294 MPa. The green compacts are sintered in a vacuum at 1428 ± 5 K for 1.08×10^4 s, and then finally annealed at 1073 K for 7.2×10^5 s. Figure 18.12 shows the U-shaped $FeSi_2$

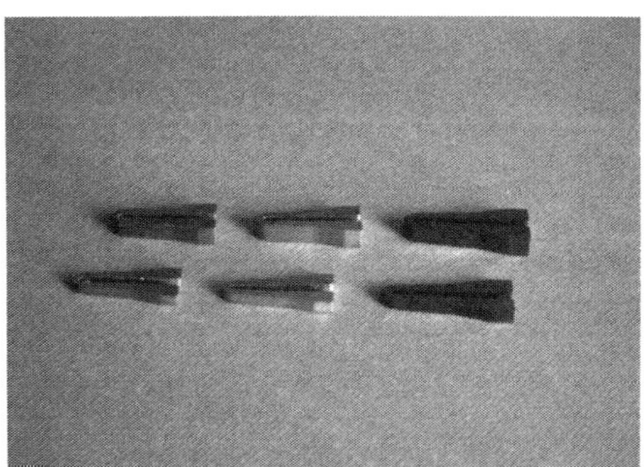

FIGURE 18.12 The U-shaped β-$FeSi_2$ thermocouple made of *p*- and *n*-type legs which were prepared by the modified cold-pressing and sintering technique. From the right, the green samples, the sintered samples, and the annealed samples.

thermocouples, sintered compacts, and green compacts. The annealing induces the phase transition for obtaining β-$FeSi_2$. The formation of the β-$FeSi_2$ phase is sensitive to the presence of the added impurities. For cobalt-doping samples (≦1.67 at%), the most effective annealing time and temperature are about 20 h and 1113 K.[14,16] For the manganese-doping samples, the annealing temperatures of the maximum formation rates are 1115 and 1135 K, and the most effective annealing times are about 4 h for 3.34 at% Mn and 40 h for the 1.67 at% Mn.[15,16] Copper-doping samples increase the rate of the eutectoid decomposition and the initial stage of the peritectoid reaction.[59,60] The effect of copper depends on the annealing temperature. Thus, the decomposition rate in the doped samples was observed to be 15 times higher at 1023 K with respect to an undoped one and was more than 30 times at 873 K.

These preparation techniques are described in detail in several papers.[13–18] The thermoelectric performance of the thermocouple was not sufficiently reproducible as it depends on the production process.

18.5 Thermoelectric Properties of β-$FeSi_2$ Thermocouple by Producing Process

18.5.1 The Effects of Raw Materials Purity

The starting materials are usually used zone-refined iron (99.99%) and a high-purity silicon single crystal (6 N). The effects of the raw materials' purity on the thermoelectric properties were investigated.[13,61] The different behaviors of the Seebeck coefficient by purity are shown in Figure 18.13. The low-purity

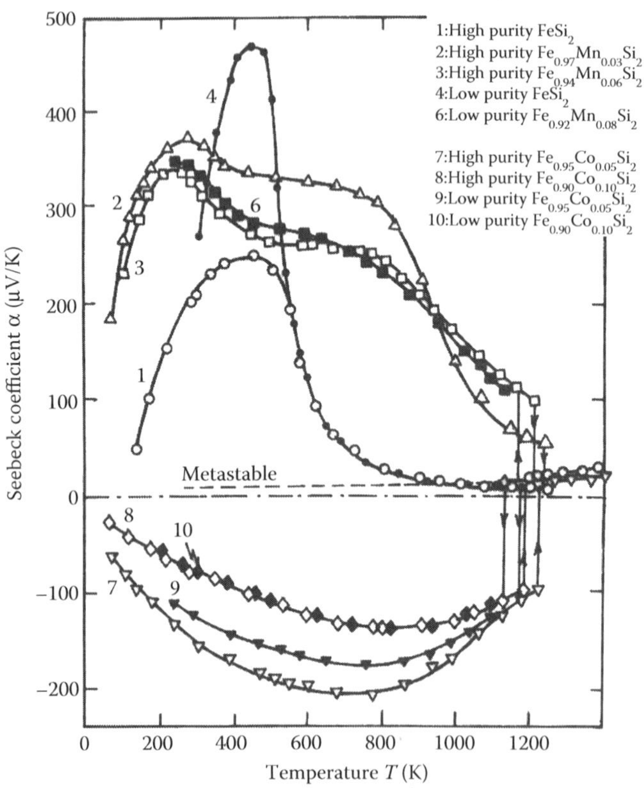

FIGURE 18.13 Temperature dependence of Seebeck coefficient for the starting materials of various purities. (From I. A. Nishida; *Thermoelectrics—Principles and Applications*, edited by M. Sakata, Realize Inc., Tokyo, 2000, 199–209. With permission.)

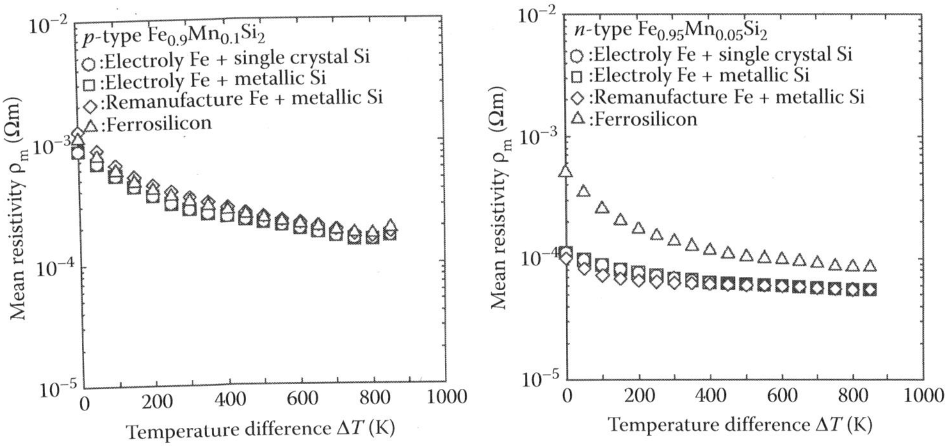

FIGURE 18.14 Mean resistivity of p-type $Fe_{0.9}Mn_{0.1}Si_2$ and n-type $Fe_{0.95}Co_{0.05}Si_2$ for the starting materials of various purities as a function of the temperature difference.

materials used electrolytic iron (99.9%) and metallic silicon (98%). The temperature dependence of the electrical resistivity for the high and low purity of the starting materials showed the same behavior, as the difference by purity cannot be identified on the resistivity. The difference was not observed on the Seebeck coefficient for the doping samples. However, the maximum value of the Seebeck coefficient for the low-purity samples was observed for a longer time compared with the high-purity samples.

The power generation properties of p-type $Fe_{0.9}Mn_{0.1}Si_2$ and n-type $Fe_{0.95}Co_{0.05}Si_2$ which use the iron and silicone of different purities are shown in Figures 18.14 and 18.15. The raw materials with the following purities were used: electrolytic iron (99.9%), remanufacture iron (99%), 50 mass%Si ferrosilicon, silicon single crystal (99.9999%), and metallic silicone (98%). Thermoelectromotive force (E_0), which used ferrosilicon of the lowest purity decreased only about 7% compared with the highest purity at a temperature difference of 800 K. The E_0 did not attain a difference in magnitude by purity. For p-type $Fe_{0.9}Mn_{0.1}Si_2$, the E_0 that used a silicon single crystal was the largest. On the other hand, the E_0 for n-type $Fe_{0.95}Co_{0.05}Si_2$ that used metal silicone is larger compared with the silicon single crystal. As shown in Figure 18.14, the

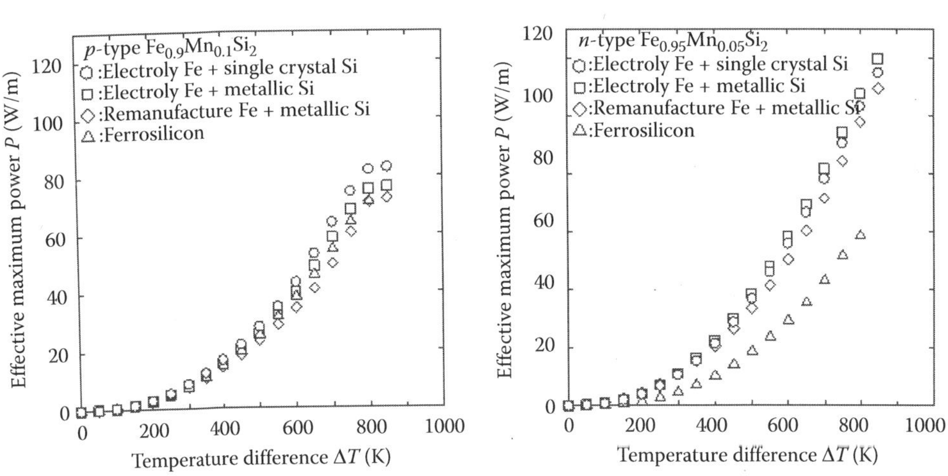

FIGURE 18.15 Effective maximum power of p-type $Fe_{0.9}Mn_{0.1}Si_2$ and n-type $Fe_{0.95}Co_{0.05}Si_2$ for the starting materials of various purities as a function of the temperature difference.

difference by raw material purity of the mean resistivity (ρ_m) was observed. The ρ_m of the sample that used electrolytic iron and a silicon single crystal was the same as that of the sample that used electrolytic iron and metallic silicone. For *p*-type $Fe_{0.9}Mn_{0.1}Si_2$, the sample that used remanufacture iron was the largest, and the sample that used ferrosilicon was second. For *n*-type $Fe_{0.95}Co_{0.05}Si_2$, the ρ_m that used ferrosilicon is the largest and is four times larger than other samples because the main impurities contained in ferrosilicon are aluminum and manganese, which act as acceptors. These recombine with an electron (donor) and increase the mean resistivity. The maximum effective power ($P = E_0^2/4\rho_m$) was estimated by using the measured thermoelectromotive force (E_0) and the mean resistivity (ρ_m). The maximum effective power was increased monotonously with the temperature differences, as shown in Figure 18.15. Except for *n*-type samples that used ferrosilicon, the *P* of the low-purity samples is lower than that of the high-purity samples at about 10% at a temperature difference of 800 K. For *n*-type samples, the sample that used electrolytic iron and metal silicon was about 3% higher than the high-purity sample. The *P* of the ferrosilicon sample decreased about 44% compared with the high-purity samples. This decreasing of the maximum effective power was a result of increasing the mean resistivity.

Thus, an industrial grade of comparatively low-purity starting materials used for thermoelectric purposes is available because the thermoelectric properties are structure insensitive.

18.5.2 The Effects of Si Content

One problem of the cold-pressing and sintering method was the variation of Si content by contamination of Fe, which cannot be avoided in the powdering process with an iron mortar and an iron ball mill. The melted ingot was crushed by a stamp mill, and then powdered to 1.1–2.2 μm in diameter by an iron ball mill. A commercially available cast-iron ball milling pot with a capacity of 0.64×10^{-3} m^3 was used, and the iron balls were of diameters 6, 13, and 16 mm with a mass of 0.7, 0.5, and 1.4 kg, respectively. Figure 18.16 shows the relationship among the ball-milling durations, the average powder particle sizes and the amount of Fe in the β-$FeSi_2$ powder.[62]

The amount *Q* of Fe-contaminant was 3.91 wt% for the ball-milling duration time of 20 h, and was 4.61 wt% for 100 h. The average powder particle sizes *d* decreased exponentially with increasing ball-milling duration *t*, while the amount *Q* of Fe-contaminant increased linearly. Therefore, the effect of the variation of Si composition by Fe-contaminant on thermoelectric properties for sintered compacts $FeSi_2$ should be investigated.

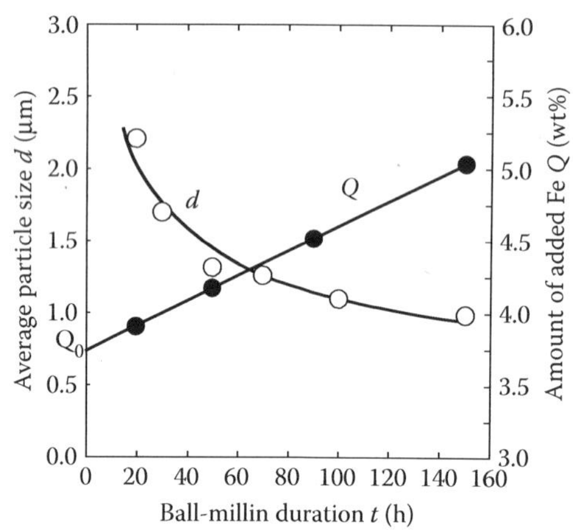

FIGURE 18.16 The effect of ball-milling duration on the average particle size and amount of added Fe for 200 g powder.

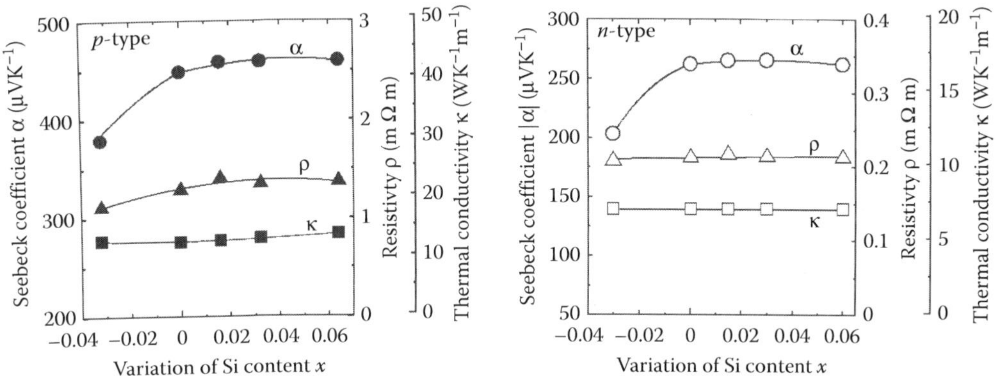

FIGURE 18.17 Thermoelectric properties for *p*-type $Fe_{0.91}Mn_{0.09}Si_{2+x}$ and *n*-type $Fe_{0.97}Co_{0.03}Si_{2+x}$ at room temperature as a function of the amount of Si variation from the stoichiometric composition.

Sintered compacts of the *p*-type $Fe_{0.91}Mn_{0.09}Si_{2+x}$ and the *n*-type $Fe_{0.97}Co_{0.03}Si_{2+x}$ with x ranging from −0.032 to 0.064 were prepared by the cold-pressing and sintering method. The solubility limits of Si for both *p*- and *n*-type specimens were determined to be $-0.032 < x < 0.015$, which is smaller than that for non-doped $FeSi_{2+x}$. The α, ρ, and thermal conductivity κ at room temperature were plotted against Si content in Figure 18.17. The absolute values of α and ρ for $x < 0$ were smaller than those of $x = 0$ (stoichiometric composition) by precipitation of ε-FeSi, while the values for $x > 0$ with free Si were independent of x. The κ was independent of x. Thermoelectromotive force and effective maximum power for the *p*-type $Fe_{0.91}Mn_{0.09}Si_{2+x}$ and the *n*-type $Fe_{0.97}Co_{0.03}Si_{2+x}$ at a temperature difference of 800 K as a function of the amount of Si variation from the stoichiometric composition are shown in Figure 18.18. The absolute values of the E_0 at a temperature difference of 800 K showed a maximum value for both *p*- and *n*-type specimens with $x = 0.03$, which were 225.0 and 195.6 mV, respectively. Both *p*- and *n*-type specimens showed maximum values 2.53×10^{-4} and 4.65×10^{-4} K^{-1} of Z at 950 and 820 K, respectively. The values were obtained for $x = 0$ (*p*-type) and $x = 0.03$ (*n*-type).

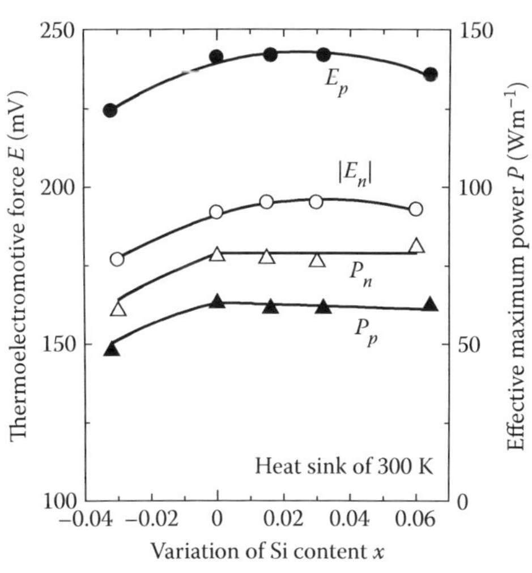

FIGURE 18.18 Thermoelectromotive force and effective maximum power for *p*-type $Fe_{0.91}Mn_{0.09}Si_{2+x}$ and *n*-type $Fe_{0.97}Co_{0.03}Si_{2+x}$ at temperature difference 800 K as a function of the amount of Si variation from the stoichiometric composition.

The important point in a weighing capacity process is adding Si more than stoichiometry composition to compensate for the contamination of Fe.

18.5.3 The Effect of Crystal Grain Size

The relationship between the ball-milling duration and the average particle size is shown in Table 18.5.[17] The mean powder size d, sintered grain size C_s, and annealed grain size C_a decrease exponentially with increasing ball-milling duration. The C_a/d of ratio for Co-doped n-type $FeSi_2$ changes from 4.7 to 6.5 with increasing d. The C_a/d for Mn-doped p-type $FeSi_2$ is greater than 10, and the maximum value is 13.0 at 1.32 μm. Mn-doped p-type $FeSi_2$ has about two times larger C_a/d than that of Co-doped n-type $FeSi_2$. This suggests that the recrystallization rate of Mn-doped $FeSi_2$ is faster than that of Co-doped ones, and that Mn accelerates the sintering of $FeSi_2$. The C_a dependences of α, ρ, and κ at room temperature for the p-type $Fe_{0.93}Mn_{0.07}Si_2$ and the n-type $Fe_{0.985}Co_{0.015}Si_2$ are shown in Figure 18.19.

The α, ρ, and κ of the p-type $FeSi_2$ increase with C_a. The maximum of α is 537 μV/K at C_a = 21.2 μm, and the minimum is 476 μV/K at C_a = 11.6 μm. The ρ shows a similar tendency to the α, as ρ increases from 1.43 mΩm at 11.6 μm to 1.66 mΩm at 24.0 μm. The κ increases linearly from 13.2 to 15.2 W/Km with increasing C_a. On the other hand, the α of Co-doped n-type $FeSi_2$ is constant (−380 μV/K) and independent of C_a. However, the ρ decreases from 0.52 to 0.43 mΩm with increasing C_a. Contrarily, the κ increases from 8.7 to 11.8 W/Km with increasing C_a. The κ shows the similar tendency for p- and n-type $FeSi_2$.

TABLE 18.5 Effect of Ball-Milling Duration on the Amount of Fe and the Average Particle Size for p-Type $Fe_{0.93}Mn_{0.07}Si_2$ and n-Type $Fe_{0.985}Co_{0.015}Si_2$

		Average Particle Size (μm)					
		Ball-Milled Powder d		Sintered Grain C_s		Annealed Grain C_a	
Ball-Milling Duration (h)	Amount of Added Fe (wt%)	p	n	p	n	p	n
20	3.91	2.21	2.12	10.6	7.7	24.0	10.0
30	4.00	1.70	1.62	9.0	6.6	21.2	9.0
50	4.17	1.32	1.42	7.4	5.4	17.1	8.2
70	4.34	1.26	1.29	5.4	4.1	14.1	7.8
100	4.61	1.10	1.10	4.2	3.7	11.6	7.1

Note: Sintered temperature was 1433 K and annealing time was 200 h at 1073 K. The amount of powder was 200 g.

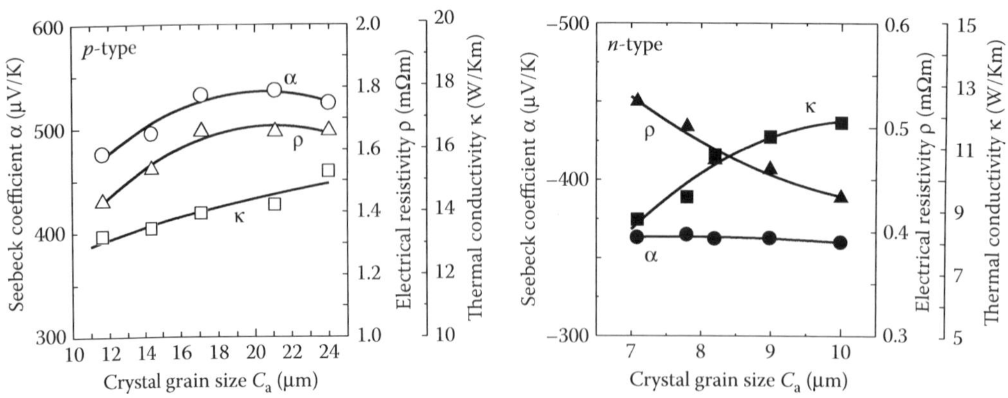

FIGURE 18.19 The mean annealed grain size Ca dependences of thermoelectric properties at room temperature for p-type $Fe_{0.93}Mn_{0.07}Si_2$ and n-type $Fe_{0.985}Co_{0.015}Si_2$.

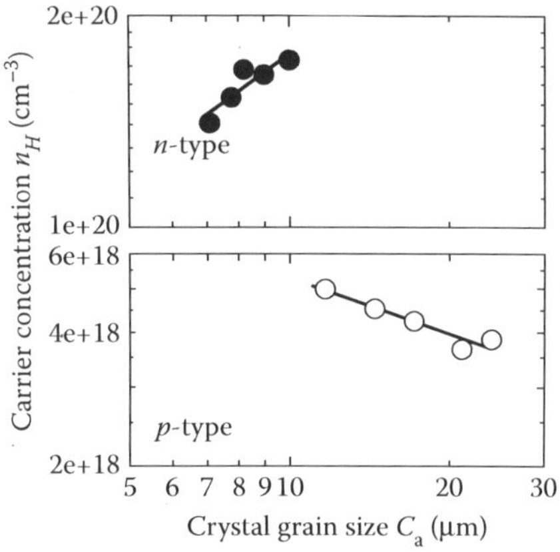

FIGURE 18.20 The mean annealed grain size Ca dependences of carrier concentration at room temperature for p-type $Fe_{0.93}Mn_{0.07}Si_2$ and n-type $Fe_{0.985}Co_{0.015}Si_2$.

The grain boundaries are smaller for the larger C_a. κ_L is decreased by increasing the grain boundaries, since phonons will be scattered by the grain boundaries. Variations of the carrier concentration as a function of C_a are shown in Figure 18.20. The carrier concentration n_H for the p-type $FeSi_2$ decreases in connection with the increase in C_a, whereas the n_H of the n-type $FeSi_2$ increases with C_a. These changes may be considered an effect of the grain boundaries. Since $FeSi_2$ has a very complicated crystal structure, there are many vacant lattice points in the grain boundary. It is considered that these vacant lattice points will catch electrons. As a result, the grain boundaries play a role of acceptors and yield holes. This effect is shown by the fact that the C_a dependencies of α and ρ are very similar to the n_H dependences. For the p-type $FeSi_2$, the increase of n_H with decreasing C_a is due to the holes created in grain boundaries. For the n-type $FeSi_2$, the decrease of n_H with decreasing C_a is due to the carrier compensation.

The C_a dependence of the thermoelectromotive force E_0 and effective maximum power P_{ef} at a temperature difference of 800 K for the p-type $Fe_{0.93}Mn_{0.07}Si_2$ and the n-type $Fe_{0.985}Co_{0.015}Si_2$ are shown in Figure 18.21. The E_0 for the p-type $FeSi_2$ shows about 250 mV at all C_a. The absolute of E_0 for the n-type $FeSi_2$, which shows the same tendency as that of p-type $FeSi_2$, is about 230 mV. The E_0 has little effect on

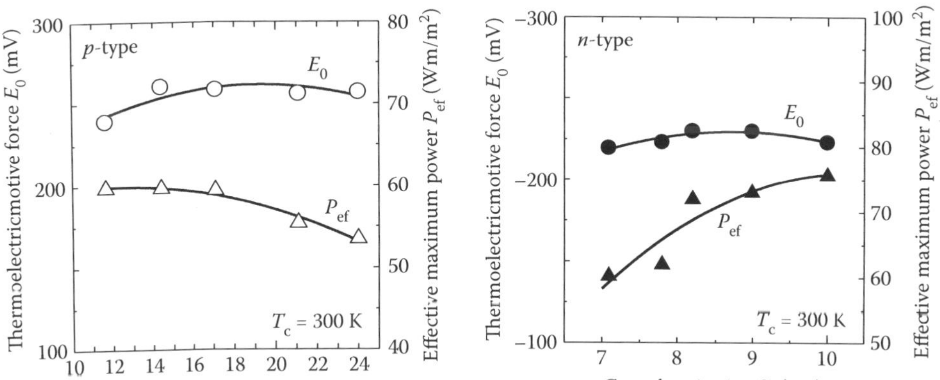

FIGURE 18.21 Thermoelectromotive force and effective maximum power for p-type $Fe_{0.93}Mn_{0.07}Si_2$ and n-type $Fe_{0.985}Co_{0.015}Si_2$ at temperature difference 800 K as a function of the mean annealed grain size C_a.

crystal grain size. P_{ef} for the *p*-type $FeSi_2$ decreases with C_a. On the other hand, P_{ef} for the *n*-type $FeSi_2$ increases with C_a. The high P_{ef} is obtained at the smallest C_a of 11.6 μm in the *p*-type $FeSi_2$ and at the largest C_a of 10 μm in the *n*-type $FeSi_2$. As a result, it was found that sintered *p*- and *n*-type $FeSi_2$ were prepared by a cold-pressing and sintering technique possessing high thermoelectric performance at ball-milled powder sizes of 1.10 and 2.12 μm, respectively.

These results suggest that the effect of crystal grain size on thermal conductivity is remarkable for $FeSi_2$, and the thermoelectric performance can consequently be improved by reducing the grain size.

18.5.4 The Effect of Density on Structural Reliability

The thermoelectric generation element (TEG) of β-$FeSi_2$ produced by the cold-pressing and sintered method is used for many civil applications in the gas flame of town gas or LPG (both contain H_2, S, C, and water vapor). In order to use TEG, high reliability and chemical stability are needed. Figure 18.22 shows the durability testing of a β-$FeSi_2$ thermocouple in a town gas flame. Here, inferior goods are samples that changed internal resistance 5% or more from the initial value. The results of continuation heating and thermal cycle durability testing are shown in Figures 18.23 and 18.24, respectively. As shown in Figure 18.23, a β-$FeSi_2$ thermocouple with 89% density ratio will deteriorate within 100 h.

FIGURE 18.22 (**See color insert.**) The durability testing of β-$FeSi_2$ thermocouples in the town gas flame.

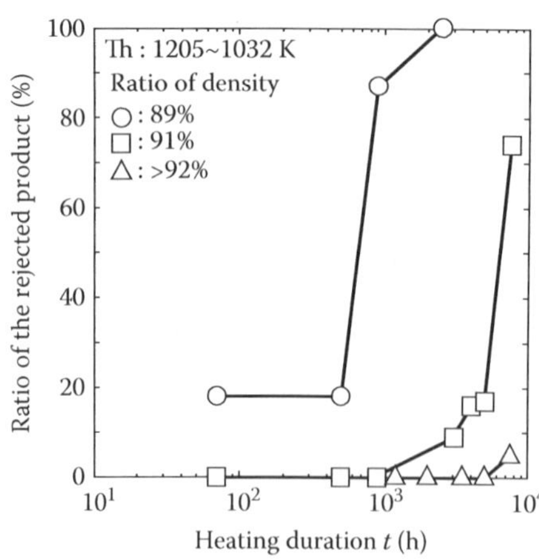

FIGURE 18.23 The ratio of rejected product for the continuation heat durability testing of β-$FeSi_2$ thermocouple in the gas flame.

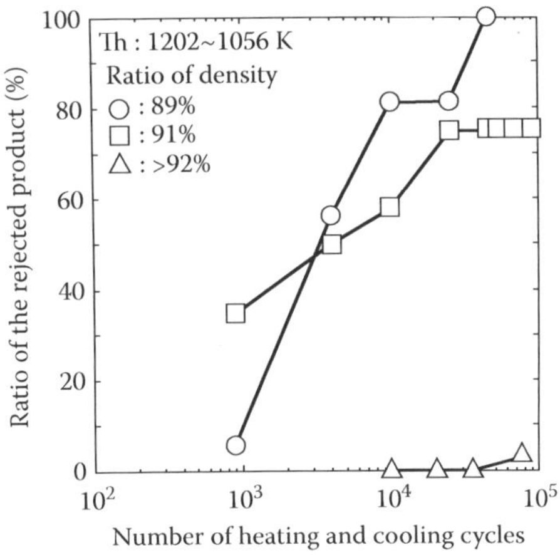

FIGURE 18.24 The ratio of rejected product for the thermal cycle durability testing of β-$FeSi_2$ thermocouple in the gas flame.

After 2500 h, all thermocouples with an 89% density ratio are the inferior goods. However, a β-$FeSi_2$ thermocouple with a density ratio above 92% has almost no degradation until 5000 h. One cycle for the thermal cycle testing shown in Figure 18.24 is 3 min of heating and 2 min of cooling. The degradation of the β-$FeSi_2$ thermocouples with density ratios of 91% and below begins at 1000 times the number of cycles. The majority will deteriorate at 4000 times. On the other hand, elements with a density ratio of 92% or more do not deteriorate at least 40,000 times. At 75,000 times, the ratio of inferior goods is only 3%. For the compacts of β-$FeSi_2$ thermocouple with high density, the sintering hole is isolated. Therefore, neither the atmosphere nor the corrosive gas can easily trespass into the interior of the material and so the corrosion inside the material is controlled. Thus, it is important to increase the density of a sintered compact.

18.5.5 Thermal Shock Resistance of β-$FeSi_2$ Thermocouple

As already described, the U-shaped $FeSi_2$ thermocouple made of *p*-type and *n*-type legs is very useful for comparatively high-temperature applications in the atmosphere. On a pilot test of a household safety application for a gas range, no change was found in the thermoelectric power and electrical resistivity after the *p–n* junction of the U-shaped $FeSi_2$ thermocouple was heated at 1200 K for 5000 h in a gas flame, and the thermocouple was proofed against more than 10^5 thermal cycles.[44] However, on the severe thermal cycles such as quenching in water, the thermal shock resistance is as poor as only one cycle as a crack initiated on the specimen surface by quenching from 1100 K.

The thermal shock resistance test was made using the following two methods: (1) A temperature difference of 750 K was given between the top and bottom ends of the specimen, and then the sample was quenched in water. This is called partial heating. (2) A specimen was homogeneously heated at 1073 K, and then quenched in water. This is called homogeneous heating. Table 18.6 shows the number of quenching cycles repeated until a crack was found on the specimen surface.[44] On the homogeneous and partial heating tests, respectively, a crack was initiated by one and two quenching cycles for the specimen without boron, while five and 21 cycles were needed for the specimen containing the boron atoms of $x = 0.01$. Such an improvement in the thermal shock resistance is more remarkable as the boron content increases. For example, even beyond the solid solubility limit of boron, a crack was present after 30 and 50 quenching cycles, respectively, during the tests mentioned above. The dissolved boron atoms improved the intergranular cleavage strength of the β-$FeSi_2$ system. By doping with Mn and Co, respectively, *p*- and

TABLE 18.6 Effect of Boron on Thermoelectric Properties and Thermal Shock Resistance of $Fe_{1-x}B_xSi_2$ Doped with Mn or Co

Composition	Dopant (%)		α (μV/K)	ρ (Ωm)	Thermal Shock Resistance[a]	
x	Co	Mn	At 300 K	At 300 K	Case 1	Case 2
0.000	0.0	0.0	−104	4.8×10^{-2}	1 ± 1	5 ± 3
0.005	0.0	0.0	−559	1.2×10^{-2}	5 ± 2	14 ± 5
0.010	0.0	0.0	−571	1.1×10^{-2}	13 ± 3	40 or more
0.020	0.0	0.0	−589	1.3×10^{-2}	20 or more	40 or more
0.030	0.0	0.0	−681	1.3×10^{-2}	20 or more	40 or more
0.040	0.0	0.0	−641	1.2×10^{-2}	20 or more	40 or more
0.010	1.0	0.0	−300	3.6×10^{-4}	13 ± 3	40 or more
0.010	0.0	2.8	+480	2.8×10^{-3}	20 or more	40 or more
0.020	2.0	0.0	−260	1.8×10^{-4}	20 or more	40 or more
0.020	0.0	3.3	+410	2.2×10^{-3}	20 or more	40 or more

Note: Case 1 and 2 are quenching cycles when a crack initiated on the specimen surface for the partial heating and for the homogeneous heating at 1073 K, respectively.
[a] Indicates the mean quenching cycles determined with every 10 specimens.

n-type semiconductors were obtained for $FeSi_2$ containing a small amount of boron. Note that the thermoelectric properties of the improved specimens were equivalent to those of pure $FeSi_2$ without boron, and therefore a $FeSi_2$ thermocouple with high thermal shock resistance has wider applications for thermoelectric generation than ever. Such a thermocouple was hardly broken when it encountered water as dew or was accidentally immersed. In addition, it requires only easy maintenance, is highly reliable, and has comparatively high thermoelectromotive force, especially when used as a fire sensor employed to detect a gas fire.

18.6 Applications of β-$FeSi_2$ Thermoelectric Generator

The thermoelectric generator (TEG) with $FeSi_2$ can generate electric power by a difference in temperature, and recover energy from heat that was previously lost. Figure 18.25 shows the thermoelectric properties of the U-shaped β-$FeSi_2$ thermocouple that weighs 0.9 g. Each leg is about 20 mm long and 2 mm thick. The open voltage E_0 of a pair of thermocouples is 0.49 V, mean resistivity *r* is 0.051 Ωcm, and the effective

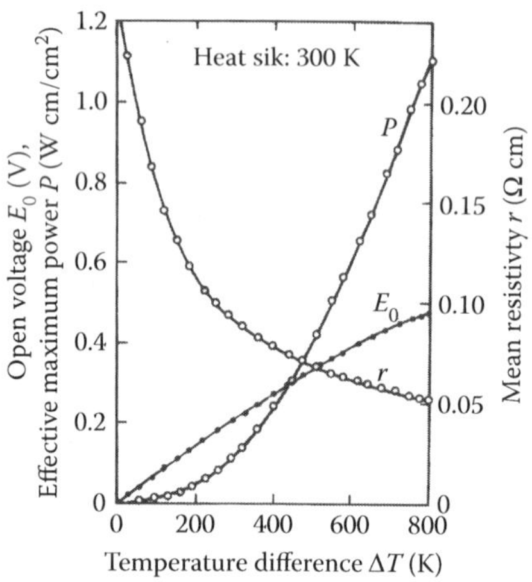

FIGURE 18.25 The power generation properties of a U-shaped β-$FeSi_2$ thermocouple with the weight of 0.9 g.

FIGURE 18.26 (**See color insert**.) The simplest thermoelectric generator system made by using a pair of β-$FeSi_2$ thermocouples. (From I. A. Nishida; *Thermoelectrics—Principles and Applications*, edited by M. Sakata, Realize Inc., Tokyo, 2000, 199–209. With permission.)

maximum power P is 1.1 Wcm/cm^2 at the temperature difference 800 K between hot and cold sides. Thus, the effective maximum power estimated from the density of $FeSi_2$ of 4.93 g/cm^3 is 223 W/Kg. A simple thermoelectric generator system using a pair of $FeSi_2$ thermocouples is shown in Figure 18.26. $FeSi_2$ has excellent thermal resistance and is a corrosion-resistant sensor inside a gas flame for going-out prevention and temperature control of gas fittings. For example, the U-shaped β-$FeSi_2$ thermocouple is used for the electrical power supply of hot-air fans, heating systems, and gas lamps. In Kobe city, Japan, the streetlights have been changed from the conventional electric lights to gas lamps because they bring harmony to the town.[63] In the gas lamp, the electric apparatus that consists of a timer, an electromagnetic valve, ignition equipment, and a flame sensor needs to be powered. Lithium batteries and a TEG are utilized as the power supplies. A TEG is used as the power supply that operates an electromagnetic valve. When the ignition time comes according to a timer, an electromagnetic valve is opened by a lithium battery. The electric power from the lithium battery will be simultaneously supplied to an ignitor. Then, sparks will occur by the spark plasma of the ignitor and the gas will be ignited. The $FeSi_2$ thermocouple is heated as the gas burns, and the power supply changes from a battery to a TEG. A valve closes when the predetermined voltage that the thermoelectromotive force of TEG needs for opening of an electromagnetic valve is not achieved. Therefore, in this system, the $FeSi_2$ thermocouple has two

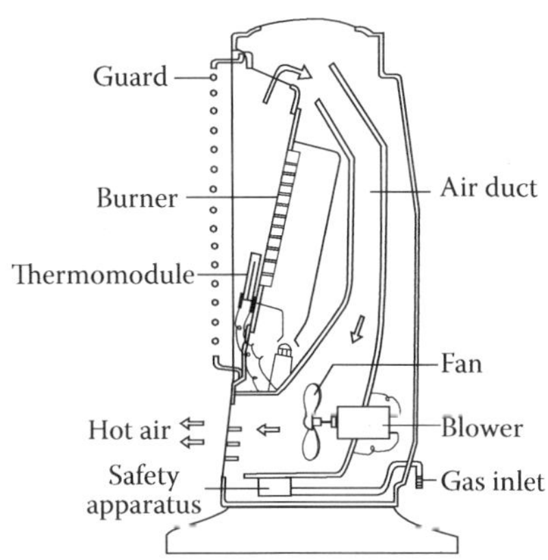

FIGURE 18.27 The appearance and structure of a cordless gas fan heater using 30 pairs of U-shaped β-$FeSi_2$ thermocouples.

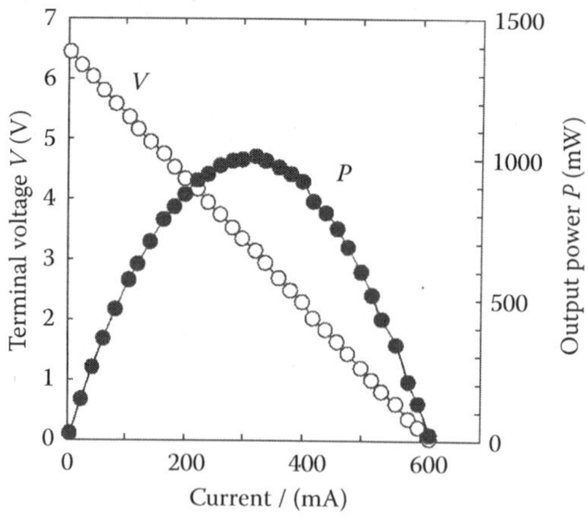

FIGURE 18.28 The load properties of the cordless fan heater when the gas cock is fully open.

functions as a going-out detection sensor and a supplier of electric power to the electromagnetic valve. Moreover, the TEG is used for going-out prevention and gas flame control in gas-heated baths. Power-saving was achieved by using the TEG without any outside electric power.

The appearance and structure of a cordless fan heater are shown in Figure 18.27.[64] The TEG of the $FeSi_2$ thermocouple has an arrangement of 30 couples (about 72 g and one couple is for going-out prevention). This is shown in Figure 18.27 around a ceramic burner and the elements are in a series of connections.[64] The arranged TEG, when heated by the gas burner, generates electric power from temperature differences. With this electric power, a fan rotates and warm air is ventilated from the stoves' lower part. This stove can change the amount of gas combustion in two steps by turning a knob so that the quantity of heat is 2000 Kcal/h when the gas is fully open and 1000 Kcal/h when it is half open. The warm air at 333 K can be ventilated at about 1 m^3/min when the gas is fully open. The load properties of the 30 pairs

FIGURE 18.29 Candle lighting fixture equipped with the portable radio that used five β-$FeSi_2$ thermocouples.

of TEG installed in the cordless fan heater when fully open are shown in Figure 18.28. When the gas was fully open, the open-voltage was $E = 6.40$ V, the maximum power was $P = 0.991$ W, and internal resistance was $R = 10.38\ \Omega$. At half open, the results were $E = 4.30$ V, $P = 0.660$ W, and $R = 7.12\ \Omega$. Because a cordless fan heater has high internal resistance compared with a general constant voltage power supply, output voltage changes remarkably with load changes. Moreover, output voltage also changes with changes of a heat source or the influences of the circumference. For this reason, output voltage stabilization of TEG was attained in reaction to the input-and-output changes by a switching regulator. The voltage variations at the time of sudden load changes can also be suppressed. However, if the load becomes large, the output voltage will decline and become unstable. When the gas is fully open, the output voltage is fixed at 6.0 V and stable to 95% of the maximum output. When changing the heat input by changing the gas heater, the output voltage is fixed and stable. A cordless fan heater that can use a TEG as a stable power supply has been made. Since moving parts are not used, a TEG installed in the cordless fan heater does not require maintenance like a regular machine.

The candle-type portable power source attached to the radio (called a candle radio), which uses five $FeSi_2$ thermocouples, is shown in Figure 18.29.[65] The hot-side p–n junctions of the five $FeSi_2$ thermocouples are positioned on the top of a conical-shaped module and are heated by the candle's flame. The heat is dissipated to the environment through the cold-side electrodes attached to radiator rings. During operation, the temperatures of hot side and cold side, internal resistance, and open terminal voltage are 680 K, 330 K, 14 Ω, and 1.45 V, respectively. The thermoelectric generation by these five $FeSi_2$ thermocouples can supply sufficient electric power to operate the candle radio (rated voltage of 1.1 V and current of 20 mA). A candle portable radio with 100 g of wax can provide over 12 h of stable power generation and illumination, and can receive AM broadcasts.

References

1. J. P. Piton and M. F. Fay; *C. R. Acad. Sci. (Paris)*, **C266**, 1968, 514–516.
2. L. Brewer, A. W. Searcy, D. H. Templeton, and C. H. Dauben; *J. Am. Ceram. Soc.*, **33**, 1960, 291.
3. R. Wandji, Y. Dusausoy, J. Protas, and B. Roques; *C. R. Seances Acad. Sci., Ser. C, V*, **269**, 1969, 907–909 (in French).
4. L. D. Ivanova, N. Kh. Abrikosova, E. I. Elagina, and V. D. Khvostikova; *Izv. Acad. Nauk. SSSR, Neorg. Mater*, **5**, 1969, 1933.
5. U. Birkholz and J. Schelm; *Phys. Stat. Sol.*, **34**, 1969, K177–K180.
6. I. Nishida; *Phys. Rev.*, **B7**, 1973, 2710–2713.
7. Y. Dusausoy, J. Protas, R. Wandji, and B. Roques; *Acta Cryst.*, **B27**(6), 1971, 1209–1218.
8. F. A. Sidorenko, P. V. Gel'd, and M. A. Shumilov; *Fiz. Met. Metalloved.*, **6**, 1960, 861–867.
9. T. Sakata, Y. Sakai, H. Yoshino, H. Fujii, and I. Nishida; *J. Less-Common Met.*, **61**, 1978, 301–308.
10. F. Wever and H. Moller; *Z. Krist.*, **75**, 1930, 362–365; *Naturwissenschaften*, **18**, 1930, 734–735.
11. L. Pauling and A. M. Soldate; *Acta Cryst.*, **1**, 1948, 212–216.
12. K. Nogi, T. Kita, and X. Q. Yan; *J. Ceram. Soc. Jpn.*, **109**(3), 2001, 265–269.
13. T. Kojima; *Phys. Stat. Sol.*, (a)**111**, 1989, 233–242.
14. T. Kojima, K. Masumoto, M. A. Okamoto, and I. Nishida; *J. Less-Common Met.*, **159**, 1990, 299–305.
15. T. Kojima, M. Sakata, and I. Nishida; *J. Less-Common Met.*, **162**, 1990, 39–49.
16. H. Kaibe, E. Ohta, M. Sakata, Y. Isoda, and I. Nishida; *J. Mater. Sci. Soc. Jpn.*, **25**, 1990, 149–158.
17. Y. Isoda, Y. Imai, and Y. Shinohara; in *Proc. 21st Int. Conf. Thermoelectrics*, edited by T. Caillat, 2002, 102–105.
18. J. Tani and H. Kido; *J. Appl. Phys.*, **86**, 1999, 464–467.
19. J. Tani and H. Kido; *J. Ceram. Soc. Jpn.*, **109**(6), 2001, 557–560.
20. M. Karakita, J. Koyama, T. Komatsu, and M. Miyamoto; *J. Jpn. Soc. Powder Powder Metallurgy*, **55**(11), 2008, 776–780.

21. M. Ito, T. Tada, and S. Hara; *Mater. Trans.*, **45**(9), 2004, 2916–2921.
22. J. Yamamoto, T. Shibayanagi, M. Maeda, and M. Naka; *Trans. JWRI*, **129**(1), 2000, 33–38.
23. M. Ae, K. Ichii, T. Oishi, N. Yoshida, and M. Ozawa; *J. Jpn. Soc. Powder Powder Metallurgy*, **45**(11), 1998, 1086–1091.
24. K. Nogi and T. Kita; *Mater. Trans.*, **42**(5), 2001, 862–869.
25. R. Wandji, Y. Dusausoy, J. Protas, and B. Roques; *C. R. Acad. Sci. Paris*, **267**, 1968, 1587.
26. Ch. Kloc, E. Arushanov, M. Wendl, H. Hohl, U. Malang, and E. Bucher; *J. Alloys Compd.*, **219**, 1995, 93–96.
27. J. F. Wang, S. Y. Ji, K. Mimura, Y. Sato, S. H. Song, H. Yamane, and M. Shimada; *Phys. Stat. Sol.*, **201**(a), 2004, 2905–2909.
28. H. Kakemoto, Y. Tsai, A.C. Beye, H. Katsumata, S. Sakuragi, Y. Makita, A. Obara et al. *Mater. Res. Soc. Symp. Proc.*, **402**, 1996, 331.
29. H. Shibata, Y. Makita, H. Kakemoto, Y. Tsai, S. Sakuragi, H. Katsumoto, A. Obara et al. in *Proc. 15th Int. Conf. Thermoelectrics*, edited by J. P. Fleurial, 1996, 62–66.
30. N. Kh. Abrikosov and L. I. Petrova; *Inorg. Matter.*, **11**, 1975, 184–186.
31. H. Udono and I. Kikuma; *Jpn. J. Appl. Phys.*, **39**, 2000, L225–L226.
32. M. Kuramoto, Y. Nose, Y. Momose, K. Saito, H. Tatsuoka, and H. Kuwabara; *J. Crys. Grow.*, **237–239**, 2000, 1981–1985.
33. H. Udono and I. Kikuma; *Jpn. J. Appl. Phys.*, **40**, 2001, 1367–1369.
34. T. Yamada, H. Morito, and H. Yamane; *Jpn. J. Appl. Phys.*, **48**, 2009, 100209.
35. G. Behr, J. Werner, G. Weise, A. Heinrich, A. Burkov, and C. Gladun; *Phys. Stat. Sol.*, **160**(2), 1997, 549–556.
36. A. Heinrich, G. Behr, H. Griessman, S. Teichert, and H. Lange; *Thermoelectric Materials—New Directions and Approaches*, edited by T. M. Tritt, M. G. Kanatzidis, H. B. Lyon and G. D. Mahan (MRS, Pittsburgh, Pennsylvania, 1997), 255–266.
37. M. Komabayashi, K. Hijikata, and S. Ido; *Jpn. J. Appl. Phys.*, **30**(29), 1991, 331–334.
38. T. Tokiai, T. Uesugi, and K. Koumoto; *J. Ceram. Soc. Jpn.*, **103**(7), 1995, 670–675.
39. M. Ito, H. Nagai, S. Katsuyama, and K. Majima; *J. Alloys Comp.*, **315**, 2001, 251–258.
40. U. Birkholz and J. Schelm; *Phys. Stat. Sol.*, 27, 1968, 413–425.
41. Y. Isoda, Y. Shinohara, Y. Imai, and I. Nishida; *Trans. IEEE Jpn.*, **116-A**, 1996, 212–217.
42. M. Umemoto; *Mater. Trans.*, **39**, 1998, 373–383.
43. H. Nagai, I. Maeda, S. Katsuyama, and K. Majima; *J. Jpn. Soc. Powder Powder Metallurgy*, **41**, 1994, 560–564.
44. Y. Isoda, Y. Shinohara, Y. Imai, I. Nishida, and O. Ohashi; *J. Jpn. Inst. Metals.*, **63**, 1999, 391–396.
45. Y. Isoda, I. Nishida, and O. Ohashi; *J. Jpn. Inst. Metals.*, **63**, 1999, 1372–1376.
46. Y. Hara, M. Tobita, S. Ohuchi, and K. Nakaoka; *Thin Solid Films* **515**, 2007, 8259–8262.
47. H. Udono, S. Takaku, and I. Kikuma; *J. Crystal Growth*, **237–239**, 2002, 1971–1975.
48. H. Udono, Y. Aoki, I. Kikuma, H. Tajima, and I. J. Ohsugi; *J. Crystal Growth*, **275**, 2004, e2003–e2007.
49. H. Udono, K. Matsumura, I. J. Osugi, and I. Kikuma; *J. Crystal Growth*, **275**, 2004, e1967–e1974.
50. H. Udono and I. Kikuma; *Jpn. J. Appl. Phys.*, **41**, 2002, L583–L585.
51. K. Irmscher, W. Gehlhoff, Y. Tomm, H. Lange, and V. Alex; *Phys. Rev. B*, **55**, 1997, 4417–4425.
52. H. Udono and I. Kikuma; *Jpn. J. Appl. Phys.*, **40**, 2001, 4164–4165.
53. H. Udono and I. Kikuma; *Mat. Sci. Semicond. Processing*, **6**, 2003, 413–416.
54. A. Heinrich, H. Griessmann, G. Behr, K. Ivanenko, J. Schumann, and H. Vinzelberg, *Thin Solid Films*, **381**, 2001, 287–295.
55. A. Heinrich, C. Gladun, A. Burkov, Y. Tomm, S. Brehme, and H. Lange; in *Proc. 14th Int. Conf. Thermoelectrics*, 1995, 259–263.
56. M. Takeda, M. Kuramitsu, and M. Yoshio; *Thin Solid Films*, **461**, 2004, 179–181.
57. H. Suzuki, H. Udono, and I. Kikuma; *Mat. Trans.*, **47**, 2006, 1428–1431.

58. H. Udono, H. Suzuki, K. Goto, S. Mashiko, M. Uchikoshi, and M. Issiki; in *Proc. 26th Int. Conf. Thermoelectrics*, 2007, 241–244.
59. I. Yamauchi, A. Suganuma, T. Okamoto, and I. Ohnaka; *J. Mater. Sci.*, **32**(17), 1997, 4603–4611.
60. I. Yamauchi, T. Okamoto, H. Ohuta, and I. Ohnaka; *J. Alloys Comp.*, **260**(1), 1997, 162–171.
61. I. A. Nishida; *Thermoelectrics—Principles and Applications*, edited by M. Sakata (Realize Inc., Tokyo, 2000), 199–209.
62. Y. Isoda, T. Ohkoshi, I. Nishida, and H. Kaibe; *J. Mater. Sci. Soc. Jpn.*, **25**, 1989, 311–319.
63. H. Kashiwaki; *Thermoelectric Energy Conversion Systems*, edited by T. Kajikawa, T. Ohta, I. A. Nishida, K. Matsuura, and K. Matsubara (Realize Inc., Tokyo, 1995), 196–199.
64. Y. Isoda; *Thermoelectric Energy Conversion Systems*, edited by T. Kajikawa, T. Ohta, I. A. Nishida, K. Matsuura, K. Matsubara (Realize Inc., Tokyo, 1995), 200–203.
65. K. Uemura, Y. Mori, T. Imai, I. A. Nishida, S. Hori, and M. Kawaguchi; in *Proc. 8th Int. Conf. Thermoelectrics Energy Conversion*, edited by H. S. Scherrer, 1989, 151–156.

19

The Deposition of Bi_2Te_3 and Sb_2Te_3 Thermoelectric Thin Films by Thermal Coevaporation and Applications in Energy Harvesting

L. M. Goncalves
University of Minho

19.1 Introduction

Bismuth, antimony, and tellurium compounds (Bi/Sb/Te) are known as the best thermoelectric materials for room-temperature operation. Despite thermoelectric devices with these materials being used for many years in macroscale dimensions (millimetres sized devices), only few attempts were made to reduce these devices to the microscale (micrometer-sized devices). The deposition of thermoelectric films was reported before using techniques like electrochemical deposition (ECD), metal–organic chemical vapor deposition (MOCVD), pulsed laser deposition, sputtering, and thermal evaporation [1–8]. Each technique has its advantages and disadvantages, and a summary can be found in Table 19.1. In the table, CVD and ECD present opposite characteristics: while CVD films present high figure of merit (*ZT*), but a low deposition rate and expensive and complicated equipment is required (specific gases are needed for the deposition), ECD is a simple process, allowing high deposition rates (tens of micrometers can be achieved) but resultant films present low *ZT*. However, ECD allows the creation of structures during the deposition process, using the LIGA process (from German "lithographie, galvanoformung, abformung," meaning lithography, electroplating, and molding). In this chapter, the deposition of Bi_2Te_3 and Sb_2Te_3 thin films by thermal coevaporation is described.

The conventional techniques used in the fabrication of macro-thermoelectric devices cannot be used in the microdevices. Few pattern techniques have been demonstrated in the fabrication of thermoelectric microdevices, with feature size below tens of micrometers. These techniques were imported from

TABLE 19.1 Resume of Deposition Techniques for Thermoelectric Films

Technique	Advantages	Disadvantages
CVD	Very good materials: High *ZT*	Low deposition rate Expensive and complicated equipment
ECD	Very high deposition rate Patterning can be done during deposition (LIGA)	Low *ZT*
Sputtering	Good deposition rate	Annealing (or substrate heating) improves *ZT*
	Average *ZT* value	Composition difficult to control (depends on power)
	Composition can be controlled with cosputtering	
Thermal coevaporation	High *ZT* Good deposition rate	Needs substrate heating Needs precise control of deposition rate and crucible temperature
	Control of film composition Simple/inexpensive equipment	

the fabrication of MEMS (micro-electromechanical systems) techniques, namely, wet-etching, lift-off [with SU-8 photoresist (PR)], reactive ion etching (RIE), and LIGA. The wet-etching patterning process assisted with UV photolithography is presented in this chapter, using thermoelectric Bi/Sb/Te films, in a planar device structure. Applications for microcooling and energy harvesting are also presented at the end of the chapter.

19.2 Thin-Film Deposition

The figure of merit of Bi/Sb/Te thermoelectric materials is always related with composition and crystalline structure of materials, despite the technique used. And these properties are related with deposition variables. Previous work [1–3] showed that the optimum composition to maximize the figure of merit is obtained with Te content in the range 60–65% (atomic). Some of the deposition techniques presented in Table 19.1 allow the composition control of the film during growth; thus the figure of merit can be improved. When evaporating Bi_2Te_3, Sb_2Te_3, or $Bi_xSb_{2-x}Te_3$ directly, the materials decompose and the final composition of the film does not match the initial composition of the evaporant due to the different vapor pressure of the elemental substances Bi, Sb, and Te. When heated, these materials decompose, the tellurium evaporates faster than bismuth or antimony, and a composition gradient is expected when thicker films are deposited. The composition differs from the surface layers (Te rich in the first fabricated layers) into the bulk and the last fabricated layers (Bi and Sb rich). A detailed study of composition along thickness was presented by Silva [1]. To overcome this problem, a composition control during film growth is necessary, mainly in thermal evaporation techniques, but also in sputtered films. Using coevaporation, the deposition rate of each element (Bi, Sb, or Te) is controlled independently, and an optimal composition can be achieved [1–4]. The same technique is also used in cosputtering [5,6].

A thermal coevaporation system is presented in Figure 19.1. A pressure below 5×10^{-6} is required in a vacuum chamber. The evaporation sources (boats) are in two molybdenum baffled boxes with volume >4 cm^3. A large volume is required in the boat to maintain deposition properties along the evaporation process, mainly if thick films are pretended. The baffled boxes are used for better stability of deposition rate, compared with typical boats. The power applied to each boat is controlled independently, using a deposition controller (e.g., IC6 thin-film deposition controller from Inficon company). The deposition rate of each boat is also measured independently, with a crystal oscillator for each evaporation source.

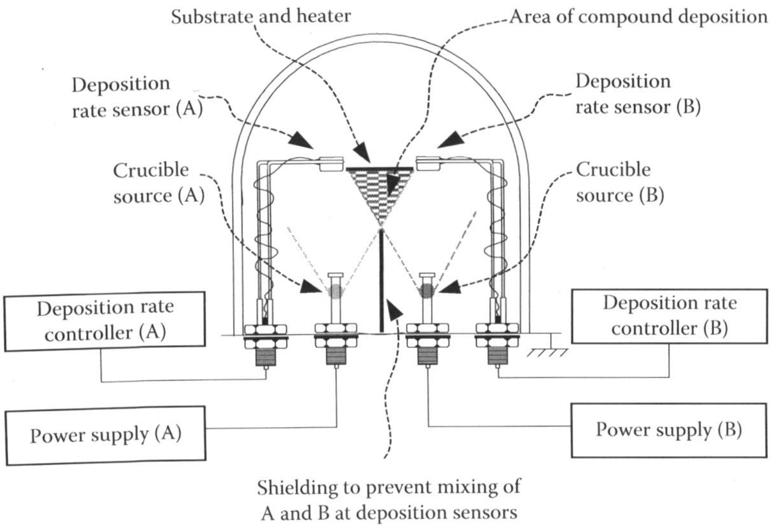

FIGURE 19.1 Vacuum chamber prepared for coevaporation.

Each crystal sensor is placed carefully in the chamber, in such way that it receives the material evaporated only from the boat it is expected to monitor. A metal sheet is placed between the two boats to partially separate the flows from the two sources, fully preventing mixing of both materials at the quartz crystals (Figure 19.1). The deposition controller [9] computes in real time the power to apply in each boat, using two proportional integral and derivative algorithms in order to maintain the deposition rate at a fixed value, different for each material. For better results, each deposition monitor should be calibrated considering its position. For this purpose, a film should be deposited in the substrate, for each material, and the measured thickness in the monitor corrected to obtain the real thickness of the film (measured with a profilometer or by SEM imaging). A rotating substrate reduces thickness and composition nonuniformity of the films. In the deposition of Bi/Sb/Te films, the substrate should be heated to temperatures (T_{sub}) in the range between 200 and 300°C. Table 19.2 compares the Seebeck coefficient (α), the resistivity (ρ), the power factor, and the figure of merit (ZT) of Bi/Sb/Te films at room temperature (300 K), deposited by different techniques.

Good thermoelectric Bi_2Te_3 films are obtained by coevaporation, with an evaporation rate of 2 Ås^{-1} at Bi source (Er_{Bi}) and evaporation rate in the range of 3–9 Ås^{-1} in the Te source (Er_{Te}). The evaporation flow rate ratio, $R_{Bi} = Er_{Te}/Er_{Bi}$, is defined as the number of atoms of Te divided by the number of atoms of Bi that arrive in unit time at the substrate during deposition. The power factor is defined as $PF = \alpha^2/\rho$, calculated using the measured Seebeck coefficient (α) and electrical resistivity (ρ). Substrates are heated to the temperature set point in the range between 190 and 270°C. All films were deposited in flexible polyimide (Kapton) substrates (Figure 19.2).

In-plane electrical resistivity, carrier concentration, and Hall mobility were measured at room temperature using the conventional four probe van der Pauw geometry in a 5 mm side square arrangement. A DC magnetic field of 80 mT was applied for Hall measurements. The Seebeck coefficient was measured by connecting one side of the film to a heated metal block at a fixed temperature and the other side to a heat sink kept at room temperature, with a temperature difference between both sides <10°C.

Figure 19.3 compares the thermoelectric properties of 1 μm Bi_2Te_3 films, as a function of R_{Bi} (different evaporation rates of tellurium (Er_{Te}), maintaining the evaporation rate of bismuth Er_{Bi} = 2 Ås^{-1}). These properties (Seebeck coefficient, resistivity, and carrier concentration) were measured at room temperature for films deposited with different R_{Bi} and different substrate temperatures (T_{sub}).

TABLE 19.2 Properties of Selected Bi_2Te_3 and $Bi_xSb_{2-x}Te_3$ Films

Material and Type		Deposition Technique	Seebeck α (μVK⁻¹)	Resistivity ρ (μΩm)	Power Factor 10^{-3} $WK^{-2}m^{-1}$	*ZT*@300 K	Reference	Obs
Bi_2Te_3	*n*	Coevaporation	−220	10.6	4.57	0.91	[7]	
Sb_2Te_3	*p*	Coevaporation	188	12.6	2.81	0.56	[10]	
Bi_2Te_3	*n*	Electrochemical	−60	10	0.36	—	[11]	
Bi_2Te_3	*n*	MOCVD	−210	12	3.7	0.74	[12]	(1)
Sb_2Te_3	*p*	MOCVD	−110	3.5	3.46	—	[12]	
Bi_2Te_3	*p*	MOCVD	190	78	0.46	0.75	[13]	(1)
Bi_2Te_3	*n*	MOCVD	−218	6.9	6.9	—	[14]	
$Bi_{0.5}Sb_{1.5}Te_3$	*p*	Flash	230	17	3.1	0.87	[15]	
$Bi_2Te_{2.72}Se_{0.3}$	*n*	Flash	−200	15	2.7	—	[16]	
$Bi_{0.5}Sb_{1.5}Te_3$	*p*	Flash	240	12	4.8	—	[16]	
$Bi_{1.8}Sb_{0.2}Te_{2.7}Se_{0.3}$	*n*	Sputtering	−235	47	1.2	—	[17]	(2)
Bi_2Te_3	*n*	CoSputtering	−160	16.3	1.6	—	[6]	(3)
$(BiSb)_2Te_3$	*p*	CoSputtering	175	12.1	2.5	—	[6]	(3)
$Bi_2Se_{0.3}Te_{2.7}$	*n*	Sputtering	−160	20	1.3	—	[18]	
$Bi_{0.5}Sb_{1.5}Te_3$	*p*	Sputtering	210	25	1.8	—	[18]	
Bi_2Te_3	*n*	CoSputtering	−55	10	0.3	—	[5]	
Bi_2Te_3	*n*	Coevaporation	−228	13.0	4.0	0.81	[19]	(1)
Sb_2Te_3	*p*	Coevaporation	171	10.4	2.8	0.53	[19]	(1)
Bi_2Te_3	*n*	Coevaporation	−228	28.3	1.8	—	[3]	
Sb_2Te_3	*p*	Coevaporation	149	12.5	1.78	—	[3]	

Obs:
(1) Z estimated by the author.
(2) Doped with CuBr.
(3) The power factor of de 3×10^{-3} $WK^{-2}m^{-1}$ and 4×10^{-3} $WK^{-2}m^{-1}$, respectively, for type *n* and type *p* was reported latter by the same authors [20] but no reference of other thermoelectric properties was found.

At each T_{sub} the maximum absolute value of the Seebeck coefficient is obtained at a value of R_{Bi} that depends on T_{sub}. Maximum PF is obtained at $R_{Bi} = 2$, 2.5, and 3.5, respectively, for $T_{sub} = 190$, 230, and 270°C. Films with compositions in Te richer than stoichiometric have lower carrier concentration ($<10^{20}$ cm^{-3}), higher mobility (>50 $cm^2V^{-1}s^{-1}$), and higher Seebeck coefficient, leading to a higher power factor. Films obtained with $R_{Bi} < 1.2$ are Bi-rich and have high carrier concentration. This results in a decreased value of power factor. High power factor correlates with a high absolute value of Seebeck coefficient, with a low carrier concentration, and with a relatively low electrical resistivity. Since the observed PFs dramatically decrease as the carrier concentration and electrical conductivity increase, one is forced

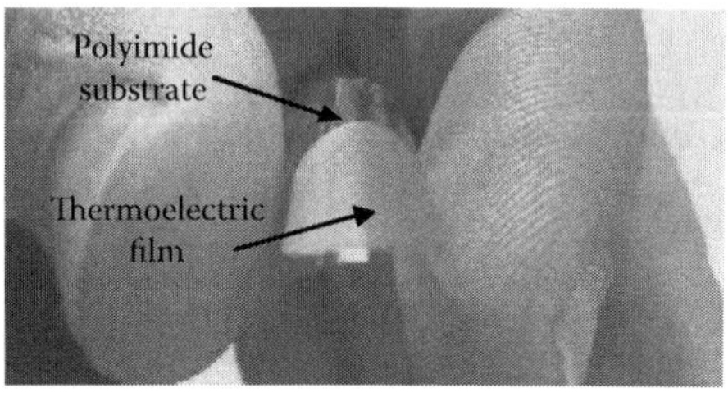

FIGURE 19.2 Thermoelectric thin film deposited on top of a flexible polyimide (Kapton) substrate. (Adapted from L.M. Goncalves et al., *Journal of Micromechanics and Microengineering*, 17, 2007, S168–S173.)

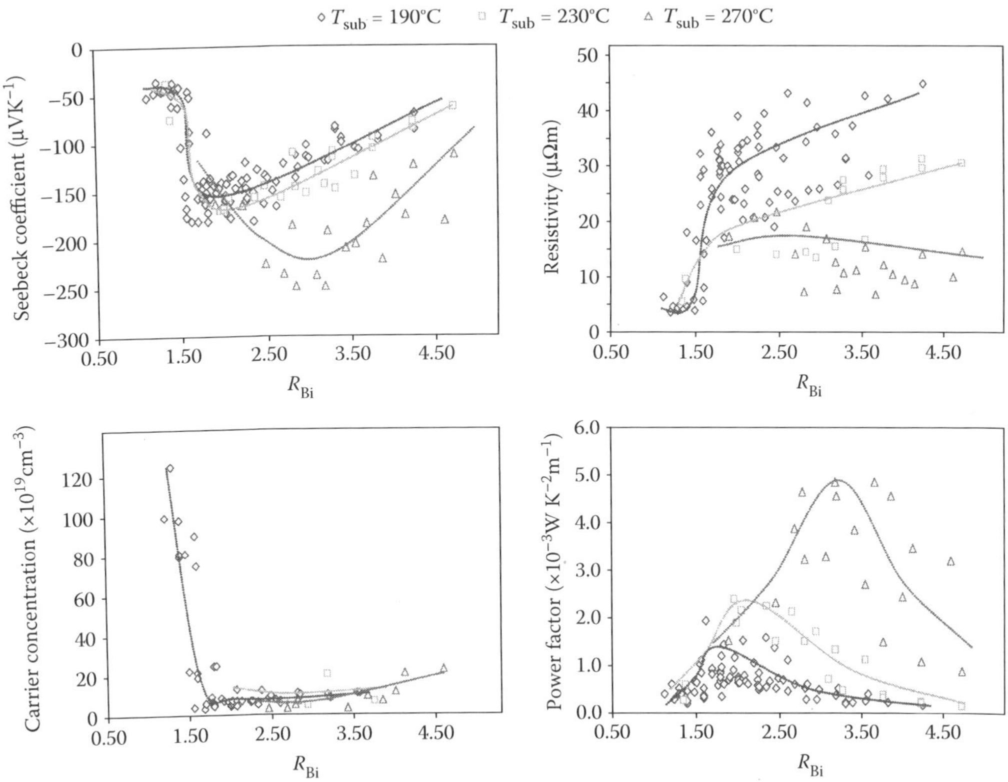

FIGURE 19.3 Seebeck coefficient, electrical resistivity, carrier concentration, and power factor of Bi_2Te_3 thin films as a function of Te/Bi evaporation flow ratio, R_{Bi} (lines are guides to the eye). (Adapted from L.M. Goncalves et al., *Thin Solid Films*, 518, 2010, 2816–2821.)

to conclude that these increments are the result of major changes in the band structure and position of the Fermi level in the density of states (DOS). These changes are also responsible for the observed decrease in α. In fact, according to the Mott theory, the Seebeck coefficient α critically depends on the derivative with respect to the energy of the DOS at the Fermi energy [7].

An equivalent behavior is also found in Sb_2Te_3 films. Figure 19.4 plots the main thermoelectric properties of 1 μm Sb_2Te_3 films, when deposited at different substrate temperatures (T_{sub} = 150, 180, and 220°C) and R_{Sb} in the range of 1.4–3.7, maintaining the Sb evaporation rate at 2 Ås^{-1}.

The value of Seebeck coefficient and the value of electrical resistivity increases as R_{Sb} is increased. The carrier concentration decreases with increase of R_{Sb}. No dependence was found on the Seebeck coefficient or the carrier concentration with the substrate temperature (T_{sub}). However, the electrical resistivity shows a strong dependence with T_{sub}. $\rho < 15$ μΩm is obtained with the substrate temperature >220°C. Films with compositions in Te richer than stoichiometric have lower carrier concentration ($n < 5 \times 10^{19}$ cm^{-3}), higher mobility ($\mu > 100$ cm^{2}V^{-1}s^{-1}), and higher Seebeck coefficient ($\alpha > 150$ μVK^{-1}), leading to a high power factor.

The influence of substrate temperature (T_{sub}) in the power factor of coevaporated Bi_2Te_3 and Sb_2Te_3 films [21] is presented in Figure 19.5. The results presented by Zou [2] and Silva [3], also for coevaporated films, are also presented in the same figure.

The chemical composition of the films and its structure were studied [23] by energy-dispersive x-ray spectroscopy. Figure 19.6 plots the power factor of coevaporated Bi_2Te_3 and Sb_2Te_3 films, as a function of composition (atomic percentage of Te atoms in composition), for films deposited at three different substrate

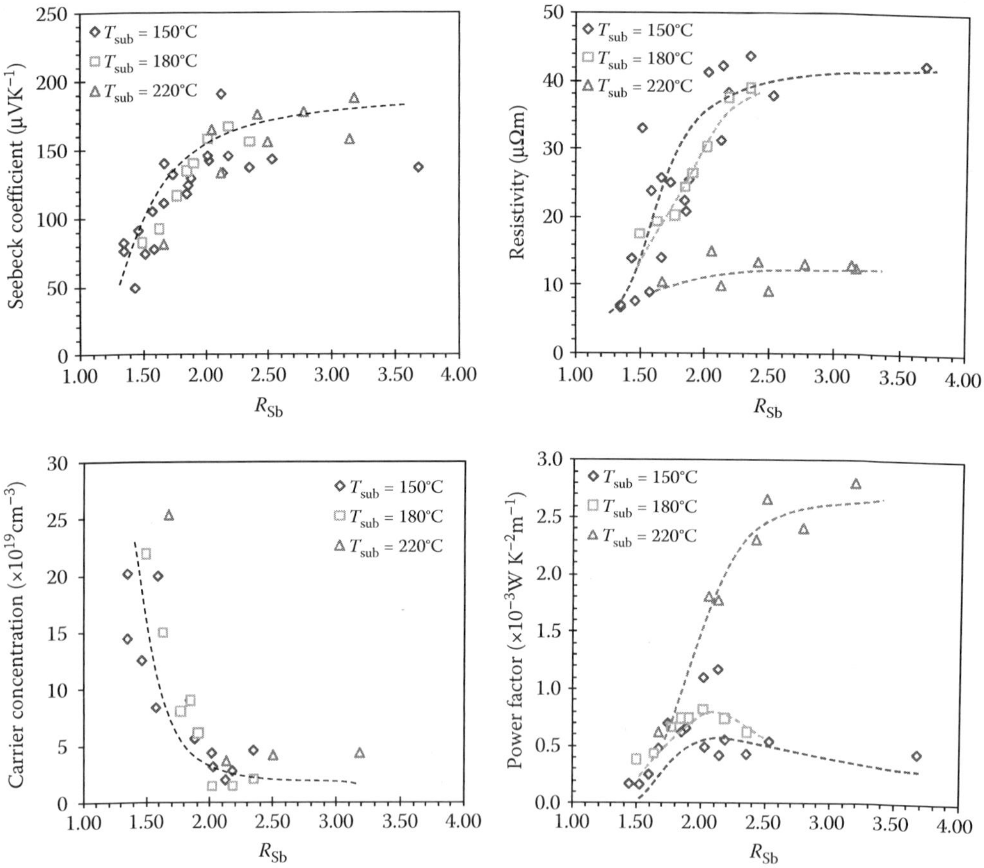

FIGURE 19.4 Seebeck coefficient, resistivity, carrier concentration, and power factor of Sb_2Te_3 thin film as a function of Te/Sb evaporation flow ratio, R_{Sb} (lines are guides for the eye). (Adapted from L.M. Goncalves et al., *Thin Solid Films*, 519, 2011, 4152–4157.)

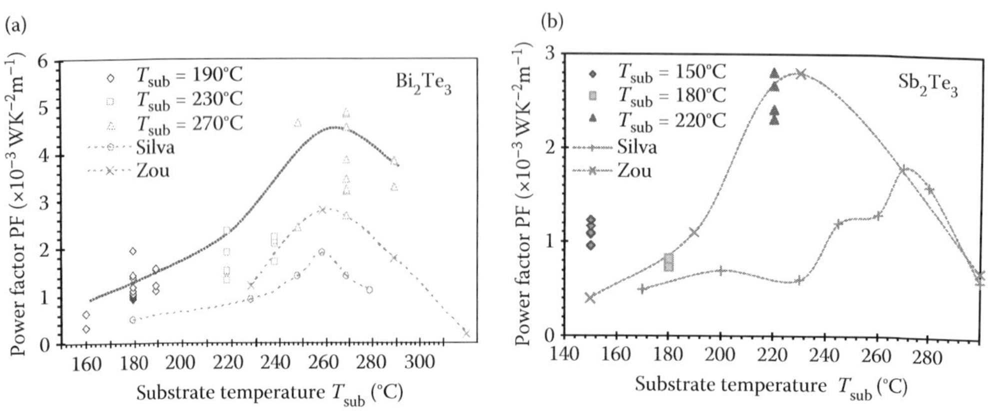

FIGURE 19.5 Power factor of selected Bi_2Te_3 (a) and Sb_2Te_3 (b) films plotted as a function of substrate temperature T_{sub} (lines are guides to the eye). Results from Zou [2] and Silva [3] are also presented. (Adapted from H. Zou, D.M. Rowe, and S.G.K. Williams, *Thin Solid Films*, 408, 2002, 270–274; L.W. da Silva, M. Kaviany, and C. Uher, *Journal of Applied Physics*, 97, 2005, 114903.)

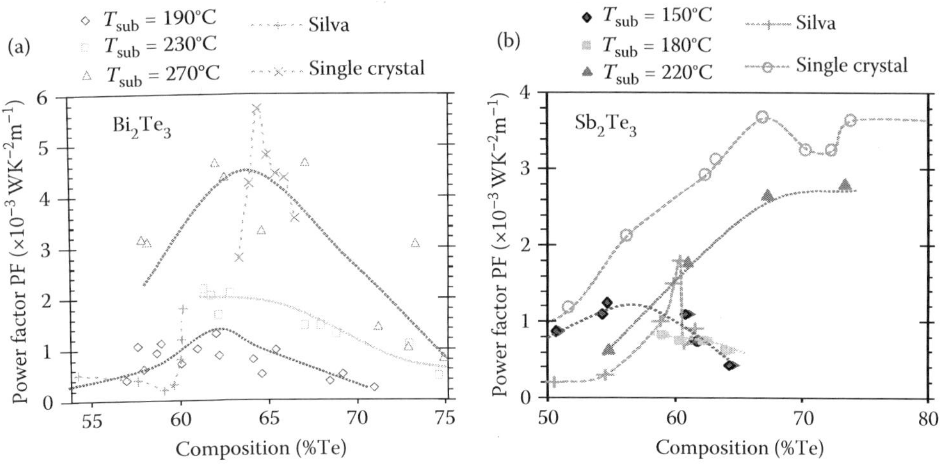

FIGURE 19.6 Power factor of Bi_2Te_3 (a) and $Sb2_2Te_3$ (b) films deposited by coevaporation plotted as a function of composition and substrate temperature during deposition (T_{sub}) (lines are guides to the eye). Results obtained in coevaporated thin films from Silva [3] and from single crystals [22] are also presented.

temperatures (T_{sub}). The power factor of films deposited by Silva [3] using the same deposition technique and the power factor of bulk single crystal of Bi_2Te_3 and Sb_2Te_3 [22] is also presented in the same figure.

In the graph of Figure 19.6, Bi_2Te_3 films present the maximum power factor at the same composition (~65% Te) as that of single-crystal Bi_2Te_3. However, the curve along the composition shows a broader peak, probably due to the polycrystalline structure of films. The power factor of Sb_2Te_3 films also presents the same behavior to those of bulk single crystal, when composition varies. The smaller power factor of Sb_2Te_3 films, compared to bulk is probably due to polycrystalline structure that promotes high electrical resistivity in interface regions.

Thermal conductivity (κ) was measured (in a direction parallel to film) on Bi_2Te_3 films (Figure 19.7) for temperatures between −20 and 110°C, using the technique described by Völklein [23]. $\kappa = 1.3$ Wm^{-1}K^{-1} was obtained at room temperature. Thermal conductivity is due to electron and phonon contributions, respectively, κ_e and κ_p, and $\kappa = \kappa_e + \kappa_p$ [24]. Electron contribution was estimated using Wiedemann–Franz law, $\kappa_e = LT/\rho_e$, where L is the Lorenz number, T the temperature, and ρ_e the electrical resistivity.

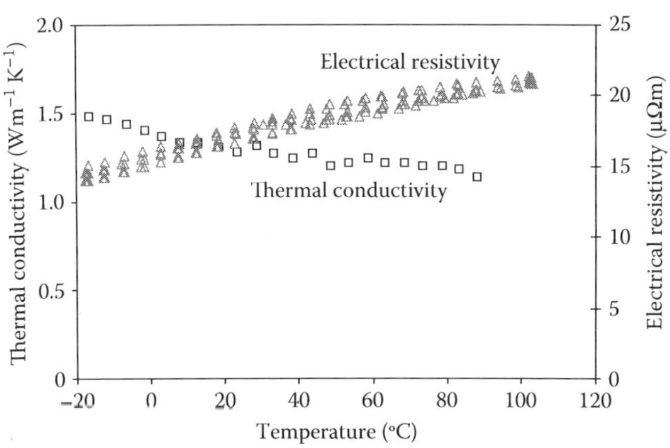

FIGURE 19.7 Thermal conductivity and electrical resistivity of a Bi_2Te_3 film, measured from −20°C to 110°C. (Adapted from L.M. Goncalves et al., *Thin Solid Films*, 518, 2010, 2816–2821.)

Considering $L = 1.5 \times 10^{-8}\ V^2K^{-2}$ at 300 K [25], $\kappa_e = 0.45\ Wm^{-1}K^{-1}$ is obtained (for a resistivity of 10 μΩm) and $\kappa_p = 0.85\ Wm^{-1}K^{-1}$. This value is smaller than $\kappa_p = 1.03\ Wm^{-1}K^{-1}$ for bulk Bi_2Te_3 [26] due to reduced lattice contribution in polycrystalline structure. Using these values and a Seebeck coefficient of 200 μVK^{-1}, a value of *ZT* near unity is obtained for the best Bi_2Te_3 thin films at room temperature.

Despite the composition, the structure of the thermoelectric thin film is also important to maximize the figure of merit, as demonstrated before [7]. Thermoelectric Bi/Sb/Te films can be amorphous or polycrystalline. Bulk material can also be found in the form of single crystal. The interface regions between crystals in a polycrystalline structure create additional electrical resistance, but also reduce thermal conductivity, in such way that films with larger crystal grain usually have a lower resistivity and higher thermal conductivity. The structure of films can be controlled by the substrate temperature during deposition or with postdeposition annealing under controlled atmosphere. Due to decomposition of Bi/Sb/Te compounds with temperature and the low vapor pressure of tellurium (compared with Bi or Sb), this heating process can also alter the composition of films. During deposition (pressure $<1 \times 10^{-5}$ mbar), Te reevaporates from substrate if temperature exceeds 200°C. This can be compensated with a higher evaporation rate of tellurium (beyond stoichiometric composition). A low deposition rate of the films also promotes the growth of crystalline films, and most important in the first film layers (<100 nm). A low growth rate (1 $Ås^{-1}$) in the first 100 nm of film, and increasing up to 6 $Ås^{-1}$, with heated substrate (200–300°C) achieves higher crystalline structure (and decreases deposition time) than a constant deposition rate of 4 $Ås^{-1}$. In films thicker than 1 μm, deposition rates can go above 6 $Ås^{-1}$ [21]. Other techniques allow faster growth of films. Böttner [6] reported a deposition rate of 5 μm/h using cosputtering, and Snyder from the jet propulsion laboratory (JPL) [8] fabricated a device with 20 μm high thermoelectric columns by ECD. Figures 19.8 and 19.9 show the diffractogram obtained by x-ray diffraction (XRD) using copper Kα radiation ($\lambda = 1.54051$ Å) of Bi_2Te_3 and Sb_2Te_3 films deposited by coevaporation, respectively.

The observed peaks in Figure 19.8 agree with the literature from powder diffraction spectra for polycrystalline Bi_2Te_3, (card 15-863 [27]) or $Bi_{0.43}Te_{0.57}$ (card 22.715 [27]) and are represented in dashed lines. Similar results are also found in the literature for the same thermoelectric materials [3,5]. The substrate temperature and the composition influences the peaks found in each film. In Figure 19.8, the two diffractograms of Bi_2Te_3 films were obtained with substrate temperatures of $T_{sub} = 160°C$ and $T_{sub} = 240°C$.

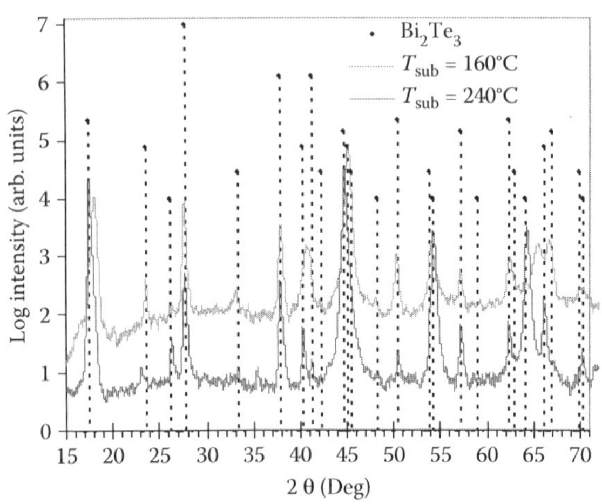

FIGURE 19.8 XRD diffractograms of Bi_2Te_3 films deposited at substrate temperatures of 160°C (top line) and 240°C (bottom line). (Adapted from L.M. Goncalves et al., *Thin Solid Films*, 518, 2010, 2816–2821.)

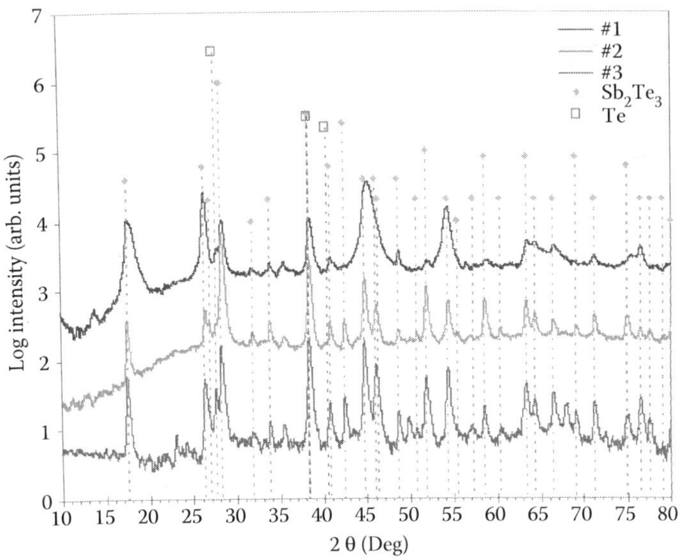

FIGURE 19.9 XRD diffractograms of Sb_2Te_3 films.

Typical diffractograms of Sb_2Te_3 films are presented in Figure 19.9 from films #1, #2, and #3, respectively, deposited at $T_{sub} = 220°C$, $T_{sub} = 150°C$, and $T_{sub} = 220°C$ and $R_{Sb} = 1.7$, $R_{Sb} = 1.7$, and $R_{Sb} = 2.5$. The planes with the higher x-ray intensities obtained with Sb_2Te_3 films deposited at $T_{sub} = 220°C$ are not the same as those from films deposited at $T_{sub} = 150°C$ or from bulk Sb_2Te_3. The substrate temperature during deposition influences both the structure and the orientation of Sb_2Te_3 thin films.

Scanning electron microscopy cross-sectional and surface images of films, deposited at optimal (substrate temperature and composition) conditions are presented in Figure 19.10. These images also reveal the polycrystalline structure of film, and grains with dimensions above 500 nm are visible [7,10]. With lower substrate temperature, the dimension of these grains is substantially reduced.

19.3 Patterning of Microdevices

The techniques used in the fabrication of microthermoelectric devices were imported from the fabrication of MEMS. The must used are wet-etching, lift-off, RIE, and LIGA. Micropelt and IPM Freiburg [6] used the RIE technique to pattern thick films of Bi/Sb/Te compounds, using PR as an etching mask, using an Oxford Instruments Bi_2Te_3 ICP etching process [28]. By other way, the JPL [8] and Institut für Halbleiter und Mikrosystemtechnik [29] fabricated microcolumns of thermoelectric (TE) materials using a LIGA process, with patterned PR and ECD deposition. Lift-off patterning was implemented by the University of Michigan [30–32]. In this technique, thermoelectric films were deposited on top of patterned SU-8 PR. The PR is then removed, also removing the TE material on top of it, defining the structures. The maximum working temperature of SU-8 PR (≈180°C) limits the substrate temperature during deposition, thus reducing the figure of merit of Bi/Sb/Te films. Shafai [33,34] has reported on the possibility of using nitric acid (HNO_3) and hydrochloric acid (HCl) diluted in water (H_2O) for etching Bi_2Te_3, but that work was not extended to full characterization of the process, and it was not applied to other tellurium compounds. Recent work from Sedky [35] also showed suspended Bi_2Te_3 microstructures fabricated by wet-etching. The use of these etchants in the wet-etching patterning of thermoelectric structures was further studied [4,21,36], using the structures of Figure 19.11. Thermoelectric Bi_2Te_3

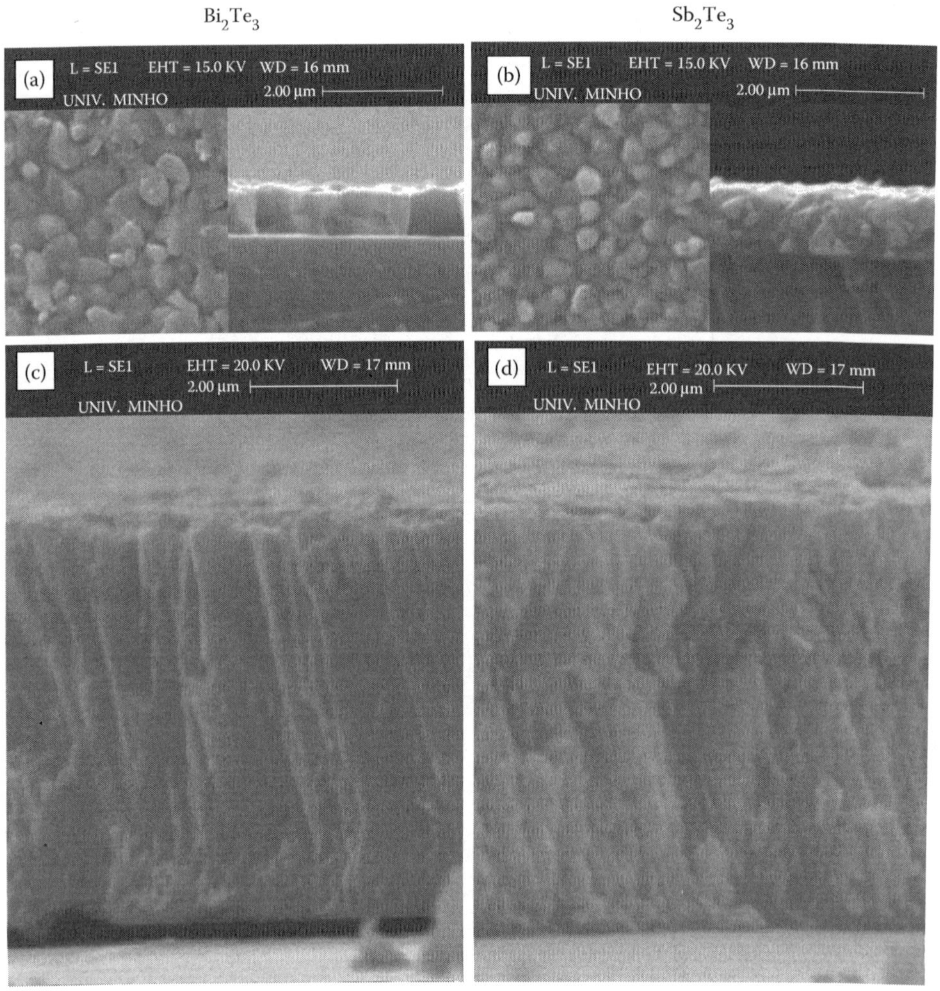

FIGURE 19.10 Surface and cross-sectional images of a Bi_2Te_3 films with $T_{sub} = 270°C$ and %Te = 62% (a and c) and Sb_2Te_3 films with $T_{sub} = 220°C$ and %Te = 70% (b and d), deposited on glass. A and B films have thickness of 1 μm and C and D films have thickness of 5 μm. (Adapted from L.M. Goncalves et al., *Thin Solid Films*, 518, 2010, 2816–2821; L.M. Goncalves et al., *Thin Solid Films*, 519, 2011, 4152–4157.)

and Sb_2Te_3 thin films (1 μm thick) were deposited on Kapton substrate. Transene's purified Kodak photoresist (PKP) negative PR was applied on the surface and test structures were patterned by wet-etching in HNO_3:HCl:H_2O etchant (pure HNO_3 and 37% HCl diluted in water). The effects of etchant composition and dilution in the etch rate are presented in Figures 19.12 and 19.13.

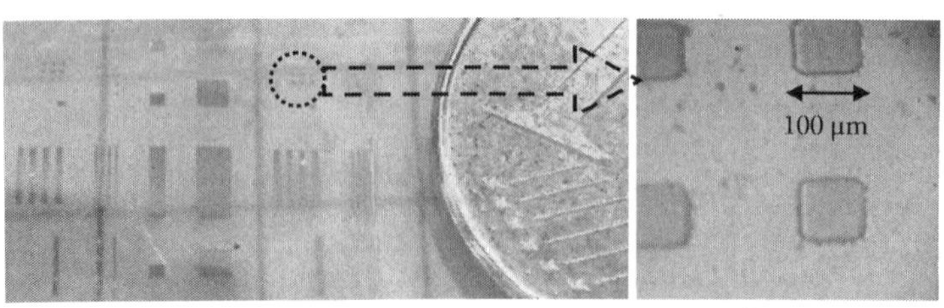

FIGURE 19.11 Test structures of thermoelectric films patterned by wet-etching.

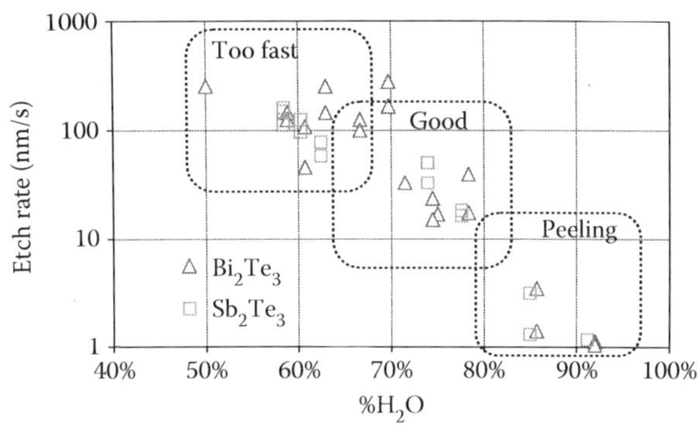

FIGURE 19.12 Etch rate of Bi_2Te_3 and Sb_2Te_3 films in 10:3 HNO_3:HCl solution, as a function of dilution in water (% in volume). (Adapted from L.M. Goncalves et al., *Journal of Micromechanics and Microengineering*, 18, 2008, 064008.)

The dilution of the etchant (% in volume) in water reduces the etch rate (Figure 19.12). Considering dilutions <70%, the etch rate is too fast (>100 nm/s) and becomes difficult to accurately control the etch time, overetching occurring frequently. On the other hand, if dilutions above 80% are used, the large time needed for the film in the solution causes peeling of the film. Dilution around 70% is recommended [36]. Figure 19.13 shows the influence of etchant composition on the etch rates. Higher percent of HCl ($\%HCl/\%HNO_3 > 0.5$) induces cracking of the film and peeling occurs. The best results are obtained with etchant composition in the range of 10:1:20 to 10:5:40 HNO_3:HCl:H_2O. The etch rate of Sb_2Te_3 films in diluted HNO_3 (solution without HCl) was more than 50 times smaller when compared with the etch rate of Bi_2Te_3 films in the same etchant. This is important in terms of the selectivity of the process in the presence of both materials, etching Bi_2Te_3 with HNO_3 while leaving Sb_2Te_3 films untouched. However, this method cannot be used when $Bi_xSb_{2-x}Te_3$ replaces Sb_2Te_3. The etching attack in this solution on $Bi_xSb_{2-x}Te_3$ is similar to that suffered by Bi_2Te_3, even for small values of x.

For chromium and aluminum films using the same etching solutions used for tellurium compounds, etch rates <0.2 nm/s were observed regardless of the etchant composition used. Both Bi_2Te_3 and Sb_2Te_3 are slightly etched (<2 nm/s) by aluminum etchant (16:1:1:2 phosphoric acid, nitric acid, acetic acid, and water) or Cr etchant (Transene 1020). This selectivity between all films and etchants allows the fabrication

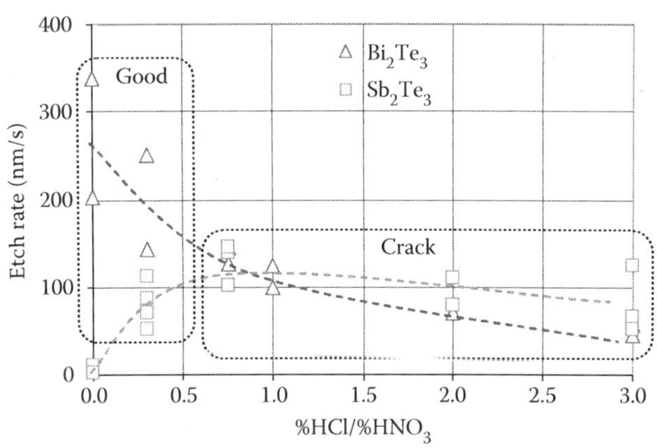

FIGURE 19.13 Etch rate of Bi_2Te_3 and Sb_2Te_3 films in $(1 - x)HNO_3:(x)HCl$ solution (diluted 70% in water, in volume). (Adapted from L.M. Goncalves et al., *Journal of Micromechanics and Microengineering*, 18, 2008, 064008.)

TABLE 19.3 Summary of Etch Rates

	Material			
Etchant	Bi_2Te_3	Sb_2Te_3	Aluminum	Nickel
Al-Transene type A	0.8 nm/s	0.5 nm/s	1–8 nm/s	<0.01 nm/s
Cr-Transene 1020	≈2 nm/s	<0.1 nm/s	—	1–4 nm/s
$3HNO_3$:1HCl (dil 70% H_20)	200 nm/s	80 nm /s	<0.2 nm/s	<0.02 nm/s
HNO_3 (dil 70% H_2O)	250 nm/s	5 nm/s	<0.01 nm/s	<0.01 nm/s

Source: Adapted from L.M. Goncalves et al., *Journal of Micromechanics and Microengineering*, 18, 2008, 064008.

of metallic contacts and thermoelectric elements by wet-etching. Table 19.3 gathers together all relevant etch rates.

Considering the etch rates and selectivity presented, the fabrication process of thermoelectric converters of Figure 19.14 can be implemented [36]. *p*-Type Sb_2Te_3 film is first deposited followed by a thin layer (100 nm) of nickel (a). The thin layers of nickel are used to avoid diffusion of thermoelectric material into the next deposited layers, promoting adhesion and avoiding large contact resistance. PR is spun and *p*-type elements are patterned by photolithography (b),(c). Nickel is etched in a chromium etchant (Transene 1020), thermoelectric film is patterned by wet-etching in HNO_3:HCl bath (d), and PR is removed. *n*-Type film is then deposited, followed by a 100 nm nickel layer (e). PR is applied and patterned by photolithography for *n*-type element definition (f),(g). Nickel is etched in a chromium etchant (Transene 1020), *n*-type film is etched in HNO_3 (h) and PR removed (i). Contacts are deposited, starting with a 100 nm layer of nickel, followed by 1 μm of aluminum (j). PR is spun and contacts patterned by photolithography (k). Nickel is etched in a chromium etchant (Transene 1020), and aluminum with a standard aluminum etchant (Transene type A). PR is removed (l). A protective layer of Si_3N_4 can also be deposited by low-temperature hot-wire chemical vapor deposition [37] and patterned if required. Figure 19.15 shows a thermoelectric device fabricated with wet-etching techniques.

19.4 Applications

Using a lateral (in-plane) configuration [38,39], thin-film techniques can be used to scale down the thermoelectric coolers and generators to microdevice dimensions. Planar thin-film technology [4,36] can be used to fabricate such devices. The conventional thermoelectric cooler, with the heat flux perpendicular to hot and cold areas, cannot be scalable to microchip dimensions, using the same fabrication methods used for macroscale devices. A new design topology [40], a thin-film planar device, as shown in Figure 19.16, has lower heat-pump capacity, but a simplified fabrication process, since all contacts are in the same plane.

Despite using figure of merit to quantify the quality of a thermoelectric material, this parameter can also be used to quantify the performance of a thermoelectric device [40]. The figure of merit of a thermoelectric device is calculated with Equation 19.1. R_{eq} and K_{eq} are the equivalent electrical resistance and thermal conductance of the device, calculated with Equations 19.2 and 19.3.

$$Z = \frac{(\alpha_p - \alpha_n)^2}{K_{eq} R_{eq}} \tag{19.1}$$

$$R_{eq} = R_e + 2R_c = \rho_n \frac{L_n}{W_n H_n} + \rho_p \frac{L_p}{W_p H_p} + 2\left(\frac{\rho_{cn}}{L_c W_n} + \frac{\rho_{cp}}{L_c W_p}\right) \tag{19.2}$$

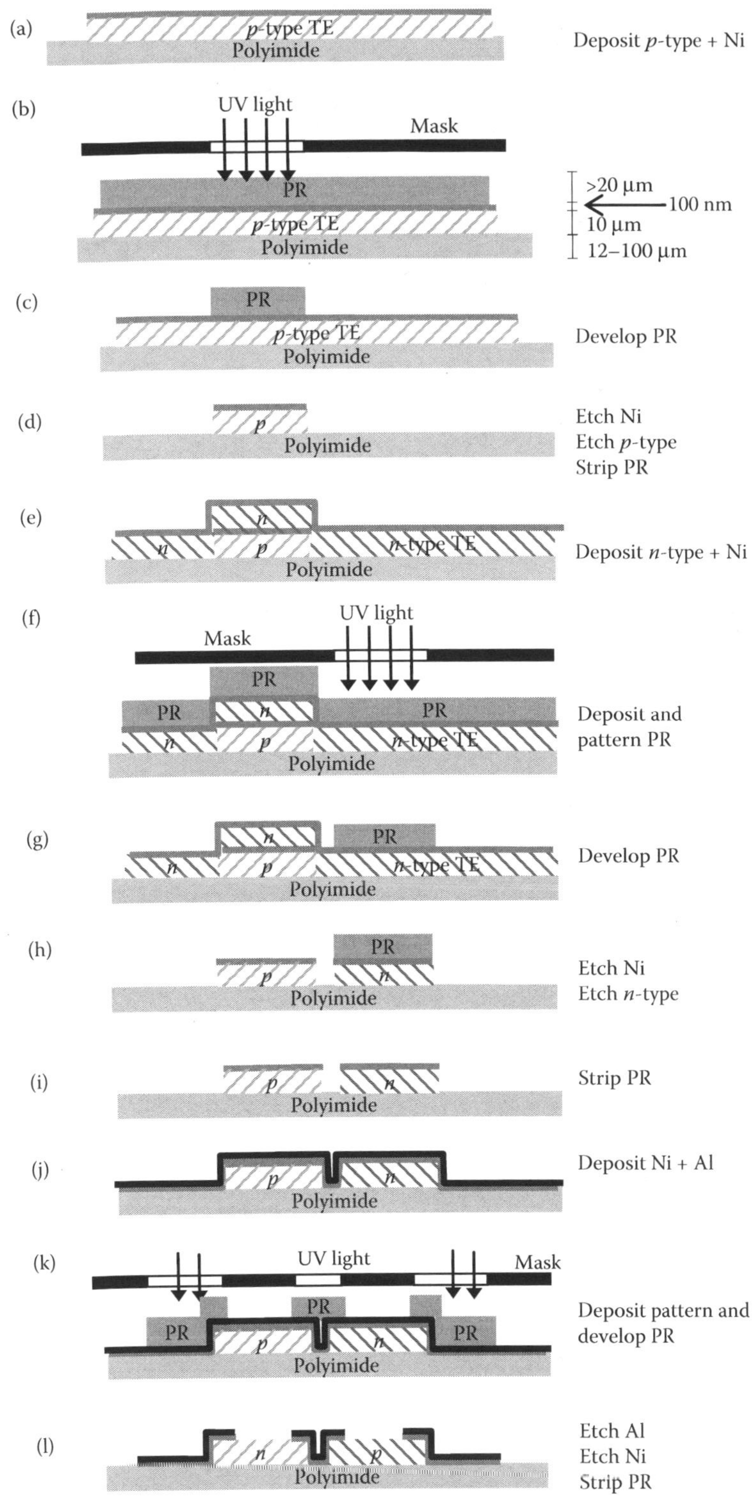

FIGURE 19.14 (**See color insert.**) Fabrication steps of thermoelectric converters. (Adapted from L.M. Goncalves et al., *Journal of Micromechanics and Microengineering*, 18, 2008, 064008.)

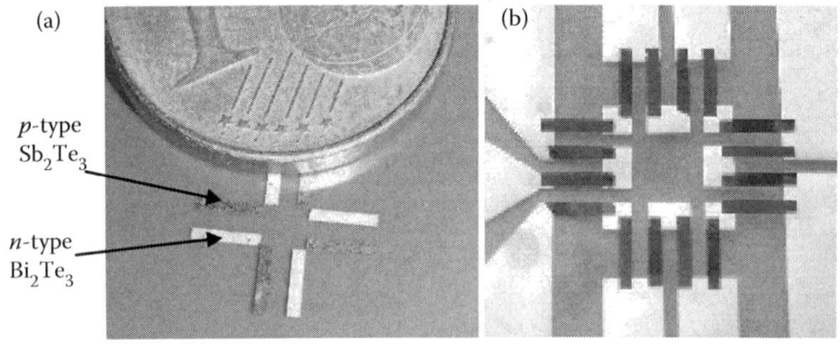

FIGURE 19.15 **(See color insert.)** (a) Photo of *n*-type and *p*-type elements, before deposition of top contacts. (b) Photo of microcooler with eight pairs of thermoelectric elements, fabricated with bottom contacts. (Adapted from L.M. Goncalves et al., *Journal of Micromechanics and Microengineering*, 18, 2008, 064008.)

$$K_{eq} = \lambda_n \frac{W_n H_n}{L_n} + \lambda_p \frac{W_p H_p}{L_p} + \lambda_m \frac{W_m H_m}{L_m} + 4\gamma W^2 \tag{19.3}$$

L_n, L_p, W_n, W_p, H_n, and H_p are, respectively, length, width, and height of thermoelectric materials (*n*- and *p*-type), L_c is the length of contacts between TE material and metal pads, α is Seebeck coefficient, ρ electrical resistivity, λ thermal conductivity, and ρ_c contact electrical resistivity. The device is supported by an isolating membrane, with thickness H_m, length L_m, width W_m, and thermal conductivity λ_m. γ represents a coefficient to include radiation and convection losses in the range $5 < \gamma < 10$ Wm^{-2}K^{-1}.

The previous equations can be used to predict the effect of reducing dimensions of a planar thermoelectric device, from millimeters to micrometers. Besides the Seebeck coefficients (α_n and α_p), the thermal conductivities (λ_n and λ_p), and electrical resistivities (ρ_n and ρ_p) of thermoelectric materials, the thermal conductivity of the supporting membrane (λ_m) and the contact resistance (ρ_c) influence the performance of planar devices. And the effect of contact resistivity is more pronounced (compared with electrical resistivities) as device dimensions get smaller.

The graph of Figure 19.17 presents the effect of scaling down the dimensions of devices in substrate plane [40]. The scale factor $f'' = 1$ (in horizontal axis) represents a device with dimensions $L = W = 1$ mm, $H = 10$ μm, and $H_m = 10$ μm. Lower values of f represent devices where dimensions (L and W) were reduced by a factor f but the same thickness (H) considered in thermoelectric materials and membrane.

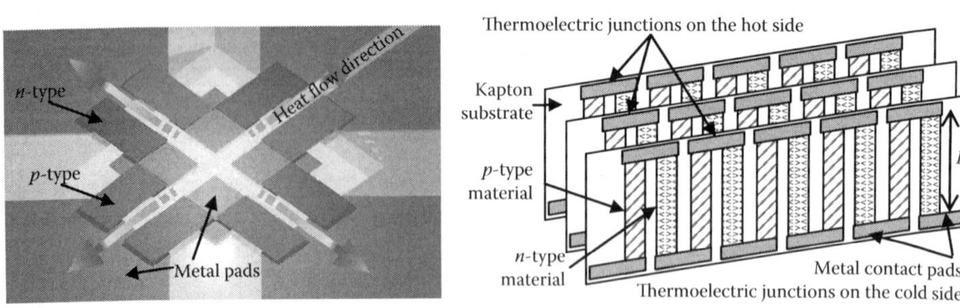

FIGURE 19.16 **(See color insert.)** Planar thermoelectric device (left) and array of thermoelectric devices, fabricated in Kapton foils.

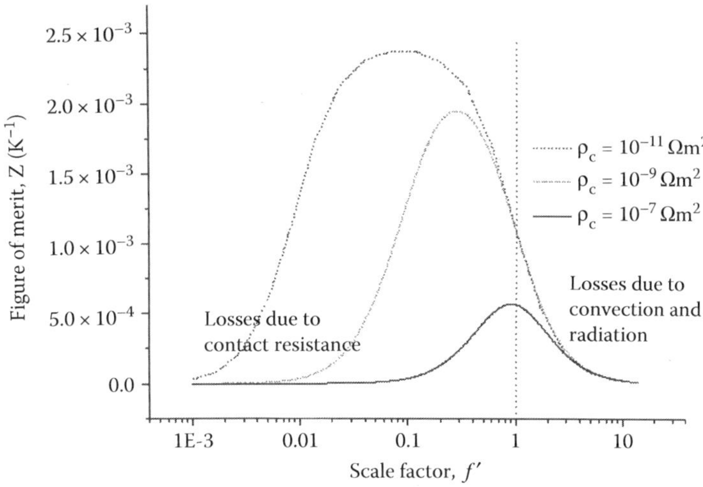

FIGURE 19.17 Effect of scaling the device in substrate plane. The height of device and support membrane are constant ($H = 10$ μm and $H_m = 10$ μm), while other dimensions are scaled by the value on horizontal axis.

Since the losses by radiation and convection are less relevant in lower-dimension devices, the figure of merit increases when the device is smaller. However, a low contact resistivity must be ensured, to keep this higher figure of merit. Considering both effects, an optimum dimension exists where the figure of merit is maximized [40].

19.4.1 Cooling Applications

A microcooler was fabricated on flexible polyimide Kapton© substrate. The polyimide substrate is excellent for thermoelectric microcoolers in the lateral configuration because of its small thermal conductivity of 0.15 $Wm^{-1}K^{-1}$ and its thermal expansion coefficient (about 20×10^{-6} K^{-1}) of the same order of magnitude as that of the thermoelectric films, thus reducing thermal stresses induced during cooling down from the temperature of deposition (200–300°C). Figure 19.15 shows a thermoelectric film deposited on top of a polyimide substrate. Since Bi_2Te_3 and Sb_2Te_3 adhesion is higher on Kapton films than on nickel metal pads, the use of top contacts process (as presented in Figure 19.14) avoids the need of depositing additional layers to promote adhesion of thermoelectric films.

The performance of the microcooler was analyzed [41] by the use of a thermal image map generated with a microscope equipped with an infrared image sensor. An image [4] under vacuum was obtained with the device excited with a 4 mA current, and cold and hot sides were clearly identified (Figure 19.18). A temperature difference of 5°C between the hot and the cold sides was measured under vacuum (10 mtorr) and 4°C in still air conditions.

From the thermal image, it is possible to conclude that thermal contact between the thermoelectric elements and the metal pads is poor since the thermal gradient on this region is quite high, compared with the gradient observed along metals or along thermoelectric elements. The 5°C hot–cold-side temperature difference obtained was lower than the value expected from the simulations. This difference is due to electrical contact resistance, which is much higher than expected. Contact resistance in this device was of the same order of magnitude as the resistance of the thermoelectric elements (2 Ω). A small rectifying barrier voltage (<80 μV) was also measured on metal–Bi_2Te_3 contacts.

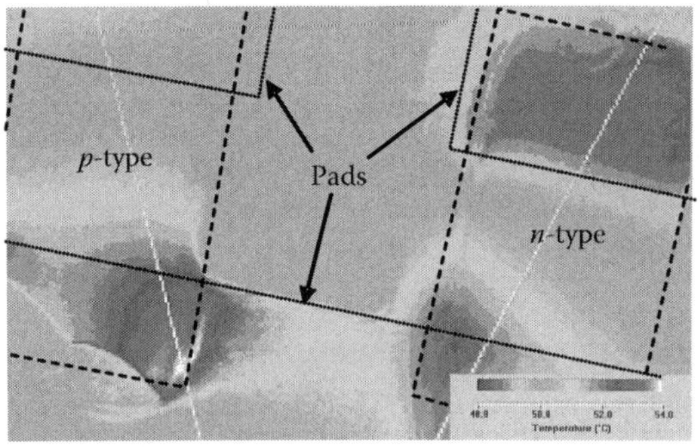

FIGURE 19.18 **(See color insert.)** Thermal image of n-type and p-type thermoelectric elements, powered with 4 mA current, under vacuum. (Adapted from L.M. Goncalves et al., *Journal of Micromechanics and Microengineering*, 17, 2007, S168–S173.)

19.4.2 Energy-Harvesting Applications

This same planar device can be used to implement a thermoelectric generator. Previous work [42–44] has demonstrated the maximum amount of thermal energy that can be removed from the human body in a wearable thermal-generator without compromising the comfort, and maximizing the thermoelectric conversion. A thermal resistance of 100–300 $KW^{-1}cm^2$ is expected in the wrist, where thermal flow can be converted with a thermo-bracelet. Also, temperatures between 27 and 36°C can be found on different parts of body. The ambient air temperature and thermal-converter to air thermal resistance also limits the maximum power available. Thermal resistance <50 $KW^{-1}cm^2$ can be achieved with a pin-heatsink. Maximum power output is obtained when the thermal resistance of the thermoelectric generator is equal to the sum of human-body and heat-sink thermal resistances [45,46]. A thermal resistance >200 $KW^{-1}cm^2$ is desirable in the thermoelectric converter. Figure 19.19 shows the open-circuit voltage and maximum power that can be obtained in a 1 cm^2 Bi_2Te_3–Sb_2Te_3 thermoelectric generator as a function of the length of the thermoelectric column (length L in Figure 19.16), in a human-body generator, when a temperature difference of 10°C (air to body) is available, with 4000 pairs of p–n junctions.

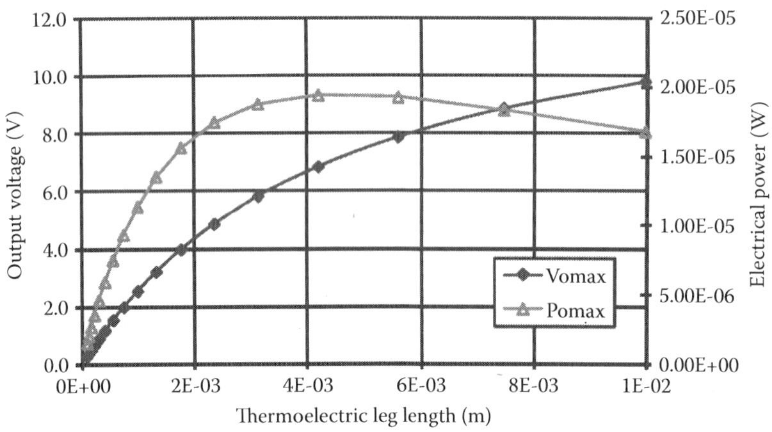

FIGURE 19.19 Open-circuit voltage and maximum power that can be obtained in a 1 cm^2 Bi_2Te_3–Sb_2Te_3 thermoelectric generator, as a function of length of the column (L), in a human-body generator, when a temperature difference of 10°C (air to body) is available.

Since each thermoelectric junction of Bi_2Te_3–Sb_2Te_3 can deliver an output voltage of 300 μVK^{-1}, more than 4000 junctions are necessary to obtain an output voltage (without load) of 10 V, under a temperature difference of 10°C, when body and heat-sink thermal resistances are considered. To obtain the desired thermal resistance of 200 $KW^{-1}cm^2$, the thermoelectric elements should have a length of about 4 mm. The array of Figure 19.16 can be used as a human body energy-harvesting device, since many junctions can be stacked in a single device and fabrication process allows the required 4 mm length. Moreover, one of the ribs of Figure 19.16 can contain a solid-state rechargeable battery and another rib the electronic circuit to extract maximum power from the thermoelectric generator and control the charge of the battery [45–48]. Since many of wireless sensors are powered in a peak basis (e.g., the transmission of data needs much more current than standby or receiving mode) and the temperature gradient could not always be present, the energy is stored in a rechargeable thin-film battery of Li-ion type (integrated in the system). Ultralow power electronics performs DC–DC rectification with a variable conversion factor (using MPPT, maximum power point tracking algorithm) and recharges the battery on optimal conditions. Electronic circuits to perform this task are commercial available [49].

Acknowledgments

The author would like to thank the Portuguese Foundation for Science and Technology (FCT/PTDC/EEA-ENE/66855/2006) for funding this work. The author also thanks Professor F. Völklein (University of Applied Sciences Wiesbaden) for thermal conductivity measurements, D.M. Rowe and Gao Min (University of Cardiff) for guidelines in this work, P. Alpuim, J.H. Correia, C. Couto, A.G. Rolo, J.P. Carmo, M.F. Silva, and J.F. Ribeiro for their help in this work and meaningful discussions.

References

1. L.W. da Silva and M. Kaviany, Miniaturized thermoelectric cooler, In *ASME Organisation ASME (ed.), Proceedings 2002 ASME International Mechanical Engineering Congress & Exhibition, IMECE02-32494*, New Orleans, LA, USA.
2. H. Zou, D.M. Rowe, and S.G.K. Williams, Peltier effect in a co-evaporated Sb_2Te_3(P)-Bi_2Te_3(N) thin film thermocouple, *Thin Solid Films*, 408, 2002, 270–274.
3. L.W. da Silva, M. Kaviany, and C. Uher, Thermoelectric performance of films in the bismuth-tellurium and antimony-tellurium systems, *Journal of Applied Physics*, 97, 2005, 114903.
4. L.M. Goncalves, J.G. Rocha, C. Couto, P. Alpuim, G. Min, D.M. Rowe, and J.H. Correia, Fabrication of flexible thermoelectric microcoolers using planar thin-film technologies, *Journal of Micromechanics and Microengineering*, 17, 2007, S168–S173.
5. D. Kim, E. Byon, G. Lee, and S. Cho, Effect of deposition temperature on the structural and thermoelectric properties of bismuth telluride thin films grown by co-sputtering, *Thin Solid Films*, 510, 2006, 148–153.
6. H. Böttner, J. Nurnus, A. Gavrikov, G. Kuhner, M. Jagle, C. Kunzel, D. Eberhard, G. Plescher, A. Schubert, and K.-H. Schlereth, New thermoelectric components using microsystem technologies, *Journal of Microelectromechanical Systems*, 13, 2004, 414–420.
7. L.M. Goncalves, C. Couto, P. Alpuim, A.G. Rolo, F. Völklein, and J.H. Correia, Optimization of thermoelectric properties on Bi_2Te_3 thin films deposited by thermal co-evaporation, *Thin Solid Films*, 518, 2010, 2816–2821.
8. G.J. Snyder, J.R. Lim, C.-K. Huang, and J.-P. Fleurial, Thermoelectric microdevice fabricated by a MEMS-like electrochemical process, *Nature Materials*, 2, 2003, 528–531.
9. L.M. Goncalves, J. Rocha, J. Correia, and C. Couto, Control of the deposition ratio of Bi_2Te_3 and Sb_2Te_3 in a vacuum evaporator for fabrication of Peltier elements, *2006 IEEE International Symposium on Industrial Electronics, IEEE*, 2006, vol. 4, pp. 2773–2777.

10. L.M. Goncalves, P. Alpuim, A.G. Rolo, and J.H. Correia, Thermal co-evaporation of Sb_2Te_3 thin-films optimized for thermoelectric application, *Thin Solid Films*, 519, 2011, 4152–4157.
11. J.R. Lim, G.J. Snyder, C.-K. Huang, J.A. Herman, M.A. Ryan, and J.-P. Fleurial, Thermoelectric microdevice fabrication process and evaluation at the Jet Propulsion Laboratory (JPL), *Twenty-First International Conference on Thermoelectrics, 2002. Proceedings ICT '02*, 2002, pp. 535–539.
12. A. Giani, A. Boulouz, F. Pascal-Delannoy, A. Foucaran, E. Charles, and A. Boyer, Growth of Bi_2Te_3 and Sb_2Te_3 thin films by MOCVD, *Materials Science and Engineering: B*, 64, 1999, 19–24.
13. A. Giani, F. Pascaldelannoy, A. Boyer, A. Foucaran, M. Gschwind, and P. Ancey, Elaboration of Bi_2Te_3 by metal organic chemical vapor deposition, *Thin Solid Films*, 303, 1997, 1–3.
14. A. Boulouz, A. Giani, F. Pascal-Delannoy, M. Boulouz, A. Foucaran, and A. Boyer, Preparation and characterization of MOCVD bismuth telluride thin films, *Journal of Crystal Growth*, 194, 1998, 336–341.
15. F. Völklein, Transport properties of flash-evaporated $(Bi_{1-x}Sb_x)_2Te_3$ films: Optimization of film properties, *Thin Solid Films*, 187, 1990, 253–262.
16. A. Foucaran, A. Sackda, A. Giani, F. Pascal-Delannoy, and A. Boyer, Flash evaporated layers of $(Bi_2Te_3–Bi_2Se_3)$(N) and $(Bi_2Te_3–Sb_2Te_3)$(P), *Materials Science and Engineering B*, 52, 1998, 154–161.
17. E. Kessler, A. Ihring, V. Baier, A. Franke, and U. Dillner, Thin-film infrared thermopile sensors with thermoelectric high-effective materials combination, *Proceedings of the 11th International Conference SENSOR, Nurnberg*, 2003, pp. 249–254.
18. M. Stordeur and I. Stark, Low power thermoelectric generator-self-sufficient energy supply for micro systems, *Proceedings ICT '97. XVI International Conference on Thermoelectrics*, IEEE, 1997, pp. 575–577.
19. H. Zou, D.M. Rowe, and G. Min, Growth of p- and n-type bismuth telluride thin films by co-evaporation, *Journal of Crystal Growth*, 222, 2001, 82–87.
20. H. Böttner, J. Nurnus, A. Schubert, and F. Volkert, New high density micro structured thermogenerators for stand alone sensor systems, *26th International Conference on Thermoelectrics*, IEEE, 2007, pp. 306–309.
21. L.M. Goncalves, P. Alpuim, and J.H. Correia, Fabrication of thermoelectric devices by applying microsystems technology, *Journal of Electronic Materials*, 39, 2010, 1516–1521.
22. H. Scherrer and S. Scherrer, Bismuth telluride, antimony telluride, and their solid solutions, *CRC Handbook of Thermoelectrics*, D.M. Rowe (ed.), CRC Press, Boca Raton, FL, 1987, pp. 211–237.
23. F. Völklein and T. Starz, Thermal conductivity of thin films-experimental methods and theoretical interpretation, *Proceedings ICT '97. XVI International Conference on Thermoelectrics*, IEEE, 1997, pp. 711–718.
24. C.M. Bhandari, Minimizing the thermal conductivity, *CRC Handbook of Thermoelectrics*, D.M. Rowe (ed.), CRC Press, Boca Raton, FL, 1987, pp. 55–65.
25. R. Venkatasubramanian, E. Siivola, T. Colpitts, and B. O'Quinn, Thin-film thermoelectric devices with high room-temperature figures of merit, *Nature*, 413, 2001, 597–602.
26. M. Stordeur, H.T. Langhammer, H. Sobotta, and V. Riede, Valence band structure of $(Bi_{1-x}Sb_x)$2Te3 single crystals, *Physica Status Solidi (B)*, 104, 1981, 513.
27. Joint Committee on Powder Diffraction Standards, *Powder Diffraction File*, ASTM, Philadelphia, PA, 1967.
28. http://www.oxford-instruments.com/products/etching-deposition-growth/processes/etch-processes/Pages/bi2te3.aspx, Oxford Intruments.
29. W. Qu, M. Ploetner, and W.J. Fischer, Microfabrication of thermoelectric generators on flexible foil substrates as a power source for autonomous microsystems, *Journal of Micromechanics and Microengineering*, 11, 2001, 146.
30. L.W. da Silva and M. Kaviany, Fabrication and measured performance of a first-generation micro-thermoelectric cooler, *Journal of Microelectromechanical Systems*, 14, 2005, 1110–1117.

31. B. Huang, C. Lawrence, A. Gross, G.-S. Hwang, N. Ghafouri, S.-W. Lee, H. Kim, C.-P. Li, C. Uher, K. Najafi, and M. Kaviany, Low-temperature characterization and micropatterning of coevaporated Bi_2Te_3 and Sb_2Te_3 films, *Journal of Applied Physics*, 104, 2008, 113710.
32. G.S. Hwang, A.J. Gross, H. Kim, S.W. Lee, N. Ghafouri, B.L. Huang, C. Lawrence, C. Uher, K. Najafi, and M. Kaviany, Micro thermoelectric cooler: Planar multistage, *International Journal of Heat and Mass Transfer*, 52, 2009, 1843–1852.
33. C. Shafai and M.J. Brett, A micro-integrated Peltier heat pump for localized on-chip temperature control, *Proceedings of 1996 Canadian Conference on Electrical and Computer Engineering*, 1996, pp. 88–91.
34. C. Shafai and M. Brett, Optimization of BiTe thin films for microintegrated Peltier heat pumps, *Journal of Vacuum Science and Technology A: Vacuum, Surfaces, and Films*, 15, 1997, 2798.
35. S. Sedky, A. Kamal, M. Yomn, H. Bakr, R. Ghannam, V. Leonov, and P. Fiorini, Bi_2Te_3 as an active material for MEMS based devices fabricated at room temperature, *Solid-State Sensors, Actuators and Microsystems Conference, 2009. TRANSDUCERS 2009. International*, IEEE, 2009, pp. 1035–1038.
36. L.M. Goncalves, C. Couto, P. Alpuim, and J.H. Correia, Thermoelectric micro converters for cooling and energy-scavenging systems, *Journal of Micromechanics and Microengineering*, 18, 2008, 064008.
37. P. Alpuim, L.M. Goncalves, E. Marins, T. Viseu, S. Ferdov, and J. Bourée, Deposition of silicon nitride thin films by hot-wire CVD at 100°C and 250°C, *Thin Solid Films*, 517, 2009, 3503–3506.
38. D. Wijngaards, S. Kong, M. Bartek, and R. Wolffenbuttel, Design and fabrication of on-chip integrated polySiGe and polySi Peltier devices, *Sensors and Actuators A: Physical*, 85, 2000, 316–323.
39. G. Min and D.M. Rowe, Cooling performance of integrated thermoelectric microcooler, *Solid-State Electronics*, 43, 1999, 923–929.
40. L.M. Goncalves, R.P. Rocha, J.P. Carmo, and J.H. Correia, Solid-state microcoolers and thermal energy harvesting microsystems, *2009 35th Annual Conference of IEEE Industrial Electronics*, Nov. 2009, pp. 4040–4044.
41. L.M. Goncalves, J. Rocha, C. Couto, P. Alpuim, and J. Correia, On-chip array of thermoelectric Peltier microcoolers, *Sensors and Actuators A: Physical*, 145–146, 2008, 75–80.
42. V. Leonov, P. Fiorini, S. Sedky, T. Torfs, and C. Van Hoof, Thermoelectric mems generators as a power supply for a body area network, *The 13th International Conference on Solid-State Sensors, Actuators and Microsystems, TRANSDUCERS '05.*, IEEE, 2005, pp. 291–294.
43. R.J.M. Vullers, R. van Schaijk, I. Doms, C. Van Hoof, and R. Mertens, Micropower energy harvesting, *Solid-State Electronics*, 53, 2009, 684–693.
44. V. Leonov and R. Vullers, Thermoelectric generators on living beings, *Proceedings of the 5th European Conference on Thermoelectrics (ECT 07)*, 2007, pp. 47–52.
45. J.P. Carmo, R.P. Rocha, A.F. Silva, L.M. Goncalves, and J.H. Correia, Integrated thin-film rechargeable battery in a thermoelectric scavenging microsystem, *2009 International Conference on Power Engineering, Energy and Electrical Drives*, Mar. 2009, pp. 359–362.
46. J.P. Carmo, L.M. Goncalves, and J.H. Correia, Thermoelectric microconverter for energy harvesting systems, *IEEE Transactions on Industrial Electronics*, 57, 2010, 861–867.
47. J.P. Carmo, L.M. Goncalves, R.F. Wolffenbuttel, and J.H. Correia, A planar thermoelectric power generator for integration in wearable microsystems, *Sensors and Actuators A: Physical*, 161, 2010, 199–204.
48. J.P. Carmo, J.F. Ribeiro, M.F. Silva, L.M. Goncalves, and J.H. Correia, Thermoelectric generator and solid-state battery for stand-alone microsystems, *Journal of Micromechanics and Microengineering*, 20, 2010, 085033.
49. http://www.cymbet.com/, Solid-State Energy Storage for Embedded Energy, Power Back-up and Energy Harvesting.

20

Thermoelectric Materials, Measurements, and Opportunities for Energy Harvesting

Patrick J. Taylor
U.S. Army Research Laboratory

20.1 Army-Related Opportunities for Energy Harvesting

Energy harvesting will be an increasingly important strategy for the U.S. Department of Defense (DoD) to address concerns regarding future power and energy. On average, the daily DoD consumption of oil exceeds 3×10^5 barrels.[1] The concern for the DoD is not just that the world's reserves of oil are finite; it is also that future military assets will be consuming dramatically more fuel, and supplying resources to DoD assets, especially those in remote duty stations, in the form of fuel and batteries is highly inefficient and alarmingly costly.

The first logistical problem is fuel supply. The base cost per gallon of JP-8 fuel is roughly $3 per gallon.[2] However, the more important metric is the "fully burdened cost" which includes the base cost, the costs of procurement, shipping, protection, delivery, and so on, and ranges from the usual $42 per gallon to possibly higher.[1] Making things worse, the fuel utilization of many DoD assets is relatively poor. One metric for comparing the fuel utilization is the common miles-per-gallon ratio, which for both the army's M-1 Abrams and Bradley-Fighting-Vehicle is <1 mile/gallon.[3] Clearly, if fuel utilization on these assets were improved, demand would be reduced, and the fully burdened costs would be lower.

Another army logistical problem is supplying batteries. The logistics of supplying batteries is worse than supplying fuel because while fuel is fully consumed, batteries are not. Fully charged batteries must be supplied with one shipment, and discharged battery casings must be removed with a second. Batteries are already very expensive assets. The logistics burden was estimated as adding >10% to the base cost of the battery.[4] However, these burdens pale in comparison with the striking human cost: on average, 1 U.S. casualty occurs for every 24 logistical supply trips,[5] so the far greater concern is the human risk for having to make trips in duplicate.

Because of these logistical problems, there are opportunities for energy harvesting with special focus on thermoelectric power generation. Thermoelectric power is of interest for these roles to support operational, intelligence, and tactical applications because thermoelectric devices can be highly covert, man-portable, ruggedized, and long-lived with very low amortized cost. Because of these desirable attributes, thermoelectric power generation is poised to significantly impact army energy needs by improving the fuel utilization on vehicles and by offering a compelling alternative to batteries.

As an example, energy harvesting can improve the usage of fuel on army vehicles. Assuming 25% of the energy in fuel goes to vehicular motion, 40% is lost to exhaust heat, 30% is transferred to the coolant, and 5% is absorbed by friction losses and drive-train parasitics.[6] If thermoelectric conversion could capture some of the energy from the exhaust and convert that into electrical energy, the fuel efficiency of vehicles may be improved by, say, 10%, and the average driving range for an individual M-1 Abrams could extend 25 miles or more.

Another example is the self-powered mobile kitchen units that use thermoelectric devices coupled to the waste heat from the stove to produce all the operational electrical power required for the mobile kitchen. For this application, batteries are not needed. Dr. Donald Pickard at the army's Natick Soldier Center has led the development of this application. For this example, the logistical supply demand for batteries to drive the electronics and control systems for the kitchen unit is eliminated entirely.

A third example is man-portable electrical power supplies. JP-8 fuel is energy-rich with a specific energy value of roughly 13,000 W h/kg. The heat released during combustion of JP-8 in a small hand-held burner can be directly converted to electrical power using thermoelectric devices. A thermoelectric device having a mere 5% efficiency could harvest the equivalent of >600 W h/kg. This equivalent specific energy far exceeds that of common batteries: a standard lithium-ion battery has <200 W h/kg, rechargeable batteries such as those based on nickel-cadmium or lead-acid chemistries have <100 W h/kg.

Overall, there are rich opportunities for thermoelectric power to improve army capabilities. Two valuable long-term goals are to show that energy harvesting using thermoelectrics can lead to improved fuel efficiency on vehicles, and can make batteries obsolete. Some of the technology needed to enable these desirable attributes is available while some fundamental research is still needed. In this present description, three focus areas of research are detailed that provide general support to advance thermoelectrics in an energy-harvesting role for the army: (1) opportunities for new thermoelectric materials, (2) developing highly accurate isothermal thermoelectric measurements, and (3) validating those measurements using device performance.

20.2 Opportunities for Developing New Materials

Thin-film heterostructures such as superlattices were proposed as an attractive route to obtain improved thermoelectric materials properties.[7] In addition to the effects from quantum confinement on the density of states, superlattices can significantly alter the phonon dispersion.[8,9] Fortunately for thermoelectrics, the practical reduction of thermal conductivity for superlattices is reported to be even greater than expected.[9] PbTe, a simple classic example of IV–VI materials suitable for thermoelectric power, can be used to develop such thin-film superlattices. However, one of the limiting issues for implementing the potential benefit of heterostructures with PbTe is that the energy gap is small, and so obtaining sufficient band offsets to demonstrate quantum-confinement at relevant temperatures is particularly challenging. Alternatively, larger band-gap materials can serve as barriers with PbTe to form heterostructures that can provide confinement, but at the expense of degraded carrier transport and reduced electrical conductivity.[10] Other complications include the need to consider thermal transport through barriers and tunneling to properly describe the properties of quantum-confined heterostructures.[11]

At MIT Lincoln Laboratory, we had significant focus on PbTe-based superlattices using molecular beam epitaxy (MBE). Our approach was to investigate heterostructure superlattice materials based on PbTe and $PbSe_{0.98}Te_{0.02}$. The band-gaps for PbTe and $PbSe_{0.98}Te_{0.02}$ are roughly the same, and the conduction band offset between PbTe and $PbSe_{0.98}Te_{0.02}$ is too small for any appreciable room-temperature quantum-confinement, so these heterostructures are more correctly termed as "nano-dot" as opposed to "quantum-dot." Because the lattice mismatch between PbTe and $PbSe_{0.98}Te_{0.02}$ exceeds 5%, the $PbSe_{0.98}Te_{0.02}$ layers grow by a Stanski–Krastanow mode and form nearly perfect self-assembled nanodot layers.[12]

One fortuitous phenomenon during (111) growth is that PbTe acts to replanarize the surface after the deposition of the nanodot layer, so that a nano-dot-superlattice (NDSL) composed of discrete $PbSe_{0.98}Te_{0.02}$ nano-dot/wetting layers, separated by flat PbTe matrix layers, is obtained. When the nanodot and planarizing layers are cyclically repeated, a superlattice heterostructure is obtained that can have essentially unlimited thickness. Prepared this way, superlattice materials have an enormous amount of internal dissimilar interface area and that is expected to effect reduction in thermal conductivity well below that from alloying alone.[13] To demonstrate the properties in as close to bulk form as possible, we investigated pseudo-bulk NDSL heterostructures having thickness roughly 100 μm, and we used BaF_2 substrates because NDSL samples can easily be removed by completely dissolving the BaF_2 in aqueous solutions.

A previously unpublished cross-sectional transmission electron microscopy (TEM) image taken from a thick NDSL using $20\bar{2}$ diffraction conditions to emphasize the $PbSe_{0.98}Te_{0.02}$ dots within the PbTe matrix is shown in Figure 20.1. The Moiré fringes are the periodic bright-and-dark parallel lines caused by the constructive-and-destructive interference between the lattice spacing of PbTe (d_1) and that of $PbSe_{0.98}Te_{0.02}$ (d_2). As given in Equation 20.1, the Moiré fringe spacing, D, is determined by d_1, d_2, and θ, the rotation angle between d_1 and d_2.

$$D = \frac{d_1 d_2}{\sqrt{\left(d_1^{\,2} + d_2^{\,2} - 2d_1 d_2 \cos(\theta)\right)}} \tag{20.1}$$

For the present NDSL case, θ is null because the dots are epitaxial, so using the respective bulk d_{220} spacings of PbTe and $PbSe_{0.98}Te_{0.02}$, D can take on any value that ranges from about 7.7 nm for complete lattice relaxation, to about 16 nm for partial relaxation, and "infinite" for the elastically strained case.[14]

One important observation from the Moiré fringe measurement here is that there is a somewhat broad, irregular distribution of $PbSe_{0.98}Te_{0.02}$ dot structures within the PbTe matrix. There is no singular D value. A second observation is that the smaller-sized, roughly 8 nm diameter, nanodots on the lower left appear to have no fringes. This means that those nanodots are coherent with the PbTe matrix and are fully strained. A third feature to note is that the larger-sized nanodots located in the center and along the left-hand side have diameter >35 nm, and measured D values of 7.6 nm which corresponds to complete relaxation. A fourth observation is that some nano-dots appear to have merged together, forming a pseudo-second phase inclusion as shown near the top left-hand side.

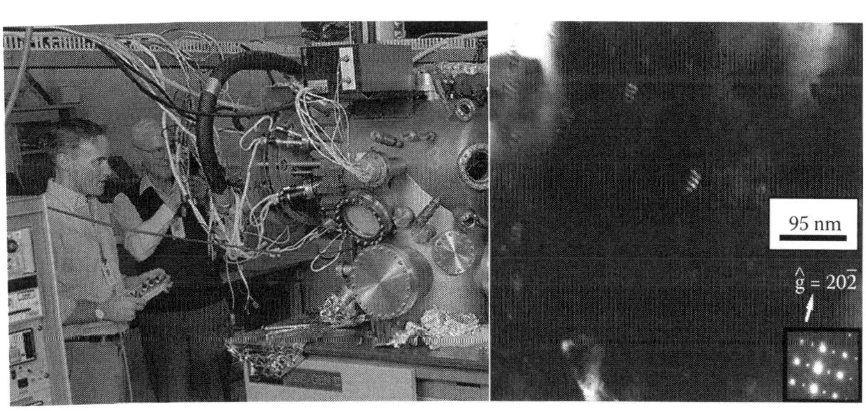

FIGURE 20.1 MBE growth of PbSeTe/PbTe NDSL and cross-sectional TEM view of NDSL.

The inherent periodicity of a basic superlattice heterostructure is well known to disrupt phonon transport.[8,9,15] However, additional periodic structure(s) beyond the basic superlattice would be capable of scattering more phonons and further disrupting phonon transport. So, the observation of wide distribution of Moiré fringes suggests two additional phonon scattering mechanisms: (1) enhanced scattering from the interatomic discontinuities and immobile strain-fields of misfit dislocations surrounding the nanodots that have relaxed (or coherency strain-field for dots that have not) and (2) Bragg interference effects between the wave nature of the phonons and the additional periodicity corresponding to the misfit spacing.

The velocity of sound waves in PbTe is 3080 m/s.[16] So, any periodic structures having a length dimension of the order of 10 nm will have destructive ($m\lambda/2$) Bragg interference throughout the terahertz and sub-terahertz range of the phonon dispersion, where m is an integer and λ is the effective phonon wavelength. Figure 20.1 shows that the relaxation of lattice strain between the nano-dots and matrix is not uniformly complete, so the Moiré fringes have periodic length dimensions ranging from 8 nm and larger. Therefore, the broad range of misfit dislocation spacings destructively interfere with broad portions of the phonon spectrum; primarily the normal longer-wavelength acoustic. The interatomic disregistry of the dislocated interfaces also disrupts the polar and nonpolar optical, phonon modes, and contribute to reduced thermal conductivity.

The strongest evidence to support enhanced phonon scattering and reduced thermal conductivity is a device result that is consistent with a reduced thermal conductivity. Such evidence was obtained and reflected in results from a single-leg cooling device using a 96 μm film and a gold wire that exhibited larger maximum temperature difference (ΔT) than a cleaved single crystal[17] of n-type $(Bi_2Te_3)_{0.90}(Sb_2Te_3)_{0.05}(Sb_2Se_3)_{0.05}$ that had a well-known ZT_{avg} of 0.9, where Z has its usual ($\alpha^2/\rho\kappa$) form and T_{avg} is the average temperature between the hot side and cold side. The experimental results are detailed fully elsewhere.[18] The salient point to make here is that because the Seebeck coefficient (α), the electrical resistivity (ρ), and device performance of the NDSL were independently known,[19] the physical device performance was used to conclude that there is no other way such a dramatic device cooling can be explained other than reduced thermal conductivity. So, one opportunity for developing new materials is to engineer or incorporate nanostructures that have (1) large internal dissimilar surface area and (2) dimensional periodicity causing destructive interference of phonons and disrupted phonon transport.

A second opportunity for new materials development is through novel bulk materials because bulk materials are generally more appropriate for practical thermoelectric devices than thin-film materials. Narsingh B. Singh and his team at Northrup-Grumman Corporation have initiated a collaboration to investigate crystalline bulk $(Tl,Bi)Te_2$ material with the objective of showing that an pseudo-binary alloy of III-telluride and V-telluride materials will possess physical properties similar to IV-telluride (PbTe) materials: Jensen and colleagues describe $(Tl,Bi)Te_2$ as a "pseudo-PbTe."[20] Past work on that material indicates that unoptimized electrical resistivity and Seebeck coefficient are interesting, but likely correspond to either heavily degenerate or semimetallic behavior.[21] There is very sparse literature data[22] on thermal conductivity of $(Tl,Bi)Te_2$, so new bulk thermal conductivity data are presented here and are used to highlight the new measurement technique for thermal conductivity.

20.3 Isothermal Thermoelectric Measurements

One of the most critical capabilities in thermoelectric materials research for energy harvesting is accurate measurement of the three principal thermoelectric properties: α, ρ, and κ. These parameters directly control the magnitudes of the important heat flows: Q_Π, the Peltier heat; Q_ρ, the Joule heat; and Q_κ, the thermally conducted heat. There are challenges associated with measuring the three parameters because, in addition to the usual electronics errors (e.g., noise, spurious voltages), the heat flows used must have reached time-invariant steady state, and there are secondary parasitic heat flows that can cause significant error. The following examples are taken from different materials to detail the materials-agnostic flexibility of the new method.

The well-known definition of the Seebeck coefficient is

$$\alpha = \lim_{\Delta T_{ss} \to 0} \frac{\Delta V_{ss}}{\Delta T_{ss}} \tag{20.2}$$

Here, we introduce a simple new method for determining the Seebeck coefficient. In more common experimental approaches, one surface of the sample is heated directly by rigid mechanical contact, and the temperature of the opposite side is held constant by a rigid heat sink. The heat imposes a temperature difference from which voltage and temperature data are used to determine the Seebeck coefficient. However, for the new technique described here, noncontact radiative heat, Q_{Rad}, is used to control the temperature of the exposed top of the sample image shown in Figure 20.2 with respect to the opposite side, shown attached to a 1 in^3 copper heat sink. In this new measurement, a range of steady-state temperature difference values, ΔT_{ss}, are imposed by noncontact thermal radiation and the corresponding steady-state open-circuit voltage data, ΔV_{ss}, are obtained. By plotting that data and determining the slope as per Equation 20.2, an accurate measurement of the Seebeck coefficient can be obtained. By using the slope of the line, most errors associated with the thermocouples (e.g., the thermocouple calibration error, the cold-junction offset) are reduced or eliminated completely. As the example in Figure 20.2 illustrates, the ΔV_{ss} data across the shown *p*-type sample of PbTe are plotted as a function of ΔT_{ss}. The slope of the line yields a Seebeck coefficient of 247.3 μV/K.

The Seebeck coefficient controls the magnitude of the Q_Π that is either absorbed or liberated at the junction held at temperature T_j, where electrical current "*I*" is passing, so accurate measurement of α is the key step to determining Q_Π: $Q_\Pi = \alpha T_j(I)$. Without changing the experimental setup, an immediate second step which depends on knowing Q_Π can yield a simultaneous measurement of both ρ and κ. To determine these two parameters, the exposed junction is cooled using negative Peltier heat ($-Q_\Pi$) by the forced passage of electrical current using a secondary, independent electrical power supply. When the negative Peltier heat exactly equals the noncontact radiative heat, then ΔT_{ss} converges exactly to zero which we term as the "special isothermal condition." For the special isothermal condition, $\Delta T_{ss} = 0$, and $Q_{Rad} \equiv -Q_\Pi$. In this case, the ΔT_{ss} along the length of the lead wires (e.g., between the sample and the meter, the thermocouples) goes to zero causing parasitic heat losses to become negligible.

Electrical resistivity using direct electrical current is usually problematic for thermoelectric materials because the electrical current causes both cooling and heating at locations of electrical contact, causing large Seebeck voltages and significant error for ρ. However, for the isothermal case, by definition there is no error from Seebeck voltage. As long as $\Delta T_{ss} = 0$, the electrical resistivity can be determined using direct electrical current. Provided that the electrical resistance from the electrical contact resistivity (R_c) is negligible compared to the resistance of the material itself, that is, $R_c \ll \rho l$, then given that the

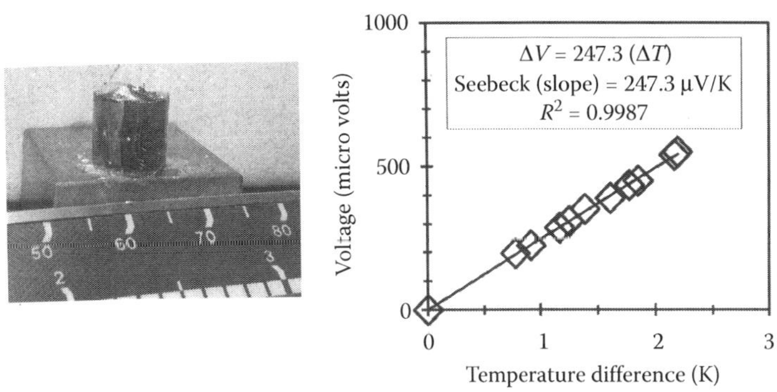

FIGURE 20.2 Sample appearance and Seebeck measurement using radiative heat.

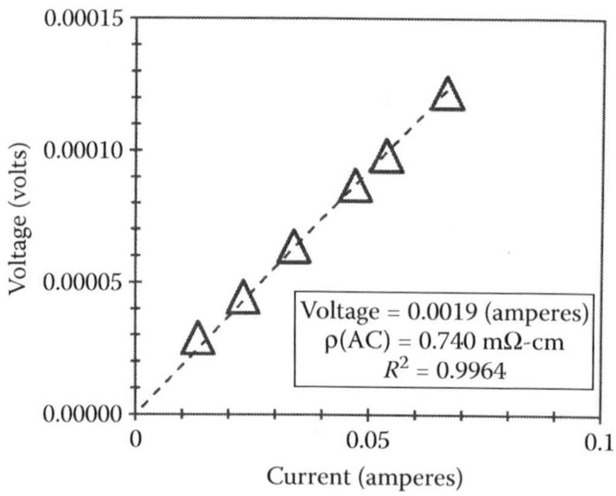

FIGURE 20.3 Comparison of two-point and four-point electrical resistivity measurements.

cross-sectional area (A) and length (l) of the sample are known, this isothermal condition yields a special two-point measurement of ρ.

As an example, the electrical resistivity of a separate n-type $(Bi_2Te_3)_{90}(Sb_2Te_3)_5(Sb_2Te_3)_5$ material using a four-point probe slope technique yielded 0.74 mΩ cm, as shown in Figure 20.3 and that measured on the exact same sample using the two-point probe yielded 0.76 mΩ cm. Those data are in agreement to better than 2%, and are not corrected for contact resistivity.

To determine κ, the electrical current, and hence cooling, is stopped, and so Q_{Rad} causes the steady-state temperature difference to rewarm back to some finite, nonzero ΔT_{ss}. The thermally conducted heat Q_κ is now exactly equal to Q_{Rad}. In this new steady-state case, κ can then be determined through Equation 20.3, the usual one-dimensional embodiment of Fourier's law:

$$Q_{Rad} = Q_\kappa = \kappa\left(\frac{A}{\ell}\right)(\Delta T_{ss}) \tag{20.3}$$

from which Q_{Rad}, ΔT_{ss}, and the (A/l) geometry ratio can yield a direct κ measurement. Present experience has been that simply averaging several single-point measurements can yield reasonably accurate data for most purposes, like quality screening. However, highly accurate κ measurements can be obtained using the sample geometry and the slope of Q_{Rad} data as a function of ΔT_{ss} values. As an example, data for the measurement of κ using this new method from samples of (Tl, Bi)Te_2 are presented in Figure 20.4. The thermal conductivity of this compound was measured using the new technique described here, and the thermal conductivity was measured using a simple average of four independent

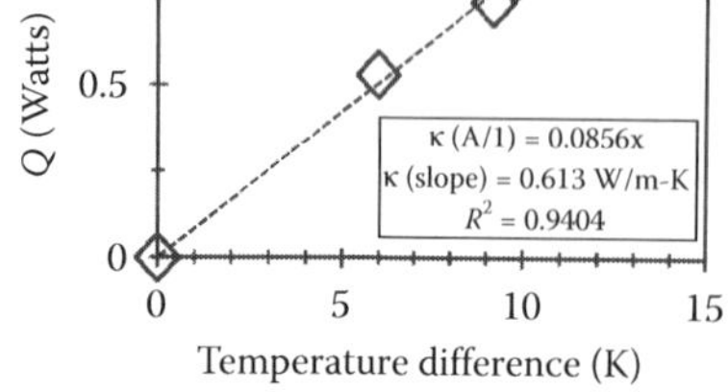

FIGURE 20.4 Measurement of thermal conductivity using the new method.

measurements: 0.606 W/m K. That measured using the slope of the $Q(\Delta T_{ss})$ data yields 0.613 W/m K. The averaged data and the slope data are in agreement to within about 1%. Note that this material is far from optimized, yet the measured thermal conductivity is at least three times lower than PbTe!

In summary, noncontact radiative heat can provide rapid, highly accurate thermoelectric measurements based on the special isothermal condition. The Seebeck coefficient is determined using a noncontact radiative heat that imposes a temperature difference. Counter-balancing the radiative heat with negative Peltier heat (cooling) yields the special isothermal condition, $\Delta T_{ss} = 0$. In the isothermal condition, two-point probe ρ data and κ data are obtained simultaneously, and the parasitics that cause error reduce to negligible levels. As a direct consequence, accurate determination of the three principal thermoelectric properties is obtained. An additional advantage is that there is no mechanical strain to cause damage/fracture, so samples can be subsequently used in devices to confirm measurements.

20.4 Validation of Measurements: Device Performance

The compelling question posed by any thermoelectric measurement is: *How does one know if it is correct?* The true test of the validity of any measurement is the performance of a device. Simple thermoelectric power generation and thermoelectric cooler devices can be constructed using the classical design and the device performance can be tested by comparison between theoretical predictions, and experimental data taken from devices. This reasoning is analogous to how device performance was used to directly demonstrate the low thermal conductivity of NDSL materials, as described earlier. Here, we review two additional test cases where experimental device performance is compared to theory: steady-state power generation, and non-steady-state Peltier cooling/heating.

For the case of simple thermoelectric power generation, a simple square module composed of two separate series-connected devices with a target T_{hot} of 500°C is considered here with the assumption that PbTe forms both the *n*- and *p*-type legs. Note that the key point for this demonstration is not to showcase novel materials. Instead, the main point is to demonstrate that device performance is the key to credible and confirmed thermoelectric measurements. For this application, *n*-type PbTe and *p*-type PbTe were initially chosen because they have adequate performance in this temperature range, and PbTe is a simple, well-established materials system with little uncertainty. Making the measurements and understanding them is the key point, while higher-efficiency materials are longer-term tasks. These materials are not optimized and it is well known that the room-temperature *n*-type ZT_{avg} can be improved to 0.52 and *p*-type PbTe to 0.46, and with those improved materials, the maximum efficiency would be about 7% for pure PbTe over $\Delta T_{ss} = 300$ K. Drs. Trivedi and Palosz at Brimrose Corporation prepared the PbTe materials by Bridgman growth using bismuth as the *n*-type dopant and thallium as the *p*-type dopant. Individual components were diced from the ingots into appropriate geometry and polished. The thermoelectric properties of the individual two *n*-type and two *p*-type components were measured before device assembly. After the final device was fully assembled and integrated onto a suitable heat sink, the measured performance was compared to theory based on individual measurements.

The equations that govern a thermoelectric power generator module's electrical power output to an impedance-matched load (P) and efficiency (Φ) depend on the open-circuit voltage (V_{oc}), the internal device resistance (R_{int}), and the heat flow in (Q_{in}) according to

$$P = \left[\frac{V_{oc}^{\ 2}}{4R_{int}}\right] \tag{20.4}$$

$$\Phi = \left[\frac{P}{Q_{in}}\right] \tag{20.5}$$

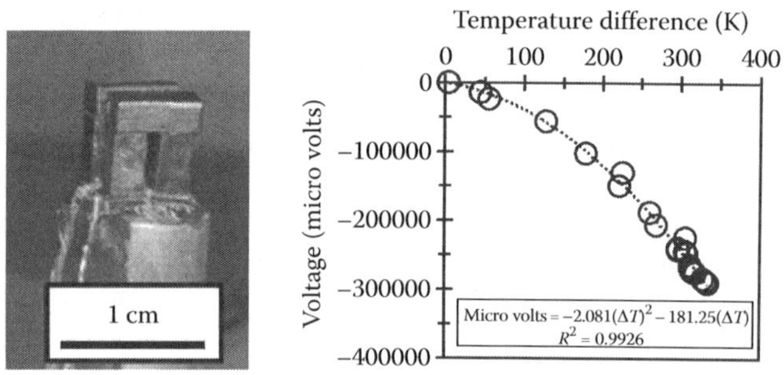

FIGURE 20.5 Comparison of experimental device performance with theoretical prediction.

Separate measurement of V_{oc}, R_{int}, and Φ allows a direct comparison between theory and experiment where V_{oc} and R_{int} are, respectively:

$$V_{oc} = (\alpha_{n1} + \alpha_{n2} - \alpha_{p1} - \alpha_{p2})\Delta T \tag{20.6}$$

$$R_{int} = (\rho_{n1} + \rho_{n2} + \rho_{p1} + \rho_{p2})\left(\frac{\ell}{A}\right) + \left(\frac{8R_c}{A}\right) \tag{20.7}$$

The assembled device shown in Figure 20.5 was mounted in a special vacuum bell jar that allows the heat sink to protrude directly out into the lab air through a feed-through, so convective cooling of the external heat sink more realistically simulates thermal management. Using a cartridge heater inside the vacuum, heat was applied to the hot side of the device and V_{oc} was measured as a function of ΔT_{ss} and a potentiometer was used to determine $1/2V_{oc}$ to obtain P and R_{int} up to $\Delta T_{ss} \geq 300$ K.

V_{oc}, R_{int}, and P are nonlinear as ΔT_{ss} increases, but near room temperature, the slope of the V_{oc} data as a function of ΔT_{ss} yields linear data that can be compared to individual measurements. In the range of V_{oc} for <100 K, the experimental device performance yields the summed absolute value of 417.56 μV/K. Comparison of the slope of the V_{oc} with ΔT_{ss} data in Figure 20.5 to the sum of the individual measurements provided in Table 20.1 yields an error of only 1.1%.

The well-known equation for Φ depends on ZT_{avg}. So, to calculate Φ, the Seebeck data can be combined with the room-temperature data which are summarized here for brevity: κ of the n-types (2.7 W/m K) and the p-types (2.2 W/m K) and ρ of the n-types (0.45 mΩ cm) and the p-types (0.50 mΩ cm). For $\Delta T_{ss} \leq 100$ K, these data yield calculated ZT_{avg} values of 0.29 and 0.38 for the n-type and p-type legs, respectively, for an overall device ZT_{avg} of 0.31, and Φ of 1.5%, as shown in Figure 20.6.

The calculation for Φ is based on measurements made assuming $\Delta T_{ss} \leq 100$ K, so the valid comparison is between calculated Φ and measured Φ for $\Delta T_{ss} \leq 100$ K. Both the calculated and measured Φ is 1.5% which confirms the calculated $ZT_{avg} = 0.31$ for near room temperature. Note that under larger ΔT_{ss},

TABLE 20.1 Individual Seebeck Coefficient Measurement Values and Their Sum

α_{n1}	−97.77 μV/K
α_{n2}	−99.23 μV/K
α_{p1}	+101.26 μV/K
α_{p2}	+114.89 μV/K
Sum	413.15 μV/K

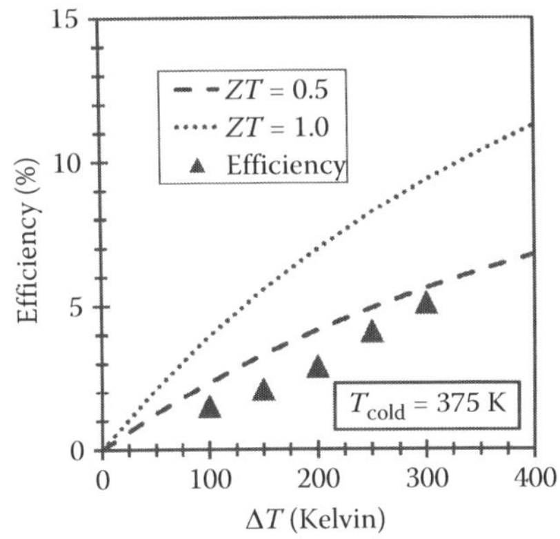

FIGURE 20.6 Comparison of experimental power output with theory and measured properties.

the output power shows quadratic increase up to the fixed $\Delta T_{ss} \leq 300$ K upper boundary condition that was fixed for this test; although the power appears to continually improve beyond that. For $\Delta T_{ss} \leq 300$ K, the measured Φ is 5% which is consistent with the well-known high-temperature increase in ZT_{avg} for PbTe; up to 0.45 for this device.

So, for this first of the three cases being considered here, the device efficiency is consistent with the thermoelectric ZT measurements. The measurement of V_{oc} agrees with the calculated V_{oc} based on the Seebeck coefficient values to within 1.1%. For $\Delta T_{ss} \leq 100$ K, the measured value of 1.5% for Φ agrees with the calculated ZT_{avg} value of 0.31. Because the measured and calculated V_{oc} based on Seebeck measurements as well as the measured and calculated device Φ are in agreement, the thermoelectric measurements are confirmed.

The second test case considered here is non-steady-state Peltier heating and Peltier cooling.[23] In the non-steady-state case, the device has temperature performance that changes with time, t, and requires knowledge of the molar amount of metal, n (moles), at the cold junction, and that metal's specific heat, C_p. A straightforward comparison between the calculated temperature and the measured temperature of the cold junction can be obtained by determining R^2, the coefficient of determination, which has its usual statistics definition. The time-dependent temperature difference $\Delta T(t)$ for either cooling or heating from the ambient temperature, T_e, depends on ΔT_{ss} according to[23]:

$$\Delta T(t) = \Delta T_{ss}\left[1 - \exp^{\frac{-\left(\frac{A}{\ell}\right)(\kappa_n+\kappa_p)-(\alpha_n-\alpha_p)I}{nC_p}}\right] \quad (20.8)$$

where

$$\Delta T_{ss} = I\left[\frac{I\left(R_c + \frac{\ell}{2A}(\rho_n + \rho_p)\right) - (\alpha_n - \alpha_p)T_e}{(\kappa_n + \kappa_p)\left(\frac{A}{\ell}\right) - (\alpha_n - \alpha_p)I}\right] \quad (20.9)$$

One device was constructed from n-type $(Bi_2Te_3)_{90}(Sb_2Te_3)_5(Sb_2Te_3)_5$ and p-type $(Sb_2Te_3)_{72}(Bi_2Te_3)_{25}(Sb_2Se_3)_3$ materials having the following independently measured properties in Table 20.2.

TABLE 20.2 Thermoelectric Measurements and Geometries for the Thermoelectric Device

Type	ρ (Ω cm)	α (μV/K)	κ (W/mK)	λ/A (1/cm)
n	0.00104	−210	0.140	0.94/0.40
p	0.00143	+243	0.132	0.96/0.37

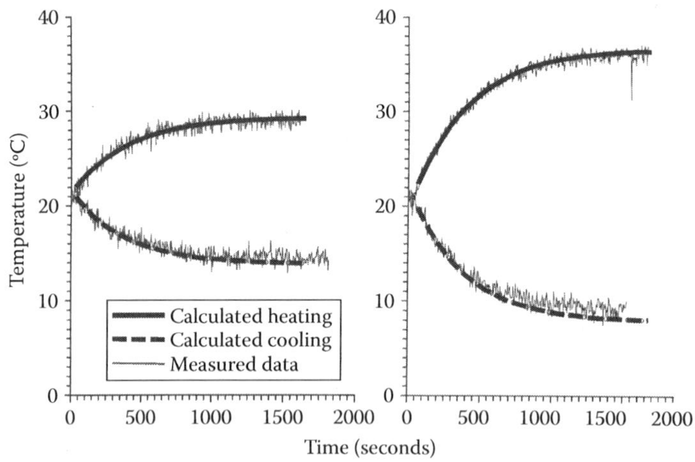

FIGURE 20.7 Comparison of experimental cooling with theory for ±8°C and ±16°C.

R_c is negligible for this device, so these properties can be substituted in the theoretical expressions for $\Delta T(t)$ and ΔT_{ss} along with the specific information about the metal cold junction: $n = 0.211$ mol and $C_p = 24.46$ J/mol-K. The device was mounted in the vacuum test stand so that convective heat flow is negligible, and the time-dependent temperature data were collected. Time starts at zero for this test when the electrical current is first switched on and supplied to the device, and ΔT_{ss} is approached after 1600 s. Two values of electrical current were used: ±0.6 and ±1.1 A. Because the materials properties are temperature dependent, the test results presented in Figure 20.7 will specifically focus on the two moderate ΔT ranges: (1) ±8°C with respect to T_e and (2) ±16°C with respect to T_e. The maximum ΔT for the device was ~70°C. For ±8°C, the experimental device measurements and theoretical predictions based on materials measurement's excellent agreement is indicated by the obtained R^2 value of 0.99. For ±16°C, excellent agreement is still obtained, but for the case of cooling, the prediction is no longer accurate to an R^2 value of 0.99, it is slightly less. So, a total temperature window of something close to ±16°C appears to be about the acceptable limit for the range where the statistical deviation between theory and experiment begins to diverge from $R^2 = 0.99$. The key point here is that the thermoelectric property measurements are confirmed because the calculated and measured performance are in agreement with $R^2 = 0.99$.

20.5 Summary

There are rich opportunities to improve military capabilities through energy harvesting using thermoelectrics. Two primary opportunities are reducing the use of batteries and improving the fuel utilization on vehicles, both of which have significant payoff to logistics and reduced human risk. There are opportunities for new nanostructured materials that have some dimensional periodicity suitable to destructively interfere with, primarily, acoustical phonons and cause reduced thermal conductivity. Bulk materials based on $(Tl,Bi)Te_2$ have shown remarkably low thermal conductivity, and presage new III–V–VI alloys with unusually low thermal conductivity and high electrical conductivity. In support of new materials development, the new isothermal technique that uses noncontact radiative heat can provide rapid, high-accuracy, nondestructive thermoelectric measurements that are compatible with subsequent device fabrication. Thermoelectric measurements are best considered confirmed and validated by comparison between the measured performance from fully assembled devices and predictions from theory.

Symbols

D	spacing of Moiré fringes
d_1	lattice spacing of matrix material
d_2	lattice spacing of nanodot material
θ	angle of lattice rotation between d_1 and d_2
m	integer
λ	phonon wavelength
α	Seebeck coefficient
ρ	electrical resistivity
κ	thermal conductivity
Q_{Π}	Peltier heat flow
Q_{ρ}	Joule heat flow
Q_{κ}	thermally conducted heat flow
Q_{Rad}	heat flow from thermal radiation
ΔV_{ss}	steady-state voltage difference
ΔT_{ss}	steady-state temperature difference
T_{j}	temperature of the thermoelectric junction
I	electrical current
R_{c}	electrical contact resistivity
A	cross-sectional area of sample
l	length of sample
P	electrical output power
V_{oc}	open-circuit voltage
R_{int}	total internal resistance of a thermoelectric device
t	time
$\Delta T(t)$	time-dependence of temperature difference
n	molar volume of metallic material at the cold junction
C_{p}	heat capacity of metallic material at the cold junction
R^2	statistical coefficient of determination
T_{hot}	temperature of the hot side of a thermoelectric device
T_{cold}	temperature of the cold side of a thermoelectric device
T_{avg}	average temperature between hot side and cold side of a thermoelectric device
Z	thermoelectric figure of merit
ZT_{avg}	dimensionless thermoelectric figure of merit

References

1. *Report of the Defense Science Board Task Force on DoD Energy Strategy*, Office of the Under Secretary of Defense for Acquisition, Technology, and Logistics, Ch. 2, p. 11, 2008.
2. W.H. McMichael, R. Maze, *Army Times*, May 18, 2008, Army Times Publishing Company, Springfield, VA, 2008.
3. M. Green, G. Stewart, *M-1 Abrams at War*, Zenith Imprint-Motorbooks Int'l, p. 117, 2005. ISBN 9780760321539.

4. T. Kiper, A. Hughley, M. McClellan, Master's thesis, Naval Postgraduate School, Monterey, CA, 2010.
5. D. Eady, S. Siegel, R. Bell, S. Dicke, AEPI Report, Sustain the mission project: Casualty factors for fuel and water resupply convoys, Final Technical Report, W74V8H-04-D-0005, p. 5, 2009.
6. K. Smith, M. Thornton, *Feasibility of Thermoelectrics for Waste Heat Recovery in Conventional Vehicles*, Technical Report NREL/TP-540-4424, p. 4, 2009.
7. T.E. Whall, E.H.C. Parker, In D.M. Rowe, ed., *Proceedings of the First European Conference on Thermoelectrics*, London: Peter Peregrinus Ltd., p. 51, 1987.
8. S. Ren, J. Dow, Thermal conductivity of superlattices, *Phys. Rev. B*, 25(6), 3750, 1982.
9. W. Capinski, H. Maris, Thermal conductivity of GaAs/AlAs superlattices, *Physica B*, 219 & 220, 699, 1996; Touzelbaev, P. Zhou, R. Venkatasubramanian, K. Goodson, Thermal characterization of Bi_2Te_3/Sb_2Te_3 superlattices, *J. Appl. Phys.*, 90(2), 763, 2001.
10. T.C. Harman, D.L. Spears, D.R. Calawa, S.H. Groves, M.P. Walsh, *Proceedings of the 16th International Conference on Thermoelectrics*, Dresden, Germany, 1997. p. 416.
11. D.A. Broido, T.L. Reineke, Effect of superlattice structure on the thermoelectric figure of merit, *Phys. Rev. B*, 51(19), 13797, 1995; T. Koga, T.C. Harman, S. Cronin, M. Dresselhaus, Mechanism of the enhanced thermoelectric power in (111)-oriented *n*-type PbTe/$Pb_{(1-x)}Eu_xTe$ multiple quantum wells, *Phys. Rev. B*, 60(20), 15, 14286, 1999.
12. G. Springholz, V. Holy, M. Pinczolits, G. Bauer, Self-organized growth of three-dimensional quantum-dot crystals with fcc-like stacking and a tunable lattice constant, *Science*, 282, 5389, 734, 1998.
13. G. Chen, M. Neagu, Thermal conductivity and heat transfer in superlattices, *Appl. Phys. Lett.* 71, 2761, 1997.
14. P.J. Taylor, T.C. Harman, Lattice misfit strain relaxation in quantum dot heterostructures, M.I.T. Lincoln Laboratory Solid-State Research Reports, SSRR 2002:4, p. 15, 2003.
15. Y. Wang, C. Liebig, X. Xu, R. Venkatasubramanian, Acoustic phonon scattering in BiTe/SbTe superlattices, *Appl. Phys. Lett.* 97, 083103, 2010.
16. D. Broido, T. Reineke, Thermoelectric transport in quantum well superlattices, *Appl. Phys. Lett.*, 70, 2834, 1997.
17. P.J. Taylor, J. Maddux, W. Jesser, F.D. Rosi, Room-temperature anisotropic, thermoelectric, and electrical properties of *n*-type $(Bi_2Te_{3})_{90}(Sb_2Te_3)_5(Sb_2Se_3)_5$ and compensated *p*-type $(Sb_2Te_3)_{72}(Bi_2Te_3)_{25}(Sb_2Se_3)_3$ semiconductor alloys, *J. Appl. Phys.*, 85(11), 7807, 1999.
18. T.C. Harman, P.J. Taylor, M.P. Walsh, B.E. LaForge, Quantum dot superlattice thermoelectric materials and devices, *Science*, 297, 2229, 2002.
19. T.C. Harman, P.J. Taylor, D. Spears, M. Walsh, Thermoelectric quantum-dot superlattices with high ZT, *J. Electr. Mater.*, 29(1), L1, 2000.
20. J. Jensen, R. Burke, D. Ernst, R. Allgaier, Structural and electrical properties of $TlBiTe_2$ and TlTe, *Phys. Rev. B*, 6(2), 319, 1972.
21. B. Yan, C. Liu, H. Zhang, C. Yam, X. Qi, T. Frauenheim, S. Zhang, Theoretical prediction of topological insulators in thallium-based III-V-VI_2 ternary chalcogenides, *Europhys. Lett*, 90, 37002, 2010.
22. K. Kurosaki, A. Kosuga, S. Yamanaka, Thermoelectric properties of $TlBiTe_2$, *J. Alloys Compd.*, 351, 14, 2003.
23. P.J. Taylor, W. Jesser, F.D. Rosi, Z. Derzko, A model for the non-steady-state temperature behaviour of thermoelectric cooling semiconductor devices, *Semicond. Sci. Technol.*, 12, 443, 1997.

21

Thermal and Thermoelectric Characterization of Individual Nanostructures and Thin Films

Li Shi
The University of Texas

21.1 Introduction

In 1993, Hicks and Dresselhaus proposed that the sharply peaked electronic density of states (DOS) in two-dimensional (2D) quantum wells and one-dimensional (1D) quantum wires can be used to increase the power factor, $P \equiv S^2\sigma$, and the thermoelectric figure of merit, $ZT \equiv S^2\sigma T/\kappa$, where S, σ, κ, and T are the Seebeck coefficient, electrical conductivity, thermal conductivity, and absolute temperature, respectively (Hicks and Dresselhaus 1993a,b). In a subsequent theoretical study, Broido and Reinecke suggested that the ZT enhancement in realistic quantum well thin-film superlattice (SL) systems can be limited by several factors, including electron tunneling through the barrier layers that modifies the DOS and limits the power factor, increased charge carrier-phonon scattering rates and reduced charge carrier mobility, and parasitic thermal conduction in the barrier layers (Broido and Reinecke 1995). They further suggested that freestanding structures such as suspended nanowires may be needed to overcome these limitations in order to obtain a large ZT. Several calculations by Lin and coworkers as well as Mingo found that $ZT > 2$ at 77 K and 1.5 at 300 K, respectively, is possible in Bi and InSb nanowires with diameter <10 nm (Lin et al. 2000; Mingo 2004, 2006). In addition, the calculation by Rabina and coworkers further suggested that $ZT > 2$ at 77 K could be achieved in $Bi_{1-x}Sb_x$ nanowires of diameter <20 nm (Rabina et al. 2001). Because the electron effective mass (m^*) is very small in Bi, InSb, and $Bi_{1-x}Sb_x$ and the de Broglie wavelength (λ_e) of electrons is proportional to $m^{*-1/2}$, the electron wavelength becomes comparable to the 10–20 nanowire diameter in these three systems, giving rise to 1D DOS.

These calculations are based on the assumption that electrons are scattered by the nanowire surface or quantum well/barrier interface specularly without reduction of the mobility, while the lattice thermal conductivity is reduced by diffuse phonon-boundary scattering. Because the electron wavelength in small-effective mass semiconductors can be >10 nm and much larger than the interface roughness, it is theoretically possible to achieve specular interface scattering of electrons. In comparison, the dominant phonon wavelength is of the order of 1 nm at room temperature, comparable to the typical interface roughness. Hence, diffuse phonon-interface scattering is expected based on Ziman's expression of the interface specularity parameter (Ziman 1962).

These theoretical predictions have motivated intense experimental studies of thin-film and nanowire structures. Although individual nanostructures and thin films cannot be used readily to fabricate Pelteir coolers or thermoelectric power generators with sufficient power for many applications, they provide model systems to understand the effects of classical and quantum size confinement on thermoelectric transport.

There have been several experimental reports of ZT enhancement in thin-film SL structures (Venkatasubramanian et al. 2001; Harman et al. 2002) and nanowires (Hochbaum et al. 2008; Boukai et al. 2008). The enhancement has been attributed mainly to the suppression of the lattice thermal conductivity instead of power factor enhancement. In addition, for the case of thin-film SLs (Venkatasubramanian et al. 2001; Harman et al. 2002), the ZT enhancement was reported for the cross-plane transport direction instead of the in-plane direction considered in earlier theoretical studies. Diffuse interface scattering of phonons is believed to be one of the main causes of the reduced lattice thermal conductivity in the cross-plane direction of thin-film SLs and along the axial direction of nanowires, although other intriguing wave behaviors such as phonon mini bands and phonon localization in thin-film SLs has also been discussed (Venkatasubramanian 2000; Chen et al. 2005).

In existing experiments with nanowire or quantum well structures, clear evidence of power factor enhancement has not been observed, partly because the presence of interface charge states can result in a depletion layer and diffuse interface scattering of charge carriers in quantum wells and nanowires. Nevertheless, the general idea of using sharply peaked electronic DOS to increase the power factor has been an accepted approach (Mahan and Sofo 1996), and was used to explain the power factor enhancement found in bulk PbTe doped with Tl, which is believed to distort and increase the local DOS near the Fermi level (Heremans et al. 2008).

The intense experimental studies of thermoelectric transport in thin film and nanowires have led to the development of a number of thermoelectric characterization methods designed specifically for thin films and nanowires. It is worth noting that the measurement of the three properties entering the ZT expression is more challenging for individual nanostructures and thin films than for bulk crystals, which already requires care to address a number of issues. This challenge can lead to large uncertainty and discrepancy in the measured properties. For example, recent measurements have found neither enhancement of the power factor (Vineis et al. 2008) nor suppression of the lattice thermal conductivity (Koh et al. 2009) in PbSeTe/PbTe nanodot thin film SLs, for which enhanced ZT was reported in an earlier report (Harman et al. 2002). Hence, independent verification of reported measurement properties of individual nanostructures and thin films would be useful.

This chapter discusses experimental methods for thermoelectric characterization of individual thin films and nanostructures. As summarized in Table 21.1, the measurement methods discussed here include four electrical methods, including the 3ω methods, suspended resistance thermometer microdevices, supported resistance thermometer microdevices, and mesa structures for cross-plane thin-film

TABLE 21.1 Thermoelectric Measurement Methods for Thin Films and Nanowires

Methods	Properties	Film ($\kappa_\perp$)	Film ($\kappa_{//}$)	Film ($\sigma_\perp, S_\perp$)	Film ($\sigma_{//}, S_{//}$)	Film ($ZT_\perp$)	Nanowire (κ)	Nanowire (σ, S)
Electrical	3ω Methods	X	X				X	
	Suspended microdevices		X		X		X	X
	Supported microdevices				X			X
	Mesa structures	X		X		X		
Optical	Thermal reflectance	X	X				X	
	Raman spectroscopy		X				X	

characterization, as well as two optical methods, namely thermal reflectance methods and Raman spectroscopy measurements. Because thin-film thermoelectric materials often consist of numerous interfaces that give rise to anisotropic thermoelectric properties, different methods have been developed for measuring thermoelectric properties along the cross-plane and in-plane directions, which are denoted in Table 21.1 by the subscript $\perp$ and //, respectively. For each method, the different variations for measuring different thin-film or nanowire properties are described, including a discussion on the error sources for some of the methods. Although some intriguing low-dimensional phenomena in thin films and nanostructures are discussed, the focus of this chapter is on the characterization methods instead of the reported measurement results.

21.2 The 3ω Method

The 3ω method is essentially an electrical resistance thermometry method for thermal conductivity measurement. Different variations of the 3ω method have been employed to measure the cross-plane and in-plane thermal conductivity of thin films, as well as the axial thermal conductivity of suspended nanowires and nanotubes.

21.2.1 Cross-Plane Thermal Conductivity Measurement of Thin Films

The 3ω method was first established by Cahill for measuring the cross-plane thermal conductivity of a thin-film sample (Cahill 1990, 2002). For this measurement, a long and thin metal line is patterned on top of the thin-film sample and serves as both an electrical heater and a resistance thermometer. If the thin film is electrically conducting, a thin dielectric film is deposited on the conducting film sample prior to the deposition of the metal line, as illustrated in Figure 21.1. During the measurement, the metal line is electrically heated with a sinusoidal current at a frequency ω, resulting in modulated electrical heating at a frequency of 2ω. The modulated heating results in a temperature oscillation at a frequency 2ω in the metal line. When the metal line of a linewidth $2b$ and a length L is deposited directly on a semi-infinite substrate of thermal diffusivity α_s, the temperature oscillation at the 2ω frequency is reduced to the following expression for the low-frequency limit of $2wb/\alpha_s \ll 1$,

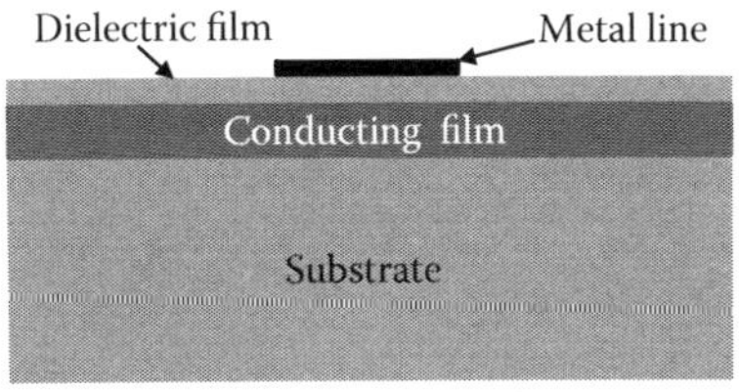

FIGURE 21.1 Cross-sectional schematic of 3ω measurement of the cross-plane thermal conductivity of a conducting thin-film sample.

$$\Delta T_{2\omega} = \frac{Q}{L\pi\kappa_s}\left[\frac{1}{2}\ln\left(\frac{\alpha_s}{b^2}\right) + \eta - \frac{1}{2}\ln(2\omega) - \frac{i\pi}{4}\right] \tag{21.1}$$

where Q is the amplitude of the electrical heating oscillation at the 2ω frequency in the heater line, κ_s is the thermal conductivity of the substrate, and η is a constant. Based on this equation, the out-of-phase component ($-i\pi Q/4L\pi\kappa_s$) of the temperature oscillation is independent of frequency. In comparison, the in-phase component is linearly proportional to $\ln\omega$, with the slope equal to $-Q/2L\pi\kappa_s$, which can be used to obtain the substrate thermal conductivity κ_s.

The metal line is usually made of a metal with a relatively large temperature coefficient of resistance, such as aluminum, gold, and platinum. For these metals, the electrical resistance (R) of the metal line shows an approximately linear relationship with the temperature within a large temperature range. Consequently, the modulated heating at a frequency 2ω causes an oscillating component of the electrical resistance at the same frequency. The voltage drop measured between the two ends of the metal line consists of an oscillating component ($V_{3\omega}$) at a frequency 3ω, which is caused by the product of the 1ω current oscillation and the 2ω resistance oscillation. This 3ω component is related to the 2ω temperature oscillation according to

$$\Delta T_{2\omega} = 2\frac{dT}{dR}\frac{R}{V_{1\omega}}V_{3\omega} \tag{21.2}$$

where $V_{1\omega}$ and $V_{3\omega}$ are the measured voltage oscillation at frequency 1ω and 3ω, respectively, R is the electrical resistance of the metal line, and dT/dR is the temperature coefficient obtained from the R versus temperature (T) curve.

To obtain the cross-plane thermal conductivity of a conducting thin film of low thermal conductivity compared to that of the substrate, the measurement is made on a sample consisting of a substrate, the conducting thin film, and a dielectric thin film between the metal line and the conducting thin film, and on a reference sample consisting of only the substrate and the thin dielectric film. When the thickness t of the conducting thin film and that of the dielectric thin film are small compared to the width ($2b$) of the heater line and the thermal penetration depth, the presence of the conducting thin film layer yields an additional frequency-independent term, $Qt/2bL\kappa_f$, in the in-phase component of the temperature oscillation of the metal line, where κ_f and t are the thermal conductivity and thickness of the thin film (Cahill et al. 1994). In this case, the cross-plane thermal conductivity can be obtained according to

$$\kappa_f = \frac{dR}{dT}\frac{Qt}{4bLR}\frac{1}{(V_{3\omega,in}/V_{1\omega,in})_{sample} - (V_{3\omega,in}/V_{1\omega,in})_{ref}} \tag{21.3}$$

where the subscripts in, sample, and ref are used to denote the in-phase component, the sample with the conducting thin film, and the reference sample, respectively. In this analysis, the thermal interface resistance between different layers is ignored. When the film thickness is small, this assumption can underestimate the cross-plane thermal conductivity. A detailed model has been developed by Borca-Tasciuc and coworkers to account for the thermal interface resistance and other factors, including thermal conductivity anisotropy, and thermal conductivity contrast between the substrate and the thin film, and finite substrate thickness, in the analysis of the 3ω measurement data (Borca-Tasciuc et al. 2001).

21.2.2 In-Plane Thermal Conductivity Measurement

The 3ω method has been extended by Ju and coworkers to obtain the in-plane thermal conductivity of a thin film by using heater lines of varying widths (Ju et al. 1999). The in-plane thermal conductivity of the film was extracted by fitting a 2D heat conduction model to the measurement results.

21.2.3 Thermal Conductivity Measurement of Individual Nanowires and Nanotubes

The 3ω method has been further developed by Lu and coworkers for measuring the axial thermal conductivity of a Pt wire (Lu et al. 2001) and a multiwalled carbon nanotube (MWCNT) bundle (Yi et al. 1999). In this measurement, the wire sample is suspended, as illustrated in Figure 21.2. With a sinusoidal electrical current at a frequency ω flowing in the suspended wire, the in-phase component of the measured 3ω component of the voltage drop along the wire is related to the thermal conductivity of the wire according to

$$V_{3\omega} \approx \frac{4I^3LR(\mathrm{d}R/\mathrm{d}T)}{\pi^4 \kappa A\sqrt{1+(2\omega\gamma)^2}} \tag{21.4}$$

where I is the root mean squared (rms) amplitude of the heating current, R is the electrical resistance of the suspended wire, κ and A are the thermal conductivity and cross-sectional area of the wire, and γ is the thermal time constant of the wire given by $\gamma \equiv L^2/\pi^2\alpha$, where L is the suspended length and α is the thermal diffusivity of the sample. In addition, the phase lag (ϕ) between the heating current oscillation and the temperature oscillation is given as

$$\tan\phi \approx \omega\gamma \tag{21.5}$$

In this method, the κ and γ can be extracted by fitting the measured $V_{3\omega}$ and ϕ at different ω values. The specific heat is obtained as

$$C = \pi^2\gamma\kappa/\rho L^2 \tag{21.6}$$

where ρ is the density.

The κ measurement is usually conducted in the low-frequency range where $\omega\gamma \ll 1$. On the other hand, for measuring γ and C, the $\omega\gamma$ product needs to be sufficiently large, and typically $\omega\gamma > 0.1$.

The 3ω method has been explored by Choi and coworkers for measuring the thermal conductivity of an individual ~1.4-μm-long and ~20-nm-diameter suspended MWCNT grown by chemical vapor deposition (CVD) (Choi et al. 2006). Because coupled electron and phonon transport can be highly nonequilibrium in an electrically heated, high-quality single-walled carbon nanotube (SWCNT), the 3ω and other self-electrical heating methods cannot be used readily for a SWCNT. To account for nonequilibrium transport in a current-carrying SWCNT, a coupled electron–phonon transport model was developed by Pop and coworkers and used to obtain the thermal conductance of a 2.6-μm-long SWCNT by fitting the current–voltage characteristics with several other parameters including the coupling constant between optical and acoustic phonons and contact thermal and electrical resistances (Pop et al. 2006).

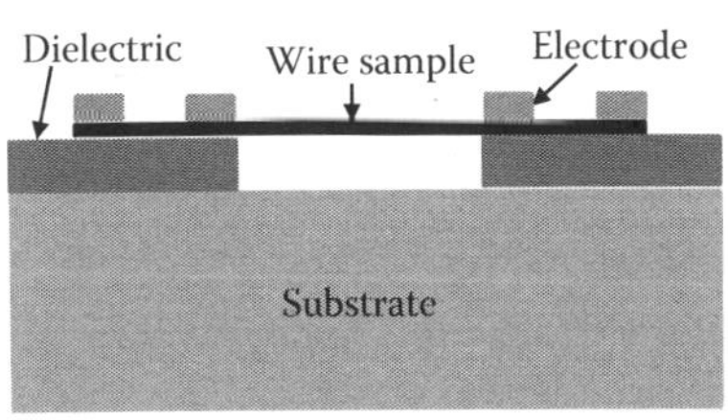

FIGURE 21.2 Schematic of 3ω measurement of the thermal conductivity of a suspended wire.

Besides the complication due to nonequilibrium transport in carbon nanotubes as well as semiconductor nanowires of high electron mobility, self-heating methods are also inconvenient for semiconductor nanowires or thin films that possess a small or varying dR/dT. To address this issue, Liu and Asheghi patterned a thin Si film into a suspended beam and employed a metal layer on top of the patterned film as a heater and resistance thermometer (Liu and Asheghi 2006). Similar approach has been used by Heron and coworkers to measure the thermal conductivity of patterned silicon nanowires (Heron et al. 2009). In this method, the thermal conductance contribution from the metal thermometer layer needs to be determined and subtracted from the measured total thermal conductance.

21.3 Suspended Resistance Thermometer Microdevices

For eliminating parasitic heat and charge transport in the substrate during in-plane thermoelectric measurements of thin films and nanostructures, different designs of suspended thermometer devices have been developed. Shi and coworkers have reported suspended platinum resistance thermometer (PRT) microdevices for measuring the in-plane or axial thermal conductivity, Seebeck coefficient, and electrical conductivity of individual thin films, nanowires, and nanotubes (Shi et al. 2003; Mavrokefalos et al. 2007). As shown in Figure 21.3, one design of such devices consists of two ~20 μm × 20 μm × 0.5 μm low-stress silicon nitride (SiN_x) membranes, each suspended by six 2-μm-wide, 200–400-μm-long SiN_x beams. One ~40-nm-thick, ~200-nm-wide Pt serpentine resistance thermometer is patterned on each SiN_x membrane, and connected to four 2-μm-wide Pt lines on four of the six SiN_x beams. Two additional Pt electrodes are patterned near the inner edge of each SiN_x membrane, and connected to two 2-μm-wide Pt lines on the remaining two of the six SiN_x beams.

For measuring the in-plane thermoelectric properties of a thin-film sample, the thin film is patterned using lithography and etching, and subsequently suspended by undercutting the substrate below the patterned thin film. A sharp tungsten probe is used to break the two ends of the suspended patterned

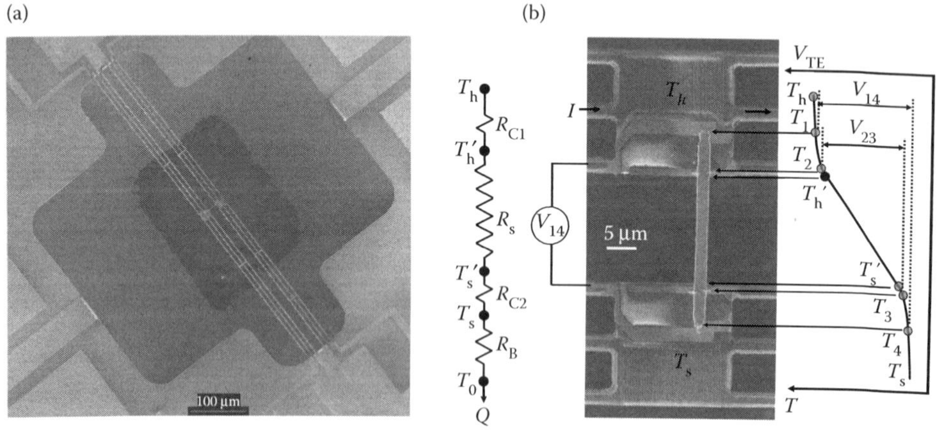

FIGURE 21.3 (a) SEM of a suspended resistance thermometer microdevice. (b) SEM of an InAlGaAs thin-film sample assembled across the two membranes of the device. The thermal resistance circuit and the temperature distribution along the sample are shown in the schematics at the left and right sides of the SEM. T_h and T_s are the temperatures of the heating (upper) and sensing (lower) membranes. T_h' and T_s' are the temperatures at the two ends of the suspended segment of the film sample. T_1, T_2, T_3, and T_4 are the temperatures at the four contacts. T_0 is the temperature of the substrate. R_s and R_B are the thermal resistances of the film sample and the six beams supporting one membrane. R_{C1} and R_{C2} are the contact thermal resistances between the sample and the heating and sensing membranes. V_{14} and V_{23} are the thermoelectric voltages (V_{TE}) measured between the two outer and between the two inner electrodes.

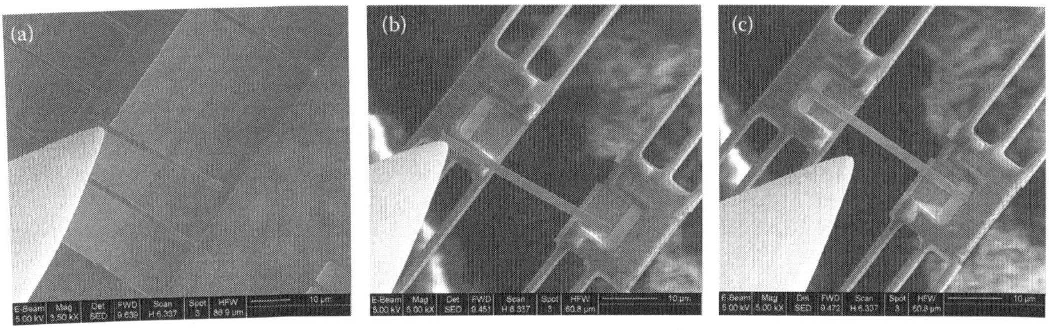

FIGURE 21.4 (a–c) SEMs of the process of assembling a thin-film sample on the suspended microdevice with the use of a tungsten probe.

thin film, pick up the film, and transfer the patterned film to bridge the two suspended membranes of the measurement device. This transfer process can be carried out with the use of a nano-manipulator stage inside a scanning electron microscope, as shown in Figure 21.4. Some skilled experimentalists have also been able to transfer nanostructures to a suspended device using a home-built micrometer stage under an optical microscope.

Several methods have been investigated for making electrical contacts between the thin-film sample and the four Pt electrodes prepatterned on the two suspended SiN_x membranes. One method is based on focused ion beam-induced deposition or focused electron beam-induced deposition of metal patterns on the contact area between the film sample and the four underlying Pt electrodes (Mavrokefalos et al. 2007). The focused electron beam-induced deposition method is preferred because of concerns on ion beam damage of the film sample, and has been used to deposit small patterns of Pt–C composite on the contact area of the thin-film sample shown in Figure 21.3 and the nanowire sample shown in Figure 21.5. One issue with both techniques is that the sample surface can be contaminated with the organometallic precursor gases used for the deposition. For addressing this issue, an *in situ* hydrogen-annealing procedure has been

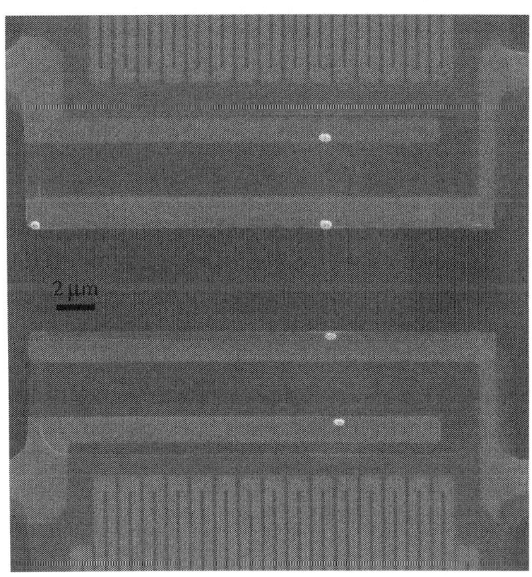

FIGURE 21.5 SEM of a nanowire trapped between the two suspended resistance thermometers of a microdevice. The four bright spots on the nanowire are Pt–C patterns deposited with the use of focused electron beam induced metal deposition.

employed for making electrical contact between a Bi_2Te_3 nanowire and the underlying Pt electrodes on the suspended device (Mavrokefalos et al. 2009). However, this annealing procedure has not been effective for other materials such as Si nanostructures with stable native oxide. With the use of a shadow mask, metal patterns can be evaporated on the contact area between the thin film sample and the underlying electrodes. This method has been recently used to make ohmic contact between an ErAs:InGaAlAs thin film and the microdevice. This method has also been used for improving the thermal contact between a porous Si thin film and the Pt electrodes of a suspended microdevice (Tang et al. 2010).

21.3.1 Two-Probe and Four-Probe Thermoelectric Measurements

The thermoelectric measurement of the thin-film sample is conducted in a variable temperature cryostat where the sample space is in high vacuum. The thermal resistance of the thin film sample is obtained as follows. When a direct current (I) is supplied to one PRT to raise the temperature (T_h) of one membrane, part of the Joule heat generated in this heating membrane, Q, is conducted through the film to the other membrane referred to as the sensing membrane, increasing its temperature (T_s), as illustrated in Figure 21.3b. Because the internal thermal resistance of each membrane is on the order of 10^5 K/W and is two orders of magnitude smaller than the thermal resistance of the six beams supporting each membrane and that of the suspended sample between the two membranes, the temperature distribution on each membrane is uniform during the heating process. The temperature uniformity has been verified in a finite-element calculation (Yu et al. 2006). The two PRTs are used to measure the temperature rises on the heating and sensing membranes at different I values, $\Delta T_h(I) \equiv T_h(I) - T_0$ and $\Delta T_s(I) \equiv T_s(I) - T_0$, respectively, where T_0 is the substrate temperature. It is worth noting that for measuring T_h, the resistance of the heating PRT can be obtained from the current–voltage (I–V) curve or by coupling a small oscillating AC excitation current into the DC heating current to measure the differential resistance. For the latter method, it is important to use a sufficiently high frequency (ω) of the AC excitation compared to the thermal time constant of the microdevice, so that the temperature modulation caused by the 1ω oscillation in the electrical heating is negligible and causes negligible errors in the obtained differential resistance measured using a lock-in amplifier at the 1ω frequency (Shi et al. 2003).

Based on the solution of the temperature distribution in each of the six beams supporting each membrane, the thermal resistance of the six beams supporting each membrane can be obtained as

$$R_B = \frac{\Delta T_h + \Delta T_s}{Q_h + Q_L} \tag{21.7}$$

where Q_h is the Joule heat dissipation in the PRT on the heating membrane, and Q_L is the Joule heat dissipation in one of the two identical Pt leads supplying the current to the heating PRT (Shi et al. 2003). This result is obtained by ignoring radiation heat loss from the microdevice to the environment. With inadequate radiation shielding, the radiation loss can become important at high and low temperatures, and cause nonlinear temperature distribution on the supporting beams. Nevertheless, with the radiation loss taken into account, a numerical analysis has shown that the error caused by radiation loss is negligible for the temperature range between 4 and 800 K, provided that the temperature coefficient of resistance of the PRT can still be measured accurately (Moore and Shi 2011).

The total thermal resistance of the sample can be obtained according to the thermal circuit in Figure 21.3b as

$$R_{total} \equiv R_{C,1} + R_S + R_{C,2} = \frac{\Delta T_h - \Delta T_s}{Q} \tag{21.8}$$

where R_S is thermal resistance of the suspended segment of the film sample, $R_{C,1}$ and $R_{C,2}$ are the contact thermal resistances between the film sample and the two membranes, and Q equals the heat conducted through the six beams of the sensing membrane to the substrate, and can be obtained as

$$Q = R_B \Delta T_s \tag{21.9}$$

One major issue with this two-probe thermal measurement is the contact thermal resistances, $R_{C,1}$ and $R_{C,2}$, which can be obtained from a four-probe thermoelectric measurement procedure (Mavrokefalos et al. 2007). According to the temperature profile schematic shown in Figure 21.3b, the contact thermal resistances can cause temperature drops, that is, $(T_h - T_h')$ and $(T_s' - T_s)$, where T_h' and T_s' are the temperatures at the two ends of the suspended segment of the film sample. While the temperature profile is linear along the suspended segment, the temperature decays approximately exponentially along the two segments in contact with the membranes. In the experiment, the four Pt contacts deposited on the thin film are used to measure the following two thermoelectric voltages generated by the film,

$$V_{14} = S(T_1 - T_4) \tag{21.10a}$$

and

$$V_{23} = S(T_2 - T_3) \tag{21.10b}$$

where V_{14} and V_{23} are the thermoelectric voltage measured between the two outer electrodes and that between the two inner electrodes, respectively, S is the Seebeck coefficient of the thin-film sample and is assumed to be constant along the thin film and much larger than that of the Pt electrodes, and T_1, T_2, T_3, and T_4 are the temperatures at the four metal contacts deposited on the thin film, as explained in Figure 21.3b.

Each thin-film segment in contact with the membrane can be treated as a fin inserted into a thermal reservoir. The thermal contact resistance between the thin film and each membrane is thus the fin resistances given as (Incropera and Dewitt 2006)

$$R_{C,i} = \frac{1}{\kappa A m \tanh(m L_{C,i})}; \quad i = 1 \text{ or } 2 \tag{21.11}$$

where κ and A are the thermal conductivity and cross-sectional area of the thin film, $L_{C,i}$ is the length of a thin-film segment in contact with one membrane, as shown in Figure 21.6, and m is defined as

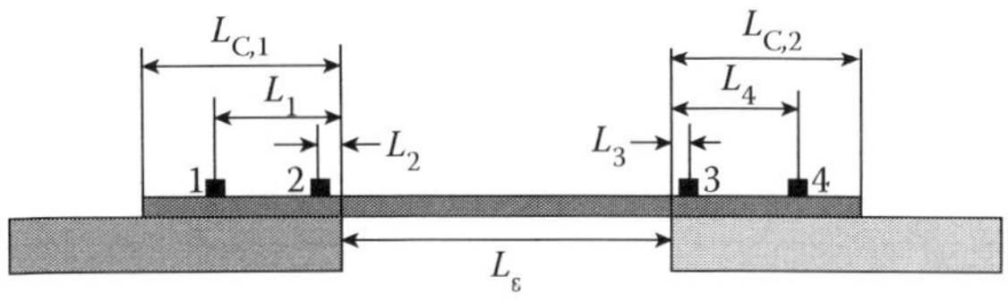

FIGURE 21.6 Schematic showing the definition of the different dimensions for the heat transfer analysis of the four-probe thermoelectric measurement results. (Reprinted with permission from Mavrokefalos, A. et al. Four-probe measurements of the in-plane thermoelectric properties of nanofilms. *Review of Scientific Instruments* 78(3):034901. Copyright 2007, American Institute of Physics.)

$$m = \sqrt{\frac{hw}{\kappa A}} \tag{21.12}$$

where h is the thermal contact conductance per unit area between the thin film and the membrane, and w is the width of the thin film.

According to the thermal resistance circuit in Figure 21.3b,

$$Q = \frac{T_h - T_h'}{R_{C,1}} = \frac{T_h' - T_s'}{R_S} = \frac{T_s' - T_s}{R_{C,2}} \tag{21.13}$$

According to the fin temperature profile (Incropera and Dewitt 2006)

$$\frac{T_h - T_i}{T_h - T_h'} = \frac{\cosh(m(L_{C,1} - L_i))}{\cosh(mL_{C,1})}; \quad i = 1 \text{ or } 2 \tag{21.14a}$$

and

$$\frac{T_i - T_s}{T_s' - T_s} = \frac{\cosh(m(L_{C,2} - L_i))}{\cosh(mL_{C,2})}; \quad i = 3 \text{ or } 4 \tag{21.14b}$$

From these equations, one can obtain

$$\beta_{14} \equiv \frac{T_1 - T_4}{T_h' - T_s'} = 1 + \frac{1 - \frac{\cosh(m(L_{C,1} - L_1))}{\cosh(mL_{C,1})}}{L_S m \tanh(mL_{C,1})} + \frac{1 - \frac{\cosh(m(L_{C,2} - L_4))}{\cosh(mL_{C,2})}}{L_S m \tanh(mL_{C,2})} \tag{21.15a}$$

$$\beta_{23} \equiv \frac{T_2 - T_3}{T_h' - T_s'} = 1 + \frac{1 - \frac{\cosh(m(L_{C,1} - L_2))}{\cosh(mL_{C,1})}}{L_S m \tanh(mL_{C,1})} + \frac{1 - \frac{\cosh(m(L_{C,2} - L_3))}{\cosh(mL_{C,2})}}{L_S m \tanh(mL_{C,2})} \tag{21.15b}$$

and

$$\beta \equiv \frac{V_{14}}{V_{23}} = \frac{T_1 - T_4}{T_2 - T_3} = \frac{\beta_{14}}{\beta_{23}} \tag{21.15c}$$

The above equation can be used to obtain the parameter m from the measured β value at each temperature. The obtained m value is used to calculate β_{23}. One can derive the following equation to calculate the ratio α defined as

$$\alpha \equiv \frac{T_h - T_s}{T_h' - T_s'} = 1 + \frac{1}{L_S m}\left(\frac{1}{\tanh(mL_{C,1})} + \frac{1}{\tanh(mL_{C,2})}\right) \tag{21.16}$$

Based on the α and β_{23} values, one can obtain the Seebeck coefficient, the sample thermal resistance, and the contact thermal resistance as

$$S = \alpha S_{23}/\beta_{23} \tag{21.17a}$$

$$R_s = R_{\text{total}}/\alpha \tag{21.17b}$$

and

$$R_C \equiv R_{\text{total}} - R_s \tag{21.17c}$$

For a number of smooth thin film and nanowire samples, the R_C/R_{total} ratio has been found to be about 10% when both the suspended and supported segments are sufficiently long and the thickness of the sample is small (Mavrokefalos et al. 2007, 2009; Zhou et al. 2007). The obtained R_s can further be used to obtain the thermal conductivity based on the dimension of the thin-film sample that can be measured by scanning electron microscopy (SEM) and atomic force microscopy.

The electrical conductivity can be measured using a four-probe configuration, where the current flows between the two outer electrodes and the voltage drop between the two inner electrodes is measured with a high-input impedance electrometer. During the measurement, Peltier cooling and heating occur at the two contacts to the film, resulting in a temperature difference (ΔT) and consequently a thermovoltage ($S\Delta T$) between the two contacts. During the I–V measurement, the temperature difference (ΔT) between the two ends of the thin-film sample can be measured simultaneously with the two PRTs, based on the assumption that the temperature is uniform on each membrane. The thermovoltage $S\Delta T$ is subtracted from the measured total voltage drop (ΔV) between the two inner contacts to obtain the IR voltage drop and the electrical conductivity. Hence, all the three in-plane transport properties entering the ZT expression can be obtained.

In addition, the presence of a through-substrate hole allows for transmission electron microscopy (TEM) characterization of the crystal structure of the sample assembled between the two membranes. This capability allows for the structure–property relationship to be established. For samples such as carbon nanotubes, TEM is one of the very few methods that can allow for the accurate determination of the cross section. Pettes and Shi have recently reported combined thermal conductance and TEM characterizations of the same carbon nanotubes (CNTs), and demonstrated the measurement of the thermal conductance, diameter, and chirality of the same SWCNT sample. For four MWCNT samples, the obtained intrinsic thermal conductivity in the range of 43–343 W m^{-1} K^{-1} of the MWCNT samples at room temperature correlates well with the axial defect concentration determined by TEM (Pettes and Shi 2009).

21.3.2 Other Suspended Resistance Thermometer Microdevices

Other designs of suspended resistance thermometer devices have also been reported. Fujii and coworkers reported a so-called T-junction sensor device for measuring the thermal conductivity of individual MWCNTs (Fujii et al. 2005). The sensor contains a suspended pattered Pt nanofilm that serves as both a heater and thermometer. An MWCNT was placed between the center of the suspended nanofilm and the Si substrate, and formed a T-shape junction with the suspended nanofilm. The average temperature of the nanofilm was measured by measuring its electrical resistance at different electrical currents. The thermal conductivity of the nanofilm was measured using a self-heating method when no nanotube was placed on the nanofilm. The self-heating measurement was repeated after the nanotube was placed on the nanofilm. The latter measurement was used to determine the thermal conductance and thermal conductivity of the nanotube. In addition, Seol and coworkers have recently reported a method for measuring the thermal conductivity of supported monolayer graphene connected to one straight and one U-shaped resistance thermometer line at each end of the graphene monolayer (Seol et al. 2010). In another method, Zink and coworkers obtained the in-plane κ of a thin film evaporated on the backside of a silicon nitride membrane device by comparing the measured thermal conductance values before and after the evaporation (Zink et al. 2005). Jain and Goodson fabricated a suspended thin-film structure with a metal heater and thermometer line at the center of the film to measure the in-plane thermal conductivity using the 3ω technique (Jain and Goodson 2008).

21.3.3 Determination of Carrier Concentration and Mobility of Nanowires

Besides the three transport properties entering the ZT expression, the Fermi Level (E_F), carrier concentration, and mobility are fundamental parameters that are critical for optimizing the ZT values. For thin-film samples, the carrier concentration and mobility can be obtained from the Hall measurement, which cannot be used for the nanowire geometry. Efforts have been made to obtain the Fermi energy or carrier concentration in nanowires and nanotubes from a field effect measurement (Durkop et al. 2003). However, extracting the carrier concentration from field effect measurements requires accurate calculation of the capacitance for the nanowire geometry. Moreover, the field effect measurement cannot be applied readily to heavily dope thermoelectric nanowires.

Recently, Moore and coworkers have used the measured Seebeck coefficient and electrical conductivity of individual nanowires to extract the Fermi level, carrier concentration and mobility based on a two-band or one-band model (Zhou et al. 2007, 2010; Seol et al. 2007; Mavrokefalos et al. 2009). In the two-band model, the Seebeck coefficient was calculated as

$$S = \frac{S_e n\mu_e + S_h p\mu_h}{n\mu_e + p\mu_h} \tag{21.18}$$

where S_e and S_h are the Seebeck coefficients of electrons and holes, respectively, n and p are the electron and hole concentrations, respectively, μ_e is electron mobility, and μ_h is hole mobility. The mean free scattering time τ is assumed to depend on the carrier energy E according to $\tau = \tau_0 E^r$, where τ_0 and r are constants. When τ is limited by either acoustic phonon scattering or boundary scattering in a nanowire, r takes a value of −1/2 (Ziman 1962). With this assumption, the electron contribution and hole contribution to S were calculated using

$$S_e = -\frac{k_B}{e}\left(\frac{\left(r+\frac{5}{2}\right)F_{r+\frac{3}{2}}(\zeta_e)}{\left(r+\frac{3}{2}\right)F_{r+\frac{1}{2}}(\zeta_e)} - \zeta_e\right) \tag{21.19}$$

and

$$S_h = \frac{k_B}{e}\left(\frac{\left(r+\frac{5}{2}\right)F_{r+\frac{3}{2}}(\zeta_h)}{\left(r+\frac{3}{2}\right)F_{r+\frac{1}{2}}(\zeta_h)} - \zeta_h\right) \tag{21.20}$$

where $F_y(\zeta) = \int_0^\infty (x^y dx)/(e^{(x-\zeta)}+1)$ is the Fermi–Dirac integral of order y, k_B is the Boltzmann constant, and the terms ζ_e and ζ_h are the reduced Fermi energies for electrons and holes.

The electron and hole concentrations can be calculated using

$$n = \frac{4\pi}{h^3}(2m_e^* k_B T)^{3/2} F_{1/2}(\zeta_e) \tag{21.21}$$

and

$$p = \frac{4\pi}{h^3}(2m_h^* k_B T)^{3/2} F_{1/2}(\zeta_h) \tag{21.22}$$

where h is Planck's constant and m_e^* and m_h^* are the electron and hole DOS effective masses, respectively.

Using the above model, E_F was extracted from the measured S data. Typically two E_F values, one in the transition regime and the other in the highly degenerate regime, can yield the same S value, as shown in the example of Figure 21.7.

The electrical conductivity σ was calculated from

$$\sigma = ne\mu_e + pe\mu_h \tag{21.23}$$

where e is the mobility elemental charge. For a Bi_2Te_3 nanowire, Mavrokefalos and coworkers found that even using the bulk values (Champness and Kipling 1966) for several different stoichiometries, the σ values calculated using the above equation with E_F in the transition regime were 2–20 times smaller than the measurement values for the nanowires. Hence, they concluded that the actual E_F should be the one found in the degenerately n-type regime. In this case, a single-band model of the electron band is sufficient for extracting the carrier concentration and mobility.

Mavrokefalos and coworkers found that the electron mobility and phonon mean free path were reduced to a similar extent in their electrodeposited Bi_2Te_3 nanowires compared to bulk crystals of the same carrier concentration (Mavrokefalos et al. 2009). In comparison, the hole mobility in $CrSi_2$ nanowires of 70–100 nm diameter has been measured to be comparable to the bulk value by Zhou and coworkers (Zhou et al. 2007). For $CrSi_2$ with a large effective mass and high carrier concentration, the mobility and charge carrier mean free path in the bulk crystal are already small although the electrical conductivity is still relatively high because of the high carrier concentration. The high carrier concentration could also screen the surface charges on a nanowire. For these reasons, the mean free path and mobility

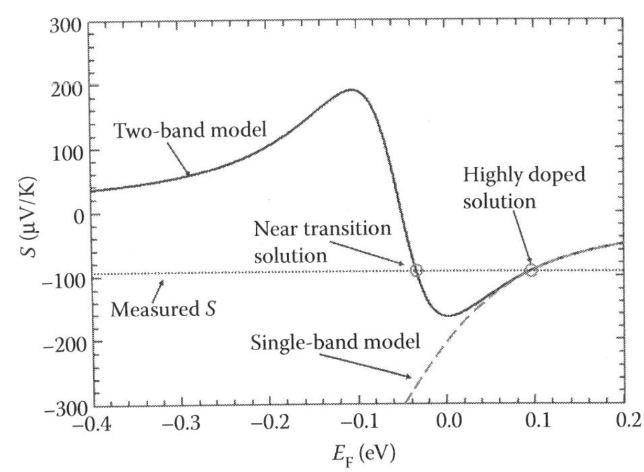

FIGURE 21.7 Calculated Seebeck coefficient of a Bi_2Te_3 nanowire as a function of the Fermi energy in comparison with the measurement result. (Reprinted with permission from Mavrokefalos, A. et al. Thermoelectric and structural characterizations of individual electrodeposited bismuth telluride nanowires. *Journal of Applied Physics* 105(10):104318. Copyright 2009, American Institute of Physics.)

are not reduced much in $CrSi_2$ nanowires with diameter larger than the charge carrier mean free path. In comparison, in a bulk semiconductor with a small effective mass, the carrier concentration is usually low, the mobility and charge carrier mean free path can be very large. Because of weak screening in the nanowire due to the low charge carrier concentration, surface charges in the nanowire can lead to reduction of the charge carrier mean free path that is long in the bulk compared to the nanowire diameter. This nonideal effect and the lack of control in the carrier concentration have obscured the predicted quantum enhancement of power factor in existing experimental efforts. Large ZT enhancement in nanowires has only been reported in etched silicon nanowires (Boukai et al. 2008; Hochbaum et al. 2008), and attributed to abnormally large suppression of the lattice thermal conductivity, whereas it was reported that the charge carrier mobility and power factor were not suppressed in the Si nanowires with a relatively large effective mass.

21.4 Supported Microdevices for Field Effect Thermoelectric Measurements

Although the suspended resistance thermometer microdevices have been used to obtain all the three thermoelectric properties of a thin-film or nanowire sample, one limitation of the method is that a gate field cannot be applied to tune the carrier concentration of the sample in order to maximize the power factor. This issue can be addressed with the measurement device shown in Figure 21.8 (Seol et al. 2007). With this device, the Seebeck coefficient and electrical conductivity of a nanowire supported on an oxidized silicon substrate can be measured with the resistance thermometer lines and electrodes patterned on top of the nanowire when a back gate voltage is applied to the doped silicon substrate. The field effect thermoelectric measurement has been demonstrated for individual carbon nanotubes (Small et al. 2003) and PbSe nanowires (Liang et al. 2009). Different heater, thermometer, electrode designs have been reported. For the design shown in Figure 21.8a, A–B is the heater line and C–D is the resistance thermometer line. For measuring the four-probe electrical resistance, the two outer electrodes (C and J) can be used as the current source and drain, and the two inner electrodes (H and I) serve as the voltage probes. For Seebeck coefficient measurement, the four-probe resistance of the thermometer between F and G can be measured using a small sinusoidal excitation current and a lock-in amplifier when the heater line is heated by a direct current. The average temperature rise (ΔT) of the thermometer line can be obtained from the measured resistance increase (ΔR) of the thermometer line.

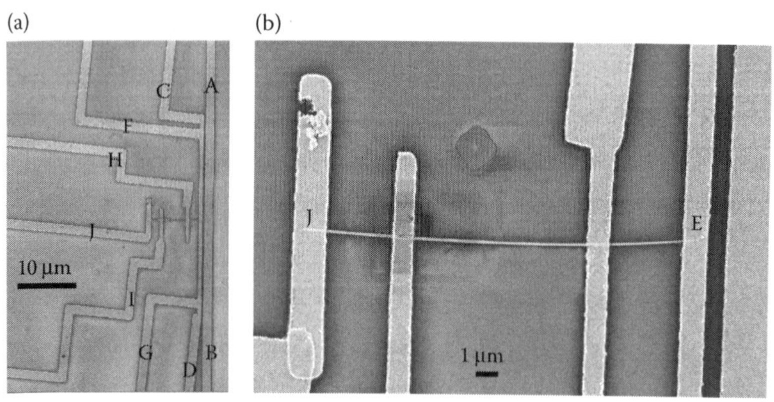

FIGURE 21.8 (a,b) SEMs of a device for measuring the Seebeck coefficient and electrical conductivity of a supported nanowire. (Reprinted with permission from Seol, J.H. et al. Measurement and analysis of thermopower and electrical conductivity of an indium antimonide nanowire from a vapor-liquid-solid method. *Journal of Applied Physics* 101(2):023706–6. Copyright 2007, American Institute of Physics.)

The thermovoltage (ΔV) between J and E in Figure 21.8b is measured and used to calculate the Seebeck coefficient as $S = \Delta V/\Delta \bar{T}$.

For the design shown in Figure 21.8, the temperature distribution on the device surface has been calculated using a finite-element method. The temperature rise ΔT_E of point E in Figure 21.8b, which was the contact point between the nanowire and the thermometer line, was found to be 16.5% larger than $\Delta \bar{T}$ (Seol et al. 2007). Three additional electrodes contacted the nanowire. The temperature rise ΔT_J at the contact between the nanowire and electrode J in Figure 21.8b was found in the thermal simulation to be about 4.8% of $\Delta \bar{T}$. Thus, the temperature difference $\Delta T_{E\text{-}J}$ between the two contact points was estimated to be about 12% larger than $\Delta \bar{T}$. This simulation result suggests that the as-obtained Seebeck coefficient $S = \Delta V/\Delta \bar{T}$ could be about 12% higher than the actual S value of $\Delta V/\Delta T_{E\text{-}J}$.

21.5 Mesa Structures for Thin Film Cross-Plane Measurements

While the 3ω method and the time domain thermal reflectance (TDTR) techniques have been widely used for measuring the cross-plane thermal conductivity, several methods have been investigated for measuring the cross-plane Seebeck coefficient, electrical conductivity, and *ZT*.

Venkatasubramanian and coworkers (Venkatasubramanian et al. 2001; Venkatasubramanian 2001) reported a the transmission line model (TLM) to measure the cross-plane electrical resistance (R) and conductivity of Bi_2Te_3/Sb_2Te_3 SL thin-film mesa structure shown in Figure 21.9. In this method, the voltage drop V between the two metal electrodes is recorded when a direct current I flows between the two electrodes. Peltier heating and cooling at the two metal-semiconductor interfaces result in a temperature difference (ΔT) between the two metal electrodes. When the current is turned off, the IR voltage drop vanishes rapidly within the dielectric relaxation time; whereas the thermovoltage $S\Delta T$ decays at a slower rate depending on the thermal time constant. Based on the two different timescales, $S\Delta T$ is determined to be the voltage drop V_0 between the two electrodes right after I is turned off, whereas the IR voltage drop is taken as the V_R value determined from Figure 21.9. The cross-plane electrical resistivity is obtained from the slope of the measured $AV_R/2I$ versus mesa height (h) curve, where A is the cross section of the area of the mesa. Moreover, when the external heat loss from the two metal electrodes is ignored, the measured thermovoltage is related to the cross-plane Seebeck coefficient and thermal conductivity of the thin film according to

$$V_0 = (S_\perp^2/\kappa_\perp)\mathrm{TI}(2h/\mathrm{A}) \qquad (21.24)$$

where T is the average temperature of the mesa. The $(S_\perp^2/\kappa_\perp)T$ term can be obtained from the slope of the measured $AV_0/2I$ versus h curve. Hence, the intrinsic thermoelectric figure of merit in the cross-plane direction can be obtained as

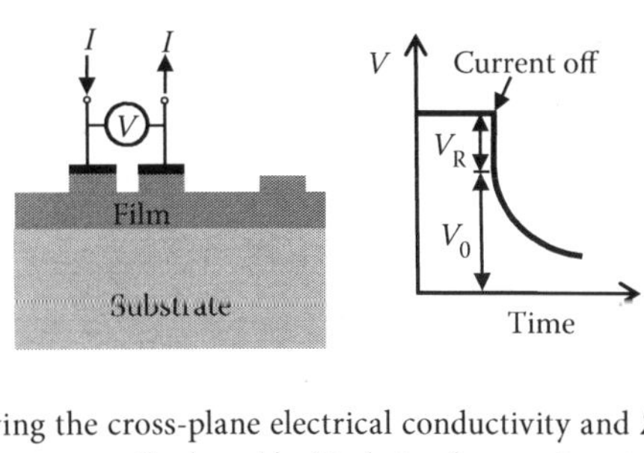

FIGURE 21.9 Schematic showing the cross-plane electrical conductivity and *ZT* measurement approach based on the TLM and the transient Harman method used by Venkatasubramanian et al. (2001).

$$ZT_{\perp} \equiv \frac{S_{\perp}^{2}\sigma_{\perp}T}{\kappa_{\perp}} = \frac{\partial(V_0/I)/\partial h}{\partial(V_R/I)/\partial h} \tag{21.25}$$

In addition, the extrinsic device figure of merit can be obtained as

$$ZT_e = V_0/V_R = ZT_{\perp}/(1+\sigma_{\perp}\rho_c/h) \tag{21.26}$$

where ρ_c is the contact electrical resistivity. The accuracy of this measurement is mainly limited by the V_0 uncertainty, which increases with decreasing film thickness because of a decreased thermal time constant.

For measuring the cross-plane S, Yang and coworkers developed a modulated heating method based on a mesa structure (Yang et al. 2002a,b). In their experiment, Si/Ge quantum dot SL films of different thickness were patterned into mesa structures on a conducting Si layer that serves as the bottom contact, as shown in Figure 21.10. A top metal contact layer was patterned on the SL film to serve as a voltage probe. A dielectric film was deposited on the top contact, followed by the patterning of an Au/Cr heater on top of the dielectric layer. When a sinusoidal current at a frequency ω was passed through the heater, a thermal wave at the 2ω frequency was generated throughout the sample. The 2ω temperature oscillation ($T_{2\omega}$) at the heater was measured using the heater as a resistance thermometer. Meanwhile, the 2ω voltage oscillation component ($V_{2\omega}$) was measured between the top and bottom electrical contacts. The measurement was conducted on two thin-film samples of different thickness. The differences $\Delta T_{2\omega}$ and $\Delta V_{2\omega}$ between the obtained $T_{2\omega}$ and $V_{2\omega}$ values for the two samples were used to obtain the cross-plane S according to

$$S = \Delta V_{2\omega}/\Delta T_{2\omega} \tag{21.27}$$

In addition, the cross-plane thermal conductivity of the SL film was obtained from the mesa thickness difference (Δt) between the two samples, the mesa area (A), the electrical heating power component ($P_{2\omega}$) at the 2ω frequency according to

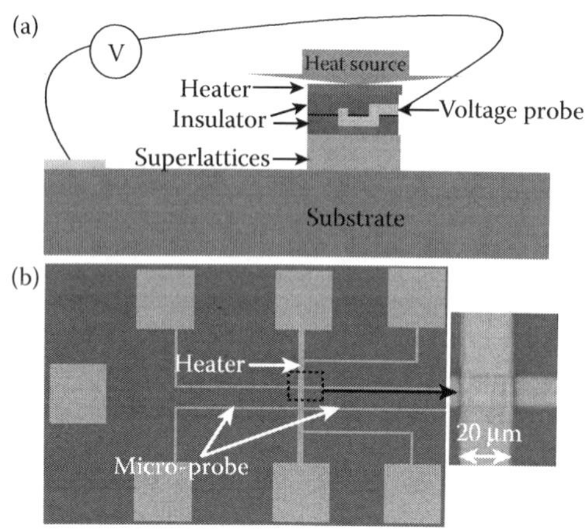

FIGURE 21.10 Cross-sectional schematic (a) and top surface photograph (b) of the device for cross-plane TE measurements by Yang et al. (2002). (Reprinted with permission from Yang, B. et al. 2002. Simultaneous measurements of Seebeck coefficient and thermal conductivity across superlattice. *Applied Physics Letters* 80(10):1758–1760. Copyright 2002. American Institute of Physics.)

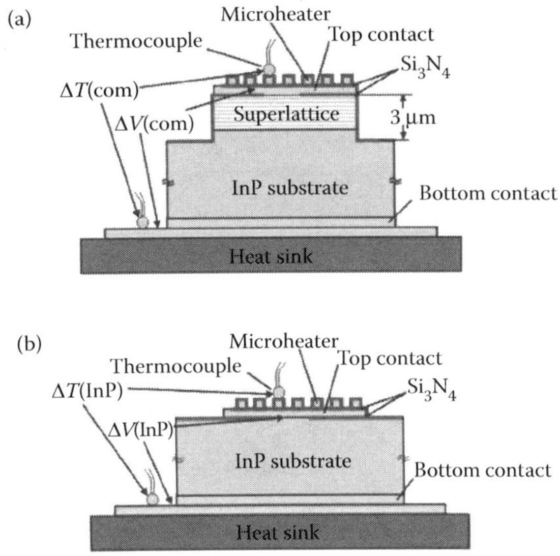

FIGURE 21.11 (a,b) Schematic of Zeng et al.'s cross-plane S measurement setup. (Reprinted with permission from Zeng, G. H. et al. Cross-plane Seebeck coefficient of ErAs : InGaAs/InGaAlAs superlattices. *Journal of Applied Physics* 101(3):0345020. Copyright 2007, American Institute of Physics.)

$$\kappa_{\perp} = P_{2\omega}\Delta t / A\Delta T_{2\omega} \tag{21.28}$$

Zeng and coworkers (Zeng et al. 2007) fabricated the mesa structure in Figure 21.11a to measure the cross-plane Seebeck coefficient of ErAs:InGaAs/InGaAlAs SL thin films. A top metal contact, an inter-metal dielectric layer made of silicon nitride, and a microheater were patterned on top of the mesa. The bottom contact was made to the back side of the InP substrate. When the microheater was heated by a direct current, the temperature drop ΔT_{com} across the mesa-substrate stack was measured using two thermocouples attached to the microheater and the bottom contact, respectively. Meanwhile, the Seebeck voltage drop ΔV_{com} across the stack was measured. Another similar test structure was made on a bare InP substrate and used as the reference device, as shown in Figure 21.11b. When the microheater was heated by a direct current, the temperature drop ΔT_{InP} and voltage drop ΔV_{InP} across the InP substrate were measured. Because there was no temperature sensor to measure the temperature at the interface between the SL and the InP substrate, a three-dimensional heat transfer model was used to calculate this temperature so as to obtain the ratio between the temperature drop (ΔT_{SL}) across the SL film and ΔT_{com}, that is, $r = \Delta T_{SL}/\Delta T_{com}$. The cross-plane S of the film was obtained from the measurement and simulation results according to

$$S = \frac{\Delta V_{com}}{r\Delta T_{com}} - \frac{(1-r)\Delta V_{InP}}{r\Delta T_{InP}} \tag{21.29}$$

21.6 Time Domain Thermal Reflectance Techniques

TDTR techniques have been increasingly employed for measuring the cross-plane thermal conductivity of thin films. In this method, an ultrafast laser pulse is used to heat a metal transducer layer deposited on a thin-film sample (Capinski and Maris 1996). A probe laser beam is used to measure the reflectance signal at different delay times. The signal-to-noise ratio is enhanced by using a lock-in amplifier to measure the modulation of the reflectance signal at the repetition rate of the heating pulse. The reflectance of the metal transducer depends approximately linearly with the temperature rise of the metal film, which in turn depends on the cross-plane thermal conductivity and interface thermal resistance. Cahill has described an approach to extracting the thermal conductivity and thermal interface resistance of a thin film from the ratio of the in-phase component and the out-of-phase component of the modulated thermal reflectance signal (Cahill 2004). In addition, his group observed that the as-measured thermal conductivity depends on the modulation frequency of the pump pulse for samples with long phonon mean free path compared to the thermal penetration depth of the modulated thermal wave (Koh and Cahill 2007). The reason is that when the thermal penetration depth is shorter than the mean free path of some long-wavelength phonons, these phonons are not thermalized within the nonuniform temperature zone, so that they do not fully contribute to heat diffusion. Consequently, the measured thermal conductivity at a very high modulation frequency is lower than that measured at a low frequency.

The TDTR method has also been employed to measure a densely packed InAs nanowire array filled with a low-thermal conductivity polymer (Persson et al. 2009). Finite-element analysis was used to extract the average thermal conductivity of the nanowires. At the low frequency limit, the finite-element analysis result was found to yield results close to that obtained from the effective medium approach. In addition, Schmidt and coworkers have developed a TDTR method to measure both the cross-plane and in-plane thermal conductivity of an anisotropic thin film by varying the heating laser beam size and modulation frequency (Schmidt et al. 2008).

21.7 Micro-Raman Spectroscopy Measurements

Raman spectroscopy methods have recently been employed to measure the temperature and temperature distribution of optically or electrically heated carbon nanotubes, silicon nanowires, and graphene. The anti-Stokes-to-Stokes ratio of a Raman peak can be used to probe the temperature of the zone center or zone boundary optical phonons (Chae et al. 2010; Faugeras et al. 2010). In addition, the Raman peak position down shifts with increasing temperature because increasing anharmonic scattering of optical phonons that are active in the Raman process (Bonini et al. 2007; Basko et al. 2009). Hence, the Raman peak shift can be used to obtain the equivalent temperature of the anharmonic scattering process. For the case of optically heated suspended graphene, the optical phonon temperature determined from the anti-Stokes-to-Stokes ratio was found to be at equilibrium with the anharmonic scattering temperature determined from the Raman G peak shift (Faugeras et al. 2010).

These Raman thermometry methods have been used to probe the thermal property of carbon nanotubes and graphene. Hsu and coworkers used the Raman G peak shift to determine the temperature at the Raman laser spot of a suspended nanotube (Hsu et al. 2008). A parabolic profile was observed when the laser spot was moved along the suspended nanotube. Because ballistic phonon transport would result in a constant temperature in the nanotube irrespective of the laser beam position, the observed parabolic temperature profile is attributed to diffusive phonon transport in these likely defective nanotube samples. The measured temperature profile could be used to obtain the ratio between the thermal contact resistance and the diffusive thermal resistance of the nanotube. However, the optical absorption by the nanotube was not measured in this work, so that the absolute values of the diffusive thermal resistance and thermal contact resistance were not obtained. To measure the optical absorption of a nanotube, Hsu and coworkers employed the Raman method to measure the temperature rise at and heat flow from the Raman laser spot of a carbon nanotube suspended between two suspended resistance thermometers, which were used to measure the temperature rises at the two ends of the nanotube during laser heating (Hsu et al. 2009). In a different work, the optical absorption of a silicon nanowire cantilever heated by a Raman laser beam was calculated from an optical absorption model, so as to obtain the thermal conductivity of the nanowire (Doerk et al. 2010).

Besides optical heating, electrically heated carbon nanotubes have also been probed using Raman thermometry. In short (~2 μm long), high-quality suspended carbon nanotube, nonequilibrium optical phonon populations were observed based on the shift of different Raman peaks, suggesting preferential coupling of hot electrons with certain optical phonon populations and a bottleneck for the hot optical phonons to be thermalized with other optical and acoustic phonons (Deshpande et al. 2009). In comparison, local thermal equilibrium among different phonon branches could be established in over 30 μm long (Li et al. 2009) or defective (Hsu et al. 2010) suspended nanotubes under an electrical bias. For these samples, a parabolic temperature profile along the nanotube can be measured and used to obtain the diffusive thermal resistance and contact thermal resistance values with the assumption of uniform electrical heating in the suspended nanotube. It is important that the measurement is conducted with the sample in vacuum, because parasitic heat loss to the surrounding gas can be appreciable especially for long or defective nanotube samples (Hsu et al. 2010).

Raman spectroscopy has also been employed to measure the thermal properties of graphene. Balandin and coworkers employed a Raman laser beam to heat a rectangular monolayer graphene flake suspended across a trench (Balandin et al. 2008). The temperature at the laser spot of the graphene was determined from the Raman G peak shift. The optical absorption of the graphene monolayer was determined by comparing the Raman peak intensity values measured on the graphene monolayer and graphite. This method obtained an optical absorption of about 6.5% per pass of the laser beam through the graphene monolayer. Based on the measured graphene temperature and optical absorption, the near-room temperature thermal conductivity of suspended graphene was obtained to be between $(4.84 \pm 0.44 \text{ to } 5.30 \pm 0.48) \times 10^3$ W m^{-1} K^{-1} (Balandin et al. 2008; Ghosh et al. 2008; Ghosh et al. 2009). However, a direct transmission measurement was used to determine the optical absorption of clean graphene monolayer to be 2.3% (Nair et al. 2008). Using this optical absorption value, Faugeras and coworkers obtained a thermal conductivity of 600 W m^{-1} K^{-1} for a graphene monolayer suspended over a 44-μm-diameter hole when optical absorption of the Raman laser beam raises the center of the monolayer to a temperature of about 660 K, which was measured from the anti-Stokes-to-Stokes ratio (Faugeras et al. 2010). They suggested that the discrepancy between the two sets of thermal conductivity values was caused by the different assumptions in the optical absorption value. At the same time, the thermal conductivity of graphene monolayer grown by CVD and suspended over 3.8 μm diameter hole was measured to decrease from $(2.5 \pm 1.1) \times 10^3$ W m^{-1} K^{-1} to $(1.4 \pm 0.5) \times 10^3$ W m^{-1} K^{-1} when the graphene temperature measured from the Raman G peak shift of the optically heated graphene increased from about 350 K to about 500 K (Cai et al. 2010). The results suggest that the discrepancy between the two other results can also be caused by different temperatures at the laser spot of the graphene. In this work, the optical absorption of the suspended graphene was measured to be 3.3 ± 1.1% by measuring the transmission through the graphene sample.

For Raman measurement of thermal transport in graphene, the circular or so-called Corbino membrane geometry matches the laser beam intensity profile and heat flow direction better than the rectangular flake geometry. For the Corbino membrane geometry, one can define the measured thermal resistance of the graphene sample as $R_m \equiv (T_m - T_0)/Q$, where T_m is the graphene temperature measured by the Raman laser beam, T_0 is the substrate temperature, and Q is the optical absorption of the graphene. The as-defined R_m contains a thermal resistance contribution from the graphene region in contact with the support, $R_c \equiv (T_1 - T_0)/Q$, and a contribution from the suspended graphene, $R_g \equiv (T_m - T_1)/Q$, where T_1 is the temperature at the edge of the suspended graphene.

In addition, the Raman measurement based on the peak position shift mainly probes the temperature profile and thermal transport of the phonons that are involved in anharmonic scattering of the Raman active optical phonons. For the case that the phonons probed by the Raman laser are thermalized locally within a scale smaller than the laser beam spot size, the as-defined R_g equals the diffusive thermal resistance ($R_{d,g}$) that is related to the thermal conductivity contribution (κ) from the diffusive phonons according to (Cai et al. 2010)

$$\kappa = \frac{\ln(R/r_0)}{2\pi t R_{d,g}}\alpha \tag{21.30}$$

where R is the radius of the hole, $t = 0.335$ nm is the graphene monolayer thickness, and r_0 is the radius of the Gaussian laser beam. Without the factor α, this equation can be obtained from the diffusive thermal resistance expression of a cylindrical wall (Incropera and Dewitt 2002). The α factor accounts for the Gaussian beam profile and is close to unity for the suspended graphene radii R range between 3 and 9 μm and the r_0 value of in the range between 170 and 300 nm. If the mean free paths of the phonons are much larger than the laser beam size so that the phonons are not thermalized locally within the laser spot, R_g would consist of an additional ballistic resistance component ($R_{b,g}$) (Chen 1996; Cai et al. 2010).

In the work by Cai and coworkers (Cai et al. 2010; Chen et al. 2011), the thermal contact resistance was calculated using the sample geometries and the values of Au-supported graphene thermal conductivity and the Au–graphene interface thermal conductance determined from a Raman measurement of the supported graphene region. The R_c value can in principle be determined with the use of a second probe laser beam to measure the temperature distribution in the circular graphene membrane when the heating laser beam is focused at the center of the graphene membrane.

With the calculated R_c subtracted from the R_m values to obtain R_g and with $R_{b,g}$ assumed to be negligible compared to $R_{d,g}$, Cai and coworkers obtained the thermal conductivity κ versus temperature T_m measured at the beam spot. The neglect of the ballistic resistance component $R_{b,g}$ could underestimate κ if the phonons are not thermalized locally within the laser spot. The as-calculated κ are shown in Figure 21.12. The rather large uncertainty in the measured thermal conductivity is due to errors in the measured temperature and optical absorption. Nevertheless, the measured thermal conductivity values of their CVD graphene samples are higher than the reported basal-plane thermal conductivity data of pyrolytic graphite samples (Slack 1962; Hooker et al. 1965; Taylor 1966), and lie in between the data reported by Balandin et al. (2008) and Faugeras et al. (2010) for mechanically exfoliated graphene samples. In addition, under a gray medium assumption where the phonon scattering mean free path is taken to be independent of phonon frequency, the phonon scattering mean free path at temperature 400 K was

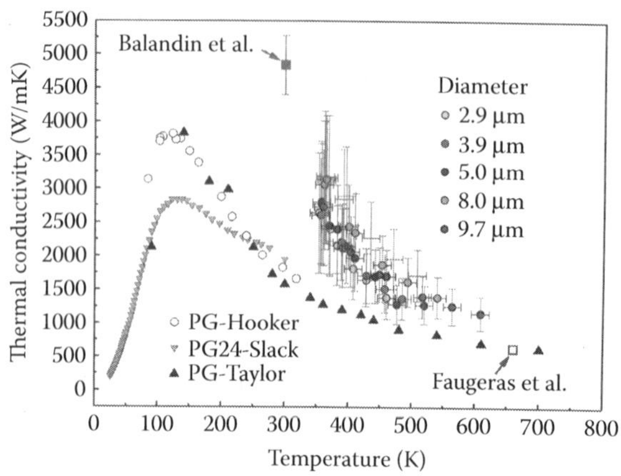

FIGURE 21.12 **(See color insert.)** Thermal conductivity of the suspended CVD graphene (Chen et al. 2011) as a function of the measured temperature of the graphene monolayer suspended in vacuum over holes of various diameters. Shown in comparison are the literature thermal conductivity data of pyrolytic graphite samples as a function of temperature (Hooker et al. 1965; Taylor 1966; Slack 1962), and the thermal conductivity data of mechanically exfoliated graphene monolayer reported in Balandin et al. (2008) and Faugeras et al. (2010). The figure was prepared by Shanshan Chen.

calculated from the thermal conductivity measured on the CVD graphene to be in the range between 330 and 400 nm, which is comparable to the laser beam diameter. Hence, the diffusive phonons are mostly thermalized within the laser beam spot so that the effect of the ballistic resistance is not expected to be large especially at higher temperatures with shorter phonon mean free paths.

21.8 Conclusion

A number of methods for characterizing the thermal and thermoelectric properties of thin films, individual nanowires, nanotubes, and graphene are reviewed in this chapter. For thin film samples, the 3ω method (Cahill 1990) and the TDTR technique (Cahill 2004) have been established for measuring the cross-plane thermal conductivity as well as the interface thermal resistance. In comparison, measurements of cross-plane Seebeck coefficient and electrical conductivity have remained a significant challenge, although different mesa structures have been fabricated to measure these two cross-plane properties as well as the cross-plane *ZT* of a thin-film sample (Venkatasubramanian et al. 2001; Yang et al. 2002; Zeng et al. 2007).

For the in-plane direction of thin films, parasitic heat loss to the substrate makes thermal conductivity measurement complicated. The 3ω method with different heater line widths has been used in conjunction with a solution of the 2D heat conduction equation to extract both the cross-plane thermal conductivity and in-plane thermal conductivity (Ju et al. 1999). To completely eliminate the heat loss to the substrate, several suspended resistance thermometer microdevices have been developed for measuring the in-plane thermal conductivity (Shi et al. 2003; Mavrokefalos et al. 2007; Jain and Goodson 2008). Similarly, for measurements of the in-plane Seebeck coefficient and electrical conductivity, the substrate can present a problem at high temperatures, where thermally generated charge carriers in some semi-insulating substrates can cause large uncertainties in the Seebeck coefficient and electrical conductivity measured on the supported thin film. This challenge has been addressed by using suspended thin-film samples including the method based on suspended resistance thermometer microdevices (Mavrokefalos et al. 2007), which obtain all the three in-plane properties entering the *ZT* expression for the same thin-film sample. The contact thermal and electrical resistances can also be determined with the use of the four-probe thermoelectric measurement procedure based on the suspended microdevices. However, this four-probe method has not been ineffective for determining the contact thermal resistance to graphene and nanotube samples, where the Seebeck coefficient of the monatomic nanostructures depends sensitively on the environment and can be different for the supported and suspended segments of the sample.

For nanowires and nanotubes, the transport properties of interest are those along the axial direction as well as at the interfaces between the nanostructures and the environment. The self-heating 3ω method has been established for measuring the thermal conductivity of individual suspended nanowires (Lu et al. 2001). Moreover, suspended resistance thermometer microdevices have been developed for direct four-probe or two-probe thermal conductance measurements of nanowires and nanotubes (Shi et al. 2003; Zhou et al. 2007). Some of these devices can also allow for the measurement of the Seebeck coefficient and electrical conductivity of the same nanowire sample used for the four-probe thermal conductivity measurements (Zhou et al. 2007), so that the *ZT* of the nanowires can be obtained. The measured Seebeck coefficient and electrical conductivity can be further used to extract the Fermi energy, carrier concentration and mobility in the nanowire sample. However, making clean ohmic electrical contact to nanowires assembled on a suspended microdevice has remained a challenge. In addition, while TEM and Raman microscopy have been employed for determining the structure and cross section of nanotubes and graphene samples measured by the suspended microdevices, accurate determination of the charge and heat-transport cross section of noncylindrical, irregular, or rough nanowire samples assembled on the microdevices has remained a challenge, which can lead to considerable uncertainty in the reported thermal and electrical conductivity values.

Microfabricated resistance thermometer devices supported on a substrate have also been developed to measure the Seebeck coefficient and electrical conductivity of individual nanowires and nanotubes as a function of a gate field (Small et al. 2003). This capability is important for tuning the Fermi energy and carrier concentration that sensitively affect the power factor in the nanostructures.

In addition, micro-Raman spectroscopy methods have been employed for measuring the thermal conductivity of individual nanotubes (Hsu et al. 2008; Li et al. 2009), silicon nanowires (Doerk et al. 2010), and graphene monolayers (Balandin et al. 2008; Faugeras et al. 2010; Cai et al. 2010). Compared to the suspended resistance thermometer microdevices, the Raman techniques can be used readily for determining the contact thermal resistance to suspended nanotube and graphene samples, but are usually subjected to relatively large measurement uncertainty because of the limited temperature sensitivity of the Raman thermometry technique, uncertainty in the optical absorption, and issues related to nonequilibrium transport of different energy carriers.

Acknowledgments

The author's research in thermal transport and thermoelectric conversion has been supported by the U.S. National Science Foundation, Office of Naval Research, and Department of Energy. Some of the results discussed in this chapter are based on collaborative research between the author and a large number of researchers. A partial list of some of the recent collaborators includes Anastassios Mavrokefalos, Feng Zhou, Arden Moore, Michael Pettes, Jae Hun Seol, Insun Jo, Shanshan Chen, Weiwei Cai, I-Kai Hsu, Steve Cronin, Rodney Ruoff, Zhen Yao, David Johnson, and Ali Shakouri.

References

Balandin, A. A., S. Ghosh, W. Z. Bao, I. Calizo, D. Teweldebrhan, F. Miao, and C. N. Lau. 2008. Superior thermal conductivity of single-layer graphene. *Nano Letters* 8(3):902–907.

Basko, D. M., S. Piscanec, and A. C. Ferrari. 2009. Electron-electron interactions and doping dependence of the two-phonon Raman intensity in graphene. *Physical Review B* 80:165413.

Bonini, N., M. Lazzeri, N. Marzari, and F. Mauri. 2007. Phonon anharmonicities in graphite and graphene. *Physical Review Letters* 99:176802.

Borca-Tasciuc, T., A. R. Kumar, and G. Chen. 2001. Data reduction in 3 omega method for thin-film thermal conductivity determination. *Review of Scientific Instruments* 72(4):2139–2147.

Boukai, A. I., Y. Bunimovich, J. Tahir-Kheli, J. K. Yu, W. A. Goddard, and J. R. Heath. 2008. Silicon nanowires as efficient thermoelectric materials. *Nature* 451(7175):168–171.

Broido, D. A. and T. L. Reinecke. 1995. Effect of superlattice structure on the thermoelectric figure of merit. *Physical Review B* 51(19):13797–13800.

Cahill, D. G. 1990. Thermal-conductivity measurement from 30 K to 750 K—The 3-omega method. *Review of Scientific Instruments* 61(2):802–808.

Cahill, D. G. 2002. Erratum: Thermal conductivity measurement from 30 to 750 K: The 3 omega method. *Review of Scientific Instruments* 61, 802, 1990: AIP.

Cahill, D. G. 2004. Analysis of heat flow in layered structures for time-domain thermoreflectance. *Review of Scientific Instruments* 75(12):5119–5122.

Cahill, D. G., M. Katiyar, and J. R. Abelson. 1994. Thermal-conductivity of alpha-Si:H thin-films. *Physical Review B* 50(9):6077–6081.

Cai, W. W., A. L. Moore, Y. W. Zhu, X. S. Li, S. S. Chen, L. Shi, and R. S. Ruoff. 2010. Thermal transport in suspended and supported monolayer graphene grown by chemical vapor deposition. *Nano Letters* 10(5):1645–1651.

Capinski, W. S. and H. J. Maris. 1996. Improved apparatus for picosecond pump-and-probe optical measurements. *Review of Scientific Instruments* 67(8):2720–2726.

Chae, D. H., B. Krauss, K. von Klitzing, and J. H. Smet. 2010. Hot phonons in an electrically biased graphene constriction. *Nano Letters* 10(2):466–471.

Champness, C. H. and A. L. Kipling. 1966. The Hall and Seebeck effects in nonstoichiometric bismuth telluride. *Canadian Journal of Physics* 44:769–788.

Chen, G. 1996. Nonlocal and nonequilibrium heat conduction in the vicinity of nanoparticles. *Journal of Heat Transfer-Transactions of the ASME* 118(3):539–545.

Chen, S., A. L. Moore, W. Cai, J. W. Suk, J. An, C. Mishra, C. Amos et al. 2011. Raman measurements of thermal transport in suspended monolayer graphene of variable sizes in vacuum and gaseous environments. *ACS Nano* 5:321–328.

Chen, Y. F., D. Y. Li, J. R. Lukes, Z. H. Ni, and M. H. Chen. 2005. Minimum superlattice thermal conductivity from molecular dynamics. *Physical Review B* 72(17):6.

Choi, T. Y., D. Poulikakos, J. Tharian, and U. Sennhauser. 2006. Measurement of the thermal conductivity of individual carbon nanotubes by the four-point three-omega method. *Nano Letters* 6(8):1589–1593.

Deshpande, V. V., S. Hsieh, A. W. Bushmaker, M. Bockrath, and S. B. Cronin. 2009. Spatially resolved temperature measurements of electrically heated carbon nanotubes. *Physical Review Letters* 102(10):4.

Doerk, G. S., C. Carraro, and R. Maboudian. 2010. Single nanowire thermal conductivity measurements by Raman thermography. *ACS Nano* 4(8):4908–4914.

Durkop, T., S. A. Getty, E. Cobas, and M. S. Fuhrer. 2003. Extraordinary mobility in semiconducting carbon nanotubes. *Nano Letters* 4(1):35–39.

Faugeras, C., B. Faugeras, M. Orlita, M. Potemski, R. R. Nair, and A. K. Geim. 2010. Thermal conductivity of graphene in corbino membrane geometry. *ACS Nano* 4(4):1889–1892.

Fujii, M., X. Zhang, H. Q. Xie, H. Ago, K. Takahashi, T. Ikuta, H. Abe, and T. Shimizu. 2005. Measuring the thermal conductivity of a single carbon nanotube. *Physical Review Letters* 95(6):065502.

Ghosh, S., I. Calizo, D. Teweldebrhan, E. P. Pokatilov, D. L. Nika, A. A. Balandin, W. Bao, F. Miao, and C. N. Lau. 2008. Extremely high thermal conductivity of graphene: Prospects for thermal management applications in nanoelectronic circuits. *Applied Physics Letters* 92(15):151911.

Ghosh, S., D. L. Nika, E. P. Pokatilov, and A. A. Balandin. 2009. Heat conduction in graphene: Experimental study and theoretical interpretation. *New Journal of Physics* 11:095012.

Harman, T. C., P. J. Taylor, M. P. Walsh, and B. E. LaForge. 2002. Quantum dot superlattice thermoelectric materials and devices. *Science* 297(5590):2229–2232.

Heremans, J. P., V. Jovovic, E. S. Toberer, A. Saramat, K. Kurosaki, A. Charoenphakdee, S. Yamanaka, and G. J. Snyder. 2008. Enhancement of thermoelectric efficiency in PbTe by distortion of the electronic density of states. *Science* 321(5888):554–557.

Heron, J. S., T. Fournier, N. Mingo, and O. Bourgeois. 2009. Mesoscopic size effects on the thermal conductance of silicon nanowire. *Nano Letters* 9(5):1861–1865.

Hicks, L. D. and M. S. Dresselhaus. 1993a. Effect of quantum-well structures on the thermoelectric figure of merit. *Physical Review B* 47(19):12727–12731.

Hicks, L. D. and M. S. Dresselhaus. 1993b. Thermoelectric figure of merit of a one-dimensional conductor. *Physical Review B* 47(24):16631–16634.

Hochbaum, A. I., R. K. Chen, R. D. Delgado, W. J. Liang, E. C. Garnett, M. Najarian, A. Majumdar, and P. D. Yang. 2008. Enhanced thermoelectric performance of rough silicon nanowires. *Nature* 451(7175):163-U5.

Hooker, C. N., A. R. Ubbelohd, and D. A. Young. 1965. Anistropy of thermal conductance in near-ideal graphene. *Proceedings of the Royal Society of London Series a-Mathematical and Physical Sciences* 284(1396):17.

Hsu, I. K., R. Kumar, A. Bushmaker, S. B. Cronin, M. T. Pettes, L. Shi, T. Brintlinger, M. S. Fuhrer, and J. Cumings. 2008. Optical measurement of thermal transport in suspended carbon nanotubes. *Applied Physics Letters* 92(6):063119.

Hsu, I. K., M. T. Pettes, M. Aykol, L. Shi, and S. B. Cronin. 2010. The effect of gas environment on electrical heating in suspended carbon nanotubes. *Journal of Applied Physics* 108(8):084307.

Hsu, I. K., M. T. Pows, A. Bushmaker, M. Aykol, L. Shi, and S. B. Cronin. 2009. Optical absorption and thermal transport of individual suspended carbon nanotube bundles. *Nano Letters* 9(2):590–594.

F. P. Incropera, and D. P. Dewitt. 2002. *Fundamentals of Heat and Mass Transfer*. 5th ed. New York: John Wiley & Sons.

F. P. Incropera and D. P. Dewitt. 2006. *Introduction to heat transfer*. 5th ed. Hoboken, NJ: Wiley.

Jain, A. and K. E. Goodson. 2008. Measurement of the thermal conductivity and heat capacity of freestanding shape memory thin films using the 3 omega method. *Journal of Heat Transfer-Transactions of the ASME* 130(10):7.

Ju, Y. S., K. Kurabayashi, and K. E. Goodson. 1999. Thermal characterization of anisotropic thin dielectric films using harmonic Joule heating. *Thin Solid Films* 339(1–2):160–164.

Koh, Y. K. and D. G. Cahill. 2007. Frequency dependence of the thermal conductivity of semiconductor alloys. *Physical Review B* 76(7):5.

Koh, Y. K., C. J. Vineis, S. D. Calawa, M. P. Walsh, and D. G. Cahill. 2009. Lattice thermal conductivity of nanostructured thermoelectric materials based on PbTe. *Applied Physics Letters* 94(15):3.

Li, Q. W., C. H. Liu, X. S. Wang, and S. S. Fan. 2009. Measuring the thermal conductivity of individual carbon nanotubes by the Raman shift method. *Nanotechnology* 20(14):145702:1–5.

Liang, W. J., A. I. Hochbaum, M. Fardy, O. Rabin, M. J. Zhang, and P. D. Yang. 2009. Field-effect modulation of seebeck coefficient in single PbSe nanowires. *Nano Letters* 9(4):1689–1693.

Lin, Y. M., X. Z. Sun, and M. S. Dresselhaus. 2000. Theoretical investigation of thermoelectric transport properties of cylindrical Bi nanowires. *Physical Review B* 62(7):4610–4623.

Liu, W. J. and M. Asheghi. 2006. Thermal conductivity measurements of ultra-thin single crystal silicon layers. *Journal of Heat Transfer-Transactions of the ASME* 128(1):75–83.

Lu, L., W. Yi, and D. L. Zhang. 2001. 3 omega method for specific heat and thermal conductivity measurements. *Review of Scientific Instruments* 72(7):2996–3003.

Mahan, G. D. and J. O. Sofo. 1996. The best thermoelectric. *Proceedings of the National Academy of Sciences of the United States of America* 93(15):7436–7439.

Mavrokefalos, A., A. L. Moore, M. T. Pettes, L. Shi, W. Wang, and X. G. Li. 2009. Thermoelectric and structural characterizations of individual electrodeposited bismuth telluride nanowires. *Journal of Applied Physics* 105(10):104318.

Mavrokefalos, A., M. T. Pettes, F. Zhou, and L. Shi. 2007. Four-probe measurements of the in-plane thermoelectric properties of nanofilms. *Review of Scientific Instruments* 78(3):034901.

Mingo, N. 2004. Thermoelectric figure of merit and maximum power factor in III-V semiconductor nanowires. *Applied Physics Letters* 84(14):2652–2654.

Mingo, N. 2006. Thermoelectric figure of merit and maximum power factor in III–V semiconductor nanowires (vol 84, pg 2652, 2004). *Applied Physics Letters* 88(14):149902.

Moore, A. L. and L. Shi. 2011. On errors in thermal conductivity measurements of suspended and supported nanowires using micro-thermometer devices from low to high temperatures. *Measurement Science and Technology* 22(1):015103.

Nair, R. R., P. Blake, A. N. Grigorenko, K. S. Novoselov, T. J. Booth, T. Stauber, N. M. R. Peres, and A. K. Geim. 2008. Fine structure constant defines visual transparency of graphene. *Science* 320(June):1308.

Persson, A. I., Y. K. Koh, D. G. Cahill, L. Samuelson, and H. Linke. 2009. Thermal conductance of InAs nanowire composites. *Nano Letters* 9(12):4484–4488.

Pettes, M. T., and L. Shi. 2009. Thermal and structural characterizations of individual single-, double-, and multi-walled carbon nanotubes. *Advanced Functional Materials* 19(24):3918–3925.

Pop, E., D. Mann, Q. Wang, K. Goodson, and H. J. Dai. 2006. Thermal conductance of an individual single-wall carbon nanotube above room temperature. *Nano Letters* 6(1):96–100.

Rabina, O., Y. M. Lin, and M. S. Dresselhaus. 2001. Anomalously high thermoelectric figure of merit in Bi1-xSbx nanowires by carrier pocket alignment. *Applied Physics Letters* 79(1):81–83.

Schmidt, A. J., X. Chen, and G. Chen. 2008. Pulse accumulation, radial heat conduction, and anisotropic thermal conductivity in pump-probe transient thermoreflectance. *Review of Scientific Instruments* 79(11):114902–114909.

Seol, J. H., I. Jo, A. L. Moore, L. Lindsay, Z. H. Aitken, M. T. Pettes, X. S. Li et al. 2010. Two-dimensional phonon transport in supported graphene. *Science* 328(5975):213–216.

Seol, J. H., A. L. Moore, S. K. Saha, F. Zhou, L. Shi, Q. L. Ye, R. Scheffler, N. Mingo, and T. Yamada. 2007. Measurement and analysis of thermopower and electrical conductivity of an indium antimonide nanowire from a vapor-liquid-solid method. *Journal of Applied Physics* 101(2):023706-6.

Shi, L., D. Y. Li, C. H. Yu, W. Y. Jang, D. Kim, Z. Yao, P. Kim, and A. Majumdar. 2003. Measuring thermal and thermoelectric properties of one-dimensional nanostructures using a microfabricated device. *Journal of Heat Transfer-Transactions of the ASME* 125(5):881–888.

Slack, G. A. 1962. Anisotropic thermal conductivity of pyrolytic graphite. *Physical Review* 127:694–701.

Small, J. P., K. M. Perez, and P. Kim. 2003. Modulation of thermoelectric power of individual carbon nanotubes. *Physical Review Letters* 91(25):4.

Tang, J. Y., H. T. Wang, D. H. Lee, M. Fardy, Z. Y. Huo, T. P. Russell, and P. D. Yang. 2010. Holey silicon as an efficient thermoelectric material. *Nano Letters* 10(10):4279–4283.

Taylor, R. 1966. The thermal conductivity of pyrolytic graphite. *Philosophical Magazine* 13:157–166.

Venkatasubramanian, R. 2000. Lattice thermal conductivity reduction and phonon localizationlike behavior in superlattice structures. *Physical Review B* 61(4):3091–3097.

Venkatasubramanian, R. 2001. Phonon blocking electron transmitting superlattice structures as advanced thin film thermoelectric materials. In *Recent Trends in Thermoelectric Materials Research iii: Semiconductors and Semimetals,* vol. 71 (Chapter 4, pp. 175–201). San Diego, CA: Academic Press.

Venkatasubramanian, R., E. Siivola, T. Colpitts, and B. O'Quinn. 2001. Thin-film thermoelectric devices with high room-temperature figures of merit. *Nature* 413(6856):597–602.

Vineis, C. J., T. C. Harman, S. D. Calawa, M. P. Walsh, R. E. Reeder, R. Singh, and A. Shakouri. 2008. Carrier concentration and temperature dependence of the electronic transport properties of epitaxial PbTe and PbTe/PbSe nanodot superlattices. *Physical Review B* 77(23):14.

Yang, B., J. L. Liu, K. L. Wang, and G. Chen. 2002a. Simultaneous measurements of Seebeck coefficient and thermal conductivity across superlattice. *Applied Physics Letters* 80(10):1758–1760.

Yang, B., W. L. Liu, J. L. Liu, K. L. Wang, and G. Chen. 2002b. Measurements of anisotropic thermoelectric properties in superlattices. *Applied Physics Letters* 81(19):3588–3590.

Yi, W., L. Lu, D. L. Zhang, Z. W. Pan, and S. S. Xie. 1999. Linear specific heat of carbon nanotubes. *Physical Review B* 59(14):R9015–R9018.

Yu, C. H., S. Saha, J. H. Zhou, L. Shi, A. M. Cassell, B. A. Cruden, Q. Ngo, and J. Li. 2006. Thermal contact resistance and thermal conductivity of a carbon nanofiber. *Journal of Heat Transfer-Transactions of the ASME* 128(3):234–239.

Zeng, G. H., J. M. O. Zide, W. Kim, J. E. Bowers, A. C. Gossard, Z. X. Bian, Y. Zhang, A. Shakouri, S. L. Singer, and A. Majumdar. 2007. Cross-plane Seebeck coefficient of ErAs : InGaAs/InGaAlAs superlattices. *Journal of Applied Physics* 101(3):034502.

Zhou, F., A. L. Moore, M. T. Pettes, Y. Lee, J. H. Seol, Q. L. Ye, L. Rabenberg, and L. Shi. 2010. Effect of growth base pressure on the thermoelectric properties of indium antimonide nanowires. *Journal of Physics D-Applied Physics* 43(2):9.

Zhou, F., J. Szczech, M. T. Pettes, A. L. Moore, S. Jin, and L. Shi. 2007. Determination of transport properties in chromium disilicide nanowires via combined thermoelectric and structural characterizations. *Nano Letters* 7:1649–1654.

Ziman, J. M. 1962. *Electrons and Phonons: The Theory of Transport Phenomena in Solids.* Oxford: Clarendon Press.

Zink, B. L., B. Revaz, J. J. Cherry, and F. Hellman. 2005. Measurement of thermal conductivity of thin films with a Si-N membrane-based microcalorimeter. *Review of Scientific Instruments* 76(2):024901.

22

Microchips and Methods for the Characterization of Thermoelectric Transport Properties of Nanostructures

Friedemann Völklein
RheinMain University of Applied Sciences

Daniel Huzel
RheinMain University of Applied Sciences

Heiko Reith
RheinMain University of Applied Sciences

Matthias Schmitt
RheinMain University of Applied Sciences

22.1 Introduction

In recent years, numerous studies on the electrical transport properties of semiconducting and metallic nanostructures have been performed because of their possible applications in future miniaturized devices, such as thermoelectric sensors or generators. In particular, the electrical transport properties of quasi one-dimensional structures such as nanowires were investigated intensively [1–3].

When the structure size is comparable to the electronic mean free path, finite-size effects and quantum size effects are expected to occur [4–6]. These effects are anticipated to also affect the thermal transport mechanisms. Besides a decrease in the thermal conductivity λ, theories predict an increase in Seebeck coefficient S and thermoelectric efficiency $z = S^2\sigma/\lambda$ [7–9] when reducing the dimensionality of the specimen (e.g., by decreasing the diameter of nanowires), where σ denotes the electrical conductivity.

Therefore, nanowires are promising candidates for new thermoelectric devices (TED) with significantly increased thermoelectric figure of merit.

Experimental verification of these predicted effects requires the simultaneous measurement of the above-mentioned transport coefficients (S, σ, λ) on single nanowires. In contrast to the extensive studies on electrical transport, only few experimental investigations of thermal conductivity and Seebeck coefficient of nanowires have been reported. For thermal conductivity experiments, any thermal bypass has to be avoided, that is, these studies must not be performed on single wires lying on a substrate but on individual suspended needles. Seebeck effect measurements with sufficiently high measuring accuracy require a temperature difference along the nanowire axis (between the electrical contacts) of few Kelvin. Therefore, at nanowires with a distance of their contact layers of the order of 10 μm, a temperature gradient of the order of 10^5 K/m has to be established between the contacts. Furthermore, the thermoelectric voltage U caused by this temperature gradient and the temperatures at the nanowire contacts (i.e., at two positions with a distance of only few micron) have to be measured precisely. Obviously, sophisticated microstructured devices for the determination of the thermoelectric transport coefficients are required [10,11]. Measurements of S, σ, and λ have been performed on assemblies of many identical nanowires of bismuth [11] and silicon [12] and, recently, on individual silicon nanowires [13].

We present new methods and microchips for the measurement of electrical and thermal conductivity and Seebeck coefficient on individual nanowires, which were electrochemically deposited by ion-track technology [14–16] and on nanostructures, that have been realized by focused electron/ion beam-induced deposition (FEBID/FIBID). Furthermore, FEM simulations for design optimization and the fabrication process for a z-microchip, which can be applied for the complete characterization of the thermoelectric efficiency of an individual nanowire, are demonstrated. With respect to applications, the thermoelectric parameters of template-embedded nanowires have also been investigated. A steady-state method for the measurement of cross-plane thermal conductivity of nanowire bundles, embedded in ion-track-etched template foils of 30 μm thickness and measurement results for template-embedded bismuth nanowires are presented.

22.2 Fabrication of Thermoelectric Nanostructures by Ion-Track Technology and FEBID/FIBID

22.2.1 Fabrication of Nanowires by Ion-Track Technology

The steps of nanowire fabrication by ion-track technology are represented in Figure 22.1. A high-energetic ion beam, consisting of 11.4 MeV/nucleon Au^{26+} or Pb^{26+} ions, strikes a 30-μm-thick polycarbonate membrane (foil) and generates latent tracks. These tracks are chemically etched in aqueous sodium hydroxide leading to the formation of nanochannels whose diameter is controlled by the etching parameters (concentration, temperature, etching time). After deposition of a metal layer on one of the membrane surfaces, the channels are electrochemically filled with Bi, $Bi_{1-x}Sb_x$ or Bi_2Te_3. Free-standing nanowires are created by dissolving the matrix in an organic solvent. Subsequently, the nanowires are detached from the back electrode by ultrasonication. After this process step, nanowires with lengths ranging from a few to 30 μm are available in the suspension. The wires are deposited on a substrate of choice (e.g., on a silicon wafer with SiO_2 film) by applying a few drops of the solvent containing the needles. The solvent evaporates within a few minutes leaving behind randomly distributed wires. By controlling the fabrication parameters, polycrystalline as well as single-crystalline wires can be created [14–16]. Figure 22.2 presents a scanning electron microscopy (SEM) image, demonstrating smooth wire contours and narrow diameter distribution.

22.2.2 Focused Electron/Ion Beam-Induced Deposition

FEBID/FIBID is a direct beam writing technique for the realization of nano- and microstructures. The schematic FEBID process is represented in Figure 22.3. Precursor molecules are emitted from a gas

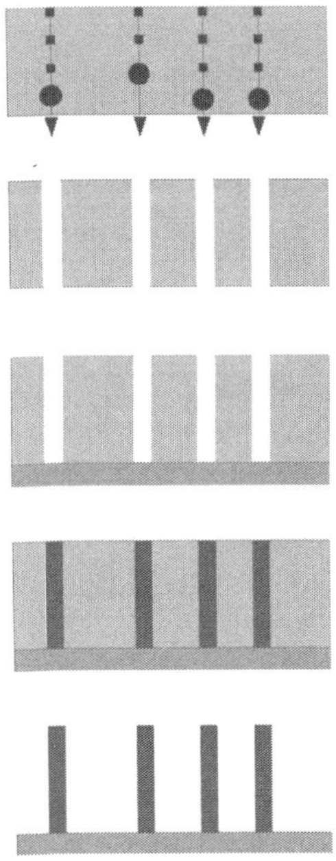

FIGURE 22.1 Flow chart of the nanowire fabrication process by ion-track technology.

injection system close to the focal point of an electron beam. The molecules are dissociated by the primary, backscattered, and secondary electrons. With a predefined pattern, the primary electron beam can be scanned over the substrate surface. Relevant process parameters for this scanning process are the distance between successive dwell points of the electron beam (pitch) and the time period over which the electron beam is held at each dwell point (dwell time). Typical pitches vary between 10 and 100 nm. The required dwell times depend on the precursor gas and the substrate material. Detailed recent reviews concerning FEBID/FIBID and related techniques are given in [17,18].

By proper selection of the precursor gas and the process parameters, different functionalities of the resulting deposits can be obtained. FEBID-generated nano-granular films are very promising substances regarding the optimization of thermoelectric material parameters due to nanodimensions.

FIGURE 22.2 SEM image of Pt nanowires after dissolving the polycarbonate foil and detachment from the back electrode.

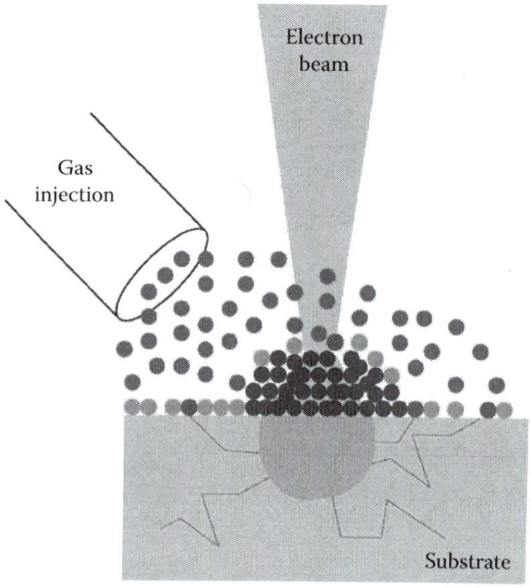

FIGURE 22.3 Schematic representation of the FEBID process: The adsorbed precursor molecules (light gray spheres) are dissociated by electron impact and a permanent deposit (black spheres) is formed in the focal area of the electron beam. The lines indicate exemplary trajectories of electrons leaving the excitation volume.

Quite generally, nanogranular metals can be considered as tunable model systems for studying the interplay of electronic correlation effects, quantum size effects, and disorder. Particularly interesting are nanogranular FEBID/FIBID structures for which an insulator-to-metal transition is observed as a function of the intergrain coupling strength. Valuable information about the electronic properties close to this transition can be obtained from Seebeck effect measurements [19].

22.3 Electrical Conductivity

22.3.1 Preparation Methods for Electrical Contacts

To perform thermoelectric measurements on nanowires and FEBID nanostructures, their integration in specific microchips is required. One important demand for this nano-micro-integration process is the formation of reliable electrical contacts, which do not influence the thermoelectric properties of the nanostructures (e.g., by diffusion processes). To obtain electrical contacts and measuring devices for nanostructures such as nanowires or FEBID/FIBID lines two different approaches are commonly used:

1. The nanostructure is (randomly) placed on a substrate and the measuring device is realized around the nanostructure by micropatterning, especially by using the technologies of surface micromachining. This approach includes the direct formation of electrical connections to the nanostructures as part of the construction of the measuring device.
2. Specific microchips are produced (preferably) by standard semiconductor batch-process fabrication and the nanostructures are placed onto the chips by particular handling and deposition techniques afterwards. Insufficient electrical connections to the prefabricated microchips have to be improved by additional measures.

The first approach has the advantage that no equipments or techniques for handling and micropositioning of the nanostructure are required. On the other hand, due to the small dimension of the samples, the patterning of the measuring device has to be performed using local microscopic lithography which makes the fabrication time consuming [19–21]. The second approach, applying separately batch-fabricated chips and afterwards putting the nanostructures onto these devices, is associated with challenging handling and positioning procedures. One method is to place droplets of a solution containing nanowires (see Section 22.2.1) onto the device and to repeat the procedure until one nanowire is (randomly) placed at the right position [13,22,23]. Another method is the direct positioning of the sample onto the microchip by using a nanomanipulator [10]. After such placing procedures, usually the electrical contacts of the nanowire to the electrical connections (e.g., thin-film contact pads) have to be improved by local deposition of FEBID/FIBID layers at the overlap of nanowire and contact pads.

22.3.1.1 Lift-Off Process

Electrical connections to individual nanowires, fabricated according to approach (1), can be realized by the lift-off technique. The preparation sequence for electrical contacts on both ends of a single nanowire is demonstrated in Figure 22.4.

Nanowires are placed randomly on the front of the (100)-oriented silicon wafer, which is deposited with an insulating SiO_2 layer on top. After spin coating of photoresist and inspection of the nanowires positions on the SiO_2 surface, a local microscopic UV exposure of the photoresist is performed by using a photolithographic micromask. Then, the exposed positive photoresist is developed and removed only at both ends of the nanowire, whereas it protects the central area of the needle. Next, a thin metal contact film is deposited by sputtering onto the wafer. Finally, this film is structured, forming contact pads at both ends of the selected nanowire by the lift-off technique. This contacting technique is used for all

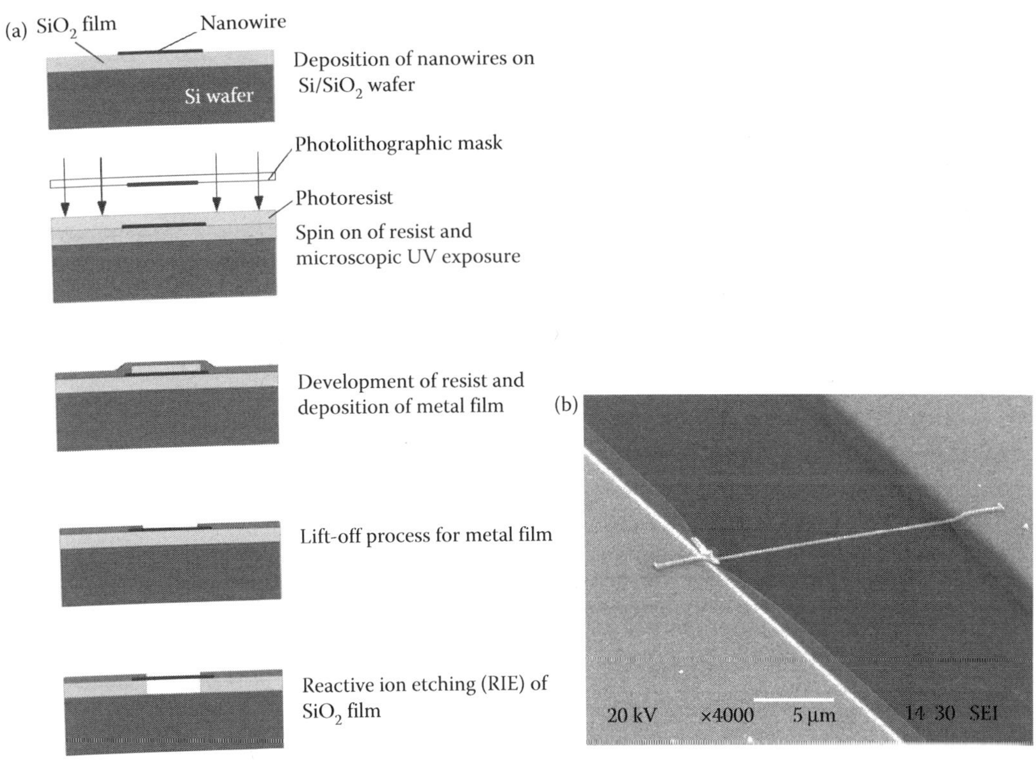

FIGURE 22.4 (a) Preparation scheme of electrical contact pads for a single nanowire performed with local microscopic photolithographic exposure of photoresist, thin-film deposition, and lift-off technique. (b) SEM image of electrical contact layers above a 20-µm-long suspended nanowire on a Si/SiO_2 substrate.

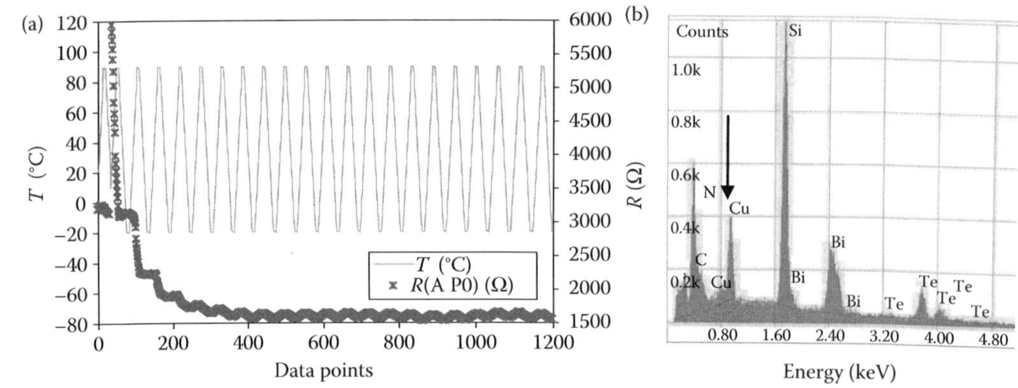

FIGURE 22.5 (a) Long-term behavior of the resistance of a Bi_2Te_3 nanowire (contacted with Cr(10 nm)/Cu(200 nm)) during thermal cycling (22 cycles with temperature oscillations between −20°C and +90°C within 3 days). (b) EDX signals after thermal cycling, detected at the center (10 μm distant from both Cr/Cu contacts) of the nanowire.

measurements of Seebeck coefficient and electrical and thermal conductivity. For the measurement of the thermal conductivity, suspended nanowires have to be prepared to avoid any thermal bypass of the substrate. Therefore, after contact preparation the SiO_2 layer below the nanowire is removed by reactive ion etching (RIE). Due to optimization of the process parameters (etching gas, pressure, bias voltage) the usually anisotropic RIE process is transformed in a partially isotropic one, removing the SiO_2 layer below the nanowire within few minutes. During the etching process, the metal contact films serve as etching mask layers. Consequently, the SiO_2 film is removed exclusively in the gap area between the contact films.

Long-term stability is a crucial parameter of thin-film contacts on nanostructures, especially for their applications at elevated temperatures. Therefore, various film materials have been tested with respect to constant ohmic contacts and annealing behavior during thermal cycling.

Figure 22.5 shows the long-term behavior of the resistance of a Bi_2Te_3 nanowire during thermal cycling (22 cycles with temperature oscillations between −20 and +90°C within 3 days). Electrical contacts were formed before by the lift-off technique, using 10 nm chromium as adhesion layer and 200 nm copper.

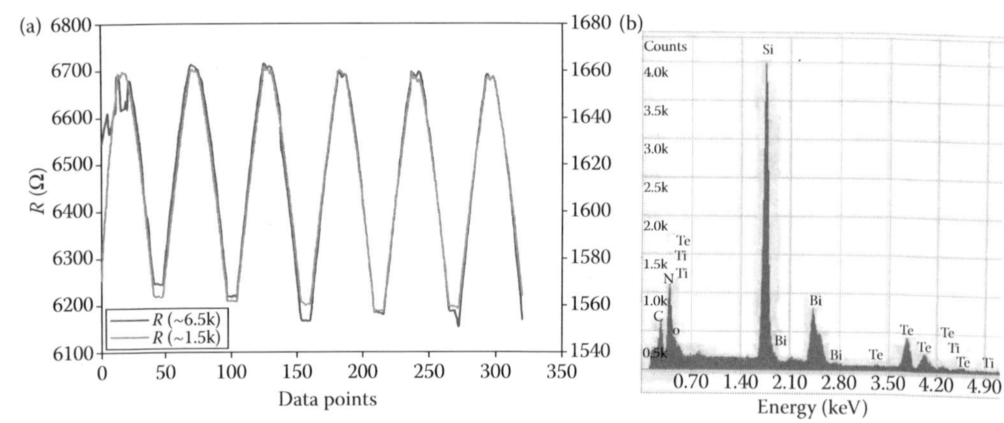

FIGURE 22.6 (a) Long-term behavior of the resistance of two Bi_2Te_3 nanowires (contacted with Ti(20 nm)/Au(150 nm)) during thermal cycling (six cycles with temperature oscillations between −20°C and +90°C within 1 day). (b) EDX signals after thermal cycling, detected at the center (10 μm distant from both Ti/Au contacts) of the nanowire, indicating the effectiveness of Ti diffusion barrier (no peaks of contact materials are detected in the suspended nanowire).

During the first cycles, a permanent decrease in the resistance from about 3300 to 1500 Ω (at the lowest temperature of −20°C) is observed. This resistance change can be attributed to surface and volume diffusion of copper, indicated by energy-dispersive x-ray (EDX) investigations. The electron beam of the EDX probe was focused on the center of the suspended nanowire, nearly 10 μm distant from both Cr/Cu contact pads. The EDX signal (Figure 22.5b) shows a strong copper peak at this position, indicating Cu diffusion with a diffusion length of >10 μm during the above-mentioned thermal cycling process. Similar results at comparable cycling conditions have been received with a contact layer sandwich of 10 nm Cr (adhesion layer) and 200 nm Ni.

Long-term stable contacts without diffusion effects can be realized by a layer sandwich of 20 nm Ti (diffusion barrier) combined with a 150-nm-thick Au film, as demonstrated in Figure 22.6. During six thermal cycling periods of two Bi_2Te_3 nanowires (Figure 22.6a) with the above-described temperature oscillations, no resistance drift or annealing occurs. Furthermore, the EDX signal indicates no peaks, which could be correlated to the contact materials Ti or Au. Therefore, all nanowires, which we have prepared for thermoelectric measurements are electrically connected to the measuring chips with Ti/Au contact pads.

22.3.1.2 Direct Writing of FEBID/FIBID Nanostructures on Measuring Devices

Figure 22.7 represents an example for the deposition of nanostructures on a prefabricated microchip according to approach (2). Here, nanogranular FEBID layers are placed by direct writing between prepared Au contact films of a sensor chip.

A very important benefit of such FEBID techniques arises from the fact that nanogranular structures can be deposited at arbitrary positions of a planar or three-dimensional measuring device or also of a specific application device. For example, in Figure 22.7, the nanogranular layer is deposited on a 10 μm thick cantilever in order to investigate its resistivity as a function of induced strain.

22.3.2 Four-Point-Probe Measurement

The lift-off technique is applied to obtain two-point-probe electrical contacts of individual nanowires. However, contact resistances may have considerable influence on the measuring results. The effect of the contact resistances can be investigated and eliminated by using a four-point-probe arrangement, which have also been realized by the previously described lift-off technique. The structure of the used lithographic micromask for four-point measurements is presented in Figure 22.8. The occurence of contact resistances can be deduced by comparing the results of two-point-probe and four-point-probe resistivity measurements. To obtain the two-point-probe resistivity ρ_{2p}, a current flow I and the corresponding

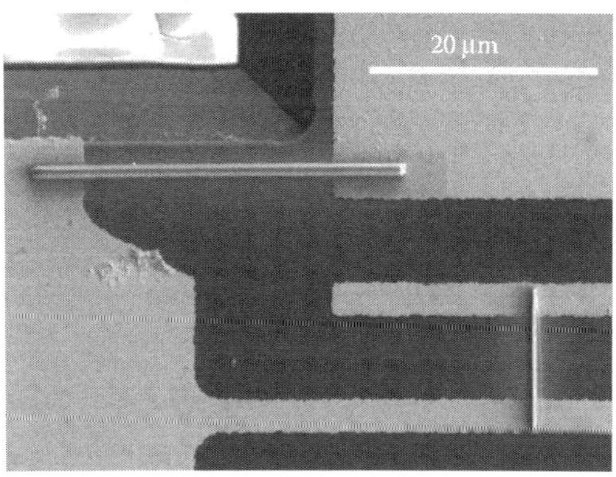

FIGURE 22.7 Nanogranular sensor element (left-to-right structure) and reference element (top-to-bottom structure) prepared by FEBID direct writing between Au contact films using the precursor $MeCpPt(Me)_3$.

FIGURE 22.8 Four-point-probe contact structure with a Ti/Au film as contact material (bright area). The gaps between the contact probes (marked with 1–4) are 5 μm.

voltage drop U_{2p} between the contacts 1 and 4 are measured. In case of negligible contact resistance, the resistivity ρ_{2p}, calculated according to

$$\rho_{2p} = \frac{U_{2p}A}{I(l_{12} + l_{23} + l_{34})} \tag{22.1}$$

represents the true resistivity of the nanowire, where A indicates its cross-sectional area and the l_{ij} denote the lengths of the wire segments between the contact probes. For the determination of the four-point-probe resistivity ρ_{4p}, the same current flow I is supplied using contacts 1 and 4, whereas the potential difference is measured between contacts 2 and 3. Without any contact resistance, the difference between the two-point-probe and four-point-probe resistivity $(\rho_{2p} - \rho_{4p})$, related to ρ_{2p},

$$\frac{\rho_{2p} - \rho_{4p}}{\rho_{2p}} = 1 - \frac{U_{4p}(l_{12} + l_{23} + l_{34})}{U_{2p}l_{23}} \tag{22.2}$$

should be close to zero.

Figure 22.9 shows an SEM image of a Bi_2Te_3 nanowire contacted by a four-point-probe arrangement. Its dimensions (length, diameter) are determined by evaluation of high-resolution SEM pictures.

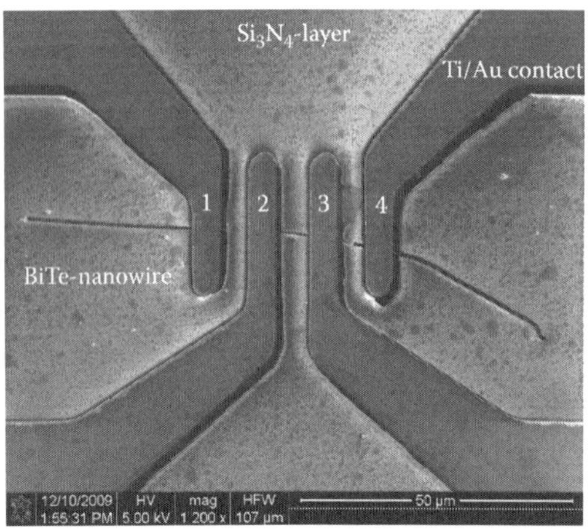

FIGURE 22.9 SEM image of a Bi_2Te_3 nanowire contacted in a four-point-probe arrangement (see labels in the picture for details).

TABLE 22.1 Lengths of Segments of a Bi_2Te_3 Nanowire between Contact Probes and Voltage Drops U_{2p} in Two-Point-Probe and U_{4p} in Four-Point-Probe Arrangement

l_{12}(μm)	l_{23}(μm)	l_{34}(μm)	U_{2p}(mV)	U_{4p}(mV)
2.73	3.87	3.55	12.7	4.7

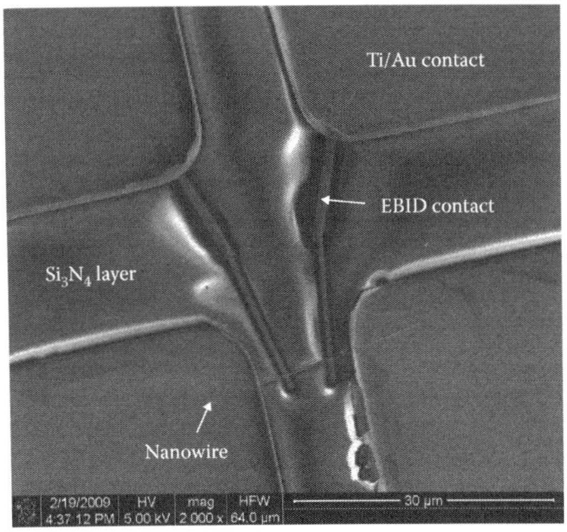

FIGURE 22.10 SEM image of a nanowire contacted in a four-point-probe arrangement with the interior potential contacts written by FEBID (see labels in the picture for details).

Table 22.1 lists the lengths of the nanowire segments between the contact probes and the corresponding voltage drops.

The relative deviation between two-point-probe and four-point-probe resistivity

$$\frac{\rho_{2p} - \rho_{4p}}{\rho_{2p}} = 1 - \frac{U_{4p}(l_{12} + l_{23} + l_{34})}{U_{2p} l_{23}} < 0.03 \quad (22.3)$$

of this nanowire is <3% and indicates that the effect of contact resistances can be neglected for our preparation technique. This may be attributed to sputter deposition of the adhesion/barrier layer (Ti) before the deposition of the conducting Au film, since sputtered particles have a very high energy (few 100 eV) and form very intimate bindings with the coated nanowire surface atoms.

Nanowires with very small dimensions cannot be observed by an optical microscope, and therefore the lift-off process by using local microscopic UV exposure for the fabrication of four-point-probe contacts cannot be applied. However, conventionally, two-point-probe contacts can be transferred to four-point-probe arrangements using FEBID for direct writing of potential contacts as indicated in Figure 22.10. Furthermore, the additional application of a focused ion beam (FIB) allows controlled doping of potential barriers (Schottky barriers) which may appear between metal–semiconductor junctions. Therefore, FIB is often used for reducing contact resistances, where the junction is doped with Ga until a metallic behavior is achieved [24]. In general, it is necessary to verify two-point-probe measurement results with appropriated four-point-probe measurements.

22.3.3 Current–Voltage Characteristics

Between metal–semiconductor junctions, potential barriers (Schottky barriers) may occur. Their appearance can be observed in the current–voltage characteristic of the contacted nanowire.

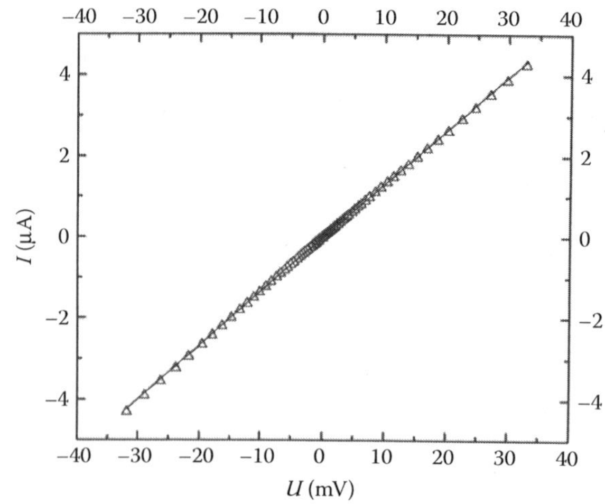

FIGURE 22.11 Current–voltage characteristic of a Bi_2Te_3 nanowire measured at room temperature.

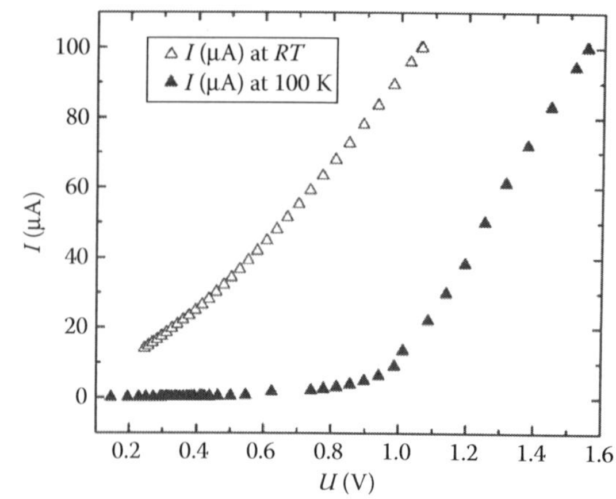

FIGURE 22.12 Current–voltage characteristics of a $Bi_{1-x}Sb_x$ nanowire ($x = 0.1$) measured at room temperature (open triangles) and at 100 K (filled triangles).

A metal-like (ohmic) contact leads to a linear current–voltage curve, as shown in Figure 22.11 for the Bi_2Te_3 nanowire with negligible contact resistance at room temperature (see Equation 22.3).

In Figure 22.12, the current–voltage characteristics of a $Bi_{1-x}Sb_x$ nanowire contacted with Ti/Au pads are shown for room temperature and 100 K. The typical current–voltage characteristic of a Schottky barrier is comparable with the I–U curve of a p–n junction. With decreasing temperature the influence of the Schottky barrier increases and leads to the observed exponentially current–voltage characteristic at $T = 100$ K. At room temperature, the effect of the Schottky barrier decreases and the current–voltage characteristic of the $Bi_{1-x}Sb_x$ nanowire is almost linear.

22.3.4 Electrical Conductivity and Temperature Coefficient of Resistance

Since the length of the nanowires is limited by the preparation technique (ion-track technology), two-point-probe ohmic contacts are preferred for our measurements (instead of four-point-probe contacts) in cases where linear current–voltage characteristics are observed. It would also be possible to contact the

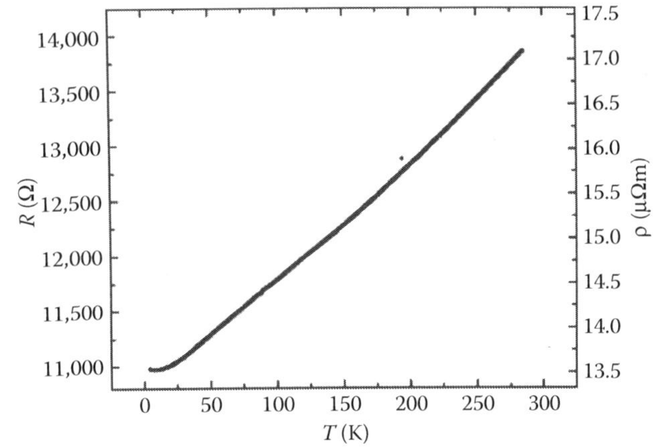

FIGURE 22.13 Resistance/resistivity as function of temperature of a Bi_2Te_3 nanowire (diameter = 180 nm, length = 18.4 μm).

nanowires by FEBID to achieve four-point-probe arrangements even for short nanowires but with much higher effort.

In Figure 22.13, the resistance and electrical resistivity of a Bi_2Te_3 nanowire with a diameter of 180 nm and a length of 18.4 μm are presented. The geometrical data were obtained by SEM. The measurements of the resistance as a function of temperature were performed in a ^{4}He cryostat. For current supply, a DC source with a constant electrical power of 10^{-8} W was used. For each measurement, the applied voltage is calculated by the given power and the resistance recorded at the previous measurement. With respect to the high current densities generated in the nanowire, especially for low resistances (metallic nanowires), preferably an AC source and lock-in technique should be used as described in [18] to prevent damages of the nanowire by electromigration. The temperature coefficient of resistance is calculated using the slope of the R/T characteristics. The slope can be obtained either by differentiation of an appropriated fit function of the resistivity data or by a partial linearization of the data points. The temperature coefficient of resistivity (TCR) presented in Figure 22.14 is calculated using a polynomial fit of the resistivity data and is related to the resistivity of the Bi_2Te_3 nanowire at room temperature.

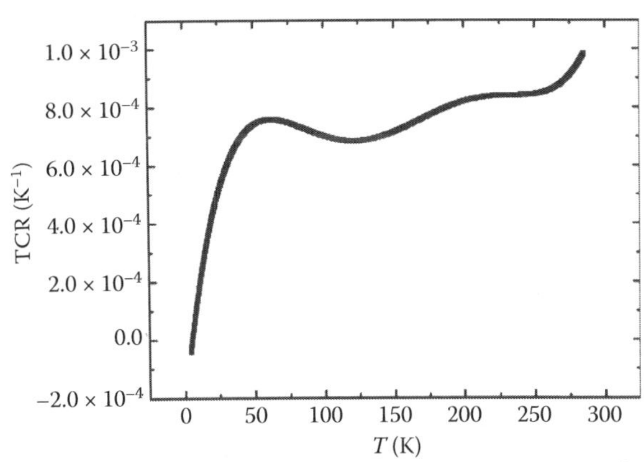

FIGURE 22.14 Temperature coefficient of resistivity of a Bi_2Te_3 nanowire (calculated slope of the curve in Figure 22.13).

22.4 Seebeck Coefficient

A specifically designed and optimized microchip is required for Seebeck effect measurements on single nanowires. For such investigations, two crucial experimental problems have to be solved:

- In order to achieve sufficient measuring accuracy, a temperature difference along the nanowire axis of at least 1 K should be realized. Thus, a temperature gradient of 10^5 K/m has to be established between the electrical contacts (assuming a contact distance of about 10 μm).
- Besides the thermoelectric voltage U caused by the temperature difference, the temperatures at the nanowire contacts (i.e., at two positions with a distance of only a few microns) also have to be measured precisely.

Micromachining and microlithographic methods were employed to realize a newly developed microchip, which solves these experimental problems.

22.4.1 Experimental Details

A 300-nm-thick Si_3N_4 membrane is prepared by micromachining in the center of a silicon chip. After low pressure chemical vapor deposition (LPCVD) of Si_3N_4 layers on both sides of a silicon wafer, square-shaped windows are opened by photolithographic patterning and RIE in the Si_3N_4 layer on the back side. Then, the exposed (100)-oriented silicon is removed by anisotropic etching in KOH solution at 80°C. The LPCVD layer is not attacked by this etchant; therefore, a square-shaped Si_3N_4 membrane of 2 × 2 mm² remains at the front side of the chip. This very thin membrane with low thermal conductivity is a crucial feature of the chip with respect to realization of the above-mentioned temperature gradient.

Prior to the membrane preparation, nanowires are placed on the front side Si_3N_4 layer of the (100)-oriented silicon wafer. After inspection of the wires and their positions on the Si_3N_4 surface, one pair of Ti/Au contact films is structured at the ends of a selected wire by lift-off technique (described in Figure 22.4). An identical second pair of contact films is patterned besides the first one as shown in Figure 22.15. After preparation of these contact layers, the opening of the back side windows for anisotropic silicon etching is established by photolithography, using a double-side exposure mask aligner. Due to double-side alignment, the correct position and orientation of the back-side windows with respect to the nanowire and

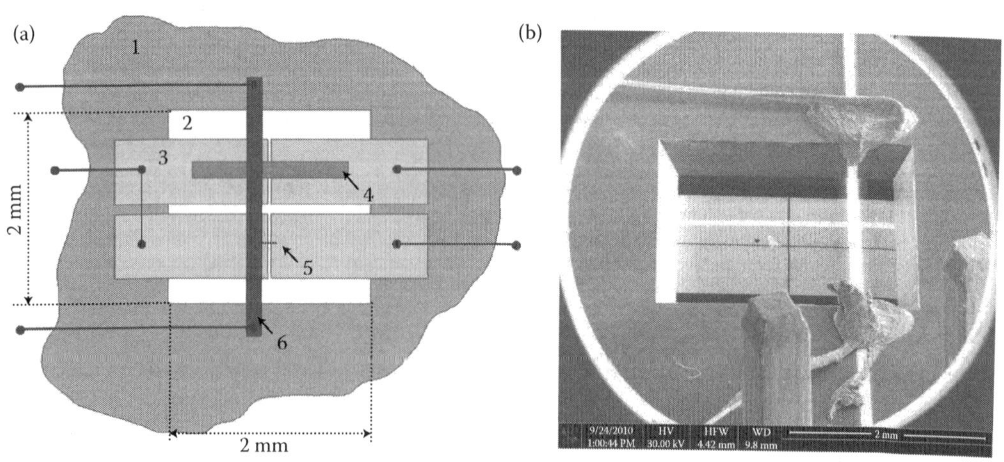

FIGURE 22.15 (a) Scheme of the microchip for Seebeck coefficient measurements with 1: surface of silicon wafer, 2: Si_3N_4 membrane, 3: Ti/Au contact pad, 4: bismuth reference layer, 5: position of the nanowire, 6: heating layer. (b) SEM picture (back side view) of the microchip, showing electrical connections to the heating layer, the Ti/Au contact pads, and Bi reference film.

contact film positions at the wafer front side can be achieved. The anisotropic etching is performed in a KOH-resistant etching box in order to protect the wafer front side against the etchant. Due to correct alignment, the single nanowire is located near the center of the membrane and the edges of the rectangular contact films are oriented parallel to the membrane edges, as shown in Figure 22.15. Next, a Seebeck reference film (thickness 200 nm) with known Seebeck coefficient is deposited on the second pair of contact films by thermal evaporation through a metal diaphragm. Polycrystalline bismuth films have been applied as reference films because their thermoelectric properties and Seebeck coefficients have been investigated in detail [25]. Furthermore, the Seebeck reference film can be deposited simultaneously on the bulk silicon rim of the chip, where its Seebeck coefficient can be measured with calibrated thermocouples.

Finally, a thin-film heater is deposited onto the membrane region of the chip in order to generate the required temperature difference. This heating film can be placed on the back side of the membrane without additional electrical insulation. Alternatively, it can also be deposited on the membrane front side, if the heater is separated from the contact films by a thin insulating layer. The contact films are electrically connected by wire bonding, where the bond contacts have to be realized outside the membrane region on the bulk silicon rim.

The heater generates a temperature difference ΔT at the gap between the metal contacts. The chip design, shown in Figure 22.15, was investigated and optimized by finite-element method (FEM) simulations. As demonstrated in the next section, an identical temperature difference ΔT on both pairs of contact films can be achieved. Therefore, ΔT at the nanostructure is determined by using the reference voltage U_R of the Seebeck reference film

$$U_R = (S_R - S_C)\Delta T \tag{22.4}$$

where S_R is the known Seebeck coefficient of the reference film ($S_R = -63$ µV/K for 200 nm Bi films at room temperature) and S_C is the Seebeck coefficient of the contact pads. For Ti/Au contact films, S_C amounts to −0.4 µV/K at room temperature. With the measured Seebeck voltage U_N of the nanowire (and the known temperature difference) its Seebeck coefficient S_N can be calculated according to

$$S_N - S_C = \frac{U_N}{\Delta T} = \frac{U_N(S_R - S_C)}{U_R} \tag{22.5}$$

22.4.2 FEM Simulation

By FEM simulations using COMSOL® multiphysics, the temperature profile in the membrane region as a function of the heating power and the position of the thin-film heater has been investigated. The temperature difference at the gap between the contact films is the most important feature of the chip design. Since the bulk silicon rim has a very high thermal capacitance and thermal conductivity compared to the membrane, it can be considered as a heat sink with constant ambient temperature T_0. Therefore, we apply the boundary condition of constant temperature $T = T_0$ at the membrane edges. Because of the very small thickness of the membrane and of all deposited films compared with their lateral dimensions, the FEM simulation can be performed with a two-dimensional model. Figure 22.16 shows the various areas of the membrane region that must be distinguished because of their different thin-film stacks. By using a two-dimensional model, all individual areas (subdomains) are modeled by a uniform model thickness d. Then, the effective thermal conductivities λ_{eff} of the subdomains have to be calculated by

$$\lambda_{eff} d = \sum_i \lambda_i d_i \tag{22.6}$$

with λ_i denoting the thermal conductivities and d_i representing the thicknesses of the individual layers of the stacks. Radiation losses of the membrane areas are included in the calculation by

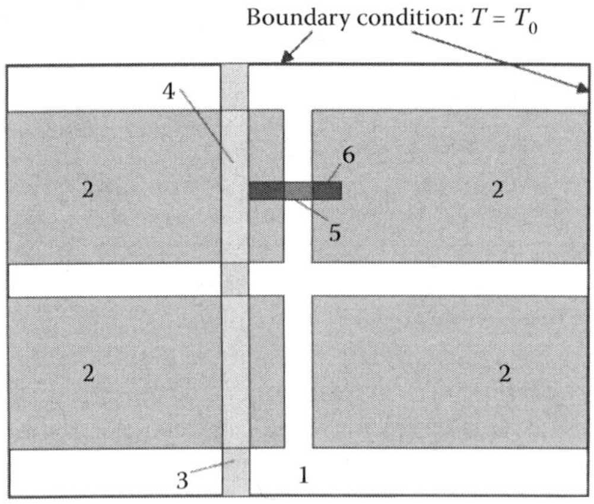

FIGURE 22.16 FEM subdomains in the membrane region, characterized by specific effective thermal conductivities λ_{eff} and emissivities ε_1, ε_2. 1: Membrane, 2: membrane + Ti/Au film, 3: membrane + Bi heating film, 4: membrane + Ti/Au film + Bi heating film, 5: membrane + Bi reference film, 6: membrane + Ti/Au film + Bi reference film.

Stefan–Boltzmann radiation law. With respect to thermal radiation, the subdomains are characterized by two emissivities ε_1 and ε_2 for the top and bottom surface, respectively.

Simulations were performed for microchips applied both in high vacuum and at atmospheric pressure. Heat losses by convection can be neglected for measurements in high vacuum at pressures $p < 10^{-5}$ mbar. Convective heat losses in atmospheric air are simulated by using a heat transfer coefficient $\alpha = 6.4$ W/m^{2}K. The temperature differences ΔT at the gaps between the contact pairs depend on the heating power density Q and on the distance between gap and thin-film heater.

Figure 22.17 demonstrates the three-dimensional temperature profile in the membrane region for a microchip with a thin-film heater placed on the back side of the membrane at a distance of 0.2 mm to the gap. A moderate heating power density $Q = 28$ W/mm^3 generates temperature differences of $\Delta T > 10$ K between the nanowire contacts for operation in high vacuum as well as at atmospheric pressure. A

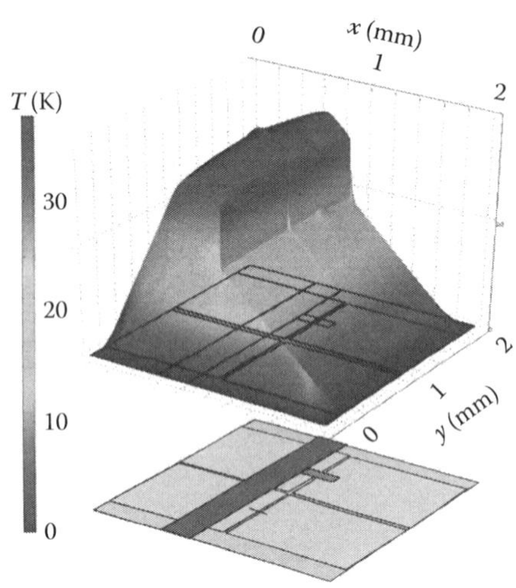

FIGURE 22.17 Simulated three-dimensional temperature profile in the membrane region for a microchip operation in high vacuum ($p < 10^{-5}$ mbar) with a thin-film heater 0.2 mm distant from the gap and a heating power density of 28 W/mm^3, indicating temperature differences $\Delta T > 10$ K for the nanowire and reference film, respectively.

nearly identical temperature difference can be achieved between the contact pairs for the Seebeck reference film (200 nm Bi), because of its small thermal conductance.

22.4.3 Results

The first experimental investigations of the Seebeck coefficient were performed on nanogranular (W)-FEBID lines [19]. Nanogranular metals may be promising thermoelectric materials because of the expected reduction of thermal conductivity. Therefore, their thermopower has been investigated with respect to thermoelectric applications. A typical FEBID layer is shown in Figure 22.7. For Seebeck effect measurements, nanogranular W-FEBID lines (length: 13 μm, thickness: 300 nm, width: 1.2 μm) are deposited on the Ti/Au contact pads of the chip. Applying a voltage of 20 mV to a thin-film heater of 50 Ω, a temperature difference of a few K is generated at the gap. By increasing the heating power, a linear increase in the thermoelectric voltage has been observed (Figure 22.18). The Seebeck coefficient is

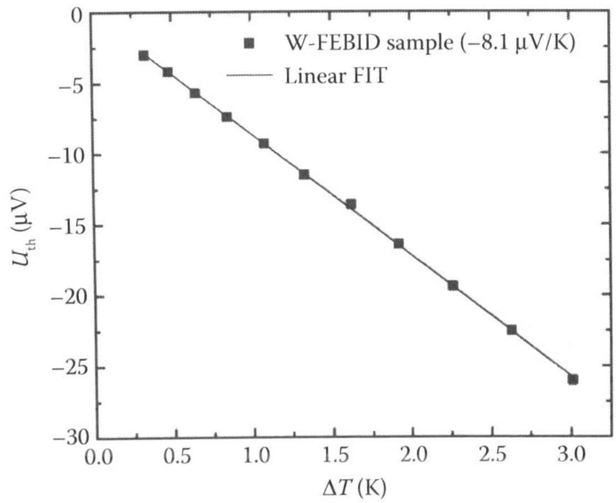

FIGURE 22.18 Thermoelectric voltage as function of temperature difference for a nanogranular W-FEBID layer (thickness: 300 nm, width: 1.2 μm, length: 13 μm) with a metal content of about 22 at.% at room temperature.

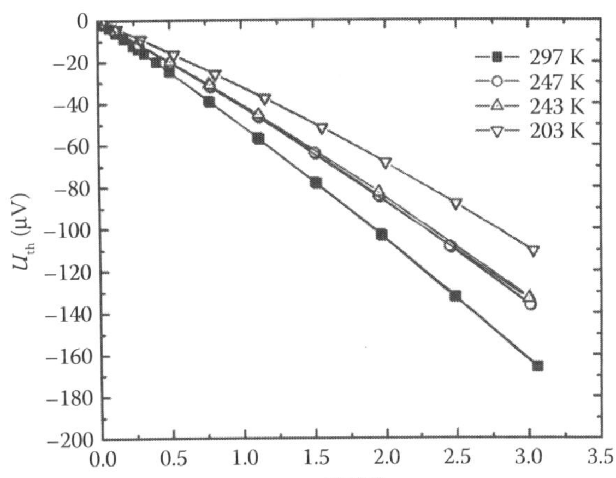

FIGURE 22.19 Thermoelectric voltage as function of temperature difference for a Bi_2Te_3 nanowire (diameter: 180 nm, length: 18 μm) at different ambient temperatures.

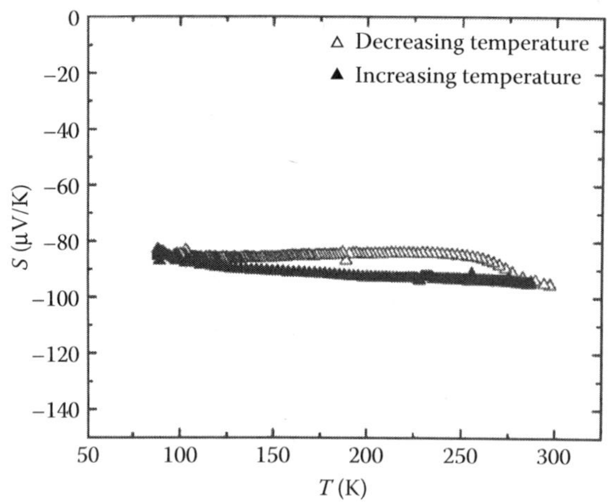

FIGURE 22.20 Seebeck coefficient of a $Bi_{1-x}Sb_x$ nanowire (diameter: 200 nm, $x = 0.1$) in the temperature range from 80 to 300 K. Data points indicate measured values during continuous temperature increase and decrease of the microchip, respectively.

calculated using the slope of the thermoelectric voltage as a function of temperature difference. A small Seebeck coefficient $S_N - S_C = -8.1$ μV/K has been determined for nano-granular W-FEBID with a metal content of about 22 at.% at room temperature, where S_C denotes the Seebeck coefficient of the contact material (−0.4 μV/K).

Figure 22.19 shows the results of Seebeck effect measurements on a Bi_2Te_3 nanowire at various ambient temperatures. The Seebeck coefficient decreases with decreasing temperature in the observed temperature range. Figure 22.20 indicates the Seebeck coefficient of a $Bi_{1-x}Sb_x$ nanowire (diameter: 200 nm, $x = 0.1$) in a temperature range from 80 to 300 K. The measured Seebeck coefficient of (−95 ± 5) μV/K at room temperature is close to the value reported for $Bi_{1-x}Sb_x$ thin films with an antimony content of about 10 at % [26,27].

22.5 Thermal Conductivity

Sophisticated microdevices have been designed and applied for the measurement of thermal conductivity of nanostructures [10–12]. Extremely fragile suspended cantilevers are prepared for measuring platforms in order to reduce the thermal conductance of the device. We present an alternative steady-state method that requires only heat sink conditions at the ohmic contacts of the nanostructure.

22.5.1 Steady-State Method for Suspended Nanowires

22.5.1.1 Experimental Details

For thermal conductivity measurements, we need a suspended nanowire to avoid thermal bypass by the substrate, which can be prepared by RIE as discussed in Section 22.3.1.1. The nanowire endings are in intimate thermal contact with the metallic contact pads and are separated from the bulk silicon wafer only by a thin Si_3N_4 layer. During electrical heating of the nanowire by an electrical current I, most of the heating power is dissipated in the suspended part of the wire, since the resistance of this part is large compared to the resistance of the contact pads. Therefore, the ends of the nanowire can be considered as heat sinks of constant temperature T_0, as demonstrated by FEM simulations (see Figure 22.21).

The measuring principle and experimental setup for the thermal conductivity measurements are illustrated in Figure 22.22. The measuring sequence starts with the determination of the nanowire

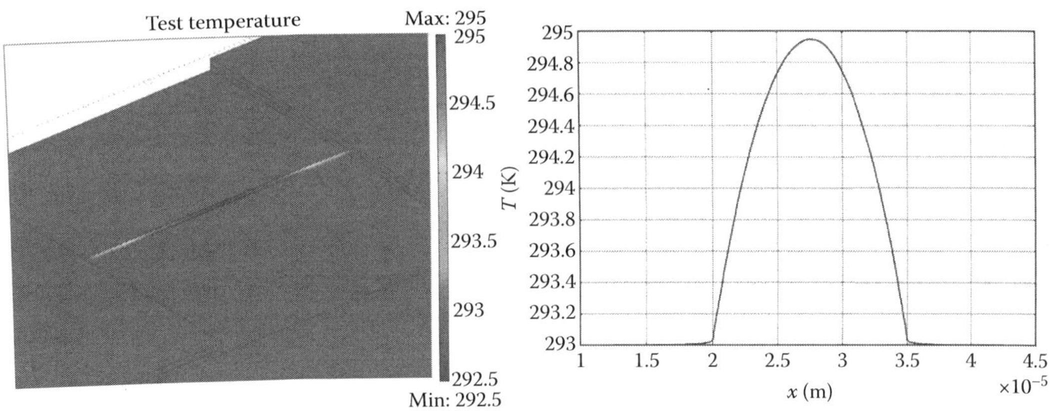

FIGURE 22.21 FEM simulation shows temperature distribution of electrically self-heated nanowire between two metallic contact pads on a Si_3N_4/Si chip (heating power $I^2R = 1$ μW, wire radius = 100 nm, wire length = 15 μm, Si_3N_4 thickness 0.3 μm). Left: Temperature profile along nanowire axis (suspended between 20–35 μm) clearly demonstrates heat sink conditions at the contacts.

resistance $R(T_0)$ at any ambient temperature T_0, using a four-point-probe technique and a low measuring current I_0. The temperature T_0 of the setup can be varied between 80 and 400 K, but, in general, measurements can also be performed at liquid He temperatures.

Because of the low I_0, the temperature increase of the nanowire due to Joule heating can be neglected. From these measurements, the slope of the resistance–temperature curve $m = dR(T_0)/dT_0$ is determined at an arbitrary T_0. Then, a heating current I is applied leading to a temperature increase of the wire which involves a rise of the electrical resistance $R(T_M) - R(T_0) = \Delta R$, where $R(T_M)$ denotes the resistance and T_M denotes the mean temperature of the heated nanowire. The usually applied heating currents lead to mean temperature increases $\Delta T_M = T_M - T_0$ of <10 K. ΔT_M can be determined experimentally by using the measured resistance increase ΔR and the slope m according to

$$\Delta T_M = \frac{\Delta R}{m} \tag{22.7}$$

On the other hand, ΔT_M is a function of the thermal conductivity λ of the nanowire. In order to find the dependence of ΔT_M on λ, the thermal heat equation has to be solved for the specific boundary

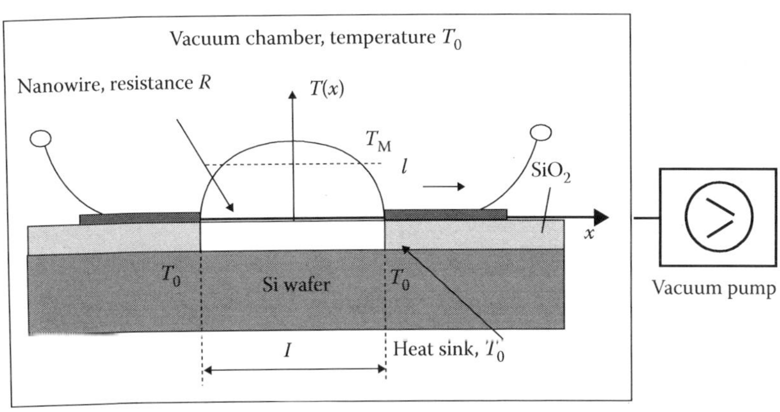

FIGURE 22.22 Schematic cross section of the experimental setup.

conditions, given by the experimental setup. Measurements are performed in a high vacuum (pressure $p < 10^{-5}$ mbar). Therefore, convective heat losses from the heated wire to the ambient can be neglected. Furthermore, radiation losses according to the Stefan–Boltzmann law can be omitted because of the small ΔT_M and the dimensions of the suspended nanowire [18]. With respect to the wire length of (typically) 10 μm compared to its small diameter of the order of 100 nm, the heat transport can be described using a one-dimensional model. A detailed derivation of the heat equation [18] leads to

$$\Delta T_M = \frac{I^2 R}{12\lambda(A/l)} \tag{22.8}$$

Combining Equations 22.7 and 22.8, we can calculate the thermal conductivity

$$\lambda = \frac{mI^2 Rl}{12A\Delta R} \tag{22.9}$$

by using the measured resistance increase ΔR, heating power I^2R, slope m of the resistance–temperature curve, and the geometrical dimensions (A, l) of the wire, which are determined by SEM investigations.

22.5.1.2 Results

For a first test of measuring principle, the thermal conductivity of Pt nanowires has been investigated. The presented measurements are performed in the temperature range from 260 to 360 K on a nanowire with 131 nm diameter and a suspended length of 14.7 μm. The values of electrical conductivity σ (Figure 22.23) are deduced from measurements of the nanowire resistance $R(T_0)$ performed with a low ac current of frequency 113 Hz. During these measurements, a very low Joule heat of 10^{-7} W is dissipated and, therefore, the mean temperature increase ΔT_M in the nanowire remains <0.3 K. Compared to the bulk value $\sigma_{bulk} = 9.6 \times 10^6\ (\Omega m)^{-1}$ at 295 K, the electrical conductivity of the Pt nanowire is reduced by a factor of 2.5.

This decrease is in good agreement with classical size effect theory [5,6]. For thermal conductivity measurements, an electrical heating power of 4×10^{-6} W was applied, causing a mean temperature increase of $\Delta T_M = 12.7$ K. Figure 22.24 shows the thermal conductivity in the temperature range 260–365 K. The bulk values (e.g., $\lambda_{bulk} = 71$ W/mK at room temperature) are more than three times larger than the measured thermal conductivities of the polycrystalline Pt nanowire. This is also attributed to size effects [28], mainly caused by grain boundary scattering of electrons [18].

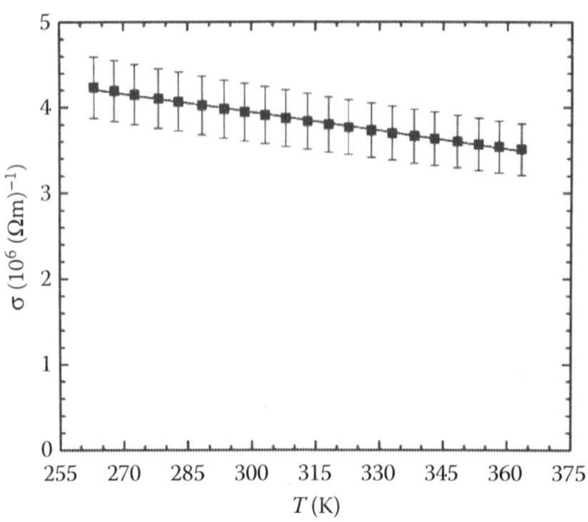

FIGURE 22.23 Electrical conductivity σ of a Pt nanowire as a function of temperature.

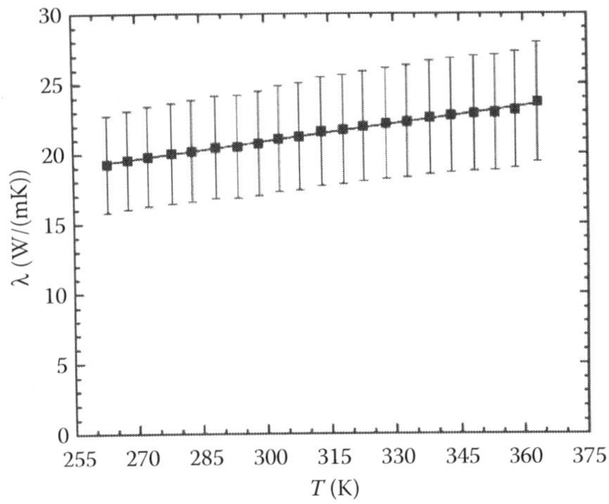

FIGURE 22.24 Thermal conductivity of a Pt nanowire as a function of temperature.

22.5.2 Cross-Plane Thermal Conductivity of Embedded Nanowires

Arrays of nanowire bundles can be applied for thermoelectric microgenerators or for arrays of thermoelectric radiation sensors. Figure 22.25 shows the fabrication sequence of an array of thermoelectric sensors, where each sensor pixel consists of a thermocouple of *p*-type and *n*-type nanowire bundles. On a substrate with high thermal conductivity (heat sink) such as silicon wafer with an electrical insulating deposition (SiO_2 or Si_3N_4), metal contact pads are structurized by microlithography (Figures 22.25 and 22.26). A polycarbonate template foil is deposited on such a substrate, which has been irradiated before by heavy ions and etched in aqueous sodium hydroxide to generate nanochannels with diameters of 100–200 nm.

Alternatively, thin polymer layers such as poly(methyl methacrylate) (PMMA), SU-8 negative resist or thick AZ® photoresists can be deposited by spin coating, irradiated with heavy ions and etched in qualified solutions. Half of the metal contact pads are connected together for selected fabrication of *n*-type nanowire bundles in a first electrodeposition process and the second group of contact pads is used in the next electrodeposition step for the realization of *p* type nanowire bundles. Then, thermocouples are formed by connecting a *p*- and *n*-type bundle by physical vapor deposition and photolithographic patterning of a metal layer. To form a radiation sensor array, these metal contact layers are coated with a broadband radiation absorbing layer with high-absorption coefficient from the visible to the infrared radiation range. Finally, the template foil can be removed by dissolution (Figure 22.26), using an organic solvent, in order to increase the sensitivity and detectivity of the array pixels. By using the presented technology, a minimum size of an individual thermocouple (pixel) of about $10 \times 20\ \mu m^2$ can be achieved. For simulation and characterization of the sensor parameters, the thermoelectric figure of merit of such template foils filled with nanowires has to be investigated.

This investigation requires the measurement of the cross-plane thermal conductivity of thin foils with thickness in the order of 10 μm. The measuring principle and experimental setup of the applied steady-state method is represented in Figure 22.27. Metallic heater/thermometer films with high-temperature coefficients of resistivity are deposited on the top surfaces of the samples and electrically insulated from them by thin dielectric layers (interfaces). Measurements are performed on two foils and results are compared in order to calculate the thermal conductivity of embedded nanowires: on a template foil after irradiation and etching (Figure 22.28) and on an identical foil (with the same thickness, nanochannel diameters, and ion-track density) filled with thermoelectric nanowires. The template foils are fixed on a substrate by an adhesive and so thermally separated from the substrate by an interface with

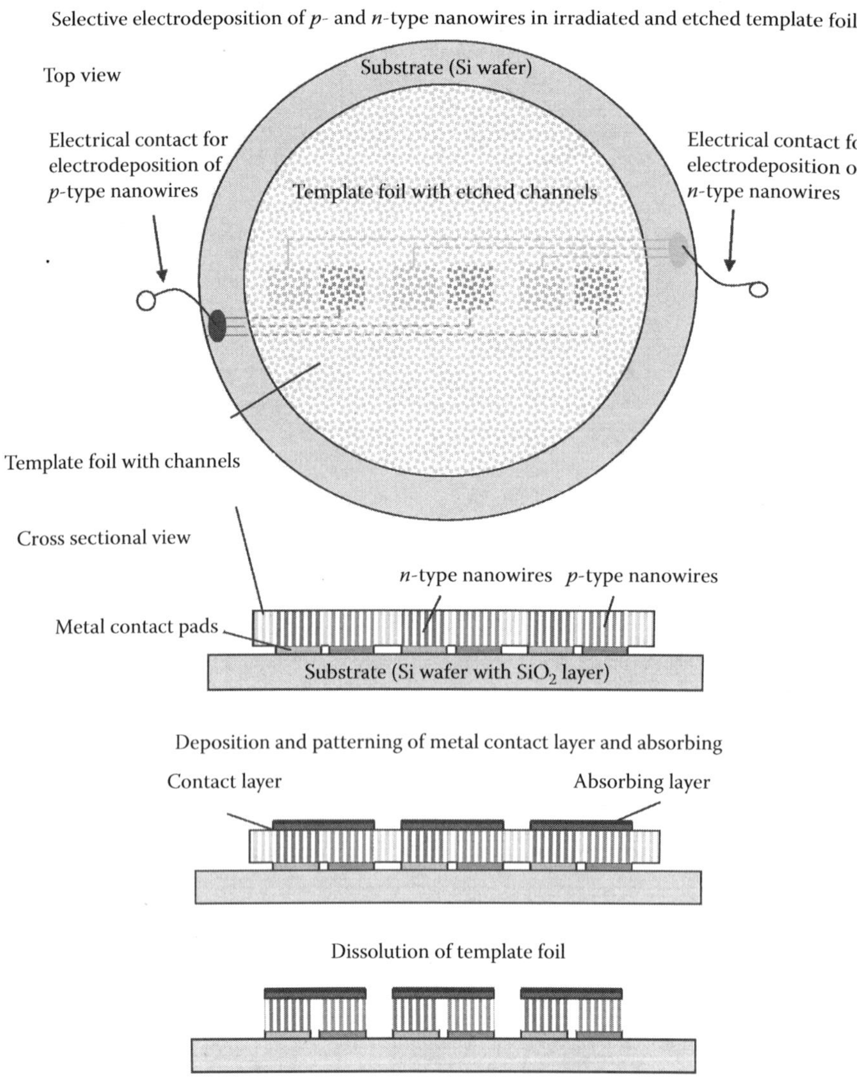

FIGURE 22.25 Fabrication process for an array of thermoelectric sensors: each pixel consists of a bundle of *p*- and *n*-type nanowires. (With kind permission of Rauber, M., GSI Helmholtz-Zentrum für Schwerionenforschung GmbH, Darmstadt, Germany.)

unknown thermal resistance. The substrate is fixed on a heat sink with temperature T_0. On the heater/thermometer films (length l, width b) on top of the two template foils with cross-plane thermal conductivities λ_1 and λ_2, the electrical heating powers N_1 and N_2 are dissipated, respectively.

This causes temperature increases $T_1 - T_0$ and $T_2 - T_0$, respectively, which are measured by the resistance increases of the heating films (thermometers). The ratios of the specific temperature increases and heating powers $(T_1 - T_0)/N_1$ and $(T_2 - T_0)/N_2$, respectively, represents the total thermal resistances R_{T1} and R_{T2} of the samples. Besides the thermal resistances of the templates $R_{TF1} = d/(\lambda_1 bl)$ and $R_{TF2} = d/(\lambda_2 bl)$, these total thermal resistances include the thermal resistances of the substrate R_{TSub}, of the interfaces R_{TInt} and of the heater/thermometer $R_H = d_H/(2\lambda_H bl)$. Usually, the thermal resistance of the heater/thermometer is very small and can be neglected since the thickness d_H of the used metallic heating films amounts to 100 nm and its thermal conductivity λ_H is >100 W/mK. For our measurements also the

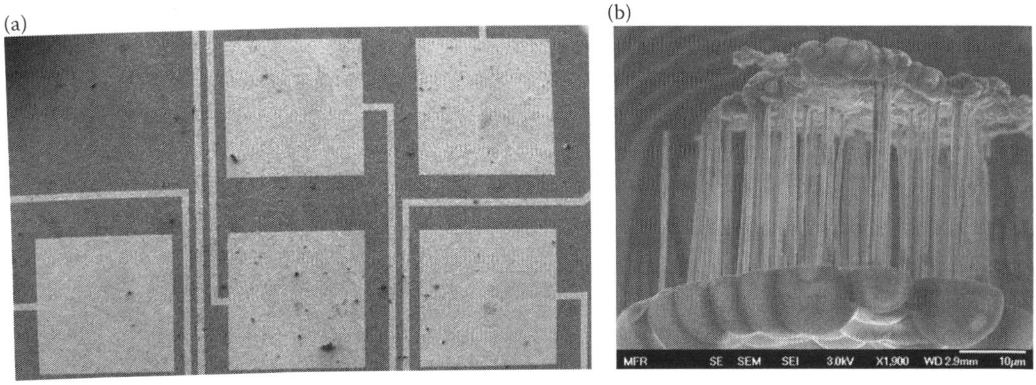

FIGURE 22.26 (a) Irradiated and etched template foil has been prepared for selected electrodeposition of p- and n-type nanowire bundles by microlithographic patterning of arrays of Au contact pads on the bottom side of the template. (b) Bundle of nanowires with metal contact layer after dissolution of the template foil [29].

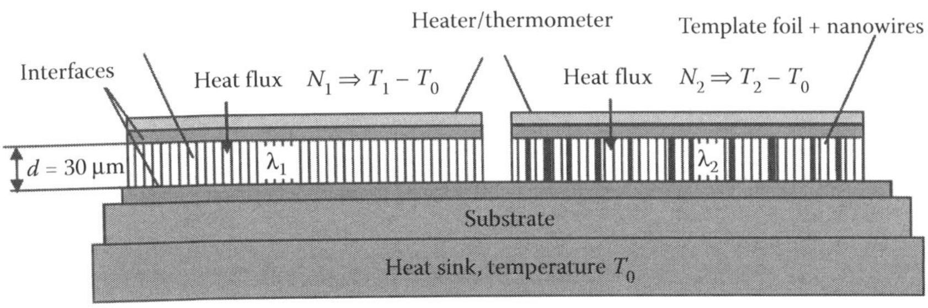

$$\frac{T_1 - T}{N_1} = R_{T1} = \left(R_{TSub} + R_{TInt} + \frac{d}{\lambda_1 bl} + \frac{d_H}{2\lambda_H bl}\right) \qquad \frac{T_2 - T_0}{N_2} = R_{T2} = \left(R_{TSub} + R_{TInt} + \frac{d}{\lambda_2 bl} + \frac{d_H}{2\lambda_H bl}\right)$$

FIGURE 22.27 Method and experimental setup (cross-sectional view) for steady-state measurements of cross-plane thermal conductivity of thin template foils (without and with embedded nanowires).

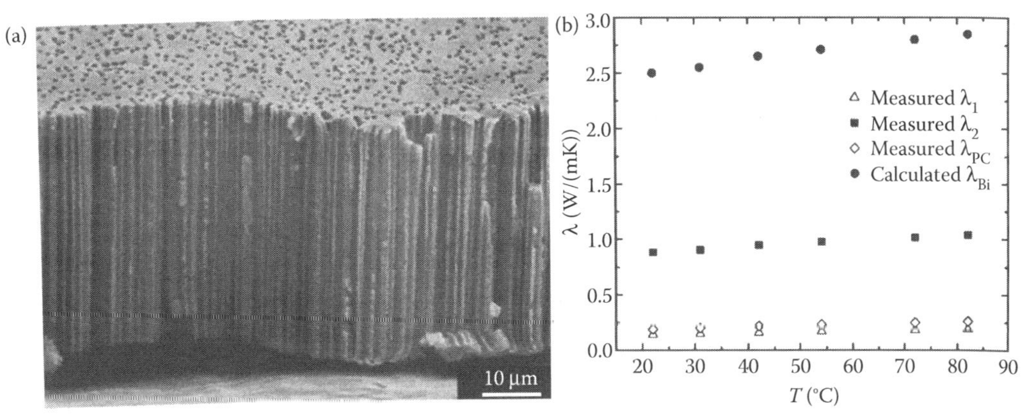

FIGURE 22.28 (a) Polycarbonate foil after irradiation with heavy ions (ion density 10^9/cm^2) and etching of nanochannels. (b) Measured thermal conductivities λ_1 and λ_2 and calculated thermal conductivities of polycarbonate template foil λ_{PC} and embedded Bi nanowires λ_{Bi}.

thermal resistance of all interfaces R_{TInt} can be neglected, since the insulation between the heater and the foil is realized by spin coating of a very thin resist layer and the interface between the foil and the substrate is formed by a thin metal-filled adhesive with high thermal conductivity. Furthermore, by proper choice of the geometrical dimensions l and b of heater and sample, the spreading thermal resistance R_{TSub} of the substrate can also be omitted. In general, the geometrical conditions $b > d$ and $b \ll l$ should be fulfilled. Then, the measured total resistance R_{T1} represents the thermal resistance R_{TF1} of the irradiated and etched polycarbonate foil and R_{T2} is equivalent to the thermal resistance R_{TF2} of the foil with embedded nanowires. Figure 22.28 shows a cross-sectional view of such a foil of about 30 μm thickness, irradiated with an ion density of $10^9/\text{cm}^2$ after etching of nanochannels with an average diameter of 200 nm. Consequently, the volume content of etched channels V_{Ch} in the foil amounts to about 30%. Figure 22.28 shows the thermal conductivity λ_1 of the polycarbonate foil with nanochannels as a function of temperature, which has been determined from the measured data of the thermal resistance R_{T1}, and the calculated thermal conductivity λ_{PC} of polycarbonate, whereby we have assumed that the nanochannels are filled with air at atmospheric pressure. Then, λ_{PC} is evaluated by

$$\lambda_{PC} = \lambda_1 \frac{(V_{PC} + V_{Ch})}{V_{PC}} - \lambda_{air} \frac{V_{Ch}}{V_{PC}} \tag{22.10}$$

where λ_{air} indicates the thermal conductivity of air at atmospheric pressure and V_{PC} is the volume content of polycarbonate in the etched foil.

Figure 22.28 also indicates the thermal conductivity λ_2 of such a polycarbonate foil filled with Bi nanowires. Taking into consideration the volume content of Bi nanowires of 30%, the thermal conductivity of embedded Bi nanowires λ_{Bi} (also shown in Figure 22.28) can be calculated using

$$\lambda_{Bi} - \lambda_{air} = (\lambda_1 - \lambda_2) \frac{(V_{PC} + V_{Ch})}{V_{Ch}} \tag{22.11}$$

Figure 22.27 demonstrates the problems associated with the measurement of the cross-plane thermal conductivity. Interfaces between substrate and template foil and between template foil and heater/thermometer may have a thermal resistance that cannot be ignored in comparison with the thermal resistance of the investigated template. However, both the presented steady-state technique and the transient 3-Omega method have to solve this experimental challenge. One advantage of the steady-state method compared to the 3-Omega method is a much lower electrical power density required for the heater/thermometer and a less expensive/complex electrical measuring setup.

22.6 Microchip for the Complete Characterization of Thermoelectric Transport Coefficients (σ, S, λ) on an Individual Nanowire (z-Chip)

Various specific and sophisticated microchips and methods for the measurement of thermoelectric transport properties have been presented in the previous sections. In the ideal case, only one chip should be applied for the complete characterization of all thermoelectric transport coefficients (σ, S, λ) on an individual nanowire. This section deals with the design (simulation) and realization of a z-chip fabricated by micromachining. Figure 22.29 shows the scheme of the z-chip and the measuring principle. By microlithographic patterning and anisotropic etching in silicon, a comb-like silicon cantilever array is fabricated. The cantilevers (length $l = 550$ μm, width $b = 16$ μm, thickness $d = 40$ μm) are separated by a distance w of about 10 μm. The top surface of the cantilevers is coated with a thin insulating Si_3N_4 film and two thin-film Ti/Au contact stripes are patterned by photolithography on each cantilever. The cantilever array is fabricated by using an silicon on insulator (SOI) wafer with a top silicon wafer of

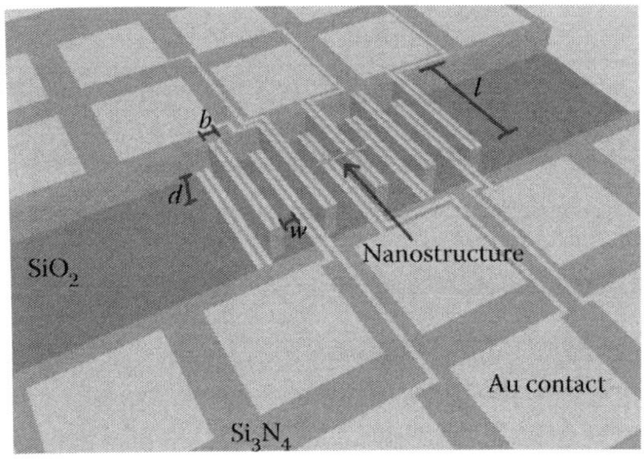

FIGURE 22.29 Scheme and measuring principle of a cantilever array for the characterization of thermoelectric efficiency z of individual nanowires (z-chip).

40 μm thickness, an insulating SiO_2 interface and a bottom silicon wafer of 300 μm thickness. After patterning of the electrical contact stripes, the cantilever array is realized by anisotropic etching of the top wafer. Finally, the cantilevers are thermally separated from each others by anisotropic etching of a cavity in the bottom wafer, where the etching process is stopped beneath the cantilevers due to the SiO_2 interface. The thermal cross-talk between the cantilevers caused by the remaining SiO_2 interface layer (300 nm thick) is very small and can be minimized by removing the SiO_2 film using RIE.

One part of the bulk silicon rim of the *z*-chip is connected with a heat sink of temperature T_1, the other part with a heat sink of temperature T_0. Because of the very small thermal cross-talk between the cantilevers and the high thermal conductivity of silicon, one half of the cantilevers have a temperature close to the heat sink temperature T_1, the other half close to T_0. Thus, two neighboring cantilevers possess a temperature difference which is proportional to the temperature difference of the heat sinks $T_1 - T_0$. Figure 22.30 shows the temperature profiles along two neighboring cantilevers, one of them connected with a heat sink of temperature $T_1 = 298$ K and the other with a heat sink of temperature $T_0 = 295$ K. Figure 22.30 demonstrates that the temperature difference between the cantilevers is about 2.4 K and nearly independent on the position.

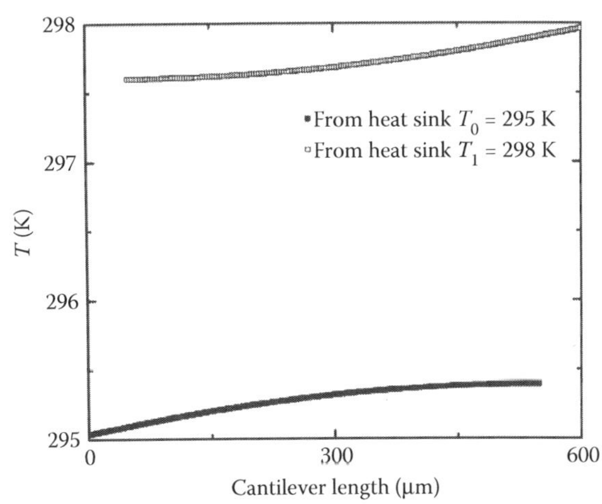

FIGURE 22.30 Temperature profile along two neighboring cantilevers connected to heat sinks of temperatures $T_1 = 298$ K and $T_0 = 295$ K, respectively.

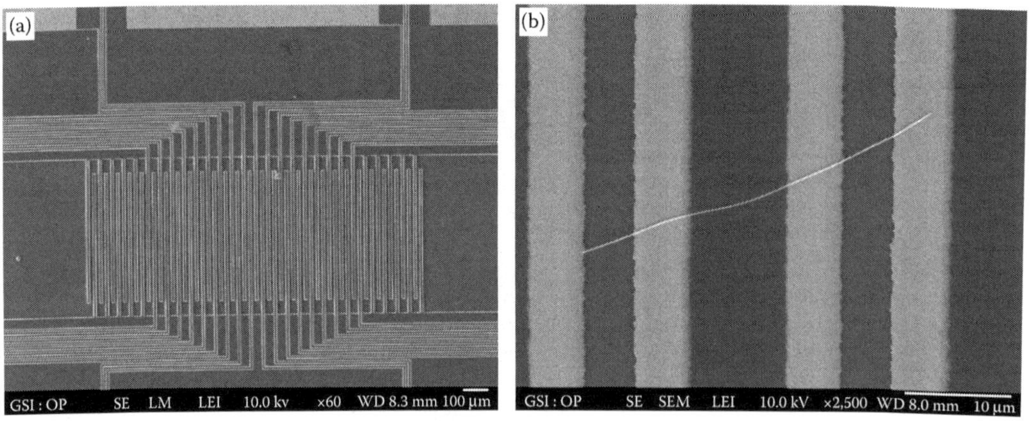

FIGURE 22.31 (a) Cantilever array with 56 cantilevers. (b) Suspended nanowire deposited and contacted on two neighboring cantilevers with thin Ti/Au metal lines.

Ion-track-fabricated nanowires are deposited on the cantilever array (Figure 22.31a) using few drops of the solvent containing the wires (see Section 22.2.1). The solvent evaporates within a few minutes leaving behind randomly distributed suspended wires (Figure 22.31b) fixed on neighboring cantilevers.

Measurements of electrical conductivity σ are performed using a four-point-probe technique, where the four thin-film Ti/Au lines of two neighboring cantilevers are applied as current and potential probes. For the measurement of the Seebeck coefficient S, a definite temperature difference between the heat sinks is generated, which leads to a specific temperature difference between neighboring cantilevers and consequently to a thermoelectric voltage, that is measured using the Ti/Au contact lines. The temperature difference between neighboring cantilevers is a definite function of the heat sink temperature difference ($T_1 - T_0$) and has been determined by separate measurements, using thin-film temperature sensors.

For measurements of thermal conductivity λ, both heat sinks are hold at the same temperature. Because of their high thermal conductivity, the cantilevers act as heat sinks of identical temperature in spite of electrical heating of the nanowire. Therefore, the steady-state method described in Section 22.5.1 can be applied. The four-point-probe technique is employed for the determination of the heating power I^2R dissipated in the nanowire and the resistance change ΔR, and λ is calculated using Equation 22.9. Thus, all transport coefficients (σ, S, λ) can be characterized on one and the same individual needle.

References

1. Dresselhaus, M.S., Lin, Y.M., Rabin, O., Jorio, A., Souza Filho, A.G., Pimenta, M.A., Saitof, R., Samsonidze, G., and Dresselhaus, G., Nanowires and nanotubes, *Materials Science and Engineering*, 23, 129, 2003.
2. Cornelius, T.W., Toimil-Molares, M.E., Neumann, R., and Karim, S., Finite-size effects in the electrical transport properties of single bismuth nanowires, *Journal of Applied Physics*, 100, 114307, 2006.
3. Toimil-Molares, M.E., Hohberger, E.M., Schaeflein, C., Blick, R.H., Neumann, R., and Trautmann, C., Electrical characterization of electrochemically grown single copper nanowires, *Applied Physics Letters*, 82, 2139, 2003.
4. Fuchs, K., The conductivity of thin metallic films according to the electron theory of metals, *Mathematical Proceedings of the Cambridge Philosophical Society*, 34, 100, 1937.
5. Sondheimer, E.H., The mean free path of electrons in metals, *Advances in Physics*, 50, 499, 1952.
6. Mayadas, A.F., and Shatzkes, M., Electrical-resistivity model for polycrystalline films: The case of arbitrary reflection at external surfaces, *Physical Review B*, 1, 1382, 1970.

7. Lin, Y.M., Sun, X., and Dresselhaus, M.S., Theoretical investigation of thermoelectric transport properties of cylindrical Bi nanowires, *Physical Review B*, 62, 4610, 2000.
8. Zou, J., and Balandin, A., Phonon heat conduction in a semiconductor nanowire, *Journal of Applied Physics*, 89, 2932, 2001.
9. Dresselhaus, M.S., and Heremans, J.P., Recent developments in low-dimensional thermoelectric materials, *Thermoelectric Handbook: Macro to Nano*, D.M. Rowe (ed.), Taylor & Francis, Boca Raton, FL, 2006.
10. Shi, L., Li, D., Yu, C., Jang, W., Kim, D., Yao, Z., Kim, P., and Majumdar, A., Measuring thermal and thermoelectric properties of one-dimensional nanostructures using a microfabricated device, *Journal of Heat Transfer*, 125, 881, 2003.
11. Boukai, A., Xu, K., and Heath, J.R., Size-dependent transport and thermoelectric properties of individual polycrystalline bismuth nanowires, *Advanced Materials*, 18, 864, 2006.
12. Boukai, A., Bunimovich, Y., Tahir-Kheli, J., Yu, J.K., Goddard III, W.A., and Heath, J.R., Silicon nanowires as efficient thermoelectric materials, *Nature*, 451, 168, 2008.
13. Hochbaum, A.I., Chen, R., Delgado, R.D., Liang, W., Garnett, E.C., Najarian, M., Majumdar, A., and Yang, P., Enhanced thermoelectric performance of rough silicon nanowires, *Nature*, 451, 163, 2008.
14. Toimil-Molares, M.E., Buschmann, V., Dobrev, D., Neumann, R., Scholz, R., Schuchert, I.U., and Vetter, J., Single-crystalline copper nanowires produced by electrochemical deposition in polymeric ion track membranes, *Advanced Materials*, 13, 62, 2001.
15. Cornelius, T.W., Brotz, J., Chtanko, N., Dobrev, D., Miehe, G., Neumann, R., and Toimil-Molares, M.E., Controlled fabrication of poly- and single-crystalline bismuth nanowires, *Nanotechnology*, 16, S246, 2005.
16. Karim, S., Toimil-Molares, M.E., Maurer, F., Miehe, G., Ensinger, W., Liu, J., Cornelius, T.W., and Neumann, R., Synthesis of gold nanowires with controlled crystallographic characteristics, *Applied Physics A*, 84, 403, 2006.
17. Utke, I., Hoffmann, P., and Melngailis, J., Gas-assisted focused electron beam and ion beam processing and fabrication, *Journal of Vacuum Science and Technology B*, 26, 1197, 2008.
18. Huth, M., Focused electron beam induced deposition—Principles and applications, *Proceedings of Beilstein Symposium Functional Nanosciences*, Bozen, Italy, 2010, pp. 193–211.
19. Voelklein, F., Reith, H., Schmitt, M.C., Huth, M., Rauber, M., and Neumann, R., Microchips for the investigation of thermal and electrical properties of individual nanowires, *Journal of Electronic Materials*, 39, 1950, 2010.
20. Voelklein, F., Schmitt, M.C., Cornelius, T.W., Picht, O., Müller, S., and Neumann, R., Microchip for the measurement of seebeck coefficients of single nanowires, *Journal of Electronic Materials*, 38, 1109, 2009.
21. Voelklein, F., Reith, H., Cornelius, T.W., Rauber, M., and Neumann, R., The experimental investigation of thermal conductivity and the Wiedemann–Franz law for single metallic nanowires, *Nanotechnology*, 20, 325706, 2009.
22. Shi, L., Yu, C., and Zhou, J., Thermal characterization and sensor applications of one-dimensional nanostructures employing microelectromechanical systems, *Journal of Physical Chemistry B*, 109, 22102, 2005.
23. Mavrokefalos, A., Moore, A.L., Pettes, M.T., Shi, L., Wang, W., and Li, X., Thermoelectric and structural characterizations of individual electrodeposited bismuth telluride nanowires, *Journal of Applied Physics*, 105, 104318, 2009.
24. Cronin, S.B., Lin, Y.M., Rabin, O., Black, M.R., Ying, J.Y., Dresselhaus, M.S., Gai, P.L., Minet, J.P., and Issi, J.P., Making electrical contacts to nanowires with a thick oxide coating, *Nanotechnology*, 13, 653, 2002.
25. Voelklein, F., and Kessler, E., Analysis of the lattice thermal conductivity of thin films by means of a modified Mayadas–Shatzkes model: The case of bismuth films, *Thin Solid Films*, 142, 169, 1986.

26. Cho, S.L., Kim, Y., DiVenere, A., Wong, G., Ketterson, J.B., and Meyer, J.R., Anisotropic Seebeck and magneto-Seebeck coefficients of Bi and $Bi_{0.92}Sb_{0.08}$ alloy thin films, *Journal of Applied Physics*, 88, 808, 2000.
27. Liu, H.J., Song, C.M., Wu, S.T., and Li, L.F., Processing method dependency of thermoelectric properties of $Bi_{85}Sb_{15}$ alloys in low temperature, *Cryogenics*, 47, 56, 2007.
28. Ouarbya, L., Tosser, A.J., and Tellier, C.R., Effects of electron scatterings on thermal conductivity of thin metal films, *Journal of Materials Science*, 16, 2287, 1981.
29. Rauber, M., GSI Helmholtz-Zentrum für Schwerionenforschung GmbH, Darmstadt, Germany.

23

Neutron Scattering Investigations of Thermoelectric Materials

Mogens Christensen
Aarhus University

This chapter looks at how neutrons can be used to investigate thermoelectric materials. The chapter is divided into two main sections related to elastic scattering, known as Bragg diffraction, and inelastic scattering describing the dynamics of atoms. Each section presents central elements of theory and gives recent examples related to host–guest structures, where neutron investigations have been undertaken.

23.1 Introduction

Thermoelectric properties are optimized through maximizing the power factor and minimizing the thermal conductivity. The thermal conductivity associated with the charge carriers is fixed through optimal doping. This leaves the lattice thermal conductivity associated with the atomic motions to be the only independent material property. In crystalline materials, lattice thermal conductivity is driven by the coherent motion of the atoms. The coherent motion can be envisioned as a wave of atomic masses moving through the material. Such mass displacement causes a propagation of heat. The heat package can be considered a particle, which is known as a phonon. The phonon is characterized by having a certain energy ($\hbar\omega$) and momentum vector (**q**). The situation resembles electromagnetic radiation propagating through vacuum, which can be described as a wave or particle, that is, the photon. The phonons are typically found to have energies ($0 < \hbar\omega < 100$ meV) and radiation with similar energies are needed to facilitate observations of energy gain or loss to the probe. Photons can be chosen to have either the energy or the momentum vector close to the values of the phonon. Scattering of visible light is known as Raman scattering and here the photons have energies of a few eV; however, the momentum vector is too short to probe the phonon momentum vector. Therefore, first-order scattering of visible light provides only information about optical modes. Photons with momentum vectors comparable to the phonons are x-rays, but they have energies in the keV range. Even though phonon dispersions can be measured using x-rays, distinguishing gain and loss of a few meV with a keV x-ray probe is experimentally difficult. Cold and thermal neutrons on the other hand have the advantage of having energies (0.1–100 meV) and the momentum vectors in the same order of magnitude as phonons.

Neutrons are the tool of choice for investigating the phonon propagation governing the lattice thermal conductivity. Neutron investigations can provide information on the microscopic level about atomic dynamics, which can help understand the observed macroscopic observations. Additionally, the neutron wavelength is comparable to interatomic distances; therefore, neutrons can probe the position of the atoms in the unit cell. The use of neutrons as probes was pioneered by Brockhouse and Shull, and their work was recognized by the 1994 Nobel Prize award in physics.[1]

23.2 Neutron Scattering

This section gives a brief introduction to basic concepts of neutron scattering. Neutrons have some special merits when compared with x-ray scattering:

1. *Energy and wavelength.* Thermal neutrons have wavelengths comparable to interatomic distances and energies comparable to atomic excitations. These virtues were highlighted in the introduction.
2. *Sensitivity toward light elements and isotopes.* Neutrons interact with atoms through nuclear forces and the interaction potential varies randomly with atomic number and isotope.
3. *Weak interaction.* The nuclear interaction between neutrons and matter is generally weak and allows spacious sample environment and probing the bulk of the sample.
4. *Magnetism.* Neutrons carry a spin which interacts with the electron spin and allows neutron experiments to extract information about magnetic ordering and magnetic spin fluctuations.

The first three points are relevant in the context of investigating thermoelectric materials. The first case—*energy and wavelength*—was outlined in the introduction and detailed examples will be given after briefly introducing the underlying theory. The second point—*sensitivity toward light elements and isotopes*—is important when comparing with x-rays, where the scattering is proportional to the atomic number. Consequently, x-rays have low contrast between neighboring elements, whereas neutrons may have an appreciable difference giving sufficient contrast to distinguish these elements. Figure 23.1 shows the scattering power of x-rays and neutrons as a function of atomic number. Figure 23.2 illustrates the neutron sensitivity to light elements and specific isotopes. The third point stated above allows investigation of thermoelectric modules in operation without the need to grind the sample. Such experiments can reveal information about stress and strain as a result of the thermal gradient across the sample.

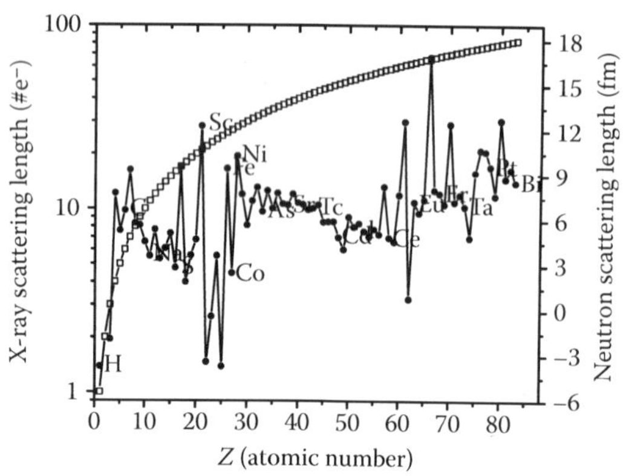

FIGURE 23.1 The x-ray and neutron scattering power as a function of atomic number. The x-ray scattering power is shown on a log-scale, but the scattering power increases linearly with the atomic number.

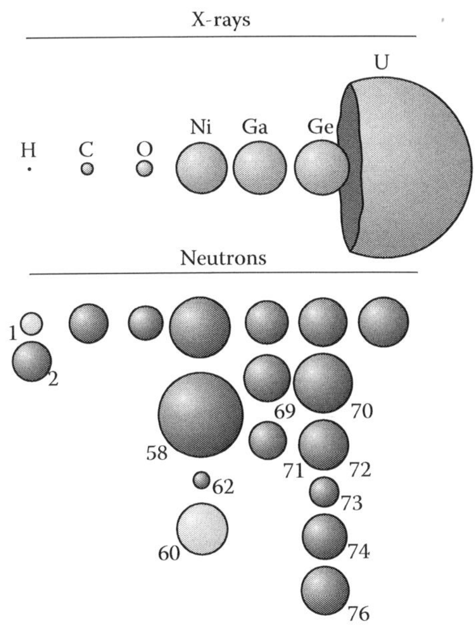

FIGURE 23.2 Comparison between the x-ray and neutron scattering power for selected elements. The radius of the sphere is proportional to the number of electrons in case of x-rays and proportional to the scattering lengths in the neutron case. The first row of spheres for neutrons shows the scattering length of the natural abundant elements, whereas the spheres below correspond to different isotopes. The light color signifies a negative scattering length.

The dominating neutron interaction with matter is scattering processes, in which the momentum and energy of the neutron can change. In the quantum mechanical formalism, the flux of neutrons scattered into an angular opening ($d\Omega$) with a certain energy range (dE_f) is given by Fermi's golden rule[2]:

$$\left.\frac{d^2\sigma}{d\Omega_f dE_f}\right|_{\lambda_i \to \lambda_f} = \frac{k_f}{k_i}\left(\frac{m_n}{2\pi\hbar^2}\right)^2 \left|\langle \mathbf{k}_f \lambda_f | V | \mathbf{k}_i \lambda_i \rangle\right|^2 \cdot (E_{\lambda,i} - E_{\lambda,f} + E_f - E_i) \tag{23.1}$$

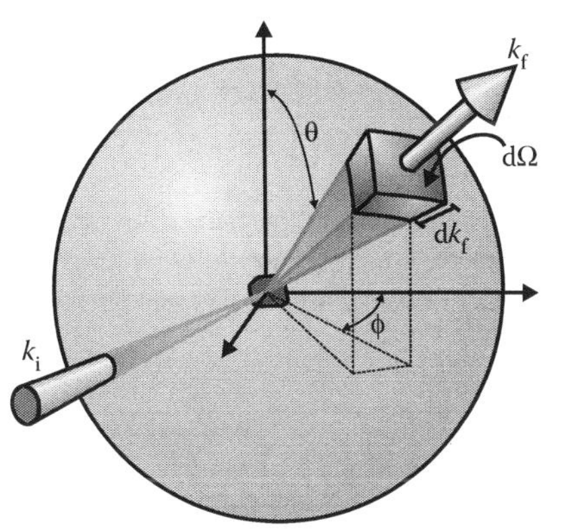

FIGURE 23.3 Illustration of Fermi's golden rule with an incoming neutron flux of k_i and a scattered beam at k_f in the angular range $d\Omega$.

where the neutron mass is given by m_n and the matrix elements $|\langle \mathbf{k}_f\lambda_f|V|\mathbf{k}_i\lambda_i\rangle|$ expressed the probability of obtaining the final state f of the neutron momentum **k** and the perturbed system λ, when the potential V operates upon the initial state i. The potential V describes the interaction between the neutron and the matter. The delta function $\delta(E_{\lambda,i} - E_{\lambda,f} + E_f - E_i)$ is energy conservation and the change in neutron energy is expressed as $\hbar\omega = E_i - E_f$. The conservation of momentum is given by the scattering vector $\mathbf{Q} = \mathbf{k}_f - \mathbf{k}_i$. Nuclear scattering is essentially a delta function in space, and the potential energy can be written as $V(r) = (2\pi\hbar^2 b_n/m_n)\delta(r)$, where b_n is the isotope-specific neutron scattering length. A neutron scattering experiment is an evaluation of the probability function described by Fermi's golden rule. The situation is illustrated in Figure 23.3.

23.3 Elastic Scattering

In an elastic scattering event, the neutron neither loses nor gains energy. Consequently, the magnitude of the neutron momentum vector is unchanged $|\mathbf{k}_i| = |\mathbf{k}_f|$. The situation can be illustrated using the Ewald's sphere concept, where the reciprocal lattice points are shown as dots and the periphery of a circle with radius k_i intersects the origin of the reciprocal lattice (see Figure 23.4a). Elastic scattering occurs when the scattering vector **Q** equals a reciprocal lattice point given by the reciprocal lattice vector $\mathbf{H}_{hkl} = h\mathbf{a}^* + k\mathbf{b}^* + l\mathbf{c}^*$, where hkl are Miller indices and $\mathbf{a}^*$, $\mathbf{b}^*$, and $\mathbf{c}^*$ are reciprocal unit cell lattice vectors. The energy terms in Fermi's golden rule can be ignored for elastic scattering and the matrix elements can be replaced by the structure factor F_{hkl}:

$$\left(\frac{d\sigma}{d\Omega}\right)_{\text{elastic}} = N\frac{(2\pi)^3}{v_0}\sum_{\mathbf{H}}\delta(\mathbf{Q} - \mathbf{H}_{hkl})\left|F_{hkl}(\mathbf{Q})\right|^2 \tag{23.2}$$

Here N is the number of unit cells and v_0 is the unit cell volume. The delta function takes care of the diffraction conditions, and F_{hkl} is known as the structure factor and contains all information related to the position and vibration of each atom in the unit cell. The structure factor is the sum over all atoms n in the unit cell with regard to the atomic site occupation fraction (sof), scattering length b_n, the atomic position $\mathbf{r}_n$, and the displacement from the equilibrium position $\mathbf{u}_n$:

$$F_{hkl} = \sum_{n} \text{sof}_n \cdot b_n \exp(i\mathbf{Q}\cdot\mathbf{r}_n)\exp\left(-\tfrac{1}{2}\langle(\mathbf{Q}\cdot\mathbf{u}_n)^2\rangle\right) \tag{23.3}$$

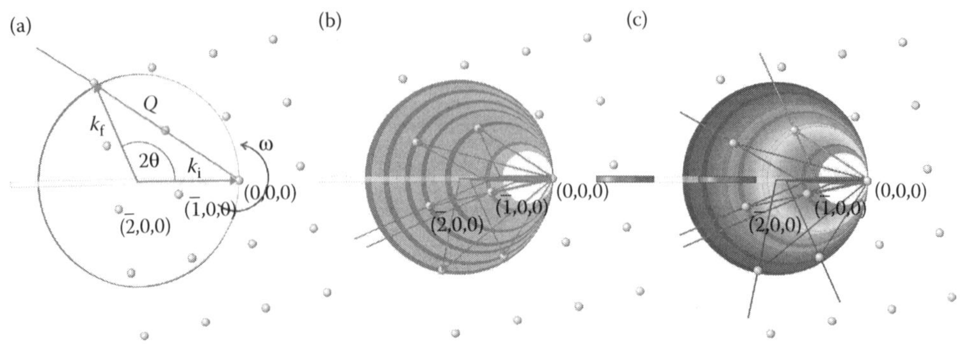

FIGURE 23.4 The Ewalds sphere and reciprocal lattice is illustrated for (a) monochromatic beam (b) white beam Laue diffraction and (c) time-of-flight Laue diffraction. The initial and final neutron momentum vectors, k_i and k_f, are shown along with the scattering vector Q in the monochromatic case along with the rotation ω. In the Laue diffraction case, it is illustrated how the reciprocal lattice vectors ($\bar{1}00$) and ($\bar{2}00$) result in the same scattering angle and they can therefore not be distinguished. For time-of-flight diffraction the different colors represent different times, which allows separation of ($\bar{1}00$) and ($\bar{2}00$) into different time channels.

The last exponential function is also known as the Debye–Waller term and it describes the atomic displacement parameter (ADP). We will look at the ADPs in further detail in Section 23.3.3 as they can provide information about Einstein and Debye temperatures of the material. The observed intensity for the Bragg reflection (*hkl*) is given as I_{hkl} and the intensities are proportional to square modulo of the structure factors as seen from Equation 23.2. Bragg reflections can be collected from either single crystalline or powder samples as described in the following subsections.

23.3.1 Single-Crystal Diffraction

In a single-crystal diffraction experiment, a sample with a single domain with known orientation is used for the collection of numerous Bragg reflections. The collected intensities are used to describe the nuclear density through refinements of the atomic coordinates and thermal vibrations. Generally, three different approaches can be employed to collect single-crystal neutron diffraction data.

Monochromatic diffraction: Monochromatic instruments use a narrow incident wavevector distribution (typically $d\mathbf{k}_i < 0.01$ Å^{-1}). Different Bragg reflections are obtained by rotating the crystal, to fulfill the condition for diffraction $\mathbf{Q} = \mathbf{H}_{hkl}$ and each reflection is uniquely defined. Monochromatic diffraction is illustrated in Figure 23.4a using the Ewald sphere concept.

Laue diffraction: In Laue diffraction experiments, a broad wavevector distribution ($d\mathbf{k}_i \sim 0.2$ Å^{-1}) is brought to scatter from the single crystal. This causes a large number of reflections to be in scattering conditions simultaneously. The scattering vector corresponding to the reciprocal lattice vector $\mathbf{H}_{hkl}$ are no longer uniquely defined, as the incident wavevector $\mathbf{k}_i$ is unknown. Software such as Lauegen or OrientExpress can uniquely index all reflections except for reflections, which are scattered into the same angular opening $d\Omega$.[3] This occurs when $\mathbf{Q} = \mathbf{H}_{hkl}$ and $2\mathbf{Q} = \mathbf{H}_{2h2k2l}$ as the scattering angle is identical in these two cases. Laue diffraction is shown in Figure 23.4b.

Time-of-flight Laue diffraction: The incident beam has a broad wavevector distribution, but the time structure is pulsed. Knowing the neutron scattering angle, length of the traversed path, and neutron time-of-flight allows unique determination of the scattering vector **Q**. In this case, **Q** and 2**Q** will still scatter into the same angular opening for $\boldsymbol{H}_{hkl}$ and $\boldsymbol{H}_{2h2k2l}$, but the events are now separated in time. The situation is shown in Figure 23.4c, where the different colors represent differences in momentum vector and consequently time-of-flight.

After the scattering experiment the recorded data are turned into reflection Miller indices and intensities. A structural model intensity I_{calc} based on Equation 23.3 is refined against the observed intensities I_{obs}. Agreement between modeled and observed intensities is obtained through variation of parameters such as atomic coordinates and thermal vibration.

23.3.2 Powder Diffraction

In a powder diffraction experiment, a large number of randomly oriented crystallites scatter a beam of neutrons. Due to the randomness of the crystallite orientations the reciprocal lattice vector can be considered a scalar $\mathbf{H}_{hkl} = |\mathbf{H}_{hkl}|$, and consequently the delta function in Equation 23.2 causes the scattering vector to describe a cone. The total cross section for each cone is described as

$$\sigma_{\text{total}}(\text{cone}) = \frac{V}{v_0^2}\frac{\lambda^3}{4\sin\theta}\sum_{Q}\left|F_{hkl}(Q)\right|^2 \tag{23.4}$$

where V is the sample volume, λ the neutron wavelength, and the structure factor $F_{hkl}(Q)$ is summed over all scattering vectors with length $|\mathbf{Q}|$. Typically, the neutron powder diffraction is collected over a broad Q range to obtain numerous structure factor intensities. The obtained scattering pattern is often treated using the Rietveld method. The Rietveld method models the scattering at each Q by the function

$$\sigma_{Q,\text{cal}} = \sigma_{Q,\text{background}} + \sum_{\phi=1}^{\#\text{phases}} S_\phi \sum_{o} L_{Q,\phi o} O_{Q,\phi o} \left|F_{hkl,\phi}(Q)\right|^2 \Omega_{Q,\phi o} \tag{23.5}$$

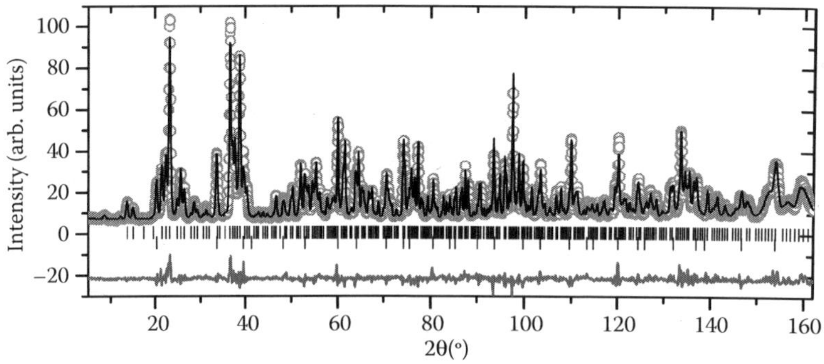

FIGURE 23.5 Rietveld refined powder diffraction pattern of $Sr_8Ga_{16}Ge_{30}$ collected at 20 K using an incoming wavelength of 1.155 Å at HRPT, PSI, Switzerland. The circles are the data points, the line is the fitted model, and the bars represent the Bragg positions. The lower line is the difference between the observed data and the calculated model.

Here, $\sigma_{background}(Q)$ is the background intensity and the first sum is over all phases present within the irradiated volume, the relative amount of each phase ϕ is given by the scale factor S_ϕ. The second summation takes care of overlapping peaks, $L_{Q\phi o}$ is the geometric Lorentz correction, $\Omega_{Q\phi o}$ describes the peak profile, and $O_{Q\phi o}$ describes eventual preferred orientation. The structure factor takes the same form as Equation 23.3. Figure 23.5 shows the observed and calculated intensity based on Rietveld refinements of $Sr_8Ga_{16}Ge_{30}$.

In addition to Rietveld refinements, powder diffraction data can be treated in a total scattering approach; this method is also known as pair distribution function (PDF) analysis. The PDF gives the radial distribution function between pairs of atoms in direct space. The method is a local probe giving information about the environment around the atoms. The direct space function is obtained by taking the Fourier transform of the powder diffraction pattern. To avoid termination ripples in direct space, the Fourier transform is taken over a large Q range (typically $0.5 > Q > 25$ Å^{-1}). To reach such high Q values it is often necessary to use time-of-flight techniques.

23.3.3 Atomic Displacement Parameters

In a single elastic scattering event, a neutron is scattered by one atom. This atom is at the given time at a specific location with in the unit cell. In the diffraction experiment, billions of neutrons are shone upon the sample and consequently, the collected data give the crystallographic time and space average of the atomic positions. The Fourier transform of the diffracted intensities gives the nuclear probability density function (pdf), not to be confused with the pair distribution function. The structure factor (Equation 23.3) describes the nuclear pdf through the atomic mean square displacement $\langle u^2 \rangle$ from the average position also known as ADPs. Commonly, an anisotropic model is used to describe the nuclear pdf, which depicts the nuclear density as an ellipsoid. The neutron interaction with the nuclei is mediated through short-range nuclear forces. Therefore, neutrons provide better details regarding the nuclear pdf compared to x-rays, which, on the other hand, can reveal information about the charge distribution. High-resolution neutron diffraction experiments allow refinement of ADPs to include anharmonic terms. Reliable refinements of ADPs are difficult because slight structural disorder, static or dynamic, may be modeled as enlarged thermal motion of the atoms. Likewise is the site occupation fraction strongly correlated to the modeled ADPs. Therefore, ADPs may act as a "sponge" sucking up uncorrected systematic errors in the experiment or errors in the refined model.

Debye approximation: Carefully extracted ADPs can provide useful information about the dynamic motion of the atoms within the crystal structure. An example of the usefulness of ADPs is the determination of the Debye temperature θ_D using the Debye model. A collection of mean square displacement $\langle u^2 \rangle$ as a function of temperature allows extraction of the Debye temperature in the monoatomic cubic lattice;[4]

$$\langle u^2 \rangle = \frac{3\hbar^2 T}{mk_B\theta_D^2}\left[\frac{T}{\theta_D}\int_0^{\theta_D/T}\frac{x}{e^x - 1}dx + \frac{\theta_D}{4T}\right] + d^2 \quad (23.6)$$

where m is the atomic mass and d^2 is added to account for possible temperature-independent disorder. In case of a polyatomic lattice, an estimate of the Debye temperature can be obtained by replacing $\langle u^2 \rangle$ with the averaged ADP weighted by site multiplicity and the mass m is weighted according to relative fraction of the different elements.

From the Debye temperature, it is possible to estimate the average velocity of sound by $v_s = k_B\theta_D/\hbar(6\pi n)^{1/3}$, where n is the number of atoms in the unit cell. In the simple kinetic approximation, the lattice contribution to the thermal conductivity can be estimated by $\kappa_L = 1/3 v_s l_p C_V$, where l_p is the phonon mean free path and C_V is the heat capacity. The Debye temperature gives an estimate of when Umklapp phonon scattering starts dominating the thermal conductivity, which is at temperatures larger than $1/2\theta_D$ as illustrated in Figure 23.6. Sales et al. suggested extracting the Debye temperatures from structures deposited in the crystallographic databases to search for materials with low thermal conductivity.[5,6]

Einstein model: An atom vibrating independently of the surrounding atoms is known as an Einstein oscillator. Guest atoms residing in oversized cage are often observed to have abnormally large thermal displacement parameters. Therefore, the guest atoms are considered to be weakly bonded and describable as an Einstein atom. The Einstein temperature reflects the eigen-frequency of the guest atom and it can be modeled from the ADPs through the Einstein expression given by

$$\langle u_{xx}^2 \rangle = \frac{\hbar^2}{2mk_B\theta_{E,xx}}\coth\left(\frac{\theta_{E,xx}}{2T}\right) + d^2 \quad (23.7)$$

Here, $\langle u_{xx}^2 \rangle$ is the displacement along xx, which is one of the main axis in the thermal ellipsoid and $\theta_{E,xx}$ is the Einstein temperature along the same direction. The guest atom mass is given by m and d^2 represents temperature-independent disorder. A low Einstein temperature corresponds to a guest atom phonon mode at low energy, which through hybridization with the acoustic phonon modes can cause

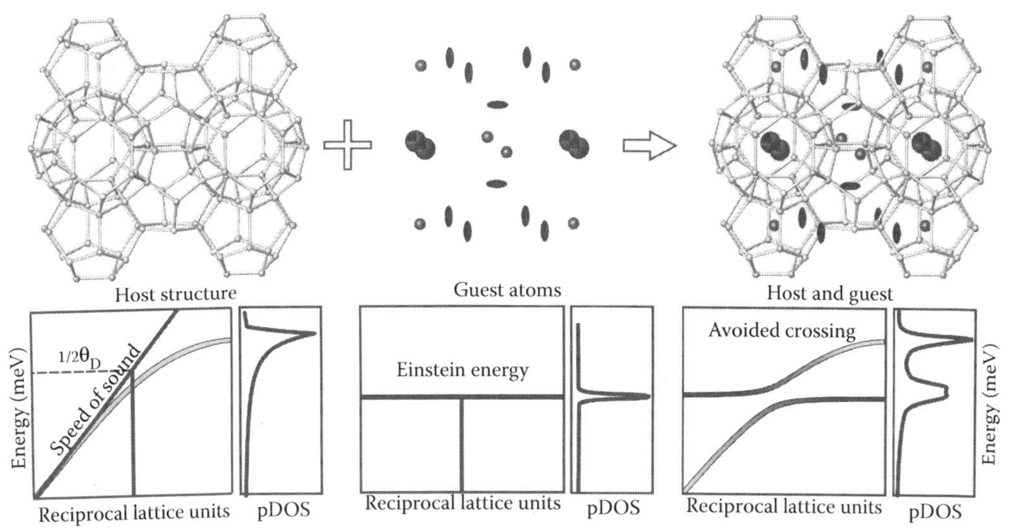

FIGURE 23.6 Classic picture of Debye and Einstein modes represented by the clathrate host structure (left) and the guest atoms isolated from the structure (middle). The combined structure is seen at the right and the resulting phonon dispersion relation and phonon density (pDOS) of states are shown.

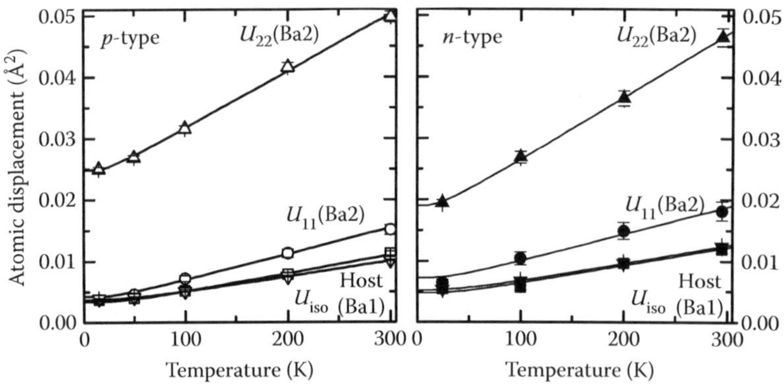

FIGURE 23.7 The atomic displacement parameters for *p*- and *n*-type $Ba_8Ga_{16}Ge_{30}$. The Ba atomic displacement parameters have been fitted with an Einstein model, whereas the Debye expression is used to fit the host structure atoms.

avoided crossing. An avoided crossing introduces flat phonon modes which lowers the group velocity, and in turn reduces the thermal conductivity.[7] A schematic illustration of the Debye and Einstein modes in a phonon dispersion relation is shown in Figure 23.6.

The idea to use ADPs to extract information relevant for thermoelectric materials was first presented by Sales and coworkers.[5,6,8,9] The usefulness was initially demonstrated for a series of skutterudite samples, where the information was compared to heat capacity.[10] The method of extracting Debye and Einstein temperatures has been extensively used for various skutterudite systems using both x-ray and neutrons as probe and on both single crystals and powder data.[6,9–12] The disorder parameter d^2 is not used or close to zero in case of the skutterudites, revealing the guest atom to be located in the center of the cage with little disorder. On the other hand, the disorder for the clathrate samples is large and the disorder parameter gives an idea about the temperature-independent off-center displacement. In the first reports of Einstein temperatures in the clathrates, isotropic temperature factors and no disorder parameter were used in the description of the guest atom in the large cage. In subsequent studies, more refined models were used and described the guest atom with anisotropic displacement parameters and by introducing the disorder parameter in Equation 23.7.[13–16] The information from the disorder term d^2 was used to reveal a correlation between the guest atom position and the site occupation fraction of the low valence host structure atoms in $Ba_8Al_{16}Ge_{30}$.[17] The guest atom disorder has a significant impact on the thermal conductivity in $Ba_8Al_{16}Ge_{30}$. Additional complexity in the atomic guest atom motion was demonstrated by Bentien et al., who used fourth-order anharmonic Gram–Charlier coefficients in the refinement of thermal vibration in $Ba_8Ga_{16}Si_{30}$.[18]

As an example, the temperature-dependent ADPs for *n*- and *p*-type $Ba_8Ga_{16}Ge_{30}$ extracted from single-crystal neutron diffraction data are shown in Figure 23.7. The disorder parameter reveals a small, but considerable, difference between the two compounds.[15] Despite the different ADPs, the Einstein and Debye temperature on the other hand was almost identical for *n*- and *p*-type $Ba_8Ga_{16}Ge_{30}$, with values of $\theta_{E,11}$(Ba2) around 60 K and the Debye temperature was around 270 K.[15]

23.3.4 Maximum Entropy Method

In the above, the nuclear pdf was modeled through the use of ADPs. The maximum entropy method (MEM) is "model free" and uses observed structure factors from either single-crystal or powder diffraction to describe the nuclear pdf. The method is based on maximum likelihood estimation and uses Bayesian statistics. For a positive additive distribution, it can be shown that the probability for a given model is $P(\text{model}) = \exp(\alpha S)$, where S is the entropy given by

$$S = \sum_N \rho_N \log \frac{\rho_N}{\tau_N} \tag{23.8}$$

where the unit cell is divided into a number of pixels N and the sum is taken over all pixels, where ρ_N is the nuclear density in pixel N, and τ_N is the density of the prior distribution (not to be confused with prior probability). The density maximization is obtained for the density with the largest entropy corresponding to the flattest and most featureless density, which simultaneously describes the data. A complication with extracting MEM nuclear densities from neutron diffraction data is the negative scattering length of some of the elements.[19] As MEM analysis is model free it cannot filter out inconsistent data; therefore, the obtained densities may contain artifacts due to systematic errors in the data. Good data quality is critical for MEM analysis.

MEM analysis of skutterudites was recently used to reveal split positions of Pr guest atom positions in $PrOs_4Sb_{12}$ at room temperature, whereas at 8 K the guest atom appears to be positioned in the center of the cage.[20] For the clathrates off-center positions was initially addressed through refinements using off-center models and Fourier difference densities.[21,22] Chakoumakos and coworkers addressed the guest atom disorder and dynamics in multitemperature single-crystal and powder neutron diffraction studies of $Sr_8Ga_{16}Ge_{30}$. Refinements of single-crystal and powder diffraction data resulted in similar residual factors for models with four-site disorder on 24*k* or 24*j*. This suggests the disorder is more complex than a simple four site off-center model.[21] The four-position split side model was used by Sales et al. to explain observed differences in the thermal conductivity between $X_8Ga_{16}Ge_{30}$ (X = Sr, Ba, and Eu).[23] Tunneling between the off-center positions in Sr and Eu, but not Ba, was taken as an explanation for the difference in thermal conductivities (see Figure 23.8).[23] Subsequent studies reported similar differences in thermal conductivity between *n*- and *p*-type of $Ba_8Ga_{16}Ge_{30}$.[24] High-resolution neutron single-crystal diffraction data revealed similar off-center nuclear densities extracted from the difference Fourier density, seriously questioning the tunneling explanation for the difference in thermal conductvity.[15] The use of MEM analysis in the clathrates is limited to studies by Bentien and coworkers, who showed disorder of barium guest atoms in both the large and the small cage in $Ba_8Ga_{16}Si_{30}$.[18]

Figure 23.9 shows different approaches toward extracting the nuclear density from neutron powder diffraction data collected on $Sr_8Ga_{16}Ge_{30}$ at 20 K. The isosurface is shown at half the maximum nuclear density. Five different models were used to refine the data: (1) on-center Sr2 with isotropic ADPs, (2) on-center Sr2 with anisotropic ADPs, (3) off-center Sr2 at 24*j* with isotropic ADPs, (4) off-center Sr2 at 24*k* with isotropic ADPs, and (5) on-center Sr2 with anharmonic ADPs. Furthermore, the nuclear density is extracted from Fourier analysis of the observed structure factor (Fobs) and by the MEM. By comparing *R*-values between the refined models it becomes clear that the isotropic on-center model is fitting the data worse than the other models. The anisotropic refinement gives lower *R*-values; however, the negative value along the U_{11} direction is unphysical. Placing the Si atom off-center gives lower *R*-values compared to the isotropic refinement and the 24*k* off-center model produces values almost as

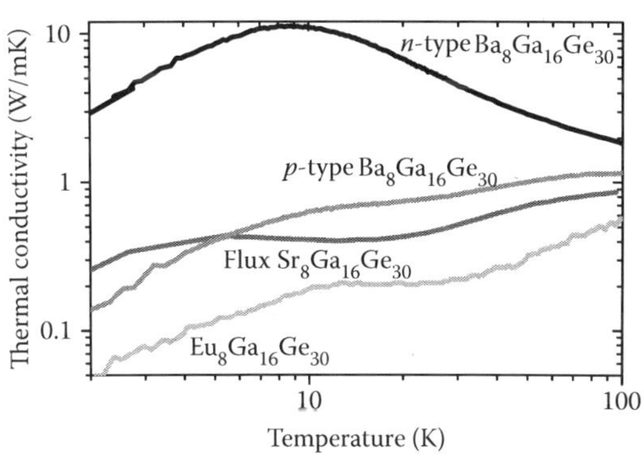

FIGURE 23.8 Thermal conductivity for different guest atom clathrates with host structures consisting of Ga/Ge. (Adapted from Bentien, A. et al. *Phys. Rev. B* 69, 045107, 2004.)

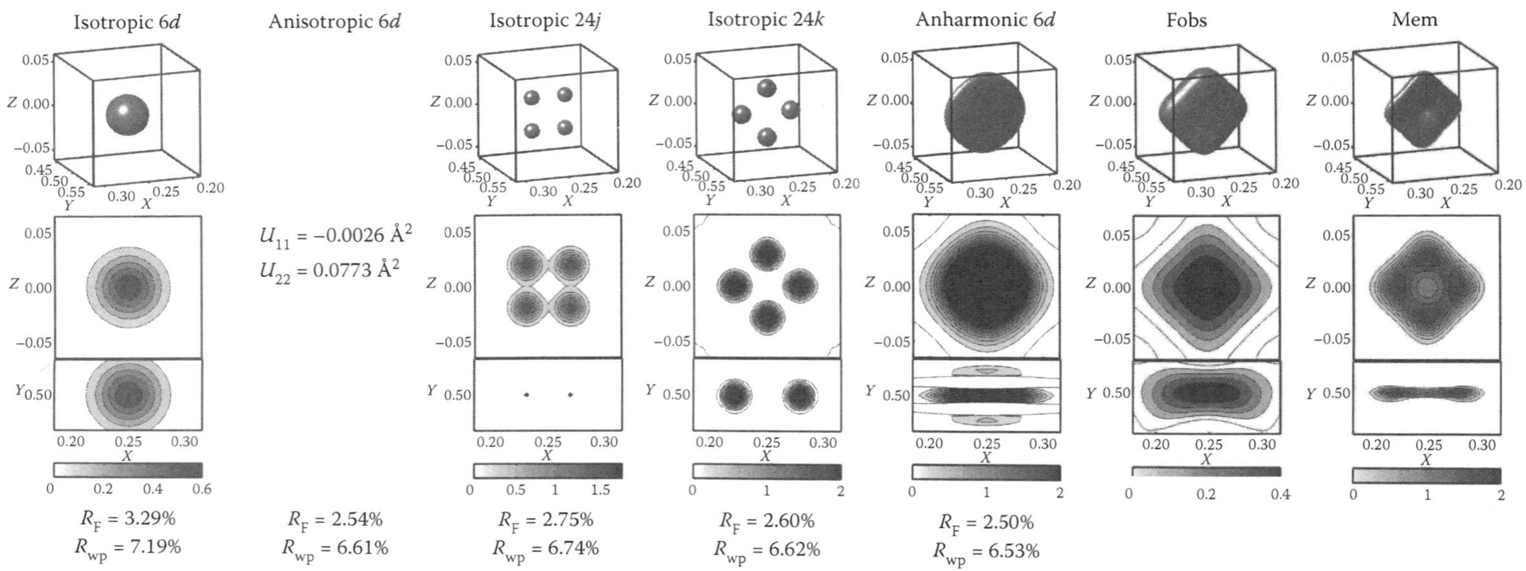

FIGURE 23.9 Different descriptions of the Sr guest atom in the large cage extracted from the powder diffraction data shown in Figure 23.5. The nuclear probability density is shown as an isosurface plot, with isosurface level of half maximum density. Contour plots show the two principal axes within the cage.

low as the anisotropic refinement. The anharmonic refinement produces the lowest R-value, which are comparable to the anisotropic R-values. When comparing the different models to the nuclear density obtained from the Fourier synthesis and MEM, it is clear that the anharmonic model is coming closest in describing the nuclear density. The MEM nuclear density reveals more features than the Fobs. The MEM nuclear density reveals that the majority of the nuclear density is torus shaped; however, there is still some density in the center of the cage and a slight buildup of density on the 24k site, explaining why the 24k model gives better R-values than the 24j site model.

In addition to the detailed refinements of the Sr2 guest atom, the powder diffraction pattern also allowed refinements of the Ga/Ge occupation on the individual host structure sites, even though the contrast is fairly small b(Ga) = 7.3 fm and b(Ge) = 8.1 fm. Ga shows a preference for the 6c site with a site occupation fraction of 61(5)%, whereas the 16i and 24k site have similar occupancies of 31(2) and 32(1)%, respectively. These results are very close to values obtained for neutron single-crystal diffraction data on p-type $Ba_8Ga_{16}Ge_{30}$, where the following occupancies were obtained: 60(4), 33(2), and 30(2)% for the 6c, 16i, and 24k site, respectively.[15]

The neutron powder diffraction example of $Sr_8Ga_{16}Ge_{30}$ concludes the introduction to elastic neutron scattering. The case of $Sr_8Ga_{16}Ge_{30}$ illustrates how neutron diffraction data can be used to give detailed information about atomic positions and how neighboring elements can be distinguished using neutrons. The example illustrates point (1) and (2) in the introduction to neutron scattering (Section 23.2). Additionally, it was shown how multitemperature diffraction data can be used to extract information about the atom dynamics through Einstein and Debye temperatures. Dynamics is the focus of the next section which deals with inelastic neutron scattering.

23.4 Inelastic Scattering

The description of inelastic scattering is in this contribution limited to coherent one-phonon scattering processes. If the neutron gains energy in the one-phonon process, it is known as *phonon absorption* (phonon annihilation) and neutron energy loss is *phonon emission* (phonon creation). The conservation of energy and momentum is fulfilled in one-phonon scattering processes. The difference between the incident and final neutron wavevector can be rewritten in terms of the reciprocal lattice vector $\mathbf{H}$ and the scattering vector $\mathbf{Q} = \mathbf{k}_f - \mathbf{k}_i = \mathbf{H} \pm \mathbf{q}$, where $-\mathbf{q}$ and $+\mathbf{q}$ correspond to phonon creation and annihilation, respectively. The differential scattering cross section for coherent one-phonon emission and absorption can be deduced from Fermi's golden rule (Equation 23.1) and it takes the form[25]:

$$\left(\frac{d^2\sigma}{d\Omega dE_f}\right)_{coh\pm 1} = \frac{k_f}{k_i}\frac{(2\pi)^3}{2V}\sum_s\sum_H \frac{1}{\omega_s}\left|\sum_m \frac{\overline{b_m}}{\sqrt{M_m}} e^{i\mathbf{Q}\cdot\mathbf{r}_m} e^{-W_m}(\mathbf{Q}\cdot\mathbf{e}_{ms})\right|^2 \times \left(\langle n_s+1\rangle\delta(\omega-\omega_s)\delta(\mathbf{Q}-\mathbf{q}-\mathbf{H}) + \langle n_s\rangle\delta(\omega+\omega_s)\delta(\mathbf{Q}+\mathbf{q}-\mathbf{H})\right) \quad (23.9)$$

The summation index s refers to normal modes in a harmonic crystal, and $\mathbf{H}$ is the sum over possible scattering vectors. The angular frequency ω_s refers to the sth normal mode in the crystal. The δ-functions manage the energy and momentum conservation. The operators $\langle n_s + 1\rangle$ and $\langle n_s\rangle$ correspond to phonon emission (neutron energy loss) and phonon absorption (neutron energy gain), respectively, and $\langle n_s\rangle = (\exp(\hbar\omega_s/k_bT) - 1)^{-1}$ is the Bose–Einstein distribution function. The scattering function $S(\mathbf{Q},\omega)$ observed in a neutron scattering experiment is proportional to the differential scattering cross section $(d^2\sigma/d\Omega dE_f)$. The squared term in Equation 23.9 is the dynamical structure factor:

$$F_{dyn}(\mathbf{Q}) = \sum_m \frac{\overline{b_m}}{\sqrt{M_m}} e^{i\mathbf{Q}\cdot\mathbf{r}_m} e^{-W_m}(\mathbf{Q}\cdot\mathbf{e}_{ms}) \quad (23.10)$$

The dynamical structure factor resembles the structure factor for elastic scattering; it deviates by including the square root of the atomic mass M_m and the phonon polarization factor $(\mathbf{Q}\cdot\mathbf{e}_{ms})$, where $\mathbf{e}_{ms}$ are the phonon polarization Eigenvectors. There is in general no simple relation between **Q** and $\mathbf{e}_{ms}$, making calculations of the dynamical structure factor difficult. The phonon polarization can be divided between two transversal modes (e_T) and the longitudinal mode (e_L). By choosing high symmetric directions for inelastic measurements, it is possible to obtain simple relations for $F_{dyn}(\mathbf{Q})$ as will be shown in Section 23.4.2.

23.4.1 Inelastic Scattering from Powders

The scattering cross section from powder samples averages over all possible scattering vectors. For powder samples dominated by incoherent scatters, the neutron-weighted generalized phonon density of states $G(\omega) = S(\omega)\hbar\omega/n_s$ can be obtained by measuring the intensity as a function of energy. For an incoherent monoatomic system, the "true" density of states can easily be extracted from $G(\omega)$, as the mass and scattering length are uniquely defined. The situation is nontrivial for polyatomic systems as the atomic contribution to each phonon branch depends on the phonon polarization ($\mathbf{Q} \cdot \mathbf{e}_{ms}$). Retrieving the phonon density of states of a system dominated by coherent scatters is even more difficult because the scattering cross section depends on the scattering vector **Q**. In other words, the measured scattering cross section for a sample dominated by coherent scatters depends on the part of reciprocal covered in the experiment. To approach the neutron-weighted generalized phonon density of state, as obtained for incoherent scatters, a large number of Brillouin zones must be covered and averaged. This is known as the incoherent approximation.[2]

Direct assignment of phonon modes to the neutron-weighted generalized phonon density of states is difficult without any prior knowledge. Flat modes have high density of states and are consequently observed as high intensities. Weakly dispersive modes arising from Einstein-like guest atom give rise to high intensities in the phonon density of states. However, acoustic modes flatten at the Brillouin zone boundary and cause so-called van Hove singularities, δ-functions folded with the instrumental resolution. Without prior knowledge the observed intensities at low energy in the phonon density of states can be assigned to either Einstein-like guest atoms or acoustic phonons at the zone boundary.

Host structure and guest atom modes can be distinguished by changing the contrast of the guest atoms, by varying the neutron scattering power of the guest atom. This method was demonstrated by V. Keppens et al. in the skutterudites XFe_4Sb_{12} (X = Ce and La).[26] Two La modes were observed at 7 and 15 meV, respectively, and compared well with values extracted from heat capacity. Different filling ratios of guest atom thallium filled skutterudites have also been used to show the presence of low energy modes caused by the guest atom.[27] The Einstein temperature extracted from ADPs obtained from diffraction experiments can likewise be used to give information about the guest atom Einstein energy. However, heat capacity and ADPs rely heavily on the specific models and are therefore biased. Nuclear resonance x-ray scattering can provide unique information about the individual element phonon modes; however, the method is limited to Mössbauer active nucleuses. This method was used to obtain information about the individual atom modes in $EuFe_4Sb_{12}$ and identified low-energy Eu modes in the phonon density of states.[28,29] Another method giving direct information about phonon modes is Raman spectroscopy. Unfortunately, the guest atom modes in the skutterudites cannot be measured by first-order Raman scattering.[30,31] Combination of inelastic neutron scattering and density functional theory has in recent years been used to identify different modes and to compare obtained intensities.[11,32,33] The experimentally obtained intensities in the neutron-weighted generalized phonon density of states $G(\omega)$ and the theoretically calculated phonon density of states cannot be directly compared. However, comparing energies can help assign the atomic modes to the observed peaks. Intensity comparison becomes possible if neutron weighting is applied to the theoretical calculations. To improve agreement between observation and theory, the theoretical data should be folded with the instrumental resolution and the instrumentally covered Q space and energy range should be taken into consideration. Multiphonon processes should also be taken into account.

Low-energy phonon modes were observed in the type I clathrates;[17,34,35] however, they could not be readily identified as guest atom or host structure acoustic modes. It was unfeasible to conduct experiments using the contrast or different filling ratio method. Heat capacity and ADPs were used to infer information about the guest atom modes, but due to the heavy model dependency, the conclusion may be incorrect. Raman spectroscopy on the other hand provides direct information about phonon modes at the zone center.[7,36–38] Therefore, by comparison of the low-energy modes in Raman and inelastic neutron scattering data, makes it possible to identify acoustic phonons and guest atom modes. Nuclear resonance scattering has provided direct evidence for the Eu guest atom modes in $Eu_8Ga_{16}Ge_{30}$.[34] Multitemperature inelastic neutron scattering has likewise been used to identify the guest atom modes in the clathrates due to mode softening upon cooling, caused by a flat-bottom U-shaped potential for the guest atom.[39] Applying pressure both chemically and physically has revealed the same effect in the clathrate system $Ba_8X_yGe_{46-y}$ ($X_y = Ni_6$, Cu_6, Zn_8, Ga_{16}).[39] Theoretical calculations of clathrates are difficult due to the disorder of the host structure. Fair agreement between experiment and theory was obtained for $Ba_8Ga_{16}Ge_{30}$ due to host structure atoms having similar number of electrons and masses giving similar force constants.[35,40] Theoretical calculations were performed on the noble metal clathrates $Ba_8X_6Ge_{40}$ (X = Cu, Ag, and Au), where the noble metals are almost exclusively found on the 6*c* site.[41] The agreement between calculation and experiment is good for the Cu sample; however, the agreement decreases with increasing mass and number of electrons. Despite lacking agreement between intensities in theoretical calculations and experimental data, identification of the phonon modes is still possible through comparison. Exceptionally good agreement between theoretical calculations and experiment was obtained by M. Koza et al. for the fully ordered clathrate $Ba_8Ge_{43}[]_3$ (M. Koza et al., private communication, 2010).

An example of inelastic scattering from a powder is shown in Figure 23.10. The sample used in this experiment was $Ba_8Ga_{16}Ge_{30}$. The top plot shows parts of the covered *Q* space and energy range obtained

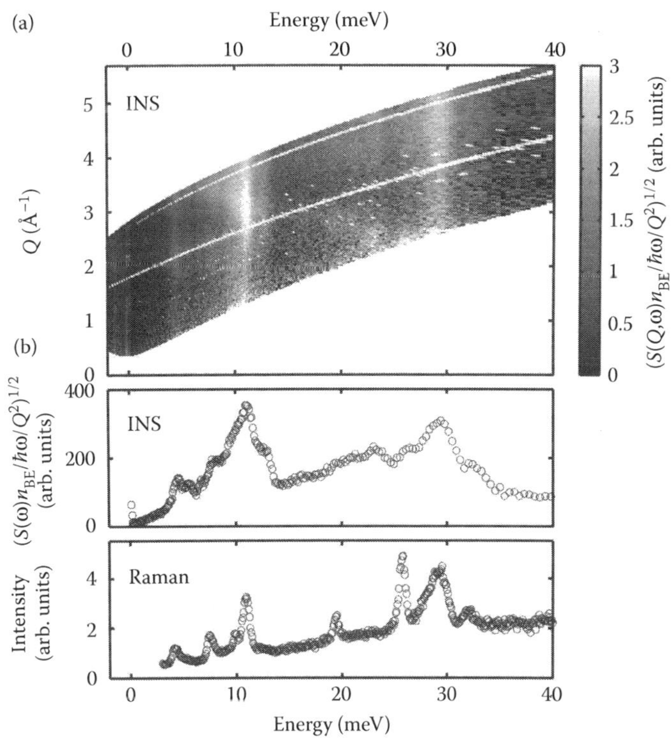

FIGURE 23.10 The $S(Q,\omega)$ map of *n*-type $Ba_8Ga_{16}Ge_{30}$ corrected for Bose–Einstein occupation statistics, energy $\hbar\omega$ and the scattering vector *Q*. (a) The $S(\omega)$ integrated over all *Q* and (b) Raman scattering from *n*-type $Ba_8Ga_{16}Ge_{30}$.

at the time-of-flight spectrometer FOCUS at PSI, Switzerland.[42] The shown intensities are the square root of $S(Q,\omega)$ corrected by Bose statistics, energy, and Q^2. The Q and energy map allows identification of strong modes at 5, 11, and 29 meV. The middle plot shows the corresponding summed intensity over the covered Q range and represents the incoherent approximation toward obtaining the generalized neutron-weighted phonon density of states. The bottom plot is the Raman spectra of $Ba_8Ga_{16}Ge_{30}$. Comparison between the inelastic neutron scattering data and Raman data reveal a first peak in both spectra at about 4.5 meV. As first-order Raman scattering is incapable of observing acoustic phonons, this first mode can be attributed to a guest atom mode. This assignment is in agreement with theoretical calculations and results from ADPs. The observed mode corresponds to the guest atom motion in the equatorial plane of the large cage.[15,40]

The powder inelastic neutron scattering from $Ba_8Ga_{16}Ge_{30}$ is an example of how neutrons can be used to reveal guest atom modes. However, the powder data cannot reveal information about phonon lifetimes. Peak broadening in the neutron-weighted generalized phonon density of states can be due to either short phonon lifetimes or dispersive behavior. Powder inelastic neutron scattering cannot be used directly to infer information about the lattice thermal conductivity.

23.4.2 Inelastic Scattering from Single Crystals

Inelastic neutron scattering from powders is limited to give information about the density of states. Powder samples cannot provide information about phonon lifetime, avoided crossing, or other details, which can be observed in the phonon dispersion relation. Inelastic neutron scattering from a single crystal can provide the full phonon dispersion with sufficient spatial resolution, including phonon lifetimes. In an inelastic single-crystal coherent scattering experiment, the scattering cross section is given by Equation 23.9. Experimentally, the crystal needs to be oriented in space to uniquely define the lattice vector $\mathbf{H}$. The scattering vector $\mathbf{Q}$ and energy change $\hbar\omega$ is defined by the instrument. Scattered neutrons can be observed when the δ-function in the cross section for coherent one-phonon scattering in Equation 23.10 is fulfilled. Additionally, the scattering vector $\mathbf{Q}$ must be chosen not to be perpendicular to the phonon polarization $\mathbf{e}_{ms}$. Actually, the phonon polarization term $(\mathbf{Q} \cdot \mathbf{e}_{ms})^2$ may be used to selectively promote one polarization and extinguish the others. At low $\mathbf{q}$, the phonon polarization $\mathbf{e}_{ms}$ is divided into transversal and longitudinal phonons, which are perpendicular and parallel to $\mathbf{q}$, respectively. Consider a cubic crystal oriented with [110] and [001] in the scattering plane. A scan at $\mathbf{H}$ = (222) along $\mathbf{q}$ = [hhh] gives only a longitudinal contribution, whereas a scan along $\mathbf{q}$ = [$hh0$] at (222) gives a contribution from both the transversal and longitudinal phonons. The situation is illustrated in Figure 23.11.

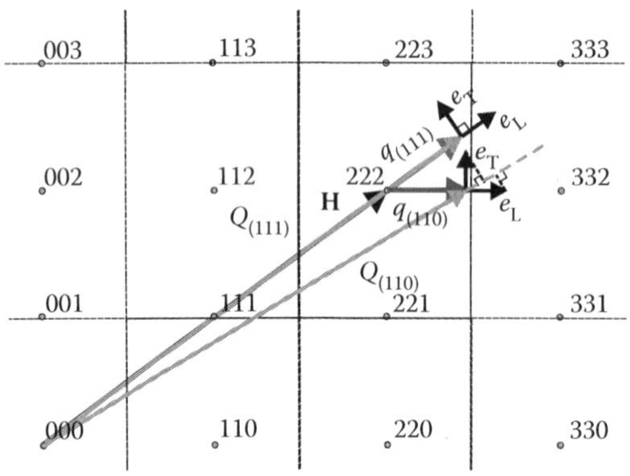

FIGURE 23.11 Reciprocal space around the (222) reflection. The phonon polarization vectors, e_T and e_L, transversal and longitudinal phonon polarization are perpendicular and parallel to the direction vector Q, respectively. The transversal polarization has an in-plane and out-of-plane component.

The main problem in measuring full phonon dispersion relations is growing single crystals of sufficient size. Even when large single crystals are obtained, the measurement is time consuming. The data are typically collected using triple-axis spectrometers, because they offer better reciprocal space resolution, compared to time-of-flight spectrometers.

Triple-axis spectroscopy has been used to investigate the phonon dispersion relation in single-crystal skutterudites. Investigations of $CeOs_4Sb_{12}$ did not reveal any features, which could be related to guest atom modes.[43,44] A study of $PrRu_4Sb_{12}$ and $PrOs_4Sb_{12}$ reveals features of a nondispersive mode from about halfway through to the end of the Brillouin zone. Furthermore, softening of the low-energy mode upon cooling was also observed; however, clear evidence for avoided crossing was not reported.[45] In a study of $CeRu_4Sb_{12}$, the authors claim to observe avoided crossing, but they do not observe the guest atom mode until halfway through the Brillouin zone.[46] Early reports on $Ba_8Ga_{16}Ge_{30}$ and $Sr_8Ga_{16}Ge_{30}$ focused on the transversal phonon modes and found guest atom energies at about 4.5 meV in good agreement with powder inelastic neutron scattering measurements.[47,48] Additionally, it was observed that the guest atom mode was strongly interacting with the acoustic phonons. High-resolution Q and energy data were measured on $Ba_8Ga_{16}Ge_{30}$ using the cold source triple-axis spectrometer, RITA-II at PSI, Switzerland and clearly reveal avoided crossing behavior between the guest atom mode and the acoustic phonons.[7] Phonon widths were estimated to extract phonon lifetimes, which turned out to be much longer than anticipated from kinetic theory. An example of the data collected on $Ba_8Ga_{16}Ge_{30}$ is shown in Figure 23.12, where the scans are made along the longitudinal direction at (222) + [*hhh*] and (330) + [*hh*0]. The avoided crossing is observed for the dispersion relation around (330), but is completely lacking in the data at (222). At (222), the intensity abruptly decreases around $h = 0.25$; this change is most likely due to sudden changes in the phonon polarization, which could be due to interactions with a guest atom mode.

The phonon widths observed, when correcting for instrumental broadening in (330) at $h = 0.18$, correspond to longitudinal and guest atom phonon lifetimes of $\tau_L = 2.6(6)$ ps and $\tau_{Guest} = 2.0(2)$ ps. In the region after the avoided crossing $h = 0.23$, the lifetimes are reduced to 1.3(6) and 1.5(2) ps, respectively. However, these lifetimes are still one order of magnitude larger than what is predicted from the simple kinetic approach $\tau_{kinetic} = 0.18$ ps.[7] Conclusively, the low thermal conductivity in clathrates are not due to scattering effects, but rather a consequence of the flat phonon modes, which causes a low phonon group velocity. The flat modes also allow for Umklapp process taking place at much lower temperatures than expected from the half-Debye temperature. The Debye temperature is approximated from the

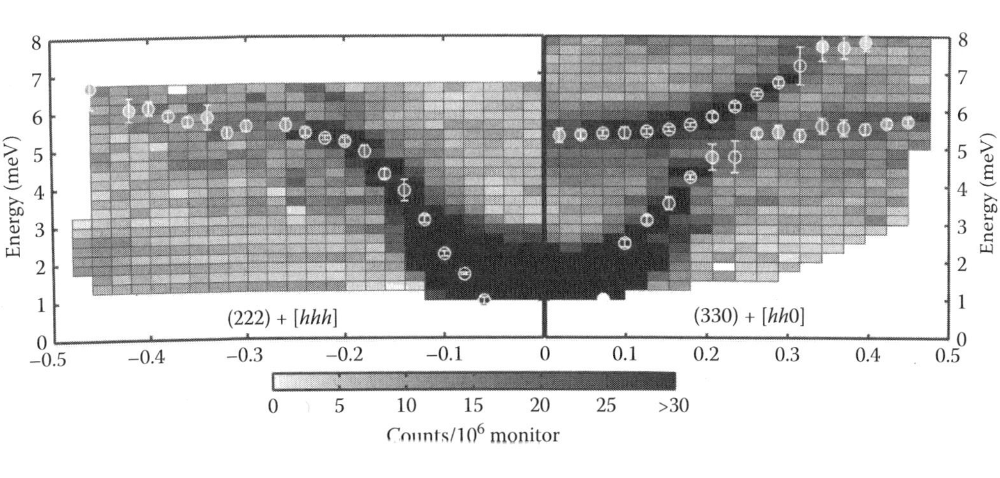

FIGURE 23.12 The phonon dispersion relation for $Ba_8Ga_{16}Ge_{30}$ measured along [*hhh*] at (222) and [*hh*0] at (330). The scattering vector and the phonon polarization are in both cases oriented to emphasize the longitudinal phonon contribution.

velocity of sound, which is extracted from the slope of the phonon dispersion.[7] Based on energies observed at half the Brillouin zone (2.5–3 meV) Umklapp scattering is possible from about 30 K, five times lower than what would be expected from half the Debye temperature $1/2\theta_D = 151$ K. The relative large number of flat phonon modes at low energy also increases the possibilities for Umklapp processes as energy and momentum must be conserved for multiphonon processes. The phonon dispersion relation for $Ba_8Ga_{16}Ge_{30}$ exemplifies how dynamic information at the microscopic level can aid in the understanding of macroscopic properties such as thermal conductivity. Neutrons have a unique role to play in the understanding of complex thermoelectric materials, especially when it comes to the lattice thermal conductivity. However, the main drawback is the large volume of single crystals necessary for carrying out experiments disclosing the phonon dispersion relation.

23.5 Concluding Remarks

This chapter has briefly reviewed neutron scattering in a thermoelectric context. An overview of neutron investigations in host–guest skutterudites and type I clathrates has been presented. The usability of neutrons toward elucidating the static and dynamic microstructure on the atomic level has been exemplified with data from clathrate research. The main drawback of the presented methods is the insufficiency of neutrons available, in terms of both neutron sources worldwide and flux on individual instruments. Consequently, it is at present necessary to work with large volume samples, which might be difficult to produce for new potential thermoelectric materials. Fortunately, these obstacles will be reduced, as new improved sources such as SNS, J-Parc, and ESS become fully operational in the near future.

Acknowledgments

Fruitful discussions with Bo B. Iversen, Kim Lefmann, Niels H. Andersen, Niels B. Christensen, Simon Johnsen, Georg Madsen, Fanni Juranyi, and Nina Lock are greatly appreciated.

References

1. Press Release: The 1994 Nobel Prize in physics. Nobelprize.org, 1994.
2. Squires, G. L. *Thermal Neutron Scattering*. (Dover, Cambridge University Press, Cambridge, UK, 1996.)
3. Campbell, J. Lauegen, an X-windows-based program for the processing of Laue diffraction data. *J. Appl. Cryst.* 28, 228–236, 1995.
4. Willis, B. T. M. and Pryor, A. W. *Thermal Vibrations in Crystallography*. (Cambridge University Press, Cambridge, UK, 1975.)
5. Sales, B. C., Chakoumakos, B. C., Mandrus, D., Sharp, J. W., Dilley, N. R., and Maple, M. B. in *Thermoelectric Materials 1998—The Next Generation Materials for Small-Scale Refrigeration and Power Generation Applications*, Vol. 545, Materials Research Society Symposium Proceedings, eds. T. M. Tritt, M. G. Kanatzidis, G. D. Mahan, and H. B. Lyon, pp. 13–21 (Materials Research Society, Warrendale, Pennsylvania, US, 1999).
6. Sales, B. C., Chakoumakos, B. C., Mandrus, D., and Sharp, J. W. Atomic displacement parameters and the lattice thermal conductivity of clathrate-like thermoelectric compounds. *J. Solid State Chem.* 146, 528–532, 1999.
7. Christensen, M., Abrahamsen, A. B., Christensen, N. B., Juranyi, F., Andersen, N. H., Lefmann, K., Andreasson, J., Bahl, C. R. H., and Iversen, B. B. Avoided crossing of rattler modes in thermoelectric materials. *Nat. Mater.* 7, 811–815, 2008.
8. Chakoumakos, B. C., Sales, B. C., Mandrus, D., and Keppens, V. Disparate atomic displacements in skutterudite-type $LaFe_3CoSb_{12}$, a model for thermoelectric behavior. *Acta Crystallogr. B, Struct. Sci.* 55, 341–347, 1999.

9. Sales, B. C., Mandrus, D. G., and Chakoumakos, B. C. *Use of Atomic Displacement Parameters in Thermoelectric Materials Research*. Vol. 70 (Academic Press Inc, Waltham, Massachusetts, US, 2001).
10. Sales, B. C., Chakoumakos, B. C., and Mandrus, D. Thermoelectric properties of thallium-filled skutterudites. *Phys. Rev. B* 61, 2475–2481, 2000.
11. Koza, M. M., Capogna, L., Leithe-Jasper, A., Rosner, H., Schnelle, W., Mutka, H., Johnson, M. R., Ritter, C., and Grin, Y. Vibrational dynamics of filled skutterudites $M_{1-x}Fe_4Sb_{12}$ (M = Ca, Sr, Ba, and Yb). *Phys. Rev. B* 81, 174302, 2010.
12. Mi, J. L., Christensen, M., Nishibori, E., Kuznetsov, V., Rowe, D. M., and Iversen, B. B. Multitemperature synchrotron powder diffraction and thermoelectric properties of the skutterudite $La_{0.1}Co_4Sb_{12}$. *J. Appl. Phys.* 107, 113507, 2010.
13. Bentien, A., Nishibori, E., Paschen, S., and Iversen, B. B. Crystal structure, atomic vibration, and disorder of the type-I thermoelectric clathrates $Ba_8Ga_{16}Si_{30}$, $Ba_8Ga_{16}Ge_{30}$, $Ba_8In_{16}Ge_{30}$ and $Sr_8Ga_{16}Ge_{30}$. *Phys. Rev. B* 71, 144107, 2005.
14. Christensen, M., Bryan, J. D., Birkedal, H., Stucky, G. D., Lebech, B., and Iversen, B. B. The nuclear and magnetic structure of $Eu_4Ga_8Ge_{16}$. *Phys. Rev. B* 68, 174428, 2003.
15. Christensen, M., Lock, N., Overgaard, J., and Iversen, B. B. Crystal structures of thermoelectric *n*-and *p*-type $Ba_8Ga_{16}Ge_{30}$ studied by single crystal, multitemperature, neutron diffraction, conventional x-ray diffraction and resonant synchrotron x-ray diffraction. *J. Am. Chem. Soc.* 128, 15657–15665, 2006.
16. Melnychenko-Koblyuk, N., Grytsiv, A., Fornasari, L., Kaldarar, H., Michor, H., Rohrbacher, F., Koza, M. et al., Ternary clathrates Ba-Zn-Ge: Phase equilibria, crystal chemistry and physical properties. *J. Phys. Condens. Matter.* 19, 216223, 2007.
17. Christensen, M., and Iversen, B. B. Host structure engineering in thermoelectric clathrates. *Chem. Mater.* 19, 4896–4905, 2007.
18. Bentien, A., Iversen, B. B., Bryan, J. D., Stucky, G. D., Palmqvist, A. E. C., Schultz, A. J., and Henning, R. W. Maximum entropy method analysis of thermal motion and disorder in thermoelectric clathrate $Ba_8Ga_{16}Si_{30}$. *J. Appl. Phys.* 91, 5694–5699, 2002.
19. Sakata, M., Uno, T., Takata, M., and Howard, C. J. Maximum-entropy-method analysis of neutron diffraction data. *J. Appl. Cryst.* 26, 159–165, 1993.
20. Kaneko, K., Metoki, N., Kimura, H., Noda, Y., Matsuda, T. D., and Kohgi, M. Visualizing rattling in $PrOs_4Sb_{12}$ by single crystal neutron diffraction and maximum-entropy analysis. *J. Phys. Soc. Jpn.* 78, 074710, 2009.
21. Chakoumakos, B. C., Sales, B. C., Mandrus, D. G., and Nolas, G. S. Structural disorder and thermal conductivity of the semiconducting clathrate $Sr_8Ga_{16}Ge_{30}$. *J. Alloys Compounds* 296, 80–86, 2000.
22. Chakoumakos, B. C., Sales, B. C., and Mandrus, D. G. Structural disorder and magnetism of the semiconducting clathrate $Eu_8Ga_{16}Ge_{30}$. *J. Alloys Compounds* 322, 127–134, 2001.
23. Sales, B. C., Chakoumakos, B. C., Jin, R., Thompson, J. R., and Mandrus, D. Structural, magnetic, thermal, and transport properties of $X_8Ga_{16}Ge_{30}$ (X = Eu, Sr, Ba) single crystals. *Phys. Rev. B* 63, 245113, 2001.
24. Bentien, A., Christensen, M., Bryan, J. D., Sanchez, A., Paschen, S., Steglich, F., Stucky, G. D., and Iversen, B. B. Thermal conductivity of thermoelectric clathrates. *Phys. Rev. B* 69, 045107, 2004.
25. Shirane, G., Shapiro, S. M., and Tranquada, J. M. *Neutron Scattering with a Triple-Axis Spectrometer: Basic Techniques*. (Cambridge University Press, Cambridge, UK, 2002.)
26. Keppens, V., Mandrus, D., Sales, B. C., Chakoumakos, B. C., Dai, P., Coldea, R., Maple, M. B., Gajewski, D. A., Freeman, E. J., and Bennington, S. Localized vibrational modes in metallic solids. *Nature* 395, 876–878, 1998.
27. Hermann, R. P., Jin, R. J., Schweika, W., Grandjean, F., Mandrus, D., Sales, B. C., and Long, G. J. Einstein oscillators in thallium filled antimony skutterudites. *Phys. Rev. Lett.* 90, 135505, 2003.
28. Long, G. J., Hermann, R. P., Grandjean, F., Alp, E. E., Sturhahn, W., Johnson, C. E., Brown, D. E., Leupold, O., and Ruffer, R. Strongly decoupled europium and iron vibrational modes in filled skutterudites. *Phys. Rev. B* 71, 140302, 2005.

29. Wille, H. C., Hermann, R. P., Sergueev, I., Leupold, O., van der Linden, P., Sales, B. C., Grandjean, F., Long, G. J., Rueffer, R., and Shvyd'ko, Y. V. Antimony vibrations in skutterudites probed by Sb-121 nuclear inelastic scattering. *Phys. Rev. B* 76, 140301 2007.
30. Ogita, N., Kondo, T., Hasegawa, T., Takasu, Y., Udagawa, M., Takeda, N., Ishikawa, K. et al., Raman scattering investigation of skutterudite compounds. *Physica B* 383, 128–129, 2006.
31. Ogita, N., Udagawa, M., Saha, S. R., Sato, H., and Sugawara, H. Raman scattering investigation of filled skutterudite PrRu4P12. *Physica B* 378–380, 206–208, 2006.
32. Feldman, J. L., Dai, P. C., Enck, T., Sales, B. C., Mandrus, D., and Singh, D. J. Lattice vibrations in La(Ce)Fe_4Sb_{12} and $CoSb_3$: Inelastic neutron scattering and theory. *Phys. Rev. B* 73, 014306, 2006.
33. Koza, M. M., Johnson, M. R., Viennois, R., Mutka, H., Girard, L., and Ravot, D. Breakdown of phonon glass paradigm in La- and Ce-filled Fe_4Sb_{12} skutterudites. *Nat. Mater.* 7, 805–810, 2008.
34. Hermann, R. P., Schweika, W., Leupold, O., Ruffer, R., Nolas, G. S., Grandjean, F., and Long, G. J. Neutron and nuclear inelastic scattering study of the Einstein oscillators in Ba-, Sr-, and Eu-filled germanium clathrates. *Phys. Rev. B* 72, 174301, 2005.
35. Christensen, M., Juranyi, F., and Iversen, B. B. The rattler effect in thermoelectric clathrates studied by inelastic neutron scattering. *Physica B* 385, 505–507, 2006.
36. Nolas, G. S., and Kendziora, C. A. Raman scattering study of Ge and Sn compounds with type-I clathrate hydrate crystal structure. *Phys. Rev. B* 62, 7157–7161, 2000.
37. Takasu, Y., Hasegawa, T., Ogita, N., Udagawa, M., Avila, M. A., Suekuni, K., and Takabatake, T. Off-center rattling and anisotropic expansion of type-I clathrates studied by Raman scattering. *Phys. Rev. Lett.* 100, 165503, 2008.
38. Takasu, Y., Hasegawa, T., Ogita, N., Udagawa, M., Avila, M. A., Suekuni, K., Ishii, I., Suzuki, T., and Takabatake, T. Dynamical properties of guest ions in the type-I clathrate compounds $X_8Ga_{16}Ge_{30}$ (X = Eu,Sr,Ba) investigated by Raman scattering. *Phys. Rev. B* 74, 174303, 2006.
39. Christensen, M., Johnsen, S., Juranyi, F., and Iversen, B. B. Clathrate guest atoms under pressure. *J. Appl. Phys.* 105, 073508, 2009.
40. Madsen, G. K. H., and Santi, G. Anharmonic lattice dynamics in type-I clathrates from first-principles calculations. *Phys. Rev. B* 72, 220301, 2005.
41. Johnsen, S., Christensen, M., Thomsen, B., Madsen, G. K. H., and Iversen, B. B. Barium dynamics in noble metal clathrates. *Phys. Rev. B* 82, 184303, 2010.
42. Janssen, S., Mesot, J., Holitzner, L., Furrer, A., and Hempelmann, R. FOCUS: A hybrid TOF-spectrometer at SINQ. *Physica B* 234–236, 1174–1176, 1997.
43. Yang, C. P., Kohgi, M., Iwasa, K., Sugawara, H., and Sato, H. Inelastic neutron scattering from $CeOs_4Sb_{12}$. *J. Phys. Soc. Jpn.* 74, 2862–2863, 2005.
44. Yang, C. P., Wang, H., Iwasa, K., Kohgi, M., Sugawara, H., and Sato, H. Lattice dynamics of the filled skutterudite $CeOs_4Sb_{12}$. *Appl. Phys. Lett.* 90, 102503, 2007.
45. Iwasa, K., Mori, Y., Lijie, H., Murakami, Y., Kohgi, M., Sugawara, H., and Sato, H. Softening phonon and relaxation mode in the filled skutterudite PrT_4Sb_{12} (T = Ru and Os). *J. Phys. Conf. Ser.* 92, 012122, 2007.
46. Lee, C. H., Hase, I., Sugawara, H., Yoshizawa, H., and Sato, H. Low-lying optical phonon modes in the filled skutterudite $CeRu_4Sb_{12}$. *J. Phys. Soc. Jpn.* 75, 123602, 2006.
47. Lee, C. H., Yoshizawa, H., Avila, M. A., Hase, I., Kihou, K., and Takabatake, T. Neutron scattering study of phonon dynamics on type-I clathrate $Ba_8Ga_{16}Ge_{30}$. *J. Phys. Conf. Ser.* 92, 012169, 2007.
48. Lee, C.-H., Yoshizawa, H., Avila, M. A., Hase, I., Kihou, K., and Takabatake, T. Phonon dynamics of type-I clathrate $Sr_8Ga_{16}Ge_{30}$ studied by inelastic neutron. *J. Phys. Soc. Jpn.* 77, 260–262, 2008.

III

Commercial

24

Patent Basics Every Researcher and Engineer Should Know

Robert M. Hansen
The Marbury Law Group, PLLC

24.1 Introduction

This chapter provides an overview of intellectual property, focusing on elements of interest to engineers and scientists in the fields of thermoelectric materials and devices. Given the arcane and ever-changing nature of intellectual property laws around the world, this chapter cannot be an exhaustive reference. Rather, this chapter provides engineers and scientists with sufficient background information regarding intellectual property to enable them to appreciate the subject, recognize when legal assistance should be obtained to protect intellectual property, and assist legal counsel in identifying and documenting intellectual property. Note that no attorney–client relationship is established between the author and the reader, and the following are not intended to be legal advice.

Intellectual property is important to engineers and scientists because much of the value of their work results in or is protected by patents, copyrights, trade secrets, and technical know-how. A great deal of intellectual labor is invested in developing new designs, discovering new phenomena, and inventing new technologies. Without laws to protect the fruits of intellectual labor, engineers, and scientists would not receive fair remuneration for their efforts and expertise.

Intellectual property laws provide the economic foundations that make the creative intellectual labors of engineers and scientists worthwhile. Companies, institutions, and governments hire engineers and scientists to create technologies and products which have value in the marketplace. Without the legal protections afforded by patent, copyright, trade secret, and similar intellectual property laws, the fruits of intellectual labors could be simply copied by others investing the time and costs required to develop new products, conceive new inventions or discover new phenomenon. Without protections to prevent such copying, the company, research center or university that invested the time and money to invent, develop new products, and create new technologies could never recover their investment, let alone properly compensate those who contributed their intellectual labors. Thus, intellectual property provides an important foundation for the compensation paid to engineers and scientists.

Economic interests in intellectual property are not limited to for-profit companies. Many private and public institutions of research and education derive substantial revenues from the licensing, sales, or other exploitations of the intellectual accomplishments of their employees. Even not-for-profit organizations that employ engineers and scientists typically have a strong economic interest in protecting the intellectual property created by their employees.

Intellectual property protections are also important to national economies. Without intellectual property protections, individuals and companies would have little incentive to invest time and money in developing new products, creating new inventions, or researching new phenomenon. Without economic incentives to create and research, citizens would spend their time on other pursuits, and over time a nation's economy would stagnate, with the nation's wealth limited to the value of its natural resources and the manual labor of its people.

This economic view of intellectual property is now widely accepted by countries around the world. Most countries have adopted intellectual property laws to protect the intellectual creations of their engineers and scientists, as well as of their artists, authors, musicians, and others who create things of value to society that could otherwise be copied. As a result, the laws underpinning intellectual property are remarkably consistent across the globe. Countries with diverse cultures and different political systems have largely agreed on relatively uniform patent, trade secret, copyright, and trademark protections. Global treaties reconciling intellectual property laws and streamlining the processes for obtaining patents and trademarks have been implemented and form a foundation of intellectual property protections in many countries.

In short, the creation of patents and copyrights, trade secrets, and industrial know-how are how engineers and scientists create the economic value that supports their salaries. Generally speaking, those who create more intellectual property will earn more money. Thus, every engineer and scientist has an interest in ensuring all of their intellectual creations are properly protected in one or more forms of intellectual property.

In addition to monetary incentives, the creation of intellectual property can yield recognition and reputation rewards. Patents are widely viewed as symbols of intellectual accomplishment. As with books and journal articles, being listed as an inventor on patents can contribute to one's professional reputation and resume. Thus, individuals are wise to ensure that their inventions and works of authorship are appropriately protected as symbols of their accomplishment and capabilities.

This chapter provides an overview of the forms of intellectual property of most importance to engineers and scientists to help you protect your intellectual creations in a manner that creates the greatest value for yourself, your employer and for society as a whole.

24.2 Types of Intellectual Property

The term "intellectual property" generally refers to the value created in human inventions and mental efforts, as opposed to value created through manual labor and following the instructions of others. Through its legal protections, intellectual property can be treated much like physical property, allowing individuals and companies to buy, sell, and license their intellectual property to others.

The metes and bounds of intellectual property are defined by the legal frameworks established by national governments. Intellectual property laws define how individuals and organizations can assert legal rights to intellectual creations. Thus, intellectual property exists primarily as legal concepts that vary somewhat from country to country depending upon the details within each country's intellectual property laws. In countries with laws that expansively protect inventions and intellectual creations, and whose court systems vigorously enforce such laws, intellectual property is highly valuable. In countries with narrow legal protections or limited legal enforcement mechanisms, intellectual property exhibits less value. Some countries also recognize the concept of "natural rights" and have enacted laws which extend extensive rights to individuals in their creative works of authorship, artwork, and design. So we must look to the laws of each country to define intellectual property, the extent of the protections that are provided, and the inherent value in such intellectual property.

As described more fully below, there are four main types of intellectual property that will be of most concern to engineers and scientists: patents; copyrights; trade secrets; and trademarks. Additionally, most industrial countries also have laws that protect other forms of intellectual property which are similar to these four main categories, including, for example, rights in industrial designs (which are similar to design patents and copyrights), rights in registered semiconductor mask works, and rights of publicity. Also, most countries have laws against industrial espionage and computer "hacking," which provide protections for intellectual property. Also many countries have legal processes to prevent Internet domain name misuse.

24.3 Patents

Patents are essentially legal grounds of limited monopolies to make, use, sell, or import inventions for a limited period of time (generally 20 years from the date that the application was first filed). Patents are only granted for inventions that are new and involve an incentive step or are not obvious in view of what was known. The patent monopoly is granted in exchange for inventors publicly disclosing their invention in a patent application which the government soon publishes for all to see. Patents have economic value because they provide legal rights to prevent others from using the claimed inventions, thereby enabling companies to exploit the invention without competition for the duration of the patent's life. Since monopolies have serious economic consequences, patents are only granted for inventions which are truly new and not merely obvious variations of what is known, only in exchange for a full public disclosure of the complete invention, and only for the limited life of the patent so long as its maintenance fees are paid. Also, the monopoly is limited to what is recited in the allowed claims, and not to everything described in the patent.

The novelty requirement may vary slightly from country to country, but generally countries examine patent applications to determine whether the claimed invention was known or would have been obvious to one fully apprised of the state of the art at the time the patent application was filed. In terminology embodied in the United States patent laws, a claim must be "new" and "not obvious" to receive patent protection. Patent claims that a country's patent office determines are not new or do not involve some inventive step (some flash of inspiration) over what was known are rejected.

The definition of "new" varies somewhat from country to country. In most countries, an invention is new only if it has not been publicly disclosed or known to others prior to the application filing date. In the United States and Brazil, an invention may still receive patent protection provided the application is filed within 1 year of a public disclosure of the invention by the inventor. However, most country patent laws do not provide such a grace period for filing an application, and instead require absolute novelty.

In the jargon of patent law, that which is known before the application filing date (or the date the invention was conceived in the United States) is referred to as "prior art." Thus, an invention is said to be new if it is not disclosed in the prior art. Publications generally encompass any book, journal, magazine or newspaper article, public advertisement, press release, Internet webpage or blog posting, graduate thesis or dissertation cataloged in a library somewhere in the world, even 5000-year-old Chinese scrolls, and any publicly released government document, such as patents and published patent applications. Further, an invention may not be "new" if it has been disclosed to anyone in the public that is not under some obligation to keep the disclosure secret (e.g., a nondisclosure agreement); that disclosure becoming prior art. Also, an invention is not "new" if it was already in use by others (except for secret uses) before the application was filed, even if such use was not published. The basic concept is that the patent monopoly cannot be applied products, methods, materials and technologies already in use and known to the public before the application filing date.

Patents may also not be granted on obvious variations or combinations of what was known in the prior art. In most countries, the patent laws require that a claim include or involve an "inventive step" by the inventor that went beyond the prior art. In the United States, this requirement is expressed differently; a claim cannot be "obvious" in view of the prior art. As may be expected, determining whether a claimed invention is "obvious" or "involves an inventive step" in view of the prior art involves a degree of judgment on the part of the examiner, which often forms the battleground of debates between inventors (or their patent attorneys) and patent offices around the world.

There are three basic types of patents that can be obtained: utility patents, design patents, and plant variety patents. Most patents are utility patents, which encompass methods, materials, machines, software programs stored on a computer-readable storage medium, chemicals, specific DNA sequences, and genetically modified plants and animals. Design patents cover only the ornamental design of an article, such as the shape of articles, containers, and visible features of a product. Design patents are similar to trademarks, with the exceptions that they are valid for a limited period of time (e.g., 14 years in the United States) and can only be obtained on novel artistic designs. Plant variety patents are awarded to those who invent or discover and asexually reproduce a distinct and new variety of plant, other than a tuber-propagated plant or a plant found in an uncultivated state. Additionally, some countries provide patent-like protections for industrial designs, such as architectural designs.

The types of inventions that are patentable vary somewhat from country to country, but most practical innovations are eligible for patent protections. In the United States, for example, with few exceptions "anything new under the sun made by man" may be patented (as explained by the U.S. Supreme Court in *Diamond v. Chakrabarty*, 447 U.S. 303, 309 (1980)). The breadth of patentable subject matter is quite extensive and encompasses most innovations that one would expect to be patentable. Without attempting to be comprehensive, clearly patentable subject matter includes new forms of materials (e.g., thermoelectric materials); practically any form of machine, mechanical device, electronics, or device with a usable function; chemicals, including the organic chemicals of life; and industrial methods, such as methods of making various chemicals or materials; and certain types of computerized computer-implemented methods as discussed below. Generally, if an invention concerns something that is tangible or involves a tangible form that involves machines, then patent protections are available for anything that is new and involves some form of inventive step (i.e., the invention is not obvious). Generally, anything that is useful, tangible, and not illegal or reprehensible may be patented.

While patents can be obtained on anything new made by man, there are some limitations that countries impose on the type of subject matter that is able to be patented. As new technologies have been developed, countries have modified their patent laws through legislation or court action to keep up. The implementation of computer-based systems has also stretched the boundaries of what is considered to be an "invention." As a result, there have been significant changes in the nature of the type of patents that can be obtained in various countries.

The question of whether an invention is patentable becomes more difficult to define in general terms when the invention involves a method, which is generally a sequence of steps that may be performed by a person or machine. In general, methods of making things, such as methods of formulating new thermoelectric materials, are generally patentable because the result is a transformation of the raw materials. However, methods that do not transform a material into a new form or are not involved in industrial processes may not be patentable in all countries. In many countries, such as typified by the European Patent Office (EPO), a method must have industrial application in order for it to be patentable. In the United States, a method that does not transform material must be necessarily tied to a machine (i.e., a machine must perform at least one of the steps of the method) in order for it to be eligible for patent protections. Also, many countries restrict the types of methods that can be patented. For example, many countries do not permit patenting of methods for treating a human patient. Prohibitions on patenting medical methods are justified on the basis that it is improper to prevent a doctor from treating a patient. In the United States there has been an evolution in the law regarding business method patents, with the current standard being that the method must be tied to a machine, such as the computer. Since the laws on patenting of methods are rapidly evolving, it is best to consult with a knowledgeable patent attorney regarding whether a method invention is suitable for patenting.

While the subject matter that is eligible for patent protections is extensive, there are limits. In particular, nontangible interventions, such as software and electronic signals, while of significant economic value and involving real inventive effort, are generally not patentable. Software and electronic signals lack a physical form. While there is debate in several countries regarding whether such intangible inventions should be patentable, there is currently general agreement that patent protections are limited to their physical manifestation. In particular, for example, software is currently claimed in the United States using words like "nontransitory computer-readable storage medium comprising ..." Thus, the CD or hard drive storing the software is claimed, not the software itself. In practice, this restriction is not particularly burdensome, but is an example of the nuance requirements imposed on patent claims. Of course, the software itself is protected from copying by copyrights.

Intangible signals, such as a particular type of wireless signal or Internet data protocol, are not patentable as such, and therefore must be covered by claims to methods tied to a computer or transmitters/receiver, or the transmission or receiver circuitry that is configured to generate or process such signals.

Another subject ineligible for patent protection in most countries is human beings. This prohibition comes up most often in the biotechnology arts where an invention might comprise some biological aspect of a human being, such as certain types of tissue genetic sequences. However, the law in this area is also rapidly evolving.

In the field of thermoelectric devices and thermoelectric materials, utility patents are most likely to be appropriate for protecting new thermoelectric compounds, methods of making thermoelectric compounds, devices implementing thermal electric compounds and components, systems using thermal electric devices, methods of using thermoelectric devices and systems, and software associated with any of the foregoing may be patented. Since thermoelectric devices and thermoelectric materials have obvious industrial and practical utility, the only standard for obtaining a patent on an invention is whether it is new and not obvious in view of what is known.

Applications for patents are filed by or in the name of the inventors of the claims. An inventor is someone who first conceived of at least one element recited in at least one claim in the patent application. Conceiving of an element of a claim means that the individual had the entire concept of the claim element in his or her mind even if some refinements were required. Patents may list single inventors or groups of inventors who worked collaboratively or independently conceived of elements recited in the claims. Indeed, many inventions are the product of a collaboration of two or more individuals working together or combining their expertise to generate something new.

Unlike technical papers which often list everyone involved in the project, the inventors listed on patents are limited to those individuals who conceived of the claimed invention. In contrast, those who merely identify a problem requiring a solution, implement or refine a concept, build prototypes, run tests or otherwise perform noninventive tasks at the direction of those who conceived the invention are not inventors.

Individual inventors can file a patent application on their own, but typically inventors assign their inventions to an employer (e.g., a company, university, research facility, etc.) that files for the application in the name of the inventors. Ownership rights in patent applications vary from country to country. Within the United States, a patent application is owned by all of the inventors equally until the inventors assign their rights to another person or company.

Generally, patent applications and patents may be freely assigned like any other property. Also, owners of patents and patent applications can license others to practice claims of a patent under terms that may be negotiated between the patent owner and the licensee.

As described in more detail below, patents grant the owner the right to exclude others via a lawsuit from making, using, selling or importing anything that infringes a patent claim. These rights to exclude others may be waived by granting a license to make, use, sell, or import things that infringe a patent claim. As with other forms of intellectual property, these patent rights are limited to the country granting the patent. Thus, a patent in one country only provides protection in that country. To protect an invention around the world, patents on the invention must be obtained in every country, or at least in countries where there is a possibility of infringement.

It is important to keep in mind that a patent is not a license or permission to practice an invention. Although an inventor may obtain a patent on a new improvement over a known compound, device, method, or system, the inventor may not be able to practice the invention without permission of those owning patents on the underlying material, device, method, or system. Further, the fact that a patent is granted provides no indication about whether the invention is free from patents owned by others. In fact, it is quite common that materials, devices, and systems are covered by a number of patents. For example, the personal computer may be covered with hundreds of patents owned by several different companies, covering details from the packaging of microchips to software operating on the processor to the exterior packaging. Thus, if a company obtains a patent on an improvement related to a personal computer, that company must obtain the permission of all other owners of patents covering various aspects of the computer before they can make, use or sell their invention without running the risk of being sued for patent infringement.

A patent is also not necessarily worth any money. Typically, only a small percentage of patents are actually worth significant money in the global marketplace. Many patents are never practiced. Some patents cover improvements of little economic value. Many patents cover a technology that is obsolete by the time the patent issues. Many patents expire before products implementing the patented technology ever reach the marketplace. Additionally, at the time a patent is filed, it is very difficult to identify whether it will be worth any money when it issues. For these reasons, many companies file large numbers of patents with the expectation that some fraction will ultimately cover valuable products and technologies.

Even if a patent does cover a valuable product or technology, it does not mean that the patent owner is automatically going to receive appropriate rewards. This is because patent holders must enforce their patents in a court of law to stop infringement if the infringer refuses to stop infringing, or purchase or license the patent. Patent enforcement litigation is notoriously expensive, with a typical patent enforcement action costing the plaintiff over two million dollars in U.S. courts in order to obtain a judgment. While prevailing in patent litigation may result in the award of royalties or license fees from an infringer, such damages may not equal the cost of litigation, particularly if the patented product, method, or system does not have significant economic value in the market. Thus, if the value of the technology covered by a patent is less than the cost of litigation, it may be very difficult to collect a large portion of its potential economic value.

Thus, patents are an imperfect tool for protecting inventions. For this reason, patents are best enforced by organizations with sufficient resources to afford litigation and combine patents together into a sizeable portfolio covering one or more products with significant market value.

If a patent covers a new product or technology that no one is infringing, it is often difficult to sell or license the patent to companies. This is because companies are often reluctant to take a risk on a product that is not already proven in the marketplace. Thus, inventors of truly new products, technologies, and materials are often frustrated the challenge of convincing companies that they should buy or license their patent. For this reason, inventors are often better rewarded if they can bring their product, material, or technology to market themselves. Once a new product, technology, or material has presented itself in the marketplace, the resulting sales, profits, or customer base are easy for companies to value for purposes of purchasing or licensing the technology.

In short, patents can be very valuable property for companies with the resources to exploit them or enforce them in litigation, although they are often of less value or usefulness to individuals.

24.4 Copyrights

Copyrights give to the owner the legal right to stop others from making or selling copies (or collect royalties from those who do make copies) of a creative expression which is fixed or recorded in some tangible medium of expression or recording. Copyrights encompass a wide range of creative endeavors, including works of authorship, musical compositions, musical recordings, photographs, films, sculptures, software programs, drawings, paintings, and so on.

Unlike patents, copyrights do not protect ideas and cannot prevent others from making, selling, or using an invention. Instead, copyrights protect the creative expression itself from being copied or reproduced by others without permission of the copyright holder. Thus, an author of a book or article describing how to make a particular thermoelectric material or device can stop others from making unauthorized copies of that book or article, but can do nothing about someone using what they learned from the author to make, use, or sell the thermoelectric material or device. Similarly, the author cannot prevent others from reading the book or article and then writing their own book or article on the same subject, provided they do not copy substantial portions.

Since copyrights are limited to protections against copying, and do create monopolies in the ideas, concepts, or designs described in the works, governments are much more liberal in granting copyrights. There is no requirement that copyrighted works be new or nonobvious; only that they are the creative works of the author (i.e., not a copy of someone else's work). Governments do not examine applications for copyrights except to confirm that the material is eligible for copyright protection. Also, in most countries, copyrights are acquired by the creator of a creative work as soon as the work is recorded in a tangible medium of expression. Thus, the very act of writing, recording, taking a photograph, drawing, or shooting a video is sufficient to create copyrights in that document, recording, photograph, drawing, or video. The registration of copyrights with government copyright offices is little more than a formality required to enforce the copyright, such as by filing a copyright infringement lawsuit.

While copyrights do not protect the underlying idea, technology, or teaching, they may nevertheless be very valuable in situations where the expression or details are of value. Obviously, authors are rewarded for their works of expression based on their creative talents and turn of phrase. Thus, while the elements of the detective novel have been around for generations, the estate of Agatha Christie continues to receive royalties from book sales because copyright protections preclude unauthorized copies that might otherwise compete in the marketplace with authorized editions. Also, copyrights provide effective protections for computer-based products because the complexity of modern software means that creating a similar or derivative work which would avoid copyright infringement involves a significant investment of time and effort.

Since nearly all works of authorship essentially automatically lead to the creation of copyrights, computer programs, laboratory notes, reports, and journal articles are some of the most common forms of intellectual property created by engineers and scientists. However, since copyrights do not require novelty examinations like patents, engineers, and scientists need not be concerned with the details of copyright law beyond understanding the copyrights of value and are essentially automatically created when the work is recorded in a tangible media. There are advantages to notifying others that you consider a particular work to be copyrighted by including a simple copyright notice that includes the "©" symbol, the name of the copyright owner, and the year or years of creation. If you suspect that a copyrighted work of yours or your organization has been infringed (i.e., copied without authorization by another) you should seek the advice of legal counsel.

24.5 Trade Secrets and Industrial Data

Another category of intellectual property important to engineers and scientists is generally known as trade secrets, although in some countries it is referred to as industrial data. Trade secrets and industrial data worthy of legal protections, by definition, are information that has economic value because it is not generally known to the public or industrial competitor. Thus, in order to be a trade secret, the information must be secret and it must have economic value because it is secret. “Secret” means that it is not known to the public, such as published in a document that is readily available, and not known to most competitors in the industry of the owner of the trade secrets. To have economic value the information must provide some competitive advantage, a head start in development, or cost saving to the possessor that is not available to competitors.

Trade secrets encompass a broad range of information defined simply by whether it provides economic value as a result of being secret. Thus, private business information like customer lists, supplier lists, pricing structures, business plans, and similar competitive information is eligible for legal protections as trade secrets. More important to engineers and scientists, trade secrets also include technical information that is not known to competitors. For example, ingredients, formulations, methods of processing, inventions that have not yet been disclosed in a patent application, product designs not yet on sale to the public or disclosed in a patent application, research plans, unpublished scientific discoveries, research and development budgets that have not been disclosed (e.g., in a company securities filing), and future product plans are all typical types of trade secret information. In short, most technical activities of engineers and scientists conducted on behalf of a for-profit organization may be eligible for trade secret protection until they are publicly disclosed in a patent application, product offering, proposal, technical paper, or other disclosure that may become known to competitors.

The legal protections afforded to trade secrets are limited to protecting against industrial espionage and unauthorized disclosure. Because of the very essence of trade secrets, once the information is publicly disclosed the information is no longer secret, and thus public disclosure extinguishes trade secrets. As a result, the enforcement of trade secrets is often an exercise in trying to herd animals back into the barn after someone has left the barn door open. An owner of a trade secret may obtain legal enforcement in the form of payments or damage caused by misappropriating or disclosing a trade secret. In cases where the misappropriated trade secrets are still not generally known, the trade secret owner may be able to obtain a court order preventing the misappropriating company from continuing to use the trade secret information. However, those competitors who innocently learn the information through public sources are then free to use that information in their business.

Trade secrets have another important limitation; unlike patents, there is no protection against others independently developing and using the same information or technology. Thus, a company maintains trade secrets in innovations that might otherwise be patentable run the risk that other companies may independently develop the same invention and may seek patent protections.

On the other hand, unlike patents, trade secrets do not expire as long as the trade secret remains a secret. A good example of a long-enduring trade secret is the formula for Coca Cola which reportedly is still kept in a safe 100 years after it was first developed. If the developer of the Coca Cola formula had applied for patent protections, such a patent would have expired long ago and that secret would have been known to everyone in the industry.

Therefore, before filing a patent application on an invention, companies generally evaluate whether the innovation could be detected or reverse engineered from a product. If not, then the innovation may be of more value to the company in the long run is maintained as a trade secret. On the other hand, if the innovation is readily discernible from a simple reverse engineering analysis, then trade secrets will be of little value and a patent application would be a better form of intellectual property.

Another form of intellectual property that is related to trade secrets is referred to as “know-how.” Know-how refers to the technical knowledge that is usually acquired by engineers and plant operators that relates to efficient and effective processes for making certain products, operating certain machines, or providing certain services. Often know-how exists in the experience of a company’s manufacturing team or laboratory staff, and is difficult to reduce to writing. While know-how is often not something that can be protected by patents or trade secrets, it nevertheless can be an important intellectual property, particularly in the context of products, methods, and services which are partially covered by patents and trade secrets. For example, it is not uncommon for license or sales agreements for patents to also address the exchange of trade secrets and know-how. For example, an agreement to license a patented process for manufacture may include license terms for disclosing related trade secrets and technical assistance to teach the know-how needed to practice the invention to the licensee's employees.

In addition to the explicit protections against trade secret's appropriation, some countries have laws governing corporate behavior that also provide protections for intellectual property. For example, laws against industrial espionage effectively make it illegal to steal corporate secrets whether they are or are not trade secrets. While trade secrets may be affected by such laws, the elements of trade secret (i.e., secrecy, economic value) do not have to be shown for such laws to be enforced. Similarly, laws against tampering with or obtaining data illegally from electronic databases (e.g., "hacking") can be leveraged to protect trade secrets and know-how.

24.6 Trademarks

Trademarks are another important form of intellectual property familiar to most people. Trademarks are important as symbols used by companies and organizations to identify themselves, their services, and products they offer to the marketplace.

Trademarks are often misunderstood by the public because, unlike patents and copyrights which are provided to protect the owner, trademarks are intended to protect consumers from being misled regarding the nature or source of products and services. Trademarks inform the purchasing public of the source of particular goods or services, and a trademark is infringed only if a consumer might be confused or misled by the same or similar mark. Thus, trademarks are linked to particular products and practices in commerce. As such, one cannot simply create an imaginative trademark and register it without being in or planning to be in business selling goods or services that will be identified by the trademark. Unlike patents and copyrights, trademarks do not expire so long as they are actively used in commerce on the registered goods or service.

Because trademarks are identifiers of organizations, goods, or services, they can acquire real value when the reputation of those organizations, goods, or services has value. For this reason, owners of trademarks often zealously guard the use of their marks to ensure they are not counterfeited, used by others or disparaged in the marketplace. For this reason, it is important to respect the trademarks of others, including identifying marks (e.g., with the "®" or "™" symbols) when the marks are used in publications and avoiding associating the trademarks of others with negative connotations or another's goods or services.

Since trademarks are the identifiers for companies, goods, and services, and not the creation of engineers and scientists, no further discussion of trademarks is included in this chapter.

24.7 Other Types of Intellectual Property

In addition to the patent, trademark, trade secret, know-how, and copyright laws, countries also provide protection for products of intellectual endeavors that do not fit neatly into these packages. For example, many countries provide protections for registered integrated circuit mask works. Falling somewhere between copyright and patent, mask works used to fabricate integrated circuits have value in the effort required to efficiently lay out the circuits and generates the corresponding photolithography masks. Registering a mask work gives the owner legal remedies against those who would copy the masks in order to fabricate copies of the integrated circuits. Another type of intellectual property protection concerns architectural designs, such as the designs of buildings and other large structures, such as bridges.

Some countries have laws protecting works of engineering design that can be readily copied, such as by making a mold of the product. For example, some states in the United States have laws protecting against the copying of boat hulls. The performance of watercraft depends upon the detailed three-dimensional contours of the hull, which requires complex fluid dynamic analysis and testing to perfect. However, a boat hull can easily be copied by using one boat to create a mold that can be used to create exact copies. Such laws are intended to protect the economic investment made by those who created the products by inhibiting competition from those who would pirate their intellectual products.

24.8 How a Patent Is Obtained

Since patents are an important form of intellectual property with which engineers and scientists are likely to be involved, it is worthwhile to describe the process by which patents are obtained. Patents begin as a patent application that is filed with a nation's patent office, such as the U.S. Patent and Trademark Office, the EPO, or the Japanese Patent Office. As described below, a patent application includes a complete description of the invention and a series of claims that define the invention for which legal protections are desired.

After patent applications have been received in a country's patent office [e.g., the United States Patent and Trademark Office (USPTO)] and confirmed to be complete, the applications wait in a queue for examination. Eventually, an examiner sees the patent application and conducts a search of the prior art, typically concentrating on patents and published patent applications from around the world. The examiner then examines the patent application to determine whether the claims are new and not obvious (or involve an inventive step) in view of the prior art found by the examiner. Typically, this evaluation begins with a search which drives off the claim language. The examiner may use keywords in the claims or look for documents that are related to the same type of technology. Quite often, the examiner is very knowledgeable of the prior art related to the invention since applications are generally assigned to examiners based upon the technology areas they examine. This practice enables examiners to develop expertise in the technologies of applications they examine.

After the examiner conducts a search, he or she will compare the claim language to the disclosure in identified prior art documents in order to determine whether the country's statutory basis for allowing a patent claim is satisfied. The examiner will reject any claims that are anticipated (i.e., not new) by a prior art document which describes or suggests every element recited in claim language. If a claim is not anticipated by a single reference, the examiner will determine whether the claims is obvious or lacks an inventive step in view of the prior art found by the examiner. The standards of obviousness or inventive step necessarily involves a judgment call, but in general, a rejection will be made when two or more references in the same technical field teach all of the elements in the claim when the references are somehow combined. These evaluations are accomplished on a claim-by-claim basis. When the examiner completes the examination, an official paper (often referred to as an "office action") is issued to the applicants or their attorneys. The office action identifies the claims that are rejected, the basis for the rejections and objections, and identifies the prior art that the examiner relied upon.

If any of the claims are rejected, the applicants have an opportunity to cancel or amend claims if needed, as well as make arguments to overcome the rejections, such as by adding limitations or elements that are not disclosed in the prior art. Typically, many if not all claims are rejected the first time the examining office reviews the claims. As an inventor, you should understand that claim rejections and amendments are parts of the process by which the patent office determines whether the claims are patentable.

Responding to a patent office action is something best left to a trained practitioner (i.e., a patent attorney or patent agent). The response to an office action typically will include amendments to the claims and arguments providing reasons why the claims should be allowed over the prior art identified in the office action. The types of arguments that will persuade an examiner are defined by the country's patent laws and thus are best left to an experienced practitioner (i.e., a patent attorney or patent agent). Office actions usually specify a period of time in which the applicant must respond. In some countries, like the United States, the office action must be responded to within a set period of time (e.g., three months in the United States), which typically can be extended with payment of fees for a limited duration. In other countries, the applicant has a set period of time in which the application must be placed in condition for allowance by overcoming all of the examiner rejections, including those that are made in the future.

When a response to the office action in the form of amendments and arguments is submitted to the patent office, the examiner will reconsider the claims and either allow them or issue another rejection. If the examiner rejects some or all of the patent claims, another office action will be issued and the applicants may have an opportunity to respond again. This rejection and response process may continue several rounds, with the procedure varying from country to country in terms of types of submissions and fees. Since patent applications are typically filed in several countries, similar procedures may be repeated in nearly all countries in which the application is filed.

Once the examiner finds that the claims are allowable, an official action will notify the applicants that the application is allowed, and the applicant may be invited to pay an issue fee for the application. A few months after the issue fees are paid, the patent will issue, which involves the public publication of the patent and issuance of a patent certificate. On the day the patent is issued, the claims can be enforced in a court of law of the issuing country.

The examination process may take anywhere from 1 to 5 years depending upon the technology and the patent office, although it may take longer, such as in Japan where applications may be pending for 7 years or more.

Within about 18 months after the patent applications are filed, the patent office will typically publish the application, which means that the document is then publicly available. Publication makes an electronic copy of the application available in public databases, many of which can be accessed via the Internet. Published applications become prior art to later-filed applications. Thus, even if a patent application does not result in an issued patent the published application can preclude others from obtaining patents on the same or a similar invention. Thus, filing patent applications creates prior art that others must overcome.

With all of the fees and legal services involved in prosecuting a patent application, it is not surprising that the process of obtaining patents can be very expensive, particularly when an applicant seeks to protect the invention in many countries.

As mentioned above, patents are enforceable only in the country issuing the patent, which requires owners of globally relevant technologies to file the same application in many countries. In order to simplify the process of global patenting, most countries in the world have entered into the Paris Convention Treaty which gives applicants a year to file in the country after filing in another country, and the Patent Cooperation Treaty (PCT) discussed below.

A PCT application is a special form of application that enables a single application to be filed in most countries around the world. A PCT application must be filed within 1 year of the filing date of a patent application in a treaty member country. The PCT application enables applicants to designate the countries of the world for filing of a national patent application. The PCT application is searched by a patent search authority (e.g., the USPTO) designated by the applicant. Results of that search are provided to the applicant and form a basis for examination of the application in the countries in which national applications are eventually filed. Filing a PCT application enables the applicant to delay filing in other countries for 30 or 31 months from the first day that the application was filed (either as the PCT application or as a preceding national application). This delay can be economically advantageous to companies since it permits them to investigate the invention for marketability and technical feasibility before having to invest in the fees and legal services associated with filing in several countries. Such expenses can be significant, especially because many countries require translation of the patent application into their official language. While the PCT application is searched and, in some cases, the claims amended, the PCT application does not result in a patent that can be enforced in any country. Instead, the PCT application simply expires after the 31 month deadline for filing national stage applications.

24.9 Contents of Patent Applications

The purpose of patents defines the information required in the patent application; the application must provide a complete teaching of how to make and use the invention so that others may readily comprehend and practice the invention after the patent expires. Thus, the patent application is a technical paper and drawings that explain the details of the invention in terms that a person knowledgeable in the science or technology can understand.

Most patent applications include a set of drawings with elements identified by reference numbers that are referred to in the written description within the application. Without trying to explain the drawing requirement for every country in the world, drawings are typically simple black-and-white line drawings. For inventions that are not amenable to drawings, such as chemical and biotechnology inventions, drawings may not be included. Inventions involving methods, such as processes of making thermoelectric materials, should include a process flow diagram showing at least each of the steps recited in the claims. Patent drawings are not works of art, and typically show simplistic or cartoon versions of the device, process or material being claimed. Since there a number of technical requirements for patent drawings, typically a draftsman prepares the final drawings for application.

The written specification is both a technical and legal paper that describes the important aspects—particularly the new and inventive aspects—of the invention. While the description of the invention needs to be complete, it does not have to explain technology that is already well known. Thus, the patent application covering a new type of thermoelectric material would not have to explain the science of thermoelectrics or known thermoelectric materials. However, it is sometimes difficult to distinguish between what is new in the invention and what is already known. When in doubt, patent attorneys will typically err on the side of including more details regarding the background of technology and the invention to ensure the application meets all disclosure requirements of any country in which it will be filed. Also, the patent laws in many countries require an explanation in the specification of the improvements that the invention makes over the prior art. Thus, drafting patent applications involves a balancing act between disclosing sufficient background information to place the invention in context, including revealing how the invention improves over what is known, and cluttering the application with unnecessary details well known in the technology.

There are few formal requirements on the format of patent applications and such requirements vary from country to country. However, the typical patent application is written in a series of sections. First, the application includes a description of the background of the invention in which known technologies may be outlined. Next, there may be a summary of the invention, which is typically a brief outline of the important aspects of the invention. Following the summary is typically a brief description of each of the drawings included in the application. This description merely explains what the drawing is, rather than what the drawing teaches. Next, the bulk of the specification provides a detailed description of the invention, including a description of the drawings and any alternative implementations and embodiments. The detailed description explains the invention in sufficient detail to enable others to practice it. Typically, the reference numerals included in the drawings are referred to in the description. The description may also explain the utility of the invention, such as an example of how it may be employed or used. The United States requires that the "best mode" be disclosed, which is the preferred or most likely the implementation of the invention known to the inventor at the time the application is filed. The specification then concludes with a set of claims and an abstract. As described below, the claims define the invention and the limits of the monopoly granted by the patent. The abstract is a brief summary, typically limited to about 150 words or less, that is printed on the face of the patent. Since the abstract is useful for those searching and screening patents, it should reflect the nature of the invention, but does not have to explain each inventive detail or the background of the invention.

If the detailed description is the heart of the specification, the claims are its soul. Claims identify the scope of the monopoly that the inventor wants to obtain if a patent is granted. The allowed claims constitute the legal rights that can be enforced in a court of law. In overview, claims must clearly articulate the scope of invention that is claimed, and with a few exceptions only need to recite those elements which are new and actually being claimed. Claims are typically written in an arcane style. There are many rules, regulations, and presumptions imposed on claim language by national patent laws, courts, and patent offices which can have significant impact on the value of a patent. For this reason, claims should always be written by an experienced practitioner, such as a patent attorney or patent agent knowledgeable about the patent laws in the country in which the application will be filed.

While patent applications are typically written by patent attorneys for patent agents, it is important that the inventor thoroughly reviews the entire application before it is filed with a patent office. In most countries, the inventor must confirm that the invention is their own and completely discloses their invention. In some countries, there are legal penalties for signing such documents without confirming that the invention is complete.

Filing of a completed patent application is accompanied by a payment of national filing fees. Such fees can be expensive in some countries, especially if the application is long or includes many claims. In addition to the cost of preparing, filing an issue fees for patents, most countries also require payment of maintenance fees throughout the life of the patent in order to keep it enforceable. Such maintenance fees can add up quickly, especially when a patent is filed in several countries. In the United States, maintenance fees are due on the 4th, 8th, and 12th year after the patent issues, but most other countries require annual maintenance fees. If a maintenance fee is not paid, the patent will go abandoned in that country, at which point it can no longer be enforced.

24.10 How to Recognize a Patentable Invention

Many engineers and scientists fail to appreciate what they have created or conceived of a patentable invention. Many believe that patents are reserved for breakthrough inventions. Also, some people believe that an invention cannot be a combination of things that are already well known. In fact, the vast majority of patents are granted on improvements to products, methods, or systems that were already known, and almost all inventions involve a combination of known things in new ways. For example, the airplane invented by the Wright brothers was the combination of a glider that had been known for years, with the gasoline engine which had been perfected in automobiles, and the propeller which had been perfected in fans and ship propellers. Yet, no one would question that the airplane was a breakthrough invention.

Additionally, just because a material is known does not mean that a patent cannot be obtained for a new and previously unknown use of the material. Patents on new uses for known drugs are quite common in medical fields because drugs developed for one purpose often turn out to have remarkable properties for treating other maladies. Viagra is a well-known example of such an invention. Those who discover new uses of known materials and develop the appropriate uses, formulations, and implementing forms for the new use may obtain patents for their discoveries just as if they had discovered the material themselves. The test for patentability, of course, is whether the claimed use is truly new (i.e., not disclosed in any previous publications, even in a footnote or speculative instruction) and that this new use is not obvious in view of what was known.

Similarly, new methods of fabricating or making known materials may also be patentable provided they are indeed new and not obvious in view of what was known. Thus, methods of making materials, such as thermoelectric materials, which improve the material's performance, manufacturing yield, or reduce costs may be eligible for patent protections.

Even when the invention involves known materials fabricated using known methods and applied in the known application, there still may be patentable subject matter if some change in the process, configuration, implementation, or method of use results in new and unexpected qualities or results. In such a situation it may be necessary to prove not only that the new formulation, implementation, or use was not previously known, but that the unexpected results were not anticipated or would have been expected by one of skill in the art. Thus, if your research yields a thermoelectric material or device which exhibits 5% or 10% greater efficiency and performance, and such increase in performance is not something one would expect, you may have a patentable invention.

Thus, the challenge is to recognize when an improvement or combination of known things in new ways rises to the level of a patentable invention. One approach to this challenge is to attempt to patent everything and let the patent office sort out the patentable from the unpatentable. However, this approach is generally impractical and unaffordable. The preparation of patent applications and invention disclosures is very time consuming, and the legal and government fees associated with filing patent applications is very expensive. For this reason, most organizations implement a review process designed to cull out those inventions which have economic value and are likely to result in an issued patent.

One reliable method for recognizing a patentable invention involves considering the effort required to conceive and perfect the idea. In the Internet era, engineers and scientists have ready access to the combined knowledge of the world. Using an Internet search engine, engineers and scientists can readily determine whether a solution to a particular problem is known in just a matter of a few hours of sorting through search results. For this reason, if a solution to a particular problem cannot be readily found on the Internet, there is a good chance that the solution, when developed, will be patentable. Also, if trained engineers or scientists cannot quickly determine a solution to a problem using their knowledge and technical skills in combination with what they find in textbooks and the Internet, it is a safe bet that the solution they eventually develop is likely in the patentable realm.

So, if you have faced a difficult technical problem that was not readily solved using ordinary technical skills and knowledge in combination with a library or Internet search—particularly if the solution required a few days of thinking and some trial and error—the solution you eventually developed probably is patentable.

Of course, not every patentable invention is worth the time and expense to patent. The costs of filing for patents should be reserved for those innovations which are likely to have significant economic value. Thus, a second test to consider before filing a patent application on a new solution to a technical problem is whether the solution is central to or enables a product that has value in the marketplace, or reduces the cost of making a valuable product.

24.11 What Not to Do When You Have Conceived an Invention

When you think you have developed or conceived of a patentable invention, you should avoid making any public disclosures until an application is filed or a decision not to patent has been made. Public disclosures that can render inventions unpatentable in many countries may be as simple as discussing it with somebody outside of the company who is not under a nondisclosure agreement; submitting your invention in technical papers for publication; making presentations at technical conferences; discussing the invention with customers and partner companies; and beta testing products with customers. Since others may be inventing in the same area of technology and other parts of your organization may be planning to disclose the invention publicly, it is also a good idea not to delay too long in filing for a patent application.

Given the subtleties of patent law and the myriad of requirements in countries around the world, the preparation of patent applications is best left to licensed patent attorneys and patent agents. However, inventors are well served by describing their invention in writing and in a series of drawings that can be turned into patent drawings. The more description and drawings the inventors can provide the attorney or agent drafting the application, the more likely the application will be drafted efficiently and include all of the important and valuable aspects of the invention. Thus, promptly providing a full description of an invention to your company or your patent attorney is the first and best step that can be taken to secure the appropriate legal protections.

24.12 How Patents Can Be Turned into Money

The monopoly right that is granted in a patent can be quite valuable. The monopoly can provide great competitive advantages for patent owners practicing the invention and using the patent rights to prevent others from competing or having a competitive advantage over the products or services covered by the patent claims. Patents can also be used to block the importation of low-cost copies. Thus, a business practicing a patent can charge higher prices for its patented products or services, thereby increasing profits.

The right to prevent others from practicing the claimed invention may also be used to induce companies to pay royalties in return for a license to practice the patent. Licensing a patent can be very rewarding when the claimed invention has a large economic potential value. Patents may be licensed exclusively (i.e., to a single company) or nonexclusively (i.e., to more than one company). Basically, a license is permission to use the technology and a promise not to sue the company for infringing the patent, which is typically made in return for payments of an upfront fee and/or royalties on sales or profits of patented products or services.

If an infringing company refuses to accept a license to a patent, the patent holder may sue for patent infringement and recover damages. The damages that may be ordered by a court will typically be similar to license royalties.

Patents may also be sold, such as to companies who would like to practice the invention in the marketplace. The value of the patent in such a situation will be the economic value of the monopoly practice of the invention by the business, which may be calculated as a net present value of a future stream of increased profits or earnings from royalty payments if the company plans to license the patent.

24.13 How Patents Are Enforced

Patents give the owner the right to sue in a court of law anyone who practices the claims of the patent without authorization. Practicing the patent includes making, using, selling, offering to sell, and importing an article of manufacture, a device, or system which satisfies every limitation recited in at least one claim. Thus, patents do not give their owners a right to practice the invention, but rather access to the country's courts to stop others from practicing the claims without paying some form of royalty.

If a patent holder believes that a company is infringing a patent, the company must follow the procedures of the country's legal system for initiating a patent infringement lawsuit. In that lawsuit, the company must prove the facts of actual use or practicing of the claims. Since patent claims are necessarily specific in nature, proving infringement is often a difficult legal challenge. Since many patents are highly practical in nature and most judges and jury members are not technically educated, proving infringement involves explaining technical details to non-technical people. As a result, patent infringement lawsuits are typically very expensive, generally costing more than two million dollars in the United States.

If the patent owner prevails in a patent infringement lawsuit, the patent owner may be compensated for the infringement in the form of a payment from the infringing entity ordered by the court. In many countries and situations, the amount of money that is paid to the patent holder will be similar to what would have been paid in royalties if a license had been negotiated in advance. The amount of damages may depend upon the value of the technology that was infringed as well as the amount and duration of infringement. In some cases and some countries, the patent holder may obtain an injunction which can be enforced to prevent the infringer from continuing to practice the patent claims.

The procedures, remedies, and standards for patent infringement lawsuits vary widely from country to country. As a result, the cost to enforce patents and the available remedies depend upon the country in which the lawsuit is brought. For this reason, the intrinsic value of patents differs from country to country.

Considering the costs of patent infringement litigation and the potential of monetary damages, patent infringement lawsuits are very often settled before the dispute sees a courtroom. A typical settlement involves the payment of royalties in return for a license, but other forms of business arrangements may form the basis of the settlement.

24.14 Trade Secrets and Know-How

Since many intellectual creations are either not sufficiently new and nonobvious to be patentable or not worth the costs to prepare a patent application, organizations typically choose to maintain many innovations as trade secrets and matters of organizational know-how. The choice between filing a patent application and maintaining the innovation as a trade secret involves weighing the risks and costs of both alternatives.

Trade secrets and know-how can be very valuable to organizations, since such knowledge provides competitive advantages and can be sold or licensed to others. Sales and licensing of trade secrets and know-how may be accomplished by contracts in which the organization agrees to teach an acquiring company the know-how and trade secrets in return for financial payments. Such contracts or agreements may require the company paying for them to also protect them from further disclosure, thereby assuring their value remains intact. Since know-how and related trade secrets are often associated with patented inventions, particularly with respect to details regarding a process of manufacturing or formulation developed as part of refining a patented invention, the sale or licensing of know-how and trade secrets are often included in patent licenses and assignment agreements. Thus, many times it is the combination of patents and enabling know-how that renders an invention economically valuable to others. Thus, trade secrets are important elements of developing value for one's employer through developing or implementing and refining inventions.

Trade secrets are typically developed in the ordinary course of research and development, as well as the competitive conduct of the business. Technical trade secrets often develop over time in the know-how gained by the technical staff regarding how products are made, processes are controlled and other technical details which are discovered after a patent application has been filed. Since such knowledge is often acquired without that feeling of inspiration or conclusion that something very new has been developed, scientists and engineers may fail to appreciate the value that has been created.

Since trade secrets acquire their value from being unknown by others (particularly competitors), it is important to implement practices and procedures which avoid inadvertent disclosure of such knowledge that would destroy its trade secret value. This is typically accomplished by implementing proprietary practices that limit access of outsiders to research and development and process information. Also, employees are typically advised not to discuss the details of their work with those outside the company. Also, policies to record observations, findings, process settings, and similar types of information in laboratory or engineering notebooks provide mechanisms for recording trade secret information that enables them to be retained and passed on to others within the company or organization.

In order to receive the legal protections afforded by trade secrets, companies and research organizations must implement procedures that protect trade secrets from disclosure. Thus, locking facilities, restricting access to facilities and the secrets and escorting visitors, locking up laboratory and engineering notebooks in a cabinet or safe, and similar procedures are all necessary to receive the benefits of trade secrets.

24.15 Conclusion

In short, intellectual property is the stock and trade of engineers and scientists. Recognizing when intellectual property has been created, appreciating its value and taking necessary steps to secure the appropriate legal protections should be part of the day-to-day activities of every professional. While many of the administrative tasks require the knowledge and services of patent attorneys or patent agents, as inventors, the process of protecting intellectual property begins with you starting from the moment of creation or inspiration.

Index

Note: n = Footnote

C

D

E

F

G

H

J

K

L

M

N

O

P

S

U

V

W

X

Z